STUDENT SOLUTIONS MANUAL
& STUDY GUIDE

to accompany

Volume 2

PHYSICS

For Scientists and Engineers

Fifth Edition

Serway • Beichner

John R. Gordon
James Madison University

Ralph McGrew
Broome Community College

Raymond A. Serway

with GOAL problem
solutions developed by

Duane Deardorff
North Carolina State University

BROOKS/COLE

THOMSON LEARNING

Australia • Canada • Mexico • Singapore • Spain
United Kingdom • United States

0-03-020973-0

**For more information about our products,
contact us at:
Thomson Learning Academic Resource Center
1-800-423-0563**

**For permission to use material from this text,
contact us by:
Phone: 1-800-730-2214
Fax: 1-800-731-2215
Web: www.thomsonrights.com**

Asia
Thomson Learning
60 Albert Complex, #15-01
Alpert Complex
Singapore 189969

Australia
Nelson Thomson Learning
102 Dodds Street
South Street
South Melbourne, Victoria 3205
Australia

Canada
Nelson Thomson Learning
1120 Birchmount Road
Toronto, Ontario M1K 5G4
Canada

Europe/Middle East/South Africa
Thomson Learning
Berkshire House
168-173 High Holborn
London WC1 V7AA
United Kingdom

Latin America
Thomson Learning
Seneca, 53
Colonia Polanco
11560 Mexico D.F.
Mexico

Spain
Paraninfo Thomson Learning
Calle/Magallanes, 25
28015 Madrid, Spain

Table of Contents

Preface

This <u>Student Solution Manual and Study Guide</u>, has been written to accompany the textbook **Physics for Scientists and Engineers**, Fifth Edition, by Raymond A. Serway and Robert J. Beichner. The purpose of this Student Solution Manual and Study Guide is to provide the students with a convenient review of the basic concepts and applications presented in the textbook, together with solutions to selected end-of-chapter problems from the textbook. This is not an attempt to rewrite the textbook in a condensed fashion. Rather, emphasis is placed upon clarifying typical troublesome points, and providing further drill for methods of problem solving.

Each chapter is divided into several parts, and every textbook chapter has a matching chapter in this book. Very often, reference is made to specific equations or figures in the textbook. Every feature of this Study Guide has been included to ensure that it serves as a useful supplement to the textbook. Most chapters contain the following components:

- **Equations and Concepts:** This represents a review of the chapter, with emphasis on highlighting important concepts, and describing important equations and formalisms.

- **Suggestions, Skills, and Strategies:** This offers hints and strategies for solving typical problems that the student will often encounter in the course. In some sections, suggestions are made concerning mathematical skills that are necessary in the analysis of problems.

- **Review Checklist:** This is a list of topics and techniques the student should master after reading the chapter and working the assigned problems.

- **Answers to Selected Conceptual Questions:** Suggested answers are provided for approximately fifteen percent of the conceptual questions.

- **Solutions to Selected End-of-Chapter Problems:** Solutions are shown for approximately forty percent of the odd-numbered problems from the text, which illustrate the important concepts of the chapter.

We sincerely hope that this Student Solution Manual and Study Guide will be useful to you in reviewing the material presented in the text, and in improving your ability to solve problems and score well on exams. We welcome any comments or suggestions which could help improve the content of this study guide in future editions; and we wish you success in your study.

John R. Gordon
Harrisonburg, Virginia

Ralph McGrew
Binghamton, New York

Raymond A. Serway
Chapel Hill, North Carolina

Acknowledgments

It is a pleasure to acknowledge the many excellent contributions of Michael Rudmin, of DSC—Publishing, LLC. to this Fifth Edition of the Student Solutions Manual and Study Guide. His graphics skills and technical expertise have combined to produce illustrations which enhance the descriptions and solutions. The attractive layout is a result of his attention to detail; and the overall appearance of the Solution Manual reflects his work on the final camera-ready copy.

Special thanks go to the professional staff at Saunders College Publishing, especially Susan Pashos and Ericka Yeoman for managing all phases of the production of the manual.

Finally, we express our appreciation to our families for their inspiration, patience, and encouragement.

Suggestions for Study

Very often we are asked "How should I study this subject, and prepare for examinations?" There is no simple answer to this question, however, we would like to offer some suggestions which may be useful to you.

1. It is essential that you understand the basic concepts and principles before attempting to solve assigned problems. This is best accomplished through a careful reading of the textbook before attending your lecture on that material, jotting down certain points which are not clear to you, taking careful notes in class, and asking questions. You should reduce memorization of material to a minimum. Memorizing sections of a text, equations, and derivations does not necessarily mean you understand the material. Perhaps the best test of your understanding of the material will be your ability to solve the problems in the text, or those given on exams.

2. Try to solve as many problems at the end of the chapter as possible. You will be able to check the accuracy of your calculations to the odd-numbered problems, since the answers to these are given at the back of the text. Furthermore, detailed solutions to approximately half of the odd-numbered problems are provided in this study guide. Many of the worked examples in the text will serve as a basis for your study.

3. The method of solving problems should be carefully planned. First, read the problem several times until you are confident you understand what is being asked. Look for key words which will help simplify the problem, and perhaps allow you to make certain assumptions. You should also pay special attention to the information provided in the problem. In many cases a simple diagram is a good starting point; and it is always a good idea to write down the given information before proceeding with a solution. After you have decided on the method you feel is appropriate for the problem, proceed with your solution. If you are having difficulty in working problems, we suggest that you again read the text and your lecture notes. It may take several readings before you are ready to solve certain problems, though the solved problems in this Study Guide should be of value to you in this regard. However, your solution to a problem does not have to look just like the one presented here. A problem can sometimes be solved in different ways, starting from different principles. If you wonder about the validity of an alternative approach, ask your instructor.

4.   After reading a chapter, you should be able to define any new quantities that were introduced, and discuss the first principles that were used to derive fundamental formulas. A review is provided in each chapter of the Study Guide for this purpose, and the marginal notes in the textbook (or the index) will help you locate these topics. You should be able to correctly associate with each physical quantity the symbol used to represent that quantity (including vector notation if appropriate) and the SI unit in which the quantity is specified. Furthermore, you should be able to express each important formula or equation in a concise and accurate prose statement.

5.   We suggest that you use this Study Guide to review the material covered in the text, and as a guide in preparing for exams. You should also use the Equations and Concepts to focus in on any points which require further study. Remember that the main purpose of this Study Guide is to improve upon the efficiency and effectiveness of your study hours and your overall understanding of physical concepts. However, it should not be regarded as a substitute for your textbook or individual study and practice in problem solving.

Problem Solving

Besides what you might expect to learn about physics concepts, another very valuable skill you should hope to take away from your physics course is the ability to solve complicated problems. The way physicists approach complex situations and break them down into manageable pieces is extremely useful. Below is a memory aid to help you easily recall the steps required for successful problem solving. When working on problems, the secret is to keep your **GOAL** in mind.

GOAL Problem Solving Steps

Gather information

The first thing to do when approaching a problem is to understand the situation. Carefully read the problem statement, looking for key phrases like "at rest," or "freely falls." What information is given? Exactly what is the question asking? Don't forget to gather information from your own experiences and common sense. What should a reasonable answer look like? You wouldn't expect to calculate the speed of an automobile to be 5×10^6 m/s. Do you know what units to expect? Are there any limiting cases you can consider? What happens when an angle approaches 0° or 90° or a mass gets huge or goes to zero? Also make sure you carefully study any drawings that accompany the problem.

Organize your approach

Once you have a really good idea of what the problem is about, you need to think about what to do next. Have you seen this type of question before? Being able to classify a problem can make it much easier to lay out a plan to solve it. You should almost always make a quick drawing of the situation. Label important events with circled letters. Indicate any known values, perhaps in a table or directly on your sketch. Some kinds of problems require specific drawings, like a free body diagram when analyzing forces. Once you've done this and have a plan of attack, it is time for the next step.

Analyze the problem

Because you have already categorized the problem, it should not be too difficult to select relevant equations that apply to this type of situation. Use algebra (and calculus if necessary) to solve for the unknown variable in terms of what is given. Substitute in the appropriate numbers, calculate the result, and round it to the proper number of significant figures.

Learn from your efforts

This is actually the most important part. Examine your numerical answer. Does it meet your expectations from the first step? What about the algebraic form of the result before you plugged in numbers? Does it make sense? (Try looking at the variables in it to see if the answer would change in a physically meaningful way if they were drastically increased or decreased or even reduced to zero.) Think about how this problem compares to others you have done. How was it similar? In what critical ways did it differ? Why was this problem even assigned? You should have learned something by doing it. Can you figure out what?

For complex problems, you may need to apply these four steps of the GOAL process recursively to subproblems. For very simple problems, you probably don't need this protocol. But when you are looking at a problem and you don't know what to do next, remember what the letters in GOAL stand for and use that as a guide.

Selected problem solutions in this solutions manual follow the GOAL problem solving strategy. These problems should serve as examples of how the GOAL strategy can be applied to nearly any physics problem.

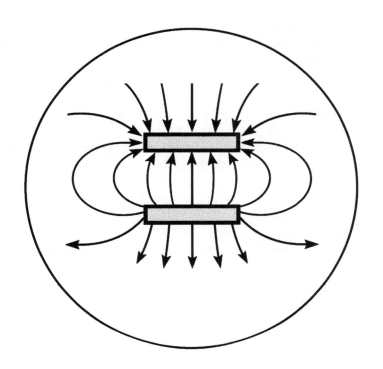

Electric Fields

ELECTRIC FIELDS

INTRODUCTION

The electromagnetic force between charged particles is one of the fundamental forces of nature. In this chapter, we begin by describing some of the basic properties of electric forces. We then discuss Coulomb's law, which is the fundamental law of force between any two charged particles. The concept of an electric field associated with a charge distribution is then introduced, and its effect on other charged particles is described. The method of using Coulomb's law to calculate electric fields of a given charge distribution is discussed, and several examples are given. Then the motion of a charged particle in a uniform electric field is discussed.

EQUATIONS AND CONCEPTS

The **magnitude** of the **electrostatic force** between two stationary point charges, q_1 and q_2, separated by a distance r is given by Coulomb's law.

$$F_e = k_e \frac{|q_1||q_2|}{r^2} \qquad (23.1)$$

$$k_e = \frac{1}{4\pi \epsilon_0}$$

$$\epsilon_0 = 8.8542 \times 10^{-12} \ \text{C}^2 / \text{N} \cdot \text{m}^2$$

In calculations, an approximate value for k_e may be used.

$$k_e = 8.99 \times 10^9 \frac{\text{N} \cdot \text{m}^2}{\text{C}^2}$$

The direction of the electrostatic force on each charge is determined from the experimental observation that like sign charges experience forces of mutual repulsion and unlike sign charges attract each other. By virtue of Newton's third law, the magnitude of the force on each of the two charges is the same regardless of the magnitude of the charges q_1 and q_2.

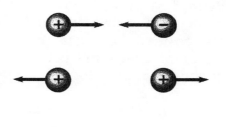

The electric force between two charges can be expressed in vector form. $\mathbf{F}_{12}$ is the force **on q_2 due to q_1** and $\hat{\mathbf{r}}$ is a unit vector directed from q_1 to q_2. Coulomb's law applies exactly only to point charges or particles.

$$\mathbf{F}_{12} = k_e \frac{q_1 q_2}{r^2} \hat{\mathbf{r}} \qquad (23.2)$$

When more than two point charges are present, the **total electrostatic force** exerted on the ith charge is the vector sum of the forces exerted on that charge by the others individually. The principle of superposition applies.

The **electric field** at any point in space is defined as the ratio of electric force per unit charge exerted on a small positive test charge placed at the point where the field is to be determined.

$$E \equiv \frac{F_e}{q_0}$$ (23.3)

The definition above together with Coulomb's law leads to an expression for calculating the **electric field a distance r from a point charge, q.** In this case the unit vector $\hat{r}$ is directed away from q and toward the point P where the field is to be calculated. The direction of the electric field is radially outward from a positive point charge and radially inward toward a negative point charge.

$$E = k_e \frac{q}{r^2}\hat{r}$$ (23.4)

The superposition principle holds when the electric field at a point is due to a number of point charges.

$$E = k_e \sum_i \frac{q_i}{r_i^2}\hat{r}_i$$ (23.5)

(vector sum)

Electric field lines are a convenient graphical representation of electric field patterns. These lines are drawn so that the electric field vector, **E**, is tangent to the electric field lines at each point.

Also, the number of lines per unit area through a surface perpendicular to the lines is proportional to the strength or magnitude of the electric field over the region. In every case, electric field lines must begin on positive charges and terminate on negative charges or at infinity; the number of lines leaving or approaching a charge is proportional to the magnitude of the charge; and no two field lines can cross.

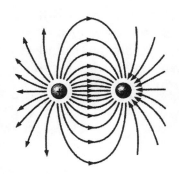

When the electric field is due to a **continuous charge distribution**, the contribution to the field by each element of charge must be integrated over the total line, surface, or volume which contains the charge.

$$\mathbf{E} = k_e \int \frac{dq}{r^2} \hat{\mathbf{r}} \qquad (23.6)$$

In order to perform the integration described above, it is convenient to represent a charge increment dq as the product of an element of length, area, or volume and the **charge density over that region**. **Note**: For those cases in which the charge is not uniformly distributed, the densities λ, σ, and ρ must be stated as functions of position.

For an element of length dx,
$$dq = \lambda dx$$

For an element of area dA,
$$dq = \sigma \, dA$$

For an element of volume dV,
$$dq = \rho \, dV$$

For uniform charge distributions the volume charge density (ρ), the surface charge density (σ), and the linear charge density (λ) can be calculated from mass properties.

$$\rho \equiv \frac{Q}{V}$$

$$\sigma \equiv \frac{Q}{A}$$

$$\lambda \equiv \frac{Q}{\ell}$$

SUGGESTIONS, SKILLS, AND STRATEGIES

PROBLEM-SOLVING STRATEGY FOR ELECTRIC FORCES AND FIELDS

- Units: When performing calculations that involve the use of the Coulomb constant k_e that appears in Coulomb's law, charges must be in coulombs and distances in meters. If they are given in other units, you must convert them to SI.

- Applying Coulomb's law to point charges: It is important to use the superposition principle properly when dealing with a collection of interacting point charges. When several charges are present, the resultant force on any one of them is found by finding the individual force that every other charge exerts on it and then finding the vector sum of all these forces. The magnitude of the force that any charged object exerts on another is given by Coulomb's law, and the direction of the force is found by noting that the forces are repulsive between like charges and attractive between unlike charges.

- Calculating the electric field of point charges: Remember that the superposition principle can also be applied to electric fields, which are also vector quantities. To find the total electric field at a given point, first calculate the electric field at the point due to each individual charge. The resultant field at the point is the vector sum of the fields due to the individual charges.

- To evaluate the electric field of a continuous charge distribution, it is convenient to employ the concept of charge density. Charge density can be written in different ways: charge per unit volume, ρ; charge per unit area, σ; or charge per unit length, λ. The total charge distribution is then subdivided into a small element of volume dV, area dA, or length dx. Each element contains an increment of charge dq(equal to $\rho\,dV$, $\sigma\,dA$, or $\lambda\,dx$). If the charge is **nonuniformly** distributed over the region, then the charge densities must be written as functions of position. For example, if the charge density along a line or long bar of length b is proportional to the distance from one end of the bar, then the linear charge density could be written as $\lambda = bx$ and the charge increment dq becomes $dq = (bx)dx$.

- Symmetry: Whenever dealing with either a distribution of point charges or a continuous charge distribution, take advantage of any symmetry in the system to simplify your calculations.

REVIEW CHECKLIST

▷ Describe the fundamental properties of electric charge and the nature of electrostatic forces between charged bodies.

▷ Use Coulomb's law to determine the net electrostatic force on a point electric charge due to a known distribution of a finite number of point charges.

▷ Calculate the electric field **E** (magnitude and direction) at a specified location in the vicinity of a group of point charges.

▷ Calculate the electric field due to a continuous charge distribution. The charge may be distributed uniformly or nonuniformly along a line, over a surface, or throughout a volume.

▷ Describe quantitatively the motion of a charged particle in a uniform electric field.

ANSWERS TO SELECTED CONCEPTUAL QUESTIONS

3. A balloon is negatively charged by rubbing and then clings to a wall. Does this mean that the wall is positively charged? Why does the balloon eventually fall?

Answer

No. The balloon induces a charge of the opposite sign in one section of the wall, and the wall surface becomes positively charged. The balloon eventually falls since its charge slowly diminishes as it "leaks" to the surroundings. Some of the charge could also be lost due to ions of opposite sign in the surrounding atmosphere which would tend to neutralize the charge.

□   □   □   □

11. Would life be different if the electron were positively charged and the proton were negatively charged? Does the choice of signs have any bearing on physical and chemical interactions? Explain.

Answer

No. Life would not be different. The character and effect of electric forces is defined by (1) the fact that there are only two types of electric charge – positive and negative, and (2) the fact that opposite charges attract, while like charges repel. The choice of signs is completely arbitrary.

As a related exercise, you might consider what would happen in a world where there were three types of electric charge, or in a world where opposite charges repelled, and like charges attracted.

□   □   □   □

SOLUTIONS TO SELECTED END-OF-CHAPTER PROBLEMS

3. Richard Feynman once said that if two persons stood at arm's length from each other and each person had 1% more electrons than protons, the force of repulsion between them would be enough to lift a "weight" equal to that of the entire Earth. Carry out an order-of-magnitude calculation to substantiate this assertion.

Solution

Suppose each person has mass 70.0 kg. In terms of elementary charges, each consists of precisely equal numbers of protons and electrons and a nearly equal number of neutrons. The electrons comprise very little of the mass, so we find the number of protons-and-neutrons in each person:

$$(70 \text{ kg})\left(\frac{1 \text{ u}}{1.66 \times 10^{-27} \text{ kg}}\right) = 4 \times 10^{28} \text{ u}$$

Of these, nearly one half, 2×10^{28}, are protons, and 1% of this is 2×10^{26}, constituting a charge of $(2 \times 10^{26})(1.60 \times 10^{-19} \text{ C}) = 3 \times 10^7 \text{ C}$. Thus, taking an "arm's length" to be half a meter, Feynman's force is

$$F = \frac{k_e q_1 q_2}{r^2} = \frac{\left(8.99 \times 10^9 \text{ N} \cdot \text{m}^2 / \text{C}^2\right)\left(3 \times 10^7 \text{ C}\right)^2}{(0.500 \text{ m})^2} \sim 10^{26} \text{ N}$$

A mass equal to the Earth's in a gravitational field of magnitude 9.80 m/s² would weigh

$$w = mg = (6 \times 10^{24} \text{ kg})(10 \text{ m} / \text{s}^2) \sim 10^{26} \text{ N}$$

Thus, the forces are of the same order of magnitude. ◊

7. Three point charges are located at the corners of an equilateral triangle as shown in Figure P23.7. Calculate the net electric force on the 7.00–μC charge.

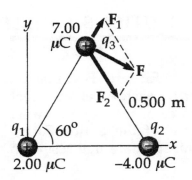

Figure P23.7
(modified)

Solution

G: Gather information. The 7.00–μC charge experiences a repulsive force $\mathbf{F}_1$ due to the 2.00–μC charge, and an attractive force $\mathbf{F}_2$ due to the –4.00–μC charge, where $F_2 = 2F_1$. If we sketch these force vectors, we find that the resultant appears to be about the same magnitude as F_2 and is directed to the right about 30.0° below the horizontal.

O: Organize your approach. We can find the net electric force by adding the two separate forces acting on the 7.00–μC charge. These individual forces can be found by applying Coulomb's law to each pair of charges.

A: Analyze the problem. The force on the 7.00–μC charge by the 2.00–μC charge is

$$\mathbf{F}_1 = k_e \frac{q_1 q_2}{r^2} \hat{\mathbf{r}}$$

$$\mathbf{F}_1 = \frac{\left(8.99 \times 10^9 \text{ N}\cdot\text{m}^2/\text{C}^2\right)\left(7.00 \times 10^{-6} \text{ C}\right)\left(2.00 \times 10^{-6} \text{ C}\right)}{(0.500 \text{ m})^2}(\cos 60°\mathbf{i} + \sin 60°\mathbf{j})$$

$$\mathbf{F}_1 = (0.252\mathbf{i} + 0.436\mathbf{j}) \text{ N}$$

Similarly, the force on the 7.00–μC by the –4.00–μC charge is

$$F_2 = k_e \frac{q_1 q_3}{r^2} \hat{r}$$

$$F_2 = -\left(8.99 \times 10^9 \ \frac{N \cdot m^2}{C^2}\right) \frac{\left(7.00 \times 10^{-6} \ C\right)\left(-4.00 \times 10^{-6} \ C\right)}{\left(0.500 \ m\right)^2} \left(\cos 60° i - \sin 60° j\right)$$

$$F_2 = \left(0.503i - 0.872j\right) N$$

Thus, the total force on the 7.00–μC, expressed as a set of components, is

$$F = F_1 + F_2 = \left(0.755i - 0.436j\right) N \qquad \lozenge$$

We can also write the total force as:

$$F = \sqrt{\left(0.755 \ N\right)^2 + \left(0.436 \ N\right)^2} \quad at \quad \tan^{-1}\left(\frac{0.436 \ N}{0.755 \ N}\right) \ below \ the \ +x \ axis$$

$$F = 0.872 \ N \ at \ 30.0° \ below \ the \ +x \ axis \qquad \lozenge$$

L: **Learn from your efforts.** Our calculated answer agrees with our initial estimate. An equivalent approach to this problem would be to find the net electric field due to the two lower charges and apply F=qE to find the force on the upper charge in this electric field.

11. What are the magnitude and direction of the electric field that will balance the weight of (a) an electron and (b) a proton? (Use the data in Table 23.1.)

Solution If the forces balance, then $m\mathbf{g} + q\mathbf{E} = 0$.

(a) For an electron:

$$\mathbf{E} = -\frac{m\mathbf{g}}{q} = \frac{-(9.11 \times 10^{-31} \text{ kg})(9.80 \text{ m}/\text{s}^2)(-\mathbf{j})}{-1.602 \times 10^{-19} \text{ C}} = (-5.57 \times 10^{-11} \text{ }\mathbf{j}) \text{ N}/\text{C} \qquad \lozenge$$

(b) For a proton:

$$\mathbf{E} = -\frac{m\mathbf{g}}{q} = \frac{-(1.67 \times 10^{-27} \text{ kg})(9.80 \text{ m}/\text{s}^2)(-\mathbf{j})}{+1.602 \times 10^{-19} \text{ C}} = (1.02 \times 10^{-7} \text{ }\mathbf{j}) \text{ N}/\text{C} \qquad \lozenge$$

17. Three equal positive charges q are at the corners of an equilateral triangle of side a, as shown in Figure P23.17. (a) Assume that the three charges together create an electric field. Find the location of a point (other than ∞) where the electric field is zero. (**Hint:** Sketch the field lines in the plane of the charges.) (b) What are the magnitude and direction of the electric field at P due to the two charges at the base?

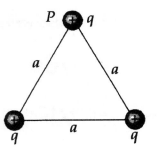

Figure P23.17

Solution

G: The electric field has the general appearance shown by the black arrows in the figure to the right. This drawing indicates that $\mathbf{E} = 0$ at the center of the triangle, since a small positive charge placed at the center of this triangle will be pushed away from each corner equally strongly. This fact could be verified by vector addition as in part (b) below. $\lozenge$

The electric field at point P should be directed upwards and about twice the magnitude of the electric field due to just one of the lower charges as shown in Figure P23.17. For part (b), we must ignore the effect of the charge at point P, because a charge cannot exert a force on itself.

O: The electric field at point P can be found by adding the electric field vectors due to each of the two lower point charges: $\mathbf{E} = \mathbf{E}_1 + \mathbf{E}_2$

A: (b) The electric field from a point charge is

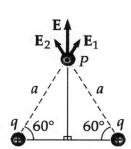

$$\mathbf{E} = k_e \frac{q}{r^2} \hat{\mathbf{r}}$$

As shown in the figure to the right,

$$\mathbf{E}_1 = k_e \frac{q}{a^2} \text{ to the right and upward at } 60°$$

$$\mathbf{E}_2 = k_e \frac{q}{a^2} \text{ to the left and upward at } 60°$$

$$\mathbf{E} = \mathbf{E}_1 + \mathbf{E}_2 = k_e \frac{q}{a^2} \left[\left(\cos 60° \mathbf{i} + \sin 60° \mathbf{j} \right) + \left(-\cos 60° \mathbf{i} + \sin 60° \mathbf{j} \right) \right]$$

$$\mathbf{E} = k_e \frac{q}{a^2} \left[2 \left(\sin 60° \mathbf{j} \right) \right] = 1.73 k_e \frac{q}{a^2} \mathbf{j} \qquad \lozenge$$

L: The net electric field at point P is indeed nearly twice the magnitude due to a single charge and is entirely vertical as expected from the symmetry of the configuration. In addition to the center of the triangle, the gray electric field lines in the figure above indicate three other points near the middle of each leg of the triangle where $\mathbf{E} = 0$, but they are more difficult to find mathematically.

27. A uniformly charged ring of radius 10.0 cm has a total charge of 75.0 μC. Find the electric field on the axis of the ring at (a) 1.00 cm, (b) 5.00 cm, (c) 30.0 cm, and (d) 100 cm from the center of the ring.

Solution

Using the result of Example 23.8:

$$E = \frac{k_e x Q}{\left(x^2 + a^2\right)^{3/2}}$$

$$E = \frac{\left(8.99 \times 10^9 \text{ N} \cdot \text{m}^2 / \text{C}^2\right)\left(75.0 \times 10^{-6} \text{ C}\right)x}{\left(x^2 + (0.100 \text{ m})^2\right)^{3/2}}$$

$$E = \frac{\left(6.75 \times 10^5 \text{ N} \cdot \text{m}^2 / \text{C}\right)x}{\left(x^2 + (0.100 \text{ m})^2\right)^{3/2}}$$

Now, using your calculator,

(a)   At $x = 0.0100$ m, $\mathbf{E} = 6.64 \times 10^6 \mathbf{i} \text{ N} / \text{C}$ ◊

(b)   At $x = 0.0500$ m, $\mathbf{E} = 2.41 \times 10^7 \mathbf{i} \text{ N} / \text{C}$ ◊

(c)   At $x = 0.300$ m, $\mathbf{E} = 6.40 \times 10^6 \mathbf{i} \text{ N} / \text{C}$ ◊

(d)   At $x = 1.00$ m, $\mathbf{E} = 6.64 \times 10^5 \mathbf{i} \text{ N} / \text{C}$ ◊

33. A uniformly charged insulating rod of length 14.0 cm is bent into the shape of a semicircle, as shown in Figure P23.33. The rod has a total charge of –7.50 μC. Find the magnitude and direction of the electric field at O, the center of the semicircle.

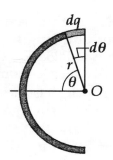

Solution

Let λ be the charge per unit length.

Then, $dq = \lambda ds = \lambda r d\theta$

Figure P23.33 (modified)

and $dE = \dfrac{kdq}{r^2}$

In component form, $E_y = 0$ (from symmetry)

$$dE_x = dE \cos \theta$$

Integrating, $E_x = \displaystyle\int dE_x = \int \dfrac{k\lambda r \cos \theta}{r^2} d\theta = \dfrac{k\lambda}{r} \int_{-\pi/2}^{\pi/2} \cos \theta \, d\theta = \dfrac{2k\lambda}{r}$

But $Q_{total} = \lambda \ell$ where $\ell = 0.140$ m   and $r = \ell / \pi$

Thus, $E_x = \dfrac{2\pi kQ}{\ell^2} = \dfrac{(2\pi)\left(8.99 \times 10^9 \text{ N} \cdot \text{m}^2 / \text{C}^2\right)(-7.50 \times 10^{-6} \text{ C})}{(0.140 \text{ m})^2}$

$$\mathbf{E} = \left(-2.16 \times 10^7 \text{ N} / \text{C}\right)\mathbf{i} \qquad \Diamond$$

35. A thin rod of length ℓ and uniform charge per unit length λ lies along the x axis, as shown in Figure P23.35. (a) Show that the electric field at P, a distance y from the rod, along the perpendicular bisector has no x component and is given by $E = 2k_e\lambda \sin \theta_0 / y$. (b) Using your result to part (a), show that the field of a rod of infinite length is $E = 2k_e\lambda / y$. (**Hint:** First calculate the field at P due to an element of length dx, which has a charge $\lambda\, dx$. Then change variables from x to θ, using the facts that $x = y \tan \theta$ and $dx = y \sec^2 \theta\, d\theta$, and integrate over θ.)

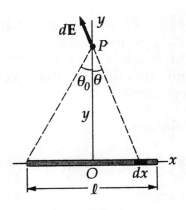

Figure P23.35
(modified)

Solution (a) The segment of rod from x to $(x + dx)$ has a charge of $\lambda\, dx$, and creates an electric field

$$d\mathbf{E} = \frac{k_e dq}{r^2}\hat{\mathbf{r}} = \frac{k_e\lambda\, dx}{y^2 + x^2} \quad \text{upward along the line from } dq \text{ to } P.$$

This bit of field has an x component

$$dE_x = \frac{k_e\lambda\, dx}{y^2 + x^2}(-\sin \theta) = \frac{-k_e\lambda x\, dx}{(y^2 + x^2)^{3/2}}$$

and a y component,

$$dE_y = \frac{k_e\lambda\, dx}{y^2 + x^2}(\cos \theta) = \frac{k_e\lambda y\, dx}{(y^2 + x^2)^{3/2}}$$

The total field has an x component

$$E_x = \int_{\text{All } q} dE_x = \int_{-\ell/2}^{\ell/2} -\frac{k_e\lambda x\, dx}{(y^2 + x^2)^{3/2}}$$

To integrate, make a change of variables to θ, such that $\quad x = y \tan \theta$
When $x = -\ell/2$, $\theta = -\theta_0$, and when $x = \ell/2$, $\theta = \theta_0$. Further,

$$(y^2 + x^2)^{3/2} = (y^2 + y^2 \tan^2 \theta)^{3/2} = y^3 \sec^3 \theta \quad \text{and} \quad dx = y \sec^2 \theta\, d\theta$$

Thus
$$E_x = \int_{-\theta_0}^{\theta_0} -\frac{k_e \lambda y (\tan \theta) y (\sec^2 \theta) \, d\theta}{y^3 \sec^3 \theta} = -\frac{k_e \lambda}{y} \int_{-\theta_0}^{\theta_0} \sin \theta \, d\theta$$

$$= +\frac{k_e \lambda}{y} \cos \theta \Big]_{-\theta_0}^{+\theta_0}$$

$$= \frac{k_e \lambda}{y} (\cos \theta_0 - \cos(-\theta_0)) = \frac{k_e \lambda}{y} (\cos \theta_0 - \cos \theta_0) = 0$$

This answer has to be zero because each segment of rod on the left produces a field whose contribution cancels out that of the corresponding segment of rod on the right. But every incremental bit of charge produces at P a contribution to the field with upward y component:

$$E_y = \int_{\text{All } q} dE_y = \int_{-\ell/2}^{\ell/2} \frac{k_e \lambda y \, dx}{(y^2 + x^2)^{3/2}}$$

Think of k_e, λ, and y as known constants. Now E_y is the unknown and x is the variable of integration, which we again change to θ, with $x = y \tan \theta$:

$$E_y = \int_{-\theta_0}^{\theta_0} \frac{k_e \lambda y (y \sec^2 \theta) \, d\theta}{y^3 \sec^3 \theta} = \frac{k_e \lambda}{y} \int_{-\theta_0}^{\theta_0} \cos \theta \, d\theta = \frac{k_e \lambda}{y} \sin \theta \Big]_{-\theta_0}^{\theta_0}$$

$$= \frac{k_e \lambda}{y} (\sin \theta_0 - \sin(-\theta_0)) = \frac{k_e \lambda}{y} (\sin \theta_0 + \sin \theta_0) = \frac{2 k_e \lambda \sin \theta_0}{y} \qquad \Diamond$$

(b)   As ℓ goes to infinity, θ_0 goes to 90° and $\sin \theta_0$ becomes 1. Then the infinite amount of charge produces a finite field at P:

$$\mathbf{E} = 0\mathbf{i} + \frac{2 k_e \lambda}{y} \mathbf{j} \qquad \Diamond$$

39. A negatively charged rod of finite length has a uniform charge per unit length. Sketch the electric field lines in a plane containing the rod.

Solution

Since the rod has negative charge, field lines point inwards. Any field line points nearly toward the center of the rod at large distances, where the rod would look like just a point charge. The lines curve to reach the rod perpendicular to its surface, where they end at equally-spaced points.

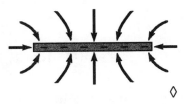

◊

43. A proton accelerates from rest in a uniform electric field of 640 N/C. At some later time, its speed is 1.20×10^6 m/s (nonrelativistic, since v is much less than the speed of light). (a) Find the acceleration of the proton. (b) How long does it take the proton to reach this speed? (c) How far has it moved in this time? (d) What is its kinetic energy at this time?

Solution

(a) $\quad a = \dfrac{F}{m} = \dfrac{qE}{m} = \dfrac{(1.602 \times 10^{-19}\ \text{C})(640\ \text{N}/\text{C})}{1.67 \times 10^{-27}\ \text{kg}} = 6.13 \times 10^{10}\ \text{m}/\text{s}^2$ ◊

(b) $\quad t = \dfrac{\Delta v}{a} = \dfrac{1.20 \times 10^6\ \text{m}/\text{s}}{6.13 \times 10^{10}\ \text{m}/\text{s}^2} = 19.5\ \mu\text{s}$ ◊

(c) $\quad x = v_0 t + \tfrac{1}{2} a t^2 = 0 + \left(\tfrac{1}{2}\right)(6.13 \times 10^{10}\ \text{m}/\text{s}^2)(19.5 \times 10^{-5}\ \text{s})^2 = 11.7\ \text{m}$ ◊

(d) $\quad K = \tfrac{1}{2} m v^2 = \tfrac{1}{2}(1.67 \times 10^{-27}\ \text{kg})(1.20 \times 10^6\ \text{m}/\text{s})^2 = 1.20 \times 10^{-15}\ \text{J}$ ◊

45. The electrons in a particle beam each have a kinetic energy K. What are the magnitude and direction of the electric field that stops these electrons in a distance of d?

Solution

G: We should expect that a larger electric field would be required to stop electrons with greater kinetic energy. Likewise, **E** must be greater for a shorter stopping distance, d. The electric field should be in the same direction as the motion of the negatively charged electrons in order to exert an opposing force that will slow them down.

O: The electrons will experience an electrostatic force $\mathbf{F} = q\mathbf{E}$. Therefore, the work done by the electric field can be equated with the initial kinetic energy since energy should be conserved.

A: The work done on the charge is $\qquad\qquad\qquad W = \mathbf{F} \cdot \mathbf{d} = q\mathbf{E} \cdot \mathbf{d}$

and $\qquad\qquad\qquad\qquad\qquad\qquad\qquad\qquad K_i + W = K_f = 0$

Assuming **v** is in the $+x$ direction, $\qquad\qquad K + (-e)\mathbf{E} \cdot d\mathbf{i} = 0$

$$e\mathbf{E} \cdot (d\mathbf{i}) = K$$

E is therefore in the direction of the electron beam: $\quad \mathbf{E} = \dfrac{K}{ed}\mathbf{i} \qquad\qquad \lozenge$

L: As expected, the electric field is proportional to K, and inversely proportional to d. The direction of the electric field is important; if it were otherwise the electron would speed up instead of slowing down! If the particles were protons instead of electrons, the electric field would need to be directed opposite to **v** in order for the particles to slow down.

47. A proton moves at 4.50×10^5 m/s in the horizontal direction. It enters a uniform vertical electric field with a magnitude of 9.60×10^3 N/C. Ignoring any gravitational effects, find (a) the time it takes the proton to travel 5.00 cm horizontally, (b) its vertical displacement after it has traveled 5.00 cm horizontally, and (c) the horizontal and vertical components of its velocity after it has traveled 5.00 cm horizontally.

Solution E is directed along the y direction, therefore, $a_x = 0$ and $x = v_{xi}t$

(a) $t = \dfrac{x}{v_{xi}} = \dfrac{0.0500 \text{ m}}{4.50 \times 10^5 \text{ m/s}} = 1.11 \times 10^{-7}$ s ◊

(b) $a_y = \dfrac{qE_y}{m} = \dfrac{(1.602 \times 10^{-19} \text{ C})(9.60 \times 10^3 \text{ N/C})}{1.67 \times 10^{-27} \text{ kg}} = 9.20 \times 10^{11}$ m/s^2

$y = v_{yi}t + \tfrac{1}{2}a_y t^2 = \left(\tfrac{1}{2}\right)(9.20 \times 10^{11} \text{ m/s}^2)(1.11 \times 10^{-7} \text{ s})^2 = 5.68$ mm ◊

(c) $v_x = v_{xi} = 4.50 \times 10^5$ m/s ◊

$v_y = v_{yi} + a_y t = 0 + (9.20 \times 10^{11} \text{ m/s}^2)(1.11 \times 10^{-7} \text{ s}) = 1.02 \times 10^5$ m/s ◊

53. A charged cork ball of mass 1.00 g is suspended on a light string in the presence of a uniform electric field, as shown in Fig. P23.53. When $\mathbf{E} = (3.00\mathbf{i} + 5.00\mathbf{j}) \times 10^5$ N/C, the ball is in equilibrium at $\theta = 37.0°$. Find (a) the charge on the ball and (b) the tension in the string.

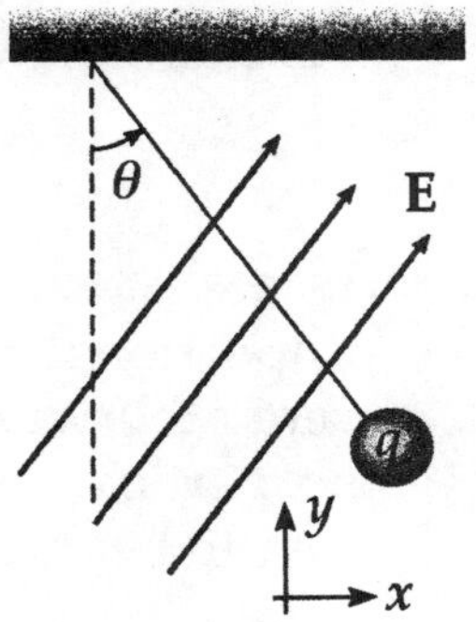

Solution

G: (a) Since the electric force must be in the same direction as **E**, the ball must be positively charged.

Figure P23.53

If we examine the free body diagram that shows the three forces acting on the ball, the sum of which must be zero, we can see that the tension is about half the magnitude of the weight.

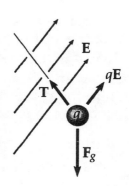

O: The tension can be found from applying Newton's second law to this statics problem (electrostatics, in this case!).   Since the force vectors are in two dimensions, we must apply $\Sigma\mathbf{F} = m\mathbf{a}$ to both the x and y directions.

A: $\Sigma\mathbf{F} = \mathbf{T} + q\mathbf{E} + \mathbf{F}_g = 0$

We are given $E_x = 3.00\times10^5\ \text{N}/\text{C}$ and $E_y = 5.00\times10^5\ \text{N}/\text{C}$

Applying Newton's Second Law in the x and y directions,

$\Sigma F_x = qE_x - T\sin 37.0° = 0$ (1)

$\Sigma F_y = qE_y + T\cos 37.0° - mg = 0$ (2)

Substitute T from Equation (1) into Equation (2):

$$q = \frac{mg}{\left(E_y + \dfrac{E_x}{\tan 37.0°}\right)} = \frac{(1.00\times10^{-3}\ \text{kg})(9.80\ \text{m}/\text{s}^2)}{\left(5.00 + \dfrac{3.00}{\tan 37.0°}\right)\times10^5\ \text{N}/\text{C}} = 1.09\times10^{-8}\ \text{C} \qquad \lozenge$$

(b) Using this result for q in Equation (1), we find that the tension is

$$T = \frac{qE_x}{\sin 37.0°} = 5.44\times10^{-3}\ \text{N} \qquad\qquad \lozenge$$

L: The tension is slightly more than half the weight of the ball ($F_g = 9.80\times10^{-3}\ \text{N}$) so our result seems reasonable based on our initial prediction.

54. A charged cork ball of mass m is suspended on a light string in the presence of a uniform electric field, as shown in Figure P23.53. When $\mathbf{E} = (A\mathbf{i} + B\mathbf{j})\,\text{N}/\text{C}$, where A and B are positive numbers, the ball is in equilibrium at the angle θ. Find (a) the charge on the ball and (b) the tension in the string.

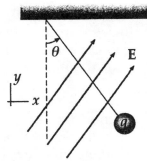

Figure P23.53

Solution

G: This is the general version of the preceding problem. The known quantities are A, B, m, g, and θ. The unknowns are q and T.

O: The approach to this problem should be the same as for the last problem, but without numbers to substitute for the variables. Likewise, we can use the free body diagram given in the solution to problem 53.

A: Again, Newton's second law gives us

$$-T\sin\theta + qA = 0 \qquad (1)$$

and

$$+T\cos\theta + qB - mg = 0 \qquad (2)$$

(a) Substituting $T = \dfrac{qA}{\sin\theta}$, into Eq. (2),

$$\frac{qA\cos\theta}{\sin\theta} + qB = mg$$

Isolating q on the left,

$$q = \frac{mg}{\left(A\cot\theta + B\right)} \qquad \lozenge$$

(b) Substituting this value into Eq. (1),

$$T = \frac{mgA}{\left(A\cos\theta + B\sin\theta\right)} \qquad \lozenge$$

L: If we had solved this general problem first, we would only need to substitute the appropriate values in the equations for q and T to find the numerical results needed for problem 53. If you find this problem more difficult than problem 53, the little list at the Gather step is useful. It shows what symbols to think of as known data, and what to consider unknown. The list is a guide for deciding what to solve for in the Analysis step, and for recognizing when we have an answer.

63. Two small spheres of mass m are suspended from strings of length ℓ that are connected at a common point. One sphere has charge Q; the other has charge $2Q$. Assume that the angles θ_1 and θ_2 that the strings make with the vertical are small. (a) How are θ_1 and θ_2 related? (b) Show that the distance r between the spheres is

$$r \cong \left(\frac{4k_eQ^2\ell}{mg}\right)^{1/3}$$

Solution

(a) The spheres have different charges, but each exerts an equal force on the other, given by $F_e = k_e(Q)(2Q)/r^2$, where r is the distance between them. Since their masses are equal, $\theta_1 = \theta_2$. ◊

(b) For equilibrium, $\Sigma F_y = 0$: $T\cos\theta - mg = 0$

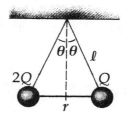

Thus $T = \dfrac{mg}{\cos\theta}$

$\Sigma F_x = 0$: $F_e - T\sin\theta = 0$

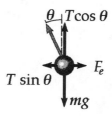

Substituting for T, $F_e = \dfrac{mg\sin\theta}{\cos\theta} = mg\tan\theta$

For small angles, $\tan\theta \cong \sin\theta = \dfrac{r}{2\ell}$

Therefore, $F_e \cong mg\dfrac{r}{2\ell}$

The force F_e is $\dfrac{k_eQ(2Q)}{r^2} \cong mg\dfrac{r}{2\ell}$

so that $4k_eQ^2\ell \cong mg\,r^3$

and $r \cong \left(\dfrac{4k_eQ^2\ell}{mg}\right)^{1/3}$ ◊

73. A negatively charged particle $-q$ is placed at the center of a uniformly charged ring, where the ring has a total positive charge Q as shown in Example 23.8. The particle, confined to move along the x axis, is displaced a **small** distance x along the axis (where $x \ll a$) and released. Show that the particle oscillates with simple harmonic motion with a frequency

$$f = \frac{1}{2\pi}\left(\frac{k_e qQ}{ma^3}\right)^{1/2}$$

Solution   From Example 23.8, the electric field at points along the x axis is

$$\mathbf{E} = \frac{k_e x Q \mathbf{i}}{(x^2 + a^2)^{3/2}}$$

The field is zero at $x = 0$, so the negative charge is in equilibrium at this point. When it is displaced by an amount x that is small compared to a,

$$\Sigma \mathbf{F} = m\frac{d^2\mathbf{x}}{dt^2}: \qquad\qquad (-q)\frac{k_e x Q \mathbf{i}}{a^3} = m\frac{d^2\mathbf{x}}{dt^2}$$

The particle's acceleration is proportional to its distance (x) from the equilibrium position and is oppositely directed, so it moves in simple harmonic motion.   ◊

Since $d^2x/dt^2 = -\omega^2 x$, $\qquad\qquad \omega = \left(\frac{k_e qQ}{ma^3}\right)^{1/2} = 2\pi f$

and $\qquad\qquad\qquad\qquad f = \frac{1}{2\pi}\left(\frac{k_e qQ}{ma^3}\right)^{1/2}$ ◊

Chapter 24

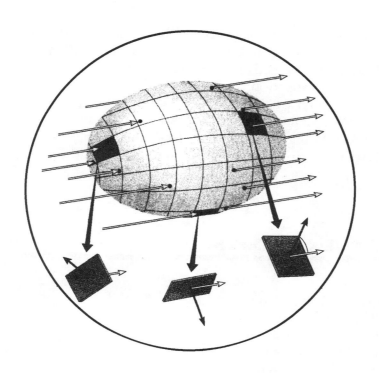

Gauss's Law

GAUSS'S LAW

INTRODUCTION

In the preceding chapter we showed how to calculate the electric field generated by a given charge distribution from Coulomb's law. This chapter describes an alternative procedure for calculating electric fields known as **Gauss's law.** This formulation is based on the fact that the fundamental electrostatic force between point charges is an inverse-square law. Although Gauss's law is a consequence of Coulomb's law, Gauss's law is much more convenient for calculating the electric field of highly symmetric charge distributions. Furthermore, Gauss's law serves as a guide for understanding more complicated problems.

EQUATIONS AND CONCEPTS

The **electric flux** is a measure of the number of electric field lines that penetrate some surface. For a **plane surface** in a **uniform field**, the flux depends on the angle between the normal to the surface and the direction of the field.

$$\Phi_E = EA\cos\theta \qquad (24.2)$$

In the case of a **general surface** in the region of a **nonuniform** field, the flux is calculated by integrating the normal component of the field over the surface in question.

$$\Phi_E = \int_{\text{surface}} \mathbf{E} \cdot d\mathbf{A} \qquad (24.3)$$

Gauss's law states that when Φ_E (Eq. 24.3) is evaluated over a **closed** surface (gaussian surface), the result equals **the net charge enclosed by the surface** divided by the constant ϵ_0. In this equation, the symbol $\oint$ indicates that the integral must be evaluated over a **closed** surface.

$$\Phi_E = \oint \mathbf{E} \cdot d\mathbf{A} = \frac{q_{in}}{\epsilon_0} \qquad (24.6)$$

The field exterior to a charged sphere of radius a is equivalent to that of a point charge located at the center of the sphere.

$$E = k_e \frac{Q}{r^2} \qquad \text{for} \quad r > a$$

At an **interior point** of a uniformly charged sphere, the electric field is proportional to the distance from the center of the sphere.

$$E = k_e \left(\frac{Q}{a^3} \right) r \qquad \text{for} \quad r < a$$

The electric field due to a uniformly charged, **nonconducting, infinite plate** is uniform everywhere.

$$E = \frac{\sigma}{2\epsilon_0} \qquad (24.8)$$

The electric field just outside the surface of a charged conductor in equilibrium can be expressed in terms of the surface charge density on the conductor. Just outside the conductor, the field is normal to the surface.

$$E_n = \frac{\sigma}{\epsilon_0} \qquad (24.9)$$

SUGGESTIONS, SKILLS, AND STRATEGIES

Gauss's law is a very powerful theorem which relates any charge distribution to the resulting electric field at any point in the vicinity of the charge. In this chapter you should learn how to apply Gauss's law to those cases in which the charge distribution has a sufficiently high degree of symmetry. As you review the examples presented in Section 24.3 of the text, observe how each of the following steps have been included in the application of the equation $\oint \mathbf{E} \cdot d\mathbf{A} = q/\epsilon_0$ to that particular situation.

- The gaussian surface should be chosen to have the **same symmetry as the charge distribution.**

- The dimensions of the surface must be such that the surface includes the point where the electric field is to be calculated.

- From the symmetry of the charge distribution, you should be able to correctly describe the direction of the electric field vector, $\mathbf{E}$, relative to the direction of an element of surface area vector, $d\mathbf{A}$, which points outward from each region of the gaussian surface.

- From the symmetry of the charge distribution, you should also be able to identify one or more portions of the closed surface (and in some cases the entire surface) over which the magnitude of $\mathbf{E}$ remains constant.

- Write $\mathbf{E} \cdot d\mathbf{A}$ as $E\, dA \cos\theta$ and separate the gaussian surface into regions such that over each region one of the following is true:

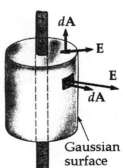

  $\mathbf{E} \perp d\mathbf{A}$ so that $E\, dA \cos\theta = 0$ (as is the case over each end of the cylindrical gaussian surface in the figure.

  $\mathbf{E} \,||\, d\mathbf{A}$ so that $E\, dA \cos\theta = E\, dA$ (as is the case over the curved portion of the cylindrical gaussian surface in the figure to the right).

  $\mathbf{E}$ and $d\mathbf{A}$ are oppositely directed so that $E\, dA \cos\theta = -E\, dA$.

  $\mathbf{E} = 0$ (as is the case over a surface inside a conductor).

- If the gaussian surface has been chosen and subdivided so that the magnitude of **E** is **constant** over those regions where $\mathbf{E} \cdot d\mathbf{A} = E\,dA$, then over each of those regions

$$\int \mathbf{E} \cdot d\mathbf{A} = E \int dA = EA$$

- The total charge enclosed by the gaussian surface is $q = \int dq$. It is often convenient to represent the charge distribution in terms of the charge density ($dq = \lambda\,dx$ for a line of charge, $dq = \sigma\,dA$ for a surface of charge, or $dq = \rho\,dV$ for a volume of charge). The integral of dq is then evaluated only over that length, area, or volume which includes that portion of the charge **inside** the gaussian surface.

- Once the left and right sides of Gauss's law have been evaluated, you can calculate the electric field on the gaussian surface, assuming the charge distribution is given in the problem. Conversely, if the electric field is known, you can calculate the charge distribution that produces the field.

REVIEW CHECKLIST

▷ Calculate the **electric** flux through a surface; in particular, find the net electric flux through a **closed** surface.

▷ Understand that a gaussian surface must be a real or imaginary **closed** surface within a conductor, a dielectric, or in space. Also remember that the net electric flux through a closed gaussian surface is equal to the net charge enclosed by the surface divided by the constant ϵ_0.

▷ Use Gauss's law to evaluate the electric field at points in the vicinity of charge distributions which exhibit spherical, cylindrical, or planar symmetry.

▷ Describe the properties which characterize an electrical conductor in electrostatic equilibrium.

ANSWERS TO SELECTED CONCEPTUAL QUESTIONS

5. If the total charge inside a closed surface is known but the distribution of the charge is unspecified, can you use Gauss's law to find the electric field? Explain.

Answer **No.** If we wish to use Gauss's law to find the electric field, we must be able to bring the electric field out of the integral. This **can only** be done in the rare case that the field is constant both in magnitude and direction. To illustrate this point, consider a sphere that contained a net charge of 100 μC. The charges could be located near the center, or they could all be grouped at the northernmost point within the sphere. In either case, the net electric flux would be the same, but the electric field would vary greatly.

□　　□　　□　　□

10. A person is placed in a large, hollow, metallic sphere that is insulated from ground. If a large charge is placed on the sphere, will the person be harmed upon touching the inside of the sphere? Explain what will happen if the person also has an initial charge whose sign is opposite that of the charge on the sphere.

Answer The metallic sphere is a good conductor, so any excess charge on the sphere will reside on the outside of the sphere. From Gauss's Law, we know that the field **inside** the sphere will then be zero. As a result, when the person touches the inside of the sphere, no charge will be exchanged between the person and the sphere, and the person will not be harmed.

What happens, then, if the person has an initial charge? Regardless of the sign of the person's initial charge, the charges in the conducting surface will redistribute themselves to maintain a net zero charge within the **conducting metal**. Thus, if the person has a 5.00 μC charge on his skin, exactly -5.00 μC will gather on the inner surface of the sphere, so that the electric field inside the metal will be zero. When the person touches the metallic sphere then, h e will receive a shock due to the charge on his own skin.

□　　□　　□　　□

SOLUTIONS TO SELECTED END-OF-CHAPTER PROBLEMS

3. A 40.0-cm-diameter loop is rotated in a uniform electric field until the position of maximum electric flux is found. The flux in this position is measured to be 5.20×10^5 N·m^2/C. What is the magnitude of the electric field?

Solution We are given $\Phi_E = 5.20 \times 10^5$ N·m^2/C and $r = 20.0$ cm $= 0.200$ m. The flux is $\Phi_E = \mathbf{E} \cdot \mathbf{A}$, and is maximum when $\mathbf{E}$ is parallel to the vector normal to the area:

$$\Phi_E = \mathbf{E} \cdot \mathbf{A} = EA\cos(0) = EA = E\left(\pi r^2\right)$$

Therefore, $E = \dfrac{\Phi_E}{\pi r^2} = \dfrac{5.20 \times 10^5 \text{ N·m}^2/\text{C}}{\pi(0.200 \text{ m})^2} = 41.4 \times 10^5$ N/C $\qquad \Diamond$

11. The following charges are located inside a submarine: 5.00 μC, –9.00 μC, 27.0 μC, and –84.0 μC. (a) Calculate the net electric flux through the submarine. (b) Is the number of electric field lines leaving the submarine greater than, equal to, or less than the number entering it?

Solution The total charge within the closed surface is

5.00 μC – 9.00 μC + 27.0 μC – 84.0 μC = –61.0 μC

so the total electric flux is

$$\Phi_E = \frac{q}{\epsilon_0} = \frac{-61.0 \times 10^{-6} \text{ C}}{(8.85 \times 10^{-12} \text{ C}^2/\text{N·m}^2)} = -6.89 \times 10^6 \text{ N·m}^2/\text{C} \qquad \Diamond$$

The minus sign means that more lines enter the surface than leave it.

15. A point charge Q is located just above the center of the flat face of a hemisphere of radius R, as shown in Figure P24.15. What is the electric flux (a) through the curved surface and (b) through the flat face?

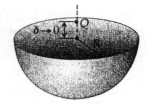

Solution

Figure P24.15

G: From Gauss's law, the flux through a sphere with a point charge in it should be Q/ϵ_0, so we should expect the electric flux through a hemisphere to be half this value: $\Phi_{\text{curved}} = Q/2\,\epsilon_0$. Since the flat section appears like an infinite plane to a point just above its surface so that half of all the field lines from the point charge are intercepted by the flat surface, the flux through this section should also equal $Q/2\,\epsilon_0$.

O: We can apply the definition of electric flux directly for part (a) and then use Gauss's law to find the flux for part (b).

A: (a) With δ very small, all points on the hemisphere are nearly at distance R from the charge, so the field everywhere on the curved surface is $k_e Q / R^2$ radially outward (normal to the surface). Therefore, the flux is this field strength times the area of half a sphere:

$$\Phi_{\text{curved}} = \int \mathbf{E} \cdot d\mathbf{A} = E_{\text{local}} A_{\text{hemisphere}}$$

$$\Phi_{\text{curved}} = \left(k_e \frac{Q}{R^2} \right)\left(\tfrac{1}{2} \right)\left(4\pi R^2 \right) = \frac{1}{4\pi\,\epsilon_0} Q(2\pi) = \frac{Q}{2\,\epsilon_0} \qquad \Diamond$$

(b) The closed surface encloses zero charge so Gauss's law gives

$$\Phi_{\text{curved}} + \Phi_{\text{flat}} = 0 \qquad \text{or} \qquad \Phi_{\text{flat}} = -\Phi_{\text{curved}} = \frac{-Q}{2\,\epsilon_0} \qquad \Diamond$$

L: The direct calculations of the electric flux agree with our predictions, except for the negative sign in part (b), which comes from the fact that the area unit vector is defined as pointing outward from an enclosed surface, and in this case, the electric field has a component in the opposite direction (down).

29. Consider a long cylindrical charge distribution of radius R with a uniform charge density ρ. Find the electric field at distance r from the axis where $r < R$.

Solution

G: According to Gauss's law, only the charge enclosed within the gaussian surface of radius r needs to be considered. The amount of charge within the gaussian surface will certainly increase as ρ and r increase, but the area of this gaussian surface will also increase, so it is difficult to predict which of these two competing factors will more strongly affect the electric field strength.

O: We can find the general equation for E from Gauss's law.

A: If ρ is positive, the field must be radially outward. Choose as the gaussian surface a cylinder of length L and radius r, contained inside the charged rod. Its volume is $\pi r^2 L$ and it encloses charge $\rho \pi r^2 L$. The circular end caps have no electric flux through them; there $\mathbf{E} \cdot d\mathbf{A} = EdA\cos 90.0° = 0$. The curved surface has $\mathbf{E} \cdot d\mathbf{A} = EdA\cos 0°$, and E must be the same strength everywhere over the curved surface.

Gauss's law, $\oint \mathbf{E} \cdot d\mathbf{A} = \dfrac{q}{\epsilon_0}$, becomes $E \int\limits_{\substack{\text{Curved} \\ \text{Surface}}} dA = \dfrac{\rho \pi r^2 L}{\epsilon_0}$

Now the lateral surface area of the cylinder is $2\pi rL$: $E(2\pi r)L = \dfrac{\rho \pi r^2 L}{\epsilon_0}$

Thus, $\mathbf{E} = \dfrac{\rho r}{2\epsilon_0}$ radially away from the cylinder axis $\lozenge$

L: As we expected, the electric field will increase as ρ increases, and we can now see that E is also proportional to r. For the region outside the cylinder ($r > R$), we should expect the electric field to decrease as r increases, just like for a line of charge.

31. Consider a thin spherical shell of radius 14.0 cm with a total charge of 32.0 μC distributed uniformly on its surface. Find the electric field (a) 10.0 cm and (b) 20.0 cm from the center of the charge distribution.

Solution (a) A gaussian sphere, radius 10.0 cm, encloses 0 charge: **E** = 0 ◊

(b) For a gaussian sphere of radius 20.0 cm, $$\oint \mathbf{E} \cdot d\,\mathbf{A} = \frac{q_{in}}{\epsilon_0}$$

The field is radially outward, and $$E\left(4\pi r^2\right) = \frac{q}{\epsilon_0}$$

$$E = \frac{k_e q}{r^2} = \frac{8.99 \times 10^9 \text{ N} \cdot \text{m}^2 \, (32.0 \times 10^{-6} \text{ C})}{\text{C}^2 \, (0.200 \text{ m})^2} = 7.19 \times 10^6 \text{ N/C}$$

So $$\mathbf{E} = (7.19 \times 10^6 \text{ N/C})\hat{\mathbf{r}}$$ ◊

---

35. A uniformly charged, straight filament 7.00 m in length has a total positive charge of 2.00 μC. An uncharged cardboard cylinder 2.00 cm in length and 10.0 cm in radius surrounds the filament at its center, with the filament as the axis of the cylinder. Using reasonable approximations, find (a) the electric field at the surface of the cylinder and (b) the total electric flux through the cylinder.

Solution The approximation in this case is that the filament length is so large when compared to the cylinder length that the "infinite line" of charge can be assumed.

(a) $$E = \frac{2k_e \lambda}{r} \quad \text{where} \quad \lambda = \frac{2.00 \times 10^{-6} \text{ C}}{7.00 \text{ m}} = 2.86 \times 10^{-7} \text{ C/m}$$

so $$E = \frac{(2)(8.99 \times 10^9 \text{ N} \cdot \text{m}^2 / \text{C})(2.86 \times 10^{-7} \text{ C/m})}{0.100 \text{ m}} = 5.14 \times 10^4 \text{ N/C}$$ ◊

(b) $\Phi_E = 2\pi rLE = 2\pi rL\left(\dfrac{2k_e\lambda}{r}\right) = 4\pi k_e\lambda L$

so $\Phi_E = 4\pi\left(8.99\times10^9\ \dfrac{N\cdot m^2}{C^2}\right)\left(2.86\times10^{-7}\ \dfrac{C}{m}\right)(0.0200\ m) = 6.46\times10^2\ \dfrac{N\cdot m^2}{C}$ ◊

---

37. A large flat sheet of charge has a charge per unit area of $9.00\ \mu C/m^2$. Find the electric field intensity just above the surface of the sheet, measured from its midpoint.

Solution For a large insulating sheet, **E** will be perpendicular to the sheet, and will have a magnitude of

$$E = \dfrac{\sigma}{2\,\epsilon_0} = 2\pi k_e\sigma = (2\pi)\left(8.99\times10^9\ \dfrac{N\cdot m^2}{C^2}\right)(9.00\times10^{-6}\ C/m^2)$$

so $\qquad E = 5.08\times10^5\ N/C$ ◊

---

39. A long, straight metal rod has a radius of 5.00 cm and a charge per unit length of 30.0 nC/m. Find the electric field (a) 3.00 cm, (b) 10.0 cm, and (c) 100 cm from the axis of the rod, where distances are measured perpendicular to the rod.

Solution (a) Inside the conductor, $\qquad E = 0$ ◊
Outside the conductor, $\quad E = 2k_e\lambda/r$

(b) At $r = 0.100$ m,

$$E = 2\dfrac{k_e\lambda}{r} = 2\dfrac{(8.99\times10^9\ N\cdot m^2/C^2)(30.0\times10^{-9}\ C/m)}{0.100\ m} = 5.40\times10^3\ N/C\ ◊$$

(c) At $r = 1.00$ m,

$$E = 2\frac{k_e \lambda}{r} = 2\frac{\left(8.99 \times 10^9 \text{ N} \cdot \text{m}^2 / \text{C}^2\right)\left(30.0 \times 10^{-9} \text{ C} / \text{m}\right)}{1.00 \text{ m}} = 540 \text{ N} / \text{C} \qquad \Diamond$$

45. A long, straight wire is surrounded by a hollow metal cylinder whose axis coincides with that of the wire. The wire has a charge per unit length of λ, and the cylinder has a net charge per unit length of 2λ. From this information, use Gauss's law to find (a) the charge per unit length on the inner and outer surfaces of the cylinder and (b) the electric field outside the cylinder, a distance r from the axis.

Solution

(a) Use a cylindrical gaussian surface S_1 within the conducting cylinder.

$E = 0;$ Thus, $\oint E_n dA = \left(\frac{1}{\epsilon_0}\right)q_{in} = 0$

and $\lambda_{inner} = -\lambda$ $\Diamond$

Also, $\lambda_{inner} + \lambda_{outer} = 2\lambda$

Thus, $\lambda_{outer} = 3\lambda$ $\Diamond$

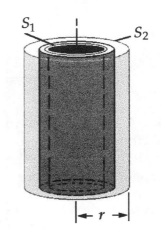

(b) For a gaussian surface S_2 outside the conducting cylinder,

$$\oint E_n dA = \left(\frac{1}{\epsilon_0}\right)q_{in}, \text{ or } E(2\pi rL) = \frac{1}{\epsilon_0}(\lambda - \lambda + 3\lambda)L, \text{ and } E = \frac{3\lambda}{2\pi \epsilon_0 r} \qquad \Diamond$$

47. A thin conducting plate 50.0 cm on a side lies in the xy plane. If a total charge of 4.00×10^{-8} C is placed on the plate, find (a) the charge density on the plate, (b) the electric field just above the plate, and (c) the electric field just below the plate.

Solution   In this problem ignore "edge" effects and assume that the total charge distributes uniformly over each side of the plate (one half the total charge on each side).

(a) $\sigma = \dfrac{q}{A} = \left(\dfrac{1}{2}\right)\dfrac{4.00 \times 10^{-8}\ C}{(0.500\ m)^2} = 8.00 \times 10^{-8}\ C/m^2$ ◊

(b)   Just above the plate,

$$E = \dfrac{\sigma}{\epsilon_0} = \dfrac{8.00 \times 10^{-8}\ C/m^2}{(8.85 \times 10^{-12}\ C^2/N \cdot m^2)} = 9.04 \times 10^3\ N/C\ \text{upward}$$ ◊

(c)   Just below the plate,

$$E = \dfrac{\sigma}{\epsilon_0} = 9.04 \times 10^3\ N/C\ \text{downward}$$ ◊

---

51. A sphere of radius R surrounds a point charge Q, located at its center. (a) Show that the electric flux through a circular cap of half-angle θ (Fig. P24.51) is

$$\Phi_E = \dfrac{Q}{2\,\epsilon_0}(1 - \cos\theta)$$

What is the flux for (b) $\theta = 90°$ and (c) $\theta = 180°$?

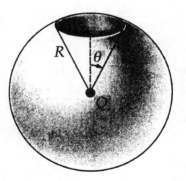

Figure P24.51

Solution

(a) The electric field of the point charge has constant strength k_eQ/R^2 over the cap and points radially outward. To find the area of the curved cap, we think of it as formed of rings, each of radius $r = R \sin \phi$, where ϕ ranges from 0 to θ. The width of each ring is $ds = R\,d\phi$, so its area is the product of its two perpendicular dimensions,

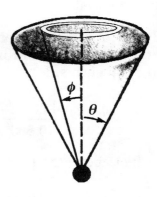

$$dA = (2\pi r)\,ds = 2\pi\,(R \sin \phi)(R\,d\phi)$$

The whole cap has an area of:

$$A = \int dA = \int_0^\theta 2\pi R^2 \sin \phi\, d\phi = 2\pi R^2(-\cos\phi)\Big]_0^\theta = 2\pi R^2(-\cos\theta + 1)$$

The flux through it is

$$\Phi_E = \int \mathbf{E}\cdot d\mathbf{A} = \int E\,dA \cos 0° = E\int dA = EA$$

$$= \frac{k_eQ}{R^2}2\pi R^2(1-\cos\theta) = \left(\frac{1}{4\pi\,\epsilon_0}\right)(2\pi Q)(1-\cos\theta) = \frac{Q}{2\,\epsilon_0}\,(1-\cos\theta) \quad \lozenge$$

(b) For $\theta = 90°$, the cap is a hemisphere and intercepts half the flux from the charge:

$$\Phi_E = \frac{Q}{2\,\epsilon_0}\,(1-\cos 90°) = \frac{Q}{2\,\epsilon_0} \qquad \lozenge$$

(c) For $\theta = 180°$, the cap is a full sphere and all the field lines go through it:

$$\Phi_E = \frac{Q}{2\,\epsilon_0}\,(1-\cos 180°) = \frac{Q}{\epsilon_0} \qquad \lozenge$$

55. A solid, insulating sphere of radius a has a uniform charge density ρ and a total charge Q. Concentric with this sphere is an uncharged, conducting hollow sphere whose inner and outer radii are b and c, as shown in Figure P24.55. (a) Find the magnitude of the electric field in the regions $r < a$, $a < r < b$, $b < r < c$, and $r > c$. (b) Determine the induced charge per unit area on the inner and outer surfaces of the hollow sphere.

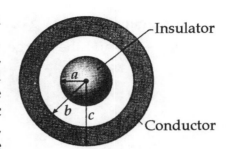

Insulator

Conductor

Figure P24.55

Solution (a) Choose as the gaussian surface a concentric sphere of radius r. The electric field will be perpendicular to its surface, and will be uniform in strength over its surface.

The sphere of radius $r < a$ encloses charge $\frac{4}{3}\rho\pi r^3$

so $\Phi_E = q/\epsilon_0$ becomes $E(4\pi r^2) = \left(\frac{4}{3}\rho\pi r^3\right)/\epsilon_0$

and $E = \rho r/3\epsilon_0$ ◊

For $a < r < b$, we have $E(4\pi r^2) = \frac{\rho 4}{3}\pi a^3 / \epsilon_0 = \dfrac{Q}{\epsilon_0}$

and $E = \dfrac{\rho a^3}{3\epsilon_0 r^2} = \dfrac{Q}{4\pi\epsilon_0 r^2}$ ◊

For $b < r < c$, we must have $E = 0$ because any nonzero field would be moving charges in the metal. ◊

Free charges did move in the metal to deposit charge $+Q_b$ on its inner surface, at radius b, leaving charge $+Q_c$ on its outer surface, at radius c. Since the shell as a whole is neutral, $Q_c + Q_b = 0$.

For $r > c$, $\Phi_E = \epsilon_0$ reads $E(4\pi r^2) = (Q + Q_c - Q_b)/\epsilon_0$

and $E = Q/\left(4\pi\epsilon_0 r^2\right)$ ◊

(b)   For a gaussian surface of radius $b < r < c$, we have $0 = (Q + Q_b)/\epsilon_0$ so $Q_b = -Q$ and the charge density on the inner surface is

$$\sigma = \frac{+Q_b}{A} = \frac{-Q}{4\pi b^2}$$ ◊

Then $Q_c = -Q_b = Q$, and the charge density on the outer surface is

$$\sigma = +\frac{Q}{4\pi c^2}$$ ◊

---

59.   Repeat the calculations for Problem 58 when both sheets have **positive** uniform charge densities of value σ.   **Note:** The new problem statement would be as follows: Two infinite, nonconducting sheets of charge are parallel to each other, as shown in Figure P24.58.   Both sheets have positive uniform charge densities σ. Calculate the value of the electric field at points (a) to the left of, (b) in between, and (c) to the right of the two sheets.

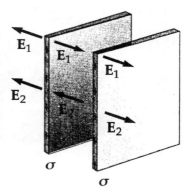

Figure P24.58
(modified)

Solution

G: When both sheets have the same charge density, a positive test charge at a point midway between them will experience the same force in opposite directions from each sheet.   Therefore, the electric field here will be zero. (We should ask: can we also conclude that the electron will experience equal and oppositely directed forces *everywhere* in the region between the plates?)

Outside the sheets the electric field will point away and should be twice the strength due to one sheet of charge, so $E = \sigma/\epsilon_0$ in these regions.

O: The principle of superposition can be applied to add the electric field vectors due to each sheet of charge.

A: For each sheet, the electric field at any point is $|\mathbf{E}| = \sigma/(2\,\epsilon_0)$ directed away from the sheet.

(a) At a point to the left of the two parallel sheets

$$\mathbf{E} = E_1(-\mathbf{i}) + E_2(-\mathbf{i}) = 2E(-\mathbf{i}) = -\frac{\sigma}{\epsilon_0}\mathbf{i} \qquad \diamond$$

(b) At a point between the two sheets

$$\mathbf{E} = E_1\mathbf{i} + E_2(-\mathbf{i}) = 0 \qquad \diamond$$

(c) At a point to the right of the two parallel sheets

$$\mathbf{E} = E_1\mathbf{i} + E_2\mathbf{i} = 2E\mathbf{i} = \frac{\sigma}{\epsilon_0}\mathbf{i} \qquad \diamond$$

L: We essentially solved this problem in the Gather information step, so it is no surprise that these results are what we expected. A better check is to confirm that the results are complementary to the case where the plates are oppositely charged (Problem 58).

---

63. A solid insulating sphere of radius R has a nonuniform charge density that varies with r according to the expression $\rho = Ar^2$, where A is a constant and $r < R$ is measured from the center of the sphere. (a) Show that the electric field outside ($r > R$) the sphere is $E = AR^5/5\epsilon_0 r^2$. (b) Show that the electric field inside ($r < R$) the sphere is $E = Ar^3/5\epsilon_0$. (**Hint:** Note that the total charge Q on the sphere is equal to the integral of $\rho\,dV$, where r extends from 0 to R; also note that the charge q within a radius $r < R$ is **less** than Q. To evaluate the integrals, note that the volume element dV for a spherical shell of radius r and thickness dr is equal to $4\pi r^2\,dr$.)

Solution (a) We call the constant A', reserving A to denote area. The whole charge of the ball is

$$Q = \int_{\text{ball}} dQ = \int_{\text{ball}} \rho dV = \int_{r=0}^{R} A'r^2 4\pi r^2 dr = 4\pi A' \frac{r^5}{5}\bigg]_0^R = \frac{4\pi A' R^5}{5}$$

To find the electric field, consider as gaussian surface a concentric sphere of radius r outside the ball of charge:

In this case, $\displaystyle\int \mathbf{E} \cdot d\mathbf{A} = \frac{Q}{\epsilon_0}$ reads $E\, A \cos 0 = \dfrac{Q}{\epsilon_0}$

Solving, $$E\left(4\pi r^2\right) = \frac{4\pi A' R^5}{5\,\epsilon_0}$$

Thus, the electric field is $$E = \frac{A' R^5}{5\,\epsilon_0\, r^2} \qquad \lozenge$$

(b) Let the gaussian sphere lie inside the ball of charge:

$$\int_{\substack{\text{sphere, radius } r}} \mathbf{E} \cdot d\mathbf{A} = \int_{\substack{\text{sphere, radius } r}} dQ/\epsilon_0$$

Now the integral becomes $$E(\cos 0)\int dA = \int \frac{\rho dV}{\epsilon_0}$$

Substituting for each term, $$E\,A = \int_0^r \frac{A'r^2\left(4\pi r^2\right) dr}{\epsilon_0}$$

and $$E\left(4\pi r^2\right) = \left(\frac{A'4\pi}{\epsilon_0}\right)\left(\frac{r^5}{5}\right)\bigg]_0^r = \frac{A'4\pi\, r^5}{5\,\epsilon_0}$$

or $$E = \frac{A'r^3}{5\epsilon_0} \qquad \lozenge$$

67. A slab of insulating material (infinite in two of its three dimensions) has a uniform positive charge density ρ. An edge view of the slab is shown in Figure P24.67. (a) Show that the electric field a distance x from its center and inside the slab is $E = \rho x/\epsilon_0$. (b) Suppose an electron of charge $-e$ and mass m_e is placed inside the slab. If it is released from rest at a distance x from the center, show that the electron exhibits simple harmonic motion with a frequency

$$f = \frac{1}{2\pi}\sqrt{\frac{\rho e}{m_e\,\epsilon_0}}$$

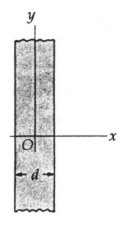

Figure P24.67

Solution

(a) The slab has left-to-right symmetry, so its field must be equal in strength at x and at $-x$. It points everywhere away from the central plane. Take as gaussian surface a rectangular box of thickness $2x$ and height and width L, centered on the $x = 0$ plane. The charge it contains is $\rho V = \rho 2xL^2$. The total flux leaving it is EL^2 through the right face, EL^2 through the left face, and zero through each of the other four sides.

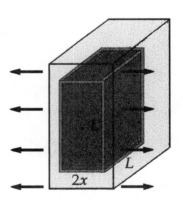

Thus Gauss's law $\displaystyle\int \mathbf{E}\cdot d\mathbf{A} = \frac{q}{\epsilon_0}$ becomes $\displaystyle 2EL^2 = \frac{\rho 2xL^2}{\epsilon_0}$

so $$E = \frac{\rho x}{\epsilon_0} \qquad \lozenge$$

(b) The electron experiences a force opposite to **E**. When displaced to $x > 0$, it experiences a restoring force to the left.

For it, $\Sigma \mathbf{F} = m_e \mathbf{a}$ reads

$$q\mathbf{E} = m_e \mathbf{a}$$

$$\frac{-e\rho x \mathbf{i}}{\epsilon_0} = m_e \mathbf{a}$$

Solving for the acceleration,

$$\mathbf{a} = -\left(\frac{e\rho}{m_e \, \epsilon_0}\right) x \mathbf{i}$$

or

$$\mathbf{a} = -\omega^2 x \mathbf{i}$$

That is, its acceleration is proportional to its displacement and oppositely directed, as is required for simple harmonic motion.

Solving for the frequency,

$$\omega^2 = \frac{e\rho}{m_e \, \epsilon_0}$$

and

$$f = \frac{\omega}{2\pi} = \frac{1}{2\pi}\sqrt{\frac{e\rho}{m_e \, \epsilon_0}} \qquad \Diamond$$

Chapter
25

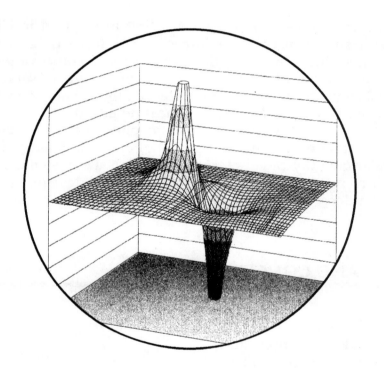

Electric Potential

Chapter 25

ELECTRIC POTENTIAL

INTRODUCTION

The concept of potential energy was first introduced in Chapter 8 in connection with such conservative forces as the force of gravity and the elastic force of a spring. By using the law of energy conservation, we were often able to avoid working directly with forces when solving various mechanical problems. In this chapter we see that the energy concept is also of great value in the study of electricity. Since the electrostatic force given by Coulomb's law is conservative, electrostatic phenomena can be described conveniently in terms of an electrical potential energy. This idea enables us to define a scalar quantity called **electric potential.** Because the potential is a scalar function of position, it offers a simpler way of describing electrostatic phenomena than does the electric field.

EQUATIONS AND CONCEPTS

The **potential difference** between two points A and B in an electric field, $\Delta V = V_B - V_A$, can be found by integrating $\mathbf{E} \cdot d\mathbf{s}$ along **any path** from A to B.

$$\Delta V = \frac{\Delta U}{q_0} = -\int_A^B \mathbf{E} \cdot d\mathbf{s} \qquad (25.3)$$

If the field is **uniform**, the potential difference depends only on the displacement d in the direction parallel to $\mathbf{E}$.

$$\Delta V = -E \int_A^B ds = -Ed \qquad (25.6)$$

$$\text{(when } \mathbf{d} \parallel \mathbf{E})$$

The **change in potential energy**, ΔU, of a charge in moving from point A to point B in an electric field depends on the sign and magnitude of the charge as well as on the change in potential, ΔV.

$$\Delta U = q_0\,\Delta V \qquad (25.7)$$

$$\Delta U = -q_0 E d \quad \text{(for } \mathbf{d} \parallel \mathbf{E} \text{ in a}$$
$$\text{uniform field)}$$

In the special case where the electric field is uniform, the change in potential energy is proportional to the distance the charge moves along a direction parallel to the electric field. Note that a positive charge loses electric potential energy when it moves in the direction of the electric field.

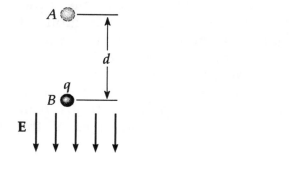

The **electric potential** at a point in the vicinity of several point charges is calculated in a manner that assumes that the potential is zero at infinity.

$$V = k_e \sum_i \frac{q_i}{r_i} \qquad (25.12)$$

The **potential energy of a pair of charges** separated by a distance r_{12} represents the work required to assemble the charges from an infinite separation. Hence, the negative of the potential energy equals the minimum work required to separate them by an infinite distance. The electric potential energy associated with a system of two charged particles is positive if the two charges have the same sign, and negative if they are of opposite sign.

$$U = k_e \frac{q_1 q_2}{r_{12}} \qquad (25.13)$$

For more than two charged particles in a system, the **total potential energy** is found by calculating U for each pair of charges and summing the terms algebraically.

$$U = \frac{k_e}{2} \sum_{i=1}^{N} \sum_{j=1}^{N} \frac{q_i q_j}{r_{ij}}$$

when $i \neq j$

If the scalar electric potential function throughout a region of space is known, then the vector electric field can be calculated from the potential function. The **components of the electric field** in rectangular coordinates are given in terms of partial derivatives of the potential.

$$E_x = -\frac{\partial V}{\partial x}$$

$$E_y = -\frac{\partial V}{\partial y}$$

$$E_z = -\frac{\partial V}{\partial z}$$

The vector expression for the electric field can be evaluated at any point $P(x,y,z)$ within the region.

$$\mathbf{E} = -\left(\frac{\partial}{\partial x}\mathbf{i} + \frac{\partial}{\partial y}\mathbf{j} + \frac{\partial}{\partial z}\mathbf{k} \right)V$$

The **potential** (relative to zero at infinity) **for a continuous charge distribution** can be calculated by integrating the contribution due to a charge element dq over the line, surface, or volume which contains all the charge. Here, as in the case of a continuous charge distribution, it is convenient to represent dq in terms of the appropriate charge density.

$$V = k_e \int \frac{dq}{r} \qquad (25.19)$$

SUGGESTIONS, SKILLS, AND STRATEGIES

The vector expressions giving the electric field **E** over a region can be obtained from the scalar function which describes the electric potential, V, over the region by using a vector differential operator called the gradient operator, ∇:

$$\nabla = -\mathbf{i}\frac{\partial}{\partial x} - \mathbf{j}\frac{\partial}{\partial y} - \mathbf{k}\frac{\partial}{\partial z}$$

$$\mathbf{E} = -\nabla V$$

This is equivalent to

$$\mathbf{E} = -\mathbf{i}\frac{\partial V}{\partial x} - \mathbf{j}\frac{\partial V}{\partial y} - \mathbf{k}\frac{\partial V}{\partial z}$$

The derivatives in the above expression are called **partial derivatives**. This means that when the derivative is taken with respect to any one coordinate, any other coordinates which appear in the expression for the potential function are treated as constants.

Since the electrostatic force is a conservative force, the work done by the electrostatic force in moving a charge q from an initial point A to a final point B depends only on the location of the two points and is independent of the path taken between A and B. When calculating potential differences using the equation

$$V_B - V_A = -\int_A^B \mathbf{E} \cdot d\mathbf{s} \tag{25.3}$$

any path between A and B may be chosen to evaluate the integral; therefore you should select a path for which the evaluation of the "line integral" in Equation 25.3 will be as convenient as possible. For example; where L, M, and N are constants, if **E** is in the form

$$\mathbf{E} = Lx\mathbf{i} + My\mathbf{j} + Nz\mathbf{k}$$

The potential is integrated as $\displaystyle\int_A^B \mathbf{E}\cdot d\mathbf{s} = \int_A^B (Lx\,dx + My\,dy + Nz\,dz)$

and Equation 25.3 becomes

$$V_B - V_A = -\left[L\int_{x_A}^{x_B} x\,dx + M\int_{y_A}^{y_B} y\,dy + N\int_{z_A}^{z_B} z\,dz \right]$$

$$V_B - V_A = \frac{L}{2}\left(x_A{}^2 - x_B{}^2\right) + \frac{M}{2}\left(y_A{}^2 - y_B{}^2\right) + \frac{N}{2}\left(z_A{}^2 - z_B{}^2\right)$$

PROBLEM-SOLVING STRATEGY

- When working problems involving electric potential, remember that potential is a **scalar quantity** (rather than a vector quantity like the electric field), so there are no components to worry about. Therefore, when using the superposition principle to evaluate the electric potential at a point due to a system of point charges, you simply take the algebraic sum of the potentials due to each charge. However, you must keep track of signs. The potential ($V = k_e q / r$) for each positive charge is positive, while the potential for each negative charge is negative.

- Only **changes** in electric potential are significant, hence the point where you choose the potential to be zero is arbitrary. When dealing with point charges or a finite-sized charge distribution, we usually define $V = 0$ to be at a point infinitely far from the charges. However, if the charge distribution itself extends to infinity, some other nearby point must be selected as the reference point.

- The electric potential at some point P due to a continuous distribution of charge can be evaluated by dividing the charge distribution into infinitesimal elements of charge dq located at a distance r from the point

P. You then treat this element as a point charge, so that the potential at P due to the element is $dV = k_e \, dq / r$. The total potential at P is obtained by integrating dV over the entire charge distribution. In performing the integration for most problems, it is necessary to express dq and r in terms of a single variable. In order to simplify the integration, it is important to give careful consideration of the geometry involved in the problem.

- Another method that can be used to obtain the potential due to a finite continuous charge distribution is to start with the definition of the potential difference given by Equation 25.3. If **E** is known or can be obtained easily (say from Gauss's law), then the line integral of $\mathbf{E} \cdot d\mathbf{s}$ can be evaluated. An example of this method is given in Example 25.8.

- Once you know the electric potential at a point, it is possible to obtain the electric field at that point by remembering that

$$E_x = -\frac{\partial V}{\partial x}, \quad E_y = -\frac{\partial V}{\partial y}, \quad \text{and} \quad E_z = -\frac{\partial V}{\partial z}$$

- Until now, we have been using the symbols V to represent the electric potential at some point and ΔV to represent the potential difference between two points. In descriptions of electrical devices, however, it is common practice to use the symbol V to represent the potential difference across the device. Hence, in this book both symbols will be used to denote potential differences, depending on the circumstances.

In practice, a variety of phrases are used to describe the potential difference between two points, the most common being "voltage." A voltage **applied** to a device or **across** a device has the same meaning as the potential difference across the device. For example, if we say that the voltage across a certain capacitor is 12 volts, we mean that the potential difference between the capacitor's plates is 12 volts.

REVIEW CHECKLIST

▷ Understand that each point in the vicinity of a charge distribution can be characterized by a scalar quantity called the electric potential, V. The values of this potential function over the region (a scalar field) are related to the values of the electrostatic field over the region (a vector field).

▷ Calculate the electric potential difference between any two points in a uniform **electric field**, and the electric potential difference between any two points in the vicinity of a **group of point charges**.

▷ Calculate the electric **potential energy** associated with a group of point charges.

▷ Calculate the electric potential due to **continuous charge distributions** of reasonable symmetry—such as a charged ring, sphere, line, or disk.

▷ Obtain an expression for the electric field (a **vector** quantity) over a region of space if the scalar electric potential function for the region is known.

▷ Calculate the work done by an external force in moving a charge q between any two points in an electric field when (a) an expression giving the field as a function of position is known, or when (b) the charge distribution (either point charges or a continuous distribution of charge) giving rise to the field is known.

ANSWERS TO SELECTED CONCEPTUAL QUESTIONS

3.   Give a physical explanation of the fact that the potential energy of a pair of like charges is positive whereas the potential energy of a pair of unlike charges is negative.

Answer   You may remember from the chapter on gravitational potential energy that potential energy of a system is defined to be positive when positive work must have been performed by an external agent.   For example, a flag has a positive potential energy relative to the ground, since positive work must be done by an external force in order to raise it from the ground to the top of the pole.

When assembling like charges from an infinite separation, it takes work to move them closer together to some distance r; therefore energy is being stored, and the potential energy is positive.

When assembling unlike charges from an infinite separation, the charges tend to accelerate towards each other, and thus energy is released as they approach a separation of distance r. Therefore, the potential energy of a pair of unlike charges is negative.

□   □   □   □

SOLUTIONS TO SELECTED END-OF-CHAPTER PROBLEMS

3.   (a) Calculate the speed of a proton that is accelerated from rest through a potential difference of 120 V.   (b) Calculate the speed of an electron that is accelerated through the same potential difference.

Solution

G: Since 120 V is only a modest potential difference, we might expect that the final speed of the particles will be substantially less than the speed of light.   We should also expect the speed of the electron to be significantly greater than the proton because, with $m_e << m_p$, an equal force on both particles will result in a much greater acceleration for the electron.

O: Conservation of energy can be applied to this problem to find the final speed from the kinetic energy of the particles. (Review this work-energy theory of motion from Chapter 8 if necessary.)

A: (a) Energy is conserved as the proton moves from high to low potential, which can be defined for this problem as moving from 120 V down to 0 V:

$$K_i + U_i + \Delta E_{nc} = K_f + U_f$$

$$0 + qV + 0 = \tfrac{1}{2}mv_p^2 + 0$$

$$(1.60 \times 10^{-19} \text{ C})(120 \text{ V})\left(\frac{1\text{J}}{1\text{ V}\cdot\text{C}}\right) = \tfrac{1}{2}(1.67 \times 10^{-27} \text{ kg})v_p^2$$

$$v_p = 1.52 \times 10^5 \text{ m/s} \qquad\qquad \lozenge$$

(b) The electron will gain speed in moving the other way, from $V_i = 0$ to $V_f = 120$ V:

$$K_i + U_i + \Delta E_{nc} = K_f + U_f$$

$$0 + 0 + 0 = \tfrac{1}{2}mv_e^2 + qV$$

$$0 = \tfrac{1}{2}(9.11 \times 10^{-31} \text{ kg})v_e^2 + (-1.60 \times 10^{-19} \text{ C})(120 \text{ J/C})$$

$$v_e = 6.49 \times 10^6 \text{ m/s} \qquad\qquad \lozenge$$

L: Both of these speeds are significantly less than the speed of light as expected, which also means that we were justified in not using the relativistic kinetic energy formula. (For precision to three significant digits, the relativistic formula is only needed if v is greater than about $0.1\,c$.)

9. An electron moving parallel to the x axis has an initial speed of 3.70×10^6 m/s at the origin. Its speed is reduced to 1.40×10^5 m/s at the point $x = 2.00$ cm. Calculate the potential difference between the origin and that point. Which point is at the higher potential?

Solution Use the work-energy theorem to equate the energy of the electron at $x = 0$ and at $x = 2.00$ cm. The unknown will be the difference in potential $V_f - V_i$. Thus, $K_i + U_i + \Delta K_{nc} = K_f + U_f$ becomes

$$\tfrac{1}{2}mv_i^2 + qV_0 + 0 = \tfrac{1}{2}mv_f^2 + qV_2 \qquad \text{or} \qquad \tfrac{1}{2}m\left(v_i^2 - v_f^2\right) = q(V_2 - V_0)$$

so $V_f - V_i = \dfrac{m(v_i^2 - v_f^2)}{2q}$

Noting that the electron's charge is negative, and evaluating the potential,

$$V_f - V_i = (9.11 \times 10^{-31}\ \text{kg})\frac{(3.70 \times 10^6\ \text{m}/\text{s})^2 - (1.40 \times 10^5\ \text{m}/\text{s})^2}{2(-1.60 \times 10^{-19}\ \text{C})} = -38.9\ \text{V} \quad \lozenge$$

The negative sign means that the 2.00-cm location is lower in potential than the origin. A positive charge would slow in free flight toward higher voltage, but the negative electron slows as it moves into lower potential. The 2.00-cm distance was unnecessary information for this problem. If the field were uniform, we could find the x-component from $\Delta V = -E_x d$. $\qquad \lozenge$

19. The Bohr model of the hydrogen atom states that the single electron can exist only in certain allowed orbits around the proton. The radius of each Bohr orbit is $r = n^2(0.0529\ \text{nm})$ where $n = 1, 2, 3, \ldots$. Calculate the electric potential energy of a hydrogen atom when the electron is in the (a) first allowed orbit, $n = 1$; (b) second allowed orbit, $n = 2$; and (c) when the electron has escaped from the atom $(r = \infty)$. Express your answers in electron volts.

Solution

G: We may remember from chemistry that the lowest energy level for hydrogen is $E_1 = -13.6$ eV, and higher energy levels can be found from $E_n = E_1 / n^2$, so that $E_2 = -3.40$ eV and $E_\infty = 0$ eV. (see section 42.2) Since these are the total energies (potential plus kinetic), the electric potential energy alone should be lower (more negative) because the kinetic energy of the electron must be positive.

O: The electric potential energy is given by $U = k_e \dfrac{q_1 q_2}{r}$

A: (a) For the first allowed Bohr orbit,

$$U = (8.99 \times 10^9 \text{ N} \cdot \text{m}^2 / \text{C}^2) \frac{(-1.60 \times 10^{-19} \text{ C})(1.60 \times 10^{-19} \text{ C})}{(0.0529 \times 10^{-9} \text{ m})}$$

$$U = -4.35 \times 10^{-18} \text{ J} = \frac{-4.35 \times 10^{-18} \text{ J}}{1.60 \times 10^{-19} \text{ J} / \text{eV}} = -27.2 \text{ eV} \qquad \Diamond$$

(b) For the second allowed orbit,

$$U = (8.99 \times 10^9 \text{ N} \cdot \text{m}^2 / \text{C}^2) \frac{(-1.60 \times 10^{-19} \text{ C})(1.60 \times 10^{-19} \text{ C})}{2^2 (0.0529 \times 10^{-9} \text{ m})}$$

$$U = -1.088 \times 10^{-18} \text{ J} = -6.80 \text{ eV} \qquad \Diamond$$

(c) When the electron is at $r = \infty$,

$$U = \left(8.99 \times 10^9 \text{ N} \cdot \text{m}^2 / \text{C}^2\right) \frac{\left(-1.60 \times 10^{-19} \text{ C}\right)\left(1.60 \times 10^{-19} \text{ C}\right)}{\infty \text{ m}} = 0 \text{ J} \qquad \Diamond$$

L: The potential energies appear to be twice the magnitude of the total energy values, so apparently the kinetic energy of the electron has the same absolute magnitude as the total energy.

21. The three charges in Figure P25.21 are at the vertices of an isosceles triangle. Calculate the electric potential at the midpoint of the base, taking $q = 7.00 \ \mu C$.

Solution   Let $q_1 = q$ and $q_2 = q_3 = -q$

The charges are at distances of

$$r_1 = \sqrt{(0.0400 \text{ m})^2 - (0.0100 \text{ m})^2} = 3.87 \times 10^{-2} \text{ m}$$

and $r_2 = r_3 = 0.0100$ m

The voltage at point P is

$$V_P = \frac{k_e q_1}{r_1} + \frac{k_e q_2}{r_2} + \frac{k_e q_3}{r_3} = k_e q\left(\frac{1}{r_1} + \frac{-1}{r_2} + \frac{-1}{r_3}\right)$$

so $V_P = \left(8.99 \times 10^9 \ \frac{\text{N} \cdot \text{m}^2}{\text{C}^2}\right)\left(7.00 \times 10^{-6} \text{ C}\right)\left(\frac{1}{0.0387 \text{ m}} - \frac{1}{0.0100 \text{ m}} - \frac{1}{0.0100 \text{ m}}\right)$

and $V_P = 11.0 \times 10^6$ V   ◊

Related Calculation   Calculate the electric field vector at the same point due to the three charges.

The separate fields of the two negative charges are in opposite directions and add to zero:

$$\mathbf{E}_P = \frac{k_e q_1}{r_1^2} \hat{\mathbf{r}}_1 = \frac{\left(8.99 \times 10^9 \ \text{N} \cdot \text{m}^2 / \text{C}^2\right)\left(7.00 \times 10^{-6} \text{ C}\right)}{\left((0.0400 \text{ m})^2 - (0.0100 \text{ m})^2\right)} \text{ down}$$

$$\mathbf{E}_P = (42.0 \times 10^6 \ \text{N} / \text{C})(-\mathbf{j})$$ ◊

23. Show that the amount of work required to assemble four identical point charges of magnitude Q at the corners of a square of side s is $5.41 k_e Q^2/s$.

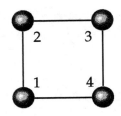

Solution

The work required equals the sum of the potential energies of the four charges. Assume that charge q_4 is positive, The work required to move q_3 into position is $U_3 = q_3 V_4$, where V_4 is the potential at q_3 due to q_4. The work required to move q_2 into position is $U_2 = q_2(V_3 + V_4)$, where V_3 and V_4 are the contribution to the potential at the location of q_2 due to q_3 and q_4, respectively. In the same manner, $U_1 = q_1(V_2 + V_3 + V_4)$. We must add up $U = qV$ contributions for all pairs:

$$U = (q_1 V_2 + q_1 V_3 + q_1 V_4) + (q_2 V_3 + q_2 V_4) + q_3 V_4$$

$$U = \frac{q_1 k_e q_2}{r_{12}} + \frac{q_1 k_e q_3}{r_{13}} + \frac{q_1 k_e q_4}{r_{14}} + \frac{q_2 k_e q_3}{r_{23}} + \frac{q_2 k_e q_4}{r_{24}} + \frac{q_3 k_e q_4}{r_{34}}$$

$$U = \frac{Q k_e Q}{s} + \frac{Q k_e Q}{s\sqrt{2}} + \frac{Q k_e Q}{s} + \frac{Q k_e Q}{s} + \frac{Q k_e Q}{s\sqrt{2}} + \frac{Q k_e Q}{s}$$

Evaluating, $\qquad U = \frac{k_e Q^2}{s}\left(4 + \frac{2}{\sqrt{2}}\right) = \frac{5.41 k_e Q^2}{s}$ $\qquad\qquad \lozenge$

---

37. Over a certain region of space, the electric potential is $V = 5x - 3x^2 y + 2yz^2$. Find the expressions for x, y, and z components of the electric field over this region. What is the magnitude of the field at the point P, which has coordinates $(1.00, 0, -2.00)$ m?

Solution   First, we find the x, y, and z components of the field; then, we evaluate them at point P.   (We assume that V is given in volts, as a function of distances in meters.)

In general,	At point P,
$E_x = -\partial V/\partial x = -5 + 6xy$	$E_x = -5 + 6(1.00\text{ m})(0.00\text{ m}) = -5.00\text{ N}/\text{C}$
$E_y = -\partial V/\partial y = 3x^2 - 2z^2$	$E_y = 3(1.00\text{ m})^2 - 2(-2.00\text{ m})^2 = -5.00\text{ N}/\text{C}$
$E_z = -\partial V/\partial z = -4yz$ ◊	$E_z = -4(0.00\text{ m})(-2.00\text{ m}) = 0.00\text{ N}/\text{C}$ ◊

and the field's magnitude

$$E = \sqrt{(-5.00\text{ N}/\text{C})^2 + (-5.00\text{ N}/\text{C})^2 + 0^2} = 7.07\text{ N}/\text{C} \qquad ◊$$

39.   It is shown in Example 25.7 that the potential at a point P a distance a above one end of a uniformly charged rod of length ℓ lying along the x axis is

$$V = \frac{k_e Q}{\ell} \ln\left(\frac{\ell + \sqrt{\ell^2 + a^2}}{a} \right)$$

Use this result to derive an expression for the y component of the electric field at P.   (**Hint:** Replace a with y.)

Solution   Replacing a with y as directed, we can differentiate the potential as shown:

$$E_y = -\frac{\partial V}{\partial y} = -\frac{k_e Q}{\ell}\frac{d}{dy}\left[\ln(\ell + \sqrt{\ell^2 + y^2}) - \ln y \right]$$

$$E_y = \frac{-k_e Q}{\ell}\left[\frac{2y/\left(2\sqrt{\ell^2 + y^2}\right)}{\ell + \sqrt{\ell^2 + y^2}} - \frac{1}{y} \right] = \frac{k_e Q}{y\sqrt{\ell^2 + y^2}} \qquad ◊$$

43. A rod of length L (Fig. P25.43) lies along the x axis with its left end at the origin and has a nonuniform charge density $\lambda = \alpha x$ (where α is a positive constant). (a) What are the units of the α? (b) Calculate the electric potential at A.

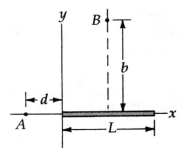

Figure P25.43

Solution

(a) As a linear charge density, λ has units of C/m. So $\alpha = \lambda/x$ must have units of C/m². ◊

(b) Consider a small segment of the rod at location x and of length dx. The amount of charge on it is $\lambda dx = (\alpha x)dx$. Its distance from A is $d+x$, so its contribution to the electric potential at A is

$$dV = k_e \frac{dq}{r} = k_e \alpha x \frac{dx}{(d+x)}$$

We must integrate these contributions for the whole rod, from $x = 0$ to $x = L$:

$$V = \int_{\text{all } q} dV = \int_0^L \frac{k_e \alpha x \, dx}{d+x}$$

To perform the integral, make a change of variables to $u = d + x$ and $du = dx$ with limits $u_{x=0} = d$ and $u_{x=L} = d + L$:

$$V = \int_d^{d+L} \frac{k_e \alpha (u-d) du}{u} = k_e \alpha \int_d^{d+L} du - k_e \alpha d \int_d^{d+L} \left(\frac{1}{u}\right) du$$

Keep track of symbols: the **unknown** is V. The values k_e, α, d, and L are **known** and constant. And x and u are variables, and will not appear in the answer. We have the answer when the unknown is expressed in terms of the d, L, and α mentioned in the problem and the universal constant k_e.

$$V = k_e \alpha u \Big|_d^{d+L} - k_e \alpha d \ln u \Big|_d^{d+L} = k_e \alpha (d+L-d) - k_e \alpha d (\ln(d+L) - \ln d)$$

$$V = k_e \alpha L - k_e \alpha d \, \ln\big((d+L)/d\big)$$ ◊

49. A spherical conductor has a radius of 14.0 cm and charge of 26.0 μC. Calculate the electric field and the electric potential at (a) $r = 10.0$ cm, (b) $r = 20.0$ cm, and (c) $r = 14.0$ cm from the center.

Solution

(a) Inside a conductor when charges are not moving, the electric field is zero and the potential is uniform, the same as on the surface.

$$\mathbf{E} = 0 \qquad \Diamond$$

$$V = \frac{k_e q}{R} = \frac{\left(8.99 \times 10^9 \ \mathrm{N \cdot m^2 / C^2}\right)\left(26.0 \times 10^{-6} \ \mathrm{C}\right)}{(0.140 \ \mathrm{m})} = 1.67 \times 10^6 \ \mathrm{V} \qquad \Diamond$$

(b) The sphere behaves like a point charge at its center when you stand outside.

$$\mathbf{E} = \frac{k_e q}{r^2}\hat{\mathbf{r}} = \frac{\left(8.99 \times 10^9 \ \mathrm{N \cdot m^2 / C^2}\right)\left(26.0 \times 10^{-6} \ \mathrm{C}\right)}{(0.200 \ \mathrm{m})^2}\hat{\mathbf{r}} = (5.84 \times 10^6 \ \mathrm{N / C})\hat{\mathbf{r}} \qquad \Diamond$$

$$V = \frac{k_e q}{r} = 1.17 \times 10^6 \ \mathrm{V} \qquad \Diamond$$

(c) $\qquad \mathbf{E} = \frac{k_e q}{r^2}\hat{\mathbf{r}} = (11.9 \times 10^6 \ \mathrm{N / C})\hat{\mathbf{r}} \qquad \Diamond$

$\qquad V = 1.67 \times 10^6 \ \mathrm{V}$ as in part (a) $\qquad \Diamond$

51. Consider a Van de Graaff generator with a 30.0-cm-diameter dome operating in dry air. (a) What is the maximum potential of the dome? (b) What is the maximum charge on the dome?

Solution

G: Van de Graaff generators produce voltages that can make your hair stand on end, somewhere on the order of about 100 kV (see the Puzzler at beginning of Chapter 25). With these high voltages, the maximum charge on the dome is probably more than typical point charge values of about $1\,\mu C$.

The maximum potential and charge will be limited by the electric field strength at which the air surrounding the dome will ionize. This critical value is determined by the **dielectric strength** of air which, from page 789 or from Table 26.1, is $E_{\text{critical}} = 3 \times 10^6\ \text{V}/\text{m}$. An electric field stronger than this will cause the air to act like a conductor instead of an insulator. This process is called dielectric breakdown and may be seen as a spark.

O: From the maximum allowed electric field, we can find the charge and potential that would create this situation. Since we are only given the diameter of the dome, we will assume that the conductor is spherical, which allows us to use the electric field and potential equations for a spherical conductor. With these equations, it will be easier to do part (b) first and use the result for part (a).

A: (b) For a spherical conductor with total charge Q, $\qquad |\mathbf{E}| = \dfrac{k_e Q}{r^2}$

$$Q = \frac{E r^2}{k_e} = \frac{\left(3.00 \times 10^6\ \text{V}/\text{m}\right)(0.150\ \text{m})^2}{8.99 \times 10^9\ \text{N} \cdot \text{m}^2 / \text{C}^2}(1\,\text{N} \cdot \text{m}/\text{V} \cdot \text{C}) = 7.51\ \mu C \qquad \Diamond$$

(a) $\quad V = \dfrac{k_e Q}{r} = \dfrac{(8.99 \times 10^9\ \text{N} \cdot \text{m}^2 / \text{C}^2)(7.51 \times 10^{-6}\ \text{C})}{0.150\ \text{m}} = 450\ \text{kV} \qquad \Diamond$

L: These calculated results seem reasonable based on our predictions. The voltage is about 4000 times larger than the 120 V found from common electrical outlets, but the charge is similar in magnitude to many of the static charge problems we have solved earlier. This implies that most of these charge configurations would have to be in a vacuum because the electric field near these point charges would be strong enough to cause sparking in air. (Example: A charged ball with $Q = 1\,\mu C$ and $r = 1\,mm$ would have an electric field near its surface of

$$E = \frac{k_e Q}{r^2} = \frac{\left(9 \times 10^9 \ N \cdot m^2 / C^2\right)\left(1 \times 10^{-6} \ C\right)}{(0.001 \ m)^2} = 9 \times 10^9 \ V / m$$

which is well beyond the dielectric breakdown of air!)

---

53. The liquid-drop model of the nucleus suggests that high-energy oscillations of certain nuclei can split the nucleus into two unequal fragments plus a few neutrons. The fragments acquire kinetic energy from their mutual Coulomb repulsion. Calculate the electric potential energy (in electron volts) of two spherical fragments from a uranium nucleus having the following charges and radii: $38e$ and 5.50×10^{-15} m; $54e$ and 6.20×10^{-15} m. Assume that the charge is distributed uniformly throughout the volume of each spherical fragment and that their surfaces are initially in contact at rest. (The electrons surrounding the nucleus can be neglected.)

Solution The problem is equivalent to finding the potential energy of a point charge $38e$ at a distance 11.7×10^{-15} m (the distance between their centers when in contact) from a point charge $54e$.

$$U = qV = k_e \frac{q_1 q_2}{r_{12}} = \left(8.99 \times 10^9 \ J \cdot m / C^2\right)\frac{(38)(1.60 \times 10^{-19} \ C)(54)(1.60 \times 10^{-19} \ C)}{(5.50 \times 10^{-15} \ m + 6.20 \times 10^{-15} \ m)}$$

$$U = 4.04 \times 10^{-11} \ J = 253 \ MeV \qquad \diamond$$

---

57. At a certain distance from a point charge, the magnitude of the electric field is 500 V/m and the electric potential is –3.00 kV. (a) What is the distance to the charge? (b) What is the magnitude of the charge?

Solution At a distance r from a point charge $V_r = \dfrac{k_e q}{r}$ and $E_r = \dfrac{k_e q}{r^2}$

Thus, $E_r = \dfrac{r V_r}{r^2} = \dfrac{V_r}{r}$

(a) $r = \dfrac{V}{E_r} = \dfrac{3000 \text{ V}}{500 \text{ N/C}} = 6.00 \ \dfrac{\text{N} \cdot \text{m/C}}{\text{N/C}} = 6.00 \text{ m}$ ◊

(b) $q = \dfrac{r V_r}{k_e} = \dfrac{(6.00 \text{ m})(-3000 \text{ V})}{8.99 \times 10^9 \text{ N} \cdot \text{m}^2 / \text{C}^2} = -2.00 \ \mu\text{C}$ ◊

---

61. Calculate the work that must be done to charge a spherical shell of radius R to a total charge Q.

Solution When the potential of the shell is V due to a charge q, the work required to add an additional increment of charge dq is

$$dW = V dq \quad \text{where} \quad V = \dfrac{k_e q}{R}$$

$$dW = \left(\dfrac{k_e q}{R}\right) dq \quad \text{and} \quad W = \dfrac{k_e}{R} \int_0^Q q \, dq$$

Therefore, $W = \left(\dfrac{k_e}{R}\right)\left(\dfrac{Q^2}{2}\right)$ ◊

---

63. From Gauss's law, the electric field set up by a uniform line of charge is

$$\mathbf{E} = \left(\frac{\lambda}{2\pi\epsilon_0 r} \right) \hat{\mathbf{r}}$$

where $\hat{\mathbf{r}}$ is a unit vector pointing radially away from the line and λ is the charge per unit length along the line. Derive an expression for the potential difference between $r = r_1$ and $r = r_2$.

Solution

In Equation 25.3, $V_2 - V_1 = \Delta V = -\int_1^2 \mathbf{E} \cdot d\mathbf{s}$; think about stepping from distance r_1 out to the larger distance r_2 away from the charged line. Then $d\mathbf{s} = dr\,\hat{\mathbf{r}}$, and we can make r the variable of integration:

$$V_2 - V_1 = -\int_{r_1}^{r_2} \frac{\lambda}{2\pi\epsilon_0 r} \hat{\mathbf{r}} \cdot dr\,\hat{\mathbf{r}}$$

with

$$\hat{\mathbf{r}} \cdot \hat{\mathbf{r}} = 1 \cdot 1 \cdot \cos 0° = 1$$

The potential difference is

$$V_2 - V_1 = -\frac{\lambda}{2\pi\,\epsilon_0} \int_{r_1}^{r_2} \frac{dr}{r} = -\frac{\lambda}{2\pi\,\epsilon_0} \ln r \Big]_{r_1}^{r_2}$$

and

$$V_2 - V_1 = -\frac{\lambda}{2\pi\,\epsilon_0}(\ln r_2 - \ln r_1) = -\frac{\lambda}{2\pi\,\epsilon_0} \ln \frac{r_2}{r_1} \quad \Diamond$$

If $r_2 > r_1$, then $V_2 - V_1$ is negative. This means the potential decreases as we move away from a positively-charged filament.

69. A dipole is located along the y axis as shown in Figure P25.69. (a) At a point P, which is far from the dipole $(r \gg a)$, the electric potential is $V = k_e(p\cos\theta)/r^2$ where $p = 2qa$. Calculate the radial component E_r and the perpendicular component, E_θ of the associated electric field. Note that $E_\theta = -(1/r)(\partial V/\partial\theta)$. Do these results seem reasonable for $\theta = 90°$ and $0°$? for $r = 0$? (b) For the dipole arrangement shown, express V in terms of rectangular coordinates using $r = (x^2 + y^2)^{1/2}$ and

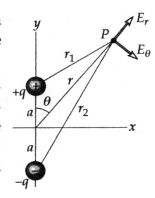

Figure P25.69

$$\cos\theta = \frac{y}{(x^2+y^2)^{1/2}}$$

Using these results and taking $r \gg a$, calculate the field components E_x and E_y.

Solution   (a)
$$E_r = -\frac{\partial V}{\partial r} = -\frac{\partial}{\partial r}\left(\frac{k_e p \cos\theta}{r^2}\right) = \frac{2k_e p \cos\theta}{r^3} = E_r \qquad \Diamond$$

In spherical coordinates, $\quad E_\theta = \frac{1}{r}\left(\frac{\partial V}{\partial\theta}\right) = \frac{1}{r}\frac{\partial}{\partial\theta}\left(\frac{k_e p\cos\theta}{r^2}\right) = \frac{k_e p\sin\theta}{r^3} = E_\theta \qquad \Diamond$

The limits of these values at $\theta = 0$ and $\theta = 90°$ correspond to an electric field that points outward from the positive charge on top, and loops to the negative charge below. This is reasonable, but it has no meaning at $r = 0$. $\Diamond$

(b)   For $r = \sqrt{x^2 + y^2}$, $\quad \cos\theta = \dfrac{y}{\sqrt{x^2+y^2}}$ and $\quad V = \dfrac{k_e p y}{\left(x^2+y^2\right)^{3/2}}$

$$E_x = -\frac{\partial V}{\partial x} = -\frac{\partial}{\partial x}\left(\frac{k_e p y}{(x^2+y^2)^{3/2}}\right) = \frac{3k_e p x y}{(x^2+y^2)^{5/2}} = E_x \qquad \Diamond$$

$$E_y = -\frac{\partial V}{\partial y} = -\frac{\partial}{\partial y}\left(\frac{k_e p y}{(x^2+y^2)^{3/2}}\right) = \frac{k_e p\left(2y^2-x^2\right)}{(x^2+y^2)^{5/2}} \qquad \Diamond$$

Chapter 26

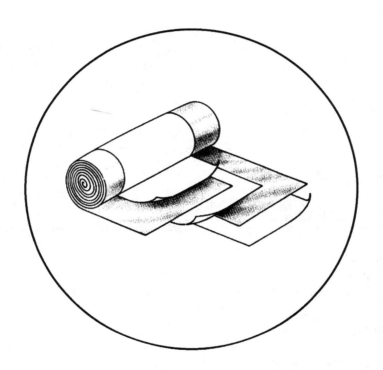

Capacitance and Dielectrics

CAPACITANCE AND DIELECTRICS

INTRODUCTION

This chapter is concerned with the properties of capacitors, devices that store charge. Capacitors are commonly used in a variety of electrical circuits. For instance, they are used (1) to tune the frequency of radio receivers, (2) as filters in power supplies, (3) to eliminate sparking in automobile ignition systems, and (4) as energy-storing devices in electronic flash units.

A capacitor basically consists of two conductors separated by an insulator. We shall see that the capacitance of a given device depends on its geometry and on the material separating the charged conductors, called a **dielectric.**

EQUATIONS AND CONCEPTS

The **capacitance** of a capacitor is defined as the ratio of the charge on either conductor (or plate) to the magnitude of the potential difference between the conductors.

$$C \equiv \frac{Q}{\Delta V} \tag{26.1}$$

For **an air-filled parallel-plate capacitor, the capacitance** is proportional to the area of the plates and inversely proportional to the separation of the plates.

$$C = \frac{\epsilon_0 A}{d} \tag{26.3}$$

When the region between the plates is completely filled by a material of dielectric constant κ, the capacitance increases by the factor κ.

$$C = \kappa \frac{\epsilon_0 A}{d} \tag{26.15}$$

68

The **equivalent capacitance** of a **parallel combination** of capacitors is larger than any individual capacitor in the group.

$$C_{eq} = C_1 + C_2 + C_3 + \ldots \qquad (26.8)$$

The **equivalent capacitance** of a **series combination** of capacitors is smaller than the smallest capacitor in the group.

$$\frac{1}{C_{eq}} = \frac{1}{C_1} + \frac{1}{C_2} + \frac{1}{C_3} + \ldots \qquad (26.10)$$

In the special case of only **t w o capacitors in series**, the equivalent capacitance is equal to the ratio of the product to the sum of their capacitance.

$$C_{eq} = \frac{C_1 C_2}{C_1 + C_2}$$

The **electrostatic energy** stored in the electrostatic field of a charged capacitor equals the work done (by a battery or other source) in charging the capacitor from $q = 0$ to $q = Q$.

$$U = \frac{Q^2}{2C} = \tfrac{1}{2} Q \Delta V = \tfrac{1}{2} C (\Delta V)^2 \qquad (26.11)$$

The **energy density** at any point in the electrostatic field of a charged capacitor is proportional to the square of the electric field intensity at that point. This is true for any electric field, not just that of a parallel plate capacitor.

$$u_E = \tfrac{1}{2} \epsilon_0 E^2 \qquad (26.13)$$

Two equal and opposite charges of magnitude q separated by a distance $2a$ constitute an electric dipole. This configuration is characterized by an **electric dipole moment, p**. The direction of vector **p** is from the negative to the positive charge.

$$p \equiv 2aq \qquad (26.16)$$

An external uniform electric field will exert a **net torque** on an electric dipole when the dipole moment makes an angle $\theta \neq 0$ with the direction of the field.

$$\tau = \mathbf{p} \times \mathbf{E} \qquad (26.18)$$

There is **potential energy** associated with the dipole-electric field system.

$$U = -\mathbf{p} \cdot \mathbf{E} \qquad (26.20)$$

SUGGESTIONS, SKILLS, AND STRATEGIES

PROBLEM-SOLVING HINTS FOR CAPACITANCE

- When analyzing a series-parallel combination of capacitors to determine the equivalent capacitance, you should make a sequence of circuit diagrams which show the successive steps in the simplification of the circuit; combine at each step those capacitors which are in simple-parallel or simple-series relationship to each other and use appropriate equations for series or parallel capacitors at each step of the simplification. At each step, you know two of the three quantities: Q, ΔV, and C. You will be able to determine the remaining quantity using the relation $Q = C\Delta V$.

- When calculating capacitance, be careful with your choice of units. To calculate capacitance in farads, make sure that distances are in meters and use the SI value of ϵ_0. When checking consistency of units, remember that the units for electric fields are newtons per coulomb (N/C) or the equivalent volts per meter (V/m).

- When two or more unequal capacitors are connected in series, they carry the same charge, but their potential differences are not the same. The capacitances add as reciprocals, and the equivalent capacitance of the combination is always less than the smallest individual capacitor.

- When two or more capacitors are connected in parallel, the potential differences across them are the same. The charge on each capacitor is proportional to its capacitance; hence, the capacitances add directly to give the equivalent capacitance of the parallel combination.

- A dielectric increases capacitance by the factor κ (the dielectric constant) because induced surface charges on the dielectric reduce the electric field inside the material from E to E/κ.

- Be careful about problems in which you may be connecting or disconnecting a battery to a capacitor. It is important to note whether modifications to the capacitor are being made while the capacitor is connected to the battery or after it is disconnected. If the capacitor remains connected to the battery, the voltage across the capacitor necessarily remains the same (equal to the battery voltage), and the charge is proportional to the capacitance, **however it may be modified** (say, by insertion of a dielectric). On the other hand, if you disconnect the capacitor from the battery before making any modifications to the capacitor, then its charge remains the same. In this case, as you vary the capacitance, the voltage across the plates changes in inverse proportion to capacitance, according to $\Delta V = Q/C$.

REVIEW CHECKLIST

▷ Use the basic definition of capacitance and the equation for finding the potential difference between two points in an electric field in order to calculate the capacitance of a capacitor for cases of relatively simple geometry—parallel plates, cylindrical, spherical.

▷ Determine the equivalent capacitance of a network of capacitors in series-parallel combination and calculate the final charge on each capacitor and the potential difference across each when a known potential is applied across the combination.

▷ Make calculations involving the relationships among potential, charge, capacitance, stored energy, and energy density for capacitors, and apply these results to the particular case of a parallel plate capacitor.

▷ Calculate the capacitance, potential difference, and stored energy of a capacitor which is partially or completely filled with a **dielectric**.

ANSWERS TO SELECTED CONCEPTUAL QUESTIONS

1. If you were asked to design a capacitor in a situation for which small size and large capacitance were required, what factors would be important in your design?

Answer You should use a dielectric filled capacitor whose dielectric constant is very large. Furthermore, you should make the dielectric as thin as possible, keeping in mind that dielectric breakdown must also be considered.

□   □   □   □

11. If the potential difference across a capacitor is doubled, by what factor does the stored energy change?

Answer Since $U = CV^2/2$, doubling V will quadruple the stored energy.

□   □   □   □

SOLUTIONS TO SELECTED END-OF-CHAPTER PROBLEMS

3.   An isolated charged conducting sphere of radius 12.0 cm creates an electric field of 4.90×10^4 N/C at a distance of 21.0 cm from its center. (a) What is its surface charge density? (b) What is its capacitance?

Solution   (a)   The electric field outside a spherical charge distribution of radius R is $E = k_e q / r^2$. Therefore, $q = E r^2 / k_e$. Since the surface charge density is $\sigma = q/A$,

$$\sigma = \frac{E r^2}{k_e 4\pi R^2} = \frac{(4.90 \times 10^4 \text{ N} / \text{C})(0.210 \text{ m})^2}{(8.99 \times 10^9 \text{ N} \cdot \text{m}^2 / \text{C}^2)(4\pi)(0.120 \text{ m})^2} = 1.33 \ \mu\text{C} / \text{m}^2 \qquad \lozenge$$

(b)   For an isolated charged sphere of radius R,

$$C = 4\pi \, \epsilon_0 \, R = (4\pi)(8.85 \times 10^{-12} \text{ C}^2 / \text{N} \cdot \text{m}^2)(0.120 \text{ m}) = 13.3 \text{ pF} \qquad \lozenge$$

7.   An air-filled capacitor consists of two parallel plates, each with area of 7.60 cm², separated by a distance of 1.80 mm. If a 20.0-V potential difference is applied to these plates, calculate (a) the electric field between the plates, (b) the surface charge density, (c) the capacitance, and (d) the charge on each plate.

Solution

(a) The potential difference between two points in a uniform electric field is $\Delta V = Ed$, so:

$$E = \frac{\Delta V}{d} = \frac{20.0 \text{ V}}{1.80 \times 10^{-3} \text{ m}} = 1.11 \times 10^4 \text{ V} / \text{m} \qquad \lozenge$$

(b) The electric field between capacitor plates is $E = \sigma/\epsilon_0$, so $\sigma = \epsilon_0 E$:

$$\sigma = \left(8.85 \times 10^{-12} \ C^2/N \cdot m^2\right)\left(1.11 \times 10^4 \ V/m\right) = 9.83 \times 10^{-8} \ C/m^2 \qquad \Diamond$$

(c) For a parallel-plate capacitor, $C = \dfrac{\epsilon_0 A}{d}$:

$$C = \frac{\left(8.85 \times 10^{-12} \ \dfrac{C^2}{N \cdot m^2}\right)\left(7.60 \times 10^{-4} \ m^2\right)}{1.80 \times 10^{-3} \ m} = 3.74 \times 10^{-12} \ F = 3.74 \ pF \qquad \Diamond$$

(d) The charge on each plate is $Q = C\Delta V$:

$$Q = \left(3.74 \times 10^{-12} \ F\right)\left(20.0 \ V\right) = 7.47 \times 10^{-11} \ C = 74.7 \ pC \qquad \Diamond$$

---

9. When a potential difference of 150 V is applied to the plates of a parallel-plate capacitor, the plates carry a surface charge density of 30.0 nC/cm². What is the spacing between the plates?

Solution We have $Q = C\Delta V$ with $C = \epsilon_0 A/d$. Thus, $Q = \epsilon_0 A\Delta V/d$. The surface charge density on each plate is the same in magnitude, so

$$\sigma = \frac{Q}{A} = \frac{\epsilon_0 \Delta V}{d}$$

Thus, $$d = \frac{\epsilon_0 \Delta V}{Q/A} = \frac{(8.85 \times 10^{-12} \ C^2/N \cdot m^2)(150 \ V)}{(30.0 \times 10^{-9} \ C/cm^2)}$$

$$d = \left(4.42 \times 10^{-2} \ \frac{V \cdot C \cdot cm^2}{N \cdot m^2}\right)\left(\frac{1 \ m^2}{10^4 \ cm^2}\right)\left(\frac{J}{V \cdot C}\right)\left(\frac{N \cdot m}{J}\right) = 4.42 \ \mu m \quad \Diamond$$

---

11. A 50.0-m length of coaxial cable has an inner conductor that has a diameter of 2.58 mm and carries a charge of 8.10 μC. The surrounding conductor has an inner diameter of 7.27 mm and a charge of –8.10 μC. (a) What is the capacitance of this cable? (b) What is the potential difference between the two conductors? Assume the region between the conductors is air.

Solution

(a) $C = \dfrac{\ell}{2k_e \ln(b/a)} = \dfrac{50.0 \text{ m}}{2(8.99 \times 10^9 \text{ N} \cdot \text{m}^2/\text{C}^2)\ln(7.27/2.58)} = 2.68 \times 10^{-9} \text{ F}$ ◊

(b) $\Delta V = \dfrac{Q}{C} = \dfrac{8.10 \times 10^{-6} \text{ C}}{2.68 \times 10^{-9} \text{ F}} = 3.02 \text{ kV}$ ◊

---

15. An air-filled spherical capacitor is constructed with inner and outer shell radii of 7.00 and 14.0 cm, respectively. (a) Calculate the capacitance of the device. (b) What potential difference between the spheres results in a charge of 4.00 μC on the capacitor?

Solution

G: Since the separation between the inner and outer shells is much larger than a typical electronic capacitor with $d \sim 0.1$ mm and capacitance in the microfarad range, we might expect the capacitance of this spherical configuration to be on the order of picofarads, (based on a factor of about 700 times larger spacing between the conductors). The potential difference should be sufficiently low to prevent sparking through the air that separates the shells.

O: The capacitance can be found from the equation for spherical shells, and the voltage can be found from $Q = C\Delta V$.

A: (a) For a spherical capacitor with inner radius a and outer radius b,

$$C = \frac{ab}{k(b-a)} = \frac{(0.0700 \text{ m})(0.140 \text{ m})}{\left(8.99 \times 10^9 \text{ N} \cdot \text{m}^2/\text{C}^2\right)(0.140 - 0.0700) \text{ m}}$$

$$C = 1.56 \times 10^{-11} \text{ F} = 15.6 \text{ pF} \qquad \Diamond$$

(b) $\quad \Delta V = \frac{Q}{C} = \frac{(4.00 \times 10^{-6} \text{ C})}{1.56 \times 10^{-11} \text{ F}} = 2.56 \times 10^5 \text{ V} = 256 \text{ kV} \qquad \Diamond$

L: The capacitance agrees with our prediction, but the voltage seems rather high. We can check this voltage by approximating the configuration as the electric field between two charged parallel plates separated by $d = 7.00$ cm, so

$$E \sim \frac{\Delta V}{d} = \frac{2.56 \times 10^5 \text{ V}}{0.0700 \text{ m}} = 3.66 \times 10^6 \text{ V} / \text{m}$$

This electric field barely exceeds the dielectric breakdown strength of air $\left(3 \times 10^6 \text{ V} / \text{m}\right)$, so it may not even be possible to place $4.00 \ \mu\text{C}$ of charge on this capacitor!

21. Four capacitors are connected as shown in Figure P26.21. (a) Find the equivalent capacitance between points a and b. (b) Calculate the charge on each capacitor if $\Delta V_{ab} = 15.0$ V.

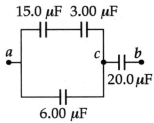

Figure P26.21

(a) We successively simplify the circuit, proceeding from the given diagram through solution figures (a) - (c).

First, the 15.0 μF and 3.00 μF in series are equivalent to

$$\frac{1}{(1/15.0\ \mu F) + (1/3.00\ \mu F)} = 2.50\ \mu F$$

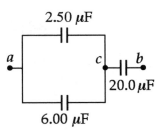

(a)

Next, 2.50 μF combines in parallel with 6.00 μF, creating an equivalent capacitance of 8.50 μF.

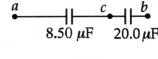

(b)

At last, 8.50 μF and 20.0 μF are in series, equivalent to

$$\frac{1}{(1/8.50\ \mu F) + (1/20.0\ \mu F)} = 5.96\ \mu F \qquad \lozenge$$

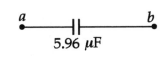

(c)

(b) We find the charge on and the voltage across each capacitor by working backwards through solution figures (c) through (a), alternately applying $Q = C\Delta V$ and $\Delta V = Q/C$ to every capacitor, real or equivalent. For the 5.96 μF capacitor, we have

$$Q = C\Delta V = (5.96\ \mu F)(15.0\ V) = 89.5\ \mu C$$

Thus, if a is higher in potential than b, just 89.5 μC flows to the right past a and past b to charge the capacitors in each picture. In (b) we have, for the 8.50 μF capacitor,

$$\Delta V_{ac} = \frac{Q}{C} = \frac{89.5 \ \mu C}{8.50 \ \mu F} = 10.5 \text{ V}$$

and for the 20.0 μF in (b), (a), and the original circuit, $Q_{20} = 89.5 \ \mu C$ ◊

$$\Delta V_{cb} = \frac{Q}{C} = \frac{89.5 \ \mu C}{20.0 \ \mu F} = 4.47 \text{ V}$$

Next, (a) is equivalent to (b), so $\Delta V_{cb} = 4.47$ V and $\Delta V_{ac} = 10.5$ V.

Thus, for the 2.50-μF and the 6.00-μF capacitors, $\Delta V = 10.5$ V

$$Q_{2.5} = C\Delta V = (2.50 \ \mu F)(10.5 \text{ V}) = 26.3 \ \mu C$$

$$Q_6 = C\Delta V = (6.00 \ \mu F)(10.5 \text{ V}) = 63.2 \ \mu C$$ ◊

Now, 26.3 μC having flowed in the upper parallel branch in (a), back in the original circuit we have

$$Q_{15} = 26.3 \ \mu C \qquad \text{and} \qquad Q_3 = 26.3 \ \mu C$$ ◊

Related Calculation   An exam problem might also ask for the voltage across each:

$$\Delta V_{15} = \frac{Q}{C} = \frac{26.3 \ \mu C}{15.0 \ \mu F} = 1.75 \text{ V}$$

and $$\Delta V_3 = \frac{Q}{C} = \frac{26.3 \ \mu C}{3.00 \ \mu F} = 8.77 \text{ V}$$ ◊

23. Consider the circuit shown in Figure P26.23, where $C_1 = 6.00 \ \mu F$, $C_2 = 3.00 \ \mu F$, and $\Delta V = 20.0$ V. Capacitor C_1 is first charged by the closing of switch S_1. Switch S_1 is then opened, and the charged capacitor is connected to the uncharged capacitor by the closing of S_2. Calculate the initial charge acquired by C_1 and the final charge on each.

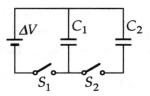

Figure P26.23

Solution

When S_1 is closed, the charge on C_1 will be

$$Q_1 = C_1 \Delta V_1 = (6.00 \ \mu F)(20.0 \ V) = 120 \ \mu C \qquad \Diamond$$

When S_1 is opened and S_2 is closed, the total charge will remain constant and be shared by the two capacitors:

$$Q_2' + Q_1' = 120 \ \mu C$$

The potential across the two capacitors will be equal.

$$\Delta V' = \frac{Q_1'}{C_1} = \frac{Q_2'}{C_2}$$

or
$$\frac{120 \ \mu C - Q_2'}{6.00 \ \mu F} = \frac{Q_2'}{3.00 \ \mu F}$$

and
$$Q_2' = 40.0 \ \mu C \qquad \Diamond$$

$$Q_1' = 120 \ \mu C - 40.0 \ \mu C = 80.0 \ \mu C \qquad \Diamond$$

27. A group of identical capacitors is connected first in series and then in parallel. The combined capacitance in parallel is 100 times larger than for the series connection. How many capacitors are in the group?

Solution

G: Since capacitors in parallel add and ones in series add as inverses, 2 capacitors in parallel would have a capacitance 4 times greater than if they were in series, and 3 capacitors would give a ratio $C_p / C_s = 9$, so maybe $n = \sqrt{C_p / C_s} = \sqrt{100} = 10$.

O: The ratio reasoning above seems like an efficient way to solve this problem, but we should check the answer with a more careful analysis based on the general relationships for series and parallel combinations of capacitors.

A: Call C the capacitance of one capacitor and n the number of capacitors. The equivalent capacitance for n capacitors in parallel is

$$C_p = C_1 + C_2 + \ldots + C_n = nC$$

The relationship for n capacitors in series is

$$\frac{1}{C_s} = \frac{1}{C_1} + \frac{1}{C_2} + \ldots + \frac{1}{C_n} = \frac{n}{C}$$

Therefore $\quad \dfrac{C_p}{C_s} = \dfrac{nC}{C/n} = n^2 \quad$ or $\quad n = \sqrt{\dfrac{C_p}{C_s}} = \sqrt{100} = 10 \qquad$ ◊

L: Our prediction appears to be correct. A qualitative reason that $C_p / C_s = n^2$ is because the amount of charge that can be stored on the capacitors increases according to the area of the plates for a parallel combination, but the total charge remains the same for a series combination.

35. A parallel-plate capacitor has a charge Q and plates of area A. Show that the force exerted on each plate by the other is $F = Q^2/2\epsilon_0 A$. (**Hint:** Let $C = \epsilon_0 A/x$ for an arbitrary plate separation x; then require that the work done in separating the two charged plates be $W = \int F\,dx$.)

Solution   The electric field in the space between the plates is

$$E = \frac{\sigma}{\epsilon_0} = \frac{Q}{A\,\epsilon_0}$$

You might think that the force on one plate is $F = QE = Q^2/A\,\epsilon_0$, but this is two times too large, because neither plate exerts a force on itself. The force **on** one plate is exerted **by** the other, through its electric field $\mathbf{E} = \sigma/2\epsilon_0 = Q/2A\,\epsilon_0$. The force on each plate is:

$$F = (Q_{\text{self}})(E_{\text{other}}) = Q^2/2A\,\epsilon_0$$

To prove this, we follow the hint, and calculate that the work done in separating the plates, which equals the potential energy stored in the charged capacitor:

$$U = \frac{1}{2}\frac{Q^2}{C} = \int F\,dx$$

From the fundamental theorem of calculus, $dU = F\,dx$, and

$$F = \frac{d}{dx}U = \frac{d}{dx}\left(\frac{Q^2}{2C}\right) = \frac{1}{2}\frac{d}{dx}\left(\frac{Q^2}{\epsilon_0\,A\,/\,x}\right)$$

Solving, $$F = \frac{1}{2}\frac{d}{dx}\left(\frac{Q^2 x}{\epsilon_0\,A}\right) = \frac{1}{2}\left(\frac{Q^2}{\epsilon_0\,A}\right) \qquad \lozenge$$

43. A commercial capacitor is constructed as in Figure 26.15a. This particular capacitor is rolled from two strips of aluminum separated by two strips of paraffin-coated paper. Each strip of foil and paper is 7.00 cm wide. The foil is 0.00400 mm thick, and the paper is 0.0250 mm thick and has a dielectric constant of 3.70. What length should the strips be if a capacitance of 9.50×10^{-8} F is desired? (Use the parallel-plate formula.)

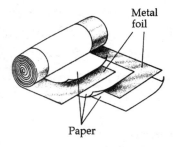

Metal foil

Paper

Figure 26.15a

Solution $C = \dfrac{\kappa \in_0 A}{d} = \dfrac{3.70(8.85 \times 10^{-12} \ C^2 / N \cdot m^2)(0.0700 \ m)L}{2.50 \times 10^{-5} \ m} = 9.50 \times 10^{-8}$ F

Solving, $L = 1.04$ m $\lozenge$

The distance between positive and negative charges is the thickness of one sheet of paper because these charges sit on adjacent surfaces of the metal foils. The electric field between the surfaces of the foil is zero and the thickness of the foil does not contribute capacitance.

---

57. A parallel-plate capacitor is constructed using a dielectric material whose dielectric constant is 3.00 and whose dielectric strength is 2.00×10^8 V/m. The desired capacitance is 0.250 μF, and the capacitor must withstand a maximum potential difference of 4000 V. Find the minimum area of the capacitor plates.

Solution $\kappa = 3.00$, $E_{max} = 2.00 \times 10^8 \ V / m = \dfrac{\Delta V_{max}}{d}$ so $d = \dfrac{\Delta V_{max}}{E_{max}}$

For $C = \dfrac{\kappa \in_0 A}{d} = 0.250 \times 10^{-6}$ F,

$A = \dfrac{Cd}{\kappa \in_0} = \dfrac{C \Delta V_{max}}{\kappa \in_0 E_{max}} = \dfrac{(0.250 \times 10^{-6} \ F)(4000 \ V)}{3.00 \left(8.85 \times 10^{-12} \ \dfrac{F}{m} \right) \left(2.00 \times 10^8 \ \dfrac{V}{m} \right)} = 0.188 \ m^2$ $\lozenge$

59. A conducting slab of a thickness d and area A is inserted into the space between the plates of a parallel-plate capacitor with spacing s and surface area A, as shown in Figure P26.59. The slab is not necessarily halfway between the capacitor plates. What is the capacitance of the system?

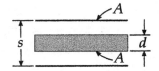

Figure P26.59

Solution

G: It is difficult to predict an exact relationship for the capacitance of this system, but we can reason that C should increase if the distance between the slab and plates were decreased (until they touched and formed a short circuit). So maybe $C \propto 1/(s-d)$. Moving the metal slab does not change the amount of charge the system can store, so the capacitance should therefore be independent of the slab position. The slab must have zero net charge, with each face of the plate holding the same magnitude of charge as the outside plates, regardless of where the slab is between the plates.

O: If the capacitor is charged with $+Q$ on the top plate and $-Q$ on the bottom plate, then free charges will move across the conducting slab to neutralize the electric field inside it, with the top face of the slab carrying charge $-Q$ and the bottom face carrying charge $+Q$. Then the capacitor and slab combination is electrically equivalent to two capacitors in series. (We are neglecting the slight fringing effect of the electric field near the edges of the capacitor.) Call x the upper gap, so that $s-d-x$ is the distance between the lower two surfaces.

A: For the upper capacitor, $\quad C_1 = \epsilon_0 \, A/x$

and the lower has $\qquad C_2 = \dfrac{\epsilon_0 \, A}{s-d-x}$

So the combination has $\quad C = \dfrac{1}{\dfrac{1}{C_1}+\dfrac{1}{C_2}} = \dfrac{1}{\dfrac{x}{\epsilon_0 \, A}+\dfrac{s-d-x}{\epsilon_0 \, A}} = \dfrac{\epsilon_0 \, A}{s-d}$ $\qquad \lozenge$

L: The equivalent capacitance is inversely proportional to $(s-d)$ as expected, and is also proportional to A. This result is the same as for the special case in Example 26.9 when the slab is just halfway between the plates; the only critical factor is the thickness of the slab relative to the plate spacing.

---

65. An isolated capacitor of unknown capacitance has been charged to a potential difference of 100 V. When the charged capacitor is then connected in parallel to an uncharged 10.0-μF capacitor, the voltage across the combination is 30.0 V. Calculate the unknown capacitance.

Solution

G: The voltage of the combination will be reduced according to the size of the added capacitance. (Example: If the unknown capacitance were $C = 10.0\ \mu F$, then $\Delta V_1 = 50.0$ V because the charge is now distributed evenly between the two capacitors.) Since the final voltage is less than half the original, we might guess that the unknown capacitor is about 5.00 μF.

O: We can use the relationships for capacitors in parallel to find the unknown capacitance, along with the requirement that the charge on the unknown capacitor must be the same as the total charge on the two capacitors in parallel.

A: We name our ignorance and call the unknown capacitance C_u. The charge originally deposited on **each** plate, + on one, − on the other, is

$$Q = C_u \Delta V = C_u (100 \text{ V})$$

Now in the new connection this same conserved charge redistributes itself between the two capacitors according to $Q = Q_1 + Q_2$.

$$Q_1 = C_u(30.0 \text{ V}) \quad \text{and} \quad Q_2 = (10.0\ \mu F)(30.0 \text{ V}) = 300\ \mu C$$

We can eliminate Q and Q_1 by substitution:

$$C_u(100 \text{ V}) = C_u(30.0 \text{ V}) + 300 \ \mu\text{C} \quad \text{so} \quad C_u = \frac{300 \ \mu\text{C}}{70.0 \text{ V}} = 4.29 \ \mu\text{F} \quad \Diamond$$

L: The calculated capacitance is close to what we expected, so our result seems reasonable. In this and other capacitance combination problems, it is important not to confuse the charge and voltage of the system with those of the individual components, especially if they have different values. Careful attention must be given to the subscripts to avoid this confusion. It is also important to not confuse the variable "C" for capacitance with the unit of charge, "C" for coulombs.

---

69. A parallel-plate capacitor of plate separation d is charged to a potential difference ΔV_0. A dielectric slab of thickness d and dielectric constant κ is introduced between the plates **while the battery remains connected to the plates.** (a) Show that the ratio of energy stored after the dielectric is introduced to the energy stored in the empty capacitor is $U / U_0 = \kappa$. Give a physical explanation for this increase in stored energy. (b) What happens to the charge on the capacitor? (Note that this situation is not the same as Example 26.7, in which the battery was removed from the circuit before the dielectric was introduced.)

Solution

(a) The capacitance changes from, say, C_0 to κC_0. The battery will maintain constant voltage across it by pumping out extra charge. The original energy is $U_0 = \frac{1}{2}C_0(\Delta V)^2$ and the final energy is $U = \frac{1}{2}\kappa C_0(\Delta V)^2$, so $U / U_0 = \kappa$. The extra energy comes from (part of the) electrical work done by the battery in separating extra charge. $\qquad \Diamond$

(b) The original charge is $Q_0 = C_0\Delta V$ and the final value is $Q = \kappa C_0\Delta V$, so the charge increases by the factor κ. $\qquad \Diamond$

---

73. The inner conductor of a coaxial cable has a radius of 0.800 mm, and the outer conductor's inside radius is 3.00 mm. The space between the conductors is filled with polyethylene, which has a dielectric constant of 2.30 and a dielectric strength of 18.0×10^6 V/m. What is the maximum potential difference that this cable can withstand?

Solution We can increase the potential difference between the core and the sheath until the electric field in the polyethylene starts to punch a hole through it, passing a spark. This will first happen at the surface of the inner conductor, where the electric field is strongest. From Example 26.2, when there is a vacuum between the conductors, the voltage between them is

$$\Delta V = |V_b - V_a| = 2 k_e \lambda \ln\left(\frac{b}{a}\right) = \frac{\lambda}{2\pi \epsilon_0} \ln\left(\frac{b}{a}\right)$$

With a dielectric, a factor $1/\kappa$ must be added,

and the equation becomes
$$\Delta V = \frac{\lambda}{2\pi \kappa \epsilon_0} \ln\left(\frac{b}{a}\right)$$

So when $E = E_{max}$ at $r = a$,
$$\frac{\lambda_{max}}{2\pi \kappa \epsilon_0} = E_{max} a$$

and
$$\Delta V_{max} = \frac{\lambda_{max}}{2\pi \kappa \epsilon_0} \ln\left(\frac{b}{a}\right) = E_{max} a \ln\left(\frac{b}{a}\right)$$

Thus, $\Delta V_{max} = \left(18.0 \times 10^6 \ \frac{V}{m}\right)\left(0.800 \times 10^{-3} \ m\right) \ln\left(\frac{3.00 \ mm}{0.800 \ mm}\right) = 19.0 \ kV$ ◊

Chapter 27

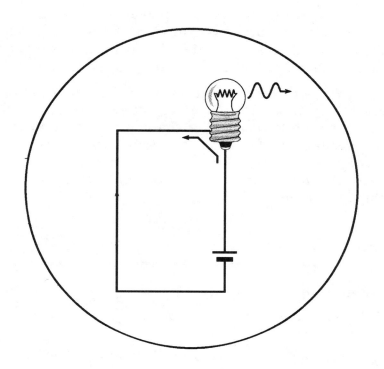

Current and Resistance

Chapter 27

CURRENT AND RESISTANCE

INTRODUCTION

Thus far our discussion of electrical phenomena has been confined to charges at rest, or electrostatics. We now consider situations involving electric charges in motion. The term **electric current,** or simply **current,** is used to describe the rate of flow of charge through some region of space.

In this chapter we first define current and current density. A microscopic description of current is given, and some of the factors that contribute to the resistance to the flow of charge in conductors are discussed. Mechanisms responsible for the electrical resistance of various materials depend on the composition of the material and on temperature. A classical model is used to describe electrical conduction in metals, and some of the limitations of this model are pointed out.

EQUATIONS AND CONCEPTS

Under the action of an electric field, electric charges will move through gases, liquids, and solid conductors. **Electric current,** I, is defined as the rate at which charge moves through a cross section of the conductor.

$$I \equiv \frac{dQ}{dt} \qquad (27.2)$$

The direction of the current is in the direction of the flow of positive charges. The SI unit of current is the **ampere** (A).

$$1\,A = 1\,C/1\,s \qquad (27.3)$$

The current in a conductor can be related to the number of mobile charge carriers per unit volume n; the charge quantity q associated with each carrier; and the **drift velocity** v_d of the carriers.

$$I_{av} = \frac{\Delta Q}{\Delta t} = nqv_d A \qquad (27.4)$$

When the surface of the cross-sectional area A is perpendicular to the direction of the current and the current density is uniform, the magnitude of the current density can be expressed as in Equation 27.5.

$$J \equiv \frac{I}{A} = nqv_d \qquad (27.5)$$

The **current density** in a conductor, **J**, is a vector quantity which is in the direction of motion for positive charge carriers. For ohmic materials (most metals), current density is proportional to the electric field in the conductor.

$$\mathbf{J} = nq\mathbf{v}_d \qquad (27.6)$$

$$\mathbf{J} = \sigma\mathbf{E} \qquad (27.7)$$

A form of Ohm's law more useful to many applications relates the potential difference across a conductor and the current in the conductor to a composite of several physical characteristics of the conductor called the **resistance**, R.

$$R \equiv \frac{\Delta V}{I} \qquad (27.8)$$

The **resistance** of a given conductor of uniform cross section depends on the length, cross-sectional area, and a characteristic property of the material of which the conductor is made. The parameter ρ is the **resistivity** of the conducting material.

$$R = \rho\frac{\ell}{A} \qquad (27.11)$$

The resistivity is the inverse of the **conductivity** and has units of ohm-meters. The unit of resistance is the ohm (Ω).

$$1\,\Omega = 1\,V/A$$

The average time between collisions with atoms of a metal is important in the description of the classical model of electronic conduction in metals. This characteristic time, denoted by τ, can be related to the drift velocity (Eq. 27.4) or the conductor's resistivity (Eq. 27.11). Here, m and q represent the mass and charge of the electron, $\mathbf{E}$ is the applied electric field, and n is the number of free electrons per unit volume.

$$\mathbf{v}_d = \frac{q\mathbf{E}}{m_e}\tau \qquad (27.14)$$

$$\rho = \frac{1}{\sigma} = \frac{m_e}{nq^2\tau} \qquad (27.17)$$

The **resistivity** and therefore the resistance of a conductor vary with temperature in an approximately linear manner. In these equations, α is the **temperature coefficient of resistance**. T_0 is a stated reference temperature (usually 20°C).

$$\rho = \rho_0\left[1 + \alpha(T - T_0)\right] \qquad (27.19)$$

$$R = R_0\left[1 + \alpha(T - T_0)\right] \qquad (27.21)$$

Power is supplied to a resistor or other current-carrying device when a potential difference is maintained between the terminals of the circuit element. The quantities can be related by Joule's law; the SI unit of power is the watt (W). For a device that obeys Ohm's law, the power dissipated can be expressed in alternative forms.

$$\mathcal{P} = I\Delta V \qquad (27.22)$$

$$\mathcal{P} = I^2 R = \frac{(\Delta V)^2}{R} \qquad (27.23)$$

SUGGESTIONS, SKILLS, AND STRATEGIES

Equation 27.11, $R = \rho \ell/A$, can be used directly to calculate the resistance of a conductor of uniform cross-sectional area and constant resistivity. For those cases in which the area, resistivity, or both vary along the length of the conductor, the resistance must be determined as an integral of dR.

The conductor is subdivided into elements of length dx over which ρ and A may be considered constant in value and the total resistance is

$$R = \int \frac{\rho}{A} dx$$

Consider, for example, the case of a truncated cone of constant resistivity, radii a and b and height h. The conductor should be subdivided into disks of thickness dx, radius r, area = πr^2 and oriented parallel to the faces of the cone as shown in the figure below. Note from the geometry that

$$x = \left(\frac{r - a}{b - a} \right) h$$

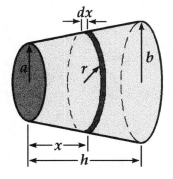

so that
$$r = \frac{x}{h}(b - a) + a$$

and
$$R = \int_0^h \frac{\rho}{\pi r^2} dx$$

The remainder of this calculation is left as a problem for you to work out (see Problem 27.70 of the text).

REVIEW CHECKLIST

▷ Define the term, electric current, in terms of rate of charge flow, and its corresponding unit of measure, the ampere. Calculate electron drift velocity, and quantity of charge passing a point in a given time interval in a specified current-carrying conductor.

▷ Determine the resistance of a conductor using Ohm's law. Also, calculate the resistance based on the physical characteristics of a conductor. Distinguish between ohmic and nonohmic conductors.

▷ Make calculations of the variation of resistance with temperature, which involves the concept of the temperature coefficient of resistivity.

▷ Use Joule's law to calculate the power dissipated in a resistor.

ANSWERS TO SELECTED CONCEPTUAL QUESTIONS

3. Two wires A and B of circular cross-section are made of the same metal and have equal lengths, but the resistance of wire A is three times greater than that of wire B. What is the ratio of their cross-sectional areas? How do their radii compare?

Answer Since $R = \rho \ell \,/\, \text{Area}$, the ratio of resistances is given by

$$R_A \,/\, R_B = \text{Area}_B \,/\, \text{Area}_A$$

Hence, the ratio of their areas is three to one; that is, the area of wire B is three times that of wire A. From the ratio of the areas, we can calculate that the radius of wire B is $\sqrt{3}$ times the radius of wire A.

□   □   □   □

9. On the basis of the atomic theory of matter, explain why the resistance of a material should increase as its temperature increases.

Answer As the temperature increases, the amplitude of atomic vibrations increases. This makes it more likely that the electrons will be scattered by atomic vibrations, and makes it more difficult for charges to move inside the conductor.

□　　□　　□　　□

14. If charges flow very slowly through a metal, why does it not require several hours for a light to turn on when you throw a switch?

Answer Individual electrons move with a small average velocity through the conductor, but as soon as the voltage is applied, electrons all along the conductor start to move. Actually, the current does not flow "immediately," but is limited by the speed of light.

□　　□　　□　　□

SOLUTIONS TO SELECTED END-OF-CHAPTER PROBLEMS

3. Suppose that the current through a conductor decreases exponentially with time according to

$$I(t) = I_o e^{-t/\tau}$$

where I_o is the initial current (at $t = 0$), and τ is a constant having dimensions of time. Consider a fixed observation point within the conductor. (a) How much charge passes this point between $t = 0$ and $t = \tau$? (b) How much charge passes this point between $t = 0$ and $t = 10\tau$? (c) How much charge passes this point between $t = 0$ and $t = \infty$?

Solution

From $I = dQ/dt$, we have $\qquad\qquad\qquad dQ = I\, dt$

From this, we derive a general integral $\qquad Q = \int_0^Q dQ = \int_0^T I\, dt$

In all three cases, define an end-time, T. $\qquad Q = \int_0^T I_o e^{-t/\tau} dt$

Integrating from time $t = 0$ to time $t = T$, $\qquad Q = \int_0^T (-I_o \tau) e^{-t/\tau}(-dt/\tau)$

$$Q = I_o \tau \left(1 - e^{-T/\tau}\right)$$

(a) If $T = \tau$ $\qquad\qquad\qquad\qquad Q = I_o \tau (1 - e^{-1}) = 0.6321\, I_o \tau \qquad \Diamond$

(b) If $T = 10\tau$, $\qquad\qquad\qquad\qquad Q = I_o \tau \left(1 - e^{-10}\right) = 0.99995 I_o \tau \qquad \Diamond$

(c) If $T = \infty$, $\qquad\qquad\qquad\qquad Q = I_o \tau \left(1 - e^{-\infty}\right) = I_o \tau \qquad\qquad \Diamond$

---

11. The electron beam emerging from a certain high-energy electron accelerator has a circular cross section of radius 1.00 mm. (a) If the beam current is 8.00 μA, what is the current density in the beam, assuming that it is uniform throughout? (b) The speed of the electrons is so close to the speed of light that their speed can be taken as $c = 3.00 \times 10^8$ m/s with negligible error. Find the electron density in the beam. (c) How long does it take for Avogadro's number of electrons to emerge from the accelerator?

Solution

(a) The current density is $\quad J = \dfrac{I}{A} = \dfrac{8.00 \times 10^{-6} \text{ A}}{\pi (1.00 \times 10^{-3} \text{ m})^2} = 2.55 \text{ A/m}^2 \quad \lozenge$

(b) From $\; J = nev_d$, we have $\quad n = \dfrac{J}{ev_d} = \dfrac{2.55 \text{ A/m}^2}{(1.60 \times 10^{-19} \text{ C})(3.00 \times 10^8 \text{ m/s})}$

$$n = 5.31 \times 10^{10} \text{ m}^{-3} \qquad \lozenge$$

(c) From $\; I = \Delta Q / \Delta t$, we have $\quad \Delta t = \dfrac{\Delta Q}{I} = \dfrac{N_A e}{I} = \dfrac{(6.02 \times 10^{23})(1.60 \times 10^{-19} \text{ C})}{8.00 \times 10^{-6} \text{ A}}$

$$\Delta t = 1.20 \times 10^{10} \text{ s} \quad \text{(or about 381 years!)} \qquad \lozenge$$

---

15. A 0.900-V potential difference is maintained across a 1.50-m length of tungsten wire that has a cross-sectional area of 0.600 mm². What is the current in the wire?

Solution

From Ohm's law, $\quad I = \dfrac{\Delta V}{R} \quad$ where $\quad R = \dfrac{\rho \ell}{A}$

Therefore, $\quad I = \dfrac{\Delta V A}{\rho \ell} = \dfrac{(0.900 \text{ V})(6.00 \times 10^{-7} \text{ m}^2)}{(5.6 \times 10^{-8} \; \Omega \cdot \text{m})(1.50 \text{ m})} = 6.43 \text{ A} \qquad \lozenge$

---

17. Suppose that you wish to fabricate a uniform wire out of 1.00 g of copper. If the wire is to have a resistance of $R = 0.500\ \Omega$, and all of the copper is to be used, what will be (a) the length and (b) the diameter of this wire?

Solution

Don't mix up symbols! Call the density ρ_d and the resistivity ρ_r. From $\rho_d = m/V$, the volume is $V = A\ell = m/\rho_d$. The resistance is $R = \rho_r \ell/A$.

(a) We can solve for L by eliminating A:

$$A = \frac{m}{\ell \rho_d} \qquad R = \frac{\rho_r \ell}{(m/\ell \rho_d)} \qquad R = \frac{\rho_r \rho_d \ell^2}{m}$$

$$\ell = \sqrt{\frac{mR}{\rho_r \rho_d}} = \sqrt{\frac{(1.00 \times 10^{-3}\ \text{kg})(0.500\ \Omega)}{(1.70 \times 10^{-8}\ \Omega \cdot \text{m})(8.92 \times 10^3\ \text{kg}/\text{m}^3)}} = 1.87\ \text{m} \qquad \Diamond$$

(b) To have a single diameter, the wire has a circular cross section:

$$A = \pi r^2 = \pi \left(\frac{d}{2}\right)^2 = \frac{m}{\ell \rho_d}$$

$$d = \sqrt{\frac{4m}{\pi \ell \rho_d}} = \sqrt{\frac{4(1.00 \times 10^{-3}\ \text{kg})}{\pi(1.82\ \text{m})(8.92 \times 10^3\ \text{kg}/\text{m}^3)}} = 2.80 \times 10^{-4}\ \text{m}$$

$$d = 0.280\ \text{mm} \qquad \Diamond$$

25. If the drift velocity of free electrons in a copper wire is 7.84×10^{-4} m/s, what is the electric field in the conductor?

Solution

G: For electrostatic cases, we learned that the electric field inside a conductor is always zero. On the other hand, if there is a current, a non-zero electric field must be maintained by a battery or other source to make the charges flow. Therefore, we might expect the electric field to be small, but definitely **not** zero.

O: The drift velocity of the electrons can be used to find the current density, which can be used with Ohm's law to find the electric field inside the conductor.

A: We first need the electron density in copper, which from Example 27.1 is $n = 8.49 \times 10^{28}$ e$^-$/m^3. The current density in this wire is then

$$J = nqv_d = (8.49 \times 10^{28} \text{ e}^-/\text{m}^3)(1.60 \times 10^{-19} \text{ C/e}^-)(7.84 \times 10^{-4} \text{ m/s})$$

$$J = 1.06 \times 10^7 \text{ A/m}^2$$

Ohm's law can be stated as $J = \sigma E = E/\rho$ where $\rho = 1.7 \times 10^{-8}$ $\Omega \cdot$m for copper, so then

$$E = \rho J = (1.70 \times 10^{-8} \text{ }\Omega \cdot \text{m})(1.06 \times 10^7 \text{ A/m}^2) = 0.181 \text{ V/m} \qquad \lozenge$$

L: This electric field is certainly smaller than typical static values outside charged objects. The direction of the electric field should be along the length of the conductor, otherwise the electrons would be forced to leave the wire! The reality is that excess charges arrange themselves on the surface of the wire to create an electric field that "steers" the free electrons to flow along the length of the wire from low to high potential (opposite the direction of a positive test charge). It is also interesting to note that when the electric field is being established it travels at the speed of light; but the drift velocity of the electrons is literally at a "snail's pace"!

31. An aluminum wire with a diameter of 0.100 mm has a uniform electric field with a magnitude of 0.200 V/m imposed along its entire length. The temperature of the wire is 50.0 °C. Assume one free electron per atom. (a) Using the information given in Table 27.1, determine the resistivity. (b) What is the current density in the wire? (c) What is the total current in the wire? (d) What is the drift speed of the conduction electrons? (e) What potential difference must exist between the ends of a 2.00-m length of the wire if the stated electric field is to be produced?

Solution The resistivity is found from $\rho = \rho_0(1 + \alpha(T - T_0))$:

(a) $\rho = \left(2.82 \times 10^{-8} \ \Omega \cdot m\right)\left(1 + \left(3.90 \times 10^{-3} \ °C^{-1}\right)(30.0 \ °C)\right) = 3.15 \times 10^{-8} \ \Omega \cdot m$ ◊

(b) $J = \sigma E = E / \rho = \dfrac{(0.200 \ V / m)}{\left(3.15 \times 10^{-8} \ \Omega \cdot m\right)}\left(1.00 \dfrac{\Omega \cdot A}{V}\right) = 6.35 \times 10^6 \ A / m^2$ ◊

(c) $J = \dfrac{I}{A} = \dfrac{I}{\pi r^2}$ $\quad I = J \pi r^2 = \left(6.35 \times 10^6 \ \dfrac{A}{m^2}\right)\pi(5.00 \times 10^{-5} \ m)^2 = 49.9 \ mA$ ◊

(d) The number-density of free electrons is given by the mass density; we assume that each atom donates one free electron:

$$n = \left(2.70 \times 10^3 \ \dfrac{kg}{m^3}\right)\left(\dfrac{1 \ mole}{26.98 \ g}\right)\left(\dfrac{10^3 \ g}{kg}\right)\left(\dfrac{6.02 \times 10^{23} \ free \ e^-}{1 \ mole}\right) = 6.02 \times 10^{28} \ \dfrac{e^-}{m^3}$$

Now $J = nqv_d$ gives:

$$v_d = \dfrac{J}{nq} = \dfrac{\left(6.35 \times 10^6 \ A / m^2\right)}{\left(6.02 \times 10^{28} \ e^- / m^3\right)\left(-1.60 \times 10^{-19} \ C / e^-\right)} = -6.59 \times 10^{-4} \ m / s$$ ◊

The sign indicates that the electrons drift opposite to the field and current.

(e) $\Delta V = E\ell = (0.200 \ V/m)(2.00 \ m) = 0.400 \ V$ ◊

33. What is the fractional change in the resistance of an iron filament when its temperature changes from 25.0 °C to 50.0 °C?

Solution $\quad R = R_0[1 + \alpha \, \Delta T] \quad$ or $\quad R - R_0 = R_0 \alpha \, \Delta T$

The fractional change in resistance is $\quad f = (R - R_0)/R_0$

Therefore, $\quad f = \dfrac{R_0 \, \alpha \, \Delta T}{R_0} = \alpha \, \Delta T = \left(5.00 \times 10^{-3} \, °C^{-1}\right)\left(50.0 \, °C - 25.0 \, °C\right) = 0.125 \quad \Diamond$

39. What is the required resistance of an immersion heater that increases the temperature of 1.50 kg of water from 10.0 °C to 50.0 °C in 10.0 min while operating at 110 V?

Solution

Assume that $\qquad E_{(thermal)} = E_{(electrical)}$

Since $\qquad E_{(thermal)} = mc \, \Delta T \quad$ and $\quad E_{(electrical)} = \mathcal{P}t = \left(\dfrac{\Delta V^2}{R}\right)t$

where $\qquad c = 4186 \, J / kg \cdot °C$

The resistance is $\quad R = \dfrac{\Delta V^2 t}{cm \, \Delta T} = \dfrac{(110 \, V)^2 (600 \, s)}{(4186 \, J / kg \cdot °C)(1.50 \, kg)(40.0 \, °C)} = 28.9 \, \Omega \quad \Diamond$

41. Suppose that a voltage surge produces 140 V for a moment. By what percentage does the power output of a 120-V, 100-W light bulb increase? (Assume that its resistance does not change.)

Solution

G: The voltage increases by about 20%, but since $\mathcal{P} = (\Delta V)^2 / R$, the power will increase as the square of the voltage:

$$\frac{\mathcal{P}_f}{\mathcal{P}_i} = \frac{(\Delta V_f)^2 / R}{(\Delta V_i)^2 / R} = \frac{(140\ \text{V})^2}{(120\ \text{V})^2} = 1.361 \text{ or a } 36.1\% \text{ increase.}$$

O: We have already found an answer to this problem by reasoning in terms of ratios, but we can also calculate the power explicitly for the bulb and compare with the original power by using Ohm's law and the equation for electrical power. To find the power, we must first find the resistance of the bulb, which should remain relatively constant during the power surge (we can check the validity of this assumption later).

A: From $\mathcal{P} = (\Delta V)^2 / R$, we find that $\qquad R = \dfrac{(\Delta V_i)^2}{\mathcal{P}} = \dfrac{(120\ \text{V})^2}{100\ \text{W}} = 144\ \Omega$

The final current is, $\qquad I_f = \dfrac{\Delta V_f}{R} = \dfrac{140\ \text{V}}{144\ \Omega} = 0.972\ \text{A}$

The power during the surge is $\qquad \mathcal{P}_f = \dfrac{(\Delta V_f)^2}{R} = \dfrac{(140\ \text{V})^2}{144\ \Omega} = 136\ \text{V}$

So the percentage increase is $\qquad \dfrac{136\ \text{W} - 100\ \text{W}}{100\ \text{W}} = 0.361 = 36.1\% \quad \lozenge$

L: Our result tells us that this 100-W light bulb momentarily acts like a 136-W light bulb, which explains why it would suddenly get brighter. Some electronic devices (like computers) are sensitive to voltage surges like this, which is the reason that **surge protectors** are recommended to protect these devices from being damaged.

In solving this problem, we assumed that the resistance of the bulb did not change during the voltage surge, but we should check this assumption. Let us assume that the filament is made of tungsten and that its resistance will change linearly with temperature according to equation 27.21. Let us further assume that the increased voltage lasts for a time long enough so that the filament comes to a new equilibrium temperature. The temperature change can be estimated from the power surge according to Stefan's law (equation 20.18), assuming that all the power loss is due to radiation. By this law, $T \propto \sqrt[4]{P}$ so that a 36% change in power should correspond to only about a 8% increase in temperature. A typical operating temperature of a white light bulb is about 3000 °C, so $\Delta T \approx 0.08(3273 \text{ °C}) = 260 \text{ °C}$. Then the increased resistance would be roughly

$$R = R_0\left(1 + \alpha(T - T_0)\right) = (144 \text{ } \Omega)\left(1 + 4.5 \times 10^{-3}(260)\right) \cong 310 \text{ } \Omega$$

It appears that the resistance could change double from 144 Ω. On the other hand, if the voltage surge lasts only a very short time, the 136 W we calculated originally accurately describes the conversion of electrical into internal energy in the filament.

47. Compute the cost per day of operating a lamp that draws 1.70 A from a 110-V line if the cost of electrical energy is $0.0600/kWh.

Solution   The power of the lamp is $\mathcal{P} = \Delta VI = U/t$, where U is the energy transformed. Then the energy you buy, in standard units, is

$$U = \Delta V I t = (110 \text{ V})(1.70 \text{ A})(1 \text{ day})\left(\frac{24 \text{ h}}{1 \text{ day}}\right)\left(\frac{3600 \text{ s}}{\text{h}}\right)\left(\frac{1 \text{ J}}{\text{V} \cdot \text{C}}\right)\left(\frac{1 \text{ C}}{\text{A} \cdot \text{s}}\right) = 16.2 \text{ MJ}$$

In kilowatt hours,

$$U = \Delta V I t = (110 \text{ V})(1.70 \text{ A})(1 \text{ day})\left(\frac{24 \text{ h}}{1 \text{ day}}\right)\left(\frac{\text{J}}{\text{V} \cdot \text{C}}\right)\left(\frac{\text{C}}{\text{A} \cdot \text{s}}\right)\left(\frac{\text{W} \cdot \text{s}}{\text{J}}\right) = 4.49 \text{ kW} \cdot \text{h}$$

So the lamp costs $(4.49 \text{ kWh})\left(\dfrac{\$0.0600}{\text{kWh}}\right) = 26.9 \text{ cents}$ ◊

---

49. A certain toaster has a heating element made of Nichrome resistance wire. When the toaster is first connected to a 120-V source of potential difference (and the wire is at a temperature of 20.0 °C), the initial current is 1.80 A. However, the current begins to decrease as the resistive element warms up. When the toaster has reached its final operating temperature, the current has dropped to 1.53 A. (a) Find the power the toaster consumes when it is at its operating temperature. (b) What is the final temperature of the heating element?

Solution

G: Most toasters are rated at about 1000 W (usually stamped on the bottom of the unit), so we might expect this one to have a similar power rating. The temperature of the heating element should be hot enough to toast bread but low enough that the nickel-chromium alloy element does not melt. (The melting point of nickel is 1455 °C, and chromium melts at 1907 °C.)

O: The power can be calculated directly by multiplying the current and the voltage. The temperature can be found from the linear conductivity equation for Nichrome, with $\alpha = 0.4 \times 10^{-3}$ °C^{-1} from Table 27.1.

A: (a) $\mathcal{P} = \Delta VI = (120 \text{ V})(1.53 \text{ A}) = 184 \text{ W}$ ◊

(b) The resistance at 20.0 °C is

$$R_0 = \frac{\Delta V}{I} = \frac{120 \text{ V}}{1.80 \text{ A}} = 66.7 \text{ }\Omega$$

At operating temperature,

$$R = \frac{120 \text{ V}}{1.53 \text{ A}} = 78.4 \text{ }\Omega$$

Neglecting thermal expansion, we have

$$R = \frac{\rho\ell}{A} = \frac{\rho_0(1 + \alpha(T - T_0))\ell}{A} = R_0(1 + \alpha(T - T_0))$$

$$T = T_0 + \frac{R/R_0 - 1}{\alpha} = 20.0 \text{ °C} + \frac{78.4 \text{ }\Omega/66.7 \text{ }\Omega - 1}{0.4 \times 10^{-3} \text{ °C}^{-1}} = 461 \text{ °C} \quad ◊$$

L: Although this toaster appears to use significantly less power than most, the temperature seems high enough to toast a piece of bread in a reasonable amount of time. In fact, the temperature of a typical 1000-W toaster would only be slightly higher because Stefan's radiation law (equation 20.18) tells us that (assuming all power is lost through radiation) $T \propto \sqrt[4]{\mathcal{P}}$, so that the temperature might be about 700 °C. In either case, the operating temperature is well below the melting point of the heating element.

55. A more general definition of the temperature coefficient of resistivity is

$$\alpha = \frac{1}{\rho}\frac{d\rho}{dT}$$

where ρ is the resistivity at temperature T. (a) Assuming that α is constant, show that

$$\rho = \rho_0 e^{\alpha(T-T_0)}$$

where ρ_0 is the resistivity at temperature T_0. (b) Using the series expansion ($e^x \cong 1+x$ for $x \ll 1$), show that the resistivity is given approximately by the expression $\rho = \rho_0[1+\alpha(T-T_0)]$ for $\alpha(T-T_0) \ll 1$.

Solution

(a) We are given $$\alpha = \frac{1}{\rho}\frac{d\rho}{dT}$$

Separating variables, $$\int_{\rho_0}^{\rho}\frac{d\rho}{\rho} = \int_{T_0}^{T}\alpha\,dT$$

Integrating both sides, $$\ln(\rho/\rho_0) = \alpha(T-T_0)$$

Thus $$\rho = \rho_0 e^{\alpha(T-T_0)}\qquad\Diamond$$

(b) From the series expansion $e^x \approx 1+x$, with x much less than 1,

$$\rho = \rho_0[1+\alpha(T-T_0)]\qquad\Diamond$$

57. An experiment is conducted to measure the electrical resistivity of Nichrome in the form of wires with different lengths and cross-sectional areas. For one set of measurements, a student uses 30 gauge wire, which has a cross-sectional area of 7.30×10^{-8} m². The student measures the potential difference across the wire and the current in the wire with a voltmeter and ammeter, respectively. For each of the measurements given in the table below taken on wires of three different lengths, calculate the resistance of the wires and the corresponding values of the resistivity. What is the average value of the resistivity, and how does it compare with the value given in Table 27.1?

ℓ (m)	V (V)	I (A)	R (Ω)	ρ ($\Omega \cdot$m)
0.540	5.22	0.500		
1.028	5.82	0.276		
1.543	5.94	0.187		

Solution For each row in the table, first find the resistance from $R = \Delta V/I$, and then find the resistivity from $R = \rho \ell/A$ or $\rho = RA/\ell$.

In the first row, $R = \dfrac{\Delta V}{I} = \dfrac{5.22 \text{ V}}{0.500 \text{ A}} = 10.4 \ \Omega$

and $\rho = \dfrac{RA}{\ell} = \dfrac{(10.4 \ \Omega)(7.30 \times 10^{-8} \text{ m}^2)}{0.540 \text{ m}} = 1.41 \times 10^{-6} \ \Omega \cdot \text{m}$

Applying this to each entry, we obtain:

ℓ (m)	R (Ω)	ρ ($\Omega \cdot$m)
0.540	10.4	1.41×10^{-6}
1.028	21.1	1.50×10^{-6}
1.543	31.8	1.50×10^{-6}

Thus the average resistivity is $\rho = 1.47 \times 10^{-6} \ \Omega \cdot \text{m}$ ◊

This differs from the tabulated $1.50 \times 10^{-6} \ \Omega \cdot \text{m}$ by 2%. The difference is accounted for by the experimental uncertainty, which we may estimate as $(1.47 - 1.41)/1.47 = 4\%$.

63. An electric car is designed to run off a bank of 12.0-V batteries with total energy storage of 2.00×10^7 J. (a) If the electric motor draws 8.00 kW, what is the current delivered to the motor? (b) If the electric motor draws 8.00 kW as the car moves at a steady speed of 20.0 m/s, how far will the car travel before it is "out of juice"?

Solution

(a) Since $\mathcal{P} = \Delta VI$
$$I = \frac{\mathcal{P}}{\Delta V} = \frac{8.00 \times 10^3 \text{ W}}{12.0 \text{ V}} = 667 \text{ A} \qquad \lozenge$$

(b) The time the car runs is
$$t = \frac{U}{\mathcal{P}} = \left(\frac{2.00 \times 10^7 \text{ J}}{8.00 \times 10^3 \text{ W}} \right) \left(\frac{1 \text{ W} \cdot \text{s}}{\text{J}} \right) = 2.50 \times 10^3 \text{ s}$$

So it moves a distance of
$$x = vt = (20.0 \text{ m / s})(2.50 \times 10^3 \text{ s}) = 50.0 \text{ km} \qquad \lozenge$$

---

69. Material with uniform resistivity ρ is formed into a wedge as shown in Figure P27.69. Show that the resistance between face A and face B of this wedge is

$$R = \rho \frac{L}{w(y_2 - y_1)} \ln\left(\frac{y_2}{y_1} \right)$$

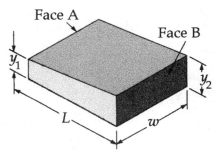

Figure P27.69

Solution

The current flows generally parallel to L. Consider a slice of the material perpendicular to this current, of thickness dx, and at distance x from face A.

Then the other dimensions of the slice are w and y, where

$$\frac{y - y_1}{x} = \frac{y_2 - y_1}{L} \quad \text{by proportion, so} \quad y = y_1 + (y_2 - y_1)\frac{x}{L}$$

The bit of resistance which this slice contributes is

$$dR = \frac{\rho \, dx}{A} = \frac{\rho \, dx}{wy} = \frac{\rho \, dx}{w(y_1 + (y_2 - y_1)x / L)}$$

and the whole resistance is that of all the slices:

$$R = \int_{x=0}^{L} dR = \int_{0}^{L} \frac{\rho \, dx}{w(y_1 + (y_2 - y_1)x / L)}$$

$$R = \frac{\rho}{w} \frac{L}{y_2 - y_1} \int_{x=0}^{L} \frac{((y_2 - y_1) / L) dx}{y_1 + (y_2 - y_1)x / L}$$

With $\quad u = y_1 + (y_2 - y_1)x / L,\quad$ this is of the form $\quad \int du / u,\quad$ so

$$R = \frac{\rho L}{w(y_2 - y_1)} \ln \left(y_1 + (y_2 - y_1)x / L \right) \Biggr]_{x=0}^{L}$$

$$R = \frac{\rho L}{w(y_2 - y_1)} (\ln y_2 - \ln y_1)$$

$$R = \frac{\rho L}{w(y_2 - y_1)} \ln \frac{y_2}{y_1} \qquad \diamond$$

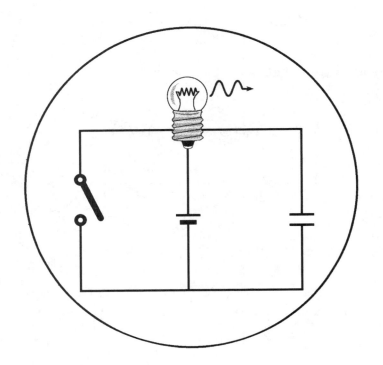

Direct Current Circuits

DIRECT CURRENT CIRCUITS

INTRODUCTION

This chapter analyzes some simple circuits whose elements include batteries, resistors, and capacitors in various combinations. Such analysis is simplified by the use of two rules known as **Kirchhoff's rules**, which follow from the laws of conservation of energy and conservation of charge. Most of the circuits are assumed to be in a **steady state**, which means that the currents are constant in magnitude and direction. We close the chapter with a discussion of circuits containing resistors and capacitors, in which current varies with time.

EQUATIONS AND CONCEPTS

When a battery is providing a current to an external circuit, the **terminal voltage** of the battery will be less than the emf due to **internal resistance** of the battery.

$$\Delta V = \mathcal{E} - Ir \tag{28.1}$$

The **current**, I, delivered by a battery in a simple **dc** circuit, depends on the value of the **emf** of the source, $\mathcal{E}$; the total **load resistance** in the circuit, R; and the **internal resistance** of the source, r.

$$I = \frac{\mathcal{E}}{R + r} \tag{28.3}$$

The total or equivalent resistance of a series combination of resistors is equal to the sum of the resistances of the individual resistors.

$$R_{eq} = R_1 + R_2 + R_3 + \ldots \tag{28.6}$$

(Series combination)

A group of resistors connected in parallel has an equivalent resistance which is less than the smallest individual value of resistance in the group.

$$\frac{1}{R_{eq}} = \frac{1}{R_1} + \frac{1}{R_2} + \frac{1}{R_3} + \ldots \qquad (28.8)$$

(Parallel combination)

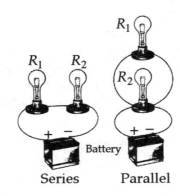

Series   Parallel

Resistors in series are connected so that they have only one common circuit point per pair; there is a common current through each resistor in the group.

Resistors in parallel are connected so that each resistor in the group has two circuit points in common with each of the other resistors; there is a common potential difference across each resistor in the group.

Many circuits which contain several resistors can be reduced to an equivalent single-loop circuit by successive step-by-step combinations of groups of resistors in series and parallel.

In the most general case, however, successive reduction is not possible and you must solve a true multiloop circuit by use of Kirchhoff's rules. Review the procedure suggested in the next section to apply Kirchhoff's rules.

When a potential difference is suddenly applied across an uncharged capacitor, the **current** I in the circuit and the **charge** q on the capacitor are functions of time with **instantaneous values** that depend on the capacitance, the resistance in the circuit. and the potential difference.

$$I(t) = \frac{\mathcal{E}}{R} e^{-t/RC} \tag{28.15}$$

$$q(t) = C\mathcal{E}\left[1 - e^{-t/RC}\right] \tag{28.14}$$

When a battery is used to charge a capacitor in series with a resistor, a quantity τ, called the time constant of the circuit, is used to describe the manner in which the charge on the capacitor varies with time. The charge on the capacitor increases from zero to 63% of its maximum value in a time interval equal to one time constant. Also, during one time constant, the charging current decreases from its initial maximum value of $I_0 = \mathcal{E}/R$ to 37% of I_0.

$$\tau = RC$$

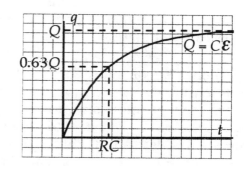

When a capacitor with an initial charge Q is discharged through a resistor, the **charge** and **current** decrease exponentially in time. In these equations, the initial current is $I_0 = Q/RC$.

$$q(t) = Qe^{-t/RC} \tag{28.17}$$

$$I(t) = -\frac{Q}{RC} e^{-t/RC} \tag{28.18}$$

where $Q/RC = I_0$

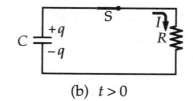

(a) $t < 0$ (b) $t > 0$

SUGGESTIONS, SKILLS, AND STRATEGIES

PROBLEM-SOLVING STRATEGY FOR RESISTORS

- Be careful with your choice of units. To calculate the resistance of a device in ohms, make sure that distances are in meters and use the SI value of ρ.

- When two or more unequal resistors are connected in **series,** they carry the same current, but the potential differences across them are not the same. The resistors add directly to give the equivalent resistance of the series combination.

- When two or more unequal resistors are connected in **parallel**, the potential differences across them are the same. Since the current is inversely proportional to the resistance, the currents through them are not the same. The equivalent resistance of a parallel combination of resistors is found through reciprocal addition, and the equivalent resistance is always **less** than the smallest individual resistor.

- A complicated circuit consisting of resistors can often be reduced to a simple circuit containing only one resistor. To do so, repeatedly examine the circuit and replace any resistors that are in series or in parallel, using the procedures outlined above. Sketch the new circuit after each set of changes has been made. Continue this process until a single equivalent resistance is found.

- If the current through, or the potential difference across, a resistor in the complicated circuit is to be identified, start with the final equivalent circuit found in the last step, and gradually work your way back through the circuits, using $V = IR$ to find the voltage drop across each equivalent resistor.

STRATEGY FOR USING KIRCHHOFF'S RULES

- First, draw the circuit diagram and assign labels and symbols to all the known and unknown quantities. You must assign **directions** to the currents in each part of the circuit. Do not be alarmed if you guess the direction of a current incorrectly; the resulting value will be negative, but **its magnitude will be correct.** Although the assignment of current directions is arbitrary, you must stick with your guess throughout as you apply Kirchhoff's rules.

- Apply the junction rule to any junction in the circuit. The junction rule may be applied as many times as a new current (one not used in a previous application) appears in the resulting equation. In general, the number of times the junction rule can be used is one fewer than the number of junction points in the circuit.

- Now apply Kirchhoff's loop rule to as many loops in the circuit as are needed to solve for the unknowns. Remember you must have as many equations as there are unknowns (I's, R's, and $\mathcal{E}$'s). In order to apply this rule, you must correctly identify the change in potential as you cross each element in traversing the closed loop. Watch out for signs!

Convenient "rules of thumb" which you may use to determine the increase or decrease in potential as you cross a resistor or seat of emf in traversing a circuit loop are illustrated in the following figure. Notice that the potential **decreases** (changes by $-IR$) when the resistor is traversed **in the direction of the current.** There is an **increase** in potential of $+IR$ if the direction of travel is **opposite** the direction of current. If a seat of emf is traversed **in the direction of the emf** (from $-$ to $+$ on the battery), the potential **increases** by $\mathcal{E}$. If the direction of travel is from $+$ to $-$, the potential **decreases** by $\mathcal{E}$ (changes by $-\mathcal{E}$).

$$\Delta V = V_b - V_a = -IR$$
$$\Delta V = V_b - V_a = IR$$
$$\Delta V = V_b - V_a = \mathcal{E}$$
$$\Delta V = V_b - V_a = -\mathcal{E}$$

• Finally, you must solve the equations simultaneously for the unknown quantities. Be careful in your algebraic steps, and check your numerical answers for consistency.

As an illustration of the use of Kirchhoff's rules, consider a three-loop circuit which has the **general form** shown in figure at the right. In this illustration, the actual circuit elements, R's and $\mathcal{E}$'s are not shown but assumed known. There are six possible different values of I in the circuit; therefore you will need six independent equations to solve for the six values of I. There are four junction points in the circuit (at points a, d, f, and h). The first rule applied at **any three** of these points will yield three equations. The circuit can be thought of as a group of three "blocks" as shown in the following figure on the right. Kirchhoff's second law, when applied to each of these loops ($abcda$, $ahfga$, and $defhd$), will yield three additional equations.

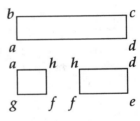

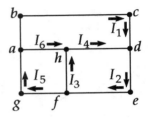

You can then solve the total of six equations simultaneously for the six values of I_1, I_2, I_3, I_4, I_5, and I_6. You can, of course, expect that the sum of the changes in potential difference around **any other closed loop** in the circuit will be zero (for example, $abcdefga$ or $ahfedcba$); however the equations found by applying Kirchhoff's second rule to these additional loops **will not be independent** of the six equations found previously.

REVIEW CHECKLIST

▷ Calculate the equivalent resistance of a group of resistors in parallel, series, or series-parallel combination.

▷ Calculate the current in a single-loop circuit and the potential difference between any two points in the circuit. Use Ohm's law to calculate the current in a circuit and the potential difference between any two points in a circuit which can be reduced to an equivalent single-loop circuit.

▷ Apply Kirchhoff's rules to solve multiloop circuits; that is, find the currents and the potential difference between any two points.

▷ Calculate the charging (discharge) current $I(t)$ and the accumulated (residual) charge $q(t)$ during charging (and discharge) of a capacitor in an *RC* circuit.

▷ Understand the circuitry and make calculations for an unknown resistance, R_x, using the ammeter-voltmeter method and the Wheatstone bridge method. Determine the value of an unknown emf, $\mathcal{E}_x$, using a potentiometer circuit.

ANSWERS TO SELECTED CONCEPTUAL QUESTIONS

3. Is the direction of current through a battery always from the negative terminal to the positive one? Explain.

Answer   No. If there is one battery in a circuit, the current inside it will be from its negative to its positive terminal. Whenever a battery is delivering electrical energy to a circuit, it will carry current in this direction. On the other hand, when another source of emf is charging the battery in question, it will have current pushed through it from its positive terminal to its negative terminal.

□   □   □   □

13. Describe what happens to the lightbulb shown in Figure Q28.13 after the switch is closed. Assume that the capacitor has a large capacitance and is initially uncharged, and assume that the light illuminates when connected directly across the battery terminals.

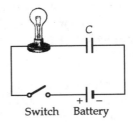

Figure Q28.13

Answer   The bulb will light up for an instant as the capacitor is being charged and there is a current in the circuit. As soon as the capacitor is fully charged, the current in the circuit will drop to zero, and the bulb will cease to glow.

□   □   □   □

18. What advantage does 120-V operation offer over 240 V? What are its disadvantages compared with 240 V?

Answer   Both 120-V and 240-V lines can deliver injurious or lethal shocks, but there is a somewhat better factor of safety with the lower voltage. To say it a different way, the insulation on a 120-V wire can be thinner. On the other hand, a 240-V line carries less current to operate a device with the same power, so the conductor itself can be thinner. Finally, as we will see in Chapter 33, the last step-down transformer can also be somewhat smaller if it has to go down only to 240 volts from the high voltage of the main power line.

□   □   □   □

SOLUTIONS TO SELECTED END-OF-CHAPTER PROBLEMS

1. A battery has an emf of 15.0 V. The terminal voltage of the battery is 11.6 V when it is delivering 20.0 W of power to an external load resistor R. (a) What is the value of R? (b) What is the internal resistance of the battery?

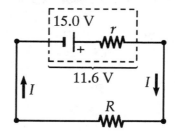

Solution

G: The internal resistance of a battery usually is less than $1\,\Omega$, with physically larger batteries having less resistance due to the larger anode and cathode areas. The voltage of this battery drops significantly (23%), when the load resistance is added, so a sizable amount of current must be drawn from the battery. If we assume that the internal resistance is about $1\,\Omega$, then the current must be about 3 A to give the 3.4 V drop across the battery's internal resistance. If this is true, then the load resistance must be about $R \approx 12\text{ V}/3\text{ A} = 4\,\Omega$.

O: We can find R exactly by using Joule's law for the power delivered to the load resistor when the voltage is 11.6 V. Then we can find the internal resistance of the battery by summing the electric potential differences around the circuit.

A: (a) Combining Joule's law, $\mathcal{P} = \Delta V I$, and the definition of resistance, $\Delta V = IR$, gives

$$R = \frac{\Delta V^2}{\mathcal{P}} = \frac{(11.6\text{ V})^2}{20.0\text{ W}} = 6.73\,\Omega \qquad \Diamond$$

(b) The electromotive force of the battery must equal the voltage drops across the resistances: $\mathcal{E} = IR + Ir$, where $I = \Delta V/R$.

$$r = \frac{\mathcal{E} - IR}{I} = \frac{(\mathcal{E} - \Delta V)R}{\Delta V} = \frac{(15.0\text{ V} - 11.6\text{ V})(6.73\,\Omega)}{11.6\text{ V}} = 1.97\,\Omega \qquad \Diamond$$

L: The resistance of the battery is larger than 1 Ω, but it is reasonable for an old battery or for a battery consisting of several small electric cells in series. The load resistance agrees reasonably well with our prediction, despite the fact that the battery's internal resistance was about twice as large as we assumed. Note that in our initial guess we did not consider the power of the load resistance; however, there is not sufficient information to accurately solve this problem without this data.

---

9. Consider the circuit shown in Figure P28.9. Find (a) the current in the 20.0-Ω resistor and (b) the potential difference between points a and b.

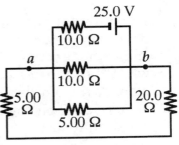

Figure P28.9

Solution If we turn the given diagram on its side, we find that it is the same as figure (a). The 20.0-Ω and 5.00-Ω resistors are in series, so the first reduction is as shown in (b). In addition, since the 10.0-Ω, 5.00-Ω, and 25.0-Ω resistors are then in parallel, we can solve for their equivalent resistance as:

$$R_{eq} = \frac{1}{\left(\dfrac{1}{10.0 \ \Omega} + \dfrac{1}{5.00 \ \Omega} + \dfrac{1}{25.0 \ \Omega} \right)} = 2.94 \ \Omega$$

This is shown in figure (c), which in turn reduces to the circuit shown in (d).

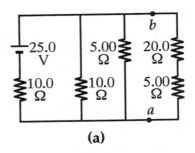

(a)

Next, we work backwards through the diagrams, applying $I = \Delta V / R$ and $\Delta V = IR$. The 12.94-Ω resistor is connected across 25.0-V, so the current through the voltage source in every diagram is

$$I = \frac{\Delta V}{R} = \frac{25.0 \ \text{V}}{12.94 \ \Omega} = 1.93 \ \text{A}$$

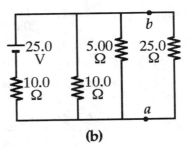

(b)

In figure (c), this 1.93 A goes through the 2.94-Ω equivalent resistor to give a voltage drop of:

$$\Delta V = IR = (1.93 \text{ A})(2.94 \text{ Ω}) = 5.68 \text{ V}$$

From figure (b), we see that this voltage drop is the same across ΔV_{ab}, the 10-Ω resistor, and the 5.00-Ω resistor.

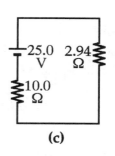
(c)

(b) Therefore, $\Delta V_{ab} = 5.68$ V ◊

Since the current through the 20-Ω resistor is also the current through the 25-Ω line ab,

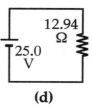
(d)

(a) $I = \dfrac{\Delta V_{ab}}{R_{ab}} = \dfrac{5.68 \text{ V}}{25 \text{ Ω}} = 0.227 \text{ A}$ ◊

---

15. Calculate the power delivered to each resistor in the circuit shown in Figure P28.15.

Solution   To find the power we must find the current in each resistor, so we find the resistance seen by the battery. The given circuit reduces as shown in the figures, since

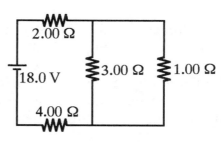
Figure P28.15

$$\frac{1}{1/1.00 \text{ Ω} + 1/3.00 \text{ Ω}} = 0.750 \text{ Ω}$$

In (b), $I = 18.0 \text{ V}/6.75 \text{ Ω} = 2.67 \text{ A}$

This is also the current in (a), so the 2.00-Ω and 4.00-Ω resistors dissipate

$\mathcal{P}_2 = I\Delta V = I^2 R = (2.67 \text{ A})^2(2.00 \text{ Ω}) = 14.2 \text{ W}$ ◊

$\mathcal{P}_4 = I^2 R = (2.67 \text{ A})^2(4.00 \text{ Ω}) = 28.4 \text{ W}$ ◊

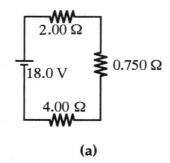
(a)

The voltage across the 0.750 Ω resistor in (a), and across both the 3.00-Ω and the 1.00-Ω resistors in Figure P28.15, is

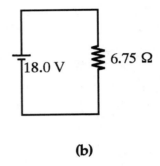

(b)

$$\Delta V = IR = (2.67 \text{ A})(0.750 \text{ Ω}) = 2.00 \text{ V}$$

Then for the 3.00-Ω resistor, $I = \dfrac{\Delta V}{R} = \dfrac{2.00 \text{ V}}{3.00 \text{ Ω}}$

and $\mathscr{P} = I\Delta V = \left(\dfrac{2.00 \text{ V}}{3.00 \text{ Ω}}\right)(2.00 \text{ V}) = 1.33 \text{ W}$ ◊

For the 1.00-Ω resistor, $I = \dfrac{2.00 \text{ V}}{1.00 \text{ Ω}}$ and $\mathscr{P} = \left(\dfrac{2.00 \text{ V}}{1.00 \text{ Ω}}\right)(2.00 \text{ V}) = 4.00 \text{ W}$ ◊

---

19. Determine the current in each branch of the circuit shown in Figure P28.19.

Solution First, we arbitrarily define the initial current directions and names, as shown in the figure below. The current rule then says that

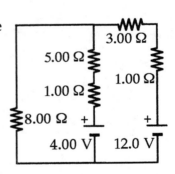

Figure P28.19

$$I_3 = I_1 + I_2 \qquad (1)$$

By the voltage rule, clockwise around the left-hand loop,

$$+ I_1(8.00 \text{ Ω}) - I_2(5.00 \text{ Ω}) - I_2(1.00 \text{ Ω}) - 4.00 \text{ V} = 0 \quad (2)$$

Clockwise around the right-hand loop, combining the 1 and 5 Ω resistors and the 1 and 3 Ω resistors,

$$4.00 \text{ V} + I_2(6.00 \text{ Ω}) + I_3(4.00 \text{ Ω}) - 12.0 \text{ V} = 0 \qquad (3)$$

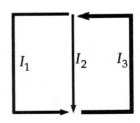

We substitute $(I_1 + I_2)$ for I_3, and reduce our three equations to:

$$(8.00\ \Omega)I_1 - (6.00\ \Omega)I_2 - 4.00\ V = 0$$

and $\quad 4.00\ V + (6.00\ \Omega)I_2 + (4.00\ \Omega)(I_1 + I_2) - 12.0\ V = 0$

Solving the top equation for I_2, $\qquad I_2 = \dfrac{(8.00\ \Omega)I_1 - 4.00\ V}{6.00\ \Omega}$

and rearranging the bottom equation, $\qquad I_1 + \dfrac{10.0\ \Omega}{4.00\ \Omega}I_2 = \dfrac{8.00\ V}{4.00\ \Omega}$

We can substitute for I_2, $\qquad I_1 + 3.33 I_1 - 1.67\ V\,/\,\Omega = 2.00\ V\,/\,\Omega$

and solve for I_1, down the 8 Ω resistor: $\quad I_1 = 0.846\ V\,/\,\Omega = 0.846\ A \qquad \Diamond$

If we were solving the three simultaneous equations by determinants or by calculator matrix inversion, we would have to do much more work to find I_2 and I_3. Our method of solving by substitution is more generally useful, and makes it easy to work out the remaining answers after the first, just as when the cat has kittens. We substitute again, putting the value for I_1 into equations we already solved for the other unknowns.

Thus, $\quad I_2 = \dfrac{(8.00\ \Omega)(0.846\ A) - 4.00\ V}{6.00\ \Omega} = 0.462\ A$ down in the middle branch $\quad \Diamond$

$\qquad I_3 = 0.846\ A + 0.462\ A = 1.31\ A$ up in the right-hand branch $\qquad \Diamond$

27. For the circuit shown in Figure P28.27, calculate (a) the current in the 2.00-Ω resistor and (b) the potential difference between points a and b.

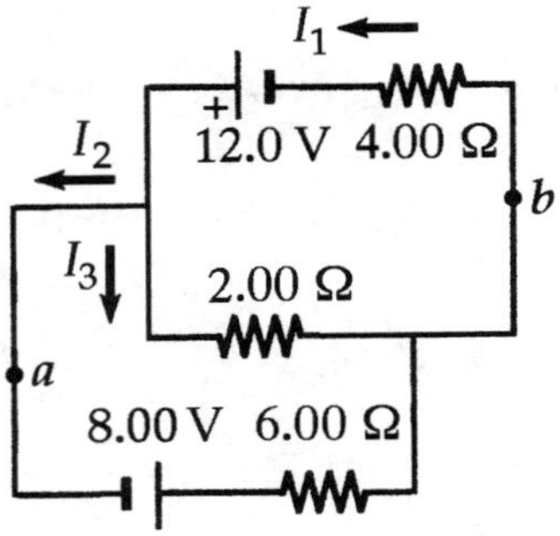

Figure P28.27

Solution Arbitrarily choose current directions as labeled in the figure to the right.

(a) From the junction point rule, we have

$$I_1 = I_2 + I_3 \qquad (1)$$

Traversing the top loop counterclockwise,

$$(12.0 \text{ V}) - (2.00 \text{ }\Omega)I_3 - (4.00 \text{ }\Omega)I_1 = 0 \qquad (2)$$

Traversing the bottom loop counterclockwise,

$$8.00 \text{ V} - (6.00 \text{ }\Omega)I_2 + (2.00 \text{ }\Omega)(I_3) = 0 \qquad (3)$$

From Equation (2),
$$I_1 = \frac{12.0 \text{ V} - (2.00 \text{ }\Omega)I_3}{4.00 \text{ }\Omega}$$

From Equation (3),
$$I_2 = \frac{(8.00 \text{ V}) + (2.00 \text{ }\Omega)I_3}{6.00 \text{ }\Omega}$$

Substituting these values into Equation (1), we find that the current in the 2.00-Ω resistor is $I_3 = 0.909$ A. ◊

(b) Through the center wire, $V_a - (0.909 \text{ A})(2.00 \text{ }\Omega) = V_b$,

Therefore, $V_a - V_b = 1.82$ V, with $V_a > V_b$ ◊

29. Consider a series RC circuit (Figure 28.16) for which $R = 1.00$ MΩ, $C = 5.00$ μF, and $\mathcal{E} = 30.0$ V. Find (a) the time constant of the circuit and (b) the maximum charge on the capacitor after the switch is closed. (c) If the switch is closed at $t = 0$, find the current in the resistor 10.0 s later.

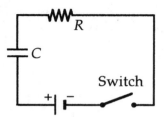

Solution

Figure 28.16

(a) $\tau = RC = (1.00 \times 10^6 \ \Omega)(5.00 \times 10^{-6} \ F) = 5.00 \ \Omega \cdot F = 5.00$ s   ◊

(b)   After a long time, the charge on the capacitor is maximum, the current is zero, and the potential across the capacitor equals the battery's emf.

$$q = Q = C\mathcal{E} = (5.00 \times 10^{-6} \ F)(30.0 \ V) = 150 \ \mu C$$ ◊

(c) $I = I_0 e^{-t/\tau}$ where $I_0 = \dfrac{\mathcal{E}}{R}$ and $\tau = RC$

$$I = \frac{\mathcal{E}}{R} e^{-t/RC} = \left(\frac{30.0 \ V}{1.00 \times 10^6 \ \Omega}\right) e^{-10.0 \ s / \left((1.00 \times 10^6 \ \Omega)(5.00 \times 10^{-6} \ F)\right)}$$

$$I = 4.06 \times 10^{-6} \ A \ = \ 4.06 \ \mu A$$ ◊

---

33. The circuit shown in Figure P28.33 has been connected for a long time. (a) What is the voltage across the capacitor? (b) If the battery is disconnected, how long does it take the capacitor to discharge to one-tenth of its initial voltage?

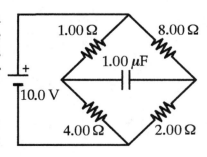

Solution   (a) After a long time the capacitor branch will carry negligible current. The current flow is as shown in Figure (a).

Figure P28.33

To find the voltage at point a, we first find the current, using the voltage rule:

$$10.0 \text{ V} - (1.00 \text{ }\Omega)I_2 - (4.00 \text{ }\Omega)I_2 = 0$$

$$I_2 = 2.00 \text{ A}$$

Similarly, $10.0 \text{ V} - (8.00 \text{ }\Omega)I_3 - (2.00 \text{ }\Omega)I_3 = 0$

$$I_3 = 1.00 \text{ A}$$

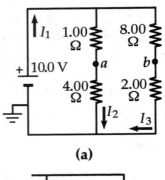

(a)

Going clockwise around the right-hand loop from point a to point b,

$$V_a + (1.00 \text{ }\Omega)(2.00 \text{ A}) - (8.00 \text{ }\Omega)(1.00 \text{ A}) = V_b$$

so $\quad \Delta V = V_a - V_b = 6.00 \text{ V} \qquad \Diamond$

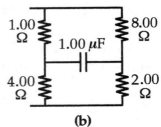

(b)

(b) We suppose the battery is pulled out leaving an open circuit. We are left with Figure (b), which can be reduced to equivalent circuits (c) and (d). From (d), we see that the capacitor discharges through a 3.60 Ω equivalent resistance. According to $q = Qe^{-t/RC}$,

$$qC = QCe^{-t/RC} \quad \text{so} \quad V = V_o e^{-t/RC}$$

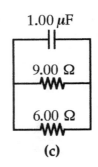

(c)

Solving, $\quad \frac{1}{10}V_o = V_o e^{-t/(3.60 \text{ }\Omega)(1.00 \text{ }\mu\text{F})}$

$$e^{-t/3.60 \text{ }\mu\text{s}} = 0.100$$

$$(-t/3.60 \text{ }\mu\text{s}) = \ln 0.100 = -2.30$$

$$t = (2.30)(3.60 \text{ }\mu\text{s}) = 8.29 \text{ }\mu\text{s} \qquad \Diamond$$

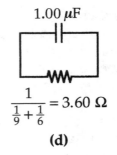

$$\frac{1}{\frac{1}{9} + \frac{1}{6}} = 3.60 \text{ }\Omega$$

(d)

39. The galvanometer described in the preceding problem can be used to measure voltages. In this case a large resistor is wired in series with the galvanometer in a way similar to that shown in Figure P28.24b This arrangement, in effect, limits the current that flows through the galvanometer when large voltages are applied. Most of the potential drop occurs across the resistor placed in series. Calculate the value of the resistor that enables the galvanometer to measure an applied voltage of 25.0 V at full-scale deflection.

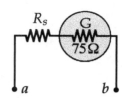

Figure 28.24b (modified)

Solution

G: The problem states that the value of the resistor must be "large" in order to limit the current through the galvanometer, so we should expect a resistance of kΩ to MΩ.

O: The unknown resistance can be found by applying the definition of resistance to the portion of the circuit shown in Figure 28.24b.

A: $\Delta V_{ab} = 25.0$ V; From Problem 38, $I = 1.50$ mA and $R_g = 75.0$ Ω. For the two resistors in series, $R_{eq} = R_s + R_g$ so the definition of resistance gives us: $\Delta V_{ab} = I(R_s + R_g)$

Therefore, $R_s = \dfrac{\Delta V_{ab}}{I} - R_g = \dfrac{25.0 \text{ V}}{1.50 \times 10^{-3} \text{A}} - 75.0 \text{ Ω} = 16.6 \text{ kΩ}$ ◊

L: The resistance is relatively large, as expected. It is important to note that some caution would be necessary if this arrangement were used to measure the voltage across a circuit with a comparable resistance. For example, if the circuit resistance was 17 kΩ, the voltmeter in this problem would cause a measurement inaccuracy of about 50%, because the meter would divert about half the current that normally would go through the resistor being measured. Problems 46 and 59 address a similar concern about measurement error when using electrical meters.

41. Assume that a galvanometer has an internal resistance of 60.0 Ω and requires a current of 0.500 mA to produce full-scale deflection. What resistance must be connected in parallel with the galvanometer if the combination is to serve as an ammeter that has a full-scale deflection for a current of 0.100 A?

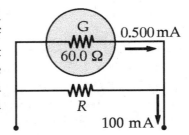

Solution

G: An ammeter reads the flow of current in a portion of a circuit; therefore it must have a low resistance so that it does not significantly alter the current that would exist without the meter. Therefore, the resistance required is probably less than 1 Ω.

O: From the values given for a full-scale reading, we can find the voltage across and the current through the shunt (parallel) resistor, and the resistance value can then be found from the definition of resistance.

A: The voltage across the galvanometer must be the same as the voltage across the shunt resistor in parallel, so when the ammeter reads full scale,

$$\Delta V = (0.500 \text{ mA})(60.0 \ \Omega) = 30.0 \text{ mV}$$

Through the shunt resistor, $I = 100 \text{ mA} - 0.500 \text{ mA} = 99.5 \text{ mA}$

Therefore, $R = \dfrac{\Delta V}{I} = \dfrac{30.0 \text{ mV}}{99.5 \text{ mA}} = 0.302 \ \Omega$ ◊

L: The shunt resistance is less than 1 Ω as expected. It is important to note that some caution would be necessary if this meter were used in a circuit that had a low resistance. For example, if the circuit resistance was 3 Ω, adding the ammeter to the circuit would reduce the current by about 10%, so the current displayed by the meter would be lower than without the meter. Problems 46 and 59 address a similar concern about measurement error when using electrical meters.

47. An electric heater is rated at 1500 W, a toaster at 750 W, and an electric grill at 1000 W. The three appliances are connected to a common 120-V circuit. (a) How much current does each draw? (b) Is a 25.0-A circuit sufficient in this situation? Explain your answer.

Solution

(a) Heater:

$$I = \frac{\mathcal{P}}{\Delta V} = \frac{1500 \text{ W}}{120 \text{ V}} \left(\frac{1\text{ J}/\text{s}}{1\text{ W}}\right)\left(\frac{1\text{ V}}{1\text{ J}/\text{C}}\right)\left(\frac{1\text{ A}}{1\text{ C}/\text{s}}\right) = 12.5 \text{ A} \quad \lozenge$$

Toaster:

$$I = \frac{750 \text{ W}}{120 \text{ V}} = 6.25 \text{ A} \qquad\qquad \lozenge$$

Grill:

$$I = \frac{1000 \text{ W}}{120 \text{ V}} = 8.33 \text{ A} \qquad\qquad \lozenge$$

(b) Together in parallel they pass current 12.5 + 6.25 + 8.33 A = 27.1 A, so 25-A wiring cannot provide power to the entire group of appliances at the same time. $\qquad\qquad \lozenge$

---

57. A battery has an emf $\mathcal{E}$ and internal resistance r. A variable resistor R is connected across the terminals of the battery. Determine the value of R such that (a) the potential difference across the terminals is a maximum, (b) the current in the circuit is a maximum, (c) the power delivered to the resistor is a maximum.

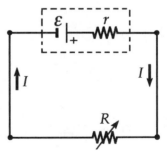

Solution

G: If we consider the limiting cases, we can imagine that the **potential** across the battery will be a maximum when $R = \infty$ (open circuit), the **current** will be a maximum when $R = 0$ (short circuit), and the **power** will be a maximum when R is somewhere between these two extremes, perhaps when $R = r$.

O: We can use the definition of resistance to find the voltage and current as functions of R, and the power equation can be differentiated with respect to R.

A: (a) The battery has a voltage $\Delta V_{\text{terminal}} = \mathcal{E} - Ir = \dfrac{\mathcal{E}R}{R+r}$

Taking the limit as $R \to \infty$, $\quad \Delta V_{\text{terminal}} \to \mathcal{E}$ ◊

(b) The circuit's current is $\quad I = \dfrac{\mathcal{E}}{R+r}$

Therefore, as $R \to 0$ $\quad I \to \dfrac{\mathcal{E}}{r}$ ◊

(c) The power delivered is $\quad \mathcal{P} = I^2 R = \dfrac{\mathcal{E}^2 R}{(R+r)^2}$

To maximize the power $\mathcal{P}$ as a function of R, we differentiate, with respect to R,

$$\frac{d\mathcal{P}}{dR} = \mathcal{E}^2 R (-2)(R+r)^{-3} + \mathcal{E}^2 (R+r)^{-2}$$

and require that $\quad \dfrac{d\mathcal{P}}{dR} = \dfrac{-2\mathcal{E}^2 R}{(R+r)^3} + \dfrac{\mathcal{E}^2}{(R+r)^2} = 0$

Then $2R = R + r$ and $\quad R = r$ ◊

L: The results agree with our predictions. Making load resistance equal to the source resistance to maximize power transfer is called impedance matching.

———————————

59. The value of a resistor R is to be determined using the ammeter-voltmeter setup shown in Figure P28.59. The ammeter has a resistance of $0.500\ \Omega$, and the voltmeter has a resistance of $20\,000\ \Omega$. Within what range of actual values of R will the measured values be correct to within 5.00% if the measurement is made using (a) the circuit shown in Figure P28.59a (b) the circuit shown in Figure P28.59b?

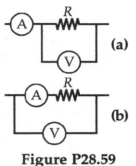

Figure P28.59

Solution

G: An ideal ammeter has zero resistance, and an ideal voltmeter has infinite resistance, so that adding the meter does not alter the current or voltage of the existing circuit. For the non-ideal meters in this problem, a low values of R will give a large voltage measurement error in circuit (b), while a large value of R will give significant current measurement error in circuit (a). We could hope that these meters yield accurate measurements in either circuit for typical resistance values of $1\ \Omega$ to $1\ \text{M}\Omega$.

O: The definition of resistance can be applied to each circuit to find the minimum and maximum current and voltage allowed within the 5.00% tolerance range.

A: (a)   In Figure P28.59a, at least a little current goes through the voltmeter, so less current flows through the resistor than the ammeter reports, and the resistance computed by dividing the voltage by the inflated ammeter reading will be too small.   Thus, we require that $\Delta V/I = 0.950R$ where I is the current through the ammeter.

Call I_R the current through the resistor; then $I - I_R$ is the current in the voltmeter. Since the resistor and the voltmeter are in parallel, the voltage across the meter equals the voltage across the resistor. Applying the definition of resistance:

$$\Delta V = I_R R = (I - I_R)(20\,000\ \Omega) \qquad \text{so} \qquad I = \frac{I_R(R + 20\,000\ \Omega)}{20\,000\ \Omega}$$

Our requirement is
$$\frac{I_R R}{\left(\dfrac{I_R(R + 20\,000\ \Omega)}{20\,000\ \Omega}\right)} \geq 0.95R$$

Solving,
$$20\,000\ \Omega \geq 0.95(R + 20\,000\ \Omega) = 0.95R + 19\,000\ \Omega$$

and
$$R \leq \frac{1000\ \Omega}{0.95} \qquad \text{or} \qquad R \leq 1.05\ \text{k}\Omega \qquad \Diamond$$

(b) If R is too small, the resistance of an ammeter in series will significantly reduce the current that would otherwise flow through R. In Figure 28.59b, the voltmeter reading is $I(0.500\ \Omega) + IR$, at least a little larger than the voltage across the resistor. So the resistance computed by dividing the inflated voltmeter reading by the ammeter reading will be too large.

We require
$$\frac{V}{I} \leq 1.05R$$

$$\frac{I(0.500\ \Omega) + IR}{I} \leq 1.05R$$

Thus,
$$0.500\ \Omega \leq 0.0500R \qquad \text{and} \qquad R \geq 10.0\ \Omega \qquad \Diamond$$

L: The range of R values seems correct since the ammeter's resistance should be less than 5% of the smallest R value ($0.500\ \Omega \leq 0.05R$ means that R should be greater than 10 Ω), and R should be less than 5% of the voltmeter's internal resistance ($R \leq 0.05 \times 20\ \text{k}\Omega = 1\ \text{k}\Omega$). Only for the restricted range between 10 ohms and 1000 ohms can we indifferently use either of the connections (a) and (b) for a reasonably accurate resistance measurement. For low values of the resistance R, circuit (a) must be used. Only circuit (b) can accurately measure a large value of R.

61. The values of the components in a simple series RC circuit containing a switch (Fig. 28.16) are $C = 1.00~\mu F$, $R = 2.00 \times 10^6~\Omega$, $\mathcal{E} = 10.0$ V. At the instant 10.0 s after the switch is closed, calculate (a) the charge on the capacitor, (b) the current in the resistor, (c) the rate at which energy is being stored in the capacitor, and (d) the rate at which energy is being delivered by the battery.

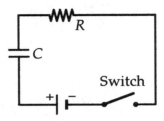

Figure 28.16

Solution

(a) $q = C\mathcal{E}\left(1 - e^{-t/RC}\right) = (1.00 \times 10^{-6}~\text{F})(10.0~\text{V})\left[1 - e^{-10.0~\text{s}/\left((2.00\times10^6~\Omega)(1.00\times10^{-6}~\text{F})\right)}\right]$

$q = 9.93 \times 10^{-6}~\text{C} = 9.93~\mu\text{C}$ ◊

(b) $I = \dfrac{dq}{dt} = \dfrac{d}{dt}\left[C\mathcal{E}\left(1 - e^{-(t/RC)}\right)\right] = \left(\dfrac{\mathcal{E}}{R}\right)e^{-(t/RC)} = \left(\dfrac{10.0~\text{V}}{2.00\times10^6~\Omega}\right)e^{-(10.0/2.00)}$

$I = 3.37 \times 10^{-8}~\text{A}$ ◊

(c) Since the energy stored in the capacitor is $U = q^2/2C$, the rate of storing energy is

$\dfrac{dU}{dt} = \dfrac{q}{C}\dfrac{dq}{dt} = \left(\dfrac{q}{C}\right)I = \left(\dfrac{9.93\times10^{-6}~\text{C}}{1.00\times10^{-6}~\text{F}}\right)(3.37\times10^{-8}~\text{A}) = 3.34\times10^{-7}~\text{W}$ ◊

(d) $\mathcal{P}_{\text{batt}} = I\mathcal{E} = (3.37\times10^{-8}~\text{A})(10.0~\text{V}) = 3.37\times10^{-7}~\text{W}$ ◊

63. Three 60.0 − W, 120.0 - V lightbulbs are connected across a 120.0 - V power source, as shown in Figure P28.63. Find (a) the total power delivered to the three bulbs and (b) the voltage across each. Assume that the resistance of each bulb conforms to Ohm's law (even though in reality the resistance increases markedly with current).

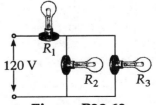

Figure P28.63

Solution   If the bulbs were all in parallel, the current in each would be

$$I = \frac{P}{\Delta V} = \frac{60.0 \text{ W}}{120 \text{ V}} = 0.500 \text{ A}$$

As they are connected, each bulb has $R = \Delta V/I = 120 \text{ V}/0.500 \text{ A} = 240 \text{ } \Omega$

so R_2 and R_3 have equivalent resistance $\dfrac{1}{1/240 \text{ } \Omega + 1/240 \text{ } \Omega} = 120 \text{ } \Omega$

The three together have net resistance $240 \text{ } \Omega + 120 \text{ } \Omega = 360 \text{ } \Omega$

The total current can be calculated as $I = \dfrac{\Delta V}{R} = \dfrac{120 \text{ V}}{360 \text{ } \Omega} = 0.333 \text{ A}$

(a) Thus the power dissipated is $\mathcal{P} = \Delta VI$: $\mathcal{P} = (120 \text{ V})(0.333 \text{ A}) = 40.0 \text{ W}$ ◊

(b) For bulb R_1, $\Delta V = IR$: $\Delta V = (0.333 \text{ A})(240 \text{ } \Omega) = 80.0 \text{ V}$ ◊

For bulb R_2 or R_3, the potential difference $\Delta V = 120 \text{ V} - 80.0 \text{ V} = 40.0 \text{ V}$ ◊

---

67. In Figure P28.67, suppose that the switch has been closed for a length of time sufficiently long for the capacitor to become fully charged. (a) Find the steady-state current in each resistor. (b) Find the charge Q on the capacitor. (c) The switch is opened at $t = 0$. Write an equation for the current I_{R_2} in R_2 as a function of time, and (d) find the time that it takes for the charge on the capacitor to fall to one-fifth its initial value.

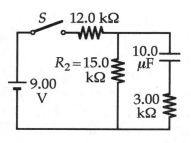

Figure P28.67

Solution   (a)   When the capacitor is fully charged, no current flows in it or in the 3.00-kΩ resistor: $I_3 = 0$ ◊

So the same current flows in resistors 1 and 2: $I_1 = I_2$, as given by the voltage rule,

$$+9.00V - (12.0 \text{ k}\Omega)I_1 - (15.0 \text{ k}\Omega)I_1 = 0.$$

$$I_1 = 9.00 \text{ V}/27.0 \text{ k}\Omega = 0.333 \text{ mA} = I_2 \qquad ◊$$

(b)   For the right-hand loop, the voltage rule gives for the capacitor voltage:

$$(+15.0 \text{ k}\Omega)(0.333 \text{ mA}) - \Delta V_c - (3.00 \text{ k}\Omega)(0) = 0$$

Since $\Delta V_c = 5.00$ V, $Q = C\Delta V_c = (10.0 \text{ }\mu\text{F})(5.00 \text{ V}) = 50.0 \text{ }\mu\text{C}$ ◊

(c)   At $t = 0$, the current in R_1 drops to zero. The capacitor, charged to 5.00 V with top plate positive, will drive the same current $I_2 = I_3$ counterclockwise around the right-hand loop. At $t = 0$, its value is given by the equation:

$$+5.00 \text{ V} - I_2(15.0 \text{ k}\Omega) - I_2(3.00 \text{ k}\Omega) = 0$$

$$I_2 = 5.00 \text{ V}/18.0 \text{ k}\Omega = 0.278 \text{ mA}$$

Thereafter, it decays as it drains the capacitor's charge, with time constant

$$R_{eq}C = (18.0 \text{ k}\Omega)(10.0 \text{ }\mu\text{F}) = 180 \text{ ms}$$

So its equation is $I_2 = (0.278 \text{ mA})e^{-t/(180 \text{ ms})}$ ◊

(d)   The charge decays according to $q = Q_0 e^{-t/RC}$

Substituting known values, $\frac{1}{5}(50.0 \text{ }\mu\text{C}) = (50.0 \text{ }\mu\text{C})e^{-t/(180 \text{ ms})}$

Solving for t, $0.200 = e^{-t/(180 \text{ ms})}$

Thus, $(-t)/180 \text{ ms} = -1.61$

and $t = (1.61)(180 \text{ ms}) = 290 \text{ ms}$ ◊

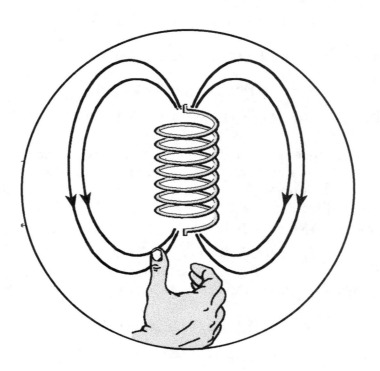

Magnetic Fields

MAGNETIC FIELDS

INTRODUCTION

Many historians of science believe that the compass, which uses a magnetic needle, was used in China as early as the 13th century B.C., its invention being of Arab or Indian origin. The early Greeks knew about magnetism as early as 800 B.C. They discovered that certain stones, now called magnetite (Fe_3O_4), attract pieces of iron. In 1269 Pierre de Maricourt mapped out the directions taken by a needle when it was placed at various points on the surface of a spherical natural magnet. He found that the directions formed lines that encircled the sphere and passed through two points diametrically opposite to each other, which he called the **poles** of the magnet. Subsequent experiments showed that every magnet, regardless of its shape, has two poles, called **north** and **south poles**, which exhibit forces on each other in a manner analogous to electric charges. That is, like poles repel each other and unlike poles attract each other.

EQUATIONS AND CONCEPTS

The magnetic field at some point in space is defined in terms of the **magnetic force** exerted on a moving positive electric charge at that point. The SI unit of the magnetic field is the tesla (T) or weber per square meter (Wb/m^2).

$$\mathbf{F}_B = q\mathbf{v} \times \mathbf{B} \tag{29.1}$$

The magnetic force will be of maximum magnitude when the charge moves along a direction perpendicular to the direction of the magnetic field. The magnitude of the magnetic force depends on the angle θ between the direction of the magnetic field and the direction of the velocity vector.

$$F_B = |q|vB\sin\theta \qquad (29.2)$$

The SI unit of magnetic field intensity is the tesla (T).

$$1\,T = 1\,\frac{N}{C\cdot m/s}$$

Equation 29.2 can be written in a form which serves to define the magnitude of the magnetic field.

$$B \equiv \frac{F_B}{|q|v\sin\theta}$$

The cgs unit of magnetic field is the gauss (G).

$$1\,T = 10^4\,G$$

In order to determine the direction of the magnetic force on a moving charge, apply the right-hand rule as desribed in **Suggestions, Skills, and Strategies.** Be sure to remember that the right-hand rule gives the force on a **positive** charge. If the charge is **negative**, then the direction of the force is reversed.

If a **straight** wire carrying a current is placed in an external magnetic field, a **magnetic force** will be exerted on the wire. To determine the direction of the force, use the right hand rule.

$$\mathbf{F}_B = I\mathbf{L} \times \mathbf{B} \qquad (29.3)$$

The magnetic force on a wire of arbitrary shape is found by integrating over the length of the wire. In these equations the direction of **L** and $d\mathbf{s}$ is that of the current.

$$\mathbf{F}_B = I\int_a^b d\mathbf{s} \times \mathbf{B} \qquad (29.5)$$

The magnitude of the magnetic force on the conductor depends on the angle between the direction of the conductor and the direction of the field.

$$F_B = BIL \sin\theta$$

The magnetic force will be maximum when the conductor is directed perpendicular to the magnetic field.

$$F_{B,\text{max}} = BIL$$

When a current loop is placed in an external magnetic field, there is a **net torque** exerted on the loop. In Equation 29.9, the area vector **A** is directed perpendicular to the area of the loop with a sense given by the right-hand rule. The magnitude of **A** is equal to the area of the loop.

$$\tau = I\mathbf{A} \times \mathbf{B} \qquad (29.9)$$

The magnitude of the torque will depend on the angle between the direction of the magnetic field and the direction of the normal (or perpendicular) to the plane of the loop.

$$\tau = IAB \sin \theta$$

The magnitude of the torque will be maximum when the magnetic field is parallel to the plane of the loop.

$$\tau_{max} = IAB \qquad (29.8)$$

The direction of rotation of the loop is such that the normal to the plane of the loop turns into a direction parallel to the magnetic field.

The torque on a current loop can also be expressed in terms of the magnetic moment of the loop.

$$\tau = \mu \times \mathbf{B} \qquad (29.11)$$

$$\text{where} \quad \mu = I\mathbf{A} \qquad (29.10)$$

When a charged particle enters the region of a uniform magnetic field with its velocity vector initially perpendicular to the direction of the field, the particle will move in a circular path in a plane perpendicular to the direction of the field. The **radius** of the circular path will be proportional to the linear momentum of the charged particle.

$$r = \frac{mv}{qB} \qquad (29.13)$$

The **angular frequency** (or cyclotron frequency) of the particle will be proportional to the ratio of charge to mass. Note that the frequency, and hence the period of rotation do **not** depend on the radius of the path.

$$\omega = \frac{qB}{m} \qquad (29.14)$$

$$T = \frac{2\pi m}{qB} \qquad (29.15)$$

There are several important applications of the motion of charged particles in a magnetic field:

Velocity Selector—When a beam of charged particles is directed into a region where uniform electric and magnetic fields are perpendicular to each other and to the initial direction of the particle beam, those particles emerging along the initial beam direction will have a common velocity. Only those particles with a velocity $v = E / B$ will get through.

$$v = \frac{E}{B} \qquad (29.17)$$

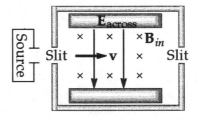

Mass Spectrometer—If an ion beam, after passing through a velocity selector, is directed perpendicularly into a second uniform magnetic field, the **ratio of charge to mass** for the isotopic species can be determined by measuring the radius of curvature of the beam in the field.

$$\frac{m}{q} = \frac{rB_0 B}{E} \qquad (29.18)$$

Velocity Selector

Cyclotron—The **maximum kinetic energy** acquired by an ion in a cyclotron depends on the radius of the dees and the intensity of the magnetic field. This relationship holds until the ion reaches relativistic energies ($\cong$ 20 MeV). This is true for ions of proton mass or greater, but is not true for electrons.

$$K = \frac{1}{2}mv^2 = \frac{q^2 B^2 R^2}{2m} \qquad (29.19)$$

When the **Hall Coefficient** ($1/nq$) is known for a calibrated sample, magnetic field strengths can be determined by accurate measurements of the **Hall voltage**, V_H. By measuring the voltage as + or –, one can find the sign of the charge carriers as well.

$$\Delta V_H = \frac{IBd}{nqA} \qquad (29.22)$$

SUGGESTIONS, SKILLS, AND STRATEGIES

REMEMBERING VECTOR SYMBOLS

To remember the symbols for vectors that point away from you and towards you, think of a three-dimensional archery arrow as it turns to point first away from you (shown on the left), and then towards you, (shown on the right). The point on the arrow is represented by a dot; the feathers are represented by an 'x'.

MAGNETIC FIELDS AND THE CROSS-PRODUCT

Equation 29.1, $\mathbf{F} = q\mathbf{v} \times \mathbf{B}$, serves as the definition of the magnetic field vector $\mathbf{B}$. The direction of the magnetic force $\mathbf{F}$ is determined by the right-hand rule for the cross product, which you have used before. This assumes that the charge q is a positive charge. If the vectors $\mathbf{v}$ and $\mathbf{B}$ are given in unit vector notation then $\mathbf{v} \times \mathbf{B}$ can be written as

$$\mathbf{v} \times \mathbf{B} = \mathbf{i}\left(v_y B_z - v_z B_y\right) + \mathbf{j}\left(v_z B_x - v_x B_z\right) + \mathbf{k}\left(v_x B_y - v_y B_x\right)$$

This means the components of the magnetic force are:

$$F_x = q\left(v_y B_z - v_z B_y\right)$$

$$F_y = q\left(v_z B_x - v_x B_z\right)$$

$$F_z = q\left(v_x B_y - v_y B_x\right)$$

USING THE RIGHT-HAND RULE

The right-hand rule is used to find the direction of the cross-product ($\mathbf{C}$) of two vectors ($\mathbf{A}$ and $\mathbf{B}$), where

$$\mathbf{A} \times \mathbf{B} = \mathbf{C}$$

There are two versions of the right-hand rule; in order to avoid confusion, you should pick the version that suits you, and use it exclusively. In either version, note that the order of $\mathbf{A}$ and $\mathbf{B}$ are very important, and should not be confused.

VERSION 1	**VERSION 2**
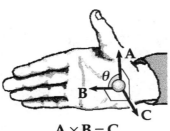	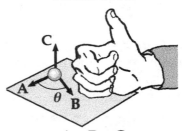
$\mathbf{A} \times \mathbf{B} = \mathbf{C}$	$\mathbf{A} \times \mathbf{B} = \mathbf{C}$
Hold your open right hand with your thumb pointing in the direction of **A** (the first named vector quantity) and your fingers pointing in the direction of **B** (the second named vector quantity). By necessity, your fingers and thumb cannot stretch farther than 180°, so if you do this, your hand will be oriented properly. The resultant vector **C** now is directed out of the palm of your hand.	Orient your hand so that your fingers point in the direction of **A** (the first named vector quantity) and then curl your fingers to point in the direction of **B** (the second named vector quantity). Note that since your fingers cannot bend farther than 180°, you may have to flip your hand upside down to do this. Your thumb now points in the direction of **C**, where **C** is perpendicular to both **A** and **B**.

To find the "curl" associated with a vector **C** by the right hand rule, let **A** and **B** be progressive elements of the curl. Thus, to find the area vector of a loop of wire, let your thumb, then your fingers point in the direction of progressive lengths of wire; the area vector then points out the palm of your hand.	To find the "curl" associated with a vector **C** by the right hand rule, simply let **A** and **B** be progressive elements of the curl. Thus, to find the area vector of a loop of wire, let your fingers curl in the direction of the current; your thumb then points in the direction of the area vector.

REVIEW CHECKLIST

▷ Use the defining equation for a magnetic field **B** to determine the magnitude and direction of the magnetic force exerted on an electric charge moving in a region where there is a magnetic field. You should understand clearly the important differences between the forces exerted on electric charges by electric fields and those forces exerted on moving electric charges by magnetic fields.

▷ Calculate the magnitude and direction of the magnetic force on a current-carrying conductor when placed in an external magnetic field. You should be able to perform such calculations for either a straight conductor or one of arbitrary shape.

▷ Determine the magnitude and direction of the torque exerted on a closed current loop in an external magnetic field. You should understand how to correctly designate the direction of the area vector corresponding to a given current loop, and to incorporate the magnetic moment of the loop into the calculation of the torque on the loop.

▷ Calculate the period and radius of the circular orbit of a charged particle moving in a uniform magnetic field. Understand the essential features of the velocity selector and mass spectrometer and make appropriate quantitative calculations regarding the operation of these instruments.

ANSWERS TO SELECTED CONCEPTUAL QUESTIONS

2. Two charged particles are projected into a region where a magnetic field is directed perpendicular to their velocities. If the charges are deflected in opposite directions, what can be said about them?

Answer We know the magnetic field is constant, and the velocity vectors are the same, but one force is the negative of the other. From $\mathbf{F} = q(\mathbf{v} \times \mathbf{B})$, we can conclude that the only thing which could cause the force to be of opposite sign is if the charges were of opposite sign.

□   □   □   □

10. Is it possible to orient a current loop in a uniform magnetic field such that the loop does not tend to rotate? Explain.

Answer Yes. If the magnetic field is perpendicular to the plane of the loop, the forces on opposite sides will be equal and opposite, but will produce no net torque.

□   □   □   □

11. How can a current loop be used to determine the presence of a magnetic field in a given region of space?

Answer The loop can be mounted free to rotate around an axis. The loop will rotate about this axis when placed in an external magnetic field for some arbitrary orientation. As the current through the loop is increased, the torque on it will increase.

□   □   □   □

SOLUTIONS TO SELECTED END-OF-CHAPTER PROBLEMS

1. Determine the initial direction of the deflection of charged particles as they enter the magnetic fields, as shown in Figure P29.1.

Solution Both versions of the right-hand rule, open-handed and curled, are shown here. Use the set of figures that you have chosen, and ignore the other set.

(a) By solution figure (a), $\mathbf{v} \times \mathbf{B}$ is (right)×(away) = up ◊

(b) By solution figure (b), $\mathbf{v} \times \mathbf{B}$ is (left)×(up) = away.

Since the charge is negative, $q\mathbf{v} \times \mathbf{B}$ is toward you ◊

(c) $\mathbf{v} \times \mathbf{B}$ is zero since the angle between $\mathbf{v}$ and $\mathbf{B}$ is 180° and sin 180° = 0. There is no deflection. ◊

(d) $\mathbf{v} \times \mathbf{B}$ is (up) × (up and right), or away from you ◊

(a) **B**$_{in}$

(b) **B**$_{up}$

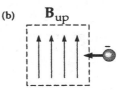

(c) **B**$_{right}$

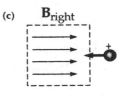

(d) **B**$_{at\ 45°}$

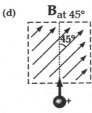

Figure P29.1

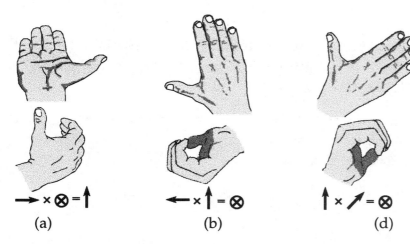

(a)   (b)   (d)

5. A proton moves in a direction perpendicular to a uniform magnetic field **B** at 1.00×10^7 m/s and experiences an acceleration of 2.00×10^{13} m/s² in the +x direction when its velocity is in the +z direction. Determine the magnitude and direction of the field.

Solution By Newton's 2nd law,

$$F = ma = (1.67 \times 10^{-27} \text{ kg})(2.00 \times 10^{13} \text{ m/s}^2)$$

The magnetic force $F = 3.34 \times 10^{-14}$ N $= qvB \sin 90°$

Rearranging, $B = \dfrac{F}{qv} = \dfrac{3.34 \times 10^{-14} \text{ N}}{(1.60 \times 10^{-19} \text{ C})(1.00 \times 10^7 \text{ m/s})} = 2.09 \times 10^{-2}$ T ◊

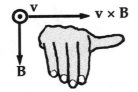

By the right-hand rule, **B** must be in the –y direction   ◊

This yields a force on the proton in the +x direction when v points in the +z direction.

---

9. A proton moving at 4.00×10^6 m/s through a magnetic field of 1.70 T experiences a magnetic force of magnitude 8.20×10^{-13} N. What is the angle between the proton's velocity and the field?

Solution Since the magnitude of the force on a moving charge in a magnetic field is $F_B = qvB \sin\theta$,

$$\theta = \sin^{-1}\left[\dfrac{F_B}{qvB}\right]$$

$$\theta = \sin^{-1}\left[\dfrac{8.20 \times 10^{-13} \text{ N}}{(1.60 \times 10^{-19} \text{ C})(4.00 \times 10^6 \text{ m/s})(1.70 \text{ T})}\right] = 48.9° \quad \text{or} \quad 131° \quad ◊$$

---

11. A proton moves with a velocity of $\mathbf{v} = (2\mathbf{i} - 4\mathbf{j} + \mathbf{k})$ m/s in a region in which the magnetic field is $\mathbf{B} = (\mathbf{i} + 2\mathbf{j} - 3\mathbf{k})$ T. What is the magnitude of the magnetic force this charge experiences?

Solution

The force on a charge is proportional to the vector product of the velocity and the magnetic field:

$$\mathbf{F}_B = q\mathbf{v} \times \mathbf{B} = \left(1.60 \times 10^{-19} \text{ C}\right)\left((2\mathbf{i} - 4\mathbf{j} + \mathbf{k}) \text{ m/s}\right) \times \left((\mathbf{i} + 2\mathbf{j} - 3\mathbf{k}) \text{ T}\right)$$

Since $1 \, \text{C} \cdot \text{m} \cdot \text{T}/\text{s} = 1 \, \text{N}$, we can write this in determinant form as:

$$\mathbf{F}_B = \left(1.60 \times 10^{-19} \text{ N}\right) \begin{vmatrix} \mathbf{i} & \mathbf{j} & \mathbf{k} \\ 2 & -4 & 1 \\ 1 & 2 & -3 \end{vmatrix}$$

Expanding the determinant as described in Equation 11.14, we have

$$\mathbf{F}_{B,x} = \left(1.60 \times 10^{-19} \text{ N}\right)[(-4)(-3) - (1)(2)]\mathbf{i}$$

$$\mathbf{F}_{B,y} = \left(1.60 \times 10^{-19} \text{ N}\right)[(1)(1) - (2)(-3)]\mathbf{j}$$

$$\mathbf{F}_{B,z} = \left(1.60 \times 10^{-19} \text{ N}\right)[(2)(2) - (1)(-4)]\mathbf{k}$$

Again in vector notation,

$$\mathbf{F}_B = \left(1.60 \times 10^{-19} \text{ N}\right)(10\mathbf{i} + 7\mathbf{j} + 8\mathbf{k}) = (16.0\mathbf{i} + 11.2\mathbf{j} + 12.8\mathbf{k}) \times 10^{-19} \text{ N}$$

$$|\mathbf{F}_B| = \left(\sqrt{16.0^2 + 11.2^2 + 12.8^2}\right) \times 10^{-19} \text{ N} = 23.4 \times 10^{-19} \text{ N} \qquad \Diamond$$

13. A wire having a mass per unit length of 0.500 g/cm carries a 2.00-A current horizontally to the south. What are the direction and magnitude of the minimum magnetic field needed to lift this wire vertically upward?

Solution

G: Since **I** = 2.00 A south, **B** must be to the east to make **F** upward according to the right-hand rule for currents in a magnetic field (see Figure 29.7). As before, in viewing the diagrams to the right, use the version of the right-hand rule you have chosen, and ignore the other version. ◊

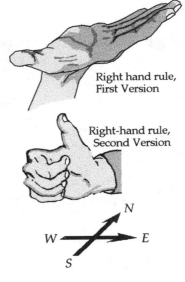

Right hand rule, First Version

Right-hand rule, Second Version

The magnitude of **B** should be significantly greater than the earth's magnetic field (~ 50 μT), since we do not typically see wires levitating when current flows through them.

O: The force on a current-carrying wire in a magnetic field is $\mathbf{F}_B = I\boldsymbol{\ell} \times \mathbf{B}$, from which we can find **B**.

A: With *I* to the south and **B** to the east, the force on the wire is simply $F_B = I\ell B \sin 90°$, which must oppose the weight of the wire, *mg*. So,

$$B = \frac{F_B}{I\ell} = \frac{mg}{I\ell} = \frac{g}{I}\left(\frac{m}{\ell}\right) = \left(\frac{9.80 \text{ m/s}^2}{2.00 \text{ A}}\right)\left(0.500 \ \frac{\text{g}}{\text{cm}}\right)\left(\frac{10^2 \text{ cm/m}}{10^3 \text{ g/kg}}\right) = 0.245 \text{ T} \quad ◊$$

L: The required magnetic field is about 5000 times stronger than the earth's magnetic field. Thus it was reasonable to ignore the earth's magnetic field in this problem. In other situations the earth's field can have a significant effect.

21. A strong magnet is placed under a horizontal conducting ring of radius r that carries current I, as shown in Figure P29.21. If the magnetic field **B** makes an angle θ with the vertical at the ring's location, what are the magnitude and direction of the resultant force on the ring?

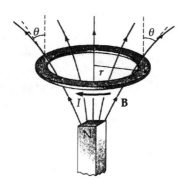

Figure P29.21

Solution

As is shown in the figure below, the magnetic force on each bit of ring is radial inward and upward, at an angle θ above the radial line, according to:

$$\left|d\mathbf{F}\right| = I\left|d\mathbf{s} \times \mathbf{B}\right| = I\,ds\,B$$

The radially inward components tend to squeeze the ring, but cancel out as forces. The upward components $\quad I\,ds\,B\sin\theta \quad$ all add to

$$\mathbf{F} = I(2\pi r)B\,\sin\theta\ \text{up} \qquad\qquad \lozenge$$

The magnetic moment of the ring is down. This problem is a model for the force on a magnetic dipole in a nonuniform magnetic field, or for the force that one magnet exerts on another magnet.

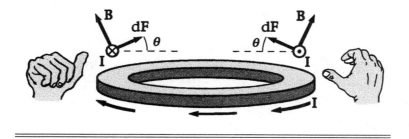

25. A rectangular loop consists of $N = 100$ closely wrapped turns and has dimensions $a = 0.400$ m and $b = 0.300$ m. The loop is hinged along the y axis, and its plane makes an angle $\theta = 30.0°$ with the x axis (Fig. P29.25). What is the magnitude of the torque exerted on the loop by a uniform magnetic field $B = 0.800$ T directed along the x axis when the current is $I = 1.20$ A in the direction shown? What is the expected direction of rotation of the loop?

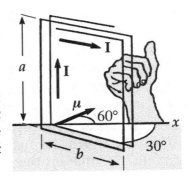

Figure P29.25

Solution The magnetic moment of the coil is $\mu = NIA$, perpendicular to its plane and making a $60°$ angle with the x axis as shown to the right. The torque on the dipole is then $\tau = \mu \times \mathbf{B} = NAIB \sin\theta$ down, having a magnitude of

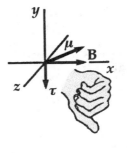

$$\tau = NBAI \sin \theta$$

$$\tau = (100)(0.800 \text{ T})(0.400 \times 0.300 \text{ m}^2)(1.20 \text{ A})\sin 60.0°$$

$$\tau = 9.98 \text{ N·m}$$

◊

Note θ is the angle between the magnetic moment and the **B** field. Loop will rotate such to align the magnetic moment with the **B** field. Looking down along the y axis, loop will rotate in the clockwise direction.

---

35. A proton (charge $+e$, mass m_p), a deuteron (charge $+e$, mass $2m_p$), and an alpha particle, (charge $+2e$, mass $4m_p$) are accelerated through a common potential difference ΔV. The particles enter a uniform magnetic field **B** with a velocity in a direction perpendicular to **B**. The proton moves in a circular path of radius r_p. Determine the values of the radii of the circular orbits for the deuteron r_d and the alpha particle r_α in terms of r_p.

Solution

G: In general, particles with greater speed, more mass, and less charge will have larger radii as they move in a circular path due to a constant magnetic force. Since the effects of mass and charge have opposite influences on the path radius, it is somewhat difficult to predict which particle will have the larger radius. However, since the mass and charge ratios of the three particles are all similar in magnitude within a factor of four, we should expect that the radii also fall within a similar range.

O: The radius of each particle's path can be found by applying Newton's second law, where the force causing the centripetal acceleration is the magnetic force: $\mathbf{F} = q\mathbf{v} \times \mathbf{B}$. The speed of the particles can be found from the kinetic energy resulting from the change in electric potential given.

A: An electric field changes the speed of each particle according to $(K+U)_i = (K+U)_f$. Therefore, assuming that the particles start from rest, we can write

$$q\Delta V = \frac{1}{2}mv^2$$

The magnetic field changes their direction as described by $\Sigma F = ma$:

$$qvB\sin 90° = \frac{mv^2}{r}; \qquad \text{thus} \qquad r = \frac{mv}{qB} = \frac{m}{qB}\sqrt{\frac{2q\Delta V}{m}} = \frac{1}{B}\sqrt{\frac{2m\Delta V}{q}}$$

For the protons,

$$r_p = \frac{1}{B}\sqrt{\frac{2m_p\Delta V}{e}}$$

For the deuterons,

$$r_d = \frac{1}{B}\sqrt{\frac{2(2m_p)\Delta V}{e}} = \sqrt{2}\,r_p \qquad\qquad \lozenge$$

For the alpha particles,

$$r_\alpha = \frac{1}{B}\sqrt{\frac{2(4m_p)\Delta V}{2e}} = \sqrt{2}\,r_p \qquad\qquad \lozenge$$

L: Somewhat surprisingly, the radii of the deuterons and alpha particles are the same and are only 41% greater than for the protons.

39. A cosmic-ray proton in interstellar space has an energy of 10.0 MeV and executes a circular orbit having a radius equal to that of Mercury's orbit around the Sun (5.80×10^{10} m). What is the magnetic field in that region of space?

Solution   Think of the proton as having accelerated through a potential difference $\Delta V = 10^7$ V. Its energy is $E = \frac{1}{2}mv^2 = e\Delta V$,

so its speed is

$$v = \sqrt{\frac{2e\Delta V}{m}}$$

Now $\Sigma F = ma$ becomes $\dfrac{mv^2}{R} = evB\sin 90°$

So $B = \dfrac{mv}{eR} = \dfrac{m}{eR}\sqrt{\dfrac{2e\Delta V}{m}} = \dfrac{1}{R}\sqrt{\dfrac{2m\Delta V}{e}}$

and $B = \dfrac{1}{5.80 \times 10^{10}\text{ m}}\sqrt{\dfrac{2(1.6727 \times 10^{-27}\text{ kg})(10^7\text{ V})}{1.60 \times 10^{-19}\text{ C}}} = 7.88 \times 10^{-12}$ T   ◊

---

45. A cyclotron designed to accelerate protons has a magnetic field with a magnitude of 0.450 T over a region of radius 1.20 m. What are (a) the cyclotron frequency and (b) the maximum speed acquired by the protons?

Solution   (a)   The cyclotron frequency is $\omega = qB/m$.

For protons, $\omega = \dfrac{(1.60 \times 10^{-19}\text{ C})(0.450\text{ T})}{1.67 \times 10^{-27}\text{ kg}} = 4.31 \times 10^7$ rad / s   ◊

(b) $R = \dfrac{mv}{Bq}$: $v = \dfrac{BqR}{m} = \dfrac{(0.450\text{ T})(1.60 \times 10^{-19}\text{ C})(1.20\text{ m})}{1.67 \times 10^{-27}\text{ kg}} = 5.17 \times 10^7$ m / s   ◊

47. The picture tube in a television uses magnetic deflection coils rather than electric deflection plates. Suppose an electron beam is accelerated through a 50.0-kV potential difference and then travels through a region of uniform magnetic field 1.00 cm wide. The screen is located 10.0 cm from the center of the coils and is 50.0 cm wide. When the field is turned off, the electron beam hits the center of the screen. What field magnitude is necessary to deflect the beam to the side of the screen? Ignore relativistic corrections.

Solution

The beam is deflected by the angle

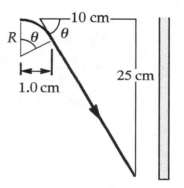

$$\theta = \tan^{-1}\frac{25.0 \text{ cm}}{10.0 \text{ cm}} = 68.2°$$

The two angles θ shown are equal because their sides are perpendicular, right side to right side and left side to left side. The radius of curvature of the electrons in the field is

$$R = \frac{1.00 \text{ cm}}{\sin 68.2°} = 1.077 \text{ cm}$$

Now $\frac{1}{2}mv^2 = q\Delta V$, so $\quad v = \sqrt{\frac{2q\Delta V}{m}} = 1.33 \times 10^8 \text{ m/s}$

$\Sigma \mathbf{F} = m\mathbf{a}$ becomes $\qquad \dfrac{mv^2}{R} = |q|vB\sin 90°$

$$B = \frac{mv}{|q|R} = \frac{(9.11 \times 10^{-31} \text{ kg})(1.33 \times 10^8 \text{ m/s})}{(1.60 \times 10^{-19} \text{ C})(1.077 \times 10^{-2} \text{ m})} = 70.1 \text{ mT} \quad \Diamond$$

51. In an experiment designed to measure the Earth's magnetic field using the Hall effect, a copper bar 0.500 cm thick is positioned along an east-west direction. If a current of 8.00 A in the conductor results in a Hall voltage of 5.10 pV, what is the magnitude of the Earth's magnetic field? (Assume that $n = 8.48 \times 10^{28}$ electrons/m³ and that the plane of the bar is rotated to be perpendicular to the direction of **B**.)

Solution

G: The Earth's magnetic field is about 50 μT (see Table 29.1), so we should expect a result of that order of magnitude.

O: The magnetic field can be found from the Hall effect voltage:

$$\Delta V_H = \frac{IB}{nqt} \qquad \text{or} \qquad B = \frac{nqt\Delta V_H}{I}$$

A: From the Hall voltage,

$$B = \frac{\left(8.48 \times 10^{28} \, \frac{e^-}{m^3}\right)\left(1.60 \times 10^{-19} \, \frac{C}{e^-}\right)(0.00500 \text{ m})\left(5.10 \times 10^{-12} \text{ V}\right)}{8.00 \text{ A}}$$

$$B = 4.32 \times 10^{-5} \text{ T} = 43.2 \, \mu\text{T} \qquad \qquad \lozenge$$

L: The calculated magnetic field is slightly less than we expected but is reasonable considering that the Earth's local magnetic field varies in both magnitude and direction.

55. Sodium melts at 99 °C. Liquid sodium, an excellent thermal conductor, is used in some nuclear reactors to cool the reactor core. The liquid sodium is moved through pipes by pumps that exploit the force on a moving charge in a magnetic field. The principle is as follows: Assume that the liquid metal is in an electrically insulating pipe having a rectangular cross section of width w and height h.   A uniform magnetic field perpendicular to the pipe affects a section of length L (Fig. P29.55). An electric current directed perpendicular to the pipe and to the magnetic field produces a current density J in the liquid sodium. (a) Explain why this arrangement produces on the liquid a force that is directed along the length of the pipe. (b) Show that the section of liquid in the magnetic field experiences a pressure increase JLB.

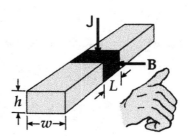

Figure 29.55

Solution

(a) By the right-hand rule, the electric current carried by the material experiences a force $I\,\mathbf{h} \times \mathbf{B}$ along the pipe.   ◊

(b) The sodium, consisting of ions and electrons, flows along the pipe transporting no net charge. But inside the section of length L, electrons drift upward to constitute downward electric current $J \cdot (\text{area}) = JLw$. The current feels magnetic force $I\,\mathbf{h} \times \mathbf{B} = JLwhB \sin 90°$. This force along the pipe axis will make the fluid move, exerting pressure

$$\frac{F}{\text{area}} = \frac{J\,LwhB}{hw} = J\,LB$$ ◊

The hand in the figure shows that the fluid moves away from you, into the page.

155

59. A positive charge $q = 3.20 \times 10^{-19}$ C moves with a velocity $\mathbf{v} = (2\mathbf{i} + 3\mathbf{j} - \mathbf{k})$ m/s through a region where both a uniform magnetic field and a uniform electric field exist. (a) What is the total force on the moving charge (in unit-vector notation) if $\mathbf{B} = (2\mathbf{i} + 4\mathbf{j} + \mathbf{k})$ T and $\mathbf{E} = (4\mathbf{i} - \mathbf{j} - 2\mathbf{k})$ V/m? (b) What angle does the force vector make with the positive x axis?

Solution The total force is the Lorentz force,

(a) $\mathbf{F}_B = q\mathbf{E} + q(\mathbf{v} \times \mathbf{B}) = q(\mathbf{E} + \mathbf{v} \times \mathbf{B})$

$\mathbf{F}_B = q[(4\mathbf{i} - \mathbf{j} - 2\mathbf{k}) \text{ V / m} + (2\mathbf{i} + 3\mathbf{j} - \mathbf{k}) \text{ m / s} \times (2\mathbf{i} + 4\mathbf{j} + \mathbf{k}) \text{ T}]$

$\mathbf{F}_B = q[(4\mathbf{i} - \mathbf{j} - 2\mathbf{k}) \text{ V / m} + (7\mathbf{i} - 4\mathbf{j} + 2\mathbf{k}) \text{ m} \cdot \text{T/s}]$

$\mathbf{F}_B = q([11\mathbf{i} - 5\mathbf{j}] \text{ V / m}) = q([11\mathbf{i} - 5\mathbf{j}] \text{ N/C})$

$\mathbf{F}_B = (3.20 \times 10^{-19} \text{ C})((11\mathbf{i} - 5\mathbf{j}) \text{ N/C}) = (3.52\mathbf{i} - 1.60\mathbf{j}) \times 10^{-18}$ N $\lozenge$

(b) $\mathbf{F}_B \cdot \mathbf{i} = F_B \cos\theta = F_x$

$\theta = \cos^{-1}\left(\dfrac{F_x}{F}\right) = \cos^{-1}\left(\dfrac{3.52}{3.87}\right) = 24.4°$ $\lozenge$

---

67. Consider an electron orbiting a proton and maintained in a fixed circular path of radius $R = 5.29 \times 10^{-11}$ m by the Coulomb force. Treating the orbiting charge as a current loop, calculate the resulting torque when the system is in a magnetic field of 0.400 T directed perpendicular to the magnetic moment of the electron.

Solution

G: Since the mass of the electron is very small ($\sim 10^{-30}$ kg), we should expect that the torque on the orbiting charge will be very small as well, perhaps $\sim 10^{-30}$ N·m.

O: The torque on a current loop that is perpendicular to a magnetic field can be found from $|\tau| = IAB \sin\theta$. The magnetic field is given, $\theta = 90°$, the area of the loop can be found from the radius of the circular path, and the current can be found from the centripetal acceleration that results from the Coulomb force that attracts the electron to proton.

A: The area of the loop is $A = \pi r^2 = \pi(5.29 \times 10^{-11} \text{ m})^2 = 8.79 \times 10^{-21} \text{ m}^2$.

If v is the speed of the electron, then the period of its circular motion will be $T = 2\pi R/v$, and the effective current due to the orbiting electron is $I = \Delta Q / \Delta t = e/T$.

Applying Newton's second law with the Coulomb force acting as the central force gives

$$\Sigma F = \frac{k_e q^2}{R^2} = \frac{mv^2}{R} \qquad \text{so that} \qquad v = q\sqrt{\frac{k_e}{mR}} \qquad \text{and} \qquad T = 2\pi\sqrt{\frac{mR^3}{q^2 k_e}}$$

$$T = 2\pi\sqrt{\frac{(9.10 \times 10^{-31} \text{ kg})(5.29 \times 10^{-11} \text{ m})^3}{(1.60 \times 10^{-19} \text{ C})^2 (8.99 \times 10^9 \text{ N} \cdot \text{m}^2/\text{C}^2)}} = 1.52 \times 10^{-16} \text{ s}$$

The torque is $|\tau| = \left(\dfrac{q}{T}\right) AB$:

$$|\tau| = \frac{1.60 \times 10^{19} \text{ C}}{1.52 \times 10^{-16} \text{ s}} (\pi)(5.29 \times 10^{-11} \text{ m})^2 (0.400 \text{ T}) = 3.70 \times 10^{-24} \text{ N} \cdot \text{m} \qquad \Diamond$$

L: The torque is certainly small, but a million times larger than we guessed. This torque will cause the atom to precess with a frequency proportional to the applied magnetic field. A similar process on the nuclear, rather than the atomic, level leads to nuclear magnetic resonance (NMR), which is used for magnetic resonance imaging (MRI) scans employed for medical diagnostic testing (see Section 44.2).

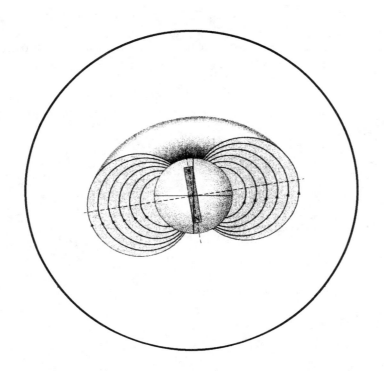

Sources of the
Magnetic Field

SOURCES OF THE MAGNETIC FIELD

INTRODUCTION

The preceding chapter treated a class of problems involving the magnetic force on a charged particle moving in a magnetic field. To complete the description of the magnetic interaction, this chapter deals with the origin of the magnetic field, namely, moving charges or electric currents. We begin by showing how to use the law of Biot and Savart to calculate the magnetic field produced at a point by a current element. Using this formalism and the superposition principle, we then calculate the total magnetic field due to a distribution of currents for several geometries. Next, we show how to determine the force between two current-carrying conductors, a calculation that leads to the definition of the ampere. We also introduce Ampère's law, which is very useful for calculating the magnetic field of highly symmetric configurations carrying steady currents. We apply Ampère's law to determine the magnetic field for several current configurations.

EQUATIONS AND CONCEPTS

The **Biot–Savart law** gives the magnetic field at a point in space due to an element of conductor $d\mathbf{s}$ which carries a current I and is at a distance r away from the point.

$$d\mathbf{B} = \frac{\mu_0}{4\pi} \frac{I\, d\mathbf{s} \times \hat{\mathbf{r}}}{r^2} \tag{30.1}$$

The permeability of free space is a constant.

$$\mu_0 = 4\pi \times 10^{-7}\ \text{T} \cdot \text{m} / \text{A}$$

The **total magnetic field** is found by integrating the Biot-Savart law expression over the entire length of the conductor.

$$\mathbf{B} = \frac{\mu_0 I}{4\pi} \int \frac{d\mathbf{s} \times \hat{\mathbf{r}}}{r^2} \qquad (30.3)$$

The magnetic field due to several important geometric arrangements of a current-carrying conductor can be calculated by use of the Biot-Savart law:

B at a distance a from a **long straight conductor**, carrying a current I.

$$B = \frac{\mu_0 I}{2\pi a} \qquad (30.5)$$

B at the center of an **arc of radius R which subtends an angle θ (in radians) at the center of the arc.**

$$B = \frac{\mu_0 I}{4\pi R} \theta \qquad (30.6)$$

B_x on the axis of a **circular loop** of radius R and at a **distance x from the plane** of the loop.

$$B_x = \frac{\mu_0 I R^2}{2\left(x^2 + R^2\right)^{3/2}} \qquad (30.7)$$

B at a distance a from a **straight wire** carrying a current I, where θ_1 and θ_2 are as shown in the figure to the right.

$$B = \frac{\mu_0 I}{4\pi a}\left(\cos\theta_1 - \cos\theta_2\right) \qquad (30.4)$$

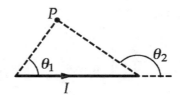

The magnitude of the **magnetic force per unit length** between very long parallel conductors depends on the distance a between the conductors and the magnitudes of the two currents.

$$\frac{F_B}{\ell} = \frac{\mu_0 I_1 I_2}{2\pi a}$$

(30.12)

If the parallel currents I_1 and I_2 are in the same direction, the force between conductors will be one of attraction. Parallel conductors carrying currents in opposite directions will repel each other. In any case, the magnitude of the forces on the two conductors will be **equal.**

Ampère's law represents a relationship between the integral of the tangential component of the magnetic field around any closed path and the total current which threads the closed path.

$$\oint \mathbf{B} \cdot d\mathbf{s} = \mu_0 I$$

(30.13)

B **inside a toroid** having N turns and at a distance r from the center of the toroid.

$$B = \frac{\mu_0 N I}{2\pi r}$$

(30.16)

B near the center of a **solenoid** of n turns per unit length.

$$B = \mu_0 \frac{N}{\ell} I = \mu_0 n I$$

(30.17)

The **magnetic flux** through a surface is the integral of the normal component of the field over the surface.

$$\Phi_B \equiv \int \mathbf{B} \cdot d\mathbf{A} \qquad (30.18)$$

The direction of the magnetic field due to current in a long wire is determined by using right-hand rule B:

> Hold the conductor in the right hand with the thumb pointing in the direction of the conventional current. The fingers will then wrap around the wire in the direction of the magnetic field lines. The magnetic field is tangent to the circular field lines at every point in the region around the conductor.

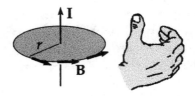

The direction of the magnetic field at the center of a current loop is perpendicular to the plane of the loop and directed in the sense given by the right-hand rule for B.

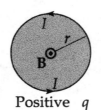

Positive q

Within a solenoid, the magnetic field is parallel to the axis of the solenoid and pointing in a sense determined by applying right-hand rule B to one of the coils.

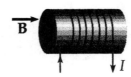

The magnetic moment μ of an orbiting electron is proportional to its orbital angular momentum L.

$$\mu = \left(\frac{e}{2m_e}\right)L \qquad (30.25)$$

$$L = 0, \hbar, 2\hbar, 3\hbar, \ldots$$

The intrinsic magnetic moment μ_{spin} associated with the spin of the electron is called the Bohr magneton.

$$\mu_B = 9.27 \times 10^{-24} \text{ J} / \text{T} \qquad (30.28)$$

The magnetic state of a substance is described by a quantity called the magnetization vector, $\mathbf{M}$. For paramagnetic and diamagnetic substances, the magnetization is proportional to the magnetic field strength.

$$\mathbf{M} = \chi\mathbf{H} \qquad (30.32)$$

In a region where the total magnetic field is due to that of a current-carrying conductor and the presence of a magnetic substance, the **total field vector B** can be expressed in terms of $\mathbf{M}$ and $\mathbf{H}$.

$$\mathbf{B} = \mu_0(\mathbf{H} + \mathbf{M}) \qquad (30.30)$$

The total field can also be expressed in terms of the permeability μ_m of the substance.

$$\mathbf{B} = \mu_m\mathbf{H} \qquad (30.33)$$

The permeability of a magnetic substance is related to its magnetic susceptibility.

$$\mu_m = \mu_0 \left(1 + \chi \right)$$

(30.34)

The magnetization of a paramagnetic substance is proportional to the applied field and inversely proportional to the absolute temperature. This is known as Curie's law.

$$M = C \frac{B_0}{T}$$

(30.35)

SUGGESTIONS, SKILLS, AND STRATEGIES

THE BIOT-SAVART LAW

It is important to remember that the **Biot-Savart law**, given by Equation 30.1,

$$d\mathbf{B} = \frac{\mu_0}{4\pi} \frac{I \, d\mathbf{s} \times \hat{\mathbf{r}}}{r^2}$$

is a **vector** expression. The unit vector $\hat{\mathbf{r}}$ is directed from the element of conductor $d\mathbf{s}$ to the point P where the magnetic field is to be calculated, and r is the distance from $d\mathbf{s}$ to point P. For the arbitrary current element shown in the figure at right, the direction of $\mathbf{B}$ at point P, as determined by the right-hand rule for the cross product, is directed **out of the plane**; while the magnetic field at point P' due to the current in the element $d\mathbf{s}$ is directed **into the plane**. In order to find the **total** magnetic field at any point due to a conductor of finite sign, you must sum up the contributions from all current elements making up the conductor. This means that the total **B** field is expressed as an integral over the entire length of the conductor:

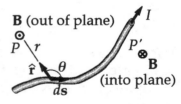

$$\mathbf{B} = \frac{\mu_0 I}{4\pi} \int \frac{d\mathbf{s} \times \hat{\mathbf{r}}}{r^2}$$

REVIEW CHECKLIST

▷ Use the Biot-Savart law to calculate the magnetic induction at a specified point in the vicinity of a current element, and by integration find the total magnetic field due to a number of important geometric arrangements. Your use of the Biot-Savart law must include a clear understanding of the **direction** of the magnetic field contribution relative to the direction of the current element which produces it and the direction of the vector which locates the point at which the field is to be calculated.

▷ Understand the basis for defining the ampere and the coulomb in terms of the magnetic force between parallel current-carrying conductors.

▷ Use Ampère's law to calculate the magnetic field due to steady current configurations which have a sufficiently high degree of symmetry such as a long straight conductor, a long solenoid, and a toroidal coil.

▷ Calculate the magnetic flux through a surface area placed in either a uniform or nonuniform magnetic field.

▷ Understand, via the generalized form of Ampère's law, that magnetic fields are produced both by **conduction currents** and by **changing electric fields**.

ANSWERS TO SELECTED CONCEPTUAL QUESTIONS

4. Explain why two parallel wires carrying currents in opposite directions repel each other.

Answer

The figure at the right will help you understand this result. The magnetic field due to wire 2 at the position of wire 1 is directed out of the paper. Hence, the magnetic force on wire 1, given by $I_1\mathbf{L}_1 \times \mathbf{B}_2$, must be directed to the left since $\mathbf{L}_1 \times \mathbf{B}_2$ is directed to the left. Likewise, you can show that the magnetic force on wire 2 due to the field of wire 1 is directed towards the right.

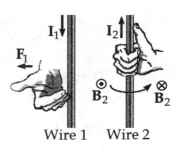

□   □   □   □

10. A hollow copper tube carries a current along its length. Why does $\mathbf{B} = 0$ inside the tube? Is $\mathbf{B}$ nonzero outside the tube?

Answer

Let us apply Ampere's circuit law to the closed path labeled 1 in this figure. Since there is no current through this path, and because of the symmetry of the configuration, we see that the magnetic field inside the tube must be zero. On the other hand, the net current through the path labeled 2 is I, the current carried by the conductor. Therefore, the field outside the tube is nonzero.

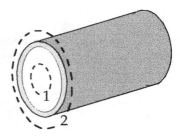

□   □   □   □

SOLUTIONS TO SELECTED END-OF-CHAPTER PROBLEMS

3. (a) A conductor in the shape of a square of edge length ℓ = 0.400 m carries a current I = 10.0 A (Fig. P30.3). Calculate the magnitude and direction of the magnetic field at the center of the square. (b) If this conductor is formed into a single circular turn and carries the same current, what is the value of the magnetic field at the center?

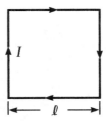

Figure P30.3

Solution

G: As shown in the diagram to the right, the magnetic field at the center is directed into the page from the clockwise current. If we consider the sides of the square to be sections of four infinite wires, then we could expect the magnetic field at the center of the square to be a little less than four times the strength of the field at a point $\ell/2$ away from an infinite wire with current I.

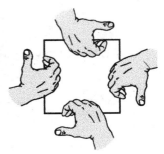

$$B < 4\frac{\mu_0 I}{2\pi a} = 4\left(\frac{\left(4\pi\times10^{-7}\ \text{T}\cdot\text{m}/\text{A}\right)(10.0\ \text{A})}{2\pi(0.200\ \text{m})}\right) = 40.0\ \mu\text{T}$$

Forming the wire into a circle should not significantly change the magnetic field at the center since the average distance of the wire from the center will not be much different.

O: Each side of the square is simply a section of a thin, straight conductor, so the solution derived from the Biot-Savart law in Example 30.1 can be applied to part (a) of this problem. For part (b), the Biot-Savart law can also be used to derive the equation for the magnetic field at the center of a circular current loop as shown in Example 30.3.

A: (a) We use Equation 30.4 for the field created by each side of the square. Each side contributes a field away from you at the center, so together they produce a magnetic field:

$$B = \frac{4\mu_0 I}{4\pi a}\left(\cos\frac{\pi}{4} - \cos\frac{3\pi}{4}\right) = \frac{4\left(4\pi\times10^{-6}\ \text{T}\cdot\text{m}/\text{A}\right)(10.0\ \text{A})}{4\pi(0.200\ \text{m})}\left(\frac{\sqrt{2}}{2} + \frac{\sqrt{2}}{2}\right)$$

so at the center of the square,

B $= 2.00\sqrt{2}\times10^{-5}$ T $= 28.3\ \mu$T   perpendicularly into the page   ◊

(b)   As in the first part of the problem, the direction of the magnetic field will be into the page. The new radius is found from the length of wire: $4\ell = 2\pi R$, so $R = 2\ell/\pi = 0.255$ m.   Equation 30.8 gives the magnetic field at the center of a circular current loop:

$$B = \frac{\mu_0 I}{2R} = \frac{\left(4\pi\times10^{-7}\ \text{T}\cdot\text{m}/\text{A}\right)(10.0\ \text{A})}{2(0.255\ \text{m})} = 2.47\times10^{-5}\ \text{T} = 24.7\ \mu\text{T} \qquad ◊$$

Caution!   If you use your calculator, it may not understand the keystrokes: [4] [×] [π] [EXP] [+/−] [7]   To get the right answer, you may need to use [4] [EXP] [+/−] [7] [×] [π].

L: The magnetic field in part (a) is less than 40μT as we predicted. Also, the magnetic fields from the square and circular loops are similar in magnitude, with the field from the circular loop being about 15% less than from the square loop.

Quick tip: A simple way to use your right hand to find the magnetic field due to a current loop is to curl the fingers of your right hand in the direction of the current. Your extended thumb will then point in the direction of the magnetic field within the loop or solenoid.

5. Determine the magnetic field at a point P located a distance x from the corner of an infinitely long wire bent at a right angle, as in Figure P30.5. The wire carries a steady current I.

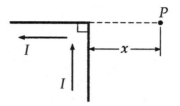

Figure P30.5

Solution The vertical section of wire constitutes one half of an infinitely long straight wire at distance x from P, so it creates a field equal to

$$B = \tfrac{1}{2}\left(\frac{\mu_o I}{2\pi x}\right)$$

Hold your right hand with extended thumb in the direction of the current; the field is away from you, into the paper. For each bit of the horizontal section of wire $d\mathbf{s}$ is to the left and $\hat{\mathbf{r}}$ is to the right, so $d\mathbf{s} \times \hat{\mathbf{r}} = 0$. The horizontal current produces zero field at P. Thus,

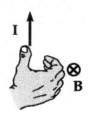

$$\mathbf{B} = \frac{\mu_o I}{4\pi x} \quad \text{into the paper} \qquad \Diamond$$

17. In Figure P30.17, the current in the long, straight wire is $I_1 = 5.00$ A and the wire lies in the plane of the rectangular loop, which carries 10.0 A. The dimensions are $c = 0.100$ m, $a = 0.150$ m, and $\ell = 0.450$ m. Find the magnitude and direction of the net force exerted on the loop by the magnetic field created by the wire.

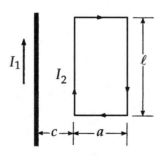

Figure P30.17

Solution

G: Even though there are forces in opposite directions on the loop, we must remember that the magnetic field is stronger near the wire than it is farther away. By symmetry the forces exerted on sides 2 and 4 (the horizontal segments of length a) are equal and opposite, and therefore cancel. The magnetic field in the plane of the loop is directed into the page to the right of I_1. By the right-hand rule, $\mathbf{F} = I\boldsymbol{\ell} \times \mathbf{B}$ is directed toward the **left** for side 1 of the loop and a smaller force is directed toward the **right** for side 3. Therefore, we should expect the net force to be to the left, possibly in the μN range for the currents and distances given.

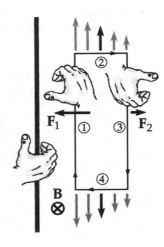

O: The magnetic force between two parallel wires can be found from Equation 30.11, which can be applied to sides 1 and 3 of the loop to find the net force resulting from these opposing force vectors.

A:
$$\mathbf{F} = \mathbf{F}_1 + \mathbf{F}_2 = \frac{\mu_o I_1 I_2 \ell}{2\pi}\left(\frac{1}{c+a} - \frac{1}{c}\right)\mathbf{i} = \frac{\mu_o I_1 I_2 \ell}{2\pi}\left(\frac{-a}{c(c+a)}\right)\mathbf{i}$$

$$\mathbf{F} = \frac{\left(4\pi \times 10^{-7} \text{ N}/\text{A}^2\right)(5.00 \text{ A})(10.0 \text{ A})(0.450 \text{ m})}{2\pi}\left(\frac{-0.150 \text{ m}}{(0.100 \text{ m})(0.250 \text{ m})}\right)\mathbf{i}$$

$$\mathbf{F} = (-2.70 \times 10^{-5}\mathbf{i}) \text{ N} \quad \text{or} \quad \mathbf{F} = 2.70 \times 10^{-5} \text{ N} \quad \text{toward the left} \qquad \lozenge$$

L: The net force is to the left and in the μN range as we expected. The symbolic representation of the net force on the loop shows that the net force would be zero if either current disappeared, if either dimension of the loop became very small ($a \to 0$ or $\ell \to 0$), or if the magnetic field were uniform ($c \to \infty$).

19. Four long, parallel conductors carry equal currents of $I = 5.00$ A. Figure P30.19 is an end view of the conductors. The direction of the current is into the page at points A and B (indicated by the crosses) and out of the page at C and D (indicated by the dots). Calculate the magnitude and direction of the magnetic field at point P, located at the center of the square of edge length of 0.200 m.

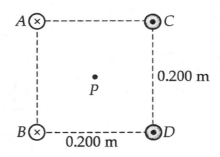

Figure P30.19

Solution

Each wire is distant from P by $(0.200 \text{ m}) \cos 45° = 0.141$ m.
Each wire produces a field at P of equal magnitude:

$$B = \frac{\mu_0 I}{2\pi a} = \frac{(2.00 \times 10^{-7} \text{ T} \cdot \text{m} / \text{A})(5.00 \text{ A})}{(0.141 \text{ m})} = 7.07 \ \mu\text{T}$$

Carrying currents away from you, the left-hand wires produce fields at P of 7.07 μT, in the following directions:

A: to the bottom and left, at 225°

B: to the bottom and right, at 315°;

Carrying currents toward you, the wires to the right also produce fields at P of 7.07 μT, in the following directions:

C: downward and to the right, at 315°

D: downward and to the left, at 225°.

The total field is then

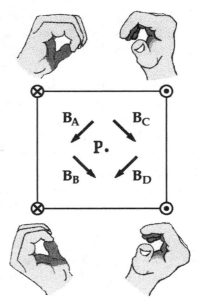

$$4(7.07 \ \mu\text{T}) \sin 45° = 20.0 \ \mu\text{T} \qquad \text{toward the bottom of the page.} \quad \Diamond$$

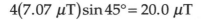

25. A packed bundle of 100 long, straight, insulated wires forms a cylinder of radius $R = 0.500$ cm. (a) If each wire carries 2.00 A, what are the magnitude and direction of the magnetic force per unit length acting on a wire located 0.200 cm from the center of the bundle? (b) Would a wire on the outer edge of the bundle experience a force greater or less than the value calculated in part (a)?

Solution

G: The force **on** one wire comes from its interaction with the magnetic field created **by** the other ninety-nine wires. According to Ampere's law, at a distance r from the center, only the wires enclosed within a radius r contribute to this net magnetic field; the other wires outside the radius produce magnetic field vectors in opposite directions that cancel out at r. Therefore, the magnetic field (and also the force on a given wire at radius r) will be greater for larger radii within the bundle, and will decrease for distances beyond the radius of the bundle, as shown in the graph to the right. Applying $\mathbf{F} = I\boldsymbol{\ell} \times \mathbf{B}$, the magnetic force on a single wire will be directed toward the center of the bundle, so that all the wires tend to attract each other.

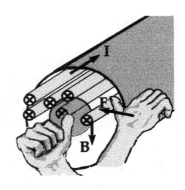

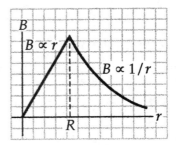

O: Using Ampere's law, we can find the magnetic field at any radius, so that the magnetic force $\mathbf{F} = I\boldsymbol{\ell} \times \mathbf{B}$ on a single wire can then be calculated.

A: (a)   Ampere's law is used to derive Equation 30.15, which we can use to find the magnetic field at $r = 0.200$ cm from the center of the cable:

$$B = \frac{\mu_o I_o r}{2\pi R^2} = \frac{\left(4\pi \times 10^{-7} \text{ T} \cdot \text{m} / \text{A}\right)(99)(2.00 \text{ A})\left(0.200 \times 10^{-2} \text{ m}\right)}{2\pi (0.500 \times 10^{-2} \text{ m})^2} = 3.17 \times 10^{-3} \text{ T}$$

This field points tangent to a circle of radius 0.200 cm and exerts a force $\mathbf{F} = I\boldsymbol{\ell} \times \mathbf{B}$ toward the center of the bundle, on the single hundredth wire:

$$\frac{F}{\ell} = IB\sin\theta = (2.00 \text{ A})\left(3.17 \times 10^{-3} \text{ T}\right)(\sin 90°) = 6.34 \text{ mN} / \text{m} \qquad \lozenge$$

(b)   As is shown above in Figure 30.12 from the text, the magnetic field increases linearly as a function of r until it reaches a maximum at the outer surface of the cable. Therefore, the force on a single wire at the outer radius $r = 5.00$ cm would be greater than at $r = 2.00$ cm by a factor of 5/2. $\qquad \lozenge$

L: We did not estimate the expected magnitude of the force, but 200 amperes is a lot of current. It would be interesting to see if the magnetic force that pulls together the individual wires in the bundle is enough to hold them against their own weight: If we assume that the insulation accounts for about half the volume of the bundle, then a single copper wire in this bundle would have a cross sectional area of about

$$(1/2)(0.01)\pi(0.500 \text{ cm})^2 = 4 \times 10^{-7} \text{ m}^2$$

with a weight per unit length of

$$\rho g A = \left(8920 \text{ kg} / \text{m}^3\right)(9.8 \text{ N} / \text{kg})\left(4 \times 10^{-7} \text{ m}^2\right) = 0.03 \text{ N} / \text{m}$$

Therefore, the outer wires experience an inward magnetic force that is about half the magnitude of their own weight. If placed on a table, this bundle of wires would form a loosely held mound without the outer sheathing to hold them together.

27. A long cylindrical conductor of radius R carries a current I, as shown in Figure P30.27. The current density J, however, is not uniform over the cross section of the conductor but is a function of the radius according to $J = br$, where b is a constant. Find an expression for the magnetic field B (a) at a distance $r_1 < R$ and (b) at a distance $r_2 > R$, measured from the axis.

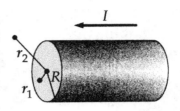

Figure P30.27

Solution   Take a circle of radius r_1 or r_2 to apply $\oint \mathbf{B} \cdot d\mathbf{s} = \mu_0 I$, where for nonuniform current density $I = \int \mathbf{J} \cdot d\mathbf{A}$. In this case $\mathbf{B}$ is parallel to $d\mathbf{s}$ and $\mathbf{J}$ is parallel to $d\mathbf{A}$, so Ampère's law gives

$$\oint B\, ds = \mu_0 \int J\, d\mathbf{A}$$

(a)   For $r_1 < R$,

$$2\pi r_1 B = \mu_0 \int_0^{r_1} br(2\pi r\, dr)$$

and

$$B = \left(\mu_0 b r_1^2\right)/3 \quad \text{(inside)} \qquad \Diamond$$

(b)   For $r_2 > R$,

$$2\pi r_2 B = \mu_0 \int_0^{R} br(2\pi r\, dr)$$

and

$$B = \left(\mu_0 b R^3\right)/3 r_2 \quad \text{(outside)} \qquad \Diamond$$

---

29.   What current is required in the windings of a long solenoid that has 1000 turns uniformly distributed over a length of 0.400 m, to produce at the center of the solenoid a magnetic field of magnitude 1.00×10^{-4} T ?

Solution   The magnetic field at the center of a coil is $B = \mu_0 \dfrac{N}{\ell} I$

So $I = \dfrac{B\ell}{\mu_0 N} = \dfrac{\left(1.00 \times 10^{-4}\ \text{T} \cdot \text{A}\right)\left(0.400\ \text{m}\right)}{\left(4\pi \times 10^{-7}\ \text{T} \cdot \text{m}\right)\left(1000\right)} = 31.8\ \text{mA} \qquad \Diamond$

---

33. A cube of edge length $\ell = $ 2.50 cm is positioned as shown in Fig. P30.33. A uniform magnetic field given by **B** = (5.00**i** + 4.00**j** + 3.00**k**) T exists throughout the region. (a) Calculate the flux through the shaded face. (b) What is the total flux through the six faces?

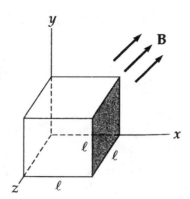

Figure P30.33

Solution   The flux is defined as $\Phi_B = \mathbf{B} \cdot \mathbf{A}$

$$\Phi_B = B_x A_x + B_y A_y + B_z A_z$$

Here, $A_y = A_z = 0$, so

(a) $\Phi_B = $ (5.00 T)(0.0250 m)^2 = 3.13 × 10^{-3} T·m^2 ◊

(b)   For a closed surface, $\oint \mathbf{B} \cdot d\mathbf{A} = 0$ so $\Phi_B = 0$ ◊

---

37. A 0.100-A current is charging a capacitor that has square plates 5.00 cm on a side. If the plate separation is 4.00 mm, find (a) the time rate of change of electric flux between the plates and (b) the displacement current between the plates.

Solution   The electric field in the space between the plates is $E = \dfrac{\sigma}{\epsilon_0} = \dfrac{Q}{\epsilon_0 A}$

The flux of this field is $\Phi_E = \mathbf{E} \cdot \mathbf{A} = \left(\dfrac{Q}{\epsilon_0 A} \right) A \cos 0° = \dfrac{Q}{\epsilon_0}$

(a)   The rate of change of flux is

$$\frac{d\Phi_E}{dt} = \frac{d}{dt} \frac{Q}{\epsilon_0} = \frac{1}{\epsilon_0} \frac{dQ}{dt} = \frac{I}{\epsilon_0} = \left(\frac{0.100 \text{ A}}{8.85 \times 10^{-12} \text{ C}^2 / \text{N} \cdot \text{m}^2} \right) \left(\frac{1 \text{ C}}{\text{A} \cdot \text{s}} \right)$$

$$\frac{d\Phi_E}{dt} = 1.13 \times 10^{10} \text{ N} \cdot \text{m}^2 / \text{C} \cdot \text{s}$$ ◊

(b)   The displacement current is defined as

$$I_d = \epsilon_0 \frac{d\Phi}{dt} = \left(8.85 \times 10^{-12} \frac{C^2}{N \cdot m^2}\right)\left(1.13 \times 10^{10} \frac{N \cdot m^2}{C \cdot s}\right) = 0.100 \ A \qquad \Diamond$$

---

39. A toroid with a mean radius of 20.0 cm and 630 turns (see Fig. 30.29) is filled with powdered steel whose magnetic susceptibility χ is 100. If the current in the windings is 3.00 A, find B (assumed uniform) inside the toroid.

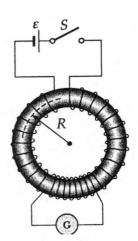

Figure 30.29

Solution

If the coil had a vacuum inside, the magnetic field would be given by Equation 30.16,

$$B = \frac{\mu_0 NI}{2\pi r}$$

That is, the magnetic field strength would be

$$H = \frac{B}{\mu_0} = \frac{NI}{2\pi r}$$

With the steel inside, $B = \mu_0(1+\chi)H = \mu_0(1+\chi)\dfrac{NI}{2\pi r}$

$$B = \frac{(4\pi \times 10^{-7} \ T \cdot m \ / \ A)(101)(630)(3.00 \ A)}{2\pi(0.200 \ m)} = 0.191 \ T \qquad \Diamond$$

---

47. The magnetic moment of the Earth is approximately 8.00×10^{22} A·m^2. (a) If this were caused by the complete magnetization of a huge iron deposit, how many unpaired electrons would this correspond to? (b) At two unpaired electrons per iron atom, how many kilograms of iron would this correspond to? (Iron has a density of 7900 kg/m^3, and approximately 8.50×10^{28} atoms/m^3.)

Solution

G: We know that most of the Earth is not iron, so if the situation described provides an accurate model, then the iron deposit must certainly be less than the mass of the Earth ($M_{Earth} = 5.98 \times 10^{24}$ kg). One mole of iron has a mass of 55.8 g and contributes $2(6.02 \times 10^{23})$ unpaired electrons, so we should expect the total unpaired electrons to be less than 10^{50}.

O: The Bohr magneton μ_B is the measured value for the magnetic moment of a single unpaired electron. Therefore, we can find the number of unpaired electrons by dividing the magnetic moment of the Earth by μ_B. We can then use the density of iron to find the mass of the iron atoms that each contribute two electrons.

A: (a) $\mu_B = \left(9.27 \times 10^{-24} \dfrac{J}{T}\right)\left(1 \dfrac{N \cdot m}{J}\right)\left(\dfrac{1\,T}{N \cdot s/C \cdot m}\right)\left(\dfrac{1\,A}{C/s}\right) = 9.27 \times 10^{-24}$ A·m^2

The number of unpaired electrons is

$$N = \frac{8.00 \times 10^{22}\ A \cdot m^2}{9.27 \times 10^{-24}\ A \cdot m^2} = 8.63 \times 10^{45}\ e^- \qquad \Diamond$$

Each iron atom has two unpaired electrons, so the number of iron atoms required is $\frac{1}{2}N = \frac{1}{2}(8.63 \times 10^{45}) = 4.31 \times 10^{45}$ iron atoms. Thus,

$$\text{(b)} \quad M_{Fe} = \frac{\left(4.31 \times 10^{45}\ atoms\right)\left(7900\ kg/m^3\right)}{8.50 \times 10^{28}\ atoms/m^3} = 4.01 \times 10^{20}\ kg \qquad \Diamond$$

L: The calculated answers seem reasonable based on the limits we expected. From the data in this problem, the iron deposit required to produce the magnetic moment would only be about 1/15 000 the mass of the Earth and would form a sphere 500 km in diameter. Although this is certainly a large amount of iron, it is much smaller than the inner core of the Earth, which is estimated to have a diameter of about 3000 km.

---

53. A very long, thin strip of metal of width w carries a current I along its length as shown in Figure P30.53. Find the magnetic field at point P in the diagram. Point P is in the plane of the strip at a distance b away from the strip.

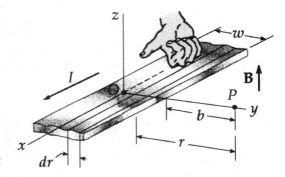

Figure P30.53

Solution   Consider a long filament of the strip which has width dr and is a distance r from point P. The magnetic field a distance r from a long conductor is

$$B = \frac{\mu_0 I}{2\pi r}$$

Thus, the field due to the thin filament is

$$dB = \frac{\mu_0 \, dI}{2\pi r} \mathbf{k} \qquad \text{where} \qquad dI = I\left(\frac{dr}{w}\right)$$

so

$$\mathbf{B} = \int_b^{b+w} \frac{\mu_0}{2\pi r}\left(I\frac{dr}{w}\right)\mathbf{k} = \frac{\mu_0 I}{2\pi w}\mathbf{k}\int_b^{b+w} \frac{dr}{r} = \frac{\mu_0 I}{2\pi w}\mathbf{k}\,\ln\!\left(1+\frac{w}{b}\right) \qquad \Diamond$$

Note: The solutions to these next two problems are set up in a dual format. Compare problem 55 and its solution with problem 56 and its solution. They contain the same steps of physical analysis and reasoning, but in problem 56 the numerical "plug and chug" steps are not included.

55. A nonconducting ring with a radius of 10.0 cm is uniformly charged with a total positive charge of 10.0 μC. The ring rotates at a constant angular speed of 20.0 rad / s about an axis through its center, perpendicular to the plane of the ring. What is the magnitude of the magnetic field on the axis of the ring, 5.00 cm from its center?

56. A nonconducting ring of radius R is uniformly charged with a total positive charge q. The ring rotates at a constant angular speed ω about an axis through its center, perpendicular to the plane of the ring. What is the magnitude of the magnetic field on the axis of the ring a distance $R/2$ from its center?

Solution

The 'static' charge on the ring constitutes an electric current as the ring rotates.

The time required for the 10.0 μC charge to go past a point is the period of rotation,

$$t = \frac{\theta}{\omega} = \frac{2\pi \text{ rad}}{20 \text{ rad / s}} = 0.314 \text{ s}$$

The current is $I = q/t$:

$$I = \frac{10.0 \times 10^{-6} \text{ C}}{0.314 \text{ s}} = 3.18 \times 10^{-5} \text{ A}$$

The time required for the charge q to go past a point is the period of rotation,

$$t = \frac{\theta}{\omega} = \frac{2\pi}{\omega}$$

The current is $I = q/t$:

$$I = \frac{q}{2\pi/\omega} = \frac{q\omega}{2\pi}$$

179

The current has the shape of a flat compact circular coil with one turn, so the magnetic field it creates is

$$B = \frac{\mu_0 I R^2}{2(x^2 + R^2)^{3/2}}$$

$$B = \frac{\mu_0 I R^2}{2(x^2 + R^2)^{3/2}}$$

Substituting numeric values for these variables,

In this case, the distance x is equal to $R/2$, so

$$B = \frac{\left(4\pi \times 10^{-7} \frac{\text{T} \cdot \text{m}}{\text{A}}\right)(3.18 \times 10^{-5} \text{ A})(0.100 \text{ m})^2}{2\left((0.0500 \text{ m})^2 + (0.100 \text{ m})^2\right)^{3/2}}$$

$$B = \frac{\mu_0 q \omega R^2}{4\pi (R^2/4 + R^2)^{3/2}}$$

$$B = \frac{4.00 \times 10^{-13} \text{ T} \cdot \text{m}^3}{2\left(1.40 \times 10^{-3} \text{ m}^3\right)} = 1.43 \times 10^{-10} \text{ T} \quad \Diamond$$

$$B = \frac{\mu_0 q \omega R^2}{4\pi (5R^2/4)^{3/2}} = \frac{2\mu_0 q \omega}{5\sqrt{5}\pi R} \quad \Diamond$$

If the ring is turning counterclockwise, the magnetic field it creates is along the axis away from its center. Observe that the ring also creates an electric field in the same direction.

---

69. A wire is formed into a square of edge length L (Fig. P.30.69). Show that when the current in the loop is I, the magnetic field at point P a distance x from the center of the square along its axis is

$$B = \frac{\mu_0 I L^2}{2\pi\left(x^2 + L^2/4\right)\sqrt{x^2 + L^2/2}}$$

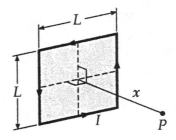

Figure P30.69

Solution   Consider the top side of the square. The distance from its center to point P is

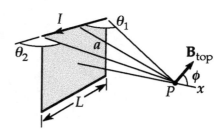

$$a = \sqrt{x^2 + (L/2)^2}$$

and Equation 30.4 describes the field it creates at point P.   The distance from one of the corners of the square to P is

$$\sqrt{x^2 + (L/2)^2 + (L/2)^2} = \sqrt{x^2 + L^2/2}$$

Thus $$\cos\theta_1 = \frac{L}{2\sqrt{x^2 + L^2/2}}$$ and $$\cos\theta_2 = -\cos\theta_1 = -\frac{L}{2\sqrt{x^2 + L^2/2}}$$

Then, $$B_{top} = \frac{\mu_0 I}{4\pi a}(\cos\theta_1 - \cos\theta_2)$$

$$B_{top} = \frac{\mu_0 I}{4\pi\sqrt{x^2 + L^2/4}}\left(\frac{L}{2\sqrt{x^2 + L^2/2}} + \frac{L}{2\sqrt{x^2 + L^2/2}}\right)$$

or $$B_{top} = \frac{\mu_0 I L}{4\pi\sqrt{x^2 + L^2/4}\sqrt{x^2 + L^2/2}}$$

The component of this field along the direction of x is $B_{top}\cos\phi = B_{top}\dfrac{L/2}{a}$

Each of the four sides of the square produces this same size field along the direction of x, with the other components adding to zero. We assemble our results, and find that the net magnetic field points away from the center of the square, with magnitude

$$B = 4B_{top}\frac{L}{2a} = \frac{\mu_0 I L^2}{2\pi\left(x^2 + L^2/4\right)\sqrt{x^2 + L^2/2}} \qquad \lozenge$$

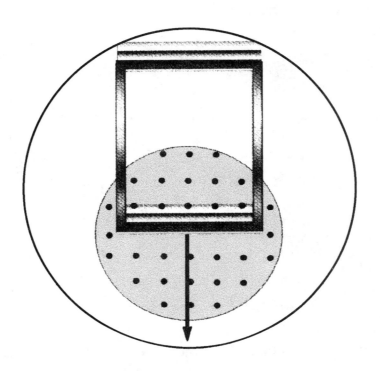

Faraday's Law

FARADAY'S LAW

INTRODUCTION

Our studies so far have been concerned with electric fields due to stationary charges and magnetic fields produced by moving charges. This chapter deals with electric fields that originate from changing magnetic fields.

Experiments conducted by Michael Faraday in England in 1831 and independently by Joseph Henry in the United States that same year showed that an electric current could be induced in a circuit by a changing magnetic field. The results of these experiments led to a basic and important law of electromagnetism known as Faraday's law of induction. This law says that the magnitude of the emf induced in a circuit equals the time rate of change of the magnetic flux through the circuit.

As we shall see, an induced emf can be produced in several ways. For instance, an induced emf and an induced current can be produced in a closed loop of wire when the wire moves into a magnetic field. We shall describe such experiments along with a number of important applications that make use of the phenomenon of electromagnetic induction.

EQUATIONS AND CONCEPTS

The total magnetic flux through a plane area, A, placed in a uniform magnetic field depends on the angle between the direction of the magnetic field and the direction perpendicular to the surface area.

$$\Phi_B \equiv B_\perp A = BA\cos\theta$$

$$\Phi_{B,\max} = BA$$

The maximum flux through the area occurs when the magnetic field is perpendicular to the plane of the surface area. When the magnetic field is parallel to the plane of the surface area, the flux through the area is zero. The unit of magnetic flux is the weber, Wb.

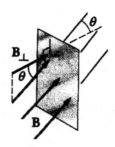

Faraday's law of induction states that the average emf induced in a circuit is proportional to the rate of change of magnetic flux through the circuit. The minus sign is included to indicate the polarity of the induced emf, which can be found by use of Lenz's law.

$$\mathcal{E} = -\frac{d\Phi_B}{dt}$$

(31.1)

Lenz's law states that the polarity of the induced emf (and the direction of the associated current in a closed circuit) produces a current whose magnetic field opposes the change in the flux through the loop. That is, the induced current tends to maintain the original flux through the circuit.

The magnetic flux threading a circuit is the integral of the normal component of the magnetic field over the area bounded by the circuit.

$$\Phi_B = \int \mathbf{B} \cdot d\mathbf{A}$$

A "motional" emf is induced in a conductor of length ℓ, moving with speed v, perpendicular to a magnetic field.

$$\mathcal{E} = -B\ell v \qquad (31.5)$$

If the moving conductor is part of a complete circuit of resistance, R, a current will be induced in the circuit.

$$I = \frac{|\mathcal{E}|}{R} = \frac{B\ell v}{R} \qquad (31.6)$$

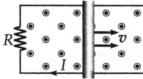

Faraday's law can be written in a more **general form** in terms of the integral of the electric field around a closed path. In this form the electric field is a **non-conservative, time-varying field**.

$$\oint \mathbf{E} \cdot d\mathbf{s} = -\frac{d\Phi_B}{dt} \qquad (31.9)$$

or

$$\oint \mathbf{E} \cdot d\mathbf{s} = -\frac{d}{dt} \int \mathbf{B} \cdot d\mathbf{A}$$

When a conducting loop of N turns and cross-sectional area, A, rotates with a constant angular velocity in a magnetic field, the emf induced in the loop will vary sinusoidally in time. For a given loop, the maximum value of the induced emf will be proportional to the angular velocity of the loop.

$$\mathcal{E} = NAB\omega\sin\omega t \qquad (31.10)$$

$$\mathcal{E}_{max} = NAB\omega \qquad (31.11)$$

Maxwell's equations as applied to free space (i.e. in the absence of any dielectric or magnetic material) are as follows (see Notes 31.7):

Gauss's law

$$\oint \mathbf{E}\cdot d\mathbf{A} = \frac{Q}{\epsilon_0} \qquad (31.12)$$

Gauss's law for magnetism

$$\oint \mathbf{B}\cdot d\mathbf{A} = 0 \qquad (31.13)$$

Faraday's law

$$\oint \mathbf{E}\cdot d\mathbf{s} = -\frac{d\Phi_B}{dt} \qquad (31.14)$$

Ampère-Maxwell law

$$\oint \mathbf{B}\cdot d\mathbf{s} = \mu_0 I + \epsilon_0\mu_0\frac{d\Phi_E}{dt} \qquad (31.15)$$

The force exerted on a particle of charge q due to the combined effect of the electric force and the magnetic field is the Lorentz force.

$$\mathbf{F} = q\mathbf{E} + q\mathbf{v}\times\mathbf{B} \qquad (31.16)$$

SUGGESTIONS, SKILLS, AND STRATEGIES

It is important to distinguish clearly between the **instantaneous value** of emf induced in a circuit and the **average value** of the emf induced in the circuit over a finite time interval.

To calculate the average induced emf, it is often useful to write Equation 31.2 as

$$\mathcal{E}_{avg} = -N\left(\frac{d\Phi_B}{dt}\right)_{avg} = -N\frac{\Delta\Phi_B}{\Delta t}$$

or

$$\mathcal{E}_{avg} = -N\left(\frac{\Phi_{B,f} - \Phi_{B,i}}{\Delta t}\right)$$

where the subscripts i and f refer to the magnetic flux through the circuit at the beginning and end of the time interval Δt. For a circuit or coil in a single plane, $\Phi_B = BA\cos\theta$, where θ is the angle between the vector normal to plane of the circuit (conducting coil) and the direction of the magnetic field.

Equation 31.3 can be used to calculate the **instantaneous value of an induced emf**. For a multi-turn loop, the induced emf is

$$\mathcal{E} = -N\frac{d}{dt}(BA\cos\theta)$$

where in a particular case B, A, θ, or any combination of those parameters can be time dependent while the others remain constant. The expression resulting from the differentiation is then evaluated using the values of B, A, and θ corresponding to the specified value.

REVIEW CHECKLIST

▷ Calculate the emf (or current) induced in a circuit when the magnetic flux through the circuit is changing in time. The variation in flux might be due to a change in (a) the area of the circuit, (b) the magnitude of the magnetic field, (c) the direction of the magnetic field, or (d) the orientation/location of the circuit in the magnetic field.

▷ Calculate the emf induced between the ends of a conducting bar as it moves through a region where there is a constant magnetic field (motional emf).

▷ Apply Lenz's law to determine the direction of an induced emf or current. You should also understand that Lenz's law is a consequence of the law of conservation of energy.

▷ Calculate the maximum and instantaneous values of the sinusoidal emf generated in a conducting loop rotating in a constant magnetic field.

▷ Calculate the electric field at various points in a charge-free region when the time variation of the magnetic field over the region is specified.

ANSWERS TO SELECTED CONCEPTUAL QUESTIONS

4. The bar shown in Figure Q31.4 moves on rails to the right with velocity **v**, and the uniform, constant magnetic field is directed out of the page. Why is the induced current clockwise? If the bar were moving to the left, what would be the direction of the induced current?

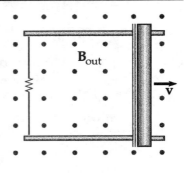

Figure Q31.4

Answer The external flux is out of the paper; as the area A enclosed by the loop increases, the external flux increases according to

$$\Phi_B = BA\cos\theta = BA$$

Under the influence of this flux, free electric charges will move in a current to create a magnetic flux inside the loop to oppose the change in flux.

In this case, the flux of the current must point into the paper in order oppose the increasing magnetic flux. By the right-hand rule (with your thumb pointing in the direction of the current along each wire), the current must be clockwise.

If the bar were moving toward the left, the area would decrease, and the flux would decrease. The **change** in the flux pointing out of the paper, then, would be negative. In order to cancel this change, the current would have to create a flux inside the loop that was positive, pointing out of the paper. This time, by the right-hand rule, we see that the current must be counterclockwise.

□   □   □   □

9. How is electrical energy produced in dams (that is, how is the energy of motion of the water converted to alternating current electricity?)

Answer As the water falls, it gains kinetic energy. It is then forced to pass through a water wheel, transferring some of its energy to the rotor of an large AC electric generator.

The rotor of the generator is supplied with a small amount of DC current, which powers electromagnets in the rotor. Because the rotor is spinning, the electromagnets then create a magnetic flux that changes with time, according to the equation $\Phi_B = BA\cos\omega t$.

Coils of wire that are placed near the magnet then experience an induced electromagnetic force according to the equation $\mathcal{E} = -N\,d\Phi_B/dt$.

Finally, a small amount of this electricity is used to supply the rotor with its DC current; the rest is sent out over power lines to supply customers with electricity.

□   □   □   □

SOLUTIONS TO SELECTED END-OF-CHAPTER PROBLEMS

5. A strong electromagnet produces a uniform field of 1.60 T over a cross-sectional area of 0.200 m². A coil having 200 turns and a total resistance of 20.0 Ω is placed around the electromagnet. The current in the electromagnet is then smoothly decreased until it reaches zero in 20.0 ms. What is the current induced in the coil?

Solution

G: A strong magnetic field turned off in a short time (20.0 ms) will produce a large emf, maybe on the order of 1 kV. With only 20.0 Ω of resistance in the coil, the induced current produced by this emf will probably be larger than 10 A but less than 1000 A.

O: According to Faraday's law, if the magnetic field is reduced uniformly, then a constant emf will be produced. The definition of resistance can be applied to find the induced current from the emf.

A: The induced voltage is

$$\mathcal{E} = -N\frac{d(\mathbf{B} \cdot \mathbf{A})}{dt} = -N\left(\frac{0 - B_i A \cos\theta}{\Delta t}\right)$$

$$\mathcal{E} = \frac{+200(1.60 \text{ T})(0.200 \text{ m}^2)(\cos 0°)}{20.0 \times 10^{-3} \text{ s}}\left(\frac{1 \text{ N·s/C·m}}{\text{T}}\right)\left(\frac{1 \text{ V·C}}{\text{N·m}}\right) = 3200 \text{ V}$$

(**Note:** The unit conversions come from $\mathbf{F} = q\mathbf{v} \times \mathbf{B}$ and $U = qV$.)

$$I = \frac{\mathcal{E}}{R} = \frac{3200 \text{ V}}{20.0 \text{ Ω}} = 160 \text{ A} \qquad \diamond$$

L: This is a large current, as we expected. The positive sign is indicative that the induced electric field is in the positive direction around the loop (as defined by the area vector for the loop).

7. An aluminum ring with a radius of 5.00 cm and resistance of 3.00×10^{-4} Ω is placed on top of a long air-core solenoid with 1000 turns per meter and a radius of 3.00 cm, as shown in Figure P31.7. Assume that the axial component of the field produced by the solenoid over the area of the end of the solenoid is one-half as strong as at the center of the solenoid. Assume that the solenoid produces negligible field outside its cross-sectional area. (a) If the current in the solenoid is increasing at a rate of 270 A/s, what is the induced current in the ring? (b) At the center of the ring, what is the magnetic field produced by the induced current in the ring? (c) What is the direction of this field?

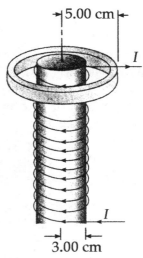

→|5.00 cm|←

I

3.00 cm

Figure P31.7

Solution So as not to confuse variables, we define the radius of the ring to be r_{ring}, and its resistance to be R.

$$\mathcal{E} = -\frac{d}{dt}(BA\cos\theta) = -\frac{d}{dt}(0.500\,\mu_o nIA\cos 0°) = -0.500\,\mu_o nA\frac{dI}{dt}$$

Note that A must be interpreted as the area of the solenoid, where the field is strong:

$$\mathcal{E} = -0.500(4\pi \times 10^{-7}\text{ T} \cdot \text{m}/\text{A})(1000\text{ turns}/\text{m})\left[\pi(0.0300\text{ m})^2\right](270\text{ A}/\text{s})$$

$$\mathcal{E} = \left(-4.80 \times 10^{-4}\ \frac{\text{T} \cdot \text{m}^2}{\text{s}}\right)\left(1\ \frac{\text{N} \cdot \text{s}}{\text{C} \cdot \text{m} \cdot \text{T}}\right)\left(1\ \frac{\text{V} \cdot \text{C}}{\text{N} \cdot \text{m}}\right) = -4.80 \times 10^{-4}\text{ V}$$

(a) $I_{ring} = \dfrac{|\mathcal{E}|}{R} = \dfrac{0.000480}{0.000300} = 1.60\text{ A}$ ◊

(b) $B_{ring} = \dfrac{\mu_o I_{ring}}{2r_{ring}} = 2.01 \times 10^{-5}\text{ T}$ ◊

(c) The coil's field points downward, and is increasing, so B_{ring} points upward. ◊

13. A long solenoid has 400 turns per meter and carries a current $I = (30.0 \text{ A})(1 - e^{-1.60t})$. Inside the solenoid and coaxial with it is a coil that has a radius of 6.00 cm and consists of a total of 250 turns of fine wire. (See Fig. P31.13) What emf is induced in the coil by the changing current?

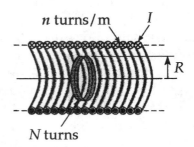

Figure P31.13

Solution The solenoid creates a magnetic field

$$B = \mu_0 n I = \left(4\pi \times 10^{-7} \text{ N} / \text{A}^2\right)(400 \text{ turns} / \text{m})(30.0 \text{ A})\left(1 - e^{-1.60t}\right)$$

$$B = \left(1.51 \times 10^{-2} \text{ N} / \text{m} \cdot \text{A}\right)\left(1 - e^{-1.60t}\right)$$

The magnetic flux through one turn of the loop is $\Phi_B = \int B \, dA \cos\theta$, but since dA refers to the area perpendicular to the flux, and the magnetic field can be estimated as constant with respect to the area dA,

$$\Phi_B = B\int dA = B\left(\pi R^2\right) = \left(1.51 \times 10^{-2} \text{ N} / \text{m} \cdot \text{A}\right)\left(1 - e^{-1.60t}\right)(\pi)(0.0600 \text{ m})^2$$

$$\Phi_B = \left(1.71 \times 10^{-4} \text{ N} \cdot \text{m} / \text{A}\right)\left(1 - e^{-1.60t}\right)$$

The emf generated in the N-turn loop is $\mathcal{E} = -N d\,\Phi_B / dt$:

$$\mathcal{E} = -(250)\left(1.71 \times 10^{-4} \frac{\text{N} \cdot \text{m}}{\text{A}}\right)\frac{d\left(1 - e^{-1.60t}\right)}{dt} = -\left(0.0426 \frac{\text{N} \cdot \text{m}}{\text{A}}\right)\left(\left(1.60 \text{ s}^{-1}\right)e^{-1.60t}\right)$$

$$\mathcal{E} = -\left(6.82 \times 10^{-2} \text{ N} \cdot \text{m} / \text{C}\right)e^{-1.60t} = -\left(6.82 \times 10^{-2} \text{ V}\right)e^{-1.60t} \qquad \Diamond$$

The minus sign indicates that the emf will produce counterclockwise current in the smaller coil, opposite to the direction of the increasing current in the solenoid.

15. A coil formed by wrapping 50.0 turns of wire in the shape of a square is positioned in a magnetic field so that the normal to the plane of the coil makes an angle of 30.0° with the direction of the field. When the magnetic field is increased uniformly from 200 μT to 600 μT in 0.400 s, an emf of 80.0 mV is induced in the coil. What is the total length of the wire?

Solution

G: If we assume that this square coil is some reasonable size between 1 cm and 1 m across, then the total length of wire would be between 2 m and 200 m.

O: The changing magnetic field will produce an emf in the coil according to Faraday's law of induction. The constant area of the coil can be found from the change in flux required to produce the emf.

A: By Faraday's law, $\mathcal{E} = -N \dfrac{d\Phi_B}{dt}$

so $$\mathcal{E} = -N \frac{d}{dt}(BA\cos\theta) = -NA\cos\theta \frac{dB}{dt}$$

For magnitudes, $\left|\overline{\mathcal{E}}\right| = NA\cos\theta\left(\dfrac{\Delta B}{\Delta t}\right)$ and $A = \dfrac{\left|\overline{\mathcal{E}}\right|}{N\cos\theta\left(\dfrac{\Delta B}{\Delta t}\right)}$

and the area is $A = \dfrac{80.0 \times 10^{-3}\ \text{V}}{50(\cos 30.0°)\left(\dfrac{600 \times 10^{-6}\ \text{T} - 200 \times 10^{-6}\ \text{T}}{0.400\ \text{s}}\right)} = 1.85\ \text{m}^2$

Each side of the coil has length $d = \sqrt{A}$, so the total length of the wire is

$$L = N(4d) = 4N\sqrt{A} = (4)(50)\sqrt{1.85\ \text{m}^2} = 272\ \text{m} \qquad \lozenge$$

193

L: The total length of wire is slightly longer than we predicted. With $d = 1.36$ m, a normal person could easily step through this large coil! As a bit of foreshadowing to a future chapter on AC circuits, an even bigger coil with more turns could be hidden in the ground below high-power transmission lines so that a significant amount of power could be "stolen" from the electric utility. There is a story of one man who did this and was arrested when investigators finally found the reason for a large power loss in the transmission lines!

---

21. Figure P31.20 shows a top view of a bar that can slide without friction. The resistor is 6.00 Ω, and a 2.50-T magnetic field is directed perpendicularly downward, into the paper. Let $\ell = 1.20$ m. (a) Calculate the applied force required to move the bar to the right at a constant speed of 2.00 m/s. (b) At what rate is energy delivered to the resistor?

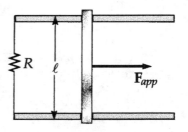

Figure P31.20

Solution

(a) At constant speed, the net force on the moving bar equals zero, or

$$\left|\mathbf{F}_{\text{app}}\right| = I\left|\boldsymbol{\ell} \times \mathbf{B}\right|$$

where the current in the bar $I = \mathcal{E}/R$ and $\mathcal{E} = B\ell v$. Therefore,

$$F_{\text{app}} = \left(\frac{B\ell v}{R}\right)\ell B = \frac{B^2\ell^2 v}{R} = \frac{(2.50 \text{ T})^2(1.20 \text{ m})^2(2.00 \text{ m}/\text{s})}{6.00 \text{ Ω}} = 3.00 \text{ N} \qquad \Diamond$$

(b) $\mathcal{P} = F_{\text{app}}v = (3.00 \text{ N})(2.00 \text{ m/s}) = 6.00 \text{ W}$ $\qquad \Diamond$

---

25. A helicopter has blades with a length of 3.00 m extending outward from a central hub and rotating at 2.00 rev/s. If the vertical component of the Earth's magnetic field is 50.0 μT, what is the emf induced between the blade tip and the center hub?

Solution We suppose the length is measured out from the hub. Following Example 31.4,

$$\mathcal{E} = \tfrac{1}{2} B \omega \ell^2 = \tfrac{1}{2}(5.00 \times 10^{-5} \text{ T})\left(\frac{2.00 \text{ rev}}{\text{s}}\right)\left(\frac{2\pi \text{ rad}}{1.00 \text{ rev}}\right)(3.00 \text{ m})^2 = 2.83 \text{ mV} \quad \Diamond$$

---

27. A rectangular coil with resistance R has N turns, each of length ℓ and width w as shown in Figure P31.27. The coil moves into a uniform magnetic field **B** with velocity **v**. What are the magnitude and direction of the resultant force on the coil (a) as it enters the magnetic field, (b) as it moves within the field, and (c) as it leaves the field?

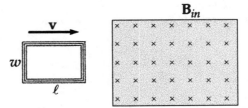

Figure P31.27

Solution (a) Call x the distance that the leading edge has penetrated into the strong field region. The flux (Bwx away from you) through the coil increases in time so the voltage, of

$$|\mathcal{E}| = N\frac{d}{dt}Bwx = NBw\frac{dx}{dt} = NBwv,$$

is induced in the coil, tending to produce counterclockwise current

$$|I| = \frac{|\mathcal{E}|}{R} = \frac{NBwv}{R}$$

The current upward in the leading edge experiences a force

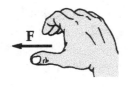

$$F = NI\, \mathbf{L} \times \mathbf{B} = N\left(\frac{NBwv}{R}\right)w\,\mathbf{j} \times B(-\mathbf{k}) = \left(\frac{N^2B^2w^2v}{R}\right)(-\mathbf{i}) \quad \lozenge$$

This retarding force models eddy-current damping.

(b) The flux through the coil is constant when it is wholly within the high-field region. The induced emf, induced current, and magnetic force are zero. ◊

(c) As the coil leaves the field, the away-from-you flux it encloses decreases. The coil carries clockwise current to make some away-from-you field of its own. Again,

$$|I| = \frac{NBwv}{R}$$

Now the trailing edge carries upward current to experience a force of

$$\mathbf{F} = \left(\frac{N^2B^2w^2v}{R}\right) \text{ to the left} \qquad \lozenge$$

---

33. A magnetic field directed into the page changes with time according to $B = (0.0300t^2 + 1.40)$ T, where t is in seconds. The field has a circular cross-section of radius $R = 2.50$ cm (see Fig. P31.32). What are the magnitude and direction of the electric field at point P_1 when $t = 3.00$ s and $r_1 = 0.0200$ m?

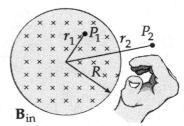

Figure P31.32

Solution   Consider a circular path of radius r_1, and integrate over that path

$$\oint \mathbf{E} \cdot d\mathbf{s} = -\frac{d\Phi_B}{dt}: \qquad E(2\pi r_1) = -\frac{d}{dt}(BA) = -A\left(\frac{dB}{dt}\right)$$

$$|E| = \frac{A}{2\pi r_1}\frac{d}{dt}(0.0300t^2 + 1.40)\text{T} = \frac{\pi r_1^2}{2\pi r_1}(0.0600t) = \frac{r_1}{2}(0.0600t)$$

At $t = 3.00$ s, $\qquad E = \left(\frac{0.0200 \text{ m}}{2}\right)(0.0600 \text{ T}/\text{s})(3.00 \text{ s}) = 1.80 \times 10^{-3} \text{ N}/\text{C}$ ◊

If there were a circle of wire of radius r_1, it would enclose increasing magnetic flux away from you. It would carry counterclockwise current to make its own magnetic field toward you, to oppose the change. Even without the wire and current, the counterclockwise electric field that would cause the current is lurking. At point P_1, it is upward and to the left, perpendicular to r_1.   ◊

---

37.   A coil of area 0.100 m² is rotating at 60.0 rev/s with the axis of rotation perpendicular to a 0.200-T magnetic field. (a) If there are 1000 turns on the coil, what is the maximum voltage induced in it? (b) What is the orientation of the coil with respect to the magnetic field, when the maximum induced voltage occurs?

Solution   (a)   For a coil rotating in a magnetic field, Equation 31.10 says that $\mathcal{E} = NBA\omega\sin\theta$. This is a maximum when $\sin\theta$ equals 1:

$$\mathcal{E}_{max} = NBA\omega$$

Thus $\mathcal{E}_{max} = (1000)(0.200 \text{ T})(0.100 \text{ m}^2)\left(60.0 \frac{\text{rev}}{\text{s}}\right)\left(2\pi \frac{\text{rad}}{\text{rev}}\right) = 7540 \text{ V}$ ◊

(b) $\mathcal{E}$ approaches $\mathcal{E}_{max}$ when $\sin\theta$ approaches 1, and $\theta = \pm\pi/2$. Therefore, at maximum emf, the plane of the coil is parallel to the field.   ◊

---

45. A conducting rectangular loop of mass M, resistance R, and dimensions w by ℓ falls from rest into a magnetic field **B** as in Figure P31.44. The loop approaches terminal speed v_t. (a) Show that $v_t = MgR/B^2w^2$. (b) Why is v_t proportional to R? (c) Why is it inversely proportional to B^2?

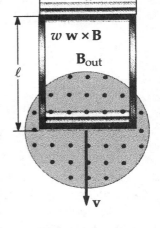

Figure P31.44

Solution   Let y represent the vertical dimension of the lower part of the loop where it is inside the strong magnetic field. As the loop falls, y increases; the loop encloses increasing flux toward you and has induced in it an emf to produce a current to make its own magnetic field away from you. This current is to the left in the bottom side of the loop, and feels an upward force in the external field.

(a)   Symbolically, the flux is $\Phi_B = BA\cos\theta = Bwy\cos 0°$

The emf is $\qquad \mathcal{E} = -N\dfrac{d}{dt}(Bwy) = -Bw\dfrac{dy}{dt} = -Bwv$

The magnitude of the current is $\qquad I = \dfrac{|\mathcal{E}|}{R} = \dfrac{Bwv}{R}$

and the force is

$$\mathbf{F}_B = I\,\mathbf{w}\times\mathbf{B} = \left(\frac{Bwv}{R}\right)wB\,\sin 90° = \frac{B^2w^2v}{R}\ \text{up}$$

At terminal speed, the loop is in equilibrium. $\Sigma F_y = 0$ becomes

$$\frac{+B^2w^2v_t}{R} - Mg = 0 \qquad\qquad \text{or} \qquad\qquad v_t = \frac{MgR}{B^2w^2} \qquad\qquad \diamond$$

(b) The emf is directly proportional to v, but the current is inversely proportional to R. A large R means a small current at a given speed, so the loop must travel faster to get $F_{B,mag}$ = weight. ◊

(c) At a given speed, the current is directly proportional to the magnetic field. But the force is proportional to the product of the current and the field. For a small B, the speed must increase to compensate for both the small B and also the current, so $v \propto B^{-2}$. ◊

---

47. A proton moves through a uniform electric field $\mathbf{E}$ = 50.0$\mathbf{j}$ V/m and a uniform magnetic field $\mathbf{B}$ = (0.200$\mathbf{i}$ + 0.300$\mathbf{j}$ + 0.400$\mathbf{k}$) T. Determine the acceleration of the proton when it has a velocity $\mathbf{v}$ = 200$\mathbf{i}$ m/s.

Solution   The combined force on the proton is the Lorentz force, as described by Equation 31.16:

$$\mathbf{F} = m\mathbf{a} = q\mathbf{E} + q\mathbf{v} \times \mathbf{B}$$

so that $\mathbf{a} = \dfrac{e}{m}[\mathbf{E} + \mathbf{v} \times \mathbf{B}]$

Taking the cross product of $\mathbf{v}$ and $\mathbf{B}$,

$$\mathbf{v} \times \mathbf{B} = \begin{vmatrix} \mathbf{i} & \mathbf{j} & \mathbf{k} \\ 200 & 0 & 0 \\ 0.200 & 0.300 & 0.400 \end{vmatrix} = -200(0.400)\mathbf{j} + 200(0.300)\mathbf{k}$$

$$\mathbf{a} = \frac{1.60 \times 10^{-19}}{1.67 \times 10^{-27}}[50.0\mathbf{j} - 80.0\mathbf{j} + 60.0\mathbf{k}] = 9.58 \times 10^{7}[-30\mathbf{j} + 60\mathbf{k}]$$

$$\mathbf{a} = \left(2.87 \times 10^{9}\right)(-\mathbf{j} + 2\mathbf{k})\ \mathrm{m/s^2} = (-2.87 \times 10^{9}\mathbf{j} + 5.74 \times 10^{9}\mathbf{k})\ \mathrm{m/s^2}$$ ◊

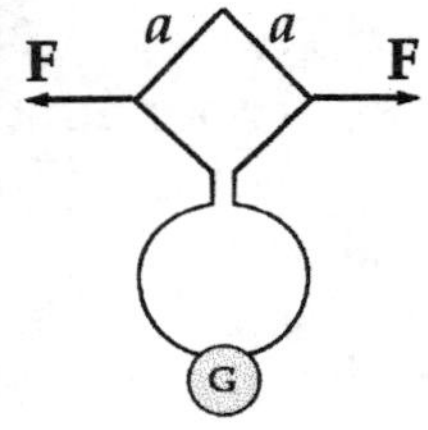

57. The plane of a square loop of wire with edge length $a = 0.200$ m is perpendicular to the Earth's magnetic field at a point where $B = 15.0$ μT, as shown in Figure P31.57. The total resistance of the loop and the wires connecting it to the galvanometer is 0.500 Ω. If the loop is suddenly collapsed by horizontal forces as shown, what total charge passes through the galvanometer?

Figure P31.57

Solution

G: For the situation described, the maximum current is probably less than 1 mA. So if the loop is closed in 0.1 s, then the total charge would be

$$Q = I\Delta t = (1 \text{ mA})(0.1 \text{ s}) = 100 \ \mu\text{C}$$

O: We do not know how quickly the loop is collapsed, but we can find the total charge by integrating the change in magnetic flux due to the change in area of the loop ($a^2 \to 0$).

A: $\quad Q = \int I dt = \int \dfrac{\mathcal{E} dt}{R} = \dfrac{1}{R}\int -\left(\dfrac{d\Phi_B}{dt}\right)dt = -\dfrac{1}{R}\int d\Phi_B = -\dfrac{1}{R}\int d(BA) = -\dfrac{B}{R}\int_{A_1=a^2}^{A_2=0} dA$

$$Q = -\dfrac{B}{R}A\Big]_{A_1=a^2}^{A_2=0} = \dfrac{Ba^2}{R} = \dfrac{(15.0\times10^{-6} \text{ T})(0.200 \text{ m})^2}{0.500 \ \Omega} = 1.20\times10^{-6} \text{ C} \qquad \lozenge$$

L: The total charge is less than the maximum charge we predicted, so the answer seems reasonable. It is interesting that this charge can be calculated without knowing either the current or the time to collapse the loop. **Note:** We ignored the internal resistance of the galvanometer. D'Arsonval galvanometers typically have an internal resistance of 50 to 100 Ω, significantly more than the resistance of the wires given in the problem. A proper solution that includes R_G would reduce the total charge by about 2 orders of magnitude ($Q \sim 0.01$ μC).

63. A rectangular coil of 60 turns, dimensions 0.100 m by 0.200 m and total resistance 10.0 Ω rotates with angular speed 30.0 rad/s about the y axis in a region where a 1.00-T magnetic field is directed along the x axis. The rotation is initiated so that the plane of the coil is perpendicular to the direction of **B** at $t = 0$. Calculate (a) the maximum induced emf in the coil, (b) the maximum rate of change of magnetic flux through the coil, (c) the induced emf at $t = 0.0500$ s, and (d) the torque exerted on the loop by the magnetic field at the instant when the emf is a maximum.

Solution Let θ represent the angle between the perpendicular to the coil and the magnetic field. Then $\theta = 0$ at $t = 0$ and $\theta = \omega t$ at all later times.

(a) $\mathcal{E} = -N\dfrac{d}{dt}(BA\cos\theta) = -NBA\dfrac{d}{dt}(\cos\omega t) = +NBA\omega\sin\omega t$

$\mathcal{E}_{max} = NBA\omega = 60(1.00 \text{ T})(0.0200 \text{ m}^2)(30.0 \text{ rad}/\text{s}) = 36.0 \text{ V}$ ◊

(b) $\dfrac{d\Phi_B}{dt} = \dfrac{d}{dt}(BA\cos\theta) = BA\omega\sin\omega t$

$\left(\dfrac{d\Phi_B}{dt}\right)_{max} = BA\omega = (1.00 \text{ T})(0.0200 \text{ m}^2)(30.0 \text{ rad}/\text{s}) = 0.600 \text{ T}\cdot\text{m}^2/\text{s}$ ◊

(c) $\mathcal{E} = NBA\omega\sin\omega t = (36.0 \text{ V})\sin\left[(30.0 \text{ rad}/\text{s})(0.0500 \text{ s})\right]$

$\mathcal{E} = (36.0 \text{ V})\sin(1.50 \text{ rad}) = (36.0 \text{ V})\sin 85.9° = 35.9 \text{ V}$

(d) $\tau = \mu \times \mathbf{B}$, so $\tau_{max} = \mu B\sin 90° = NIAB = N\mathcal{E}_{max}\dfrac{AB}{R}$

The emf is maximum when $\theta = 90°$,

and $\tau_{max} = 60(36.0 \text{ V})\dfrac{(0.0200 \text{ m}^2)(1.00 \text{ T})}{10.0 \text{ Ω}} = 4.32 \text{ N}\cdot\text{m}$ ◊

69. The magnetic flux threading a metal ring varies with time t according to $\Phi_B = 3(at^3 - bt^2)$ T·m², with $a = 2.00$ s⁻³ and $b = 6.00$ s⁻². The resistance of the ring is 3.00 Ω. Determine the maximum current induced in the ring during the interval from $t = 0$ to $t = 2.00$ s.

Solution   Substituting given values, $\Phi_B = \left(6.00t^3 - 18.0t^2\right) \text{T} \cdot \text{m}^2$

Therefore, the induced emf is $\mathcal{E} = -\dfrac{d\Phi_B}{dt} = -18.0t^2 + 36.0t$

The maximum $\mathcal{E}$ occurs when $\dfrac{d\mathcal{E}}{dt} = -36.0t + 36.0 = 0$

This yields the time of $\mathcal{E}_{max}$ and I_{max}: $t = 1.00$ s

Thus, maximum current at this time is $I_{max} = \dfrac{\mathcal{E}}{R} = \dfrac{(-18.0 + 36.0)\text{V}}{3.00\ \Omega} = 6.00$ A ◊

73. A long, straight wire carries a current $I = I_{max} \sin(\omega t + \phi)$ and lies in the plane of a rectangular coil of N turns of wire, as shown in Figure P31.9. The quantities I_{max}, ω, and ϕ are all constants. Determine the emf induced in the coil by the magnetic field created by the current in the straight wire. Assume $I_{max} = 50.0$ A, $\omega = 200\pi$ s⁻¹, $N = 100$, $h = w = 5.00$ cm, and $L = 20.0$ cm.

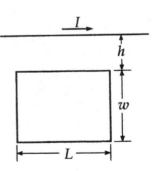

Figure P31.9

Solution   The coil is the boundary of a rectangular area. The magnetic field produced by the current in the straight wire is perpendicular to the plane of the area at all points. The magnitude of the field is
$B = \mu_0 I / 2\pi r$

Thus the flux is $N\Phi_B = \dfrac{\mu_0 NL}{2\pi} I \displaystyle\int_h^{h+w} \dfrac{dr}{r} = \dfrac{\mu_0 NL}{2\pi} I_{max} \ln\left(\dfrac{h+w}{h}\right)\sin(\omega t + \phi)$

Finally, the induced emf is in absolute value

$$|\mathcal{E}| = N\dfrac{d\,\Phi_B}{dt} = \dfrac{\mu_0 NL}{2\pi} I_{max}\omega \ln\left(\dfrac{h+w}{h}\right)\cos(\omega t + \phi)$$

$$|\mathcal{E}| = \left(4\pi \times 10^{-7}\,\dfrac{T\cdot m}{A}\right)\dfrac{(100)(0.200\ m)}{2\pi}(50.0\ A)(200\pi\ s^{-1})\ln\left(\dfrac{10.0\ cm}{5.00\ cm}\right)\cos(\omega t + \phi)$$

$$|\mathcal{E}| = (87.1\ mV)\cos(200\pi t + \phi) \qquad\qquad\qquad \Diamond$$

Note: The term $\sin(\omega t + \phi)$ in the expression for the current in the straight wire does not change appreciably when ωt changes by 0.1 rad or less. Thus, the current does not change appreciably during a time interval

$$t < 0.100/\left(200\pi\ s^{-1}\right) = 1.59 \times 10^{-4}\ s$$

We define a critical length, $c\,t = (3.00 \times 10^8\ m/s)(1.59 \times 10^{-4}\ s) = 4.77 \times 10^4\ m$

equal to the distance to which field changes could be propagated during an interval of 1.59×10^{-4} s. This length is so much larger than any dimension of the loop or its distance from the wire that, although we consider the straight wire to be infinitely long, we can also safely ignore the field propagation effects in the vicinity of the loop. Moreover, the phase angle can be considered to be constant along the wire in the vicinity of the loop. If the frequency ω were much larger, say, $200\pi \times 10^5$ s^{-1}, the corresponding critical length would be only 48 cm. In this situation, propagation effects would be important and the above expression of $\mathcal{E}$ would require modification. As a "rule of thumb," we can consider field propagation effects for circuits of laboratory size to be negligible for frequencies, $f = \omega/2\pi$, that are less than about 10^6 Hz.

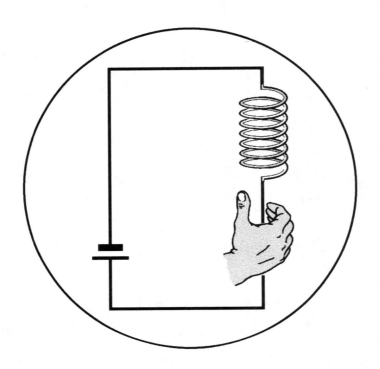

Inductance

INDUCTANCE

INTRODUCTION

In the previous chapter, we saw that currents and emfs are induced in a circuit when the magnetic flux through the circuit changes with time. This electromagnetic induction has some practical consequences, which we describe in this chapter. First, we describe an effect known as **self-induction**, in which a time-varying current in a conductor induces in the conductor an emf that opposes the external emf that set up the current. Self-induction is the basis of the **inductor**, an electrical element that plays an important role in circuits that use time-varying currents. We discuss the energy stored in the magnetic field of an inductor and the energy density associated with a magnetic field.

Next, we study how an emf is induced in a circuit as a result of a changing magnetic flux produced by an external circuit, which is the basic principle of **mutual induction**. Finally, we examine the characteristics of circuits containing inductors, resistors, and capacitors in various combinations.

EQUATIONS AND CONCEPTS

When the current in a coil changes in time, a self-induced emf is present in the coil. The inductance, L, is a measure of the opposition of the coil to a change in the current.

$$\varepsilon_L = -N\frac{d\Phi_B}{dt} = -L\frac{dI}{dt} \qquad (32.1)$$

The inductance of a given device, for example a coil, depends on its physical makeup — diameter, number of turns, type of material on which the wire is wound, and other geometric parameters.

A circuit element which has a large inductance is called an inductor. The SI unit of inductance is the henry, H. A rate of change of current of 1 ampere per second in an inductor of 1 henry will produce a self-induced emf of 1 volt.

$$1\,\mathrm{H} = 1\,\frac{\mathrm{V \cdot s}}{\mathrm{A}} = 1\,\Omega \cdot \mathrm{s}$$

A coil, solenoid, toroid, coaxial cable, or other conducting device is characterized by a parameter called its **inductance**, L. The inductance can be calculated, knowing the current and magnetic flux.

$$L = \frac{N\Phi_B}{I} \tag{32.2}$$

The **inductance** of a particular circuit element can also be expressed as the ratio of the induced emf to the time rate of change of current in the circuit.

$$L = -\frac{\mathcal{E}_L}{dI\,/\,dt} \tag{32.3}$$

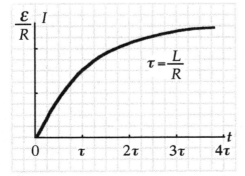

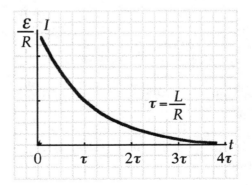

If the switch in the series circuit shown (which contains a battery, resistor, and inductor) is closed in position 1 at time $t = 0$, **current in the circuit** will increase in a characteristic fashion toward a maximum value of $\mathcal{E}/R$. This is shown in the first graph, above.

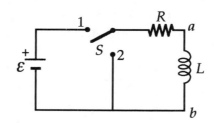

$$I = \frac{\mathcal{E}}{R}(1 - e^{-t/\tau}) \qquad\qquad (32.7)$$

Let the switch in the circuit shown in the figure be at position 1 with the current at its maximum value $I_0 = \mathcal{E}/R$. If the switch is thrown to position 2 at $t = 0$, the current will decay exponentially with time. The curve of the decay is shown in the second graph, above.

$$I = \frac{\mathcal{E}}{R}e^{-t/\tau} = I_0 e^{-t/\tau} \qquad\qquad (32.10)$$

where $\dfrac{\mathcal{E}}{R} = I_0$

In Equation 32.7 and Equation 32.10, the constant in the exponent is called the **time constant** of the circuit, τ.

$$\tau = L/R \qquad\qquad (32.8)$$

The value of the time constant is the time required for the current to reach 63% of its final value.

The **stored energy** U_L in the magnetic field of an inductor is proportional to the square of the current in the inductor.

$$U_L = \tfrac{1}{2}LI^2 \qquad (32.12)$$

It is often useful to express the energy in a magnetic field as **energy density** u_B; that is, energy per unit volume.

$$u_B = \frac{U_L}{A\ell} = \frac{B^2}{2\mu_0} \qquad (32.14)$$

When two coils are nearby, an emf is produced by mutual induction in one coil which is proportional to the rate of change of current in the other. The **mutual inductance** M depends on the coil geometries and the relative positions of the coils.

$$\varepsilon_1 = -M\frac{dI_2}{dt}$$

$$\varepsilon_2 = -M\frac{dI_1}{dt}$$

The situation in which two closely wound coils are adjacent to each other can be used to define the **mutual inductance,** M. In this expression Φ_{12} is the magnetic flux through coil 2 due to the current in coil 1.   The unit of mutual inductance is the henry, H. Note that M is a **shared property of the pair of coils.**

$$M_{12} \equiv \frac{N_2\Phi_{12}}{I_1} \qquad (32.15)$$

The figure shows a charged capacitor which can be connected by switch S to an inductor in a circuit with zero resistance. When the switch S is closed, the circuit exhibits a continual transfer of energy (back and forth) between the electric field of the capacitor and the magnetic field of the inductor. The process occurs at an angular frequency ω, called the **frequency of oscillation of the circuit**.

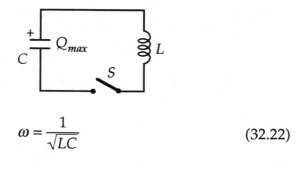

$$\omega = \frac{1}{\sqrt{LC}} \qquad (32.22)$$

The **total energy** of this circuit can be expressed in terms of the charge on the capacitor and the current in the inductor at some arbitrary time (Equation 32.18). Alternatively, the total energy can be written in terms of the maximum charge on the capacitor and the maximum current in the inductor at a specified time t after the switch is closed (Eq. 32.26).

$$U = U_C + U_L = \frac{Q^2}{2C} + \tfrac{1}{2}LI^2 \qquad (32.18)$$

$$U = \frac{Q_{max}^2}{2C}\cos^2\omega t + \frac{LI_{max}^2}{2}\sin^2\omega t \qquad (32.26)$$

The **charge** on the capacitor and the **current** in the inductor vary sinusoidally in time and are 90° out of phase with each other. If the initial conditions are such that $I = 0$ and $Q = Q_{max}$ at $t = 0$, the charge and current vary with time according to Equations 32.24 and 32.25.

$$Q = Q_{max}\cos\omega t \qquad (32.24)$$

$$I = -\omega Q_{max}\sin\omega t = -I_{max}\sin\omega t \qquad (32.25)$$

SUGGESTIONS, SKILLS, AND STRATEGIES

Equation 32.2, $L = N\Phi_B/I$ and Equation 32.12, $U_L = \frac{1}{2}LI^2$, provide two different approaches for the calculation of the inductance L of a particular device.

In order to use Equation 32.2 to calculate L, take the following steps:

- Assume a current I to exist in the conductor for which you wish to calculate L (coil, solenoid, coaxial cable, or other device).

- Calculate the magnetic flux through the appropriate cross section using $\Phi_B = \int \mathbf{B} \cdot d\mathbf{A}$. Remember that in many cases, $\mathbf{B}$ will not be uniform over the area.

- Calculate L directly from the defining Equation 32.2.

In order to use Equation 32.13 to calculate L, take the following steps:

- Assume a current I in the conductor.

- Find an expression for $\mathbf{B}$ for the magnetic field produced by I.

- Use Equation 32.14, $u_B = B^2/2\mu_0$, and integrate this value of u_B over the appropriate volume to find the total energy stored in the magnetic field of the inductor $U_L = \int u \, dV$.

- Substitute this value of U_L into Equation 32.12 and solve for L.

REVIEW CHECKLIST

▷ Calculate the inductance of a device of suitable geometry.

▷ Calculate the magnitude and direction of the self-induced emf in a circuit containing one or more inductive elements when the current changes with time.

▷ Determine instantaneous values of the current in an *LR* circuit while the current is either increasing or decreasing with time.

▷ Calculate the total magnetic energy stored in a magnetic field. You should be able to perform this calculation if (1) you are given the values of the inductance of the device with which the field is associated and the current in the circuit, or (2) given the value of the magnetic field throughout the region of space in which the magnetic field exists. In the latter case, you must integrate the expression for the energy density u_B over an appropriate volume.

▷ Calculate the emf induced by mutual inductance in one winding due to a time-varying current in a nearby inductor.

ANSWERS TO SELECTED CONCEPTUAL QUESTIONS

2. The current in a circuit containing a coil, resistor, and battery reaches a constant value. Does the coil have an inductance? Does the coil affect the value of the current?

Answer The coil has an inductance regardless of the nature of the current in the circuit. Inductance depends only on the coil geometry and its construction. Since the current is constant, the self-induced emf in the coil is zero, and the coil does not affect the steady-state current. (We assume the resistance of the coil is negligible.)

□   □   □   □

12. In the *LC* circuit shown in Figure 32.15, the charge on the capacitor is sometimes zero, even though current is in the circuit. How is this possible?

Answer When the capacitor is fully discharged, the current in the circuit is a maximum. The inductance of the coil is making the current continue to flow. At this time the magnetic field of the coil contains all the energy that was originally stored in the charged capacitor.

□   □   □   □

SOLUTIONS TO SELECTED END-OF-CHAPTER PROBLEMS

3. A 2.00-H inductor carries a steady current of 0.500 A. When the switch in the circuit is thrown open, the current is effectively zero in 10.0 ms. What is the average induced emf in the inductor during this time?

Solution $\mathcal{E}_L = -L\dfrac{dI}{dt}$

$$\mathcal{E}_{L,ave} = -L\frac{I_f - I_0}{t} = (-2.00 \text{ H})\left(\frac{0 - 0.500 \text{ A}}{1.00 \times 10^{-2} \text{ s}}\right)\left(1 \frac{\text{V} \cdot \text{s}/\text{A}}{\text{H}}\right) = +100 \text{ V} \quad \Diamond$$

7. A 10.0-mH inductor carries a current $I = I_{max} \sin \omega t$, with $I_{max} = 5.00$ A and $\omega/2\pi = 60.0$ Hz. What is the back emf as a function of time?

Solution $\mathcal{E}_{back} = -\mathcal{E}_L = L\dfrac{dI}{dt} = L\dfrac{d}{dt}(I_0 \sin \omega t)$

$$\mathcal{E}_{back} = L\omega I_0 \cos \omega t = (0.0100 \text{ H})(120 \pi \text{ s}^{-1})(5.00 \text{ A})\cos(120\pi t)$$

$$\mathcal{E}_{back} = (18.8 \text{ V})\cos(377t) \qquad \Diamond$$

9. An inductor in the form of a solenoid contains 420 turns, is 16.0 cm in length, and has a cross-sectional area of 3.00 cm². What uniform rate of decrease of current through the inductor induces an emf of 175 μV?

Solution The inductance is $$L = \frac{\mu_0 N^2 A}{\ell}$$

and the emf is as a function of time is $$\mathcal{E}_L = -L\frac{dI}{dt}$$

Therefore, rearranging terms, $$\frac{dI}{dt} = \frac{-\mathcal{E}_L}{L} = \frac{-\mathcal{E}_L \ell}{\mu_0 N^2 A}$$

Substituting the given values, we find that

$$\frac{dI}{dt} = \frac{\left(-175 \times 10^{-6} \text{ V}\right)(0.160 \text{ m})}{\left(4\pi \times 10^{-7} \text{ N / A}^2\right)(420)^2\left(3.00 \times 10^{-4} \text{ m}^2\right)} = -0.421 \text{ A / s} \qquad \lozenge$$

17. A 12.0-V battery is about to be connected to a series circuit containing a 10.0-Ω resistor and a 2.00-H inductor. How long will it take the current to reach (a) 50.0% and (b) 90.0% of its final value?

Solution

G: The time constant for this circuit is $\tau = L/R = 0.2$ s, which means that in 0.2 s, the current will reach $1/e = 63\%$ of its final value, as shown in the graph to the right. We can see from this graph that the time to reach 50% of I_{max} should be slightly less than the time constant, perhaps about 0.15 s, and the time to reach $0.9 I_{max}$ should be about $2.5\tau = 0.5$ s.

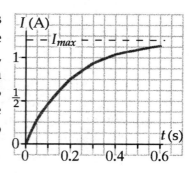

O: The exact times can be found from the equation that describes the rising current in the above graph and gives the current as a function of time for a known emf, resistance, and time constant.

A: At time t after connecting the circuit, $I(t) = \dfrac{\mathcal{E}(1 - e^{-t/\tau})}{R}$

where, after a long time, the current $I_{max} = \dfrac{\mathcal{E}(1 - e^{-\infty})}{R} = \dfrac{\mathcal{E}}{R}$

At 50% of this maximum value, $I(t) = 0.500 I_{max}$

So, $(0.500)\dfrac{\mathcal{E}}{R} = \dfrac{\mathcal{E}(1 - e^{t/0.200 \text{ s}})}{R}$

This then yields $0.500 = 1 - e^{-t/0.200 \text{ s}}$

Isolating the constants on the right, $\ln(e^{-t/2.00 \text{ s}}) = \ln(0.500)$

and solving for t, $-\dfrac{t}{0.200 \text{ s}} = -0.693$

Thus, to reach 50% of I_{max} takes: $t = 0.139$ s   ◊

(b) Similarly, to reach 90% of I_{max}, $0.900 = 1 - e^{-t/\tau}$

so $t = -\tau \ln(1 - 0.900)$ or $t = -(0.200 \text{ s})\ln(0.100) = 0.461$ s   ◊

L: The calculated times agree reasonably well with our predictions. We must be careful to avoid confusing the equation for the rising current with the similar equation for the falling current. Checking our answers against predictions is a safe way to prevent such mistakes.

21. For the *RL* circuit shown in Figure P32.19, let $L = 3.00$ H, $R = 8.00\ \Omega$, and $\mathcal{E} = 36.0$ V. (a) Calculate the ratio of the potential difference across the resistor to that across the inductor when $I = 2.00$ A. (b) Calculate the voltage across the inductor when $I = 4.50$ A.

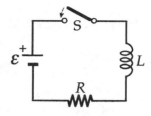

Solution

Figure P32.19

G: The voltage across the resistor is proportional to the current, $\Delta V_R = IR$, while the voltage across the inductor is proportional to the **rate of change** in the current, $\mathcal{E}_L = -L\,dI/dt$. When the switch is first closed, the voltage across the inductor will be large as it opposes the sudden change in current. As the current approaches its steady state value, the voltage across the resistor increases and the inductor's emf decreases. The maximum current will be $\mathcal{E}/R = 4.50$ A, so when $I = 2.00$ A, the resistor and inductor will share similar voltages at this mid-range current, but when $I = 4.50$ A, the entire circuit voltage will be across the resistor, and the voltage across the inductor will be zero.

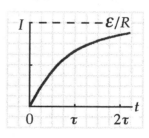

O: We can use the definition of resistance to calculate the voltage across the resistor for each current. We will find the voltage across the inductor by using Kirchhoff's loop rule.

A: (a) When $I = 2.00$ A, the voltage across the resistor is

$$\Delta V_R = IR = (2.00\ \text{A})(8.00\ \Omega) = 16.0\ \text{V}$$

Kirchhoff's loop rule tells us that the sum of the changes in potential around the loop must be zero:

$$\mathcal{E} - \Delta V_R - \mathcal{E}_L = 36.0\ \text{V} - 16.0\ \text{V} - \mathcal{E}_L = 0$$

so $\quad \mathcal{E}_L = 20.0$ V $\quad$ and $\qquad\qquad \dfrac{V_R}{\mathcal{E}_L} = \dfrac{16.0 \text{ V}}{20.0 \text{ V}} = 0.800$ $\qquad\qquad$ ◊

(b) Similarly, for $I = 4.50$ A, $\quad \Delta V_R = IR = (4.50 \text{ A})(8.00 \ \Omega) = 36.0$ V

$$\mathcal{E} - \Delta V_R - \mathcal{E}_L = 36.0 \text{ V} - 36.0 \text{ V} - \mathcal{E}_L = 0$$

So $\qquad\qquad\qquad\qquad \mathcal{E}_L = 0$ $\qquad\qquad$ ◊

L: We see that when $I = 2.00$ A, $\Delta V_R < \mathcal{E}_L$, but they are similar in magnitude as expected. Also as predicted, the voltage across the inductor goes to zero when the current reaches its maximum value. A worthwhile exercise would be to consider the ratio of these voltages for several different times after the switch is reopened.

---

29. A 140-mH inductor and a 4.90-Ω resistor are connected with a switch to a 6.00-V battery as shown in Figure P32.29. (a) If the switch is thrown to the left (connecting the battery), how much time elapses before the current reaches 220 mA? (b) What is the current in the inductor 10.0 s after the switch is closed? (c) Now the switch is quickly thrown from A to B. How much time elapses before the current falls to 160 mA?

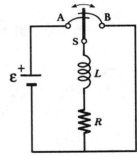

Figure P32.29

Solution The general *LR* equation is achieved by combining equations 32.7 and 32.8:

(a) $\qquad I = \dfrac{\mathcal{E}(1 - e^{-Rt/L})}{R} \qquad$ or $\qquad 0.220 \text{ A} = \dfrac{6.00 \text{ V}}{4.90 \ \Omega}\left(1 - e^{-(4.90 \ \Omega)t/0.140 \text{ H}}\right)$

$\qquad\qquad 0.180 = 1 - e^{-\left(35.0 \text{ s}^{-1}\right)t} \qquad$ so $\qquad e^{-\left(35.0 \text{ s}^{-1}\right)t} = 1.22$

Thus, $t = \dfrac{\ln 1.22}{35.0 \text{ s}^{-1}} = 5.66 \text{ ms}$ ◊

Again referring to the general equation,

(b) $I = \dfrac{6.00 \text{ V}}{4.90 \ \Omega}\left(1 - e^{\left(-35.0 \text{ s}^{-1}\right)(10.0 \text{ s})}\right) = 1.22 \text{ A}$ ◊

(c) $0.160 \text{ A} = (1.22 \text{ A}) \, e^{(-4.90 \ \Omega)t/0.140 \text{ H}}$

$7.65 = e^{\left(35.0 \text{ s}^{-1}\right)t}$ and $t = \dfrac{\ln 7.65}{35.0 \text{ s}^{-1}} = 58.1 \text{ ms}$ ◊

───────────────────────────

33. An air-core solenoid with 68 turns is 8.00 cm long and has a diameter of 1.20 cm. How much energy is stored in its magnetic field when it carries a current of 0.770 A?

Solution

For a solenoid of length ℓ, $L = \dfrac{\mu_0 N^2 A}{\ell}$

Thus, since $U_L = \tfrac{1}{2}LI^2$, $U_L = \dfrac{\mu_0 N^2 A I^2}{2\ell}$

$$U_L = \dfrac{\left(4\pi \times 10^{-7} \text{ N/A}^2\right)(68)^2 \, \pi (6.00 \times 10^{-3} \text{ m})^2 (0.770 \text{ A})^2}{2(0.0800 \text{ m})} = 2.44 \times 10^{-6} \text{ J} \quad ◊$$

───────────────────────────

35. On a clear day, there is a 100-V/m vertical electric field near the Earth's surface. At the same place, the Earth's magnetic field has a magnitude of 0.500×10^{-4} T. Compute the energy density of the two fields.

Solution

$$u_E = \frac{\epsilon_0 E^2}{2} = \frac{\left(8.85 \times 10^{-12}\, C^2/N \cdot m^2\right)\left(100\ N/C\right)^2\left(1\, J/N \cdot m\right)}{2} = 44.2\ nJ/m^3 \qquad \Diamond$$

$$u_B = \frac{B^2}{2\mu_0} = \frac{\left(5.00 \times 10^{-5}\ T\right)^2}{2\left(4\pi \times 10^{-7}\ T \cdot m/A\right)} = 995 \times 10^{-6}\ T \cdot A/m$$

$$u_B = \left(995 \times 10^{-6}\ T \cdot A/m\right)\left(1\ \frac{N \cdot s}{T \cdot C \cdot m}\right)\left(1\ \frac{J}{N \cdot m}\right) = 995\ \mu J/m^3 \qquad \Diamond$$

Magnetic energy density is 22500 times greater than that in the electric field.

---

43. Two solenoids A and B, spaced close to each other and sharing the same cylindrical axis, have 400 and 700 turns, respectively. A current of 3.50 A in coil A produces a flux of $300\ \mu T \cdot m^2$ at the center of A and a flux of $90.0\ \mu T \cdot m^2$ through each turn of B. (a) Calculate the mutual inductance of the two solenoids. (b) What is the self-inductance of A? (c) What emf is induced in B when the current in A increases at the rate of 0.500 A/s?

Solution (a) $\quad M_{12} = \dfrac{N_2 \Phi_{12}}{I_1} = \dfrac{700(90 \times 10^{-6}\ Wb)}{3.50\ A} = 18.0\ mH \qquad \Diamond$

(b) $\quad L = \dfrac{N\Phi_B}{I} = \dfrac{400(300 \times 10^{-6}\ Wb)}{3.50\ A} = 34.3\ mH \qquad \Diamond$

(c) $\quad \mathcal{E}_2 = -M_{12}\dfrac{dI_1}{dt} = -\left(18.0 \times 10^{-3}\ H\right)(0.500\ A/s) = -9.00\ mV \qquad \Diamond$

---

51. A fixed inductance $L = 1.05 \ \mu H$ is used in series with a variable capacitor in the tuning section of a radio. What capacitance tunes the circuit to the signal from a station broadcasting at 6.30 MHz?

Solution

G: It is difficult to predict a value for the capacitance without doing the calculations, but we might expect a typical value in the μF or pF range.

O: We want the resonance frequency of the circuit to match the broadcasting frequency, and for a simple *RLC* circuit, the resonance frequency only depends on the magnitudes of the inductance and capacitance.

A: The resonance frequency is $\qquad f_0 = \dfrac{1}{2\pi\sqrt{LC}}$

Thus, $\quad C = \dfrac{1}{(2\pi f_0)^2 L} = \dfrac{1}{\left[(2\pi)(6.30 \times 10^6 \ \text{Hz})\right]^2 (1.05 \times 10^{-6} \ \text{H})} = 608 \ \text{pF} \quad \lozenge$

L: This is indeed a typical capacitance, so our calculation appears reasonable. However, you probably would not hear any familiar music on this broadcast frequency. The frequency range for FM radio broadcasting is 88.0 – 108.0 MHz, and AM radio is 535 – 1605 kHz. The 6.30 MHz frequency falls in the Maritime Mobile SSB Radiotelephone range, so you might hear a ship captain instead of Top 40 tunes! This and other information about the radio frequency spectrum can be found on the National Telecommunications and Information Administration (NTIA) website, which at the time of this printing was at http://www.ntia.doc.gov/osmhome/allochrt.html

55. An *LC* circuit like that in Figure 32.14 consists of a 3.30-H inductor and an 840-pF capacitor, initially carrying a 105-μC charge. At $t = 0$ the switch is thrown closed. Compute the following quantities at $t = 2.00$ ms: (a) the energy stored in the capacitor, (b) the energy stored in the inductor, and (c) the total energy in the circuit.

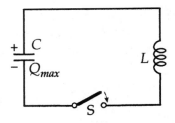

Figure 32.14

Solution   At $t = 0$ the capacitor charge is at its maximum value, so $\phi = 0$ in

$$Q = Q_{max} \cos (\omega t + \phi) = Q_{max} \cos \left(\frac{t}{\sqrt{LC}} \right)$$

$$Q = \left(105 \times 10^{-6} \text{ C} \right) \cos \left(\frac{2.00 \times 10^{-3} \text{ s}}{\sqrt{(3.30 \text{ H})(840 \times 10^{-12} \text{ F})}} \right)$$

$$Q = \left(105 \times 10^{-6} \text{ C} \right) (\cos 38.0 \text{ rad}) = 1.01 \times 10^{-4} \text{ C}$$

(a) $$U_C = \frac{Q^2}{2C} = \frac{(1.01 \times 10^{-4} \text{ C})^2}{2(840 \times 10^{-12} \text{ F})} = 6.03 \text{ J}$$ ◊

(c)   The constant total energy is that originally of the capacitor:

$$U = \frac{Q_{max}^2}{2C} = \frac{(1.05 \times 10^{-4} \text{ C})^2}{2(840 \times 10^{-12} \text{ F})} = 6.56 \text{ J}$$ ◊

(b) $$U_L = 6.56 \text{ J} - 6.03 \text{ J} = 0.529 \text{ J}$$ ◊

We could also find this from

$$\tfrac{1}{2} L I^2 = \tfrac{1}{2} L \left(\frac{d}{dt} Q_{max} \cos \omega t \right)^2 = \tfrac{1}{2} L Q_{max}^2 \omega^2 \sin^2 \omega t$$ ◊

57. Consider an LC circuit in which $L = 500$ mH and $C = 0.100\ \mu F$. (a) What is the resonant frequency ω_0? (b) If a resistance of 1.00 kΩ is introduced into this circuit, what is the frequency of the (damped) oscillations? (c) What is the percent difference between the two frequencies?

Solution

(a) $\omega_0 = \dfrac{1}{\sqrt{LC}} = \dfrac{1}{\sqrt{(0.50\ \text{H})(1.00\times10^{-7}\ \text{F})}} = 4.47\times10^3\ \text{rad/s}$ ◊

(b) $\omega_d = \sqrt{\dfrac{1}{LC} - \left(\dfrac{R}{2L}\right)^2} = \sqrt{\left(\dfrac{1}{(0.500\ \text{H})(1.00\times10^{-7}\ \text{F})}\right) - \left(\dfrac{1.00\times10^3\ \Omega}{(2)(0.500\ \text{H})}\right)^2}$

$\omega_d = 4.36\times10^3\ \text{rad/s}$ ◊

(c) $\dfrac{\Delta\omega}{\omega_0} = \dfrac{4.47 - 4.36}{4.47} = 0.0253 = 2.53\%$ ◊

Thus, the damped frequency is 2.53% lower than the undamped frequency.

---

69. At $t = 0$, the switch in Figure P32.69 is thrown closed. Using Kirchhoff's laws for the instantaneous currents and voltages in this two-loop circuit, show that the current in the inductor is

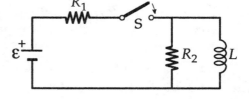

Figure P32.69

$$I(t) = \dfrac{\mathcal{E}}{R_1}[1 - e^{-(R'/L)t}]$$

where $R' = R_1 R_2/(R_1 + R_2)$.

Solution   Call I the downward current through the inductor and I_2 the downward current through R_2. Then $I + I_2$ is the current in R_1.

Left-hand loop:

$$\mathcal{E} - (I + I_2)R_1 - I_2 R_2 = 0$$

Outside loop:

$$\mathcal{E} - (I + I_2)R_1 - L\frac{dI}{dt} = 0$$

Eliminate I_2, obtaining

$$I\underbrace{\left(\frac{R_1 R_2}{R_1 + R_2}\right)}_{R'} + L\frac{dI}{dt} = \underbrace{\left(\frac{R_2}{R_1 + R_2}\right)\mathcal{E}}_{\mathcal{E}'}$$

Thus,

$$\mathcal{E}' - IR' - L\frac{dI}{dt} = 0$$

This is of the same form as Equation 32.6, so the reasoning on page 1018 and 1019 of the text shows that the solution is the same form as Equation 32.7,

$$I = \frac{\mathcal{E}'}{R'}(1 - e^{-R't/L})$$

with $\dfrac{\mathcal{E}'}{R'} = \dfrac{\mathcal{E}R_2 / (R_1 + R_2)}{R_1 R_2 / (R_1 + R_2)} = \dfrac{\mathcal{E}}{R_1}$: $I(t) = \dfrac{\mathcal{E}}{R_1}[1 - e^{-(R't/L)}]$ ◊

---

71. In Figure P32.71, the switch is closed at $t < 0$, and steady-state conditions are established. The switch is thrown open at $t = 0$. (a) Find the initial voltage $\mathcal{E}_0$ across L just after $t = 0$. Which end of the coil is at the higher potential: a or b? (b) Make freehand graphs of the currents in R_1 and in R_2 as a function of time, treating the steady-state directions as positive. Show values before and after $t = 0$. (c) How long after $t = 0$ does the current in R_2 have the value 2.00 mA?

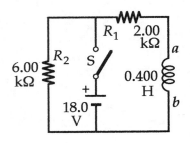

Figure P32.71

Solution   Before time $t = 0$, the current downward in R_2 is equal to $18.0 \text{ V}/6.00 \text{ k}\Omega = 3.00 \text{ mA}$, and the current is clockwise in R_1 and the coil is equal to $18.0 \text{ V}/2.00 \text{ k}\Omega = 9.00 \text{ mA}$.

(a)   Just after $t = 0$, the current in the outer loop is 9.00 mA clockwise but decreasing. Noting that $\mathcal{E}_0 = V_b - V_a$, the outer loop gives (clockwise):

$$V_b - (9.00 \text{ mA})(6.00 \text{ k}\Omega) - (9.00 \text{ mA})(2.00 \text{ k}\Omega) = V_a$$

$$\mathcal{E}_0 = V_b - V_a = 72.0 \text{ V} \quad \lozenge \qquad \text{Thus, point } b \text{ is at the higher potential.} \quad \lozenge$$

(b)   The currents in R_1 and R_2 are shown below. Just after $t = 0$, the current in R_1 decreases from an initial value of 9.00 mA according to $I = I_0 e^{-Rt/L}$. Taking the downward direction as positive in each resistor, the current decreases from +9.00 mA (downward) to zero in R_1. In R_2 the current goes from +3.00 mA (downward) to -9.00 mA (upward) and then decreases in magnitude to zero. $\qquad \lozenge$

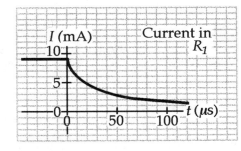
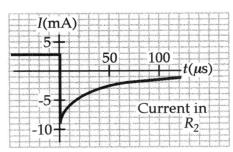

(c) $I = \dfrac{\mathcal{E}}{R} e^{-Rt/L} \qquad 2.00 \times 10^{-3} \text{ A} = \dfrac{18.0 \text{ V}}{2.00 \text{ k}\Omega} e^{(-8.00 \text{ k}\Omega)t/0.400 \text{ H}}$

Resistance R_1 establishes the original value of the current in the outer loop, but the series combination of R_1 and R_2 establishes the decay constant.

$$t = \frac{0.400 \text{ H}}{8.00 \text{ k}\Omega} \ln 4.50 = 75.2 \ \mu s \qquad \qquad \lozenge$$

73. To prevent damage from arcing in an electric motor, a discharge resistor is sometimes placed in parallel with the armature. If the motor is suddenly unplugged while running, this resistor limits the voltage that appears across the armature coils. Consider a 12.0-V dc motor with an armature that has a resistance of 7.50 Ω and an inductance of 450 mH. Assume that the back emf in the armature coils is 10.0 V when the motor is running at normal speed. (The equivalent circuit for the armature is shown in Figure P32.73.) Calculate the maximum resistance R that limits the voltage across the armature to 80.0 V when the motor is unplugged.

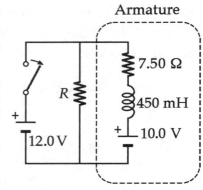

Figure P32.73

Solution

G: We should expect R to be significantly greater than the resistance of the armature coil, for otherwise a large portion of the source current would be diverted through R and much of the total power would be wasted on heating this discharge resistor.

O: When the motor is unplugged, the 10-V back emf will still exist for a short while because the motor's inertia will tend to keep it spinning. Now the circuit is reduced to a simple series loop with an emf, inductor, and two resistors. The current that was flowing through the armature coil must now flow through the discharge resistor, which will create a voltage across R that we wish to limit to 80 V. As time passes, the current will be reduced by the opposing back emf, and as the motor slows down, the back emf will be reduced to zero, and the current will stop.

Chapter 32

A: The steady-state coil current when the switch is closed is found from applying Kirchhoff's loop rule to the outer loop:

$$+12.0 \text{ V} - I(7.50 \text{ }\Omega) - 10.0 \text{ V} = 0$$

so

$$I = \frac{2.00 \text{ V}}{7.50 \text{ }\Omega} = 0.267 \text{ A}$$

We then require that

$$\Delta V_R = 80.0 \text{ V} = (0.267 \text{ A})R$$

so

$$R = \frac{\Delta V_R}{I} = \frac{80.0 \text{ V}}{0.267 \text{ A}} = 300 \text{ }\Omega \qquad \Diamond$$

L: As we expected, this discharge resistance is considerably greater than the coil's resistance. Note that while the motor is running, the discharge resistor turns $\mathcal{P} = (12 \text{ V})^2/300 \text{ }\Omega = 0.48 \text{ W}$ of power into heat (or wastes 0.48 W). The source delivers power at the rate of about $\mathcal{P} = IV = [0.267 \text{ A} + (12 \text{ V} / 300 \text{ }\Omega)](12 \text{ V}) = 3.68 \text{ W}$, so the discharge resistor wastes about 13% of the total power. To give a sense of perspective, this 4-W motor could lift a 40-N weight at the rate of 0.1 m/s.

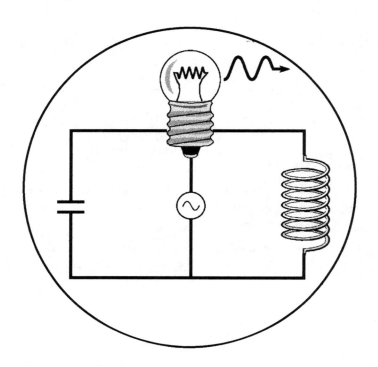

Alternating-Current
Circuits

ALTERNATING-CURRENT CIRCUITS

INTRODUCTION

In this chapter, we describe alternating current (ac) circuits. We investigate the characteristics of circuits containing familiar elements and driven by a sinusoidal voltage. Our discussion is limited to simple series circuits containing resistors, inductors, and capacitors, and we find that the ac current in each element is also sinusoidal but not necessarily in phase with the applied voltage. We conclude the chapter with two sections concerning the characteristics of RC filters, transformers, and power transmission.

EQUATIONS AND CONCEPTS

A simple series alternating current circuit with a sinusoidal source of emf is shown in the figure to the right. The rectangle ▭ used here represents the circuit element(s) which, in a particular case, may be

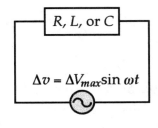

a resistor R,
a capacitor C,
an inductor L, or
some combination of the
   above components

The **applied sinusoidal voltage** of frequency ω has a maximum value of ΔV_{max}.

$$\Delta v = \Delta v_R = \Delta V_{max} \sin \omega t \qquad (33.1)$$

When the circuit element is a **resistor** of value R, the current and voltage across the resistor are **in phase**, and I_{max} is the maximum current.

$$i_R = \frac{\Delta v_R}{R} = I_{max} \sin \omega t \qquad (33.2)$$

$$\Delta v_R = I_{max} R \sin \omega t \qquad (33.3)$$

When the circuit element is an **inductor** of value L, the **current lags the voltage** across the inductor by 90°.

$$-\Delta v_L = L \frac{di}{dt} = \Delta V_{max} \sin \omega t \qquad (33.6)$$

$$i_L = \frac{\Delta V_{max}}{\omega L} \sin\left(\omega t - \frac{\pi}{2}\right) \qquad (33.8)$$

When the circuit element is a **capacitor** of value C, the **current leads the voltage** across the capacitor by 90°.

$$\Delta v = \Delta v_C = \Delta V_{max} \sin \omega t \qquad (33.12)$$

$$i_C = \omega C \Delta V_{max} \sin\left(\omega t + \frac{\pi}{2}\right) \qquad (33.15)$$

The maximum value of the current (or current amplitude) through each element is proportional to the amplitude of the ac voltage across the element. In the case of an inductor and a capacitor, the maximum value of the current depends also on the angular frequency of the source of emf.

Resistor:

$$I_{max} = \frac{\Delta V_{max}}{R}$$

Inductor:

$$I_{max} = \frac{\Delta V_{max}}{\omega L} = \frac{\Delta V_{max}}{X_L} \qquad (33.9)$$

Capacitor:

$$I_{max} = \omega C \Delta V_{max} = \frac{\Delta V_{max}}{X_C} \qquad (33.16)$$

In the general case, the ac circuit will contain a resistor, inductor, and capacitor in series with a sinusoidally varying voltage source.

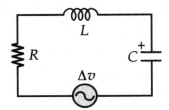

Since at any instant, the current has the same value (amplitude and phase) at every point in the circuit, it is convenient to express the phase relationship between the current and **instantaneous voltage** drops across R, L, and C relative to **the common current phase.**

$$\Delta v_R = I_{max}R \sin \omega t \qquad (33.19)$$

$$= \Delta V_R \sin \omega t$$

$$\Delta v_L = I_{max}X_L \sin\left(\omega t + \frac{\pi}{2}\right) \qquad (33.20)$$

$$= \Delta V_L \cos \omega t$$

$$\Delta v_C = I_{max}X_C \sin\left(\omega t - \frac{\pi}{2}\right) \qquad (33.21)$$

$$= -\Delta V_C \cos \omega t$$

Compare Equations 33.19, 33.20, and 33.21 to Equations 33.3, 33.8, and 33.15. You should convince yourself that **these two sets of equations express the same phase relationship between current and voltage.**

The **maximum voltage** across each circuit element can be written in the form of Ohm's law.

$$\Delta V_R = I_{max} R$$

$$\Delta V_L = I_{max} X_L$$

$$\Delta V_C = I_{max} X_C$$

In the previous set of equations, R is of course the resistance while X_L and X_C represent the **inductive reactance** and the **capacitive reactance** respectively. The reactances are frequency dependent. The inductive reactance increases with increasing frequency, while the capacitive reactance decreases with increasing frequency.

$$X_L = \omega L \qquad (33.10)$$

$$X_C = \frac{1}{\omega C} \qquad \text{where } \omega = 2\pi f \qquad (33.17)$$

The **maximum current** in the circuit depends on the angular frequency ω of the source of emf, as well as the values of ΔV_{max}, R, L, and C.

$$I_{max} = \frac{\Delta V_{max}}{\sqrt{R^2 + (X_L - X_C)^2}}$$

It is useful to define an operating parameter of the circuit called the **impedance**, Z, defined by Equation 33.23. ΔV_{max} and I_{max} can be related in the form of Ohm's law, Equation 33.24. This requires that Z have the SI unit of ohm (Ω).

$$Z \equiv \sqrt{R^2 + (X_L - X_C)^2} \qquad (33.23)$$

$$\Delta V_{max} = I_{max} Z \qquad (33.24)$$

The applied voltage (across the source) and the current in the circuit will differ in phase by some angle ϕ, the **phase angle** of the circuit, given by Equation 33.25.

$$\phi = \tan^{-1}\left(\frac{X_L - X_C}{R}\right) \tag{33.25}$$

The **average power** delivered by a generator (source of emf) to an RLC series circuit is dissipated as heat in the resistor (there is zero power loss in ideal inductors and capacitors) and is directly proportional to $\cos \phi$, where ϕ is the phase angle. The quantity $\cos \phi$ is called the **power factor** of the circuit. Since ϕ is frequency dependent (see Equation 33.25), the power factor also depends on frequency.

$$\mathcal{P}_{av} = I_{rms}\Delta V_{rms} \cos \phi \tag{33.29}$$

or

$$\mathcal{P}_{av} = I_{rms}^2 R \tag{33.30}$$

When measuring values of current and voltage in ac circuits, it is customary to use instruments which respond to **root-mean-square** (rms) values of these quantities rather than their maximum or instantaneous values.

$$\Delta V_{rms} = \frac{\Delta V_{max}}{\sqrt{2}}$$

$$I_{rms} = \frac{I_{max}}{\sqrt{2}}$$

You should notice that ΔV_{rms} and I_{rms} are in the same ratio as ΔV_{max} and I_{max}. Compare this equation to Equation 33.24.

$$\Delta V_{rms} = I_{rms} Z$$

From Equation 33.23 you can see that when the inductive reactance X_L equals the capacitive reactance X_C, the impedance of the circuit has its **minimum** value. Under these conditions, $Z = R$, and the current in the circuit will have its **maximum** value.

The condition that $Z = R$ occurs for a characteristic frequency of the circuit called the **resonance frequency**, ω_0, given by Equation 33.33. This is obtained from the condition $X_L = X_C$, where $X_L = \omega L$ and $X_C = 1/\omega C$.

$$\omega_0 = \frac{1}{\sqrt{LC}} \tag{33.33}$$

A simple transformer consists of a primary coil of N_1 turns and a secondary coil of N_2 turns wound on a common core. In Equation 33.39, ΔV_1 represents the **voltage across the primary** (input voltage), while ΔV_2 represents the **voltage across the secondary** (output voltage). Likewise, I_1 and I_2 represent the currents in the primary and secondary circuits. In the **ideal** transformer, the ratio of voltages is equal to the ratio of turns, and the ratio of currents is equal to the inverse of the ratio of turns.

$$\Delta V_2 = \frac{N_2}{N_1} \Delta V_1 \tag{33.39}$$

$$I_1 \Delta V_1 = I_2 \Delta V_2 \tag{33.40}$$

SUGGESTIONS, SKILLS, AND STRATEGIES

The **phasor diagram** is a very useful technique to use in the analysis of ac RLC circuits. In such a diagram, each of the rotating quantities ΔV_R, ΔV_L, ΔV_C, and I_{max} is represented by a separate phasor (rotating vector). A 'phasor' diagram which describes the ac circuit of figure (a) below is shown in figure (b). Each phasor has a length which is proportional to the magnitude of the voltage or current which it represents and rotates counterclockwise about the common origin with an angular frequency which equals the angular frequency of the alternating source, ω. The direction of the phasor which represents the current in the circuit is used as the **reference** direction to establish the correct phase differences among the phasors, which represent the voltage drops across the resistor, inductor, and capacitor. The **instantaneous values** Δv_R, Δv_L, Δv_C, and i are given by the **projection onto the vertical axis of the corresponding phasor**.

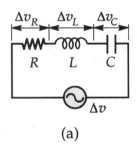

(a)

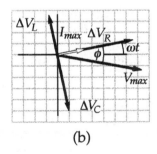

(b)

Consider the phasor diagram in figure (b) above, where the maximum voltage across the resistor, ΔV_R, is greater than the maximum voltage across the inductor, ΔV_L At the instant shown, the **instantaneous** value of the voltage across the inductor is greater than that across the resistor. Also notice that as time increases and the phasors rotate counterclockwise, maintaining their constant relative phase, ΔV_R, ΔV_L, and ΔV_C (the voltage amplitudes) will remain constant in magnitude but the instantaneous values Δv_R, Δv_L, and Δv_C will vary sinusoidally with time. For the case shown in figure (b), the phase angle ϕ is negative (this is because $X_C > X_L$ and therefore $\Delta V_C > \Delta V_L$); hence, the current in the circuit **leads** the applied voltage in phase.

The maximum voltage across each element in the circuit is the product of I_{max} and the resistance or reactance of that component. It is possible, therefore, to construct an **impedance triangle** for any series circuit as shown in the figure to the right.

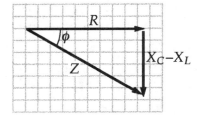

The following procedures are recommended when solving alternating current problems:

- The first step in analyzing alternating current circuits is to calculate as many of the unknown quantities such as X_L and X_C as possible. (Note that when calculating X_C, the capacitance should be expressed in farads, rather than, say, microfarads.)

- Apply the equation $\Delta V = IZ$ to that portion of the circuit of interest. That is, if you want to know the voltage drop across the combination of an inductor and a resistor, the equation reduces to $\Delta V = I\sqrt{R^2 + X_L{}^2}$.

REVIEW CHECKLIST

▷ Apply the formulas that give the reactance values in an ac circuit as a function of (i) capacitance, (ii) inductance, and (iii) frequency. Interpret the meaning of the terms **phase angle** and **power factor** in an ac circuit.

▷ Given an *RLC* series circuit in which values of resistance, inductance, capacitance, and the characteristics of the generator (source of emf) are known, calculate: (i) the instantaneous and rms voltage drop across each component, (ii) the instantaneous and rms current in the circuit, (iii) the phase angle by which the current leads or lags the voltage, (iv) the power expended in the circuit, and (v) the resonance frequency of the circuit.

▷ Understand the use of phasor diagrams for the description and analysis of ac circuits. Sketch circuit diagrams for high-pass and low-pass filter circuits and make calculations of the ratio of output to input voltage in each case.

▷ Understand the manner in which step-up and step-down transformers are used in the process of transmitting electrical power over large distances; and make calculations of primary to secondary voltage and current ratios for an ideal transformer.

ANSWERS TO SELECTED CONCEPTUAL QUESTIONS

5. Does the phase angle depend on frequency? What is the phase angle when the inductive reactance equals the capacitive reactance?

Answer Yes. Since the phase angle is a function of the reactance, which depends on frequency, it must be frequency dependent. The phase angle is zero when the inductive reactance equals the capacitive reactance.

□   □   □   □

11. Will a transformer operate if a battery is used for the input voltage across the primary? Explain.

Answer No. A voltage can only be induced in the secondary coil if the flux through the core changes in time.

□   □   □   □

SOLUTIONS TO SELECTED END-OF-CHAPTER PROBLEMS

5. The current in the circuit shown in Figure 33.1 equals 60.0% of the peak current at $t = 7.00$ ms. What is the smallest frequency of the generator that gives this current?

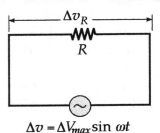

$\Delta v = \Delta V_{max} \sin \omega t$

Figure 33.1

Solution $\quad i = \dfrac{\Delta v}{R} = \left(\dfrac{\Delta V_{max}}{R} \right) \sin \omega t$

$\dfrac{0.600 \Delta V_{max}}{R} = \left(\dfrac{\Delta V_{max}}{R} \right) \sin \left[\omega \, (7.00 \text{ ms}) \right]$

$0.600 = \sin \left[\omega \, (7.00 \text{ ms}) \right]$

$\omega \, (7.00 \text{ ms}) = 36.9° = 0.644 \text{ rad}$

Thus, $\qquad \omega = 91.9 \text{ rad / s} \qquad$ and $\qquad f = \dfrac{\omega}{2\pi} = 14.6 \text{ cycle / s} \qquad \Diamond$

---

9. In a purely inductive ac circuit, such as that shown in Figure 33.4, $\Delta V_{max} = 100$ V. (a) If the maximum current is 7.50 A at 50.0 Hz, what is the inductance L? (b) At what angular frequency ω is the maximum current 2.50 A?

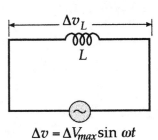

$\Delta v = \Delta V_{max} \sin \omega t$

Figure 33.4

Solution $\quad I_{max} = \Delta V_{max} / X_L$

(a) $\qquad X_L = \dfrac{\Delta V_{max}}{I_{max}} = \dfrac{100 \text{ V}}{7.50 \text{ A}} = 13.3 \ \Omega = \omega L$

$L = \dfrac{13.3 \ \Omega}{\omega} = \dfrac{13.3 \ \Omega}{2\pi \left(50.0 \text{ s}^{-1} \right)} = 42.4 \text{ mH} \qquad \Diamond$

(b) $\qquad X_L = \dfrac{\Delta V_{max}}{I_{max}} = \dfrac{100 \text{ V}}{2.50 \text{ A}} = 40.0 \ \Omega = \omega L$

$\qquad\qquad \omega = \dfrac{40.0 \ \Omega}{42.4 \text{ mH}} = 942 \text{ rad / s}$ ◊

A frequency 3 times higher makes the inductive reactance 3 times larger.

---

11. For the circuit shown in Figure 33.4 (above), $\Delta V_{max} = 80.0 \text{ V}$, $\omega = 65.0\pi \text{ rad / s}$, and $L = 70.0 \text{ mH}$. Calculate the current in the inductor at $t = 15.5 \text{ ms}$.

Solution $\qquad X_L = \omega L = \left(65.0\pi \text{ s}^{-1}\right)\left(70.0 \times 10^{-3} \ \dfrac{\text{V} \cdot \text{s}}{\text{A}}\right) = 14.3 \ \Omega$

$\qquad\qquad I_{max} = \dfrac{\Delta V_{max}}{X_L} = \dfrac{80.0 \text{ V}}{14.3 \ \Omega} = 5.60 \text{ A}$

$\qquad\qquad I = -I_{max} \cos \omega t = -(5.60 \text{ A})\cos\left[\left(65\pi \text{ s}^{-1}\right)(0.0155 \text{ s})\right]$

$\qquad\qquad I = -(5.60 \text{ A})\cos(3.17 \text{ rad}) = +5.60 \text{ A}$ ◊

---

17. What maximum current is delivered by an ac generator with $\Delta V_{max} = 48.0 \text{ V}$ and $f = 90.0 \text{ Hz}$ when connected across a $3.70 - \mu\text{F}$ capacitor?

Solution $\qquad I_{max} = \dfrac{\Delta V_{max}}{X_C} = \Delta V_{max}\omega C = \Delta V_{max}(2\pi f C)$

$\qquad\qquad I_{max} = (48.0 \text{ V})(2\pi)(90.0 \text{ Hz})(3.70 \times 10^{-6} \text{ F}) = 0.100 \text{ A} = 100 \text{ mA}$ ◊

---

19. An inductor $(L = 400 \text{ mH})$, a capacitor $(C = 4.43 \ \mu\text{F})$, and a resistor $(R = 500 \ \Omega)$ are connected in series. A 50.0-Hz ac generator produces a peak current of 250 mA in the circuit. (a) Calculate the required peak voltage ΔV_{max}. (b) Determine the angle by which the current leads or lags the applied voltage.

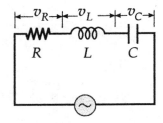

Solution

We first find the impedance of the capacitor and the inductor:

(a) $$X_L = \omega L = 2\pi(50.0 \text{ Hz})(400 \times 10^{-3} \text{ H}) = 126 \ \Omega$$

and $$X_C = \frac{1}{\omega C} = \frac{1}{(2\pi)(50.0 \text{ Hz})(4.43 \times 10^{-6} \text{ F})} = 719 \ \Omega$$

Then, we substitute these values into the equation for a series LRC circuit:

$$\Delta V_{max} = I_{max} Z = I_{max}\sqrt{R^2 + (X_L - X_C)^2}$$

Thus, $$Z = \sqrt{(500 \ \Omega)^2 + (126 \ \Omega - 719 \ \Omega)^2} = 776 \ \Omega$$

and $$\Delta V_{max} = I_{max} Z = (0.250 \text{ A})(776 \ \Omega) = 194 \text{ V} \qquad \Diamond$$

(b) $$\tan\phi = \frac{X_L - X_C}{R} \quad \text{so} \quad \phi = \tan^{-1}\left(\frac{126 - 719}{500}\right) = -49.9° \qquad \Diamond$$

The current **leads** the voltage by 49.9°.

23. An RLC circuit consists of a 150-Ω resistor, a $21.0-\mu F$ capacitor, and a $460-mH$ inductor, connected in series with a 120-V, 60.0-Hz power supply. (a) What is the phase angle between the current and the applied voltage? (b) Which reaches its maximum earlier, the current or the voltage?

Solution The reactance of the inductor is

$$X_L = \omega L = 2\pi f L = 2\pi(60.0 \text{ s}^{-1})(0.460 \text{ H}) = 173 \ \Omega$$

The reactance of the capacitor is

$$X_C = \frac{1}{\omega C} = \frac{1}{2\pi f C} = \frac{1}{2\pi(60.0 \text{ s}^{-1})(21.0 \times 10^{-6} \text{ F})} = 126 \ \Omega$$

(a) $\quad \tan \phi = \dfrac{X_L - X_C}{R} = \dfrac{173 \ \Omega - 126 \ \Omega}{150 \ \Omega} = 0.304, \quad$ so $\quad \phi = 0.304 \text{ rad} = 17.4° \quad \lozenge$

(b) $\quad$ Since $X_L > X_C$, ϕ is positive, so Δv leads the current. $\hfill \lozenge$

---

29. An ac voltage of the form $\Delta v = (100 \text{ V})\sin(1000 \ t)$ is applied to a series RLC circuit. If $R = 400 \ \Omega$, $C = 5.00 \ \mu F$, and $L = 0.500 \text{ H}$, what is the average power delivered to the circuit?

Solution

G: Comparing $\Delta v = (100 \text{ V})\sin(1000 \ t)$ with $\Delta v = \Delta V_{max} \sin \omega t$, we see that
$$\Delta V_{max} = 100 \text{ V} \quad \text{and} \quad \omega = 1000 \text{ s}^{-1}$$

Only the resistor takes electric energy out of the circuit, but the capacitor and inductor will impede the current flow and therefore reduce the voltage across the resistor. Because of this impedance, the average power dissipated by the resistor must be less than the maximum power from the source:
$$\mathcal{P}_{max} = \frac{(\Delta V_{max})^2}{R} = \frac{(100 \text{ V})^2}{(400 \ \Omega)} = 25.0 \text{ W}$$

O: The average power dissipated by the resistor can be found from $\mathcal{P}_{av} = I_{rms}^2 R$, where $I_{rms} = \Delta V_{rms} / Z$.

A:
$$\Delta V_{rms} = \frac{100}{\sqrt{2}} = 70.7 \text{ V}$$

In order to calculate the impedance, we first need the capacitive and inductive reactances:

$$X_C = \frac{1}{\omega C} = \frac{1}{(1000 \text{ s}^{-1})(5.00 \times 10^{-6} \text{ F})} = 200 \text{ }\Omega$$

$$X_L = \omega L = (1000 \text{ s}^{-1})(0.500 \text{ H}) = 500 \text{ }\Omega$$

Then,
$$Z = \sqrt{R^2 + (X_L - X_C)^2}$$

$$Z = \sqrt{(400 \text{ }\Omega)^2 + (500 \text{ }\Omega - 200 \text{ }\Omega)^2} = 500 \text{ }\Omega$$

$$I_{rms} = \frac{\Delta V_{rms}}{Z} = \frac{70.7 \text{ V}}{500 \text{ }\Omega} = 0.141 \text{ A}$$

$$\mathcal{P}_{av} = I_{rms}^2 R = (0.141 \text{ A})^2 (400 \text{ }\Omega) = 8.00 \text{ W} \qquad \lozenge$$

L: The power dissipated by the resistor is less than 25 W, so our answer appears to be reasonable. As with other *RLC* circuits, the power will be maximized at the resonance frequency where $X_L = X_C$ so that $Z = R$. Then the average power dissipated will simply be the 25 W we calculated first.

31. In a certain series RLC circuit, $I_{rms} = 9.00$ A, $\Delta V_{rms} = 180$ V, and the current leads the voltage by 37.0°. (a) What is the total resistance of the circuit? (b) What is the magnitude of the reactance of the circuit $(X_L - X_C)$?

Solution The power is $\mathcal{P}_{av} = I_{rms}\Delta V_{rms} \cos\phi = I_{rms}^2 R$

(a) Therefore, $\quad\quad\quad\quad R = \dfrac{\Delta V_{rms} \cos\phi}{I_{rms}} = \dfrac{(180 \text{ V})\cos(-37.0°)}{9.00 \text{ A}} = 16.0 \; \Omega \quad \Diamond$

(b) $\tan\phi = \dfrac{X_L - X_C}{R}:\quad\quad X_L - X_C = R \tan\phi = (16.0 \; \Omega) \tan(-37.0°) = -12.0 \; \Omega$

or $\quad\quad\quad\quad\quad\quad\quad\quad |X_L - X_C| = 12.0 \; \Omega \quad\quad\quad\quad\quad\quad\quad\quad\quad\quad\quad\quad\quad\quad \Diamond$

---

37. An RLC circuit is used in a radio to tune in to an FM station broadcasting at 99.7 MHz. The resistance in the circuit is 12.0 Ω, and the inductance is 1.40 μH. What capacitance should be used?

Solution The circuit is to be in resonance when $\quad\quad \omega L = \dfrac{1}{\omega C}$

$$C = \dfrac{1}{\omega^2 L} = \dfrac{1}{4\pi^2 f^2 L} = \dfrac{1}{4\pi^2 (99.7 \text{ MHz})^2 (1.40 \; \mu\text{V} \cdot \text{s / A})} = 1.82 \text{ pF} \quad \Diamond$$

---

43. A transformer has $N_1 = 350$ turns and $N_2 = 2000$ turns. If the input voltage is $\Delta v(t) = (170 \text{ V})\cos\omega t$, what rms voltage is developed across the secondary coil?

Solution $\quad \Delta V_{1,rms} = \dfrac{170 \text{ V}}{\sqrt{2}} = 120$ V

$$\Delta V_{2,rms} = \dfrac{N_2}{N_1}\Delta V_{1,rms} = \dfrac{2000}{350}(120 \text{ V}) = 687 \text{ V} \quad\quad\quad\quad\quad\quad \Diamond$$

---

49. The RC high-pass filter shown in Figure 33.22 has a resistance $R = 0.500\ \Omega$. (a) What capacitance gives an output signal that has one-half the amplitude of a 300-Hz input signal? (b) What is the gain $(\Delta V_{out} / \Delta V_{in})$ for a 600-Hz signal?

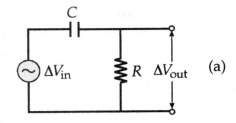

(a)

Solution

G: It is difficult to estimate the capacitance required without actually calculating it, but we might expect a typical value in the μF to pF range. The nature of a high-pass filter is to yield a larger gain at higher frequencies, so if this circuit is

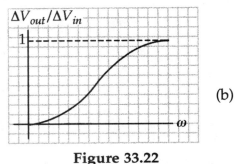

(b)

Figure 33.22

designed to have a gain of 0.5 at 300 Hz, then it should have a higher gain at 600 Hz, apparently near 1.0 based on the graph above.

O: The output voltage of this circuit is taken across the resistor, but the input sees the impedance of the resistor and the capacitor. Therefore, the gain will be the ratio of the resistance to the impedance.

A: $\dfrac{\Delta V_{out}}{\Delta V_{in}} = \dfrac{R}{\sqrt{R^2 + (1/\omega C)^2}}$

(a)   When $\Delta V_{out} / \Delta V_{in} = 0.500$, solving for C gives

$$C = \frac{1}{\omega R \sqrt{\left(\dfrac{\Delta V_{in}}{\Delta V_{out}}\right)^2 - 1}} = \frac{1}{(2\pi)(300\ \text{Hz})(0.500\ \Omega)\sqrt{(2.00)^2 - 1}} = 613\ \mu\text{F} \qquad \Diamond$$

(b)   At 600 Hz, we have $\omega = (2\pi\,\mathrm{rad})(600\ \mathrm{s}^{-1})$, so

$$\frac{\Delta V_{out}}{\Delta V_{in}} = \frac{0.500\ \Omega}{\sqrt{(0.500\ \Omega)^2 + \left(\dfrac{1}{(1200\pi\ \mathrm{rad}/\mathrm{s})(613\ \mu F)}\right)^2}} = 0.756 \qquad \Diamond$$

L: The capacitance value seems reasonable, but the gain is considerably less than we expected. Based on our calculation, we can modify the above graph to more accurately represent the characteristics of this high-pass filter. If this were an audio filter, it would reduce low frequency "humming" sounds while allowing high pitch sounds to pass through. A low pass filter would be needed to reduce high frequency "static" noise.

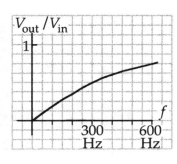

55. A series RLC circuit consists of an 8.00-Ω resistor, a 5.00-μF capacitor, and a 50.0-mH inductor. A variable frequency source applies an emf of 400 V (rms) across the combination. Determine the power delivered to the circuit when the frequency is equal to one half the resonance frequency.

Solution

G: Maximum power is delivered at the resonance frequency, and the power delivered at other frequencies depends on the quality factor, Q. For the relatively small resistance in this circuit, we could expect a high $Q = \omega_0 L/R$. So at half the resonant frequency, the power should be a small fraction of the maximum power,

$$\mathscr{P}_{av,\,max} = \Delta V_{rms}^2 / R = (400\ \mathrm{V})^2 / 8\ \Omega = 20\ \mathrm{kW}.$$

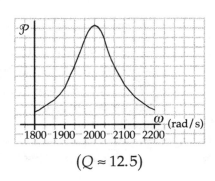

$(Q \approx 12.5)$

O: We must first calculate the resonance frequency in order to find half this frequency. Then the power delivered by the source must equal the power taken out by the resistor. This power can be found from $\mathcal{P}_{av} = I_{rms}^2 R$ where $I_{rms} = \Delta V_{rms} / Z$.

A: The resonance frequency is

$$f_0 = \frac{1}{2\pi\sqrt{LC}} = \frac{1}{2\pi\sqrt{(0.0500 \text{ H})(5.00 \times 10^{-6} \text{ F})}} = 318 \text{ Hz}$$

The operating frequency is $f = f_0 / 2 = 159 \text{ Hz}$. We can calculate the impedance at this frequency:

$$X_L = 2\pi f L = 2\pi(159 \text{ Hz})(0.0500 \text{ H}) = 50.0 \ \Omega$$

$$X_C = \frac{1}{2\pi f C} = \frac{1}{2\pi(159 \text{ Hz})(5.00 \times 10^{-6} \text{ F})} = 200 \ \Omega$$

$$Z = \sqrt{R^2 + (X_L - X_C)^2} = \sqrt{8.00^2 + (50.0 - 200)^2} \ \Omega = 150 \ \Omega$$

So,
$$I_{rms} = \frac{\Delta V_{rms}}{Z} = \frac{400 \text{ V}}{150 \ \Omega} = 2.66 \text{ A}$$

The power delivered by the source is the power dissipated by the resistor:

$$\mathcal{P}_{av} = I_{rms}^2 R = (2.66 \text{ A})^2 (8.00 \ \Omega) = 56.7 \text{ W} \qquad \Diamond$$

L: This power is only about 0.3% of the 20 kW peak power delivered at the resonance frequency. The significant reduction in power for frequencies away from resonance is a consequence of the relatively high Q-factor of about 12.5 for this circuit. A high Q is beneficial if, for example, you want to listen to your favorite radio station that broadcasts at 101.5 MHz, and you do not want to receive the signal from another local station that broadcasts at 101.9 MHz.

63. Consider a series RLC circuit having the following circuit parameters: $R = 200\ \Omega$, $L = 663$ mH, and $C = 26.5\ \mu F$. The applied voltage has an amplitude of 50.0 V and a frequency of 60.0 Hz. Find the following amplitudes: (a) the current I_{max}, including its phase constant ϕ relative to the applied voltage Δv; (b) the voltage ΔV_R across the resistor and its phase relative to the current; (c) the voltage ΔV_C across the capacitor and its phase relative to the current; and (d) the voltage ΔV_L across the inductor and its phase relative to the current.

Solution We identify that $R = 200\ \Omega$, $L = 663$ mH, $C = 26.5\ \mu F$, $\omega = 377$ rad/s, and $\Delta V_{max} = 50.0$ V

So $\omega L = 250\ \Omega$, and $\left(\dfrac{1}{\omega C}\right) = 100\ \Omega$

The impedance is

$$Z = \sqrt{R^2 + \left(\omega L - \frac{1}{\omega C}\right)^2} = \sqrt{(200\ \Omega)^2 + (250\ \Omega - 100\ \Omega)^2} = 250\ \Omega$$

(a) $\quad I_{max} = \dfrac{\Delta V_{max}}{Z} = \dfrac{50.0\ V}{250\ \Omega} = 0.200$ A $\hfill \lozenge$

$\quad\quad \phi = \tan^{-1}\left(\dfrac{X_L - X_C}{R}\right) = 36.8°$ $\quad\quad$ with Δv leading i $\hfill \lozenge$

(b) $\quad \Delta V_R = I_{max}R = 40.0$ V at $\phi = 0°$ $\hfill \lozenge$

(c) $\quad \Delta V_C = I_{max}X_C = (0.200\ A)(100\ \Omega) = 20.0$ V at $\phi = -90.0°$ $\hfill \lozenge$

(d) $\quad \Delta V_L = I_{max}X_L = (0.200\ A)(250\ \Omega) = 50.0$ V at $\phi = 90.0°$ $\hfill \lozenge$

65. Example 28.2 showed that maximum power is transferred when the internal resistance of a dc source is equal to the resistance of the load. A transformer may be used to provide maximum power transfer between two ac circuits that have different impedances. (a) Show that the ratio of turns N_1/N_2 needed to meet this condition is

$$\frac{N_1}{N_2} = \sqrt{\frac{Z_1}{Z_2}}$$

(b) Suppose you want to use a transformer as an impedance-matching device between an audio amplifier that has an output impedance of 8.00 kΩ and a speaker that has an input impedance of 8.00 Ω. What should your N_1/N_2 ratio be?

Solution $\quad \dfrac{N_1}{N_2} = \dfrac{\Delta V_1}{\Delta V_2}$

with $\quad Z_1 = \dfrac{\Delta V_1}{I_1} \quad$ and $\quad Z_2 = \dfrac{\Delta V_2}{I_2}$

Thus, $\quad \dfrac{N_1}{N_2} = \dfrac{Z_1 I_1}{Z_2 I_2}$

(a) Since $\dfrac{I_1}{I_2} = \dfrac{N_2}{N_1}$ we find $\dfrac{N_1}{N_2} = \sqrt{\dfrac{Z_1}{Z_2}}$ ◊

(b) $\dfrac{N_1}{N_2} = \sqrt{\dfrac{8000\ \Omega}{8.00\ \Omega}} = 31.6$ ◊

70. A series RLC circuit in which $R = 1.00$ Ω, $L = 1.00$ mH, and $C = 1.00$ nF is connected to an ac generator delivering 1.00 V (rms). Make a precise graph of the power delivered to the circuit as a function of the frequency, and verify that the full width of the resonance peak at half-maximum is $R/2\pi L$.

Solution

At resonance, $$\omega = \frac{1}{\sqrt{LC}} = \frac{1}{\sqrt{\left(1.00 \times 10^{-3} \text{ H}\right)\left(1.00 \times 10^{-9} \text{ F}\right)}}$$

so $$\omega = 1.00 \times 10^6 \text{ rad / s}$$

At that point $$Z = R = 1.00 \ \Omega$$

and $$I = \frac{1.00 \text{ V}}{1.00 \ \Omega} = 1.00 \text{ A}$$

and the power is $$I^2R = (1.00 \text{ A})^2(1.00 \ \Omega) = 1.00 \text{ W}$$

We compute the power at some other angular frequencies. Thus,

ω, 10^6 rad/s	ωL, Ω	$1/\omega C$, Ω	Z, Ω	I, A	$I^2R = \mathcal{P}$, W
0.9990	999	1001	2.24	0.447	0.19984
0.9994	999.4	1000.6	1.56	0.640	0.40969
0.9995	999.5	1000.5	1.41	0.707	0.49988
0.9996	999.6	1000.4	1.28	0.781	0.60966
0.9998	999.8	1000.2	1.08	0.928	0.86205
1	1000	1000	1	1	1
1.0002	1000.2	999.8	1.08	0.928	0.86209
1.0004	1000.4	999.6	1.28	0.781	0.60985
1.0005	1000.5	999.5	1.41	0.707	0.50012
1.0006	1000.6	999.4	1.56	0.640	0.40998
1.001	1001	999	2.24	0.447	0.20016

The angular frequencies giving half the maximum power are

$$0.9995 \times 10^6 \text{ rad/s}$$

and $1.0005 \times 10^6 \text{ rad/s},$

so the full width at half the maximum is

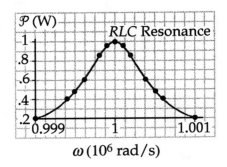

RLC Resonance

$$\Delta\omega = (1.0005 - 0.9995) \times 10^6 \text{ rad / s}$$

$$= 1.00 \times 10^3 \text{ rad / s}$$

Since $\Delta\omega = 2\pi \Delta f,$ $\Delta f = 159 \text{ Hz}$

and $\dfrac{R}{2\pi L} = \dfrac{1.00 \ \Omega}{2\pi \left(1.00 \times 10^{-3} \text{ H}\right)} = 159 \text{ Hz}$ (They agree.)   ◊

Chapter
34

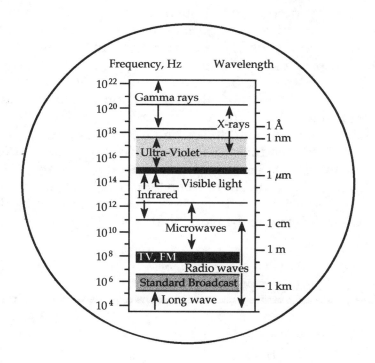

Frequency, Hz   Wavelength

10^{22}	Gamma rays	
10^{20}	X-rays	
10^{18}		1 Å
	Ultra-Violet	1 nm
10^{16}		
10^{14}	Visible light	1 μm
	Infrared	
10^{12}		
10^{10}	Microwaves	1 cm
10^{8}	TV, FM	1 m
	Radio waves	
10^{6}	Standard Broadcast	1 km
10^{4}	Long wave	

Electromagnetic Waves

ELECTROMAGNETIC WAVES

INTRODUCTION

The waves described in Chapters 16, 17, and 18 are mechanical waves. By definition, mechanical disturbances, such as sound waves, water waves, and waves on a string, require the presence of a medium. This chapter is concerned with the properties of electromagnetic waves that (unlike mechanical waves) can propagate through empty space.

In Section 31.7 we gave a brief description of Maxwell's equations, which form the theoretical basis of all electromagnetic phenomena. The consequences of Maxwell's equations are far-reaching and very dramatic for the history of physics. One of them, the Ampere-Maxwell law, predicts that a time-varying electric field produces a magnetic field just as a time-varying magnetic field produces an electric field (Faraday's law). From this generalization, Maxwell introduced the concept of displacement current, a new source of a magnetic field. Thus, Maxwell's theory provided the final important link between electric and magnetic fields.

EQUATIONS AND CONCEPTS

Maxwell's equations are the fundamental laws governing the behavior of electric and magnetic fields. Electromagnetic waves are a natural consequence of these laws. You should notice that the integrals in Eq. 34.1 and 34.2 are **surface integrals** in which the normal components of electric and magnetic fields are integrated over a **closed surface.**

$$\oint \mathbf{E} \cdot d\mathbf{A} = \frac{Q}{\epsilon_0} \qquad (34.1)$$

$$\oint \mathbf{B} \cdot d\mathbf{A} = 0 \qquad (34.2)$$

Unlike the first two of Maxwell's equations, Equations 34.3 and 34.4 involve line integrals in which the tangential components of electric and magnetic fields are integrated around a **closed path**.

$$\oint \mathbf{E} \cdot d\mathbf{s} = -\frac{d\Phi_B}{dt} \tag{34.3}$$

$$\text{where } \Phi_B = \int \mathbf{B} \cdot d\mathbf{A}$$

$$\oint \mathbf{B} \cdot d\mathbf{s} = \mu_0 I + \mu_0 \,\epsilon_0 \frac{d\Phi_E}{dt} \tag{34.4}$$

$$\text{where } \Phi_E = \int \mathbf{E} \cdot d\mathbf{A}$$

Both **E** and **B** satisfy a differential equation which has the form of the general wave equation. These are the wave equations for electromagnetic waves in free space (where $Q = 0$ and $I = 0$). As stated here, they represent linearly polarized waves traveling with a speed c.

$$\frac{\partial^2 E}{\partial x^2} = \mu_0 \,\epsilon_0 \frac{\partial^2 E}{\partial t^2} \tag{34.8}$$

$$\frac{\partial^2 B}{\partial x^2} = \mu_0 \,\epsilon_0 \frac{\partial^2 B}{\partial t^2} \tag{34.9}$$

$$c = \frac{1}{\sqrt{\mu_0 \,\epsilon_0}} \tag{34.10}$$

The electric and magnetic fields vary in position and time as **sinusoidal transverse waves**. Their planes of vibration are perpendicular to each other and perpendicular to the direction of propagation.

$$E = E_{max} \cos(kx - \omega t) \tag{34.11}$$

$$B = B_{max} \cos(kx - \omega t) \tag{34.12}$$

The ratio of the magnitude of the electric field to the magnitude of the magnetic field is constant and equal to the speed of light c.

$$\frac{E_{max}}{B_{max}} = \frac{E}{B} = c \tag{34.13}$$

The **Poynting vector S** describes the energy flow associated with an electromagnetic wave. The direction of **S** is along the direction of propagation and the magnitude of **S** is the rate at which electromagnetic energy crosses a unit surface area perpendicular to the direction of **S**.

$$\mathbf{S} \equiv \frac{1}{\mu_0} \mathbf{E} \times \mathbf{B} \qquad (34.18)$$

The **wave intensity** is the time average of the magnitude of the Poynting vector. E_{max} and B_{max} are the **maximum values** of the field magnitudes.

$$I = S_{av} = \frac{E_{max}B_{max}}{2\mu_0} = \frac{E_{max}^2}{2\mu_0 c} = \frac{cB_{max}^2}{2\mu_0} \qquad (34.20)$$

The electric and magnetic fields have **equal instantaneous energy densities.**

$$u_B = u_E = \tfrac{1}{2}\epsilon_0 E^2 = \frac{B^2}{2\mu_0}$$

The total instantaneous energy density u is proportional to E^2 and B^2 while the **total average energy density** is proportional to E_{max}^2 and B_{max}^2. The average energy density is also proportional to the wave intensity.

$$u = \epsilon_0 E^2 = B^2/\mu_0$$

$$u_{av} = \tfrac{1}{2}\epsilon_0 E_{max}^2 = \frac{B_{max}^2}{2\mu_0} \qquad (34.21)$$

$$I = S_{av} = cu_{av} \qquad (34.22)$$

The **linear momentum p delivered to an absorbing surface** by an electromagnetic wave at normal incidence depends on the fraction of the total energy absorbed.

$$p = \frac{U}{c} \quad \left(\begin{array}{c}\text{complete}\\\text{absorption}\end{array}\right) \qquad (34.23)$$

$$p = \frac{2U}{c} \quad \left(\begin{array}{c}\text{complete}\\\text{reflection}\end{array}\right) \qquad (34.25)$$

An absorbing surface (at normal incidence) will experience a **radiation pressure** P which depends on the magnitude of the Poynting vector and the degree of absorption.

$$P = \frac{S}{c} \qquad \left(\begin{array}{c}\text{Perfectly} \\ \text{absorbing surface}\end{array}\right) \qquad (34.24)$$

$$P = \frac{2S}{c} \qquad \left(\begin{array}{c}\text{Perfectly} \\ \text{reflecting surface}\end{array}\right) \qquad (34.26)$$

The **magnetic field** due to an infinite current sheet in the y-z plane is in the x-z plane and varies in a sinusoidal fashion according to Equation 34.27. Note that the electromagnetic wave associated with this sheet of current propagates along the x axis as a linearly polarized wave. For small values of x, the magnetic field is independent of x.

$$B_z = \frac{\mu_0 J_{max}}{2} \cos(kx - \omega t) \qquad (34.27)$$

$$B_z = \frac{\mu_0}{2} J_{max} \cos \omega t$$

$$(\text{for } x \cong 0)$$

A **radiated electric field** vibrates in the y-z plane according to Equation 34.28 and has the same space and time variations as the accompanying magnetic field.

$$E_y = \frac{\mu_0 J_{max} c}{2} \cos(kx - \omega t) \qquad (34.28)$$

There is an outgoing electromagnetic wave on each side of the infinite sheet. In each direction, the **rate of energy emission per unit area** (average intensity) is equal to the average of the Poynting vector.

$$I = S_{av} = \frac{\mu_0 J_{max}^2 c}{8} \qquad (34.30)$$

REVIEW CHECKLIST

▷ Describe the essential features of the apparatus and procedure used by Hertz in his experiments leading to the discovery and understanding of the source and nature of electromagnetic waves.

▷ For a properly described plane electromagnetic wave, calculate the values for the Poynting vector (magnitude), wave intensity, and instantaneous and average energy densities.

▷ Calculate the radiation pressure on a surface and the linear momentum delivered to a surface by an electromagnetic wave.

▷ Using the geometry of the infinite current sheet as an example, describe the relative directions and the space and time dependencies of the radiated electric and magnetic fields.

▷ Understand the production of electromagnetic waves and radiation of energy by an oscillating dipole. Use a diagram to show the relative directions for **E**, **B**, and **S** and account for the intensity of the radiated wave at points near the dipole and at distant points.

ANSWERS TO SELECTED CONCEPTUAL QUESTIONS

9. If you charge a comb by running it through your hair and then hold the comb next to a bar magnet, do the electric and magnetic fields that are produced constitute an electromagnetic wave?

Answer No. Charge on the comb creates an electric field and the bar magnet sets up a magnetic field. However, at any point these fields are constant in magnitude and direction. In an electromagnetic wave, the electric and magnetic fields must both be changing in order to create each other.

□   □   □   □

16. What does a radio wave do to the charges in the receiving antenna to provide a signal in your car radio?

Answer   Consider a typical metal rod antenna for a car radio. The rod detects the electric field portion of the carrier wave. Variations in the amplitude of the carrier wave cause the electrons in the rod to vibrate with amplitudes emulating those of the carrier wave. Likewise, for frequency modulation, the variations of the frequency of the carrier wave cause constant-amplitude vibrations of the electrons in the rod but at frequencies that imitate those of the carrier.

□   □   □   □

22. Suppose a creature from another planet had eyes that were sensitive to infrared radiation. Describe what the creature would see if it looked around the room you are now in. That is, what would be bright and what would be dim?

Answer   Light bulbs and the toaster glow brightly in the infrared. Somewhat fainter are the back of the refrigerator and the back of the television set, while the TV screen is dark. The pipes under the sink show the same weak glow as the walls until you turn on the faucets. Then the pipe on the right gets darker while that on the left develops a rich gleam that quickly runs up along its length. The food on your plate shines; so does human skin, the same color for all races. Clothing is dark as a rule, but your seat glows like a monkey's rump when you get up from a chair, and you leave a patch of the same glow on your chair. Your face appears lit from within, like a jack-o-lantern; your nostrils and openings of your ear canals are bright; brighter still are the pupils of your eyes.

□   □   □   □

SOLUTIONS TO SELECTED END-OF-CHAPTER PROBLEMS

5. Figure 34.3 shows a plane electromagnetic sinusoidal wave propagating in the x direction. The wavelength is 50.0 m, and the electric field vibrates in the xy plane with an amplitude of 22.0 V/m. Calculate (a) the sinusoidal frequency and (b) the magnitude and direction of **B** when the electric field has its maximum value in the negative y direction. (c) Write an expression for B in the form

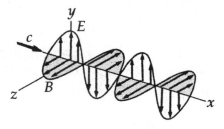

Figure 34.3a

$$B = B_{max} \cos(kx - \omega t)$$

with numerical values for B_{max}, k, and ω.

Solution

(a) $c = f\lambda$ or $f = \dfrac{c}{\lambda} = \dfrac{3.00 \times 10^8 \text{ m/s}}{50.0 \text{ m}} = 6.00 \times 10^6 \text{ Hz}$ ◊

(b) $c = \dfrac{E}{B}$ or $B = \dfrac{E}{c} = \dfrac{22.0 \text{ V/m}}{3.00 \times 10^8 \text{ m/s}} = 7.33 \times 10^{-8} \text{ T} = 73.3 \text{ nT}$ ◊

B is directed along **negative z direction** when **E** is in the negative y direction; therefore, $\mathbf{S} = \mathbf{E} \times \mathbf{B}/\mu_0$ will propagate in the direction of $(-\mathbf{j}) \times (-\mathbf{k}) = +\mathbf{i}$.

(c) $\qquad\qquad\qquad B = B_{max} \cos(kx - \omega t)$

We identify $\qquad k = \dfrac{2\pi}{\lambda} = \dfrac{2\pi}{50.0 \text{ m}} = 0.126 \text{ m}^{-1}$

and $\qquad\qquad \omega = 2\pi f = (2\pi \text{ rad})(6.00 \times 10^6 \text{ Hz}) = 3.77 \times 10^7 \text{ rad/s}$

Substituting, $\quad B = (73.3 \text{ nT}) \cos\big((0.126 \text{ rad/m})x - (3.77 \times 10^7 \text{ rad/s})t\big)$ ◊

7. In SI units, the electric field in an electromagnetic wave is described by

$$E_y = 100 \sin (1.00 \times 10^7 x - \omega t)$$

Find (a) the amplitude of the corresponding magnetic field, (b) the wavelength λ, and (c) the frequency f.

Solution

(a) $$B_{max} = \frac{E_{max}}{c} = \frac{100 \text{ V} / \text{m}}{3.00 \times 10^8 \text{ m} / \text{s}} = 3.33 \times 10^{-7} \text{ T} \qquad \Diamond$$

(b) We compare the given wave function with $y = A \sin(kx - \omega t)$ to see that

$$k = 1.00 \times 10^7 \text{ m}^{-1}$$

and $$\lambda = \frac{2\pi}{k} = \frac{2\pi}{1.00 \times 10^7 \text{ m}^{-1}}$$

$$\lambda = 6.28 \times 10^{-7} \text{ m} \qquad \Diamond$$

(c) $$f = \frac{c}{\lambda} = \frac{3.00 \times 10^8 \text{ m} / \text{s}}{6.28 \times 10^{-7} \text{ m}}$$

$$f = 4.77 \times 10^{14} \text{ Hz} \qquad \Diamond$$

13. What is the average magnitude of the Poynting vector 5.00 miles from a radio transmitter broadcasting isotropically with an average power of 250 kW?

Solution

G: As the distance from the source is increased, the power per unit area will decrease, so at a distance of 5 miles from the source, the power per unit area will be a small fraction of the Poynting vector near the source.

O: The Poynting vector is the power per unit area, where A is the surface area of a sphere with a 5-mile radius.

A: The Poynting vector is

$$S_{av} = \frac{\mathscr{P}}{A} = \frac{\mathscr{P}}{4\pi r^2}$$

In meters,

$$r = (5.00 \text{ mi})(1609 \text{ m / mi}) = 8045 \text{ m}$$

and the magnitude is

$$S = \frac{250 \times 10^3 \text{ W}}{(4\pi)(8045)^2} = 3.07 \times 10^{-4} \text{ W / m}^2 \quad \lozenge$$

L: The magnitude of the Poynting vector ten meters from the source is 199 W/m^2, on the order of a million times larger than it is 5 miles away! It is surprising to realize how little power is actually received by a radio (at the 5-mile distance, the signal would only be about 30 nW, assuming a receiving area of about 1 cm^2).

15. A community plans to build a facility to convert solar radiation to electric power. They require 1.00 MW of power, and the system to be installed has an efficiency of 30.0% (that is, 30.0% of the solar energy incident on the surface is converted to electrical energy). What must be the effective area of a perfectly absorbing surface used in such an installation, assuming a constant intensity of 1000 W/m²?

Solution

At 30.0% efficiency, the power $\mathcal{P} = 0.300SA$

$$A = \frac{\mathcal{P}}{0.300S} = \frac{1.00 \times 10^6 \text{ W}}{(0.300)(1000 \text{ W/m})} = 3330 \text{ m}^2 \quad \Diamond$$

This area is approximately $A = 0.750$ acres $\hspace{3cm} \Diamond$

---

17. The filament of an incandescent lamp has a 150-Ω resistance and carries a direct current of 1.00 A. The filament is 8.00 cm long and 0.900 mm in radius. (a) Calculate the Poynting vector at the surface of the filament. (b) Find the magnitude of the electric and magnetic fields at the surface of the filament.

Solution   In this problem, the Poynting vector does not represent the light radiated or the heat convected away from the filament. Rather, it represents the energy flow in the static electric and magnetic fields created in the surrounding empty space, by the power supply pushing current through the filament.

The rate at which the resistor converts electromagnetic energy into heat is

$$\mathcal{P} = I^2R = 150 \text{ W}$$

and the surface area is

$$A = 2\pi rL = 2\pi\left(0.900\times 10^{-3}\ \text{m}\right)\left(80.0\times 10^{-3}\ \text{m}\right)$$

$$A = 4.52\times 10^{-4}\ \text{m}^2$$

(a) The Poynting vector is $S = \dfrac{\mathcal{P}}{A} = 3.32\times 10^5\ \text{W/m}^2$ (radially inward)   ◊

(b) At the filament, $B = \mu_o \dfrac{I}{2\pi r} = \dfrac{\mu_o(1)}{2\pi(0.900\times 10^{-3})} = 2.22\times 10^{-4}\ \text{T}$ ◊

$$E = \frac{\Delta V}{\Delta x} = \frac{IR}{L} = \frac{150\ \text{V}}{0.0800\ \text{m}} = 1880\ \text{V}/\text{m}$$ ◊

Note: We could also calculate the Poynting vector from

$$S = \frac{EB}{\mu_o} = 3.32\times 10^5\ \text{W}/\text{m}^2$$ ◊

_____

25. A radio wave transmits 25.0 W/m² of power per unit area. A flat surface of area A is perpendicular to the direction of propagation of the wave. Calculate the radiation pressure on it if the surface is a perfect absorber.

Solution   For complete absorption,

$$P = \frac{S}{c} = \frac{25.0\ \text{W}/\text{m}^2}{3.00\times 10^8\ \text{m}/\text{s}} = 8.33\times 10^{-8}\ \text{N}/\text{m}^2$$ ◊

_____

29. A 15.0-mW helium-neon laser (λ = 632.8 nm) emits a beam of circular cross section with a diameter of 2.00 mm. (a) Find the maximum electric field in the beam. (b) What total energy is contained in a 1.00-m length of the beam? (c) Find the momentum carried by a 1.00-m length of the beam.

Solution The intensity of the light is the average magnitude of the Poynting vector:

$$I = \frac{\mathcal{P}}{\pi r^2} = \frac{E_{max}^2}{2\mu_o c}$$

(a) The maximum electric field is $E_{max} = \sqrt{\dfrac{\mathcal{P}(2\mu_o c)}{\pi r^2}} = 1.90 \times 10^3 \text{ N / C}$ ◊

(b) The power being 15.0 mW means that 15.0 mJ passes through a cross section of the beam in one second. This energy is uniformly spread through a beam length of 3.00×10^8 m, since that is how far the front end of the energy travels in one second. Thus, the energy in just a one-meter length is

$$\frac{15.0 \times 10^{-3} \text{ J / s}}{3.00 \times 10^8 \text{ m / s}}(1.00 \text{ m}) = 5.00 \times 10^{-11} \text{ J}$$ ◊

(c) The linear momentum carried by a 1.00 m length of the beam is the momentum that would be received by an absorbing surface, under complete absorption:

$$p = \frac{U}{c} = \frac{5.00 \times 10^{-11} \text{ J}}{3.00 \times 10^8 \text{ m / s}} = 1.67 \times 10^{-19} \text{ kg} \cdot \text{m / s}$$ ◊

33. A rectangular surface of dimensions 120 cm × 40.0 cm is parallel to and 4.40 m away from a much larger conducting sheet in which a sinusoidally varying surface current exists that has a maximum value of 10.0 A/m. (a) Calculate the average power incident on the smaller sheet. (b) What power per unit area is radiated by the larger sheet?

Solution   Solving part (b) first, we have

(b) $S_{av} = \dfrac{\mu_0 J_{max}{}^2 c}{8}$

$S_{av} = \dfrac{(4\pi \times 10^{-7} \text{ N}/\text{A}^2)(10.0 \text{ A}/\text{m})^2(3.00 \times 10^8 \text{ m}/\text{s})}{8} = 4.71 \text{ kW}/\text{m}^2$ ◊

(a) $\mathcal{P} = S_{av} A = \left(\dfrac{\mu_0 J_{max}{}^2 c}{8} \right) A$

$\mathcal{P} = \left(4.71 \text{ kW}/\text{m}^2 \right)(1.20 \text{ m})(0.400 \text{ m}) = 2.26 \times 10^3 \text{ W} = 2.26 \text{ kW}$ ◊

35. Two radio-transmitting antennas are separated by half the broadcast wavelength and are driven in phase with each other. In which directions are (a) the strongest and (b) the weakest signals radiated?

Solution

G: The strength of the radiated signal will be a function of the location around the two antennas and will depend on the interference of the waves.

O: A diagram helps to visualize this situation. The two antennas are driven in phase, which means that they both create maximum electric field strength at the same time, as shown in the diagram. The radio EM waves travel radially outwards from the antennas, and the received signal will be the vector sum of the two waves.

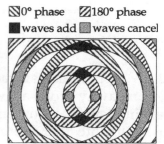

A: (a) Along the perpendicular bisector of the line joining the antennas, the distance is the same to both transmitting antennas. The transmitters oscillate in phase, so along this line the two signals will be received in phase, constructively interfering to produce a maximum signal strength that is twice the amplitude of one transmitter. ◊

(b) Along the extended line joining the sources, the wave from the more distant antenna must travel one-half wavelength farther, so the waves are received 180° out of phase. They interfere destructively to produce the weakest signal with zero amplitude. ◊

L: Radio stations may use an antenna array to direct the radiated signal toward a highly-populated region and reduce the signal strength delivered to a sparsely-populated area.

---

43. What are the wavelengths of electromagnetic waves in free space that have frequencies of (a) 5.00×10^{19} Hz and (b) 4.00×10^9 Hz?

Solution

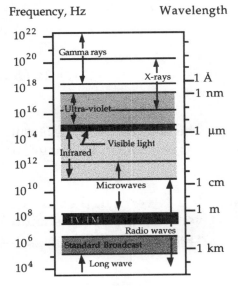

(a) $\lambda = \dfrac{c}{f} = \dfrac{3.00 \times 10^8 \text{ m/s}}{5.00 \times 10^{19} \text{ s}^{-1}} = 6.00 \text{ pm}$ ◊

This would be called an x-ray if it were emitted when an inner electron in an atom loses energy, or an electron in a vacuum tube. It would be called a gamma ray if it were radiated by an atomic nucleus.

(b) $\lambda = \dfrac{c}{f} = \dfrac{3.00 \times 10^8 \text{ m/s}}{4.00 \times 10^9 \text{ s}^{-1}} = 7.50 \text{ cm}$ ◊

By Figure 34.17, this finger-length wave is called a radio wave or a microwave.

Figure 34.17

---

51. In the absence of cable input or a satellite dish, a television set can use a dipole receiving antenna for VHF channels and a loop antenna for UHF channels (see Figure Q34.7). The UHF antenna produces an emf from the changing magnetic flux through the loop. The TV station broadcasts a signal with a frequency f, and the signal has an electric field amplitude E_{max} and magnetic field amplitude B_{max} at the location of the receiving antenna. (a) Using Faraday's law, derive an expression for the amplitude of the emf that appears in a single-turn circular loop antenna with a radius r, which is small compared to the wavelength of the wave. (b) If the electric field in the signal points vertically, what should be the orientation of the loop for best reception?

Solution

We can approximate the magnetic field as uniform over the area of the loop while it oscillates in time as $B = B_{max} \cos \omega t$. The induced voltage is

$$\mathcal{E} = -\frac{d\,\Phi_B}{dt} = -\frac{d}{dt}(BA \cos \theta) = -A\,\frac{d}{dt}(B_{max} \cos \omega t \cos \theta)$$

$$\mathcal{E} = AB_{max}\omega(\sin \omega t \cos \theta)$$

$$\mathcal{E}(t) = 2\pi f B_{max} A \sin 2\pi f t \cos \theta = 2\pi^2 r^2 f B_{max} \cos \theta \sin 2\pi f t$$

(a)   The amplitude of this emf is $\mathcal{E}_{max} = 2\pi^2 r^2 f B_{max} \cos \theta$, where θ is the angle between the magnetic field and the normal to the loop.   ◊

(b)   If **E** is vertical, then **B** is horizontal, so the plane of the loop should be vertical; the plane should contain the line of sight to the transmitter. This will make $\theta = 0°$, so $\cos \theta$ takes on its maximum value.   ◊

53. A dish antenna having a diameter of 20.0 m receives (at normal incidence) a radio signal from a distant source, as shown in Figure P34.53. The radio signal is a continuous sinusoidal wave with amplitude $E_{max} = 0.200 \ \mu V/m$. Assume the antenna absorbs all the radiation that falls on the dish. (a) What is the amplitude of the magnetic field in this wave? (b) What is the intensity of the radiation received by this antenna? (c) What power received by the antenna? (d) What force is exerted on the antenna by the radio waves?

Figure P34.53

Solution (a) $B_{max} = E_{max}/c = 6.67 \times 10^{-16} \ T$ ◊

(b) $S_{av} = E_{max}^2/2\mu_0 c = 5.31 \times 10^{-17} \ W/m^2$ ◊

(c) $\mathcal{P}_{av} = S_{av}A = 1.67 \times 10^{-14} \ W$ ◊

(Do not confuse this power with the expression for pressure $P = S/c$, which we use to find the force.)

(d) $F = PA = (S_{av}/c)A = 5.56 \times 10^{-23} \ N$ (~3000 Hydrogen atoms' weight!) ◊

---

57. In 1965, Arno Penzias and Robert Wilson discovered the cosmic microwave radiation that was left over from the Big Bang expansion of the Universe. Suppose the energy density of this background radiation is equal to $4.00 \times 10^{-14} \ J/m^3$. Determine the corresponding electric-field amplitude.

Solution $u_{av} = \frac{1}{2} \epsilon_0 E_{max}^2$ (Eq. 34.21)

$$E_{max} = \sqrt{\frac{2u_{av}}{\epsilon_0}} = \sqrt{\frac{2\left(4.00 \times 10^{-14} \ N/m^2\right)}{8.85 \times 10^{-12} \ C^2/N \cdot m^2}} = 95.1 \ mV/m \qquad ◊$$

59. A linearly polarized microwave with a wavelength of 1.50 cm is directed along the positive x axis. The electric field vector has a maximum value of 175 V/m and vibrates in the xy plane. (a) Assume that the magnetic-field component of the wave can be written in the form $B = B_{max} \sin(kx - \omega t)$ and give values for B_{max}, k, and ω. Also, determine in which plane the magnetic field vector vibrates. (b) Calculate the magnitude of the Poynting vector for this wave. (c) What maximum radiation pressure would this wave exert if directed at normal incidence onto a perfectly reflecting sheet? (d) What maximum acceleration would be imparted to a 500-g sheet (perfectly reflecting and at normal incidence) of dimensions 1.00 m × 0.750 m?

Solution $\qquad B = B_{max} \sin(kx - \omega t)$

(a) $\quad B_{max} = \dfrac{E_{max}}{c} = \dfrac{175 \text{ V} / \text{m}}{3.00 \times 10^8 \text{ m} / \text{s}} = 5.83 \times 10^{-7} \text{ T}$ $\qquad\qquad\qquad\quad \Diamond$

$\quad k = \dfrac{2\pi}{\lambda} = \dfrac{2\pi}{0.015 \text{ m}} = 419 \text{ m}^{-1}$ $\qquad\qquad\qquad\qquad\qquad\quad \Diamond$

$\quad \omega = kc = (419 \text{ m}^{-1})(3.00 \times 10^8 \text{ m} / \text{s}) = 1.26 \times 10^{11} \text{ rad} / \text{s}$ $\qquad \Diamond$

The magnetic field is in the z direction so that $\mathbf{S}_{av} = \dfrac{1}{\mu_o} \mathbf{E} \times \mathbf{B}$ can be in the

direction of $\hat{\jmath} \times \hat{k} = \hat{\imath}$. $\qquad\qquad\qquad\qquad\qquad\qquad\qquad\qquad\qquad\qquad\quad \Diamond$

(b) $\quad S_{av} = \dfrac{E_{max} B_{max}}{2\mu_o} = \dfrac{(175 \text{ V} / \text{m})(5.83 \times 10^{-7} \text{ T})}{2 \times 4\pi \times 10^{-7} \text{ N} / \text{A}^2} = 40.6 \text{ W} / \text{m}^2$ $\qquad \Diamond$

(c) For perfect reflection,

$\quad P = \dfrac{2S}{c} = \dfrac{(2)(40.6 \text{ W} / \text{m}^2)}{3.00 \times 10^8 \text{ m} / \text{s}} = 2.71 \times 10^{-7} \text{ N} / \text{m}^2$ $\qquad\qquad\qquad \Diamond$

(d) $\quad a = \dfrac{F}{m} = \dfrac{PA}{m} = \dfrac{(2.71 \times 10^{-7} \text{ N} / \text{m}^2)(0.750 \text{ m}^2)}{0.500 \text{ kg}} = 4.06 \times 10^{-7} \text{ m} / \text{s}^2$ $\qquad \Diamond$

61. An astronaut, stranded in space 10.0 m from his spacecraft and at rest relative to it, has a mass (including equipment) of 110 kg. Since he has a 100-W light source that forms a directed beam, he decides to use the beam as a photon rocket to propel himself continuously toward the spacecraft. (a) Calculate how long it takes him to reach the spacecraft by this method. (b) Suppose, instead, he decides to throw the light source away in a direction opposite the spacecraft. If the mass of the light source has a mass of 3.00 kg and, after being thrown, moves at 12.0 m/s **relative to the recoiling astronaut**, how long does it take for the astronaut to reach the spacecraft?

Solution

G: Based on our everyday experience, the force exerted by photons is too small to feel, so it may take a very long time (maybe days!) for the astronaut to travel 10 m with his "photon rocket." Using the momentum of the thrown light seems like a better solution, but it will still take a while (maybe a few minutes) for the astronaut to reach the spacecraft because his mass is so much larger than the mass of the light source.

O: In part (a), the radiation pressure can be used to find the force that accelerates the astronaut toward the spacecraft. In part (b), the principle of conservation of momentum can be applied to find the time required to travel the 10 m.

A : (a) Light exerts on the astronaut a pressure $P = F/A = S/c$, and a force of

$$F = \frac{SA}{c} = \frac{\mathcal{P}}{c} = \frac{100 \text{ J}/\text{s}}{3.00 \times 10^8 \text{ m}/\text{s}} = 3.33 \times 10^{-7} \text{ N}$$

By Newton's 2nd law, $\qquad a = \dfrac{F}{m} = \dfrac{3.33 \times 10^{-7} \text{ N}}{110 \text{ kg}} = 3.03 \times 10^{-9} \text{ m}/\text{s}^2$

This acceleration is constant, so the distance traveled is $x = \frac{1}{2}at^2$, and the amount of time it travels is

$$t = \sqrt{\frac{2x}{a}} = \sqrt{\frac{2(10.0 \text{ m})}{3.03 \times 10^{-9} \text{ m/s}^2}} = 8.12 \times 10^4 \text{ s} = 22.6 \text{ h} \qquad \Diamond$$

(b) Because there are no external forces, the momentum of the astronaut before throwing the light is the same as afterwards when the now 107-kg astronaut is moving at speed v towards the spacecraft and the light is moving away from the spacecraft at $(12.0 \text{ m/s} - v)$. Thus, $\mathbf{p}_i = \mathbf{p}_f$ gives

$$0 - (107 \text{ kg})v - (3.00 \text{ kg})(12.0 \text{ m/s} - v)$$

$$0 - (107 \text{ kg})v - (36.0 \text{ kg} \cdot \text{m/s}) + (3.00 \text{ kg})v$$

$$v = \frac{36.0}{110} = 0.327 \text{ m/s}$$

$$t = \frac{x}{v} = \frac{10.0 \text{ m}}{0.327 \text{ m/s}} = 30.6 \text{ s} \qquad \Diamond$$

L: Throwing the light away is certainly a more expedient way to reach the spacecraft, but there is not much chance of retrieving the lamp unless it has a very long cord. How long would the cord need to be, and does its length depend on how hard the astronaut throws the lamp? (You should verify that the minimum cord length is 367 m, independent of the speed that the lamp is thrown.)

Chapter
35

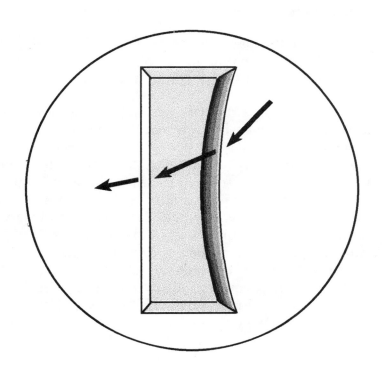

The Nature of Light and
The Laws of Geometric
Optics

THE NATURE OF LIGHT AND THE LAWS OF GEOMETRIC OPTICS

INTRODUCTION

The chief architect of the particle theory of light was Newton. With this theory he provided simple explanations of some known experimental facts concerning the nature of light, namely the laws of reflection and refraction.

In 1678 a Dutch physicist and astronomer, Christian Huygens (1629 - 1695), showed that a wave theory of light could also explain the laws of reflection and refraction. The wave theory did not receive immediate acceptance for several reasons. All the waves known at the time (sound, water, and so on) traveled through some sort of medium, but light from the Sun could travel to Earth through empty space. Furthermore, it was argued that if light were some form of wave, it would bend around obstacles; hence, we should be able to see around corners. It is now known that light does indeed bend around the edges of objects. This phenomenon, known as **diffraction**, is not easy to observe because light waves have such short wavelengths. For more than a century most scientists rejected the wave theory and adhered to Newton's particle theory. This was, for the most part, due to Newton's great reputation as a scientist.

EQUATIONS AND CONCEPTS

Consider the situation in the figures to the right, in which a light ray is incident obliquely on a smooth, planar surface which forms the boundary between two transparent media of different optical densities.

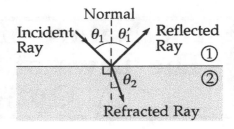

A portion of each incoming ray will be reflected back into the original medium, while the remaining fraction will be transmitted into the second medium. For light waves, the direction of each wave's motion is perpendicular to its wave front.

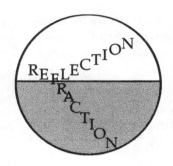

The energy of a photon is proportional to the frequency of the associated electromagnetic wave.

$$E = hf \qquad (35.1)$$

The ratio between the energy and the frequency is known as Planck's constant.

$$h = 6.63 \times 10^{-34} \text{ J} \cdot \text{s}$$

The law of reflection states that the angle of incidence (the angle measured between the incident ray and the normal line) equals the angle of reflection (the angle measured between the reflected ray and the normal line).

$$\theta_1' = \theta_1 \qquad (35.2)$$

This is one form of the statement of Snell's law. The angle of refraction (measured relative to the normal line) depends on the angle of incidence, and also on the ratio of the speeds of light in the two media on either side of the refracting surface.

$$\frac{\sin \theta_2}{\sin \theta_1} = \frac{v_2}{v_1} = \text{constant} \qquad (35.3)$$

This is the most widely used and most practical form of Snell's law. This equation involves a parameter called the index of refraction, the value of which is characteristic of a particular medium. The index of refraction is defined in Equation 35.4 and Equation 35.7.

$$n_1 \sin \theta_1 = n_2 \sin \theta_2 \qquad (35.8)$$

Each transparent medium is characterized by a dimensionless number, the index of refraction, which equals the ratio of the speed of light in the medium to the speed of light in vacuum.

$$n \equiv \frac{\text{Speed of light in vacuum}}{\text{Speed of light in a medium}} = \frac{c}{v} \qquad (35.4)$$

The frequency of a wave is characteristic of the source. Therefore, as light travels from one medium into another of different index of refraction, the frequency remains constant but the wavelength changes. The index of refraction of a given medium can be expressed as the ratio of the wavelength of light in vacuum to the wavelength in that medium.

$$n = \frac{\lambda}{\lambda_n} \qquad (35.7)$$

For angles of incidence equal to or greater than the critical angle, the incident ray will be totally internally reflected back into the first medium.

$$\sin \theta_c = \frac{n_2}{n_1} \qquad (\text{for } n_1 > n_2) \qquad (35.10)$$

Total internal reflection is possible only when a light ray is directed from a medium of high index of refraction into a medium of lower index of refraction.

REVIEW CHECKLIST

▷ Understand Huygens' principle and the use of this technique to construct the subsequent position and shape of a given wave front.

▷ Describe the methods used by Roemer and Fizeau for the measurement of c and make calculations using sets of typical values for the quantities involved.

▷ Determine the directions of the reflected and refracted rays when a light ray is incident obliquely on the interface between two optical media.

▷ Understand the manner in which Fermat's principle of least time can be used as a basis of a derivation of the laws of reflection and refraction.

▷ Understand the conditions under which total internal reflection can occur in a medium and determine the critical angle for a given pair of adjacent media.

ANSWERS TO SELECTED CONCEPTUAL QUESTIONS

4. As light travels from one medium to another, does the wavelength of the light change? Does the frequency change? Does the speed change? Explain.

Answer You can think about this based upon the fact that the light wave must be continuous. Therefore, the frequency of the light must remain constant, since any one crest coming up to the interface must create one crest in the new medium. However, the speed of light does vary in different media, as described by Snell's Law. In addition, since the speed of the light wave is equal to the light wave's frequency times its wavelength, and the frequency does not change, we can be sure that the wavelength must change. The speed decreases upon entering the new medium; therefore the wavelength decreases as well.

□   □   □   □

11. Explain why a diamond sparkles more than a glass crystal of the same shape and size.

Answer Diamond has a larger index of refraction than glass, and consequently has a smaller critical angle for internal reflection. This results in a greater amount of light being internally reflected, and more light being emitted within the critical angles.

□   □   □   □

15. Why do astronomers looking at distant galaxies talk about looking backward in time?

Answer Light travels through a vacuum at a speed of 300 000 km per second. Thus, an image we see from a distant star or galaxy must have been generated some time ago.

For example, the star Altair is 16 light-years away; if we look at an image of Altair today, we know only what was happening 16 years ago.

This may not initially seem significant, but astronomers who look at other galaxies can gain an idea of what galaxies looked like when they were significantly younger. Thus, it actually makes sense to speak of "looking backward in time."

□   □   □   □

SOLUTIONS TO SELECTED END-OF-CHAPTER PROBLEMS

3. In an experiment to measure the speed of light using the apparatus of Fizeau (see Fig. P35.2), the distance between light source and mirror was 11.45 km and the wheel had 720 notches. The experimentally determined value of c was 2.998×10^8 m/s. Calculate the minimum angular speed of the wheel for this experiment.

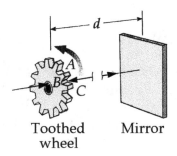

Figure P35.2

Solution

The distance down to the mirror and back is

$$2 \times 11.45 \times 10^3 \text{ m} = 2.290 \times 10^4 \text{ m}$$

The round-trip time is expected to be

$$t = \frac{x}{v} = \frac{2.290 \times 10^4 \text{ m}}{2.998 \times 10^8 \text{ m/s}} = 7.64 \times 10^{-5} \text{ s}$$

In this time the wheel should turn by $\frac{1}{2}(1/720)$ rev, to move a tooth into the place of a notch. Its angular speed should be

$$\omega = \frac{\theta}{t} = \frac{1}{2}\left(\frac{1 \text{ rev}}{720}\right)\left(\frac{1}{7.64 \times 10^{-5} \text{ s}}\right) = 9.09 \text{ rev/s} \qquad \Diamond$$

Higher angular speeds must be available so that the experimenter can home in on the special dark setting from both sides. The demonstration is most convincing if the wheel turns at twice this angular speed, to replace a notch with another notch during the light's travel time, restoring the returning light to full brightness.

7.   An underwater scuba diver sees the Sun at an apparent angle of 45.0°
from the vertical. What is the actual direction of the Sun?

Solution

G: The sunlight refracts as it enters the water from the
air. Because the water has a higher index of
refraction, the light slows down and bends toward
the vertical line that is normal to the interface.
Therefore, the elevation angle of the Sun above the
water will be less than 45° as shown in the diagram
to the right, even though it appears to the diver that
the sun is 45° above the horizon.

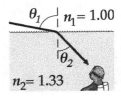

O: We can use Snell's law of refraction to find the precise angle of
incidence.

A: Snell's law is: $n_1 \sin\theta_1 = n_2 \sin\theta_2$

which gives $\sin\theta_1 = 1.33 \sin 45.0°$

$\sin\theta_1 = (1.333)(0.707) = 0.943$

The sunlight is at $\theta_1 = 70.5°$ to the vertical,
so the Sun is 19.5° above the horizon.   ◊

L: The calculated result agrees with our prediction. When applying Snell's
law, it is easy to mix up the index values and to confuse angles-with-
the-normal and angles-with-the-surface.   Making a sketch and a
prediction as we did here helps avoid careless mistakes.

11. A ray of light strikes a flat block of glass ($n = 1.50$) of thickness 2.00 cm at an angle of 30.0° with the normal. Trace the light beam through the glass, and find the angles of incidence and refraction at each surface.

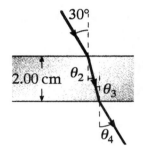

Solution

At entry, $\qquad n_1 \sin\theta_1 = n_2 \sin\theta_2$

$$(1.00)\sin 30° = (1.50)\sin\theta_2$$

and $\qquad \theta_2 = \sin^{-1}\left(\dfrac{0.500}{1.50}\right) = 19.5°$ $\qquad\qquad \Diamond$

To do geometrical optics, you must remember some geometry. The surfaces of entry and exit are parallel so their normals are parallel. Then angle θ_2 of refraction at entry and the angle θ_3 of incidence at exit are alternate interior angles formed by the ray as a transversal cutting parallel lines.

So $\qquad\qquad \theta_3 = \theta_2 = 19.5°$ $\qquad\qquad\qquad\qquad \Diamond$

At the exit, $\qquad n_2 \sin\theta_3 = n_1 \sin\theta_4$

$$(1.50)\sin 19.5° = (1.00)\sin\theta_4 \qquad \text{and} \qquad \theta_4 = 30.0° \qquad \Diamond$$

The exiting ray in air is parallel to the original ray in air.

27. A prism that has an apex angle of 50.0° is made of cubic zirconia, with $n = 2.20$. What is its angle of minimum deviation?

Solution   From Equation 35.9,

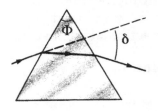

$$n = \frac{\sin\left(\dfrac{\Phi + \delta_{min}}{2}\right)}{\sin\left(\dfrac{\Phi}{2}\right)}$$

Solving for δ_{min}, $\delta_{min} = 2\,\sin^{-1}\left(n\sin\dfrac{\Phi}{2}\right) - \Phi$

$$\delta_{min} = 2\,\sin^{-1}(2.20\,\sin\,25.0°) - 50.0° = 86.8° \qquad \lozenge$$

---

29.   The index of refraction for violet light in silica flint glass is 1.66, and that for red light is 1.62. What is the angular dispersion of visible light passing through a prism of apex angle 60.0° if the angle of incidence is 50.0°? (See Fig. P35.29.)

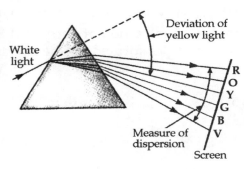

Figure P35.29

Solution   The drawing at the right follows a monochromatic ray through the prism. From the geometry, observe that

$\beta = 90.0° - \alpha$ and $\gamma = 180.0° - 60.0° - \beta$.

Combining these equations gives
$\gamma = 180.0° - 60.0° - 90.0° + \alpha = 30.0° + \alpha$

Also, observe that $\theta = 90.0° - \gamma = 90.0° - 30.0° - \alpha$

or $\theta = 60.0° - \alpha$

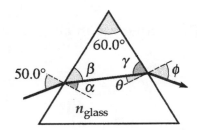

Note that in the previous equation, α is the angle of refraction at the first surface, and θ is the angle of incidence at the second surface.

For red light, the index of refraction of the glass is $n_{glass} = 1.62$. Applying Snell's law to the refraction of red light at the first surface,

$$\sin \alpha_R = \left(\frac{n_{air}}{n_{glass}} \right) \sin 50.0° = \left(\frac{1.00}{1.62} \right) \sin 50.0° = 0.473 \quad \text{or} \quad \alpha_R = 28.22°$$

Thus, $\theta_R = 60° - \alpha_R = 60° - 28.22° = 31.78°$ is the angle of incidence at the second surface. Then, applying Snell's law to the refraction at the second surface gives

$$\sin \phi_R = \left(\frac{n_{glass}}{n_{air}} \right) \sin \theta_R = \left(\frac{1.62}{1.00} \right) \sin 31.8° = 0.853$$

The angle of refraction of the red light at the second surface is therefore $\phi_R = 58.56°$. Similarly, the index of refraction of the glass for violet light is $n_{glass} = 1.66$ and application of Snell's law at the first surface gives

$$\sin \alpha_V = \left(\frac{1.00}{1.66} \right) \sin 50.0° = 0.461 \quad \text{or} \quad \alpha_V = 27.48°$$

Then, $\theta_V = 60° - 27.48° = 32.52°$, and application of Snell's law at the second surface for the violet light gives $\sin \phi_V = (1.66 / 1.00) \sin 32.52° = 0.892$.

The angle of refraction of the violet light at the second surface is then $\phi_V = 63.17°$. The total angular dispersion of visible light incident on this prism at an angle of incidence of 50.0° is then

$$\Delta \phi = \phi_V - \phi_R = 63.17° - 58.56° = 4.61° \qquad \lozenge$$

31. A triangular glass prism with an apex angle 60.0° has an index of refraction $n = 1.50$. What is the smallest angle of incidence θ_1 for which a light ray can emerge from the other side?

32. A triangular glass prism with an apex angle of Φ has an index of refraction n. What is the smallest angle of incidence θ_1 for which a light ray can emerge from the other side?

Solution

Call the angles of incidence and refraction, at the surfaces of entry and exit, θ_1, θ_2, θ_2', and θ_3, in order as shown. The apex angle ϕ is the angle between the surfaces of entry and exit. The ray in the glass forms a triangle with these surfaces, in which the interior angles must add to 180°. Thus, with

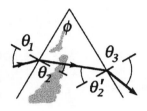

$$\phi = 60.0°,$$

$$(90 - \theta_2) + 60 + (90 - \theta_2') = 180$$

so $\theta_2 + \theta_2' = 60.0°$ (1)

$$\phi = \Phi,$$

$$(90 - \theta_2) + \Phi + (90 - \theta_2') = 180$$

so $\theta_2 + \theta_2' = \Phi$ (1)

which is a general rule for light going through prisms. At the first refraction, Snell's law gives

$$\sin \theta_1 = 1.50 \sin \theta_2$$ (2)

which is a general rule for light going through prisms. At the first interface between the air and prism,

$$\sin \theta_1 = n \sin \theta_2$$ (2)

At the second boundary, we want total internal reflection:

$$1.50 \sin \theta_2' = 1.00 \sin 90° = 1.00$$

or $\theta_2' = \sin^{-1}(1.00 / 1.50) = 41.8°$

At the second interface, we want total internal reflection:

$$n \sin \theta_2' = \sin 90° = 1$$

or $\theta_2' = \sin^{-1}(1/n)$

Now by equation (1) above,

$$\theta_2 = 60.0° - 41.8° = 18.2°$$

Now by equation (1) above,

$$\theta_2 = \Phi - \theta_2' = \Phi - \sin^{-1}(1/n)$$

while by equation (2), we find that

$$\theta_1 = \sin^{-1}(1.50 \sin 18.2°)$$

So $\theta_1 = 27.9°$ ◊

Note that in the case of this problem, the numeric solution eliminated several of the analytic steps of the variable solution, and thus was much shorter. Whether a given problem is best solved analytically or numerically depends on the complexity of the problem.

---

while by equation (2), we find that

$$\theta_1 = \sin^{-1}\left(n \sin\left(\Phi - \sin^{-1}(1/n)\right)\right)$$

Looking at the drawing of a right triangle, remember the identities that

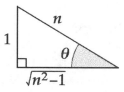

$$\sin(\alpha - \beta) = \sin\alpha\cos\beta - \cos\alpha\sin\beta$$

$$\cos\left(\sin^{-1}(1/n)\right) = \sqrt{n^2 - 1}/n$$

θ_1 therefore simplifies to

$$\theta_1 = \sin^{-1}\left(\cos\Phi + \left(\sqrt{n^2-1}\right)\sin\Phi\right)$$ ◊

---

37. Consider a common mirage formed by superheated air just above a roadway. A truck driver whose eyes are 2.00 m above the road, where $n = 1.0003$, looks forward. She perceives the illusion of a patch of water ahead on the road, where her line of sight makes an angle of 1.20° below the horizontal. Find the index of refraction of the air just above the road surface. (**Hint:** Treat this problem as a problem in total internal reflection.)

Solution   Think of the air as in two discrete layers, the first medium being cooler air with $n_1 = 1.0003$ and the second medium being hot air with a lower index, which reflects light from the sky by total internal reflection. Use $n_1 \sin\theta_1 \geq n_2 \sin 90°$:

$$1.0003 \sin 88.8° \geq n_2 \qquad \text{or} \qquad n_2 \leq 1.00008 \qquad ◊$$

47. A small underwater pool light 1.00 m below the surface. The light emerging from the water forms a circle on the water's surface. What is the diameter of this circle?

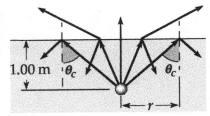

Solution

G: Only the light that is directed upwards and hits the water's surface at less than the critical angle will be transmitted to the air so that someone outside cold see it. The light that hits the surface farther from the center at an angle greater than θ_c will be totally reflected within the water, unable to be seen from the outside. From the diagram to the right, the diameter of this circle of light appears to be about 2 m.

O: We can apply Snell's law to find the critical angle, and the diameter can then be found from the geometry.

A: The critical angle is found when the refracted ray just grazes the surface ($\theta_2 = 90°$). The index of refraction of water is $n_2 = 1.33$, and $n_1 = 1.00$ for air, so

$$n_1 \sin\theta_c = n_2 \sin 90° \quad \text{gives} \quad \theta_c = \sin^{-1}\left(\frac{1}{1.333}\right) = \sin^{-1}(0.750) = 48.6°$$

The radius then satisfies $$\tan\theta_c = \frac{r}{(1.00 \text{ m})}$$

So the diameter is $$d = 2r = 2(1.00 \text{ m})\tan 48.6° = 2.27 \text{ m} \quad \Diamond$$

L: Only the light rays within a 97.2° cone above the lamp escape the water and can be seen by an outside observer (Note: this angle does not depend on the depth of the light source). The path of a light ray is always reversible, so if a person were located beneath the water, they could see the whole hemisphere above the water surface within this cone; this is a good experiment to try the next time you go swimming!

53. A hiker stands on a mountain peak near sunset and observes a rainbow caused by water droplets in the air 8.00 km away. The valley is 2.00 km below the mountain peak and entirely flat. What fraction of the complete circular arc of the rainbow is visible to the hiker? (See Figure 35.25.)

Solution Horizontal light rays from the setting Sun pass above the hiker. The light rays are twice refracted and once reflected, as in Figure 35.24 of the text, by just the certain special raindrops at 40.0° to 42.0° from the hiker's shadow, and reach the hiker as the rainbow.

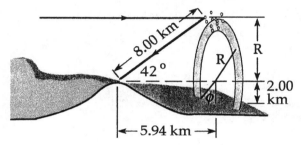

The hiker sees a greater percentage of the violet inner edge, so we consider the red outer edge. The radius R of the circle of droplets is

$$R = (8.00 \text{ km})(\sin 42.0°) = 5.35 \text{ km}$$

Then the angle ϕ, between the vertical and the radius where the bow touches the ground, is given by

Figure 35.24

$$\cos \phi = \frac{2.00 \text{ km}}{R} = \frac{2.00 \text{ km}}{5.35 \text{ km}} = 0.374 \qquad \text{or} \qquad \phi = 68.1°$$

The angle filled by the visible bow is $360° - (2 \times 68.1°) = 224°$, so the visible bow is

$$224°/360° = 62.2\% \text{ of a circle} \qquad\qquad ◊$$

This striking view motivated Charles Wilson's 1906 invention of the cloud chamber, a standard tool of nuclear physics. Look for a full-circle rainbow around your shadow when you fly in an airplane.

55. A laser beam strikes one end of a slab of material, as in Figure P35.55. The index of refraction of the slab is 1.48. Determine the number of internal reflections of the beam before it emerges from the opposite end of the slab.

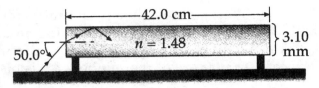

Figure P35.55

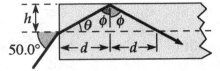

Solution   As the beam enters the end of the slab, the angle of refraction is found from Snell's law:

$$\sin \theta = \frac{n_{\text{air}}}{n_{\text{slab}}} \sin 50.0° = \left(\frac{1.00}{1.48}\right) \sin 50.0° = 0.518 \qquad \text{and} \qquad \theta = 31.2°$$

Note that the two normal lines are perpendicular to each other. Using the right triangle having these lines as two of its sides, observe that the angle of incidence at the top of the slab is $\phi = 90.0° - \theta = 58.8°$. The critical angle as the light tries to go from the slab back into air is

$$\theta_c = \arcsin\left(\frac{n_{\text{air}}}{n_{\text{slab}}}\right) = \arcsin\left(\frac{1.00}{1.48}\right) = 42.5°$$

Since $\phi > \theta_c$, total internal reflection will indeed occur at the top and bottom of the slab. The distance the beam travels down the length of the slab for each reflection is $2d$, where d is the base of the right triangle shown in the sketch. Given that the altitude of the triangle, h, is one half the thickness of the slab,

$$d = \frac{h}{\tan \theta} = \frac{(3.10 \text{ mm})/2}{\tan \theta} \qquad \text{and} \qquad 2d = \frac{3.10 \times 10^{-1} \text{ cm}}{\tan 31.2°} = 5.12 \times 10^{-1} \text{ cm}$$

The number of internal reflections made before reaching the opposite end of the slab is then

$$N = \frac{\text{length of slab}}{2d} = \frac{42.0 \text{ cm}}{5.12 \times 10^{-1} \text{ cm}} = 82 \text{ reflections} \qquad \diamond$$

59. The light beam shown in Figure P35.59 strikes surface 2 at the critical angle. Determine the angle of incidence θ_1.

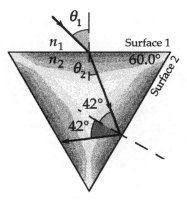

Figure P35.59

Solution

G: From the diagram it appears that the angle of incidence is about 40°.

O: We can find θ_1 by applying Snell's law at the first interface where the light is refracted. At surface 2, knowing that the 42.0° angle of reflection is the critical angle, we can work backwards to find θ_1.

A: Define n_1 to be the index of refraction of the surrounding medium and n_2 to be that for the prism material. We can use the critical angle of 42.0° to find the ratio n_2/n_1:

$$n_2 \sin 42.0° = n_1 \sin 90.0°$$

So,
$$\frac{n_2}{n_1} = \frac{1}{\sin 42.0°} = 1.49$$

Call the angle of refraction θ_2 at the surface 1. The ray inside the prism forms a triangle with surfaces 1 and 2, so the sum of the interior angles of this triangle must be 180°. Thus,

$$(90.0° - \theta_2) + 60.0° + (90.0° - 42.0°) = 180°$$

Therefore, $\theta_2 = 18.0°$

Applying Snell's law at surface 1, $n_1 \sin \theta_1 = n_2 \sin 18.0°$

$$\sin \theta_1 = (n_2/n_1)\sin \theta_2 = (1.49)\sin 18.0°$$

$$\theta_1 = 27.5° \qquad \qquad \lozenge$$

L: The result is a bit less than the 40.0° we expected, but this is probably because the figure is not drawn to scale. This problem was a bit tricky because it required four key concepts (refraction, reflection, critical angle, and geometry) in order to find the solution. One practical extension of this problem is to consider what would happen to the exiting light if the angle of incidence were varied slightly. Would all the light still be reflected off surface 2, or would some light be refracted and pass through this second surface?

---

61. A light ray of wavelength 589 nm is incident at an angle θ on the top surface of a block of polystyrene, as shown in Figure P35.61. (a) Find the maximum value of θ for which the refracted ray undergoes total internal reflection at the left vertical face of the block. Repeat the calculation for the case in which the polystyrene block is immersed in (b) water and (c) carbon disulfide.

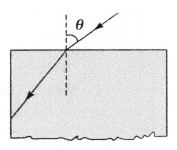

Figure P35.61

Solution

The index of refraction (for 589 nm light) of each material is listed below:

Air:	1.00
Water:	1.33
Polystyrene:	1.49
Carbon disulfide:	1.63

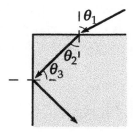

(a) For polystyrene **surrounded by air**,

internal reflection requires $\theta_3 = \sin^{-1}(1/1.49) = 42.2°$

Then from the geometry, $\qquad\qquad\qquad\qquad\qquad \theta_2 = 90.0° - \theta_3 = 47.8°$

From Snell's law, $\qquad\qquad\qquad\qquad\qquad \sin\theta_1 = (1.49)\sin 47.8°$

There is no solution: no angle has a sine grater than one. Thus, the real maximum value for θ_1 is 90.0°: total internal reflection always occurs. ◊

(b) For polystyrene **surrounded by water,** $\qquad \theta_3 = \sin^{-1}\left(\dfrac{1.33}{1.49}\right) = 63.2°$

and $\qquad\qquad\qquad\qquad\qquad\qquad\qquad \theta_2 = 26.8°$

From Snell's law, $n_1\sin\theta_1 = n_2\sin\theta_2$: $\qquad 1.33\sin\theta_1 = 1.49\sin 26.8°$

and $\qquad\qquad\qquad\qquad\qquad\qquad\qquad \theta_1 = 30.3° \qquad\qquad\qquad\qquad ◊$

(c) **This is not possible** since the beam is initially traveling in a medium of lower index of refraction.

---

63. A shallow glass dish is 4.00 cm wide at the bottom, as shown in Figure P35.63. When an observer's eye is positioned as shown, the observer sees the edge of the bottom of the empty dish. When this dish is filled with water, the observer sees the center of the bottom of the dish. Find the height of the dish.

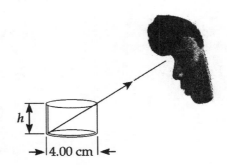

Figure P35.63

Chapter 35

Solution Refer to the diagram showing the two light rays (refracted and unrefracted). After passing the rim of the dish, the refracted ray travels half the horizontal distance of the unrefracted ray. Writing this in equation form and squaring both sides,

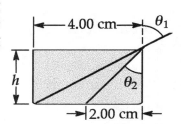

$$\tan\theta_1 = \frac{4.00 \text{ cm}}{h} \quad \text{and} \quad \tan\theta_2 = \frac{2.00 \text{ cm}}{h}$$

so that

$$\tan^2\theta_1 = \left(2.00\tan\theta_2\right)^2 = 4.00\tan^2\theta_2$$

In terms of sines, this is

$$\frac{\sin^2\theta_1}{(1-\sin^2\theta_1)} = 4.00\frac{\sin^2\theta_2}{(1-\sin^2\theta_2)} \qquad (1)$$

Snell's law in this case is: $n_1\sin\theta_1 = n_2\sin\theta_2$ or $\sin\theta_1 = 1.333\ \sin\theta_2$

Squaring both sides, $\sin^2\theta_1 = 1.777\ \sin^2\theta_2$ (2)

Substituting (2) into (1),

$$\frac{1.777\ \sin^2\theta_2}{(1-1.777\ \sin^2\theta_2)} = 4.00\frac{\sin^2\theta_2}{(1-\sin^2\theta_2)}$$

Defining $x = \sin^2\theta_2$,

$$\frac{0.444}{(1-1.777x)} = \frac{1}{(1-x)}$$

Solving for x, $0.444 - 0.444x = 1 - 1.777x$ and $x = 0.417$

From x we solve for θ_2: $\theta_2 = \sin^{-1}\sqrt{0.417} = 40.2°$

and

$$h = \frac{(2.00 \text{ cm})}{\tan\theta_2} = \frac{(2.00 \text{ cm})}{\tan(40.2°)} = 2.37 \text{ cm} \qquad \lozenge$$

65. Derive the law of reflection (Eq. 35.2) from Fermat's principle of least time. (See the procedure outlined in Section 35.9 for the derivation of the law of refraction from Fermat's principle.)

Solution

To derive the law of **reflection**, locate point O so that the time of travel from point A to point B will be minimum. Let $c + d = R$, so that we can replace d with $R - c$. Since the light travels through the same medium for both paths, you only need to minimize the distance traveled, D.

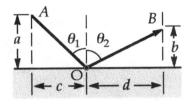

$$D = \sqrt{a^2 + c^2} + \sqrt{b^2 + (R - c)^2}$$

Require that $\dfrac{dD}{dc} = 0$:

$$\frac{2c}{\sqrt{a^2 + c^2}} - \frac{2(R - c)}{\sqrt{b^2 + (R - c)^2}} = 0$$

By inspecting the triangles of the diagram, it follows that

$$\sin\theta_1 = \frac{c}{\sqrt{a^2 + b^2}} \qquad \text{and} \qquad \sin\theta_2 = \frac{R - c}{\sqrt{b^2 + (R - c)^2}}$$

Substituting these values, $\qquad \sin\theta_1 = \sin\theta_2$

Therefore, $\qquad\qquad\qquad \theta_1 = \theta_2 \qquad\qquad \lozenge$

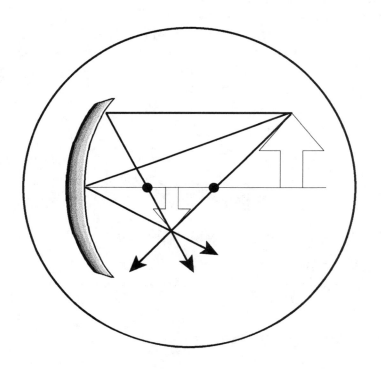

Geometric Optics

GEOMETRIC OPTICS

INTRODUCTION

This chapter is concerned with the images formed when spherical waves fall on flat and spherical surfaces. We find that images can be formed by reflection or by refraction and that mirrors and lenses work because of this reflection and refraction. Such devices, commonly used in optical instruments and systems, are described in detail. In this chapter, we continue to use the ray approximation and to assume that light travels in straight lines, both steps valid because here we are studying the field called **geometric optics**. In subsequent chapters, we shall concern ourselves with interference and diffraction effects, or the field of **wave optics.**

EQUATIONS AND CONCEPTS

The **mirror equation** is used to locate the position of an image formed by reflection of paraxial rays. The focal point of a spherical mirror is located midway between the center of curvature and the vertex of the mirror.

$$\frac{1}{p} + \frac{1}{q} = \frac{1}{f} \tag{36.6}$$

$$f = \frac{R}{2} \tag{36.5}$$

In using the equations related to the image-forming properties of spherical mirrors, spherical refracting surfaces, and thin lenses, you must be very careful to use the correct algebraic sign for each physical quantity. The sign conventions appropriate for the equation forms stated here are summarized in the SKILLS section.

The **lateral magnification** of a spherical mirror can be stated either as a ratio of image size to object size or in terms of the ratio of image distance to object distance.

$$M = \frac{h'}{h} = -\frac{q}{p} \qquad (36.2)$$

A magnified image of an object can be formed by a single spherical refracting surface of radius R which separates two media whose indices of refraction are n_1 and n_2.

$$\frac{n_1}{p} + \frac{n_2}{q} = \frac{n_2 - n_1}{R} \qquad (36.8)$$

A special case is that of the virtual image formed by a **planar refracting surface** $(R = \infty)$.

$$\frac{n_1}{p} = -\frac{n_2}{q}$$

$$q = -\frac{n_2}{n_1} p \qquad (36.9)$$

The **thin lens** is an important component in many optical instruments. The location of the image formed by a given object is determined by the characteristic properties of the lens (index of refraction n and radii of curvature R). If the lens is surrounded by a medium other than air, the index of refraction given in Equation 36.10 must be the **index of the lens relative to the surrounding medium.**

$$\frac{1}{p} + \frac{1}{q} = (n - 1)\left(\frac{1}{R_1} - \frac{1}{R_2}\right) \qquad (36.10)$$

$$\frac{1}{p} + \frac{1}{q} = \frac{1}{f} \qquad (36.12)$$

This means, for example, that a hollow biconvex lens ("air lens"), if immersed in water, would have a **negative** focal length. The lateral magnification of a thin lens has the same form as that of a spherical mirror (Eq. 36.2).

Thin lenses are often used in combination. A special case occurs when two thin lenses are in contact. The **focal length of the combination** given by Equation 36.13 will be **less** than that of either lens individually.

$$\frac{1}{f} = \frac{1}{f_1} + \frac{1}{f_2}$$ (36.13)

Camera: The **light intensity** I incident on the film per unit area is inversely proportional to the square of a ratio of the diameter of the lens to its focal length. The **f-number** equals the ratio of the focal length to the lens diameter.

$$I \propto \frac{1}{(f/D)^2} \propto \frac{1}{(f-\text{number})^2}$$ (36.15)

$$f-\text{number} \equiv \frac{f}{D}$$ (36.14)

Eye: The power of a lens in diopters is the reciprocal of the focal length measured in meters (including the correct algebraic sign).

$$P = \frac{1}{f}$$

Simple magnifier: When an object is at the near point (25 cm), the angle subtended by the object at the eye is θ. When a converging lens of focal length f is placed between the eye and object, an image which subtends an angle θ_0 can be formed at the near point. The magnifying power or **angular magnification** of the lens can be expressed in alternate forms. The angular magnification of a simple magnifier is minimized when the object is at the focal point of the lens.

$$m \equiv \frac{\theta}{\theta_0} \tag{36.16}$$

$$m_{max} = 1 + \frac{25.0 \text{ cm}}{f} \tag{36.18}$$

$$m_{min} = \frac{25.0 \text{ cm}}{f} \tag{36.19}$$

Compound microscope: This instrument contains an objective lens of short focal length f_0 and an eye piece of focal length f_e. The two lenses are separated by a distance L. When an object is located just beyond the focal point of the objective, the two lenses in combination form an enlarged, virtual and inverted image of lateral magnification M.

$$M = M_1 m_e = -\frac{L}{f_0}\left(\frac{25.0 \text{ cm}}{f_e}\right) \tag{36.20}$$

Astronomical telescope: Two converging lenses are separated by a distance equal to the sum of their focal lengths. The angular magnification is equal to the ratio of the two focal lengths.

$$m = -\frac{f_0}{f_e} \tag{36.21}$$

SUGGESTIONS, SKILLS, AND STRATEGIES

A major portion of this chapter is devoted to the development and presentation of equations which can be used to determine the location and nature of images formed by various optical components acting either singly or in combination. It is essential that these equations be used with the correct algebraic sign associated with each quantity involved. You must understand clearly the sign conventions for mirrors, refracting surfaces, and lenses. The following discussion represents a review of these sign conventions.

SIGN CONVENTIONS FOR MIRRORS

Equation: $\dfrac{1}{p}+\dfrac{1}{q}=\dfrac{1}{f}=\dfrac{2}{R}$ $\qquad M=\dfrac{h'}{h}=-\dfrac{q}{p}$

The front side of the mirror is the region on which light rays are incident and reflected.

p is + if the object is in front of the mirror (real object).
p is − if the object is in back of the mirror (virtual object).

q is + if the image is in front of the mirror (real image).
q is − if the image is in back of the mirror (virtual image).

Both f and R are + if the center of curvature is in front of the mirror　　(concave mirror).

Both f and R are − if the center of curvature is in back of the mirror　　(convex mirror).

If M is positive, the image is upright.
If M is negative, the image is inverted.

You should check the sign conventions as stated against the situations described in Figure 23.1.

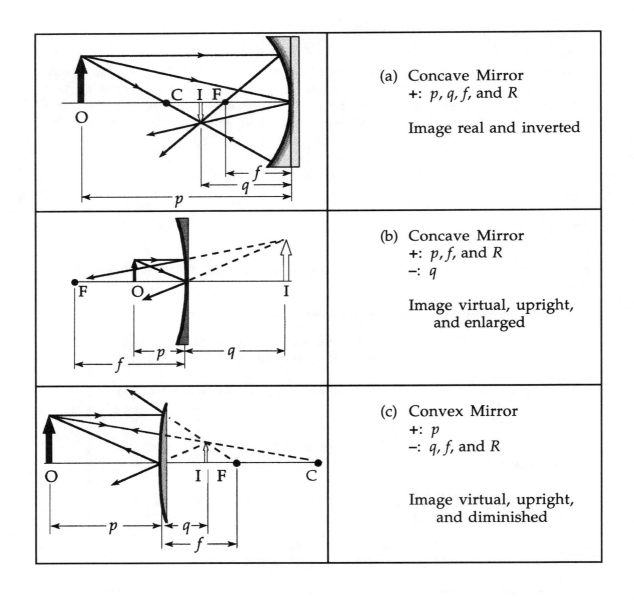

(a) Concave Mirror
+: p, q, f, and R

Image real and inverted

(b) Concave Mirror
+: p, f, and R
−: q

Image virtual, upright, and enlarged

(c) Convex Mirror
+: p
−: q, f, and R

Image virtual, upright, and diminished

SIGN CONVENTIONS FOR REFRACTING SURFACES

Equations: $\dfrac{n_1}{p} + \dfrac{n_2}{q} = \dfrac{n_2 - n_1}{R}$

In the following table, the **front** side of the surface is the side **from which the light is incident.**

> p is + if the object is in front of the surface (real object).
> p is − if the object is in back of the surface (virtual object).
> q is + if the image is in back of the surface (real image).
> q is − if the image is in front of the surface (virtual image).
> R is + if the center of curvature is in back of the surface.
> R is − if the center of curvature is in front of the surface.
>
> n_1 refers to the index of the medium on the side of the interface from which the light comes.
>
> n_2 is the index of the medium into which the light is transmitted after refraction at the interface.

Review the above sign conventions for the situations shown in Figure 23.2.

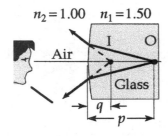

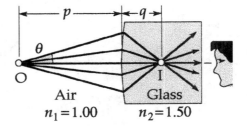

$p\ +$ (real object)

$q\ -$ (virtual image)

$R\ -$ (concave to incident light)

n_1 and n_2 as shown

$p\ +$ (real object)

$q\ +$ (real image)

$R\ +$ (convex to incident light)

n_1 and n_2 as shown

These figures describe sign conventions for refracting surfaces. In the first case, the object appears closer than it actually is. In the second case, an object beyond the glass appears to be in the glass.

SIGN CONVENTIONS FOR THIN LENSES

Equations: $$\frac{1}{p}+\frac{1}{q}=\frac{1}{f}=(n-1)\left(\frac{1}{R_1}-\frac{1}{R_2}\right) \qquad M=\frac{h'}{h}=-\frac{q}{p}$$

In the following table, the **front** of the lens is the **side from which the light is incident.**

> p is + if the object is in front of the lens.
> p is − if the object is in back of the lens.
>
> q is + if the image is in back of the lens.
> q is − if the image is in front of the lens.
>
> f is + if the lens is thickest in the center.
> f is − if the lens is thickest at the edges.
>
> R_1 and R_2 are + if the center of curvature is in back of the lens.
>
> R_1 and R_2 are − if the center of curvature is in front of the lens.

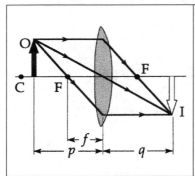	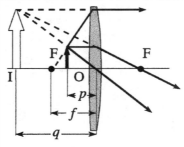	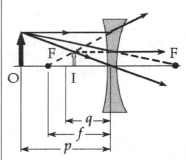
(a) Double-Convex (Converging)	(b) Double-Convex (Converging)	(c) Double-Concave (Diverging)
+: $p, q, f,$ and R_1	+: $p, f,$ and R_1	+: p, R_2
−: R_2	−: q, R_2	−: q, f, R_1
Image real, inverted	Image virtual, upright, enlarged	Image virtual, upright, diminished

REVIEW CHECKLIST

▷ Identify the following properties which characterize an image formed by a lens or mirror system with respect to an object: position, magnification, orientation (i.e. inverted, upright or right-left reversal) and whether real or virtual.

▷ Understand the relationship of the algebraic signs associated with calculated quantities to the nature of the image and object: real or virtual, upright or inverted.

▷ Calculate the location of the image of a specified object as formed by a plane mirror, spherical mirror, plane refracting surface, spherical refracting surface, thin lens, or a combination of two or more of these devices. Determine the magnification and character of the image in each case.

▷ Construct ray diagrams to determine the location and nature of the image of a given object when the geometrical characteristics of the optical device (lens or mirror) are known.

ANSWERS TO SELECTED CONCEPTUAL QUESTIONS

12. Why do some automobile mirrors have printed on them the statement "Objects in mirror are closer than they appear"?

Answer The mirror would probably be a convex mirror. This type would have been selected because the designers wish to give the driver a wide field of view, and a virtual, upright image for all distances.

□ □ □ □

14. Explain why a fish in a spherical goldfish bowl appears larger than it really is.

Answer

A spherical goldfish bowl acts as a single refracting surface; the magnification of any object in the bowl can be calculated from Equations 36.8 and 36.9 to be

$$M = 1 + \left(\frac{n_w - 1}{R}\right)q = 1 + \frac{0.33q}{R}$$

where R is negative.

Since the image distance q is also always negative, this gives a magnification that is always greater than 1. Therefore, the image of the goldfish will always be larger than the object.

□   □   □   □

15. Lenses used in eyeglasses, whether converging or diverging, are always designed such that the middle of the lens curves away from the eye, like the center lenses of Figure 36.26a and b. Why?

Answer

With the meniscus design, when you direct your gaze near the outer circumference of the lens you receive a ray that has passed through glass with more nearly parallel surfaces of entry and exit. Thus, the lens minimally distorts the direction to the object you are looking at. If you wear glasses, you can demonstrate this by turning them around and looking through them the wrong way, maximizing the distortion.

□   □   □   □

SOLUTIONS TO SELECTED END-OF-CHAPTER PROBLEMS

3. Determine the minimum height of a vertical flat mirror in which a person 5'10" in height can see his or her full image. (A ray diagram would be helpful.)

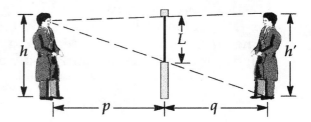

Solution

G: A diagram with the optical rays that create the image of the person is shown above and to the right. From this diagram, it appears that the mirror only needs to be about half the height of the person.

O: The required height of the mirror can be found from the mirror equation, where this flat mirror is described by $R = \infty$, $f = \infty$, and $1/f = 0$.

A: The general mirror equation is

$$\frac{1}{p} + \frac{1}{q} = \frac{1}{f}, \quad \text{so with } f = \infty, \quad q = -p$$

Thus, the image is as far behind the mirror as the person is in front. The magnification is then

$$M = \frac{-q}{p} = 1 = \frac{h'}{h}, \quad \text{so} \quad h' = h = 70.0 \text{ in.}$$

The required height of the mirror is defined by the triangle from the person's eyes to the top and bottom of the image, as shown. From the geometry of the similar triangles, we see that the length of the mirror must be:

$$L = h'\left(\frac{p}{p-q}\right) = h'\left(\frac{p}{2p}\right) = \frac{h'}{2} = \frac{70.0 \text{ in}}{2} = 35.0 \text{ in}$$

Thus, the mirror must be at least 35.0 in high. ◊

L: Our result agrees with our prediction from the ray diagram. Evidently, a full-length mirror only needs to be a half-height mirror! On a practical note, the vertical positioning of such a mirror is also important for the person to be able to view his or her full image. To allow for some variation in positioning and viewing by persons of different heights, most full-length mirrors are about 5′ in length.

---

9. A spherical convex mirror has a radius of curvature of 40.0 cm. Determine the position of the virtual image and magnification (a) for an object distance of 30.0 cm and (b) for an object distance of 60.0 cm. (c) Are the images upright or inverted?

Solution

The convex mirror is described by $f = R/2 = -40.0 \text{ cm}/2 = -20.0 \text{ cm}$

(a) $\dfrac{1}{p} + \dfrac{1}{q} = \dfrac{1}{f}$, so $\dfrac{1}{30.0 \text{ cm}} + \dfrac{1}{q} = \dfrac{1}{-20.0 \text{ cm}}$ and $q = -12.0 \text{ cm}$ ◊

$$M = \frac{-q}{p} = -\frac{-12.0 \text{ cm}}{30.0 \text{ cm}} = +0.400$$ ◊

(c) The principal ray diagram is an essential complement to the numerical description of the image. Draw the rays into this diagram; use the diagram to see that the image is behind the mirror, upright, virtual and diminished.

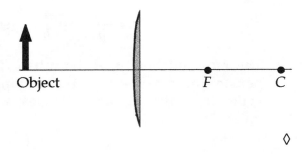

◊

(b) $\dfrac{1}{60.0\ \text{cm}} + \dfrac{1}{q} = -\dfrac{1}{20.0\ \text{cm}}$

so $q = -15.0\ \text{cm}$ ◊

$M = \dfrac{-q}{p} = -\dfrac{-15.0\ \text{cm}}{60.0\ \text{cm}} = +0.250$ ◊

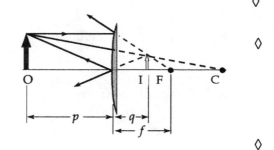

(c) The image is behind the mirror, upright, virtual, and diminished. ◊

---

11. A concave mirror has a radius of curvature of 60.0 cm. Calculate the image position and magnification of an object placed in front of the mirror (a) at a distance of 90.0 cm and (b) at a distance of 20.0 cm. (c) In each case, draw ray diagrams to obtain the image characteristics.

Solution

G: It is always a good idea to first draw a ray diagram for any optics problem. This gives a qualitative sense of how the image appears relative to the object. From the ray diagrams below, we see that when the object is 90 cm from the mirror, the image will be real, inverted, diminished, and located about 45 cm in front of the mirror, midway between the center of curvature and the focal point. When the object is 20 cm from the mirror, the image is be virtual, upright, magnified, and located about 50 cm behind the mirror.

O: The mirror equation can be used to find precise quantitative values.

A: (a)   The mirror equation is applied using the sign conventions listed in **Suggestions, Skills, and Strategies**:

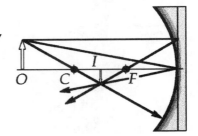

$$\frac{1}{p}+\frac{1}{q}=\frac{2}{R} \quad \text{or} \quad \frac{1}{90.0 \text{ cm}}+\frac{1}{q}=\frac{2}{60.0 \text{ cm}}$$

$$\frac{1}{q}=\frac{2}{60.0 \text{ cm}}-\frac{1}{90.0 \text{ cm}}=0.0222 \text{ cm}^{-1}$$

q = 45.0 cm (real, in front of the mirror)   ◊

$$M=\frac{-q}{p}=-\frac{45.0 \text{ cm}}{90.0 \text{ cm}}=-0.500 \text{ (inverted)}$$ ◊

(b)   We again use the mirror equation:

$$\frac{1}{p}+\frac{1}{q}=\frac{2}{R} \quad \text{or} \quad \frac{1}{20.0 \text{ cm}}+\frac{1}{q}=\frac{2}{60.0 \text{ cm}}$$

$$\frac{1}{q}=\frac{2}{60.0 \text{ cm}}-\frac{1}{20.0 \text{ cm}}=-0.0167 \text{ cm}^{-1}$$

q = −60.0 cm (virtual, behind the mirror)   ◊

$$M=-\frac{q}{p}=-\frac{-60.0 \text{ cm}}{20.0 \text{ cm}}=3.00 \text{ (upright)}$$ ◊

L: The calculated image characteristics agree well with our predictions. It is easy to miss a minus sign or to make a computational mistake when using the mirror-lens equation, so the qualitative values obtained from the ray diagrams are useful for a check on the reasonableness of the calculated values.

13. A spherical mirror is to be used to form, on a screen 5.00 m from the object, an image five times the size of the object. (a) Describe the type of mirror required. (b) Where should the mirror be positioned relative to the object?

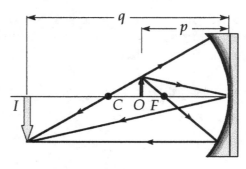

Solution

We are given that

$$q = (p + 5.00 \text{ m})$$

The image must be real, so

$$M = -5.00 = -q/p$$

or

$$q = 5.00p$$

(b) The distance of the mirror from the object is

$$p + 5.00 = 5.00p$$

thus

$$p = 1.25 \text{ m} \qquad \Diamond$$

(a) Applying Equation 26.6,

$$\frac{1}{f} = \frac{1}{p} + \frac{1}{q} = \frac{1}{1.25 \text{ m}} + \frac{1}{6.25 \text{ m}}$$

so the focal length of the mirror is

$$f = 1.04 \text{ m}$$

Noting that the image is real, inverted and enlarged, we conclude that the mirror is concave, and must have a radius of curvature of

$$R = 2f = 2.08 \text{ m} \qquad \Diamond$$

---

19. A cubical block of ice 50.0 cm on a side is placed on a level floor over a speck of dust. Find the location of the image of the speck if the index of refraction of ice is 1.309.

Solution The upper surface of the block is a single refracting surface with zero curvature, and with infinite radius of curvature.

$$\frac{n_1}{p}+\frac{n_2}{q}=\frac{n_2-n_1}{R} \quad \text{or} \quad \frac{1.309}{50.0\text{ cm}}+\frac{1}{q}=\frac{1.00-1.309}{\infty}=0$$

$$q=\frac{-50.0\text{ cm}}{1.309}=-38.2\text{ cm}$$

The speck of dust on the floor appears to be 38.2 cm below the upper surface of the ice. The image is virtual, upright, and actual size. ◊

---

21. A glass sphere (n = 1.50) with a radius of 15.0 cm has a tiny air bubble 5.00 cm above its center. The sphere is viewed looking down along the extended radius containing the bubble. What is the apparent depth of the bubble below the surface of the sphere?

Solution

The image is virtual, upright, and enlarged.

From Equation 36.8, $$\frac{n_1}{p}+\frac{n_2}{q}=\frac{n_2-n_1}{R}$$

We solve for q to find $$q=\frac{n_2Rp}{p(n_2-n_1)-n_1R}$$

In this case, $n_1 = 1.50$ $n_2 = 1.00$ $p = 10.0$ cm

So the apparent depth $$q=\frac{(1.00)(-15.0\text{ cm})(10.0\text{ cm})}{(10.0\text{ cm})(1.00-1.50)-(1.50)(-15.0\text{ cm})}=-8.57\text{ cm}$$ ◊

27. The left face of a biconvex lens has a radius of curvature of magnitude 12.0 cm, and the right face has a radius of curvature of magnitude 18.0 cm. The index of refraction of the glass is 1.44. (a) Calculate the focal length of the lens. (b) Calculate the focal length if the radii of curvature of the two faces are interchanged.

Solution

G: Since this is a biconvex lens, the center is thicker than the edges, and the lens will tend to converge incident light rays. Therefore it has a positive focal length. Exchanging the radii of curvature amounts to turning the lens around so the light enters the opposite side first. However, this does not change the fact that the center of the lens is still thicker than the edges, so we should not expect the focal length of the lens to be different (assuming the thin-lens approximation is valid).

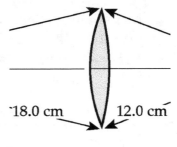

18.0 cm   12.0 cm

O: The lens makers' equation can be used to find the focal length of this lens.

A: The centers of curvature of the lens surfaces are on opposite sides, so the second surface has a negative radius:

(a) $$\frac{1}{f} = (n-1)\left(\frac{1}{R_1} - \frac{1}{R_2}\right) = (1.44 - 1.00)\left(\frac{1}{12.0 \text{ cm}} - \frac{1}{-18.0 \text{ cm}}\right)$$

$f = 16.4$ cm ◊

(b) $$\frac{1}{f} = (0.440)\left(\frac{1}{18.0 \text{ cm}} - \frac{1}{-12.0 \text{ cm}}\right)$$

$f = 16.4$ cm ◊

L: As expected, reversing the orientation of the lens does not change what it does to the light, as long as the lens is relatively thin (variations may be noticed with a thick lens). The fact that light rays can be traced forward or backward through an optical system is sometimes referred to as the **principle of reversibility**.   We can see that the focal length of this biconvex lens is about the same magnitude as the average radius of curvature. A few approximations, useful as checks, are that a symmetric biconvex lens with radii of magnitude R will have focal length $f \approx R$; a plano-convex lens with radius R will have $f \approx R/2$; and a symmetric biconcave lens has $f \approx -R$. These approximations apply when the lens has $n \approx 1.5$, which is typical of many types of clear glass and plastic.

---

31. The nickel's image in Figure P36.31 has twice the diameter of the nickel and is 2.84 cm from the lens. Determine the focal length of the lens.

Solution Looking through the lens, you see the image beyond the lens.   Therefore, the image is virtual, with $q = -2.84$ cm.

Now, $M = \dfrac{h'}{h} = 2 = -\dfrac{q}{p}$

Figure P36.31

so $p = -\dfrac{q}{2} = 1.42$ cm

Thus, $f = \left(\dfrac{1}{p} + \dfrac{1}{q} \right)^{-1}$

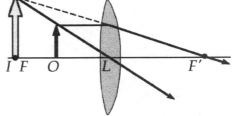

and $f = \left(\dfrac{1}{1.42 \text{ cm}} + \dfrac{1}{-2.84 \text{ cm}} \right)^{-1} = 2.84$ cm   ◊

37. An object is positioned 20.0 cm to the left of a diverging lens with focal length $f = -32.0$ cm. Determine (a) the location and (b) the magnification of the image. (c) Construct a ray diagram for this arrangement.

Solution $\quad \dfrac{1}{p} + \dfrac{1}{q} = \dfrac{1}{f} \quad$ or $\quad \dfrac{1}{20.0 \text{ cm}} + \dfrac{1}{q} = \dfrac{1}{-32.0 \text{ cm}}$

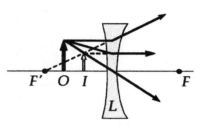

(a) So $\quad q = -\left(\dfrac{1}{20.0 \text{ cm}} + \dfrac{1}{32.0 \text{ cm}} \right)^{-1} = -12.3 \text{ cm}$ ◊

(b) $\qquad M = -\dfrac{q}{p} = -\dfrac{-12.3 \text{ cm}}{20.0 \text{ cm}} = 0.615$ ◊

(c) The image is virtual, upright, diminished. ◊

---

45. A nearsighted person cannot see objects clearly beyond 25.0 cm (her far point.) If she has no astigmatism and contact lenses are prescribed for her, what power and type of lens are required to correct her vision?

Solution The lens should take parallel light rays from a very distant object ($p = \infty$) and make them diverge from a virtual image at the woman's far point, which is 25.0 cm beyond the lens, at $q = -25.0$ cm. Thus,

$$\frac{1}{f} = \frac{1}{p} + \frac{1}{q} = \frac{1}{\infty} + \frac{1}{-25.0 \text{ cm}}$$

Hence, the power of the lens is $\qquad P = \dfrac{1}{f} = -\dfrac{1}{0.250 \text{ m}} = -4.00 \text{ diopter}$

This is a diverging lens. ◊

---

51. The Yerkes refracting telescope has a 1.00-m diameter objective lens with a focal length of 20.0 m. Assume that it is used with an eyepiece that has a focal length of 2.50 cm. (a) Determine the magnification of the planet Mars as seen through this telescope. (b) Are the Martian polar caps right side up or upside down?

Solution

(a) The angular magnification is $\qquad m = -\dfrac{f_0}{f_e} = \dfrac{-2.00 \text{ m}}{0.0250 \text{ m}} = -800$ ◊

(b) The minus sign means the image is inverted relative to the object. ◊

---

61. A parallel beam of light enters a glass hemisphere perpendicular to the flat face, as shown in Figure P36.61. The radius is $R = 6.00$ cm, and the index of refraction is $n = 1.560$. Determine the point at which the beam is focused. (Assume paraxial rays.)

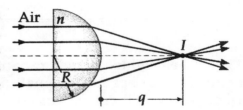

Solution

Figure P36.61

A hemisphere is too thick to be described as a thin lens. The light is undeviated on entry into the flat face. Because of this, we instead consider the light's exit from the second surface, for which $R = -6.00$ cm. The incident rays are parallel, so $p = \infty$.

Then $\qquad \dfrac{n_1}{p} + \dfrac{n_2}{q} = \dfrac{n_2 - n_1}{R}$ becomes $0 + \dfrac{1}{q} = \dfrac{(1 - 1.56)}{-6.00 \text{ cm}}$

and $\qquad q = 10.7$ cm ◊

---

63. An object is placed 12.0 cm to the left of a diverging lens with a focal length of –6.00 cm. A converging lens with a focal length of 12.0 cm is placed a distance d to the right of the diverging lens. Find the distance d that corresponds to a final image at infinity. Draw a ray diagram for this case.

Solution From Eq. 36.12, $\quad q_1 = \dfrac{f_1 p_1}{p_1 - f_1} = \dfrac{(-6.00 \text{ cm})(12.0 \text{ cm})}{12.0 \text{ cm} - (-6.00 \text{ cm})} = -4.00 \text{ cm}$

When we require that $q_2 \to \infty$, Equation 36.12 becomes $p_2 = f_2 = 12.0$ cm. Since the object for the converging lens must be 12.0 cm to its left, and since this is the image for the diverging lens which is 4.00 cm to **its** left, the two lenses must be separated by 8.00 cm. Mathematically,

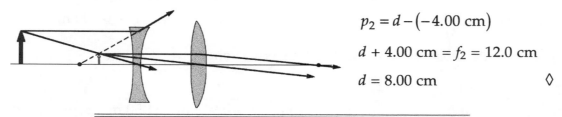

$$p_2 = d - (-4.00 \text{ cm})$$

$$d + 4.00 \text{ cm} = f_2 = 12.0 \text{ cm}$$

$$d = 8.00 \text{ cm} \qquad \Diamond$$

65. The disk of the Sun subtends an angle of 0.533° at the Earth. What are the position and diameter of the solar image formed by a concave spherical mirror of radius 3.00 m?

Solution For the mirror, $f = R/2 = +1.50$ m. In addition, because the distance to the Sun is so much larger than any other figures, we can take $p = \infty$.

The mirror equation, $\dfrac{1}{p} + \dfrac{1}{q} = \dfrac{1}{f}$ then gives $q = f = 1.50 \text{ m} \qquad \Diamond$

Now, in $M = -q/p = h'/h$, the magnification is nearly zero, but we can be more precise: the definition of radian measure means that h/p is the angular diameter of the object. Thus the image diameter is

$$h' = -hq/p = (-0.533°)(\pi \text{ rad} / 180°)(1.50 \text{ m}) = -0.140 \text{ m} = -1.40 \text{ cm} \qquad \Diamond$$

67. In a darkened room, a burning candle is placed 1.50 m from a white wall. A lens is placed between candle and wall at a location that causes a larger, inverted image to form on the wall. When the lens is moved 90.0 cm toward the wall, another image of the candle is formed. Find (a) the two object distances that produce the images stated above and (b) the focal length of the lens. (c) Characterize the second image.

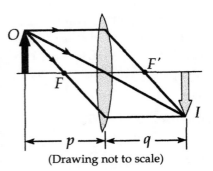

(Drawing not to scale)

Solution   Originally, $p_1 + q_1 = 1.50 \text{ m}$

In the final situation, $p_2 = p_1 + 0.900 \text{ m}$ and $q_2 = q_1 - 0.900 \text{ m}$

We substitute these values into the lens equation.

$$\frac{1}{p_1} + \frac{1}{q_1} = \frac{1}{p_2} + \frac{1}{q_2}: \qquad \frac{1}{p_1} + \frac{1}{1.50 \text{ m} - p_1} = \frac{1}{p_1 + 0.900 \text{ m}} + \frac{1}{(1.50 \text{ m} - p_1) - 9.00 \text{ m}}$$

Adding the fractions, $\dfrac{1.50 \text{ m} - p_1 + p_1}{p_1(1.50 \text{ m} - p_1)} = \dfrac{0.600 \text{ m} - p_1 + p_1 + 0.900 \text{ m}}{(p_1 + 0.900 \text{ m})(0.600 \text{ m} - p_1)}$

Simplified, this is $p_1(1.50 \text{ m} - p_1) = (p_1 + 0.900 \text{ m})(0.600 \text{ m} - p_1)$

(a)   Thus, $p_1 = \dfrac{0.540 \text{ m}^2}{1.80 \text{ m}} = 0.300 \text{ m}$

and $p_2 = p_1 + 0.900 \text{ m} = 1.20 \text{ m}$ ◊

(b) $\dfrac{1}{f} = \dfrac{1}{0.300 \text{ m}} + \dfrac{1}{1.50 \text{ m} - 0.300 \text{ m}}$ and $f = 0.240 \text{ m}$ ◊

(c)   The second image is real, inverted, and diminished, with $M = -q_2/p_2 = -0.250$ ◊

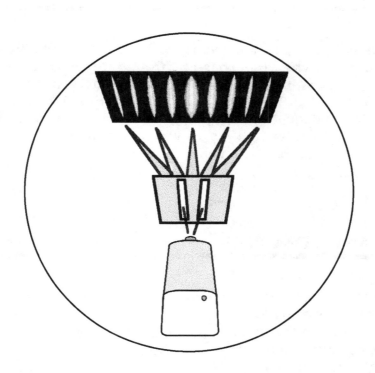

INTERFERENCE OF
LIGHT WAVES

INTERFERENCE OF LIGHT WAVES

INTRODUCTION

In the previous chapter on geometric optics, we used light rays to examine what happens when light passes through a lens or reflects from a mirror. The next two chapters are concerned with wave optics, which deals with the interference, diffraction, and polarization of light. These phenomena cannot be adequately explained with the ray optics of Chapter 36, but we describe how treating light as waves rather than as rays leads to a satisfying description of such phenomena.

EQUATIONS AND CONCEPTS

In the arrangement used for the Young's double-slit experiment, two slits separated by a distance, d, serve as monochromatic coherent sources. The light intensity at any point on the screen is the resultant of light reaching the screen from both slits. Also, light from the two slits reaching any point on the screen (except the center) travel unequal path lengths. This difference in length of path is called the path difference.

$$\delta = r_2 - r_1 = d\sin\theta \qquad (37.1)$$

$$y = L\tan\theta \approx L\sin\theta \qquad (37.4)$$

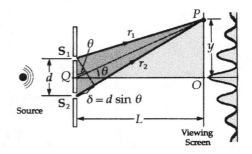

Bright fringes (constructive interference) will appear at points on the screen for which the path difference is equal to an integral multiple of the wavelength. The positions of bright fringes can also be located by calculating their vertical distance from the center of the screen (y). In each case, the number m is called the order number of the fringe. The central bright fringe ($\theta = 0$, $m = 0$) is called the zeroth-order maximum.

$$\delta = d\sin\theta = m\lambda \quad \text{(constructive)} \quad (37.2)$$

$$(m = 0, \pm 1, \pm 2, \ldots)$$

$$y_{\text{bright}} \cong \frac{\lambda L}{d} m \quad \text{(for small } \theta) \quad (37.5)$$

Destructive interference (dark fringes) will occur at points on the screen which correspond to path differences of an odd multiple of half wavelengths. For these points, waves which leave the two slits in phase arrive at the screen 180° out of phase.

$$\delta = d\sin\theta = \left(m + \tfrac{1}{2}\right)\lambda \quad \text{(destructive)} \quad (37.3)$$

$$(m = 0, \pm 1, \pm 2, \ldots)$$

$$y_{\text{dark}} \cong \frac{\lambda L}{d}\left(m + \tfrac{1}{2}\right) \quad \text{(for small } \theta) \quad (37.6)$$

Two waves which leave the slits initially in phase arrive at the screen **out of phase** by an amount which depends on the path difference.

$$\phi = \frac{2\pi}{\lambda}\delta = \frac{2\pi}{\lambda}d\sin\theta \quad (37.8)$$

Two waves initially in phase and of equal amplitude (E_0) will produce a resultant amplitude at some point on the screen which depends on the phase difference (and therefore on the path difference).

$$E_p = 2E_0 \cos\left(\frac{\phi}{2}\right)\sin\left(\omega t + \frac{\phi}{2}\right) \quad (37.10)$$

The average light intensity (I) at any point P on the screen is proportional to the square of the amplitude of the resultant wave. **The average intensity can be written:**

- as a function of phase difference ϕ;

$$I = I_{max}\cos^2\left(\frac{\phi}{2}\right)$$ (37.11)

- as a function of the angle (θ) subtended by the screen point at the source midpoint; or

$$I = I_{max}\cos^2\left(\frac{\pi d\sin\theta}{\lambda}\right)$$ (37.12)

- as a function of the vertical distance (y) from the center of the screen.

$$I \cong I_{max}\cos^2\left(\frac{\pi d}{\lambda L}y\right)$$ (for small θ) (37.13)

The intensity pattern observed on the screen will vary as the number of equally spaced sources is increased; however, the **positions of the principal maxima remain the same.**

In thin-film interference (shown on the next page), the wavelength of light in the film, λ_n, is not the same as the wavelength in the surrounding medium.

$$\lambda_n = \frac{\lambda}{n}$$ (37.14)

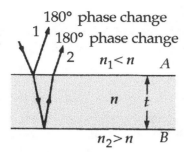

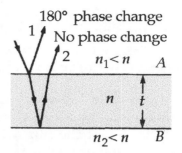

The conditions for interference of light reflected perpendicularly from a thin film can be stated in terms of the thickness (t) and index of refraction of the film. When there is only **one** phase change (at either the top or bottom surface of the film), the Equations 37.16 and 37.17 are the conditions for constructive and destructive interference, respectively. When there are either **two** or **no** phase changes, the roles of equations 37.16 and 37,17 are reversed.

Constructive interference, with one phase change:

$$2nt = \left(m + \tfrac{1}{2}\right)\lambda \quad (m = 0, 1, 2 \ldots) \quad (37.16)$$

Destructive interference, with one phase change:

$$2nt = m\lambda \qquad (m = 0, 1, 2 \ldots) \quad (37.17)$$

SUGGESTIONS, SKILLS, AND STRATEGIES

The following features should be kept in mind while working thin-film interference problems:

- Identify the thin film from which interference effects are being observed.

- The type of interference that occurs in a specific problem is determined by the phase relationship between that portion of the wave reflected at the upper surface of the film and that portion reflected at the lower surface of the film.

- Phase differences between the two portions of the wave occur because of differences in the distances traveled by the two portions and by phase changes occurring upon reflection.

The wave reflected from the lower surface of the film has to travel a distance equal to twice the thickness of the film before it returns to the upper surface of the film where it interferes with that portion of the wave reflected at the upper surface. When **distances alone are considered**, if this extra distance is equal to an integral multiple of λ the interference will be constructive; if the extra distance equals $\frac{1}{2}\lambda$, $\frac{3}{2}\lambda$, and so forth, destructive interference will occur.

Reflections may change the results above which are based solely on the distance traveled. When a wave traveling in a particular medium reflects off a surface having a higher index of refraction than the one it is in, a 180° phase shift occurs. This has the same effect as if the wave lost $\frac{1}{2}\lambda$. These losses must be considered in addition to those losses that occur because of the extra distance one wave travels over another.

- When distance and phase changes upon reflection are both taken into account, the interference will be constructive if the waves are out of phase by an integral multiple of λ Destructive interference will occur when the phase difference is $\frac{1}{2}\lambda$, $\frac{3}{2}\lambda$, and so forth.

The technique of phasor addition offers a convenient alternative to the algebraic method for finding the resultant wave amplitude at some point on a screen. This is especially true when a large number of waves are to be combined. The method of phasor addition is outlined in the following steps and is illustrated in Figure 37.3 for the case of two equal amplitude waves differing in phase by an angle of ϕ.

- Draw the phasors representing the waves end to end. The angle between successive phasors is equal to the phase angle between the waves from successive source slits. The length of each phasor is proportional to the magnitude of the wave it represents.

- The resultant wave is the vector sum (vector from the tail of the first phasor to the head of the last one) of the individual phasors.

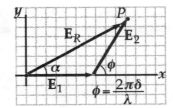

- The phase angle (α) of the resultant wave is the angle between the direction of the resultant phasor and the direction of the first phasor.

REVIEW CHECKLIST

▷ Describe Young's double-slit experiment to demonstrate the wave nature of light. Account for the phase difference between light waves from the two sources as they arrive at a given point on the screen. State the conditions for constructive and destructive interference in terms of each of the following: path difference, phase difference, distance from the center of the screen, and angle subtended by the observation point at the source mid-point.

▷ Outline the manner in which the superposition principle leads to the correct expression for the intensity distribution on a distant screen due to two coherent sources of equal intensity.

▷ Describe the use of the phasor diagram method to determine the amplitude and phase of the wave which is the resultant of two or three coherent sources.

▷ Account for the conditions of constructive and destructive interference in thin films considering both path difference and any expected phase changes due to reflection.

ANSWERS TO SELECTED CONCEPTUAL QUESTIONS

1. What is the necessary condition on the path length difference between two waves that interfere (a) constructively and (b) destructively?

Answer (a) Two waves interfere constructively if their path difference is either zero or some integral multiple of the wavelength; that is, if the path difference equals $m\lambda$. (b) Two waves interfere destructively if their path difference is an odd multiple of one-half of a wavelength; that is, where m is any integer, the path difference equals $\left(m + \frac{1}{2}\right)\lambda$.

□　　□　　□　　□

4. In Young's double-slit experiment, why do we use monochromatic light? If white light is used, how would the pattern change?

Answer Every color produces its own pattern, with a spacing between the maximums that is characteristic of the wavelength. With several colors, the patterns are superimposed, and it can become difficult to pick out a single maximum. Using monochromatic light can eliminate this problem.

With white light, the central maximum is white. However, the side maximums will occur at different angles for different wavelengths, resulting in a display of the colors of the rainbow for the first side maximum. As the angle increases further, secondary maximums will be resolved and overlap in patterns too difficult to guess without calculation. At larger angles, the light soon starts mixing to white again, though it is often so faint that you would call it gray.

□　　□　　□　　□

SOLUTIONS TO SELECTED END-OF-CHAPTER PROBLEMS

3. Two radio antennas separated by 300 m as shown in Figure P37.3, simultaneously broadcast identical signals at the same wavelength. A radio in a car traveling due north receives the signals. (a) If the car is at the position of the second maximum, what is the wavelength of the signals? (b) How much farther must the car travel to encounter the next minimum in reception? (**Note:** Do not use the small-angle approximation in this problem.)

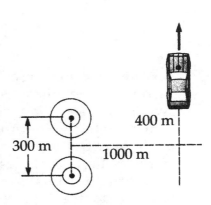

Figure P37.3

Solution

With the conditions given, the small angle approximation **does not work well.** That is, sin θ, tan θ, and θ are significantly different. Instead, the approach to be used is outlined below.

(a) At the $m = 2$ maximum, $\tan\theta = \dfrac{400 \text{ m}}{1000 \text{ m}} = 0.400$

and $\theta = 21.8°$ so $\lambda = \dfrac{d\sin\theta}{m} = \dfrac{(300 \text{ m})\sin 21.8°}{2} = 55.7 \text{ m}$ ◊

(b) The next minimum encountered is the $m = 2$ minimum, and at that point, $d\sin\theta = \left(m + \frac{1}{2}\right)\lambda = \frac{5}{2}\lambda$.

Solving for the angle, $\sin\theta = \dfrac{5\lambda}{2d} = \dfrac{5(55.7 \text{ m})}{2(300 \text{ m})} = 0.464$

so $\theta = 27.7°$ and $y = (1000 \text{ m})\tan 27.7° = 524 \text{ m}$

Therefore, the car must travel an additional 124 m.   ◊

5. Young's double-slit experiment is performed with 589-nm light and a slit-to-screen distance of 2.00 m. The tenth interference minimum is observed 7.26 mm from the central maximum. Determine the spacing of the slits.

Solution

G: For the situation described, the observed interference pattern is very narrow, (the minima are less than 1 mm apart when the screen is 2 m away). In fact, the minima and maxima are so close together that it would probably be difficult to resolve adjacent maxima, so the pattern might look like a solid blur to the naked eye. Since the angular spacing of the pattern is inversely proportional to the slit width, we should expect that for this narrow pattern, the space between the slits will be larger than the typical fraction of a millimeter, and certainly much greater than the wavelength of the light ($d \gg \lambda$ = 589 nm).

O: Since we are given the location of the tenth minimum for this interference pattern, we should use the equation for **destructive interference** from a double slit. The figure for Problem 7 shows the critical variables for this problem.

A: In the equation, $d\sin\theta = \left(m + \frac{1}{2}\right)\lambda$, the first minimum is described by $m = 0$ and the tenth by $m = 9$.

So, $\qquad \sin\theta = \dfrac{\lambda}{d}\left(9 + \dfrac{1}{2}\right)$

Also, $\tan\theta = y/L$, but for small θ, $\sin\theta \approx \tan\theta$. Thus, the distance between the slits is,

$$d = \frac{9.5\lambda}{\sin\theta} = \frac{9.5\lambda L}{y} = \frac{9.5(5890 \times 10^{-10} \text{ m})(2.00 \text{ m})}{7.26 \times 10^{-3} \text{ m}} = 1.54 \times 10^{-3} \text{ m}$$

$d = 1.54$ mm $\qquad\qquad\qquad\qquad\qquad\qquad\qquad\qquad\qquad\qquad\qquad$ ◊

L: The spacing between the slits is relatively large, as we expected (about 3 000 times greater than the wavelength of the light). In order to more clearly distinguish between maxima and minima, the pattern could be expanded by increasing the distance to the screen. However, as L is increased, the overall pattern would be less bright as the light expands over a larger area, so that beyond some distance, the light would be too dim to see.

7.   A pair of narrow, parallel slits separated by 0.250 mm is illuminated by green light (λ = 546.1 nm). The interference pattern is observed on a screen 1.20 m away from the plane of the slits. Calculate the distance (a) from the central maximum to the first bright region on either side of the central maximum and (b) between the first and second dark bands.

Solution

G: The spacing between adjacent maxima and minima should be fairly uniform across the pattern as long as the width of the pattern is much less than the distance to the screen (so that the small angle approximation is valid). The separation between fringes should be at least a millimeter if the pattern can be easily observed with a naked eye.

O: The bright regions are areas of constructive interference and the dark bands are destructive interference, so the corresponding double-slit equations will be used to find the y distances.

It can be confusing to keep track of four different symbols for distances. Three are shown in the drawing to the right. Note that:

> y is the unknown distance from the bright central maximum (m = 0) to another maximum or minimum on either side of the center of the interference pattern.

> λ is the wavelength of the light, determined by the source.

A: (a) For **very small** θ $\sin\theta \approx \tan\theta$ and $\tan\theta = y/L$

and the equation for constructive interference

$$\sin\theta = m\lambda/d \qquad \text{(Eq. 37.2)}$$

becomes $\qquad\qquad y_{\text{bright}} \approx (\lambda L/d)m \qquad \text{(Eq. 37.5)}$

Substituting values, $\qquad y_{\text{bright}} = \dfrac{(546\times10^{-9}\text{ m})(1.20\text{ m})}{0.250\times10^{-3}\text{ m}}(1) = 2.62\text{ mm}$ ◊

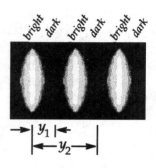

(b) If you have trouble remembering whether Equation 37.5 or Equation 37.6 applies to a given situation, you can instead remember that the first bright band is in the center, and dark bands are halfway between bright bands. Thus, Equation 37.5 describes them all, with $m = 0, 1, 2 \ldots$ for bright bands, and with $m = 0.5, 1.5, 2.5 \ldots$ for dark bands. The dark band version of Eq. 37.5 is simply Eq. 37.6:

$$y_{\text{dark}} = \frac{\lambda L}{d}\left(m + \tfrac{1}{2}\right)$$

$$\Delta y_{\text{dark}} = \left(1 + \tfrac{1}{2}\right)\frac{\lambda L}{d} - \left(0 + \tfrac{1}{2}\right)\frac{\lambda L}{d} = \frac{\lambda L}{d} = 2.62\text{ mm}$$ ◊

L: This spacing is large enough for easy resolution of adjacent fringes. The distance between minima is the same as the distance between maxima. We expected this equality since the angles are small:

$$\theta = (2.62\text{ mm})/(1.20\text{ m}) = 0.00218\text{ rad} = 0.125°$$

When the angular spacing exceeds about 3°, then $\sin\theta$ differs from $\tan\theta$ when written to three significant figures.

11. In Figure 37.4 let L = 1.20 m and d = 0.120 mm, and assume that the slit system is illuminated with monochromatic 500-nm light. Calculate the phase difference between the two wavefronts arriving at point P when (a) $\theta = 0.500°$ and (b) $y = 5.00$ mm. (c) What is the value of θ for which the phase difference is 0.333 rad? (d) What is the value of θ for which the path difference is $\lambda / 4$?

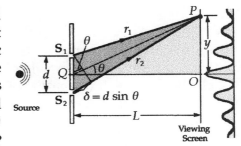

Figure 37.4

Solution

(a) The path difference is $\delta = d\sin\theta = \left(0.120 \times 10^{-3}\ \text{m}\right)\sin 0.500° = 1.05 \times 10^{-6}\ \text{m}$

The phase difference is $\phi = \dfrac{2\pi\delta}{\lambda} = \dfrac{2\pi\left(1.05 \times 10^{-6}\ \text{m}\right)}{\left(500 \times 10^{-9}\ \text{m}\right)} = 13.2$ rad $\lozenge$

This is equivalent to $\phi = 0.593$ rad $= 34.0°$ $\lozenge$

(b) $\tan\theta = \dfrac{y}{L} = \dfrac{5.00 \times 10^{-3}\ \text{m}}{1.20\ \text{m}} \cong \sin\theta$

$\phi = \dfrac{2\pi d\sin\theta}{\lambda} = \dfrac{2\pi\left(0.120 \times 10^{-3}\ \text{m}\right)\left(4.17 \times 10^{-3}\right)}{500 \times 10^{-9}\ \text{m}} = 2\pi$ rad $= 0$ $\lozenge$

(c) $\phi = \dfrac{2\pi d\sin\theta}{\lambda}$: $\theta = \sin^{-1}\left(\dfrac{\lambda\phi}{2\pi d}\right) = \sin^{-1}\left[\dfrac{\left(500 \times 10^{-9}\ \text{m}\right)\left(0.333\right)}{2\pi\left(1.20 \times 10^{-4}\ \text{m}\right)}\right] = 0.0127°$ $\lozenge$

(d) $\dfrac{\lambda}{4} = d\sin\theta$: $\theta = \sin^{-1}\left(\dfrac{\lambda}{4d}\right) = \sin^{-1}\left[\dfrac{\left(500 \times 10^{-9}\ \text{m}\right)}{4\left(1.20 \times 10^{-4}\ \text{m}\right)}\right] = 0.0597°$ $\lozenge$

15. In Figure 37.4, let L = 120 cm and d = 0.250 cm. The slits are illuminated with coherent 600-nm light. Calculate the distance y above the central maximum for which the average intensity on the screen is 75.0% of the maximum.

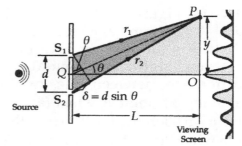

Solution

For small θ,

$$I = I_{max} \cos^2\left(\frac{\pi d \sin\theta}{\lambda}\right)$$

From the drawing, $\sin\theta \cong \dfrac{y}{L}$:

$$y = \frac{\lambda L}{\pi d}\cos^{-1}\sqrt{I_{av}/I_{max}}$$

In addition, since $I = 0.750 I_{max}$, we can substitute a value for each variable:

$$y = \frac{\left(6.00 \times 10^{-7}\ \text{m}\right)\left(1.20\ \text{m}\right)}{\pi\left(2.50 \times 10^{-3}\ \text{m}\right)}\cos^{-1}\sqrt{0.750} = 0.0480\ \text{mm} \qquad \Diamond$$

17. Two narrow parallel slits separated by 0.850 mm are illuminated by 600-nm light, and the viewing screen is 2.80 m away from the slits. (a) What is the phase difference between the two interfering waves on a screen at a point 2.50 mm from the central bright fringe? (b) What is the ratio of the intensity at this point to the intensity at the center of a bright fringe?

Solution

G: It is difficult to accurately predict the relative intensity at the point of interest without actually doing the calculation. The waves from each slit could meet in phase ($\phi = 0$) to produce a bright spot of **constructive interference**, out of phase ($\phi = 180°$) to produce a dark region of **destructive interference**, or most likely the phase difference will be somewhere between these extremes, $0 < \phi < 180°$, so that the relative intensity will be $0 < I/I_{max} < 1$.

O: The phase angle depends on the path difference of the waves according to Equation 37.8. This phase difference is used to find the average intensity at the point of interest. Then the relative intensity is simply this intensity divided by the maximum intensity.

A: (a) Using the variables shown in the diagram for problem 7 we have,

$$\phi = \frac{2\pi d}{\lambda}\sin\theta = \frac{2\pi d}{\lambda}\left(\frac{y}{\sqrt{y^2+L^2}}\right) \cong \frac{2\pi yd}{\lambda L}$$

$$\phi = \frac{2\pi\left(0.850\times10^{-3}\text{ m}\right)(0.00250\text{ m})}{\left(600\times10^{-9}\text{ m}\right)(2.80\text{ m})} = 7.95\text{ rad} = 2\pi + 1.66\text{ rad} = 95.4°\ \lozenge$$

(b) $$\frac{I}{I_{max}} = \frac{\cos^2\left(\dfrac{\pi d}{\lambda}\sin\theta\right)}{\cos^2\left(\dfrac{\pi d}{\lambda}\sin\theta_{max}\right)} = \frac{\cos^2\left(\dfrac{\phi}{2}\right)}{\cos^2(m\pi)} = \cos^2\left(\frac{\phi}{2}\right)$$

$$\frac{I}{I_{max}} = \cos^2\left(\frac{95.4°}{2}\right) = 0.453 \qquad\qquad \lozenge$$

L: It appears that at this point, the waves show **partial interference** so that the combination is about half the brightness found at the central maximum. We should remember that the equations used in this solution do not account for the diffraction caused by the finite width of each slit. This diffraction effect creates an "envelope" that diminishes in intensity away from the central maximum as shown by the dotted line in Figures 37.13 and P37.60. Therefore, the relative intensity at y = 2.50 mm will actually be slightly less than 0.452.

21. Determine the resultant of the two waves

$$E_1 = 6.00 \sin(100\pi t)$$

and $$E_2 = 8.00 \sin (100\pi t + \pi/2).$$

Solution

Let the x axis lie along E_1 in phase space. Its component form is then

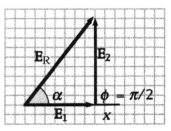

$$\mathbf{E}_1 = 6.00\mathbf{i} + 0\mathbf{j}$$

The components of $\mathbf{E}_2$ are

$$\mathbf{E}_2 = 8.00 \, \cos\!\left(\frac{\pi}{2}\right)\!\mathbf{i} + 8.00 \, \sin\!\left(\frac{\pi}{2}\right)\!\mathbf{j} = 8.00\mathbf{j}$$

The resultant is $\mathbf{E}_R = \mathbf{E}_1 + \mathbf{E}_2 = 6.00\mathbf{i} + 8.00\mathbf{j}$

with amplitude $\sqrt{6.00^2 + 8.00^2} = 10.0$

and phase $\alpha = \tan^{-1}\!\left(\dfrac{8.00}{6.00}\right) = 0.927$ rad

Thus, $E_R = 10.0 \sin (100\pi t + 0.927)$ ◊

27. Consider N coherent sources described by

$$E_1 = E_0 \sin(\omega t + \phi), \qquad\qquad E_2 = E_0 \sin(\omega t + 2\phi),$$

$$E_3 = E_0 \sin(\omega t + 3\phi), \ldots, \qquad\qquad E_N = E_0 \sin(\omega t + N\phi).$$

Find the minimum value of ϕ for which $E_R = E_1 + E_2 + E_3 + \ldots + E_N$ is zero.

Solution

Take $\qquad \phi = 360°/N$

and $\qquad E_R = \Sigma E_0 \sin(\omega t + N\phi) = 0$

where N defines the number of coherent sources.

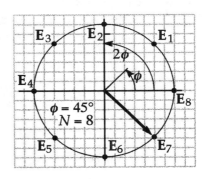

In essence, the set of electric field components completes a full 360° circle, returning to zero.

In this situation, when the vectors are added, the sources are symmetric about $E_R = 0$, and thus sum to zero.

Alternatively, one can note that the vectors must create a regular polygon when summed, in order to return to the origin. (The figures to the right may help.)

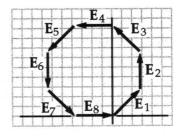

If N has an integer divisor m, then the choice $\phi = 360°/m$ will also make all the phasors add to zero.

◊

29. An oil film (n = 1.45) floating on water is illuminated by white light at normal incidence. The film is 280 nm thick. Find (a) the dominant observed color in the reflected light and (b) the dominant color in the transmitted light. Explain your reasoning.

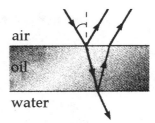

air

oil

water

Solution

The light reflected from the top of the oil film undergoes phase reversal. Since 1.45 > 1.33, the light reflected from the bottom undergoes no reversal. For constructive interference of reflected light, we then have

$$2nt = \left(m + \tfrac{1}{2}\right)\lambda \qquad \text{or} \qquad \lambda_m = \frac{2nt}{\left(m + \tfrac{1}{2}\right)} = \frac{2(1.45)(280 \text{ nm})}{\left(m + \tfrac{1}{2}\right)}$$

(a) Substituting for m, we have
$m = 0$: $\quad \lambda_0$ = 1624 nm (infrared)
$m = 1$: $\quad \lambda_1$ = 541 nm (green)
$m = 2$: $\quad \lambda_2$ = 325 nm (ultraviolet)

Both infrared and ultraviolet light are invisible to the human eye, so the dominant color is green. ◊

(b) The reflected light contains little red and violet, so these colors make it through to be in the transmitted beam.

According to the condition for destructive interference, $2nt = m\lambda$

For $\quad m = 1, \quad \lambda$ = 812 nm (near infrared)
$\quad m = 2, \quad \lambda$ = 406 nm (violet) ◊

37. An air wedge is formed between two glass plates separated at one edge by a very fine wire as shown in Figure P37.37. When the wedge is illuminated from above by 600-nm light, 30 dark fringes are observed. Calculate the radius of the wire.

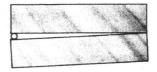

Figure P37.37

Solution

G: The radius of the wire is probably less than 0.1 mm since it is described as a "very fine wire."

O: Light reflecting from the bottom surface of the top plate undergoes no phase shift, while light reflecting from the top surface of the bottom plate is shifted by π, and also has to travel an extra distance $2t$, where t is the thickness of the air wedge.

For destructive interference, $2t = m\lambda$ $\quad (m = 0, 1, 2, 3, \ldots)$

The first dark fringe appears where $m = 0$ at the line of contact between the plates. The 30th dark fringe gives for the diameter of the wire $2t = 29\lambda$, and $t = 14.5\lambda$.

A: $\quad r = \dfrac{t}{2} = 7.25\lambda = 7.25\left(600 \times 10^{-9} \text{ m}\right) = 4.35 \ \mu\text{m}$ $\qquad\qquad \Diamond$

L: This wire is not only less than 0.1 mm; it is even thinner than a typical human hair ($\sim 50 \ \mu$m).

41. Mirror M_1 in Figure 37.22 is displaced a distance ΔL. During this displacement, 250 fringe reversals (formation of successive dark or bright bands) are counted. The light being used has a wavelength of 632.8 nm. Calculate the displacement ΔL.

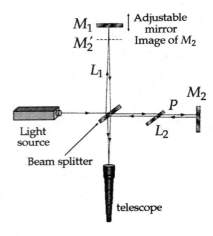

Figure 37.22

Solution

When the mirror on one arm is displaced by ΔL, the path difference increases by $2\Delta L$. A shift resulting in the formation of successive dark (or bright) fringes requires a path length change of one-half wavelength. Therefore, since in this case $m = 250$,

$$2\Delta L = m\lambda/2 \qquad \text{and} \qquad \Delta L = \frac{m\lambda}{4} = \frac{(250)(6.328 \times 10^{-7} \text{ m})}{4} = 3.955 \times 10^{-5} \text{ m} \quad \Diamond$$

51. Astronomers observed a 60.0-MHz radio source both directly and by reflection from the sea. If the receiving dish is 20.0 m above sea level, what is the angle of the radio source above the horizon at first maximum?

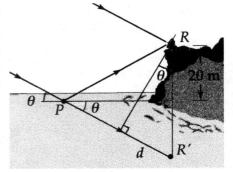

Solution One radio wave reaches the receiver R directly from the distant source at an angle θ above the horizontal.

The other wave undergoes phase reversal as it reflects from the water at P.

Constructive interference first occurs for a path difference of $\quad d = \lambda/2 \qquad$ (1)

The angles θ in the figure are equal because they each form part of a right triangle with a shared angle at R'. It is equally far from P to R as from P to R', the mirror image of the telescope.

So the path difference is

$$d = 2(20.0 \text{ m}) \sin \theta = (40.0 \text{ m}) \sin \theta$$

The wavelength is

$$\lambda = \frac{c}{f} = \frac{3.00 \times 10^8 \text{ m/s}}{60.0 \times 10^6 \text{ Hz}} = 5.00 \text{ m}$$

Substituting for d and λ in Eq. (1),

$$(40.0 \text{ m}) \sin \theta = \frac{5.00 \text{ m}}{2}$$

Solving for the angle θ,

$$\theta = \sin^{-1}\left(\frac{5.00 \text{ m}}{80.0 \text{ m}}\right) = 3.58° \qquad \Diamond$$

---

53. Measurements are made of the intensity distribution in a Young's interference pattern (see Fig. 37.6). At a particular value of y, it is found that $I/I_{max} = 0.810$ when 600-nm light is used. What wavelength of light should be used if the relative intensity at the same location is to be reduced to 64.0%?

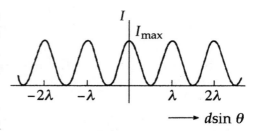

Figure 37.6

Solution

From Equation 37.13,

$$\frac{I_1}{I_{max}} = \cos^2\left(\frac{\pi y d}{\lambda_1 L}\right) = 0.810$$

Therefore,

$$\frac{\pi y d}{L} = \lambda_1 \cos^{-1}\sqrt{0.810} = (600 \text{ nm})\cos^{-1}(0.900) = 271 \text{ nm}$$

For the 2nd wavelength,

$$\frac{I_2}{I_{max}} = \cos^2\left(\frac{\pi y d}{\lambda_2 L}\right) = 0.640$$

and

$$\lambda_2 = \frac{\pi y d / L}{\cos^{-1}\sqrt{0.640}}$$

But this is simply

$$\lambda_2 = \frac{271 \text{ nm}}{\cos^{-1}\sqrt{0.640}} = 421 \text{ nm} \qquad \Diamond$$

---

57. The condition for constructive interference by reflection from a thin film in air as developed in Section 37.6 assumes nearly normal incidence. Show that if the light is incident on the film at a nonzero angle ϕ_1 (relative to the normal), then the condition for constructive interference is $2nt\cos\theta_2 = \left(m+\frac{1}{2}\right)\lambda$, where θ_2 is the angle of refraction.

Solution

In the figure, before they head off together in parallel paths, one beam travels distance y and undergoes phase reversal on reflection; the other beam travels distance $2x$ inside the film, where its wavelength is λ/n. This makes its optical path length $2nx$; the condition for constructive interference is

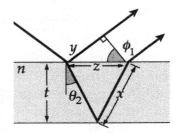

$$2nx - y = \left(m+\tfrac{1}{2}\right)\lambda$$

From the figure we have $\quad \cos\theta_2 = \dfrac{t}{x}, \quad \tan\theta_2 = \dfrac{z}{2t}, \quad$ and $\quad \sin\phi_1 = \dfrac{y}{z}$

Therefore, using Snell's law $\quad y = z\sin\phi_1 = (\sin\phi_1)2t\tan\theta_2 = \dfrac{(n\sin\theta_2)2t\sin\theta_2}{\cos\theta_2}$

Then the condition for constructive interference becomes

$$\frac{2tn}{\cos\theta_2} - \frac{2nt\sin^2\theta_2}{\cos\theta_2} = \left(m+\tfrac{1}{2}\right)\lambda$$

Since $\quad 1 - \sin^2\theta_2 = \cos^2\theta_2, \qquad 2nt\cos\theta_2 = \left(m+\tfrac{1}{2}\right)\lambda \qquad\qquad\qquad ◊$

61. Consider the double-slit arrangement shown in Figure P37.60, where the slit separation is d and the slit to screen distance is L. A sheet of transparent plastic having an index of refraction n and thickness t is placed over the upper slit. As a result, the central maximum of the interference pattern moves upward a distance y'. Find y'.

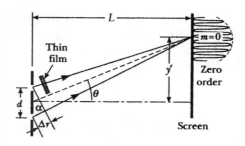

Figure P37.60

Solution

G: Since the film shifts the pattern upward, we should expect y' to be proportional to n, t, and L.

O: The film increases the optical path length of the light passing through the upper slit, so the physical distance of this path must be shorter for the waves to meet in phase ($\phi = 0$) to produce the central maximum. Thus, the added distance Δr traveled by the light from the lower slit must introduce a phase difference equal to that introduced by the plastic film.

A: First calculate the additional phase difference due to the plastic. Recall that the relation between phase difference and path difference is $\phi = 2\pi\delta/\lambda$. The presence of plastic affects this by changing the wavelength of the light, so that the phase change of the light in air and plastic, as it travels over the thickness t is

$$\phi_{air} = \frac{2\pi t}{\lambda_{air}} \qquad \text{and} \qquad \phi_{plastic} = \frac{2\pi t}{\lambda_{air}/n}$$

Thus, plastic causes an additional phase change of

$$\Delta\phi = \frac{2\pi t}{\lambda_{air}}(n-1)$$

335

Next, in order to interfere constructively, we must calculate the additional distance that the light from the bottom slit must travel.

$$\Delta r = \frac{\Delta\phi\,\lambda_{air}}{2\pi} = t(n-1)$$

In the small angle approximation we can write $\Delta r = y'd/L$, so

$$y' = \frac{t(n-1)L}{d} \qquad\qquad \Diamond$$

L: As expected, y' is proportional to t and L. It increases with increasing n, being proportional to $(n-1)$. It is also inversely proportional to the slit separation d, which makes sense since slits that are closer together make a wider interference pattern.

Chapter
38

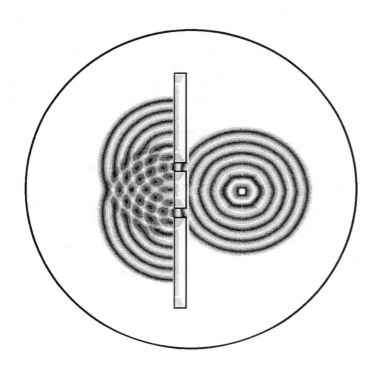

Diffraction and
Polarization

DIFFRACTION AND POLARIZATION

INTRODUCTION

When light waves pass through a small aperture, a diffraction pattern is observed rather than a sharp spot of light, showing that light spreads beyond the aperture into regions where a shadow would be expected if light traveled in straight lines. Other waves, such as sound waves and water waves, also have this property of being able to bend around corners. This phenomenon, known as diffraction, can be regarded as interference from a great number of coherent wave sources. In other words, diffraction and interference are basically equivalent.

In Chapter 34, we learned that electromagnetic waves are transverse. That is, the electric and magnetic field vectors are perpendicular to the direction of propagation. In this chapter, we see that under certain conditions, light waves can be polarized in various ways, such as by passing light through polarizing sheets.

EQUATIONS AND CONCEPTS

In single-slit diffraction, the total phase difference β between the first and last phasors will depend on the angle θ which determines the direction to an arbitrary point on the screen.

$$\beta = \frac{2\pi}{\lambda} a \sin \theta \qquad (38.3)$$

The resultant phasor will be zero (the condition for destructive interference) when β is an integer multiple of 2π This leads to a general **condition for destructive interference** in terms of θ.

$$\sin \theta = m \frac{\lambda}{a} \quad \text{(Destructive interference)} \qquad (38.1)$$

$$(m = \pm 1, \ \pm 2, \ \pm 3, \ldots)$$

338

The **intensity** at any point on the screen is given in terms of the intensity I_{max} at $\theta = 0$ where β is given by Equation 38.3.

$$I = I_{max}\left[\frac{\sin(\beta/2)}{\beta/2}\right]^2 \qquad (38.4)$$

Rayleigh's criterion states the condition for the resolution of two images due to nearby sources. For a slit, the angular separation between the sources must be greater than the ratio of the wavelength to slit width. In the case of a **circular aperture**, the minimum angular separation depends on D, the diameter of the aperture (or lens).

$$\theta_{min} = \frac{\lambda}{a} \qquad \text{(slit)} \qquad (38.8)$$

$$\theta_{min} = 1.22\frac{\lambda}{D} \qquad \text{(circular aperture)} \qquad (38.9)$$

A grating of equally spaced parallel slits (separated by a distance d) will produce an interference pattern in which there is a series of maxima for each wavelength. Maxima due to wavelengths of different value comprise a spectral order denoted by order number m.

$$d\sin\theta = m\lambda \qquad (38.10)$$

$$(m = 0, 1, 2, 3, \ldots)$$

The **resolving power** of a grating increases as the **number of lines illuminated** is increased and is proportional to the order in which the spectrum is observed.

$$R = \frac{\lambda}{\lambda_2 - \lambda_1} = \frac{\lambda}{\Delta\lambda} \qquad (38.11)$$

$$R = Nm \qquad (38.12)$$

When the order number m is substituted from Equation 38.10, $d\sin\theta = m\lambda$, into Equation 38.12, it becomes clear that **the resolution depends on the width of the grating** (Nd). It should also be noted that the angular width $\Delta\theta$ of a spectral line formed by a diffraction grating is inversely proportional to the width of the grating. This statement can be written as $\Delta\theta \propto 1/Nd$.

$$R = \frac{Nd}{\lambda}\sin\theta$$

In the 0th order (central maximum) all wavelengths are indistinguishable. If in a particular order ($m > 0$) $R = 10\,000$, the grating will produce a spectrum in which wavelengths differing in value by 1 part in 10 000 can be resolved.

Bragg's law gives the conditions for **constructive interference** of x-rays reflected from the parallel planes of a crystalline solid separated by a distance d. θ **is the angle between the incident beam and the surface.**

$$2d\sin\theta = m\lambda \qquad (38.13)$$
$$(m = 1, 2, 3 \ldots)$$

When polarized light is incident on a polarizing film, the fraction of the incident intensity transmitted through the film depends on the angle between the transmission axis of the polarizer and the electric field vector of the incident light. This is known as **Malus's law.**

$$I = I_{max}\cos^2\theta \qquad (38.14)$$

Brewster's law gives the polarizing angle for a particular surface. The **polarizing angle** θ_p is the angle of incidence for which the reflected beam is **completely polarized.**

$$n = \tan\theta_p \qquad\qquad (38.15)$$

REVIEW CHECKLIST

▷ Determine the positions of the maxima and minima in a single-slit diffraction pattern and calculate the intensities of the secondary maxima relative to the intensity of the central maximum. Determine the positions of the principal maxima in the interference pattern of a diffraction grating.

▷ Determine whether or not two sources under a given set of conditions are resolvable as defined by Rayleigh's criterion.

▷ Understand what is meant by the resolving power and the dispersion of a grating, and calculate the resolving power of a grating under specified conditions.

▷ Describe the technique of x-ray diffraction and make calculations of the lattice spacing using Bragg's law.

▷ Understand how the state of polarization of a light beam can be determined by use of a polarizer-analyzer combination. Describe qualitatively the polarization of light by selective absorption, reflection, scattering, and double refraction. Also, make appropriate calculations using Malus's law and Brewster's law.

ANSWERS TO SELECTED CONCEPTUAL QUESTIONS

1. Why can you hear around corners but not see around them?

Answer Audible sound has wavelengths on the order of meters or centimeters, while visible light has wavelengths on the order of half a micrometer. In this world of breadbox-size objects, λ is comparable to the object size for sound, and sound diffracts around walls and through doorways. But λ / a is much smaller for visible light passing ordinary-size objects or apertures, so light diffracts only through very small angles.

Another way of answering this question would be as follows. We can see by a small angle around a small obstacle or around the edge of a small opening. The side fringes in Figure 38.6 and the Arago spot in the center of the penny in Figure 38.3 (both in the textbook) show this diffraction. Conversely, we cannot always hear around corners. Out-of-doors, away from reflecting surfaces, have someone a few meters distant face away from you and whisper. The high-frequency, short-wavelength, information-carrying components of the sound do not diffract around his head enough for you to understand his words.

□   □   □   □

4. Describe the change in width of the central maximum of the single-slit diffraction pattern as the width of the slit is made narrower.

Answer Equation 38.1 describes the angles at which you get destructive interference; from it, we can obtain an estimate of the width of the central maximum. For small angles, the equation can be rewritten as $\theta_m = \sin^{-1}(m\lambda / a) \cong m\lambda / a$. Thus, as the width of the slit a decreases, the angle of the first destructive interference θ_1 grows, and the width of the central maximum grows as well.

□   □   □   □

SOLUTIONS TO SELECTED END-OF-CHAPTER PROBLEMS

3. A screen is placed 50.0 cm from a single slit, which is illuminated with 690-nm light. If the distance between the first and third minima in the diffraction pattern is 3.00 mm, what is the width of the slit?

Solution In the equation for single-slit diffraction minima at small angles,

$$y/L \cong \sin\theta = m\lambda/a$$

take differences between the first and third minima, to see that

$$\Delta y/L = \Delta m\lambda/a \quad \text{with} \quad \Delta y = 3.00 \times 10^{-3} \text{ m, and} \quad \Delta m = 3 - 1 = 2$$

The width of the slit is then

$$a = \frac{\lambda L \Delta m}{\Delta y} = \frac{\left(690 \times 10^{-9} \text{ m}\right)(0.500 \text{ m})(2)}{3.00 \times 10^{-3} \text{ m}} = 2.30 \times 10^{-4} \text{ m} \qquad \Diamond$$

7. A diffraction pattern is formed on a screen 120 cm away from a 0.400-mm-wide slit. Monochromatic 546.1-nm light is used. Calculate the fractional intensity I/I_{max} at a point on the screen 4.10 mm from the center of the principle maximum.

Solution $\sin\theta \cong \dfrac{y}{L} = \dfrac{4.10 \times 10^{-3} \text{ m}}{1.20 \text{ m}} = 3.417 \times 10^{-3}$

$$\frac{\beta}{2} = \frac{\pi a \sin\theta}{\lambda} = \frac{\pi\left(4.00 \times 10^{-4} \text{ m}\right)\left(3.417 \times 10^{-3} \text{ m}\right)}{546.1 \times 10^{-9} \text{ m}} = 7.863 \text{ rad}$$

$$\frac{I}{I_{max}} = \left[\frac{\sin\beta/2}{\beta/2}\right]^2 = \left[\frac{\sin(7.863 \text{ rad})}{7.863 \text{ rad}}\right]^2 = 1.62 \times 10^{-2} \qquad \Diamond$$

13. A helium-neon laser emits light that has a wavelength of 632.8 nm. The circular aperture through which the beam emerges has a diameter of 0.500 cm. Estimate the diameter of the beam 10.0 km from the laser.

Solution

G: A typical laser pointer makes a spot about 5 cm in diameter at 100 m, so the spot size at 10 km would be about 100 times bigger, or about 5 m across. Assuming that this HeNe laser is similar, we could expect a comparable beam diameter.

O: We assume that the light is parallel and not diverging as it passes through and fills the circular aperture. However, as the light passes through the circular aperture, it will spread from diffraction according to Equation 38.9.

A: The beam spreads into a cone of half-angle

$$\theta_{min} = 1.22\frac{\lambda}{D} = 1.22\frac{\left(632.8\times10^{-9}\text{ m}\right)}{(0.00500\text{ m})} = 1.54\times10^{-4}\text{ rad}$$

The radius of the beam ten kilometers away is, from the definition of radian measure,

$$r_{beam} = \theta_{min}\left(1.00\times10^{4}\text{ m}\right) = 1.54\text{ m}$$

and its diameter is $\quad d_{beam} = 2r_{beam} = 3.09\text{ m}$ ◊

L: The beam is several meters across as expected, and is about 600 times larger than the laser aperture. Since most HeNe lasers are low power units in the mW range, the beam at this range would be so spread out that it would be too dim to see on a screen.

15. The Impressionist painter Georges Seurat created paintings with an enormous number of dots of pure pigment each of which was approximately 2.00 mm in diameter. The idea was to have colors such as red and green next to each other to form a scintillating canvas (Fig. P38.15). Outside what distance would one be unable to discern individual dots on the canvas? (Assume $\lambda = 500$ nm and that the pupil diameter is 4.00 mm.)

Solution   We will assume that the dots are just touching and do not overlap so that the distance between their centers is 2.00 mm. By Rayleigh's criterion, two dots separated center-to-center by 2.00 mm would be seen to overlap when

$$\theta_{min} = \frac{d}{L} = 1.22\frac{\lambda}{D}$$

with $d = 2.00$ mm, $\lambda = 500$ nm,   and $D = 4.00$ mm

Thus, $L = \dfrac{Dd}{1.22\lambda} = \dfrac{\left(4.00\times10^{-3}\text{ m}\right)\left(2.00\times10^{-3}\text{ m}\right)}{1.22\left(500\times10^{-9}\text{ m}\right)} = 13.1$ m   ◊

23. White light is spread out into its spectral components by a diffraction grating. If the grating has 2000 lines per centimeter, at what angle does red light of wavelength 640 nm appear in first order?

Solution   The grating spacing is $d = \dfrac{1.00\times10^{-2}\text{ m}}{2000} = 5.00\times10^{-6}$ m

The light is deflected according to $\sin\theta = \dfrac{m\lambda}{d} = \dfrac{1\left(640\times10^{-9}\text{ m}\right)}{5.00\times10^{-6}\text{ m}} = 0.128$

at an angle of $\theta = 7.35°$ ◊

25. The hydrogen spectrum has a red line at 656 nm and a violet line at 434 nm. What is the angular separation between two spectral lines obtained with a diffraction grating that has 4500 lines/cm?

Solution

G: Most diffraction gratings yield several spectral orders within the 180° viewing range, which means that the angle between red and violet lines is probably 10° to 30°.

O: The angular separation is the difference between the angles corresponding to the red and violet wavelengths for each visible spectral order according to the diffraction grating equation, $d \sin \theta = m \lambda$.

A: The grating spacing is $\quad d = \left(1.00 \times 10^{-2} \text{ m}\right)/4500 \text{ lines} = 2.22 \times 10^{-6} \text{ m}$

In the first-order spectrum ($m = 1$), the angles of diffraction are given by $\sin \theta = \lambda/d$:

$$\sin \theta_{1r} = \frac{656 \times 10^{-9} \text{ m}}{2.22 \times 10^{-6} \text{ m}} = 0.295 \quad \text{so} \quad \theta_{1r} = 17.17°$$

$$\sin \theta_{1v} = \frac{434 \times 10^{-9} \text{ m}}{2.22 \times 10^{-6} \text{ m}} = 0.195 \quad \text{so} \quad \theta_{1v} = 11.26°$$

The angular separation is $\quad \Delta\theta_1 = \theta_{1r} - \theta_{1v} = 17.17° - 11.26° = 5.91° \quad \lozenge$

In the 2nd-order ($m = 2$) $\qquad \Delta\theta_2 = \sin^{-1}\left(\dfrac{2\lambda_r}{d}\right) - \sin^{-1}\left(\dfrac{2\lambda_v}{d}\right) = 13.2° \quad \lozenge$

In the third order ($m = 3$), $\qquad \Delta\theta_3 = \sin^{-1}\left(\dfrac{3\lambda_r}{d}\right) - \sin^{-1}\left(\dfrac{3\lambda_v}{d}\right) = 26.5° \quad \lozenge$

Examining the fourth order, we find the red line is not visible:

$$\theta_{4r} = \sin^{-1}(4\lambda_r / d) = \sin^{-1}(1.18) \text{ does not exist} \quad \lozenge$$

L: The full spectrum is visible in the first 3 orders with this diffraction grating, and the fourth is partially visible. We can also see that the pattern is dispersed more for higher spectral orders so that the angular separation between the red and blue lines increases as *m* increases. It is also worth noting that the spectral orders can overlap (as is the case for the second and third order spectra above), which makes the pattern look confusing if you do not know what you are looking for.

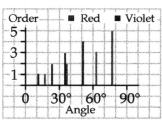

---

29. A diffraction grating of length 4.00 cm has been ruled with 3000 grooves per centimeter. (a) What is the resolving power of this grating in the first three orders? (b) If two monochromatic waves incident on this grating have a mean wavelength of 400 nm, what is their wavelength separation if they are just resolved in the third order?

Solution We assume the two distances are measured in the same direction.

From Eq. 38.12, $R = mN$

where $N = (3000 \text{ lines / cm})(4.00 \text{ cm}) = 12000 \text{ lines}$

(a) In the 1st order, $R = (1)(12000 \text{ lines}) = 12000$ ◊

In the 2nd order, $R = (2)(12000 \text{ lines}) = 24000$ ◊

In the 3rd order, $R = (3)(12000 \text{ lines}) = 36000$ ◊

(b) From Eq. 38.11, $R = \lambda / \Delta\lambda$

In the 3rd order, $\Delta\lambda = \dfrac{\lambda}{R} = \dfrac{4.00 \times 10^{-7} \text{ m}}{3.60 \times 10^{4}} = 1.11 \times 10^{-11} \text{ m} = 0.0111 \text{ nm}$ ◊

---

31. A source emits 531.62-nm and 531.81-nm light. (a) What minimum number of lines is required for a grating that resolves the two wavelengths in the first-order spectrum? (b) Determine the slit spacing for a grating 1.32 cm wide that has the required minimum number of lines.

Solution The resolving power of the diffraction grating is $Nm = \lambda/\Delta\lambda$.

(a) Resolving the first-order spectrum, $N(1) = \dfrac{531.7 \text{ nm}}{0.190 \text{ nm}} = 2800 \text{ lines}$ ◊

(b) The slits are spaced at intervals of $\dfrac{1.32\times10^{-2} \text{ m}}{2800} = 4.72 \ \mu\text{m}$ ◊

33. A grating with 250 lines/mm is used with an incandescent light source. Assume that the visible spectrum ranges in wavelength from 400 to 700 nm. In how many orders can one see (a) the entire visible spectrum and (b) the short-wavelength region?

Solution The grating spacing is $d = 1.00 \text{ mm}/250 = 4.00\times10^{-6} \text{ m}$

(a) In each order of interference m, red light diffracts at a larger angle than the other colors with shorter wavelengths. We find the largest number m satisfying

$$d\sin\theta = m\lambda \qquad \text{with} \qquad \lambda = 700 \text{ nm};$$

With $\sin\theta$ having its largest value

$$\left(4.00\times10^{-6} \text{ m}\right)\sin 90° = m\left(700\times10^{-9} \text{ m}\right)$$

so that $m = 5.71$

Thus red light cannot be seen in the 6th order, and the full visible spectrum appears in only five orders. ◊

(b) Now consider light at the boundary between violet and ultraviolet.

$d \sin \theta = m\lambda$ becomes $\left(4.00 \times 10^{-6}\ m\right) \sin 90.0° = m\left(400 \times 10^{-9}\ m\right)$

and $m = 10$ ◊

---

37. If the interplanar spacing of NaCl is 0.281 nm, what is the predicted angle at which 0.140-nm x-rays are diffracted in a first-order maximum?

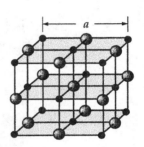

Solution

The atomic planes in this crystal are shown in Figure 38.24 of the text. The diffraction they produce is described by the Bragg condition, that

Figure 38.24

$$2d \sin \theta = m\lambda$$

Solving for θ, $\sin \theta = \dfrac{m\lambda}{2d} = \dfrac{1\left(0.140 \times 10^{-9}\ m\right)}{2\left(0.281 \times 10^{-9}\ m\right)} = 0.249$

The maximum occurs at $\theta = 14.4°$ ◊

---

43. Plane-polarized light is incident on a single polarizing disk with the direction of E_0 parallel to the direction of the transmission axis. Through what angle should the disk be rotated so that the intensity in the transmitted beam is reduced by a factor of (a) 3.00, (b) 5.00, (c) 10.0?

Solution   We start from $\theta = 0$, in $I = I_{max} \cos^2 \theta$.

(a)   For $I = I_{max}/3.00$, $\cos\theta = \dfrac{1}{\sqrt{3.00}}$ and $\theta = 54.7°$ ◊

(b)   For $I = I_{max}/5.00$, $\cos\theta = \dfrac{1}{\sqrt{5.00}}$ and $\theta = 63.4°$ ◊

(c)   For $I = I_{max}/10.0$, $\cos\theta = \dfrac{1}{\sqrt{10.0}}$ and $\theta = 71.6°$ ◊

---

45.   The critical angle for total internal reflection for sapphire surrounded by air is 34.4°. Calculate the polarizing angle for sapphire.

Solution $n = \tan\theta_p$

$\theta_p = \tan^{-1} n$

and $\sin\theta_c = \dfrac{1}{n}$ $n = \dfrac{1}{\sin\theta_c}$

Therefore, $\theta_p = \tan^{-1}\left(\dfrac{1}{\sin\theta_c}\right) = \tan^{-1}\left(\dfrac{1}{\sin 34.4°}\right) = 60.5°$ ◊

---

57.   Light of wavelength 500 nm is incident normally on a diffraction grating. If the third-order maximum of the diffraction pattern is observed at 32.0°, (a) what is the number of rulings per centimeter for the grating? (b) Determine the total number of primary maxima that can be observed in this situation.

Solution

G: The diffraction pattern described in this problem seems to be similar to previous problems that have diffraction gratings with 2 000 to 5 000 lines/mm. With the third-order maximum at 32°, there are probably 5 or 6 maxima on each side of the central bright fringe, for a total of 11 or 13 primary maxima.

O: The diffraction grating equation can be used to find the grating spacing and the angles of the other maxima that should be visible within the 180° viewing range.

A: (a) Use Equation 38.10, $d\sin\theta = m\lambda$

$$d = \frac{m\lambda}{\sin\theta} = \frac{(3)(5.00\times10^{-7}\ \text{m})}{\sin(32.0°)} = 2.83\times10^{-6}\ \text{m}$$

Thus, the line spacing is $\frac{1}{d} = 3.534\times10^{5}$ lines $/$ m $= 3530$ lines $/$ cm ◊

(b)

$$\sin\theta = m\left(\frac{\lambda}{d}\right) = \frac{m(5.00\times10^{-7}\ \text{m})}{2.83\times10^{-6}\ \text{m}} = m(0.177)$$

For $\sin\theta \le 1$, we require that $m(1.77) \le 1$ or $m \le 5.65$. Since m must be an integer, its maximum value is really 5. Therefore, the total number of maxima is $2m+1=11$ ◊

L: The results agree with our predictions, and apparently there are 5 maxima on either side of the central maximum. If more maxima were desired, a grating with **fewer** lines/cm would be required; however, this would reduce the ability to resolve the difference between lines that appear close together.

61. Suppose that the single slit in Figure 38.6 is 6.00 cm wide and in front of a microwave source operating at 7.50 GHz. (a) Calculate the angle subtended by the first minimum in the diffraction pattern. (b) What is the relative intensity I/I_{max} at $\theta = 15.0°$? (c) Consider the case when there are two such sources, separated laterally by 20 cm, behind the slit. What must the maximum distance between the plane of the sources and the slit be if the diffraction patterns are to be resolved? (In this case, the approximation $\sin\theta \approx \tan\theta$ is not valid because of the relatively small value of a/λ.)

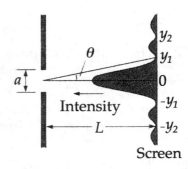

Figure 38.6

Solution (a) From Eq. 38.1, $\theta = \sin^{-1}(m\lambda/a)$

In this case, $m = 1$, and
$$\lambda = \frac{c}{f} = \frac{(3.00 \times 10^8 \text{ m/s})}{(7.50 \times 10^9 \text{ s}^{-1})} = 0.0400 \text{ m}$$

so
$$\theta = \sin^{-1}(0.666) = 41.8° \qquad \Diamond$$

(b) From Equation 38.4,
$$\frac{I}{I_{max}} = \left(\frac{\sin(\beta/2)}{\beta/2}\right)^2 \quad \text{where} \quad \beta = \frac{2\pi a \sin\theta}{\lambda}$$

When $\theta = 15.0°$, $\beta = 2.44$ rad and $I/I_{max} = 0.593$ $\qquad \Diamond$

(c) Let L' be the maximum distance between the plane of the two sources and the slit. The minimum angle subtended by the two sources at the slit is $\theta = 41.81°$, and the half-angle between the sources is $\alpha = \theta/2 = 20.91°$. From the figure,

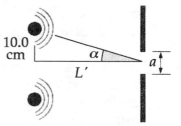

$$\tan\alpha = 10.0 \text{ cm}/L'$$

so
$$L' = \frac{10.0 \text{ cm}}{\tan 20.91°} = 26.2 \text{ cm} \qquad \Diamond$$

67. Another method to solve the equation $\phi = \sqrt{2} \sin \phi$ in Problem 66 is to use a calculator, guess a first value of ϕ, see if it fits, and continue to update your estimate until the equation balances. How many steps (iterations) does this take?

Solution   We can list each trial as we try to home in on the solution to $\phi = \sqrt{2} \sin \phi$ by narrowing the range in which it must lie:

ϕ	$\sqrt{2} \sin \phi$	
1	1.19	bigger than ϕ
2	1.29	smaller than ϕ
1.5	1.41	smaller
1.4	1.394	
1.39	1.391	bigger
1.395	1.392	
1.392	1.3917	smaller
1.3915	1.39154	bigger
1.39152	1.39155	bigger
1.3916	1.391568	smaller
1.39158	1.391563	
1.39157	1.391560	
1.39156	1.391558	
1.391559	1.3915578	
1.391558	1.3915575	
1.391557	1.3915573	
1.3915574	1.3915574	

We get the answer to seven digits after 17 steps. Clever guessing, like just using the value of $\sqrt{2} \sin \phi$ as the next guess for ϕ, could reduce this to around 13 steps. ◊

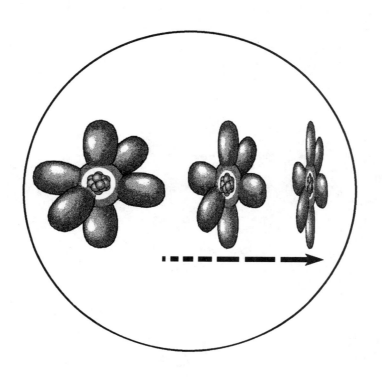

Chapter 39

Relativity

RELATIVITY

INTRODUCTION

Most of our everyday experiences and observations have to do with objects that move at speeds much less than that of light. Newtonian mechanics and early ideas on space and time were formulated to describe the motion of such objects. This formalism is very successful in describing a wide range of phenomena that occur at low speeds. It fails, however, when applied to particles whose speeds approach that of light. For example, it is possible to accelerate an electron to a speed of $0.99c$ (where c is the speed of light) by using a potential difference of several million volts. According to Newtonian mechanics, if the potential difference is increased by a factor of 4, the electron speed should jump to $1.98c$. However, experiments show that the speed of the electron always remains less than the speed of light, regardless of the size of the accelerating voltage. Thus, we can see that the validity of Newtonian mechanics is limited to low-velocity interactions.

In 1905, at the age of only 26, Einstein published his special theory of relativity. With this theory, experimental observations can be correctly predicted over the range of speeds from $v = 0$ to speeds approaching the speed of light. Newtonian mechanics, which was accepted for over 200 years, is in fact a special case of Einstein's theory. This chapter gives an introduction to the special theory of relativity, with emphasis on some of its consequences.

EQUATIONS AND CONCEPTS

In order to explain the motion of particles moving at speeds approaching the speed of light, one must use the **Lorentz coordinate transformation equations**. These equations represent the transformation between any two inertial frames in **relative** motion with velocity v in the x direction.

$$t' = \gamma\left(t - \frac{v}{c^2}x\right)$$

$$x' = \gamma(x - vt)$$

$$y' = y$$

$$z' = z$$

where

$$\gamma \equiv \frac{1}{\sqrt{1 - v^2/c^2}}$$

(39.11)

Lorentz velocity transformation equations relate the observed velocity u' in the moving (S') frame to the measured velocity u in the S frame, and the relative velocity of S' with respect to S.

$$u'_x = \frac{u_x - v}{1 - u_x v/c^2} \qquad (39.16)$$

$$u'_y = \frac{u_y}{\gamma\left(1 - u_x v/c^2\right)} \qquad (39.17)$$

$$u'_z = \frac{u_z}{\gamma\left(1 - u_x v/c^2\right)}$$

Consider a single clock in the S' frame in which an observer in this frame measures the time interval of an event Δt_p (say the time it takes a light pulse to travel from a source to a mirror and back to the source). The time interval Δt for this event as measured by an observer in S is greater than Δt_p, since the observer in S must make clock readings at two different positions. That is, **a moving clock appears to run slower than an identical clock at rest with respect to the observer.** This is known as time dilation.

$$\Delta t = \frac{\Delta t_p}{\sqrt{1 - v^2/c^2}} = \gamma t_p \qquad (39.7)$$

If an object moves along the x axis with a speed v, and has a length L_p as measured in the moving frame, the length L as measured by a stationary observer is shorter than L_p by the factor $1/\gamma$. This is called **length contraction**, and the length L_p measured in the reference frame in which the object is at rest is called the **proper length**.

$$L = \frac{L_p}{\gamma} = L_p\left(1 - v^2/c^2\right)^{1/2} \qquad (39.9)$$

The fact that observers in the S and S' frames do not reach the same conclusions regarding such measurements can be understood by recognizing that simultaneity is not an absolute concept. That is, **Two events that are simultaneous in one reference frame are not simultaneous in another reference frame that is in motion relative to the first.**

The relativistic equation for the **momentum** of a particle of mass m moving with a speed u satisfies the conditions that (1) momentum is conserved in all collisions and (2) p approaches the classical expression as u approaches 0. That is, as $u \rightarrow 0, p \rightarrow mu$.

$$\mathbf{p} \equiv \frac{m\mathbf{u}}{\sqrt{1-u^2/c^2}} = \gamma m\mathbf{u} \qquad (39.19)$$

In the relativistic equation for the **kinetic energy** of a particle of mass m moving with a speed u, the term mc^2 is called the **rest energy.**

$$K = \frac{mc^2}{\sqrt{1-\dfrac{u^2}{c^2}}} - mc^2 = \gamma mc^2 - mc^2 \qquad (39.23)$$

The **total energy** E of a particle is the sum of the kinetic energy and the rest energy. This expression shows that **mass and energy are equivalent concepts**; that is, mass is a form of energy.

$$E = \frac{mc^2}{\sqrt{1-u^2/c^2}} \qquad (39.25)$$

When the momentum or energy of a particle is known (rather than the speed), it is useful to have an expression relating the total energy and relativistic momentum.

$$E^2 = p^2c^2 + (mc^2)^2 \qquad (39.26)$$

There is an exact expression relating energy and momentum for particles which have zero mass (e.g. photons). These zero-mass particles always travel at the speed of light.

$$E = pc \qquad (39.27)$$

A convenient energy unit to use to express the energies of electrons and other subatomic particles is the electron volt (eV).

$$1 \text{ eV} = 1.60 \times 10^{-19} \text{ J}$$

REVIEW CHECKLIST

▷ State Einstein's two postulates of the special theory of relativity.

▷ Understand the Michelson-Morley experiment, its objectives, results, and the significance of its outcome.

▷ Understand the idea of simultaneity, and the fact that simultaneity is not an absolute concept. That is, two events which are simultaneous in one reference frame are not simultaneous when viewed from a second frame moving with respect to the first.

▷ Make calculations using the equations for time dilation, length contraction, and relativistic mass.

▷ State the correct relativistic expressions for the momentum, kinetic energy, and total energy of a particle. Make calculations using these equations.

ANSWERS TO SELECTED CONCEPTUAL QUESTIONS

7. List some ways our day-to-day lives would change if the speed of light were only 50 m/s.

Answer

For a wonderful fictional exploration of this question, get a "Mr. Tompkins" book by George Gamow. All of the relativity effects would be obvious in our lives. Time dilation and length contraction would both occur. Driving home in a hurry, you would push on the gas pedal not to increase your speed very much, but to make the blocks shorter. Big Doppler shifts in wave frequencies would make red lights look green as you approached, and make car horns and radios useless. High-speed transportation would be both very expensive, requiring huge fuel purchases, as well as dangerous, since a speeding car could knock down a building. When you got home, hungry for lunch, you would find that you had missed dinner; there would be a five-day delay in transit when you watched the Olympics in Australia on live TV. Finally, we would not be able to see the Milky Way, since the fireball of the Big Bang would surround us at the distance of Rigel, or Deneb.

□   □   □   □

8. Give a physical argument that shows that it is impossible to accelerate an object of mass m to the speed of light, even if it has a continuous force acting on it.

Answer

As an object approaches the speed of light, its energy approaches infinity. Hence, it would take an infinite amount of work to accelerate the object to the speed of light under the action of a continuous force or it would take an infinitely large force.

□   □   □   □

SOLUTIONS TO SELECTED END-OF-CHAPTER PROBLEMS

3. In a laboratory frame of reference, an observer notes that Newton's second law is valid. Show that it is also valid for an observer moving at a constant speed, small compared with the speed of light, relative to the laboratory frame.

Solution

The first observer watches some object accelerate under applied forces. Call the instantaneous velocity of the object u_x. The second observer has constant velocity v relative to the first, and measures the object to have velocity

$$u'_x = u_x - v$$

The acceleration is $\qquad \dfrac{du'_x}{dt} = \dfrac{du_x}{dt} - 0$

This is the same as that measured by the first observer. In this nonrelativistic case, they measure also the same mass and forces; so the second observer also confirms that

$$\Sigma F = ma. \qquad\qquad\qquad \Diamond$$

---

9. An atomic clock moves at 1000 km/h for 1 h as measured by an identical clock on Earth. How many nanoseconds slow will the moving clock be at the end of the 1-h interval?

Solution This problem is slightly more difficult than most, for the simple reason that your calculator probably cannot hold enough decimal places to yield an accurate answer.

However, we can bypass the difficulty by noting the approximation:

$$\sqrt{1-\frac{v^2}{c^2}} \cong 1-\frac{v^2}{2c^2}$$

Squaring both sides will show that when v/c is small, these two terms are equivalent.

Solving for v/c,

$$\frac{v}{c} = \left(\frac{1000 \times 10^3 \text{ m/h}}{3.00 \times 10^8 \text{ m/s}}\right)\left(\frac{1 \text{ h}}{3600 \text{ s}}\right) = 9.26 \times 10^{-7}$$

From Equation 9.7,

$$\Delta t = \gamma \Delta t_p = \frac{\Delta t_p}{\sqrt{1-v^2/c^2}}$$

Our approximation yields

$$\Delta t_p = \left(\sqrt{1-\frac{v^2}{c^2}}\right)\Delta t \cong \left(1-\frac{v^2}{2c^2}\right)\Delta t$$

and

$$\Delta t - \Delta t_p = \frac{v^2}{2c^2}\Delta t$$

Substituting,

$$\Delta t - \Delta t_p = \frac{\left(9.26 \times 10^{-7}\right)^2}{2}(3600 \text{ s})$$

Thus the time lag of the moving clock is

$$\Delta t - \Delta t_p = 1.54 \times 10^{-9} \text{ s} = 1.54 \text{ ns} \qquad \Diamond$$

---

11. A spaceship with a proper length of 300 m takes 0.750 μs seconds to pass an Earth observer. Determine its speed as measured by the Earth observer.

Solution

G: We should first determine if the spaceship is traveling at a relativistic speed: classically, $v = (300\text{m})/(0.750\ \mu s) = 4.00 \times 10^8$ m/s, which is faster than the speed of light (impossible)! Quite clearly, the relativistic correction must be used to find the correct speed of the spaceship, which we can guess will be close to the speed of light.

O: We can use the contracted length equation to find the speed of the spaceship in terms of the proper length and the time. The time of $0.750\ \mu s$ is the **proper time** measured by the Earth observer, because it is the time interval between two events that she sees as happening at the same point in space. The two events are the passage of the front end of the spaceship over her stopwatch, and the passage of the back end of the ship.

A: $L = L_p / \gamma$, with $L = v\Delta t$:

$$v\Delta t = L_p\left(1 - v^2/c^2\right)^{1/2}$$

Squaring both sides,

$$v^2\Delta t^2 = L_p{}^2\left(1 - v^2/c^2\right)$$

$$v^2c^2 = L_p{}^2c^2/\Delta t^2 - v^2L_p{}^2/\Delta t^2$$

Solving for the velocity,

$$v = \frac{cL_p/\Delta t}{\sqrt{c^2 + L_p{}^2/\Delta t^2}}$$

So

$$v = \frac{\left(3.00 \times 10^8\right)(300\ \text{m})\big/\left(0.750 \times 10^{-6}\ \text{s}\right)}{\sqrt{\left(3.00 \times 10^8\right)^2 + (300\ \text{m})^2\big/\left(0.750 \times 10^{-6}\ \text{s}\right)^2}} = 2.40 \times 10^8\ \text{m/s}\quad \Diamond$$

L: The spaceship is traveling at $0.8c$. We can also verify that the general equation for the speed reduces to the classical relation $v = L_p/\Delta t$ when the time is relatively large.

23. Two jets of material from the center of a radio galaxy fly away in opposite directions. Both jets move at 0.750c relative to the galaxy. Determine the speed of one jet relative to the other.

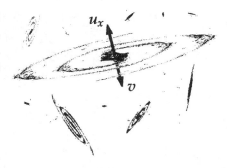

Solution

Take the galaxy as the unmoving frame. Arbitrarily define the jet moving upwards to be the object, and the jet moving downwards to be the "moving" frame.

$$u_x = 0.750\ c \qquad\qquad v = -0.750\ c$$

Thus, $\quad u'_x = \dfrac{u_x - v}{1 - \dfrac{u_x v}{c^2}} = \dfrac{0.750c - (-0.750c)}{1 - \dfrac{(0.750c)(-0.750c)}{c^2}} = \dfrac{1.50c}{1 + 0.750^2} = 0.960c$ ◊

---

31. An unstable particle at rest breaks into two fragments of **unequal** mass. The rest mass of the lighter fragment is 2.50×10^{-28} kg, and that of the heavier fragment is 1.67×10^{-27} kg. If the lighter fragment has a speed of 0.893c after the breakup, what is the speed of the heavier fragment?

Solution

G: The heavier fragment should have a speed less than that of the lighter piece since the momentum of the system must be conserved. However, due to the relativistic factor, the ratio of the speeds will not equal the simple ratio of the particle masses, which would give a speed of 0.134c for the heavier particle.

O: Relativistic momentum of the system must be conserved. For the total momentum to be zero after the fission, as it was before, $\mathbf{p_1 + p_2} = 0$, where we will refer to the lighter particle with the subscript '1', and to the heavier particle with the subscript '2.'

A:
$$\gamma_2 m_2 v_2 + \gamma_1 m_1 v_1 = 0$$

$$\gamma_2 m_2 v_2 + \left(\frac{2.50 \times 10^{-28} \text{ kg}}{\sqrt{1 - 0.893^2}}\right)(0.893c) = 0$$

Rearranging,
$$\left(\frac{1.67 \times 10^{-27} \text{ kg}}{\sqrt{1 - v_2^2/c^2}}\right)\frac{v_2}{c} = -4.96 \times 10^{-28} \text{ kg}$$

Squaring both sides,
$$\left(2.79 \times 10^{-54} \text{ kg}\right)\left(\frac{v_2}{c}\right)^2 = \left(2.46 \times 10^{-55} \text{ kg}\right)\left(1 - \frac{v_2^2}{c^2}\right)$$

and
$$v_2 = -0.285c$$

We choose the negative sign only to mean that the two particles must move in opposite directions. The speed, then, is $|v_2| = 0.285c$ ◊

L: The speed of the heavier particle is less than the lighter particle, as expected. We can also see that for this situation, the relativistic speed of the heavier particle is about twice as great as was predicted by a simple non-relativistic calculation.

33. Find the momentum of a proton in MeV/c units if its total energy is twice its rest energy.

Solution This problem has several steps. First, we solve for the rest energy of the proton; then we use that in the equation for total energy, to find the speed of the proton. Last, we substitute the speed and rest energy into the relativistic momentum equation to obtain the momentum.

For a proton, $mc^2 = \left(1.67 \times 10^{-27} \text{ kg}\right)\left(2.998 \times 10^8 \text{ m / s}\right)^2 = 1.50 \times 10^{-10}$ J

In MeV, $mc^2 = \left(1.50 \times 10^{-10} \text{ J}\right)\left(\dfrac{1 \text{ eV}}{1.60 \times 10^{-19} \text{ kg} \cdot \text{m}^2/\text{s}^2}\right) = 938$ MeV

Total energy is $\gamma mc^2 = 2mc^2$ so that $\gamma = 2$

and $\dfrac{1}{\sqrt{1 - v^2 / c^2}} = 2$ and $v / c = 0.866$

Momentum is $p = \gamma mv = \dfrac{\left(\gamma mc^2\right)}{c}(v / c)$

So in this case $p = \dfrac{2(938 \text{ MeV})}{c}(0.866) = 1620$ MeV $/ c$ ◊

---

35. A proton moves at 0.950c. Calculate its (a) rest energy, (b) total energy, and (c) kinetic energy.

Solution At $v = 0.950c$, $\gamma = \dfrac{1}{\sqrt{1 - v^2 / c^2}} = \dfrac{1}{\sqrt{1 - 0.950^2}} = 3.20$

(a) $E_R = mc^2 = (1.67 \times 10^{-27} \text{ kg})(2.998 \times 10^8 \text{ m / s})^2 = 1.50 \times 10^{-10}$ J $= 938$ MeV ◊

(b) $E = \gamma mc^2 = \gamma E_R = (3.20)(938 \text{ MeV}) = 3.00$ GeV ◊

(c) $K = E - E_R = 3.01 \text{ GeV} - 938 \text{ MeV} = 2.06$ GeV ◊

---

39. Show that the energy-momentum relationship $E^2 = p^2c^2 + (mc^2)^2$ follows from the expressions $E = \gamma mc^2$ and $p = \gamma mu$.

Solution $\qquad\qquad E = \gamma mc^2 \qquad\qquad p = \gamma mu$

Squaring both equations, $\qquad E^2 = \left(\gamma mc^2\right)^2 \qquad p^2 = \left(\gamma mu\right)^2$

Multiplying the second equation by
c^2, and subtracting it from the first, $\quad E^2 - p^2c^2 = (\gamma mc^2)^2 - (\gamma mu)^2 c^2$

$$E^2 - p^2c^2 = \gamma^2\left(\left(mc^2\right)\left(mc^2\right) - \left(mc^2\right)\left(mu^2\right)\right)$$

Extracting the $\left(mc^2\right)$ terms, $\qquad E^2 - p^2c^2 = \gamma^2\left(mc^2\right)^2\left(1 - \frac{u^2}{c^2}\right)$

and applying the definition of γ, $\quad E^2 - p^2c^2 = \left(1 - \frac{u^2}{c^2}\right)^{-1}\left(mc^2\right)^2\left(1 - \frac{u^2}{c^2}\right)$

The γ terms drop out, leaving $\qquad E^2 - p^2c^2 = \left(mc^2\right)^2 \qquad\qquad \lozenge$

---

43. A pion at rest ($m_\pi = 270m_e$) decays to a muon ($m_\mu = 206m_e$) and an antineutrino ($m_{\bar{\nu}} \cong 0$). The reaction is written $\pi^- \to \mu^- + \bar{\nu}$. Find the kinetic energy of the muon and the antineutrino in electron volts. (**Hint:** Relativistic momentum is conserved.)

Solution By conservation of energy, $\qquad\qquad m_\pi c^2 = \gamma m_\mu c^2 + |p_\nu|c$

By conservation of momentum, $\qquad\qquad p_\nu = -p_\mu = -\gamma m_\mu v$

Chapter 39

Substituting the second equation into the first, $m_\pi c^2 = \gamma m_\mu c^2 + \gamma m_\mu vc$

Simplified, this equation then reads $m_\pi = m_\mu(\gamma + \gamma v/c)$

Substituting for the masses, $270 m_e = (206 m_e)\left(\gamma + \dfrac{\gamma v}{c}\right)$

where $m_e c^2 = 0.511\ \text{MeV}$.

Numerically, $\dfrac{270 m_e}{206 m_e} = 1.31 = \dfrac{1 + v/c}{\sqrt{1 - (v/c)^2}}$

Solving for v/c with the quadratic formula, $\dfrac{v}{c} = \dfrac{270^2 - 206^2}{270^2 + 206^2} = 0.264$

Therefore, $\gamma = \dfrac{1}{\sqrt{1 - v^2/c^2}} = 1.0368$

and $K_\mu = (0.0368)(206 \times 0.511\ \text{MeV}) = 3.88\ \text{MeV}$ ◊

$K_{\bar\nu} = (270 \times 0.511\ \text{MeV}) - (206 \times 0.511\ \text{MeV} + 3.88\ \text{MeV}) = 28.8\ \text{MeV}$ ◊

49. The power output of the Sun is 3.77×10^{26} W. How much mass is converted to energy in the Sun each second?

Solution From $E_R = mc^2$, we have

$$m = \dfrac{E_R}{c^2} = \dfrac{3.77 \times 10^{26}\ \text{J}}{(3.00 \times 10^8\ \text{m/s})^2} = 4.19 \times 10^9\ \text{kg}$$ ◊

53. The cosmic rays of highest energy are protons, which have kinetic energy on the order of 10^{13} MeV. (a) How long would it take a proton of this energy to travel across the Milky Way galaxy, having a diameter on the order of ~10^5 light-years, as measured in the proton's frame? (b) From the point of view of the proton, how many kilometers across is the galaxy?

Solution

G: We can guess that the energetic cosmic rays will be traveling close to the speed of light, so the time it takes a proton to traverse the Milky Way will be much less in the proton's frame than 10^5 years. The galaxy will also appear smaller to the high-speed protons than the galaxy's proper diameter of 10^5 light-years.

O: The kinetic energy of the protons can be used to determine the relativistic γ-factor, which can then be applied to the time dilation and length contraction equations to find the time and distance in the proton's frame of reference.

A: The relativistic kinetic energy of a proton is $K = (\gamma - 1)mc^2 = 10^{13}$ MeV
Its rest energy is

$$mc^2 = \left(1.67 \times 10^{-27} \text{ kg}\right)\left(2.998 \times 10^8 \, \frac{\text{m}}{\text{s}}\right)^2 \left(\frac{1 \text{ eV}}{1.60 \times 10^{-19} \text{ kg} \cdot \text{m}^2/\text{s}^2}\right) = 938 \text{ MeV}$$

So 10^{13} MeV $= (\gamma - 1)(938 \text{ MeV})$, and therefore $\gamma = 1.07 \times 10^{10}$

The proton's speed in the galaxy's reference frame can be found from
$\gamma = 1/\sqrt{1 - v^2/c^2}$: $\quad 1 - v^2/c^2 = 8.80 \times 10^{-21}$

and $\quad v = c\sqrt{1 - 8.80 \times 10^{-21}} = \left(1 - 4.40 \times 10^{-21}\right)c \approx 3.00 \times 10^8$ m / s

The proton's speed is nearly as large as the speed of light. In the galaxy frame, the traversal time is

$$\Delta t = x / v = 10^5 \text{ light-years} / c = 10^5 \text{ years}$$

(a) This is dilated from the proper time measured in the proton's frame. The proper time is found from $\Delta t = \gamma \Delta t_p$:

$$\Delta t_p = \Delta t / \gamma = 10^5 \text{ yr}/1.07 \times 10^{10} = 9.38 \times 10^{-6} \text{ years} = 296 \text{ s}$$

$$\Delta t_p \sim \text{ a few hundred seconds} \qquad \Diamond$$

(b) The proton sees the galaxy moving by at a speed nearly equal to c, passing in 296 s:

$$\Delta L_p = v \Delta t_p = \left(3.00 \times 10^8\right)(296 \text{ s}) = 8.88 \times 10^7 \text{ km} \sim 10^8 \text{ km} \qquad \Diamond$$

$$\Delta L_p = \left(8.88 \times 10^{10} \text{ m}\right)\left(9.46 \times 10^{15} \text{ m} / \text{ly}\right) = 9.39 \times 10^{-6} \text{ ly} \sim 10^{-5} \text{ ly}$$

L: The results agree with our predictions, although we may not have guessed that the protons would be traveling so close to the speed of light! The calculated results should be rounded to zero significant figures since we were given order of magnitude data. We should also note that the relative speed of motion v and the value of γ are the same in both the proton and galaxy reference frames.

---

55. The net nuclear fusion reaction inside the Sun can be written as $4^1\text{H} \rightarrow {}^4\text{He} + \Delta E$. If the rest energy of each Hydrogen atom is 938.78 MeV and the rest energy of the Helium-4 atom is 3728.4 MeV, what is the percentage of the starting mass that is released as energy?

Solution The original rest energy is $E_R = 4(938.78 \text{ MeV}) = 3755.12 \text{ MeV}$

The energy given off is $\Delta E = (3755.12 - 3728.4) \text{ MeV} = 26.7 \text{ MeV}$

The fractional energy released is $\dfrac{\Delta E}{E_R} = \dfrac{26.7 \text{ MeV}}{3755 \text{ MeV}} \times 100\% = 0.712\% \qquad \Diamond$

---

59. Spaceship I, which contains students taking a physics exam, approaches the Earth with a speed of 0.600c (relative to Earth), while spaceship II, which contains professors proctoring the exam, moves at 0.280c (relative to Earth) directly toward the students. If the professors stop the exam after 50.0 min have passed on their clock, how long does the exam last as measured by (a) the students and (b) an observer on Earth?

Solution

Suppose that in the Earth frame, the students are moving to the right ($u_x = 0.600c$), and the professors are moving to the left, at a speed of ($v = -0.280c$).

In the professor's frame,
the Earth moves to the right at $\qquad\qquad u_e = 0.280c$,

and the students move to the right at $\qquad u'_x = \dfrac{u_x - v}{1 - u_x v / c^2}$

$$u'_x = \frac{0.600c - (-0.280c)}{1 - (0.600c)(-0.280c) / c^2} = 0.753c$$

The professors measure 50 minutes on a clock at rest in their frame: they measure proper time and everyone else sees longer, dilated time intervals.

(a) For the students, $\qquad \Delta t = \gamma \Delta t_p = \dfrac{50.0 \text{ min}}{\sqrt{1 - (0.753)^2}} = 76.0 \text{ min} \qquad\qquad \Diamond$

(b) On Earth, $\qquad \Delta t = \gamma \Delta t_p = \dfrac{50 \text{ min}}{\sqrt{1 - (0.280)^2}} = 52 \text{ min, 5 sec} \qquad\qquad \Diamond$

61. A supertrain (proper length, 100 m) travels at a speed of $0.950c$ as it passes through a tunnel (proper length, 50.0 m). As seen by a trackside observer, is the train ever completely within the tunnel? If so, with how much space to spare?

Solution

The observer sees the proper length of the tunnel, 50.0 m, but sees the train Lorentz-contracted to length

$$L = L_p\sqrt{1 - v^2/c^2} = (100 \text{ m})\sqrt{1 - (0.950c)^2} = 31.2 \text{ m}$$

This is shorter than the tunnel by 18.8 m, so it is completely within the tunnel. ◊

_____

63. A charged particle moves along a straight line in a uniform electric field E with a speed of u. If the motion and the electric field are both in the x direction, (a) show that the acceleration of the charge q in the x direction is given by

$$a = \frac{du}{dt} = \frac{qE}{m}\left(1 - \frac{u^2}{c^2}\right)^{3/2}$$

(b) Discuss the significance of the dependence of the acceleration on the speed. (c) If the particle starts from rest at $x = 0$ at $t = 0$, how would you proceed to find the speed of the particle and its position after a time t has elapsed?

Solution (a) At any speed, the **momentum** of the particle is given by

$$p = \gamma mu = \frac{mu}{\sqrt{1 - u^2/c^2}}$$

Since $F = qE = \dfrac{dp}{dt}$,

$$qE = \frac{d}{dt}\left(mu\left(1 - u^2/c^2\right)^{-1/2}\right)$$

$$qE = m\left(1 - \frac{u^2}{c^2}\right)^{-1/2}\frac{du}{dt} + \tfrac{1}{2}mv\left(1 - \frac{u^2}{c^2}\right)^{-3/2}\left(\frac{2u}{c^2}\right)\frac{du}{dt}$$

Simplifying, we find that

$$\frac{qE}{m} = \frac{du}{dt}\left(1 - \frac{u^2}{c^2}\right)^{-3/2}$$

and

$$a = \frac{du}{dt} = \frac{qE}{m}\left(1 - \frac{u^2}{c^2}\right)^{3/2} \qquad \Diamond$$

(b) As $u \to c$, we see that $a \to 0$. The particle thus never attains the speed of light.

(c) Taking the acceleration equation, isolating the velocity terms and integrating,

$$\int_0^u \left(1 - \frac{u^2}{c^2}\right)^{-3/2} du = \int_0^t \frac{qE}{m} dt$$

$$u = \frac{qEct}{\sqrt{m^2c^2 + q^2E^2t^2}} = \frac{dx}{dt}$$

$$x = \int_0^x dx = qEc\int_0^t \frac{t\, dt}{\sqrt{m^2c^2 + q^2E^2t^2}} = \frac{c}{qE}\left(\sqrt{m^2c^2 + q^2E^2t^2} - mc\right) \qquad \Diamond$$

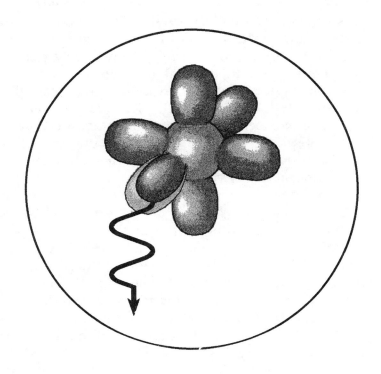

INTRODUCTION TO QUANTUM PHYSICS

INTRODUCTION TO QUANTUM PHYSICS

INTRODUCTION

In the previous chapter, we noted that Newtonian mechanics must be replaced by Einstein's special theory of relativity when dealing with particles whose speeds are comparable to the speed of light. Although many problems were indeed resolved by the theory of relativity in the early part of the 20th century, many experimental and theoretical problems remained unsolved. Attempts to explain the behavior of matter on the atomic level with the laws of classical physics were totally unsuccessful. Various phenomena, such as blackbody radiation, the photoelectric effect, and the emission of sharp spectra lines by atoms in a gas discharge tube, could not be understood within the framework of classical physics. We shall describe these phenomena because of their importance in subsequent developments.

Another revolution took place in physics between 1900 and 1930. This was the era of a new and more general formulation called **quantum mechanics.** This new approach was highly successful in explaining the behavior of atoms, molecules, and nuclei. As with relativity, the quantum theory requires a modification of our ideas concerning the physical world.

An extensive study of quantum theory is certainly beyond the scope of this book. This chapter is simply an introduction to the underlying ideas of quantum theory and the wave-particle nature of matter. We also discuss some simple applications of quantum theory, including the photoelectric effect, the Compton effect, and x-rays.

EQUATIONS AND CONCEPTS

A **black body** is an ideal body that absorbs all radiation incident on it. Any body at a temperature T emits thermal radiation which is characterized by the properties of the body and its temperature.

As the temperature of a blackbody increases, the radiation intensity increases, while the peak of the distribution shifts to shorter wavelengths.

The Wien displacement law properly describes the peak in the spectrum of radiation emitted by a blackbody.

$$\lambda_{max}T = 2.898 \times 10^{-3} \text{ m} \cdot \text{K} \qquad (40.1)$$

Vibrating molecules are characterized by discrete energy levels called quantum states. The positive integer n is called a quantum number.

$$E_n = nhf \qquad (40.5)$$

$$h = 6.626 \times 10^{-34} \text{ J} \cdot \text{s} \qquad (40.4)$$

Molecules emit or absorb energy in discrete units of light energy called quanta. The energy of a quantum or photon corresponds to the energy difference between adjacent quantum states.

$$E = hf \qquad (40.6)$$

When light is incident on certain metallic surfaces, electrons can be emitted from the surfaces. This is the photoelectric effect, and was discovered by Hertz. One cannot explain many of the features of the photoelectric effect using classical concepts. In 1905, Einstein provided a successful explanation of the photoelectric effect by extending Planck's quantum concept to include electromagnetic fields. In his model, Einstein assumed that light consists of a stream of particles called **photons** whose energy is given by $E = hf$, where h is Planck's constant and f is their frequency.

The maximum kinetic energy of an ejected photoelectron depends on the work function of the metal, ϕ, which is typically a few eV. This is in excellent agreement with experimental results, including the prediction of a cutoff (threshold) wavelength above which no photoelectric effect is observed.

$$K_{max} = hf - \phi \tag{40.8}$$

$$\lambda_c = \frac{hc}{\phi} \tag{40.9}$$

The Compton effect involves the scattering of an x-ray by an electron. The scattered x-ray undergoes a change in wavelength called the Compton shift, which cannot be explained using classical concepts.

$$\lambda' - \lambda_0 = \frac{h}{m_e c}(1 - \cos\theta) \tag{40.10}$$

By treating the x-ray as a photon (the quantum concept), the scattering process between the photon and electron predicts a shift in photon (x-ray) wavelength, where θ is the angle between the incident and scattered x-ray and m_e is the mass of the electron. The formula is in excellent agreement with experimental results.

The quantity $h/(m_e c)$ is called the Compton wavelength.

$$\frac{h}{m_e c} = 0.00243 \text{ nm}$$

When an atomic electron undergoes a transition from one allowed orbit to another, the frequency of the emitted photon is proportional to the difference in energies of the initial and final states.

$$E_i - E_f = hf \qquad (40.18)$$

The Bohr model of the atom assumed that the electron orbited the nucleus in a circular path under the influence of the Coulomb force of attraction. Also, the angular momentum of the electron about the nucleus must be quantized in units of $nh/(2\pi)$.

$$m_e vr = n\hbar \qquad (40.19)$$

$$n = 1, 2, 3, \ldots$$

While in one of the allowed orbits or stationary states (determined by quantization of the orbital angular momentum), the electron does not radiate energy.

The total energy of the hydrogen atom $(K + U)$ depends on the radius of the allowed orbit of the electron.

$$E = -\frac{k_e e^2}{2r} \qquad (40.22)$$

The electron can exist only in certain allowed orbits, the radii of which can be expressed in terms of the Bohr radius, a_0.

$$r_n = \frac{n^2 \hbar^2}{m_e k_e e^2} \qquad (40.23)$$

$$n = 1, 2, 3, \ldots$$

The **Bohr radius** corresponds to $n = 1$.

$$a_0 = \frac{\hbar^2}{m_e k_e e^2} = 0.0529 \text{ nm} \qquad (40.24)$$

Any orbit radius in the hydrogen atom can then be expressed in terms of the Bohr radius.

$$r_n = n^2 a_0 = n^2 (0.0529 \text{ nm}) \qquad (40.25)$$

When numerical values of the constants are used, the energy level values can be expressed in units of electron volts (eV).

$$E_n = -\frac{13.606}{n^2} \text{ eV} \qquad (40.27)$$

$$n = 1, 2, 3, \ldots$$

The lowest energy state or ground state corresponds to the principal quantum number $n = 1$. The absolute value of the ground state energy is equal to the ionization energy of the atom. The energy level approaches $E = 0$ as r approaches infinity.

A photon of frequency, f, (and wavelength λ) is emitted when an electron undergoes a transition from an initial energy level to a final lower level.

$$f = \frac{E_i - E_f}{h} \qquad (40.28)$$

$$\frac{1}{\lambda} = R_H \left(\frac{1}{n_f^2} - \frac{1}{n_i^2} \right) \qquad (40.30)$$

$$\text{where} \quad R_H = \frac{k_e e^2}{2 a_0 h c}$$

The lines observed in the hydrogen spectrum can be arranged into series corresponding to assigned values of principal quantum numbers of the initial and final states.

For the Lyman series,
$$n_f = 1; \quad n_i = 2, 3, 4, \ldots$$

For the Balmer series (as illustrated)
$$n_f = 2; \quad n_i = 3, 4, 5, \ldots$$

For the Paschen series,
$$n_f = 3; \quad n_i = 4, 5, 6, \ldots$$

For the Brackett series,
$$n_f = 4; \quad n_i = 5, 6, 7, \ldots$$

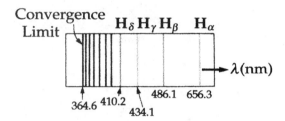

In the case of very large values of the principal quantum number, the energy differences between adjacent levels approach zero and essentially a continuous range (as opposed to a quantized set) of energy values of the emitted photon is possible. In this limit of large quantum numbers, the classical model is reasonable.

According to the de Broglie hypothesis, a material particle moves as a wave with a wavelength λ that is inversely proportional to its momentum.

$$\lambda = \frac{h}{mv} \tag{40.33}$$

REVIEW CHECKLIST

▷ Describe the formula for blackbody radiation proposed by Planck, and the assumption made in deriving this formula.

▷ Describe the Einstein model for the photoelectric effect, and the predictions of the fundamental photoelectric effect equation for the maximum kinetic energy of photoelectrons. Recognize that Einstein's model of the photoelectric effect involves the photon concept ($E = hf$), and the fact that the basic features of the photoelectric effect are consistent with this model.

▷ Describe the Compton effect (the scattering of x-rays by electrons) and be able to use the formula for the Compton shift. Recognize that the Compton effect can only be explained using the photon concept.

▷ State the basic postulates of the Bohr model of the hydrogen atom.

▷ Sketch an energy level diagram for hydrogen (include assignment of values of the principle quantum number, n), show transitions corresponding to spectral lines in the several known series, and make calculations of the wavelength values.

ANSWERS TO SELECTED CONCEPTUAL QUESTIONS

3. If the photoelectric effect is observed for one metal, can you conclude that the effect will also be observed for another metal under the same conditions? Explain.

Answer No. Suppose that the incident light frequency at which you first observed the photoelectric effect is above the cutoff frequency of the first metal, but less than the cutoff frequency of the second metal. In that case, the photoelectric effect would not be observed at all in the second metal.

□   □   □   □

13. An x-ray photon is scattered by an electron. What happens to the frequency of the scattered photon relative to that of the incident photon?

Answer The x-ray photon transfers some of its energy to the electron. Thus, its energy, and therefore its frequency, must be decreased.

□   □   □   □

14. Why does the existence of a cutoff frequency in the photoelectric effect favor a particle theory for light rather than a wave theory?

Answer Wave theory predicts that the photoelectric effect should occur at any frequency, provided that the light intensity is high enough. As is implied by the question, this is in contradiction to experimental results.

□   □   □   □

SOLUTIONS TO SELECTED END-OF-CHAPTER PROBLEMS

1. The human eye is most sensitive to 560-nm light. What is the temperature of a black body that would radiate most intensely at this wavelength?

Solution We use Wien's law: $\lambda_{max}T = 2.898 \times 10^{-3} \text{ m} \cdot \text{K}$

Thus, $$T = \frac{2.90 \text{ mm} \cdot \text{K}}{560 \times 10^{-6} \text{ mm}} = 5180 \text{ K} \qquad \lozenge$$

Related Information: This is close to the temperature of the surface of the Sun (which acts as a pretty good black body). Living things on Earth evolved to be sensitive to electromagnetic waves near this wavelength because there is such a lot of it bouncing around, carrying information.

7. Calculate the energy, in electron volts, of a photon the frequency of which is (a) 620 THz, (b) 3.10 GHz, (c) 46.0 MHz. (d) Determine the corresponding wavelengths for these photons and state the classification of each on the electromagnetic spectrum.

Solution $E = hf$

(a) $E = \left(6.63 \times 10^{-34} \text{ J} \cdot \text{s}\right)\left(6.20 \times 10^{14} \text{ Hz}\right) = 4.11 \times 10^{-19} \text{ J} = 2.57 \text{ eV}$ ◊

(b) $E = \left(6.63 \times 10^{-34} \text{ J} \cdot \text{s}\right)\left(3.10 \times 10^9 \text{ Hz}\right) = 2.06 \times 10^{-24} \text{ J} = 12.8 \text{ }\mu\text{eV}$ ◊

(c) $E = \left(6.63 \times 10^{-34} \text{ J} \cdot \text{s}\right)\left(46.0 \times 10^6 \text{ Hz}\right) = 3.05 \times 10^{-26} \text{ J} = 1.91 \times 10^{-7} \text{eV}$ ◊

(d) $\lambda_a = c/f = \left(3.00 \times 10^8 \text{ m}/\text{s}\right)\Big/\left(6.20 \times 10^{14} \text{ s}^{-1}\right) = 4.84 \times 10^{-7} \text{ m} = 484 \text{ nm}$ ◊

$\lambda_b = \left(3.00 \times 10^8 \text{ m}/\text{s}\right)\Big/\left(3.10 \times 10^9 \text{ s}^{-1}\right) = 0.0968 \text{ m} = 9.68 \text{ cm}$ ◊

$\lambda_c = \left(3.00 \times 10^8 \text{ m}/\text{s}\right)\Big/\left(46.0 \times 10^6 \text{ s}^{-1}\right) = 6.52 \text{ m}$ ◊

These wavelengths correspond, respectively, to blue light, microwave radiation, and radio waves in the public LO band. ◊

---

9. An FM radio transmitter has a power output of 150 kW and operates at a frequency of 99.7 MHz. How many photons per second does the transmitter emit?

Solution Each photon has an energy

$$E = hf = \left(6.63 \times 10^{-34} \text{ J} \cdot \text{s}\right)\left(99.7 \times 10^6 \text{ s}^{-1}\right) = 6.61 \times 10^{-26} \text{ J}$$

The number of photons per second is the power divided by the energy per photon:

$$R = \frac{\mathcal{P}}{E} = \frac{150 \times 10^3 \text{ J}/\text{s}}{6.61 \times 10^{-26} \text{ J}} = 2.27 \times 10^{30} \text{ photons}/\text{s}$$ ◊

---

19. Two light sources are used in a photoelectric experiment to determine the work function for a particular metal surface. When green light from a mercury lamp ($\lambda = 546.1$ nm) is used, a retarding potential of 0.376 V reduces the photocurrent to zero. (a) Based on this measurement, what is the work function for this metal? (b) What stopping potential would be observed when using the yellow light from a helium discharge tube ($\lambda = 587.5$ nm)?

Solution

G: According to Table 40.1, the work function for most metals is on the order of a few eV, so this metal is probably similar. We can expect the stopping potential for the yellow light to be slightly lower than 0.376 V since the yellow light has a longer wavelength (lower frequency) and therefore less energy than the green light.

O: In this photoelectric experiment, the green light has sufficient energy hf to overcome the work function of the metal ϕ so that the ejected electrons have a maximum kinetic energy of 0.376 eV. With this information, we can use the photoelectric effect equation to find the work function, which can then be used to find the stopping potential for the less energetic yellow light.

A: (a) Einstein's photoelectric effect equation is $K_{max} = hf - \phi$, and the energy required to raise an electron through a 1 V potential is 1 eV, so that $K_{max} = eV_s = 0.376$ eV.

The energy of a photon from the mercury lamp is:

$$hf = \frac{hc}{\lambda} = \frac{\left(4.14 \times 10^{-15} \text{ eV} \cdot \text{s}\right)\left(3.00 \times 10^8 \text{ m/s}\right)}{546.1 \times 10^{-9} \text{ m}} = 2.27 \text{ eV}$$

Therefore, the work function for this metal is:

$$\phi = hf - K_{max} = 2.27 \text{ eV} - (0.376 \text{ eV}) = 1.90 \text{ eV} \qquad \lozenge$$

(b) For the yellow light, $\lambda = 587.5$ nm, and

$$hf = \frac{hc}{\lambda} = \frac{\left(4.14 \times 10^{-15} \text{ eV} \cdot \text{s}\right)\left(3.00 \times 10^8 \text{ m/s}\right)}{587.5 \times 10^{-9} \text{ m}} = 2.11 \text{ eV}$$

Therefore,

$$K_{max} = hf - \phi = 2.11 \text{ eV} - 1.89 \text{ eV} = 0.216 \text{ eV}, \quad \text{so} \quad V_s = 0.216 \text{ V} \qquad \lozenge$$

L: The work function for this metal is lower than we expected, and does not correspond with any of the values in Table 40.1. Further examination in the **CRC Handbook of Chemistry and Physics** reveals that all of the metal elements have work functions between 2 and 6 eV. However, a single metal's work function may vary by about 1 eV depending on impurities in the metal, so it is just barely possible that a metal might have a work function of 1.89 eV.

The stopping potential for the yellow light is indeed lower than for the green light as we expected. An interesting calculation is to find the wavelength for the lowest energy light that will eject electrons from this metal. That threshold wavelength for $K_{max} = 0$ is 658 nm, which is red light in the visible portion of the electromagnetic spectrum.)

---

27. A 0.00160-nm photon scatters from a free electron. For what (photon) scattering angle does the recoiling electron have kinetic energy equal to the energy of the scattered photon?

Solution

The energy of the incoming photon is

$$E_o = \frac{hc}{\lambda} = \frac{\left(6.63 \times 10^{-34} \text{ J} \cdot \text{s}\right)\left(3.00 \times 10^8 \text{ m/s}\right)}{0.00160 \times 10^{-9} \text{ m}} = 1.24 \times 10^{-13} \text{ J}$$

Since the outgoing photon and the electron each have half of this energy in kinetic form,

$$E' = 6.22 \times 10^{-14} \text{ J} \quad \text{and} \quad \lambda' = \frac{hc}{E'} = 3.20 \times 10^{-12} \text{ m}$$

The shift in wavelength is

$$\Delta\lambda = \lambda' - \lambda = 1.60 \times 10^{-12} \text{ m}$$

But by Equation 40.10,

$$\Delta\lambda = \lambda_c(1 - \cos\theta)$$

so

$$\cos\theta = 1 - \frac{\Delta\lambda}{\lambda_c} = 1 - \frac{1.60 \times 10^{-12} \text{ m}}{0.00243 \times 10^{-9} \text{ m}} = 0.342 \quad \text{and} \quad \theta = 70.0° \quad \Diamond$$

37. (a) What value of n is associated with the 94.96-nm line in the Lyman hydrogen series? (b) Could this wavelength be associated with the Paschen or Brackett series?

Solution   Our equation is

$$\frac{1}{\lambda} = R_H\left(\frac{1}{n_f^2} - \frac{1}{n_i^2}\right)$$

where

$$R_H = 1.097 \times 10^7 \text{ m}^{-1}$$

and for the Lyman series,

$$n_f = 1, \text{ and } n_i = 2, 3, 4, \ldots$$

Substituting the given values,

$$\frac{1}{94.96 \times 10^{-9} \text{ m}} = \left(1.097 \times 10^{-9} \text{ m}^{-1}\right)\left(1 - \frac{1}{n_i^2}\right)$$

Solving for n_i,

$$n_i = 5 \qquad \Diamond$$

(b)   By Figure 40.17, spectral lines in the Balmer, Paschen, and Brackett series all have much longer wavelengths, since much smaller energy losses put the atom into energy levels 2, 3, or 4. $\qquad \Diamond$

41. A hydrogen atom is in its first excited state $(n = 2)$. Using the Bohr theory of the atom, calculate (a) the radius of the orbit, (b) the linear momentum of the electron, (c) the angular momentum of the electron, (d) the kinetic energy, (e) the potential energy, and (f) the total energy.

Solution We note, during our calculations, that the nominal velocity of the electron is less than 1% of the speed of light; therefore, we do not need to use relativistic equations.

(a) By Bohr's model, $r_n = n^2(a_0) = 2^2(0.0529 \text{ nm}) = 2.12 \times 10^{-10} \text{ m}$ ◊

(b) Since $m_e vr = n\hbar$, $p = m_e v = \dfrac{n\hbar}{r} = \dfrac{2\left(1.0546 \times 10^{-34} \text{ J·s}\right)}{2.12 \times 10^{-10} \text{ m}}$

$p = 9.97 \times 10^{-25} \text{ kg·m/s}$ ◊

(c) $\mathbf{L} = \mathbf{r} \times \mathbf{p}$ becomes $L = rp = n\hbar = 2.11 \times 10^{-34} \text{ J·s}$ ◊

(d) Next, $v = p/m_e$: $v = \dfrac{p}{m_e} = \dfrac{9.97 \times 10^{-25} \text{ kg·m/s}}{9.11 \times 10^{-31} \text{ kg}} = 1.09 \times 10^6 \text{ m/s}$

So $K = \frac{1}{2} m_e v^2$: $K = \dfrac{(9.11 \times 10^{-31} \text{ kg})(1.09 \times 10^6 \text{ m/s})^2}{2}$

$K = \dfrac{5.45 \times 10^{-19} \text{ J}}{1.60 \times 10^{-19} \text{ J/eV}} = 3.40 \text{ eV}$ ◊

(e) By Eq. 25.13,: $U = -\dfrac{k_e e^2}{r} = -\dfrac{(8.99 \times 10^9 \text{ N·m}^2/\text{C}^2)(1.60 \times 10^{-19} \text{ C})^2}{2.12 \times 10^{-10} \text{ m}}$

and $U = -1.09 \times 10^{-18} \text{ J} = -6.80 \text{ eV}$ ◊

(f) Thus, $E = K + U = -5.45 \times 10^{-19} \text{ J} = -3.40 \text{ eV}$ ◊

The total energy ought to be equal to minus the kinetic energy: our answer confirms this, but we needed to use more than 3 significant figures for some constants. For an accurate answer from data with fewer significant figures, we could calculate the total energy first:

$$E = \frac{-k_e e^2}{2r} = \frac{-(8.99 \times 10^9 \text{ N} \cdot \text{m}^2/\text{C}^2)(1.60 \times 10^{-19} \text{ C})^2}{2(2.12 \times 10^{-10} \text{ m})} = -5.45 \times 10^{-19} \text{ J} \qquad \Diamond$$

$$K = E - U = 5.45 \times 10^{-19} \text{ J} = 3.40 \text{ eV} \qquad \Diamond$$

43. A photon is emitted as a hydrogen atom undergoes a transition from the $n = 6$ state to the $n = 2$ state. Calculate (a) the energy, (b) the wavelength, and (c) the frequency of the emitted photon.

Solution By conservation of energy, the energy of the photon is equal to the energy lost in the transition.

(a) From Equation 40.27, $\qquad E_6 - E_2 = (-13.6 \text{ eV})\left(\frac{1}{6^2} - \frac{1}{2^2}\right) = 3.02 \text{ eV} \qquad \Diamond$

(b) $\qquad \frac{1}{\lambda} = R_H\left(\frac{1}{n_f^2} - \frac{1}{n_i^2}\right) = (1.097 \times 10^7 \text{ m}^{-1})\left(\frac{1}{4} - \frac{1}{36}\right)$

$\qquad \lambda = 410 \text{ nm} \qquad \Diamond$

(This is the deep violet spectral line farthest to the left in the hydrogen spectrum of the text's Figure 40.13)

(c) $\qquad f = c/\lambda = (3.00 \times 10^8 \text{ m/s})/(410 \times 10^{-9} \text{ m}) = 7.32 \times 10^{14} \text{ Hz} \qquad \Diamond$

49. (a) Construct an energy level diagram for the He+ ion, for which $Z = 2$.
(b) What is the ionization energy for He+?

Solution

From Equation 40.32,

$$E_n = -\frac{k_e e^2 Z^2}{2a_0 n^2} = \frac{(-13.6 \text{ eV})(2)^2}{n^2}$$

For the lowest energy level, $n = 1$

and $\qquad E_1 = -54.4 \text{ eV}$

For the first excited state, $n = 2$

and $\qquad E_2 = -13.6 \text{ eV}$

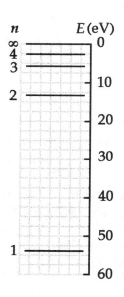

(a) The diagram is shown on the left. ◊

(b) The ionization energy is the energy required to lift the electron from the ground state to the point of breaking free from the nucleus, or from $n = 1$ to $n = \infty$, from energy -54.4 eV to 0 eV. The answer then, is 54.4 eV. ◊

57. The nucleus of an atom is on the order of 10^{-14} m in diameter. For an electron to be confined to a nucleus, its de Broglie wavelength would have to be of this order of magnitude or smaller. (a) What would be the kinetic energy of an electron confined to this region? (b) On the basis of this result, would you expect to find an electron in a nucleus? Explain.

Solution

G: The de Broglie wavelength of a normal ground-state orbiting electron is on the order 10^{-10} m (the diameter of a hydrogen atom), so with a shorter wavelength, the electron would have more kinetic energy if confined inside the nucleus. If the kinetic energy is much greater than the potential energy from its attraction with the positive nucleus, then the electron will escape from its electrostatic potential well.

O: If we try to calculate the velocity of the electron from the de Broglie wavelength, we find that

$$v = \frac{h}{m_e \lambda} = \frac{6.63 \times 10^{-34} \text{ J} \cdot \text{s}}{\left(9.11 \times 10^{-31} \text{ kg}\right)\left(10^{-14} \text{ m}\right)} = 7.27 \times 10^{10} \text{ m/s}$$

which is not possible since it exceeds the speed of light. Therefore, we must use the relativistic energy expression to find the kinetic energy of this fast-moving electron.

A: (a) The relativistic kinetic energy of a particle is $K = E - mc^2$, where $E^2 = (pc)^2 + \left(mc^2\right)^2$, and the momentum is $p = h/\lambda$:

$$p = \frac{6.63 \times 10^{-34} \text{ J} \cdot \text{s}}{10^{-14} \text{ m}} = 6.63 \times 10^{-20} \text{ N} \cdot \text{s}$$

$$E = \sqrt{\left(1.99 \times 10^{-11} \text{ J}\right)^2 + \left(8.19 \times 10^{-14} \text{ J}\right)^2} = 1.99 \times 10^{-11} \text{ J}$$

$$K = E - mc^2 = \frac{1.99 \times 10^{-11} \text{ J} - 8.19 \times 10^{-14} \text{ J}}{1.60 \times 10^{-19} \text{ J/eV}} = 124 \text{ MeV} \sim 100 \text{ MeV} \qquad \lozenge$$

(b) The electrostatic potential energy of the electron 10^{-14} m away from a positive proton is $U = -k_e e^2/r$:

$$U = -\frac{\left(8.99 \times 10^9 \ \frac{N \cdot m^2}{C^2}\right)\left(1.60 \times 10^{-19} \ C\right)^2}{10^{-14} \ m} = -2.30 \times 10^{-14} \ J \sim -0.1 \ eV \ \Diamond$$

L: Since the kinetic energy is nearly 1000 times greater than the potential energy, the electron would immediately escape the proton's attraction and would not be confined to the nucleus. $\Diamond$

It is also interesting to notice in the above calculations that the rest energy of the electron is negligible compared to the momentum contribution to the total energy.

65. The table below shows data obtained in a photoelectric experiment. (a) Using these data, make a graph similar to Figure 40.9 that plots as a straight line. From the graph, determine (b) an experimental value for Planck's constant (in joule-seconds) and (c) the work function (in electron volts) for the surface. (Two significant figures for each answer are sufficient.)

Wavelength (nm)	Maximum Kinetic Energy of Photoelectrons (eV)
588	0.67
505	0.98
445	1.35
399	1.63

Solution   Convert each wavelength to a frequency using the relation $\lambda f = c$, where c is the speed of light:

$$\lambda_1 = 588 \times 10^{-9} \text{ m} \qquad f_1 = 5.10 \times 10^{14} \text{ Hz}$$

$$\lambda_2 = 505 \times 10^{-9} \text{ m} \qquad f_2 = 5.94 \times 10^{14} \text{ Hz}$$

$$\lambda_3 = 445 \times 10^{-9} \text{ m} \qquad f_3 = 6.74 \times 10^{14} \text{ Hz}$$

$$\lambda_4 = 399 \times 10^{-9} \text{ m} \qquad f_4 = 7.52 \times 10^{14} \text{ Hz}$$

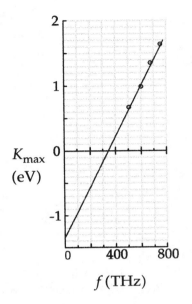

(a)   Plot each point on an energy vs. frequency graph, as shown at the right.   Extend a line through the set of 4 points, as far as the -y intercept.

(b) Our basic equation is $K_{max} = hf - \phi$. Therefore, Planck's constant should be equal to the slope of the linear K-f graph, which can be found from a least-squares fit or from reading the graph as:

$$h_{exp} = \frac{\text{Rise}}{\text{Run}} = \frac{1.25 \text{ eV} - 0.25 \text{ eV}}{6.5 \times 10^{14} \text{ Hz} - 4.0 \times 10^{14} \text{ Hz}} = 4.0 \times 10^{-15} \text{ eV} \cdot \text{s} = 6.4 \times 10^{-34} \text{ J} \cdot \text{s} \quad \lozenge$$

From the scatter of the data points on the graph, we estimate the uncertainty of the slope to be about 3%. Thus we choose to show two significant figures in writing the experimental value of Planck's constant.

(c)   From the linear equation $K_{max} = hf - \phi$, the work function for the metal surface is the negative of the y-intercept of the graph, so $\phi_{exp} = -(-1.4 \text{ eV}) = 1.4 \text{ eV}$. $\lozenge$

Based on the range of slopes that appear to fit the data, the estimated uncertainty of the work function is 5%.

67. Positronium is a hydrogen-like atom consisting of a positron (a positively charged electron) and an electron revolving around each other. Using the Bohr model, find the allowed radii (relative to the center of mass of the two particles) and the allowed energies of the system.

Solution

G: Since we are told that positronium is like hydrogen, we might expect the allowed radii and energy levels to be about the same as for hydrogen: $r = a_0 n^2 = (5.29 \times 10^{-11} \text{ m})n^2$ and $E_n = (-13.6 \text{ eV})/n^2$.

O: Similar to the textbook calculations for hydrogen, we can use the quantization of angular momentum of positronium to find the allowed radii and energy levels.

A: Let r represent the distance between the electron and the positron. The two move in a circle of radius $r/2$ around their center of mass with opposite velocities. The total angular momentum is quantized according to

$$L_n = \frac{mvr}{2} + \frac{mvr}{2} = n\hbar, \quad \text{where} \quad n = 1, 2, 3, \ldots$$

For each particle, $\Sigma F = ma$ expands to $\dfrac{k_e e^2}{r^2} = \dfrac{mv^2}{r/2}$

We can eliminate $v = \dfrac{n\hbar}{mr}$ to find $\dfrac{k_e e^2}{r} = \dfrac{2mn^2\hbar}{m^2 r^2}$

So the separation distances are $r = \dfrac{2n^2\hbar^2}{mk_e e^2} = 2a_0 n^2 = (1.06 \times 10^{-10} \text{ m})\, n^2$ ◊

The orbital radii are $r/2 = a_0 n^2$, the same as for the electron in hydrogen.

The energy can be calculated from $\qquad E = K + U = \frac{1}{2}mv^2 + \frac{1}{2}mv^2 - \dfrac{k_e e^2}{r}$

Since $mv^2 = \dfrac{k_e e^2}{2r}$, $\qquad E = \dfrac{k_e e^2}{2r} - \dfrac{k_e e^2}{r} = -\dfrac{k_e e^2}{2r} = \dfrac{-k_e e^2}{4a_0 n^2} = -\dfrac{6.80\ \text{eV}}{n^2}$ $\qquad \lozenge$

L: It appears that the allowed radii for positronium are twice as large as for hydrogen, while the energy levels are half as big. One way to explain this is that in a hydrogen atom, the proton is much more massive than the electron, so the proton remains nearly stationary with essentially no kinetic energy. However, in positronium, the positron and electron have the same mass and therefore both have kinetic energy that separates them from each other and reduces their total energy compared with hydrogen.

---

69. An example of the correspondence principle: Use Bohr's model of the hydrogen atom to show that when the electron moves from the state n to the state $n - 1$, the frequency of the emitted light is

$$f = \frac{2\pi^2 m_e k_e^2 e^4}{h^3}\left[\frac{2n-1}{(n-1)^2 n^2}\right]$$

Show that as $n \to \infty$, this expression varies as $1/n^3$ and reduces to the classical frequency one expects the atom to emit. (**Hint:** To calculate the classical frequency, note that the frequency of revolution is $v/2\pi r$, where r is given by Eq. 40.25.)

Solution

Combining Eq. 40.22 and 40.23, $E_n = -\dfrac{m_e k_e^2 e^4}{2n^2 \hbar^2}$

Because $\hbar = h / 2\pi$, $hf = \Delta E = \dfrac{4\pi^2 m_e k_e^2 e^4}{2h^2}\left(\dfrac{1}{(n-1)^2} - \dfrac{1}{n^2}\right)$

which reduces to $f = \dfrac{2\pi^2 m_e k_e^2 e^4}{h^3}\left(\dfrac{2n-1}{(n-1)^2 n^2}\right)$ ◊

As $n \to \infty$, the '-1' terms lose importance and drop out, leaving the right-hand factor equal to

$$\lim_{n \to \infty}\left(\dfrac{2n-1}{(n-1)^2 n^2}\right) = \dfrac{2n}{n^4} = \dfrac{2}{n^3}$$ ◊

Thus the frequency becomes $f = \dfrac{4\pi^2 m_e k_e^2 e^4}{h^3}\left(\dfrac{1}{n^3}\right)$

But in the Bohr model, $v^2 = \dfrac{k_e e^2}{m_e r}$ and $r = \dfrac{n^2 h^2}{4\pi^2 m_e k_e e^2}$

and the classical frequency is $f = \dfrac{v}{2\pi r} = \dfrac{1}{2\pi r^{3/2}}\sqrt{\dfrac{k_e e^2}{m_e}}$

or $f = \dfrac{1}{2\pi}\left(\sqrt{\dfrac{k_e e^2}{m_e}}\right)\left(\dfrac{4\pi^2 m_e k_e e^2}{n^2 h^2}\right)^{3/2}$

But this is identical to $f = \dfrac{4\pi^2 m_e k_e^2 e^4}{h^3}\left(\dfrac{1}{n^3}\right)$ ◊

73. The total power per unit area radiated by a black body at a temperature T is the area under the $I(\lambda, T)$ versus λ curve, as shown in Figure 40.3. (a) Show that this power per unit area is

$$\int_0^\infty I(\lambda, T)d\lambda = \sigma T^4$$

where $I(\lambda, T)$ is given by Planck's radiation law and σ is a constant independent of T. This result is known as the Stefan-Boltzman law (see Section 20.7). To carry out the integration, you should make the change of variable $x = hc/\lambda k_B T$ and use the fact that

$$\int_0^\infty \left(\frac{x^3 dx}{e^x - 1}\right) = \frac{\pi^4}{15}$$

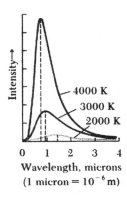

Figure 40.3

(b) Show that the Stefan-Boltzman constant σ has the value

$$\sigma = \frac{2\pi^5 k_B^4}{15c^2 h^3} = 5.67 \times 10^{-8} \frac{W}{m^2 \cdot K^4}$$

Solution   In order to make the suggested substitution, we find λ and $d\lambda$:

$$x = \frac{hc}{\lambda k_B T} \qquad \lambda = \frac{hc}{xk_B T} \qquad d\lambda = -\frac{hc\, dx}{x^2 k_B T}$$

We also note that the limits of integration change from $\lambda = (0, \infty)$ to $x = (\infty, 0)$. Substituting these variables into the integral, the intensity of the blackbody radiation is:

$$\int_0^\infty I(\lambda, T)d\lambda = \int_0^\infty \frac{2\pi hc^2}{\lambda^5 \left(e^{hc/\lambda k_B T} - 1\right)}d\lambda = \int_\infty^0 -\left(\frac{2\pi hc^2}{e^x - 1}\right)\left(\frac{x^5 k_B^5 T^5}{h^5 c^5}\right)\left(\frac{hc\, dx}{x^2 k_B T}\right)$$

$$\int_0^\infty I(\lambda, T)d\lambda = \frac{2\pi k_B^4 T^4}{h^3 c^2}\int_\infty^0 -\frac{x^3}{e^x - 1}dx = \frac{2\pi k_B^4 T^4}{h^3 c^2}\int_0^\infty \frac{x^3}{e^x - 1}dx$$

But the integral is $\pi^4/15$, so

$$\int_0^\infty I(\lambda,T)d\lambda = \frac{2\pi^5 k_B{}^4 T^4}{15 h^3 c^2} = \sigma T^4 \qquad \lozenge$$

with

$$\sigma = \frac{2\pi^5 k_B{}^4}{15 c^2 h^3} = 5.67 \times 10^{-8} \text{ W}/\text{m}^2 \cdot \text{K}^4 \qquad \lozenge$$

---

79. Show that the ratio of the Compton wavelength λ_C to the de Broglie wavelength $\lambda = h/p$ for a relativistic electron is

$$\frac{\lambda_C}{\lambda} = \left[\left(\frac{E}{mc^2} \right)^2 - 1 \right]^{1/2}$$

where E is the total energy of the electron and m is its mass.

Solution   From the definition of the Compton wavelength, $\lambda_C = h/mc$

Taking the ratio of the Compton wavelength to the de Broglie wavelength,

$$\frac{\lambda_C{}^2}{\lambda^2} = \frac{p^2}{(mc)^2}$$

From Eq. 39.26, the relativistic momentum is $p^2 = \dfrac{E^2 - m^2 c^4}{c^2}$

Substituting and simplifying,

$$\frac{\lambda_C{}^2}{\lambda^2} = \frac{\left(E^2 - m^2 c^4 \right)}{\left(mc^2 \right)^2}$$

Finally, $\dfrac{\lambda_C{}^2}{\lambda^2} = \left(\dfrac{E}{mc^2} \right)^2 - 1$ and $\dfrac{\lambda_C}{\lambda} = \sqrt{\left(\dfrac{E}{mc^2} \right)^2 - 1}$ $\lozenge$

Chapter
41

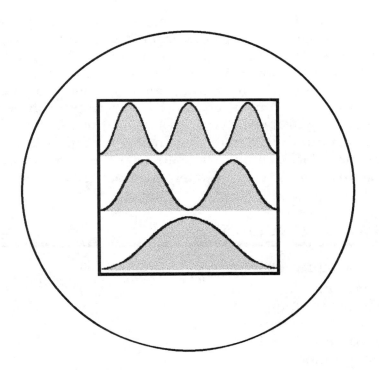

QUANTUM
MECHANICS

QUANTUM MECHANICS

INTRODUCTION

In this chapter, we introduce quantum mechanics, a theory for explaining atomic structure. This scheme, developed from 1925 to 1926 by Schrödinger, Heisenberg, and others, addresses the limitations of the Bohr model and lets us understand a host of phenomena involving atoms, molecules, nuclei, and solids. We begin by describing wave-particle duality, wherein a particle can be viewed as having wave-like properties, and its wavelength can be calculated if its momentum is known. Next, we describe some of the basic features of the formalism of quantum mechanics and their application to simple, one-dimensional systems. For example, we shall treat the problem of a particle confined to a potential well having infinitely high barriers.

EQUATIONS AND CONCEPTS

Matter waves can be represented by a wave function, ψ, which, in general, depends on position and time. In the expression for the part of the wave function which depends only on position, k is the wave number. Equation 41.3 is the wave function for a free particle.

$$\psi(x) = A\sin\left(\frac{2\pi x}{\lambda}\right) = A\sin(kx) \qquad (41.3)$$

The normalization condition is a statement of the requirement that the particle exists at some point (along the x axis in the one-dimensional case) at all times.

$$\int_{-\infty}^{\infty} |\psi|^2 \, dx = 1 \qquad (41.4)$$

Although it is not possible to specify the position of a particle exactly, it is possible to calculate the probability P_{ab} of finding the particle within an interval $a \leq x \leq b$.

$$P_{ab} = \int_a^b |\psi|^2 \, dx \qquad (41.5)$$

The probability of measuring a certain value for the position of a particle is called the expectation value of the coordinate x.

$$\langle x \rangle \equiv \int_{-\infty}^{\infty} x |\psi|^2 \, dx \qquad (41.6)$$

The uncertainty principle states that if a measurement of position is made with a precision Δx and a **simultaneous** measurement of momentum is made with a precision Δp, then the product of the two uncertainties can never be smaller than a number of the order of h.

$$\Delta x \Delta p_x \geq \frac{\hbar}{2} \qquad (41.1)$$

The allowed wave functions for a particle in a rigid box of width L are sinusoidal.

$$\psi(x) = A \sin\left(\frac{n\pi x}{L}\right) \qquad (41.8)$$

$$n = 1, 2, 3, \ldots$$

The energy of a particle in a box is quantized and the least energy which the particle can have is called the zero-point energy.

$$E_n = \left(\frac{h^2}{8mL^2}\right) n^2 \qquad (41.9)$$

$$n = 1, 2, 3, \ldots$$

The time independent Schrödinger equation for a bound system (total energy E is constant) allows in principle the determination of the wave functions and energies of the allowed states if the potential energy function is known.

$$\frac{\partial^2 \psi}{\partial x^2} = -\frac{2m}{\hbar^2}(E - U)\psi \qquad (41.12)$$

The Schrödinger equation can be written for a harmonic oscillator of total energy, E.

$$\frac{d^2\psi}{dx^2} = -\left[\left(\frac{2mE}{\hbar^2}\right) - \left(\frac{m\omega}{\hbar}\right)^2 x^2\right]\psi \qquad (41.19)$$

The energy levels of the harmonic oscillator are quantized.

$$E_n = \left(n + \frac{1}{2}\right)\hbar\omega$$

$$n = 0, 1, 2, \ldots$$

REVIEW CHECKLIST

▷ Discuss the wave properties of particles, the de Broglie wavelength concept, and the dual nature of both matter and light.

▷ Describe the concept of wave function for the representation of matter waves and state in equation form the normalization condition and expectation value of the coordinate.

▷ Discuss the manner in which the uncertainty principle makes possible a better understanding of the dual wave-particle nature of light and matter.

▷ State the time-independent form of the Schrödinger equation for a bound system of total energy, E, and discuss the required conditions on the wave function.

▷ Describe the allowed wave functions and energy levels for a particle in a one-dimensional box.

ANSWERS TO SELECTED CONCEPTUAL QUESTIONS

3. If matter has a wave nature, why is this wave-like characteristic not observable in our daily experience?

Answer For any object that we can perceive directly, the de Broglie wavelength $\lambda = h / mv$ is too small to be measured by any means; therefore, no wavelike characteristics can be observed. The object will not diffract noticeably when it goes through an aperture. It will not show resolvable interference maxima and minima when it goes through two openings. It will not show resolvable nodes and antinodes if it is in resonance.

□ □ □ □

8. Discuss the relationship between zero-point energy and the uncertainty principle.

Answer Consider a particle bound to a restricted region of space. If its minimum energy were zero, then the particle could have zero momentum and zero uncertainty in its momentum. At the same time, the uncertainty in its position would not be infinite, but equal to the width of the region. In such a case, the uncertainty product $\Delta x \Delta p_x$ would be zero, violating the uncertainty principle. This contradiction proves that the minimum energy of the particle is not zero.

□ □ □ □

SOLUTIONS TO SELECTED END-OF-CHAPTER PROBLEMS

1.   Neutrons traveling at 0.400 m/s are directed through a double slit having a 1.00-mm separation. An array of detectors is placed 10.0 m from the slit. (a) What is the de Broglie wavelength of the neutrons? (b) How far off axis is the first zero-intensity point on the detector array? (c) When a neutron reaches a detector, can we say which slit the neutron passed through? Explain.

Solution

(a) $\lambda = h/mv$, so
$$\lambda = \frac{6.63 \times 10^{-34} \text{ J} \cdot \text{s}}{(1.67 \times 10^{-27} \text{ kg})(0.400 \text{ m/s})} = 9.93 \times 10^{-7} \text{ m} \quad \lozenge$$

(b)   The condition for destructive interference in a multiple-slit experiment is $d \sin\theta = \left(m + \frac{1}{2}\right)\lambda$ with $m = 0$ for the first minimum. Then,

$$\theta = \sin^{-1}\left(\frac{\lambda}{2d}\right) = 0.0284°$$

Since $\dfrac{y}{L} = \tan\theta$, $y = L\tan\theta = (10.0 \text{ m})(\tan 0.0284°) = 0.00496 \text{ m} \quad \lozenge$

(c)   We cannot say the neutron passed through one slit. We can only say it passed through the pair of slits, as a water wave does to produce an interference pattern.

5. The resolving power of a microscope depends on the wavelength used. If one wished to "see" an atom, a resolution of approximately 1.00×10^{-11} m would be required. (a) If electrons are used (in an electron microscope), what minimum kinetic energy is required for the electrons? (b) If photons are used, what minimum photon energy is needed to obtain the required resolution?

Solution (a) Since the de Broglie wavelength is $\lambda = \dfrac{h}{p}$,

$$p_e = \frac{h}{\lambda} = \frac{6.63 \times 10^{-34} \text{ J} \cdot \text{s}}{1.00 \times 10^{-11} \text{ m}} = 6.63 \times 10^{-23} \text{ kg} \cdot \text{m / s}$$

$$K_e = \frac{p_e^2}{2m_e} = \frac{(6.63 \times 10^{-23} \text{ kg} \cdot \text{m / s})^2}{2(9.11 \times 10^{-31} \text{ kg})} = 2.41 \times 10^{-15} \text{ J} = 15.1 \text{ keV} \qquad \lozenge$$

For better accuracy, you can use the relativistic equation

$$\left(m_e c^2 + K\right)^2 = p^2 c^2 + m_e^2 c^4 \qquad \text{to find that} \qquad K = 14.9 \text{ keV} \qquad \lozenge$$

(b) For photons:

$$E = hf = \frac{hc}{\lambda} = \frac{(6.63 \times 10^{-34} \text{ J} \cdot \text{s})(3.00 \times 10^8 \text{ m / s})}{1.00 \times 10^{-11} \text{ m}} = 1.99 \times 10^{-14} \text{ J} = 124 \text{ keV} \qquad \lozenge$$

For the photon, this wavelength $\lambda = 0.100$ Å is in the x-ray range of the electromagnetic spectrum.

---

9. An electron ($m_e = 9.11 \times 10^{-31}$ kg) and a bullet ($m = 0.0200$ kg) each have a speed of 500 m/s, accurate to within 0.0100%. Within what limits could we determine the position of the objects?

Solution

G: It seems reasonable that a tiny particle like an electron could be located within a more narrow region than a bigger object like a bullet, but we often find that the realm of the very small does not obey common sense.

O: Heisenberg's uncertainty principle can be used to find the uncertainty in position from the uncertainty in the momentum.

A: The uncertainty principle states: $\Delta x \Delta p_x \geq \hbar/2$ where $\Delta p_x = m \Delta v$ and $\hbar = h/2\pi$.

Both the electron and bullet have a velocity uncertainty,

$$\Delta v = (0.000100)(500 \text{ m / s}) = 0.0500 \text{ m / s}$$

For the electron, the minimum uncertainty in position is

$$\Delta x = \frac{h}{4\pi m \Delta v} = \frac{6.63 \times 10^{-34} \text{ J} \cdot \text{s}}{4\pi \left(9.11 \times 10^{-31} \text{ kg}\right)(0.0500 \text{ m / s})} = 1.16 \text{ mm} \qquad \lozenge$$

For the bullet,

$$\Delta x = \frac{h}{4\pi m \Delta v} = \frac{6.63 \times 10^{-34} \text{ J} \cdot \text{s}}{4\pi (0.0200 \text{ kg})(0.0500 \text{ m / s})} = 5.28 \times 10^{-32} \text{ m} \qquad \lozenge$$

L: Our intuition did not serve us well here, since the position of the center of the larger bullet can be determined much more precisely than the electron. Quantum mechanics describes all objects, but the quantum fuzziness in position is too small to observe for the bullet, yet large for the small-mass electron.

15. A free electron has a wave function $\psi(x) = A \sin(5.00 \times 10^{10}\, x)$ where x is in meters. Find (a) the de Broglie wavelength, (b) the linear momentum, and (c) the kinetic energy in electron volts.

Solution

(a) The wave function $\psi(x) = A \sin(5.00 \times 10^{10}\, x)$ will go through one full cycle between $x_1 = 0$ and $(5.00 \times 10^{10})x_2 = 2\pi$. The wavelength is then

$$\lambda = x_2 - x_1 = \frac{2\pi}{5.00 \times 10^{10}\ \text{m}^{-1}} = 1.26 \times 10^{-10}\ \text{m} \qquad \Diamond$$

To say the same thing, we can inspect $A \sin((5.00 \times 10^{10}\ \text{m}^{-1})x)$ to see that the wave number is $k = 5.00 \times 10^{10}\ \text{m}^{-1}$.

(b) Since $\lambda = \dfrac{h}{p}$, the momentum is

$$p = \frac{h}{\lambda} = \frac{6.63 \times 10^{-34}\ \text{J} \cdot \text{s}}{1.26 \times 10^{-10}\ \text{m}} = 5.28 \times 10^{-24}\ \text{kg} \cdot \text{m}\, /\, \text{s} \qquad \Diamond$$

(c) The electron's kinetic energy is

$$K = \tfrac{1}{2} m_e v^2 = \frac{p^2}{2 m_e}$$

Solving, $\qquad K = \dfrac{(5.28 \times 10^{-24}\ \text{kg} \cdot \text{m}\, /\, \text{s})^2}{2(9.11 \times 10^{-31}\ \text{kg})} \left(\dfrac{1\ \text{eV}}{1.60 \times 10^{-19}\ \text{J}} \right) = 95.5\ \text{eV} \qquad \Diamond$

17. An electron is contained in a one-dimensional box of width 0.100 nm. (a) Draw an energy-level diagram for the electron for levels up to $n = 4$. (b) Find the wavelengths of all photons that can be emitted by the electron in making transitions that will eventually get it from the $n = 4$ state to the $n = 1$ state (by all spontaneous paths).

Solution   (a) We can draw a diagram that parallels our treatment of standing mechanical waves. In each state, we measure the distance d from one node to another (N to N), and base our solution upon that:

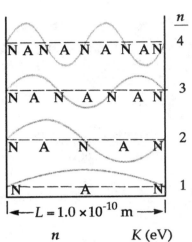

Since $d_{N \text{ to } N} = \dfrac{\lambda}{2}$ and $\lambda = \dfrac{h}{p}$, $\qquad p = \dfrac{h}{\lambda} = \dfrac{h}{2d}$

Next, $K = \dfrac{p^2}{2m_e} = \dfrac{h^2}{8m_e d^2} = \dfrac{1}{d^2}\left(\dfrac{\left(6.63 \times 10^{-34}\ \text{J} \cdot \text{s}\right)^2}{8\left(9.11 \times 10^{-31}\ \text{kg}\right)}\right)$

So $\qquad K = \dfrac{6.03 \times 10^{-38}\ \text{J} \cdot \text{m}^2}{d^2} = \dfrac{3.77 \times 10^{-19}\ \text{eV} \cdot \text{m}^2}{d^2}$

In state 1, $\quad d = 1.00 \times 10^{-10}$ m $\qquad K_1 = 37.7$ eV

In state 2, $\quad d = 5.00 \times 10^{-11}$ m $\qquad K_2 = 151$ eV

In state 3, $\quad d = 3.33 \times 10^{-11}$ m $\qquad K_3 = 339$ eV

In state 4, $\quad d = 2.50 \times 10^{-11}$ m $\qquad K_4 = 603$ eV

These energy levels are shown to the right.

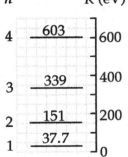

(b) When the charged, massive electron inside the box makes a downward transition from one energy level to another, a chargeless, massless photon comes out of the box, carrying the difference in energy, ΔE. Its wavelength is

$$\lambda = \dfrac{c}{f} = \dfrac{hc}{\Delta E} = \dfrac{\left(6.63 \times 10^{-34}\ \text{J} \cdot \text{s}\right)\left(3.00 \times 10^8\ \text{m}/\text{s}\right)}{\Delta E\left(1.602 \times 10^{-19}\ \text{J}/\text{eV}\right)} = \dfrac{1.24 \times 10^{-6}\ \text{eV} \cdot \text{m}}{\Delta E}$$

Transition	$4 \to 3$	$4 \to 2$	$4 \to 1$	$3 \to 2$	$3 \to 1$	$2 \to 1$
ΔE (eV)	264	452	565	188	302	113
Wavelength (nm)	4.71	2.75	2.20	6.60	4.12	11.0

The wavelengths of light released for each transition are given in the table above. ◊

21. The nuclear potential energy that binds protons and neutrons in a nucleus is often approximated by a square well. Imagine a proton confined in an infinitely high square well of width 10.0 fm, a typical nuclear diameter. Calculate the wavelength and energy associated with the photon emitted when the proton moves from the $n = 2$ state to the ground state. In what region of the electromagnetic spectrum does this wavelength belong?

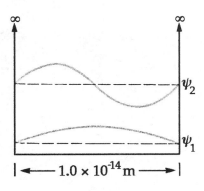

Solution

G: Nuclear radiation from nucleon transitions is usually in the form of high energy gamma rays with short wavelengths.

O: The energy of the particle can be obtained from the wavelengths of the standing waves corresponding to each level. The transition between energy levels will result in the emission of a photon with this energy difference.

A: At level 1, the node-to-node distance of the standing wave is 1.00×10^{-14} m, so the wavelength is twice this distance: $h/p = 2.00 \times 10^{-14}$ m. The proton's kinetic energy is

$$K = \frac{1}{2}mv^2 = \frac{p^2}{2m} = \frac{h^2}{2m\lambda^2} = \frac{\left(6.63 \times 10^{-34} \text{ J} \cdot \text{s}\right)^2}{2\left(1.67 \times 10^{-27} \text{ kg}\right)\left(2.00 \times 10^{-14} \text{ m}\right)^2}$$

$$K = \frac{3.29 \times 10^{-13} \text{ J}}{1.60 \times 10^{-19} \text{ J/eV}} = 2.06 \times 10^6 \text{ eV} = 2.06 \text{ MeV}$$

In the first excited state, level 2, the node-to-node distance is two times smaller than in state 1. The momentum is two times larger and the energy is four times larger: $K = 8.23$ MeV.

The proton has mass, has charge, moves slowly compared to light in a standing-wave state, and stays inside the nucleus. When it falls from level 2 to level 1, its energy change is 2.06 MeV – 8.23 MeV = –6.17 MeV. Therefore, we know that a photon (a traveling wave with no mass and no charge) is emitted at the speed of light, and that it has an energy of +6.17 MeV ◊

Its frequency is

$$f = \frac{E}{h} = \frac{\left(6.17 \times 10^6 \text{ eV}\right)\left(1.60 \times 10^{-19} \text{ J/eV}\right)}{6.63 \times 10^{-34} \text{ J} \cdot \text{s}} = 1.49 \times 10^{21} \text{ Hz}$$

and its wavelength is $\quad \lambda = \frac{c}{f} = \frac{3.00 \times 10^8 \text{ m/s}}{1.49 \times 10^{21} \text{ s}^{-1}} = 2.02 \times 10^{-13} \text{ m} \quad$ ◊

This is a gamma ray, according to Figure 34.17. ◊

L: The radiated photons are energetic gamma rays as we expected for a nuclear transition. In the above calculations, we assumed that the proton was not relativistic ($v < 0.1c$), but we should check this assumption for the highest energy state we examined ($n = 2$):

$$v = \sqrt{\frac{2K}{m}} = \sqrt{\frac{2\left(8.23 \times 10^6 \text{ eV}\right)\left(1.60 \times 10^{-19} \text{ J/eV}\right)}{1.67 \times 10^{-27} \text{ kg}}} = 3.97 \times 10^7 \text{ m/s} = 0.133c$$

This appears to be a borderline case where we should probably use relativistic equations, but our classical treatment should give reasonable results, within about $(0.133)^2 = 1\%$ accuracy.

23. Use the particle-in-a-box model to calculate the first three energy levels of a neutron trapped in a nucleus of diameter 20.0 fm. Do the energy-level differences have a realistic order of magnitude?

Solution $E_n = \dfrac{h^2 n^2}{8mL^2}$

$$E_1 = \dfrac{(6.626 \times 10^{-34} \text{ J} \cdot \text{s})^2 (1)^2}{8(1.67 \times 10^{-27} \text{ kg})(2.00 \times 10^{-14} \text{ m})^2} = 8.21 \times 10^{-14} \text{ J} = 0.513 \text{ MeV} \quad \lozenge$$

$$E_2 = 4E_1 = 2.05 \text{ MeV} \qquad \text{and} \qquad E_3 = 9E_1 = 4.62 \text{ MeV} \qquad \lozenge$$

Yes, the energy differences are of the order of 1 MeV, which is a typical energy for a γ-ray photon, as emitted by a nucleus in an excited state.

---

31. Show that the wave function $\psi = Ae^{i(kx-\omega t)}$ is a solution to the Schrödinger equation (Eq. 41.12), where $k = 2\pi/\lambda$ and $U = 0$.

Solution From $\qquad \psi = A\, e^{i(kx-\omega t)},$ $\qquad\qquad\qquad\qquad$ (1)

we evaluate $\qquad \dfrac{d\psi}{dx} = ikAe^{i(kx-\omega t)} \quad$ and $\quad \dfrac{d^2\psi}{dx^2} = -k^2 Ae^{i(kx-\omega t)}$ $\quad$ (2)

We substitute Equations (1) and (2) into the Schrödinger equation, so that

$$\dfrac{d^2\psi}{dx^2} = -\dfrac{2m}{\hbar^2}(E-U)\psi \qquad\qquad \text{(Eq. 41.12)}$$

becomes $\qquad -k^2 Ae^{i(kx-\omega t)} = -\dfrac{2m}{\hbar^2}(K)Ae^{i(kx-\omega t)}$ $\qquad\qquad$ (3)

The wave function $\psi = Ae^{i(kx-\omega t)}$ is a solution to the Schrödinger equation if Equation (3) is true. Both sides depend on A, x, and t in the same way, so we can cancel several factors, and determine that we have a solution if

$$k^2 = 2mK/\hbar^2$$

But this is true for a nonrelativistic particle with mass, since

$$\frac{2m}{\hbar^2}K = \frac{2m}{(h/2\pi)^2}\left(\tfrac{1}{2}mv^2\right) = \frac{4\pi^2 m^2 v^2}{h^2} = \left(\frac{2\pi p}{h}\right)^2 = \left(\frac{2\pi}{\lambda}\right)^2 = k^2$$

Therefore, the given wave function does satisfy Equation 41.12. ◊

---

35. Suppose a particle is trapped in its ground state in a box that has infinitely high walls (see Fig. 41.11). Now suppose the left-hand wall is suddenly lowered to a finite height and width. (a) Qualitatively sketch the wave function for the particle a short time later. (b) If the box has a width L, what is the wavelength of the wave that penetrates the barrier?

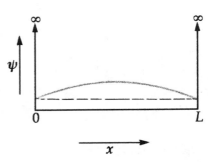

Solution Before the wall is lowered, the wave function looks the wave in the figure to the right.

Figure 41.11a

(ground state only)

(a) If the left-hand wall is shrunk to finite height and finite thickness, the wave function will look much the same in region I. It will show exponential decay in region II and a small-amplitude traveling wave in region III as in the second figure, shown below. ◊

(b) Nothing changes the particle's energy; so it has the same momentum and wavelength as before, namely $2d_{NN} = 2L$. ◊

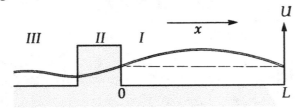

37. An electron with kinetic energy $E = 5.00$ eV is incident on a barrier with thickness $L = 0.200$ nm and height $U = 10.0$ eV (Fig. P41.37). What is the probability that the electron (a) will tunnel through the barrier and (b) will be reflected?

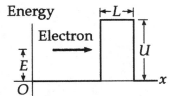

Solution   **Figure P41.37**

G: Since the barrier energy is higher than the kinetic energy of the electron, transmission is not likely, but should be possible since the barrier is not infinitely high or thick.

O: The probability of transmission is found from the transmission coefficient equation 41.18.

A: The transmission coefficient is

$$C = \frac{\sqrt{2m(U-E)}}{\hbar} = \frac{\sqrt{2(9.11\times10^{-31} \text{ kg})(10.0 \text{ eV} - 5.00 \text{ eV})(1.60\times10^{-19} \text{ J/eV})}}{6.63\times10^{-34} \text{ J}\cdot\text{s}/2\pi}$$

$$C = 1.14\times10^{10} \text{ m}^{-1}$$

(a) The probability of transmission is

$$T \approx e^{-2CL} = e^{-2(1.14\times10^{10} \text{ m}^{-1})(2.00\times10^{-10} \text{ m})} = e^{-4.58} = 0.0103 \qquad \lozenge$$

(b) If the electron does not tunnel, it is reflected, with probability

$$1 - 0.0103 = 0.990 \qquad \lozenge$$

L: Our expectation was correct; there is only a 1% chance that the electron will penetrate the barrier. This tunneling probability would be greater if the barrier were thinner, shorter, or if the kinetic energy of the electron were greater.

49. An electron is represented by the time-independent wave function

$$\psi(x) = \begin{cases} Ae^{-\alpha x} & \text{for } x > 0 \\ Ae^{+\alpha x} & \text{for } x < 0 \end{cases}$$

(a) Sketch the wave function as a function of x. (b) Sketch the probability that the electron is found between x and $x + dx$. (c) Why do you suppose this is a physically reasonable wave function? (d) Normalize the wave function. (e) Determine the probability of finding the electron somewhere in the range

$$x_1 = -1/2\alpha \qquad \text{to} \qquad x_2 = 1/2\alpha$$

Solution

(a)

(b) $P(x \rightarrow x + dx) = |\psi|^2 dx = A^2 e^{-2\alpha x} dx$

for $x > 0$ ◊

$P(x \rightarrow x + dx) = |\psi|^2 dx = A^2 e^{+2\alpha x} dx$

for $x < 0$ ◊

(c) This might be reasonable since (1) ψ is continuous, (2) $\psi \rightarrow 0$ as $x \rightarrow \infty$, and (3) The waveform mimics an electron bound at $x = 0$. ◊

(d) As ψ is symmetric, $\displaystyle\int_{-\infty}^{\infty} |\psi|^2 dx = 2\int_{0}^{\infty} |\psi|^2 dx = 1$

or $\displaystyle 2A^2 \int_{0}^{\infty} e^{-2\alpha x} dx = 1$

Integrating, $\displaystyle \left[\frac{2A^2}{(-2\alpha)}\right]\left[e^{-\infty} - e^0\right] = 1$ gives $A = \sqrt{\alpha}$ ◊

(e) $\displaystyle P_{\left(\frac{-1}{2\alpha}\right) \rightarrow \left(\frac{1}{2\alpha}\right)} = 2\int_{0}^{1/2\alpha} (\sqrt{\alpha})^2 e^{-2\alpha x} dx = \left[\frac{2\alpha}{(-2\alpha)}\right]\left[e^{-2\alpha/2\alpha} - 1\right] = \left[1 - e^{-1}\right] = 0.632$ ◊

51. Particles incident from the left are confronted with a step potential energy shown in Figure P41.50. The step has a height U, and the particles have energy $E = 2U$. Classically, all the particles would pass into the region of higher potential energy at the right. However, according to quantum mechanics, a fraction of the particles are reflected at the barrier. Use the result of Problem 50 to determine the fraction of the incident particles that are reflected. (This situation is analogous to the partial reflection and transmission of light striking an interface between two different media.)

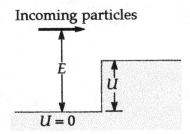

Figure P41.50

Solution   The reflection coefficient R, found from Problem 50, is

$$R = \frac{(k_1 - k_2)^2}{(k_1 + k_2)^2} \qquad \text{or} \qquad R = \frac{(1 - k_2/k_1)^2}{(1 + k_2/k_1)^2}$$

where $k_1 = 2\pi/\lambda_1$ and $k_2 = 2\pi/\lambda_2$ are the angular wave numbers for the incident and transmitted particles, respectively. Taking K to be the kinetic energy of the particle,

$$k = \frac{p}{\hbar} = \frac{\sqrt{2mK}}{\hbar}$$

For the incident particles, $K = E = 2U$: $k_1\hbar = \sqrt{2mE} = \sqrt{4mU}$

For the transmitted particles, $K = E - U = U$: $k_2\hbar = \sqrt{2mU}$

Therefore, the ratio of the wave numbers is $\dfrac{k_2}{k_1} = \dfrac{k_2\hbar}{k_1\hbar} = \dfrac{\sqrt{2mU}}{\sqrt{4mU}} = \dfrac{1}{\sqrt{2}}$

Solving for R, $R = \dfrac{(1 - 1/\sqrt{2})^2}{(1 + 1/\sqrt{2})^2} = \left[\dfrac{(\sqrt{2} - 1)}{(\sqrt{2} + 1)}\right]^2 = 0.0294$ ◊

57. For a particle described by a wave function $\psi(x)$, the expectation value of a physical quantity $f(x)$ associated with the particle is defined by

$$\langle f(x) \rangle \equiv \int_{-\infty}^{\infty} f(x)|\psi|^2\, dx$$

For a particle in a one-dimensional box extending from $x = 0$ to $x = L$, show that

$$\langle x^2 \rangle = \frac{L^2}{3} - \frac{L^2}{2n^2\pi^2}$$

Solution   We note that $\psi_n(x) = A\sin\left(\dfrac{n\pi x}{L}\right)$

where $A = \sqrt{2/L}$

Substituting x^2 for $f(x)$, $\langle x^2 \rangle = \int_{-\infty}^{\infty} x^2 |\psi|^2\, dx$

Integrating, $\langle x^2 \rangle = \left(\dfrac{2}{L}\right)\int_0^L x^2 \sin^2\left(\dfrac{n\pi x}{L}\right)dx = \dfrac{L^2}{3} - \dfrac{L^2}{2n^2\pi^2}$ ◊

---

59. A particle has a wave function

$$\psi(x) = \begin{cases} \sqrt{2/a}\; e^{-x/a} & \text{for } x > 0 \\ 0 & \text{for } x < 0 \end{cases}$$

(a) Find and sketch the probability density. (b) Find the probability that the particle will be at any point where $x < 0$. (c) Show that ψ is normalized and then find the probability that the particle will be found between $x = 0$ and $x = a$.

Chapter 11

Solution

(a) $|\psi|^2 = \begin{cases} \dfrac{2}{a}e^{-2x/a} & \text{for } x > 0 \\ 0 & \text{for } x < 0 \end{cases}$ ◊

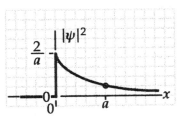

(b) The particle has zero probability of being at any point where $x < 0$. ◊

(c) For normalization, $\displaystyle\int_{\text{all } x} |\psi|^2 dx = 1$

$$0 + \int_0^\infty \frac{2}{a}e^{-2x/a}\,dx = -\int_0^\infty e^{-2x/a}(-2dx/a) = -e^{-2x/a}\Big|_0^\infty = -[0-1] = 1$$

Thus, $\sqrt{2/a}$ was the right normalization coefficient for ψ in the first place. Probability of finding the particle in $0 < x < a$ is

$$\int_0^a \frac{2}{a}e^{-2x/a}\,dx = -e^{-2x/a}\Big|_0^a = -[e^{-2}-1] = 0.865 \qquad ◊$$

---

61. An electron of momentum p is at a distance r from a stationary proton. The electron has kinetic energy $K = p^2/2m_e$, potential energy $U = -2k_e e^2/r$, and total energy $E = K + U$. If the electron is bound to the proton to form a hydrogen atom, its average position is at the proton, but the uncertainty in its position is approximately equal to the radius r of its orbit. The electron's average vector momentum is zero, but its average squared momentum is approximately equal to the squared uncertainty in its momentum as given by the uncertainty principle. Treating the atom as a one-dimensional system, (a) estimate the uncertainty in the electron's momentum in terms of r. (b) Estimate the electron's kinetic, potential, and total energies in terms of r. (c) The actual value of r is the one that **minimizes the total energy**, resulting in a stable atom. Find the value of r and the resulting total energy. Compare your answer with the predictions of the Bohr theory.

Chapter 41

Solution

(a) $\Delta x \Delta p \geq \dfrac{\hbar}{2}$ so if $\Delta x = r,$ $\Delta p \geq \dfrac{\hbar}{2r}$ ◊

(b)   Suppose that $p = \dfrac{\hbar}{r}$

Then $K = \dfrac{p^2}{2m} = \dfrac{\hbar^2}{2mr^2}$ ◊

$$U = -\dfrac{k_e e^2}{r}$$ ◊

$$E = \dfrac{\hbar^2}{2mr^2} - \dfrac{k_e e^2}{r}$$ ◊

(c)   To minimize E, $\dfrac{dE}{dr} = -\dfrac{\hbar^2}{mr^3} + \dfrac{k_e e^2}{r^2} = 0$ or $r = \dfrac{\hbar^2}{mk_e e^2}$ = Bohr radius

Then, $E = \dfrac{\hbar^2}{2m}\left(\dfrac{mk_e e^2}{\hbar^2}\right) - k_e e^2\left(\dfrac{mk_e e^2}{\hbar^2}\right) = -\dfrac{mk_e^2 e^4}{2\hbar^2} = -13.6\text{ eV}$ ◊

This is the same value as that predicted by the Bohr theory for the ground state of hydrogen.

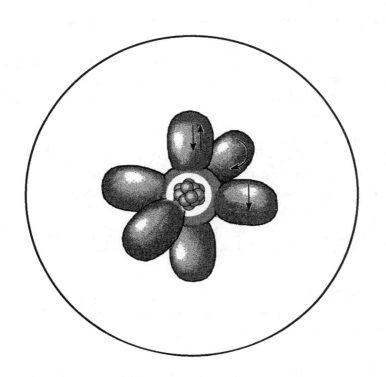

ATOMIC PHYSICS

ATOMIC PHYSICS

INTRODUCTION

In Chapter 41, we introduced some of the basic concepts and techniques used in quantum mechanics, along with their applications to various simple systems. This chapter deals with the application of quantum mechanics to the real world of atomic structure.

A large portion of this chapter is an application of quantum mechanics to the study of the hydrogen atom. Understanding the hydrogen atom, the simplest atomic system, is especially important for several reasons:

- Much of what is learned about the hydrogen atom with its single electron can be extended to such single-electron ions as He^+ and Li^{2+}.

- The hydrogen atom is an ideal system for performing precise tests of theory against experiment and for improving our overall understanding of atomic structure.

- The quantum numbers used to characterize the allowed states of hydrogen can be used to describe the allowed states of more complex atoms. This enables us to understand the periodic table of the elements, one of the greatest triumphs of quantum mechanics.

- The basic ideas about atomic structure must be well understood before we attempt to deal with the complexities of molecular structures and the electronic structure of solids.

EQUATIONS AND CONCEPTS

The **potential energy** of the hydrogen atom depends on the radius of the allowed orbit of the electron.

$$U(r) = -k_e \frac{e^2}{r} \qquad (42.1)$$

When numerical values of the constants are used, the energy level values for hydrogen can be expressed in units of electron volts (eV).

$$E_n = -\frac{13.606}{n^2} \text{ eV} \qquad (42.2)$$

$$(\text{for } n = 1, 2, 3, \ldots)$$

The lowest energy state or ground state corresponds to the principal quantum number $n = 1$. The energy level approaches $E = 0$ as n approaches infinity.

The simplest wave function for hydrogen is the one which describes the 1s state and depends only on the radial distance r. The parameter a_0 is the Bohr radius.

$$\psi_{1s}(r) = \frac{1}{\sqrt{\pi a_0{}^3}} e^{-r/a_0} \qquad (42.3)$$

$$a_0 = 0.0529 \text{ nm}$$

The radial probability density for the 1s state of hydrogen is defined as the probability of finding the electron in a spherical shell of radius r and thickness dr.

$$P_{1s}(r) = \left(\frac{4r^2}{a_0^3}\right)e^{-2r/a_0}$$ (42.6)

The value of the quantum number ℓ determines the magnitude of the electron's **orbital angular momentum**, L.

$$L = \sqrt{\ell(\ell+1)}\,\hbar$$ (42.9)

$$(\ell = 0, 1, 2, 3, \ldots, n-1)$$

When an atom is placed in an external magnetic field, the projection (or component) of the orbital angular momentum L_z along the direction of the magnetic field is quantized.

$$L_z = m_\ell \hbar$$ (42.10)

$$(m_\ell = -\ell, -\ell+1, -\ell+2, \ldots \ell)$$

If an orbiting electron is placed in a weak magnetic field directed along the z axis, the projection of the angular momentum vector of the electron (**L**) along the z axis can have only discrete values.

$$\cos\theta = \frac{m_\ell}{\sqrt{\ell(\ell+1)}}$$ (42.11)

In addition to orbital angular momentum, the electron has an intrinsic angular momentum or **spin angular momentum** S, as if due to spinning on its axis. This spin angular momentum is described by a single quantum number s whose value can only be 1/2.

$$S = \sqrt{s(s+1)}\,\hbar = \frac{\sqrt{3}}{2}\hbar$$ (42.12)

The spin angular momentum is space quantized with respect to the direction of an external magnetic field (along the z direction). The spin magnetic quantum number m_s can have values of $\pm 1/2$.

$$S_z = m_s \hbar = \pm \frac{1}{2} \hbar \qquad (42.13)$$

The **spin magnetic moment** μ_{spin} is related to the spin angular momentum.

$$\mu_{spin} = -\frac{e}{m_e} \mathbf{S} \qquad (42.14)$$

The shielding effect of the nuclear charge by inner-core electrons must be taken into account when calculating allowed energy levels of multielectron atoms. The atomic number is replaced by an effective atomic number Z_{eff}, which depends on the values of n and ℓ.

$$E_n = -\frac{13.6 Z_{eff}^2}{n^2} eV \qquad (42.18)$$

Z_{eff} is the atomic number, modified by the number of shielding electrons: for K-shell electrons, $Z_{eff} = Z - 1$; For an M shell electron, $Z_{eff} = Z - 9$. (See Example 42.7 in the text.)

$$Z_{eff} = Z - n_s$$

where

n_s = # of shielding electrons

SUGGESTIONS, SKILLS, AND STRATEGIES

After reading the chapter in your text, review the significance and set of allowed values for each of the quantum numbers used to describe the various electronic states of electrons in an atom.

In addition to the principal quantum number n (which can range from 1 to ∞), other quantum numbers are necessary to specify completely the possible energy levels in the hydrogen atom and also in more complex atoms.

All energy states with the same principal quantum number, n, form a shell. These shells are identified by the spectroscopic notation K, L, M, . . . corresponding to $n = 1, 2, 3, ...$

The orbital quantum number ℓ [which can range from 0 to $(n - 1)$], determines the allowed value of orbital angular momentum. All energy states having the same values of n and ℓ form a subshell. The letter designations $s, p, d, f, . . .$ correspond to values of $\ell = 0, 1, 2, 3, ...$

The magnetic orbital quantum number m_ℓ (which can range from $-\ell$ to ℓ) determines the possible orientations of the electron's orbital angular momentum vector in the presence of an external magnetic field.

The spin magnetic quantum number m_s, can have only two values, $m_s = -\frac{1}{2}$ and $m_s = \frac{1}{2}$, which in turn correspond to the two possible directions of the electron's intrinsic spin.

REVIEW CHECKLIST

▷ Understand the significance of the wave function and the associated radial probability density for the ground state of hydrogen.

▷ For each of the quantum numbers, n, ℓ (the orbital quantum number), m_ℓ (the orbital magnetic quantum number), and m_s (the spin magnetic quantum number):

  • qualitatively describe what each implies concerning atomic structure;
  • state the allowed values which may be assigned to each, and the number of allowed states which may exist in a particular atom corresponding to each quantum number.

▷ Associate the customary shell and subshell spectroscopic notations with allowed combinations of quantum numbers n and ℓ. Calculate the possible values of the orbital angular momentum, L, corresponding to a given value of the principal quantum number.

▷ Describe how allowed values of the magnetic orbital quantum number, m_ℓ, may lead to a restriction on the orientation of the orbital angular momentum vector in an external magnetic field. Find the allowed values for L_z (the component of the angular momentum along the direction of an external magnetic field) for a given value of L.

▷ State the Pauli exclusion principal and describe its relevance to the periodic table of the elements. Show how the exclusion principal leads to the known electronic ground state configuration of the light elements.

ANSWERS TO SELECTED CONCEPTUAL QUESTIONS

5.   Could the Stern-Gerlach experiment be performed with ions rather than neutral atoms? Explain.

Answer   Practically speaking, the answer would be no. Since ions have a net charge, the magnetic force $q\mathbf{v} \times \mathbf{B}$ would deflect the beam, making it very difficult to separate the ions with different orientations of their magnetic moments.

□   □   □   □

7.   Discuss some of the consequences of the exclusion principle.

Answer   If the Pauli exclusion principle were not valid, the elements and their chemical behavior would be grossly different because every electron would end up in the lowest energy level of the atom. All matter would therefore be nearly alike in its chemistry and composition, since the shell structures of each element would be identical. Most materials would have a much higher density, and the spectra of atoms and molecules would be very simple, resulting in the existence of less color in the world.

□   □   □   □

SOLUTIONS TO SELECTED END-OF-CHAPTER PROBLEMS

5. A general expression for the energy levels of one-electron atoms and ions is

$$E_n = -\left(\frac{\mu k_e^2 q_1^2 q_2^2}{2\hbar^2 n^2}\right)$$

where k_e is the Coulomb constant, q_1 and q_2 are the charges of the two particles, and μ is the reduced mass, given by $\mu = m_1 m_2 / (m_1 + m_2)$. In Problem 4 we found that the wavelength for the $n = 3$ to $n = 2$ transition of the hydrogen atom is 656.3 nm (visible red light). What are the wavelengths for this same transition in (a) positronium, which consists of an electron and a positron, and (b) singly ionized helium? (**Note:** A positron is a positively charged electron.)

Solution

G: The reduced mass of positronium is **less** than hydrogen, so the photon energy will be **less** for positronium than for hydrogen. This means that the wavelength of the emitted photon will be **longer** than 656.3 nm. On the other hand, helium has about the same reduced mass but more charge than hydrogen, so its transition energy will be **larger**, corresponding to a wavelength **shorter** than 656.3 nm.

O: All the factors in the above equation are constant for this problem except for the reduced mass and the nuclear charge. Therefore, the wavelength corresponding to the energy difference for the transition can be found simply from the ratio of mass and charge variables.

A: For hydrogen,

$$\mu = \frac{m_p m_e}{m_p + m_e} \approx m_e$$

The photon energy is

$$\Delta E = E_3 - E_2$$

Its wavelength is $\lambda = 656.3$ nm, where

$$\lambda = \frac{c}{f} = \frac{hc}{\Delta E}$$

(a)   For positronium,

$$\mu = \frac{m_e m_e}{m_e + m_e} = \frac{m_e}{2}$$

so the energy of each level is one half as large as in hydrogen, which we could call "protonium." The photon energy is inversely proportional to its wavelength, so for positronium,

$$\lambda_{32} = 2(656.3 \text{ nm}) = 1313 \text{ nm} \quad \text{(in the infrared region)} \qquad \lozenge$$

(b)   For He⁺, $\mu \approx m_e$, $q_1 = e$, and $q_2 = 2e$, so the transition energy is $2^2 = 4$ times larger than hydrogen. Then,

$$\lambda_{32} = \left(\frac{656}{4}\right) \text{ nm} = 164 \text{ nm} \quad \text{(in the ultraviolet region)} \qquad \lozenge$$

L: As expected, the wavelengths for positronium and helium are respectively larger and smaller than for hydrogen. Other energy transitions should have wavelength shifts consistent with this pattern. It is important to remember that the reduced mass is not the total mass, but is generally close in magnitude to the smaller mass of the system (hence the name **reduced** mass).

11. Show that the $1s$ wave function for an electron in hydrogen,

$$\psi_{1s}(r) = \frac{1}{\sqrt{\pi a_o^{\,3}}} e^{-r/a_o}$$

satisfies the radially symmetric Schrödinger equation,

$$-\frac{\hbar^2}{2m}\left(\frac{d^2\psi}{dr^2} + \frac{2}{r}\frac{d\psi}{dr}\right) - \frac{k_e e^2}{r}\psi = E\psi$$

Solution   To solve this problem, we substitute the wave function and its derivatives into the Schrödinger equation, and then start simplifying the resulting equation. If we find that the resulting equation is true, then we know that the Schrödinger equation is satisfied.

$$\frac{d\psi}{dr} = \frac{1}{\sqrt{\pi a_o^{\,3}}}\frac{d}{dr}\left(e^{-r/a_o}\right) = -\frac{1}{\sqrt{\pi a_o^{\,3}}}\left(\frac{1}{a_o}\right)e^{-r/a_o} = -\frac{\psi}{a_o}$$

Likewise,
$$\frac{d^2\psi}{dr^2} = -\frac{1}{\sqrt{\pi a_o^{\,5}}}\frac{d}{dr}e^{-r/a_o} = \frac{1}{\sqrt{\pi a_o^{\,7}}}e^{-r/a_o} = \frac{1}{a_o^{\,2}}\psi$$

Substituting these two terms into the Schrödinger equation,

$$-\frac{\hbar^2}{2m_e}\left(\frac{1}{a_o^{\,2}} - \frac{2}{a_o r}\right)\psi - \frac{k_e e^2}{r}\psi = E\psi$$

and noting that $E = -\dfrac{k_e e^2}{2a_o}$, $-\dfrac{\hbar^2}{2m_e}\left(\dfrac{1}{a_o^{\,2}} - \dfrac{2}{a_o r}\right)\psi - \dfrac{k_e e^2}{r}\psi = -\dfrac{k_e e^2}{2a_0}\psi$

Dividing the entire equation by ψ, and moving the last term on the left-hand side to the right, we find that the equation is true if $\hbar^2 = k_e m_e e^2 a_0$. But we know that this is indeed the case, from the definition of the Bohr radius. Therefore, the Schrödinger equation is satisfied.   ◊

17. How many sets of quantum numbers are possible for an electron for which (a) $n = 1$, (b) $n = 2$, (c) $n = 3$, (d) $n = 4$, and (e) $n = 5$? Check your results to show that they agree with the general rule that the number of sets of quantum numbers is equal to $2n^2$.

Solution (a) For $n = 1$, $\ell = 0$,

$m_\ell = 0$, and $m_s = \pm\frac{1}{2}$:

n	ℓ	m_ℓ	m_s
1	0	0	$-\frac{1}{2}$
1	0	0	$+\frac{1}{2}$

This yields $2n^2 = 2(1)^2 = 2$ sets   ◊

(b) For $n = 2$, we have

n	ℓ	m_ℓ	m_s
2	0	0	$\pm\frac{1}{2}$
2	1	-1	$\pm\frac{1}{2}$
2	1	0	$\pm\frac{1}{2}$
2	1	1	$\pm\frac{1}{2}$

This yields $2n^2 = 2(2)^2 = 8$ sets   ◊

Note that the number is twice the number of m_ℓ values. Also, for each ℓ there are $(2\ell + 1)$ different m_ℓ values. Finally, ℓ can take on values ranging from 0 to $n - 1$. So the general expression is

$$s = \sum_{0}^{n-1} 2(2\ell + 1)$$

The series is an arithmetic progression $2 + 6 + 10 + 14$, the sum of which is

$$s = \frac{n}{2}[2a + (n-1)d] \quad \text{where } a = 2, \ d = 4$$

$$s = \frac{n}{2}[4 + (n-1)4] = 2n^2$$

(c) $n = 3$: $2(1) + 2(3) + 2(5) = 2 + 6 + 10 = 18$ $2n^2 = 2(3)^2 = 18$ ◊

(d) $n = 4$: $2(1) + 2(3) + 2(5) + 2(7) = 32$ $2n^2 = 2(4)^2 = 32$ ◊

(e) $n = 5$: $32 + 2(9) = 32 + 18 = 50$ $2n^2 = 2(5)^2 = 50$ ◊

---

23. The ρ-meson has a charge of $-e$, a spin quantum number of 1, and a mass of 1507 times that of the electron. Imagine that the electrons in atoms were replaced by ρ-mesons, and list the possible sets of quantum numbers for ρ-mesons in the 3d subshell.

Solution   The 3d subshell has $\ell = 2$, and $n = 3$. Also, we have $s = 1$. Therefore, we can have $n = 3$, $\ell = 2$, $m_\ell = -2, -1, 0, 1, 2$, $s = 1$, and $m_s = -1, 0, 1$, leading to the following table:

n	ℓ	m_ℓ	s	m_s
3	2	−2	1	−1
3	2	−2	1	0
3	2	−2	1	+1
3	2	−1	1	−1
3	2	−1	1	0
3	2	−1	1	+1
3	2	0	1	−1
3	2	0	1	0
3	2	0	1	+1
3	2	+1	1	−1
3	2	+1	1	0
3	2	+1	1	+1
3	2	+2	1	−1
3	2	+2	1	0
3	2	+2	1	+1

---

29. (a) Scanning through Table 42.4 in order of increasing atomic number, note that the electrons fill the subshells in such a way that the subshells with the lowest values of $n + \ell$ are filled first. If two subshells have the same value of $n + \ell$, the one with the lower value of n is filled first. Using these two rules, write the order in which the subshells are filled through $n + \ell = 7$. (b) Predict the chemical valence for the elements that have atomic numbers 15, 47, and 86, and compare your predictions with the actual valences.

Solution

(a)

$n + \ell$	1	2	3	4	5	6	7
subshell	$1s$	$2s$	$2p, 3s$	$3p, 4s$	$3d, 4p, 5s$	$4d, 5p, 6s$	$4f, 5d, 6p, 7s$

(b) $Z = 15$: Filled subshells: $1s, 2s, 2p, 3s$ (12 e$^-$)

Valence subshell: 3 electrons in $3p$ subshell

Prediction: Valence +3 or –5

Element is phosphorus: Valence +3 or –5

$Z = 47$: Filled subshells: $1s, 2s, 2p, 3s, 3p, 4s, 3d, 4p, 5s$ (38 e$^-$)

Outer subshell: 9 electrons in $4d$ subshell

Prediction: Valence –1

Element is silver,   Valence +1   (Prediction fails)

$Z = 86$: Filled shells: $1s, 2s, 2p, 3s, 3p, 4s, 3d, 4p,$ $5s, 4d, 5p, 6s, 4f, 5d, 6p$

Outer subshell: Full

Prediction: Inert gas

Element is Radon,   Inert gas (Prediction works)

35. Use the method illustrated in Example 42.7 to calculate the wavelength of the x-ray emitted from a molybdenum target ($Z = 42$) when an electron moves from the L shell ($n = 2$) to the K shell ($n = 1$).

Solution   Following Example 42.7, suppose the electron is originally in the L shell with just one other electron in the K shell between it and the nucleus, so it moves in a field of effective charge $(42 - 1)e$. Its energy is then $E_L = -(42 - 1)^2\ 13.6$ eV$/4$. In its final state we estimate the screened charge holding it in orbit as again $(42 - 1)e$, so its energy is $E_K = -(42 - 1)^2\ 13.6$ eV. The photon energy emitted is the difference.

$$E_\gamma = \tfrac{3}{4}(42 - 1)^2 (13.6 \text{ eV}) = 1.71 \times 10^4 \text{ eV} = 2.74 \times 10^{-15} \text{ J}$$

Then $f = E/h = 4.14 \times 10^{18}$ Hz   and $\lambda = c/f = 0.725$ Å   ◊

---

43. A ruby laser delivers a 10.0-ns pulse of 1.00 MW average power. If the photons have a wavelength of 694.3 nm, how many are contained in the pulse?

Solution

G: Lasers generally produce concentrated beams that are bright (except for IR or UV lasers that produce invisible beams). Since our eyes can detect light levels as low as a few photons, there are probably at least 1000 photons in each pulse.

O: From the pulse width and average power, we can find the energy delivered by each pulse. The number of photons can then be found by dividing the pulse energy by the energy of each photon, which is determined from the photon wavelength.

A: The energy in each pulse is

$$E = \mathscr{P}t = (1.00 \times 10^6 \text{ W})(1.00 \times 10^{-8} \text{ s}) = 1.00 \times 10^{-2} \text{ J}$$

The energy of each photon is

$$E_\gamma = hf = \frac{hc}{\lambda} = \frac{(6.626 \times 10^{-34})(3.00 \times 10^8)}{694.3 \times 10^{-9}} \text{ J} = 2.86 \times 10^{-19} \text{ J}$$

So $N = \dfrac{E}{E_\gamma} = \dfrac{1.00 \times 10^{-2} \text{ J}}{2.86 \times 10^{-19} \text{ J/photon}} = 3.49 \times 10^{16}$ photons   ◊

L: With 10^{16} photons/pulse, this laser beam should produce a bright red spot when the light reflects from a surface, even though the time between pulses is generally much longer than the width of each pulse. For comparison, this laser produces more photons in a single ten-nanosecond pulse than a typical 5 mW helium-neon laser produces over a full second (about 1.6×10^{16} photons/second).

========

49. Show that the average value of r for the 1s state of hydrogen has the value $3a_0/2$. (**Hint:** Use Eq. 42.6.)

Solution   The average (expectation) value of r is $\langle r \rangle = \int_0^\infty r P_{1s}(r)\,dr$

where $P_{1s}(r) = \left(4r^2/a_o{}^3\right)e^{-2r/a_o}$: $\langle r \rangle = \dfrac{4}{a_o{}^3} \int_0^\infty r^3 e^{-2r/a_o}\,dr$

Letting $x = 2r/a_o$, we find: $\langle r \rangle = \dfrac{1}{4}a_o \int_0^\infty x^3 e^{-x}\,dx$

Integrating by parts gives $\langle r \rangle = \dfrac{3}{2}a_o$ ◊

========

51. Suppose a hydrogen atom is in the 2s state. Taking $r = a_0$, calculate values for (a) $\psi_{2s}(a_0)$, (b) $|\psi_{2s}(a_0)|^2$, and (c) $P_{2s}(a_0)$. (**Hint:** Use Eq. 42.7.)

Solution The wave function for the 2s state is given by Equation 42.7:

$$\psi_{2s}(r) = \frac{1}{4\sqrt{2\pi}}\left(\frac{1}{a_0}\right)^{3/2}\left[2 - \frac{r}{a_0}\right]e^{-r/2a_0}$$

(a) Taking $r = a_0 = 0.529 \times 10^{-10}$ m, we find

$$\psi_{2s}(a_0) = \frac{1}{4\sqrt{2\pi}}\left(\frac{1}{0.529 \times 10^{-10} \text{ m}}\right)^{3/2}(2-1)e^{-1/2} = 1.57 \times 10^{14} \text{ m}^{-3/2} \qquad \lozenge$$

(b) $|\psi_{2s}(a_0)|^2 = \left(1.57 \times 10^{14} \text{ m}^{-3/2}\right)^2 = 2.47 \times 10^{28} \text{ m}^{-3} \qquad \lozenge$

(c) Using Equation 42.5 and the results to (b) gives

$$P_{2s}(a_0) = 4\pi a_0{}^2 |\psi_{2s}(a_0)|^2 = 8.69 \times 10^8 \text{ m}^{-1} \qquad \lozenge$$

---

55. (a) Show that the most probable radial position for an electron in the 2s state of hydrogen is $r = 5.236a_0$. (b) Show that the wave function given by Equation 42.7 is normalized.

Solution The wave function for an electron in the 2s state is

$$\psi_{2s}(r) = \tfrac{1}{4}\left(2\pi a_0{}^3\right)^{-1/2}\left(2 - \frac{r}{a_0}\right)e^{-r/2a_0}$$

By Equation 42.5, taking $x = r / a_0$, the radial probability distribution function is

$$P(r) = 4\pi r^2 |\psi|^2 = \frac{1}{8}\left(\frac{r^2}{a_0^3}\right)\left(2 - \frac{r}{a_0}\right)^2 e^{-r/a_0}$$

$$P(x) = \frac{1}{8}\left(\frac{x^2}{a_0}\right)(2-x)^2 e^{-x}$$

(a) This probability function is a minimum or maximum at $dP(x)/dx = 0$. Taking the derivitive of $P(x)$, therefore, we find that

$$\frac{dP(x)}{dx} = \left(\frac{(2-x)xe^{-x}}{8a_0}\right)(x^2 - 6x + 4)$$

Setting this equal to zero, we find that maximums and minimums occur at $x = r / a_0 = 3 \pm \sqrt{5}$, $x = \infty$, $x = 0$, and $x = 2$. Substituting each of these values in to $P(x)$, we find that the maximum value occurs at $x = 3 + \sqrt{5} = 5.236$.   ◊

(b)   The probability function will be normalized if the integral of the probability function from $r = 0$ to $r = \infty$ is equal to 1. We make the same change of variables $x = r / a_0$, as before, but we also note that $dr = a_0 dx$. Therefore, the integral becomes

$$\int_0^\infty P(r)dr = \int_0^\infty \frac{1}{8}(x^2)(2-x)^2 e^{-x}dx = \int_0^\infty \frac{1}{8}(4x^2 - 4x^3 + x^4)e^{-x}dx$$

Using a table of integrals or integrating by parts repeatedly, we find, as desired,

$$\int_0^\infty P(r)dr = -\frac{1}{8}(x^4 + 4x^2 + 8x + 8)e^{-x}\Big|_0^\infty = 1$$ ◊

57. An electron in chromium moves from the $n = 2$ state to the $n = 1$ state without emitting a photon. Instead, the excess energy is transferred to an outer electron (one in the $n = 4$ state), which is then ejected by the atom. (This is called an Auger [pronounced 'ohjay'] process, and the ejected electron is referred to as an Auger electron.) Use the Bohr theory to find the kinetic energy of the Auger electron.

Solution The chromium atom with nuclear charge $Z = 24$ starts with one vacancy in the $n = 1$ shell, perhaps produced by the absorption of an x-ray which ionized the atom. An electron from the $n = 2$ shell tumbles down to fill the vacancy. We suppose that this electron is shielded from the electric field of the full nuclear charge by the one K-shell electron originally below it. Its change in energy is

$$\Delta E = -(Z-1)^2 (13.6 \text{ eV})\left(\frac{1}{1^2} - \frac{1}{2^2}\right) = -5.40 \text{ keV}$$

Then +5.40 keV can be transferred to the single 4s electron. Suppose that it is shielded by the 22 electrons in the K, L, and M shells. To break the outermost electron out of the atom, producing a Cr^{2+} ion, requires an energy investment of

$$E_{\text{ionize}} = \frac{(Z-22)^2(13.6 \text{ eV})}{4^2} = \frac{2^2(13.6 \text{ eV})}{16} = 1.70 \text{ eV}$$

As evidence that this relatively tiny amount of energy can still be the right order of magnitude, note that the (first) ionization energy for neutral chromium is tabulated as 6.76 eV. Then the remaining energy that can appear as kinetic energy is

$$K = \Delta E - E_{\text{ionize}} = 5395.8 \text{ eV} - 1.7 \text{ eV} = 5.39 \text{ keV} \qquad \lozenge$$

Because of conservation of momentum and the tiny mass of the electron compared to that of the Cr^{2+} ion, almost all of this kinetic energy will belong to the electron.

61. For hydrogen in the 1s state, what is the probability of finding the electron farther than $2.50\,a_o$ from the nucleus?

Solution

G: From the graph shown in Figure 42.8, it appears that the probability of finding the electron beyond $2.5\,a_0$ is about 20%.

O: The precise probability can be found by integrating the 1s radial probability distribution function from $r = 2.50\,a_o$ to ∞.

A: The general radial probability distribution function is $P(r) = 4\pi r^2 |\psi|^2$

With $\psi_{1s} = (\pi a_0^3)^{-1/2} e^{-r/a_0}$ it is $P(r) = 4r^2 a_0^{-3} e^{-2r/a_0}$

The required probability is then $P = \int\limits_{2.50a_0}^{\infty} P(r)dr = \int\limits_{2.50a_0}^{\infty} \frac{4r^2}{a_0^3} e^{-2r/a_0} dr$

Let $z = 2r/a_0$ and $dz = 2dr/a_0$: $P = \frac{1}{2}\int\limits_{5.00}^{\infty} z^2 e^{-z} dz$

Performing this integration by parts, $P = -\frac{1}{2}\left(z^2 + 2z + 2\right)e^{-z}\Big]_{5.00}^{\infty}$

$P = -\frac{1}{2}(0) + \frac{1}{2}(25.0 + 10.0 + 2.00)e^{-5.00} = \left(\frac{37}{2}\right)(0.00674) = 0.125$ ◊

L: The probability of 12.5% is less than the 20% we estimated, but close enough to be a reasonable result. In comparing the 1s probability density function with the others in Figure 42.8, it appears that the ground state is the most narrow, indicating that a 1s electron will probably be found in the narrow range of 0 to 4 Bohr radii, and most likely at $r = a_0$.

63. According to classical physics, a charge e moving with an acceleration a radiates at a rate

$$\frac{dE}{dt} = -\frac{1}{6\pi\epsilon_0}\frac{e^2 a^2}{c^3}$$

(a) Show that an electron in a classical hydrogen atom (see Fig. 42.3) spirals into the nucleus at a rate

$$\frac{dr}{dt} = -\frac{e^4}{12\pi^2 \epsilon_0^2 r^2 m_e^2 c^3}$$

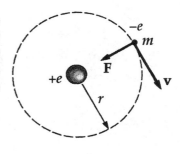

Figure 42.3

(b) Find the time it takes the electron to reach $r = 0$, starting from $r_o = 2.00 \times 10^{-10}$ m.

Solution (a) According to a classical model, the electron moving in a circular orbit about the proton in the hydrogen atom experiences a force $k_e e^2 / r^2$; and from Newton's second law, $F = ma$, its acceleration is $k_e e^2 / m_e r^2$. Using the fact that the Coulomb constant $k_e = 1/4\pi\epsilon_0$,

$$a = \frac{v^2}{r} = \frac{k_e e^2}{m_e r^2} = \frac{e^2}{4\pi\epsilon_0 m_e r^2} \qquad (1)$$

From the Bohr model of the atom (Chapter 40), we can write the total energy of the atom as

$$E = -\frac{k_e e^2}{2r} = -\frac{e^2}{8\pi\epsilon_0 r} \quad \text{so that} \quad \frac{dE}{dt} = \frac{e^2}{8\pi\epsilon_0 r^2}\frac{dr}{dt} = -\frac{1}{6\pi\epsilon_0}\frac{e^2 a^2}{c^3} \qquad (2)$$

Substituting (1) into (2) for a, solving for dr/dt, and simplifying gives

$$\frac{dr}{dt} = -\frac{4r^2}{3c^3}\left(\frac{e^2}{4\pi\epsilon_0 m_e r^2}\right)^2 = -\frac{e^4}{12\pi^2 \epsilon_0^2 r^2 m_e^2 c^3} \qquad \lozenge$$

(b) We can express dr/dt in the simpler form: $\dfrac{dr}{dt} = -\dfrac{A}{r^2} = -\dfrac{3.15 \times 10^{-21}}{r^2}$

Thus, $-\displaystyle\int_{2\times 10^{-10}\ m}^{0} r^2 dr = 3.15 \times 10^{-21} \int_{0}^{T} dt$

and $T = \left(3.17 \times 10^{20}\right) \dfrac{r^3}{3}\Bigg|_{0}^{2\times 10^{-10}\ m} = 8.46 \times 10^{-10}\ s = 0.846\ ns$ ◊

We know that atoms 'last' much longer than 0.8 ns; thus, classical physics does not hold (fortunately) for atomic systems.

===

65. (Review Problem) In the technique known as electron spin resonance (ESR), a sample containing unpaired electrons is placed in a magnetic field. Consider the simplest situation, in which only one electron is present and therefore only two energy states are possible, corresponding to $m_s = \pm 1/2$. In ESR, the absorption of a photon causes the electron's spin magnetic moment to flip from a lower energy state to a higher energy state. (The lower energy state corresponds to the case in which the magnetic moment μ_{spin} is aligned with the magnetic field, and the higher energy state corresponds to the case where μ_{spin} is aligned opposite the field.) What is the photon frequency required to excite an ESR transition in a 0.350-T magnetic field?

Solution   As in Sec. 29.3, the magnetic moment feels torque $\tau = \mu \times \mathbf{B}$ in an external field. In turning it from alignment with the field to the opposite direction, the field does work according to Eq. 10.22,

$$W = \int dW = \int_{0}^{180^\circ} \tau\, d\theta = \int_{0}^{\pi} \mu B \sin\theta\, d\theta = -\mu B \cos\theta\big|_{0}^{\pi} = 2\mu B$$

To make the electron flip, the photon must carry energy $\Delta E = 2\mu_B B = hf$:

Therefore, $f = \dfrac{2\mu_B B}{h} = \dfrac{2\left(9.27 \times 10^{-24}\right)(0.350\ \text{T})}{6.63 \times 10^{-34}\ \text{J}\cdot\text{s}} = 9.79 \times 10^9\ \text{Hz}$ ◊

67. A dimensionless number that often appears in atomic physics is the fine-structure constant $\alpha = k_e e^2/\hbar c$, where k_e is the Coulomb constant. (a) Obtain a numerical value for $1/\alpha$. (b) In scattering experiments, the electron size is taken to be the classical electron radius, $r_e = k_e e^2/m_e c^2$. In terms of α, what is the ratio of the Compton wavelength (Section 40.3), $\lambda_C = h/m_e c$, to the classical electron radius? (c) In terms of α, what is the ratio of the Bohr radius, a_0, to the Compton wavelength? (d) In terms of α, what is the ratio of the Rydberg wavelength, $1/R_H$, to the Bohr radius (Section 40.5)?

Solution

(a) $\dfrac{1}{\alpha} = \dfrac{\hbar c}{k_e e^2} = \dfrac{(1.05457 \times 10^{-34} \text{ J} \cdot \text{s})(2.997925 \times 10^8 \text{ m/s})}{(8.9875 \times 10^9 \text{ N} \cdot \text{m}^2/\text{C}^2)(1.60219 \times 10^{-19} \text{ C})^2} = 137.034$ ◊

(b) $\dfrac{\lambda_c}{r_e} = \dfrac{h/mc}{k_e e^2/mc^2} = \dfrac{hc}{k_e e^2} = \dfrac{2\pi}{\alpha}$ ◊

(c) $\dfrac{a_0}{\lambda_c} = \dfrac{\hbar^2/mk_e e^2}{h/mc} = \left(\dfrac{1}{2\pi}\right)\dfrac{\hbar c}{k_e e^2} = \dfrac{1}{2\pi\alpha}$ ◊

(d) $\dfrac{1}{R_H a_0} = \left(\dfrac{4\pi c \hbar^3}{mk_e^2 e^4}\right)\left(\dfrac{mk_e e^2}{\hbar^2}\right) = 4\pi\left(\dfrac{\hbar c}{k_e e^2}\right) = \dfrac{4\pi}{\alpha}$ ◊

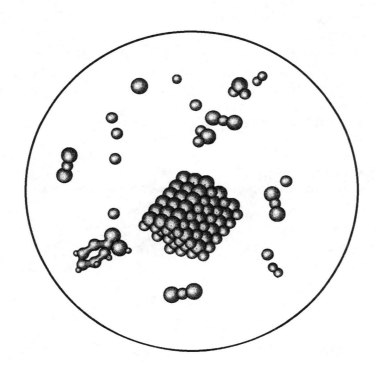

Molecules and Solids

MOLECULES AND SOLIDS

INTRODUCTION

In this chapter, we study the aggregates of atoms known as molecules. First we describe the bonding mechanisms in molecules, the various modes of molecular excitation, and the radiation emitted or absorbed by molecules. We then take the next logical step and show how molecules combine to form solids. Then, by examining their electronic distributions, we explain the differences between insulating, metallic, and semiconducting crystals. The chapter includes discussions of semiconducting junctions and several semiconductor devices, and concludes with treatment of superconductors.

EQUATIONS AND CONCEPTS

The rotational energy of a molecule is quantized and depends on the value of the moment of inertia.

$$E_{\text{rot}} = \frac{\hbar^2}{2I} J(J+1) \tag{43.6}$$

$$(J = 0, 1, 2 \ldots)$$

For a diatomic molecule, the moment of inertia I can be written in terms of the reduced mass, μ.

$$I = \mu r^2 \tag{43.3}$$

$$\mu = \frac{m_1 m_2}{m_1 + m_2} \tag{43.4}$$

The vibrational energy for a diatomic molecule is quantized and is characterized by the vibrational quantum number, v. The selection rule for allowed vibrational transitions is given by $\Delta v \pm 1$.

$$E_{\text{vib}} = \left(v + \tfrac{1}{2}\right) \frac{h}{2\pi} \sqrt{\frac{k}{\mu}} \tag{43.10}$$

$$(v = 0, 1, 2 \ldots)$$

The energy difference between successive vibrational levels is hf, where f is the frequency of vibration.

$$\Delta E_{vib} = hf \qquad (43.11)$$

The ionic cohesive energy U_0 of a solid represents the energy necessary to separate the solid into a collection of positive and negative ions. In this expression, r_0 is the equilibrium ion separation and α, the Madelung constant, has a value which is characteristic of a specific crystal structure. The parameter m is a small integer.

$$U_0 = -\alpha k_e \frac{e^2}{r_0}\left(1 - \frac{1}{m}\right) \qquad (43.17)$$

The Fermi-Dirac distribution function $f(E)$ gives the probability of finding an electron in a particular energy state, E.

$$f(E) = \frac{1}{e^{(E-E_F)/k_BT} + 1} \qquad (43.18)$$

In the Fermi-Dirac distribution function, E_F is called the Fermi energy and is a function of the total number of electrons per unit volume, n_e.

$$E_F(0) = \frac{h^2}{2m_e}\left(\frac{3n_e}{8\pi}\right)^{2/3} \qquad (43.25)$$

In thermal equilibrium, the number of electrons per unit volume with energy between E and $E+dE$ is the product of the probability of finding an electron in a state and the density of states.

$$n_e = \int_0^\infty N(E)dE = C\int_0^\infty \frac{E^{1/2}\,dE}{e^{(E-E_F)/k_BT} + 1} \qquad (43.23)$$

$$\text{where} \quad C = \frac{8\sqrt{2}\pi m_e^{3/2}}{h^3} \qquad (43.21)$$

REVIEW CHECKLIST

▷ Understand the essential bonding mechanisms involved in ionic, covalent, hydrogen, and van der Waals bonding.

▷ Describe in terms of appropriate quantum numbers the allowed energy levels associated with rotational and vibrational motions of molecules. Use the selection rules to determine the separation between adjacent energy levels.

▷ Discuss the free-electron theory of metals including the significance of the Fermi-Dirac distribution function, Fermi energy, and particle distribution by energy interval.

▷ Use the band theory of solids as a basis for a qualitative discussion of the mechanisms for conduction in metals, insulators, and semiconductors.

▷ Describe a *p-n* junction and the diffusion of electrons and holes through the junction. Discuss the fabrication and function of a junction diode and junction transistor.

ANSWERS TO SELECTED CONCEPTUAL QUESTIONS

5. The resistivity of metals increases with increasing temperature, whereas the resistivity of an intrinsic semiconductor decreases with increasing temperature. Explain.

Answer First consider electric conduction in a metal. The number of conduction electrons is essentially fixed. They conduct electricity by having drift motion in an applied electric field superposed on their random thermal motion. At higher temperature, the ion cores vibrate more and scatter more effeciently the conduction electrons flying among them. The mean time between collisions is reduced. The electrons have time to develop only a lower drift speed. The electric current is reduced, so we see the resistivity increasing with temperature.

Now consider an intrinsic semiconductor. At absolute zero its valence band is full and its conduction band is empty. It is an insulator, with very high resistivity. As the temperature increases, more electrons are promoted to the conduction band, leaving holes in the valence band. Then both electrons and holes move in response to an applied electric field. Thus we see the resistivity decreasing as temperature goes up.

□   □   □   □

15. Which is easier to excite in a diatomic molecule, rotational or vibrational motion?

Answer Rotation of a diatomic molecule involves less energy than vibration. Absorption of microwave photons, of frequency ~10^{11} Hz, excites rotational motion, while absorption of infrared photons, of frequency ~10^{13} Hz, excites vibration in typical simple molecules.

□   □   □   □

SOLUTIONS TO SELECTED END-OF-CHAPTER PROBLEMS

1. **Review Problem.** A K^+ ion and a Cl^- ion are separated by a distance of 5.00×10^{-10} m. Assuming that the two ions act like point charges, determine (a) the force each ion exerts on the other and (b) the potential energy of attraction in electron volts.

Solution The force of each ion is directed towards the other:

(a) $F = \dfrac{k_e |q_1||q_2|}{r^2} = \dfrac{(8.99 \times 10^9 \ \text{N} \cdot \text{m}^2/\text{C}^2)(1.60 \times 10^{-19} \ \text{C})^2}{(5.00 \times 10^{-10} \ \text{m})^2} = 9.21 \times 10^{-10} \ \text{N}$ ◊

(b) $U = \dfrac{k_e q_1 q_2}{r} = \dfrac{(1.60 \times 10^{-19} \ \text{C})(-1.60 \times 10^{-19} \ \text{C})}{5.00 \times 10^{-10} \ \text{m}} \left(8.99 \times 10^9 \ \dfrac{\text{N} \cdot \text{m}^2}{\text{C}^2} \right)$

$U = -4.60 \times 10^{-19} \ \text{J} = -2.88 \ \text{eV}$ ◊

7.   An HCl molecule is excited to its first rotational-energy level, corresponding to $J = 1$. If the distance between its nuclei is 0.1275 nm, what is the angular speed of the molecule about its center of mass?

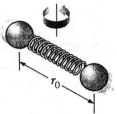

Solution

G: For a system as small as a molecule, we can expect the angular speed to be much faster than the few rad/s typical of everyday objects we encounter.

O: The rotational energy is given by the angular momentum quantum number, J. The angular speed can be calculated from this kinetic rotational energy and the moment of inertia of this one-dimensional molecule.

A: For the HCl molecule in the $J = 1$ rotational energy level, we are given $r_0 = 0.1275$ nm.

$$E_{rot} = \frac{\hbar^2}{2I}J(J + 1)$$

With $J = 1$, $E_{rot} = \frac{\hbar^2}{I} = \frac{1}{2}I\omega^2$ and $\omega = \sqrt{\frac{2\hbar^2}{I^2}} = \sqrt{2}\frac{\hbar}{I}$

The moment of inertia of the molecule is given by:

$$I = \mu r_0{}^2 = \left(\frac{m_1 m_2}{m_1 + m_2}\right)r_0{}^2 = \left[\frac{(1\ u)(35\ u)}{1\ u\ +\ 35\ u}\right]r_0{}^2 = (0.972\ u)r_0{}^2$$

$$I = (0.972\ u)(1.66 \times 10^{-27}\ kg\ /\ u)(1.275 \times 10^{-10}\ m)^2 = 2.62 \times 10^{-47}\ kg \cdot m^2$$

Therefore, $\omega = \sqrt{2}\dfrac{\hbar}{I} = \sqrt{2}\left(\dfrac{1.055 \times 10^{-34}\ J \cdot s}{2.62 \times 10^{-47}\ kg \cdot m^2}\right) = 5.69 \times 10^{12}$ rad / s ◊

L: This angular speed is more than a billion times faster than the spin rate of a music CD (200 to 500 revolutions per minute, or $\omega = 20$ rad / s to 50 rad / s).

---

11. If the effective force constant of a vibrating HCl molecule is $k = 480$ N / m, find the energy difference between the ground state and the first excited vibrational level.

Solution

The reduced mass of the pair of atoms is

$$\mu = \frac{m_1 m_2}{m_1 + m_2} = \frac{(1.01 \text{ u})(35.5 \text{ u})}{1.01 \text{ u} + 35.5 \text{ u}} \left(1.66 \times 10^{-27} \text{ kg / u}\right) = 1.63 \times 10^{-27} \text{ kg}$$

The energy difference between adjacent vibration states is

$$\Delta E_{\text{vib}} = \frac{h}{2\pi}\sqrt{\frac{k}{\mu}} = \left(\frac{6.63 \times 10^{-34} \text{ J} \cdot \text{s}}{2\pi}\right)\sqrt{\frac{480 \text{ N / m}}{1.63 \times 10^{-27} \text{ kg}}} = 5.72 \times 10^{-20} \text{ J} = 0.358 \text{ eV} \quad \Diamond$$

---

21. Consider a one-dimensional chain of alternating positive and negative ions. Show that the potential energy of an ion in this hypothetical crystal is

$$U(r) = -k_e \alpha \frac{e^2}{r}$$

where the Madelung constant is $\alpha = 2 \ln 2$ and r is the interionic spacing. [**Hint:** Use the series expansion for $\ln(1 + x)$.]

Solution

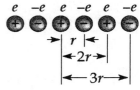

The total potential energy is obtained by summing over all pairs of interactions:

$$U = \sum_{i \neq j} k_e \frac{q_i q_j}{r_{ij}} = -k_e \left[\frac{e^2}{r} + \frac{e^2}{r} - \frac{e^2}{2r} - \frac{e^2}{2r} + \frac{e^2}{3r} + \frac{e^2}{3r} - \frac{e^2}{4r} - \frac{e^2}{4r} + \cdots \right]$$

$$U = -2 k_e \frac{e^2}{r} \left[1 - \frac{1}{2} + \frac{1}{3} - \frac{1}{4} + \cdots \right]$$

But $\quad \ln(1+x) = x - \dfrac{x^2}{2} + \dfrac{x^3}{3} - \dfrac{x^4}{4} + \cdots$

Therefore, $x = 1$ for our series, and $\quad U = -2 \ln(2) k_e \dfrac{e^2}{r} = -\alpha k_e \dfrac{e^2}{r}$ ◊

---

27. Calculate the energy of a conduction electron in silver at 800 K if the probability of finding an electron in that state is 0.950. The Fermi energy is 5.48 eV at this temperature.

Solution

G: Since there is a 95% probability of finding the electron in this state, its energy should be slightly less than the Fermi energy, as indicated by the graph in Figure 43.21.

O: The electron energy can be found from the Fermi-Dirac distribution function.

A: Taking $E_F = 5.48$ eV for silver at 800 K, and given $f(E) = 0.950$, we find

$$f(E) = \frac{1}{e^{(E-E_F)/k_B T} + 1} = 0.950$$

$$e^{(E-E_F)/k_B T} = \frac{1}{0.950} - 1 = 0.05263$$

$$\frac{E - E_F}{k_B T} = \ln(0.05263) = -2.944$$

$$E - E_F = -2.944 k_B T = -2.944 (1.38 \times 10^{-23} \text{ J / K})(800 \text{ K})$$

$$E = E_F - 3.25 \times 10^{-20} \text{ J} = 5.48 \text{ eV} - 0.203 \text{ eV} = 5.28 \text{ eV} \qquad \lozenge$$

L: As expected, the energy of the electron is slightly less than the Fermi energy, which is about 5 eV for most metals. There is very little probability of finding an electron significantly above the Fermi energy in a metal.

---

29. Show that the average kinetic energy of a conduction electron in a metal at 0 K is $E_{av} = \frac{3}{5} E_F$. (**Hint:** In general, the average kinetic energy is

$$E_{av} = \frac{1}{n_e} \int EN(E)\, dE$$

where n_e is the density of electrons, and $N(E)dE$ is given by Equation 43.22, and the integral is over all possible values of the energy.)

Solution
$$E_{av} = \frac{1}{n_e} \int_0^\infty EN(E)\,dE$$

where
$$N(E) = \frac{CE^{1/2}}{e^{(E-E_F)/k_BT}+1} = Cf(E)E^{1/2}$$

But at $T = 0$,
$$f(E) = 0 \quad \text{for} \quad E > E_F$$

Also,
$$f(E) = 1 \quad \text{for} \quad E < E_F$$

So we can take $N(E) = CE^{1/2}$:
$$E_{av} = \frac{1}{n_e} \int_0^{E_F} CE^{3/2}\,dE = \frac{2C}{5n_e}E_F^{5/2}$$

But from Equation 43.24,
$$\frac{C}{n_e} = \frac{3}{2}E_F^{-3/2}$$

so
$$E_{av} = \left(\frac{2}{5}\right)\left(\frac{3}{2}\right)\left(E_F^{-3/2}\right)E_F^{5/2} = \left(\frac{3}{5}\right)E_F \qquad \Diamond$$

---

31. (a) Consider a system of electrons confined to a three-dimensional box. Calculate the ratio of the number of allowed energy levels at 8.50 eV to the number at 7.00 eV. (b) Copper has a Fermi energy of 7.0 eV at 300 K. Calculate the ratio of the number of occupied levels at an energy of 8.50 eV to the number at the Fermi energy. Compare your answer with that obtained in part (a).

Solution The density of states at the energy E is $g(E) = CE^{1/2}$. Hence, the required ratio is

(a)
$$R = \frac{g(8.50 \text{ eV})}{g(7.00 \text{ eV})} = \frac{C(8.50)^{1/2}}{C(7.00)^{1/2}} = 1.10 \qquad \Diamond$$

(b) From Equation 43.22 we see that the number of occupied states having energy E is

$$N(E) = \frac{CE^{1/2}}{e^{(E-E_F)/k_BT}+1}$$

Hence, the required ratio is

$$R = \frac{N(8.50 \text{ eV})}{N(7.00 \text{ eV})} = \sqrt{\frac{8.50}{7.00}}\left[\frac{e^{(7.00-7.00)/k_BT}+1}{e^{(8.50-7.00)/k_BT}+1}\right]$$

At $T = 300$ K, $k_BT = 0.0259$ eV: $\quad R = (1.10)\left(\frac{2}{e^{1.50/0.0259}+1}\right) = 1.55 \times 10^{-25}$ ◊

Comparing this result with (a), we conclude that very few states with $E > E_F$ are occupied.

---

35. Most solar radiation has a wavelength of $1\,\mu$m or less. What energy gap should the material in a solar cell have in order to absorb this radiation? Is silicon appropriate (see Table 43.5)?

Solution

G: Since most photovoltaic solar cells are made of silicon, this semiconductor seems to be an appropriate material for these devices.

O: To absorb the longest-wavelength photons, the energy gap should be no larger than the photon energy.

A: The minimum photon energy is

$$hf = \frac{hc}{\lambda} = \frac{(6.63 \times 10^{-34} \text{ J} \cdot \text{s})(3.00 \times 10^8 \text{ m/s})}{10^{-6} \text{ m}}\left(\frac{1 \text{ eV}}{1.60 \times 10^{-19} \text{ J}}\right) = 1.24 \text{ eV}$$

Therefore, the energy gap in the absorbing material should be smaller than 1.24 eV.
◊

L: So silicon, with gap of 1.14 eV < 1.24 eV, is an appropriate material for absorbing solar radiation.
◊

---

41. Determine the current generated in a superconducting ring of niobium metal 2.00 cm in diameter if a 0.0200-T magnetic field in a direction perpendicular to the ring is suddenly decreased to zero. The inductance of the ring is 3.10×10^{-8} H.

Solution

G: The resistance of a superconductor is zero, so the current is limited only by the change in magnetic flux and self-inductance. Therefore, unusually large currents (greater than 100 A) are possible.

O: The change in magnetic field through the ring will induce an emf according to Faraday's law of induction. Since we do not know how fast the magnetic field is changing, we must use the ring's inductance and the geometry of the ring to calculate the magnetic flux, which can then be used to find the current.

A: From Faraday's law (Eq. 31.1), we have

$$|\varepsilon| = \frac{\Delta \Phi_B}{\Delta t} = A\frac{\Delta B}{\Delta t} = L\frac{\Delta I}{\Delta t}$$

or $\quad \Delta I = \dfrac{A \Delta B}{L} = \dfrac{\pi(0.0100 \text{ m})^2(0.0200 \text{ T})}{3.10 \times 10^{-8} \text{ H}} = 203 \text{ A}$ $\qquad \Diamond$

The current is directed so as to produce its own magnetic field in the direction of the original field.

L: This induced current should remain constant as long as the ring is superconducting. If the ring failed to be a superconductor (e.g. if it warmed above the critical temperature), the metal would have a non-zero resistance, and the current would quickly drop to zero. It is interesting to note that we were able to calculate the current in the ring without knowing the emf. In order to calculate the emf, we would need to know how quickly the magnetic field goes to zero.

47. Show that the ionic cohesive energy of an ionically bonded solid is given by Equation 43.17. (**Hint:** Start with Equation 43.16, and note that $dU/dr = 0$ at $r = r_0$.)

Solution   The total potential energy is given by Equation 43.16:

$$U_{total} = -\alpha k_e \frac{e^2}{r} + \frac{B}{r^m}$$

The potential energy has its minimum value U_0 when $r = r_0$, where r_0 is the equilibrium spacing. At this point, the slope of the curve U versus r is **zero**. That is,

$$\left.\frac{dU}{dr}\right|_{r=r_0} = 0 \quad \text{with} \quad \frac{dU}{dr} = \frac{d}{dr}\left(-\alpha k_e \frac{e^2}{r} + \frac{B}{r^m}\right) = \alpha k_e \frac{e^2}{r^2} - \frac{mB}{r^{m+1}}$$

Taking $r = r_0$ and setting this equal to zero, we find $\alpha k_e \dfrac{e^2}{r_0^2} - \dfrac{mB}{r_0^{m+1}} = 0$

Therefore, $B = \alpha \dfrac{k_e e^2}{m} r_0^{m-1}$

Substituting this value of B into U_{total} gives

$$U_0 = -\alpha k_e \frac{e^2}{r_0} + \alpha \frac{k_e e^2}{m} r_0^{m-1}\left(\frac{1}{r^m}\right) = -\alpha \frac{k_e e^2}{r_0}\left(1 - \frac{1}{m}\right) \qquad \lozenge$$

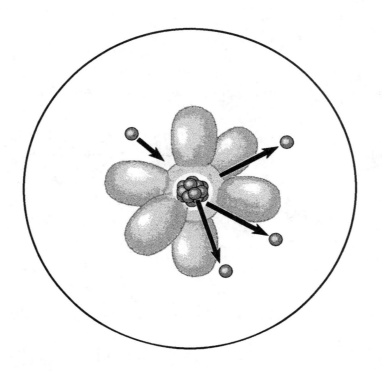

Nuclear Structure

NUCLEAR STRUCTURE

INTRODUCTION

In this chapter we discuss the properties and structure of the atomic nucleus. We start by describing the basic properties of nuclei, and then discuss nuclear forces and binding energy, nuclear models, and the phenomenon of radioactivity. We also discuss nuclear reactions and the various processes by which nuclei decay.

EQUATIONS AND CONCEPTS

Most nuclei are nearly spherical in shape and have an average radius which is proportional to the cube root of the mass number or total number of nucleons. This means that the volume is proportional to A and that all nuclei have nearly the same density.

$$r = r_0 A^{1/3} \qquad (44.1)$$

The nucleus has an angular momentum and a corresponding nuclear magnetic moment associated with it. The nuclear magnetic moment is measured in terms of a unit of moment called the nuclear magneton μ_n.

$$\mu_n \equiv \frac{e\hbar}{2m_p} = 5.05 \times 10^{-27} \text{ J / T} \qquad (44.3)$$

The binding energy of any nucleus can be calculated in terms of the mass of a neutral hydrogen atom, the mass of a neutron, and the atomic mass of the associated compound nucleus.

$$E_b(\text{MeV}) = (Zm_H + Nm_n - M_A)\left(931.494\frac{\text{MeV}}{\text{u}}\right) \quad (44.4)$$

The formula for binding energy is semiempirical, and is based on the liquid drop model of the nucleus.

$$E_b = C_1A - C_2A^{2/3} - C_3\frac{Z(Z-1)}{A^{1/3}} - C_4\frac{(N-Z)^2}{A} \quad (44.5)$$

The number of radioactive nuclei in a given sample which undergo decay during a time interval Δt depends on the number of nuclei present. The number of decays depends also on the decay constant, λ, which is characteristic of a particular isotope.

$$\frac{dN}{dt} = -\lambda N \quad (44.6)$$

The number of nuclei in a radioactive sample decreases exponentially with time. The plot of number of nuclei N versus elapsed time t is called a decay curve.

$$N = N_0e^{-\lambda t} \quad (44.7)$$

454

The decay rate R or activity of a sample of radioactive nuclei is defined as the number of decays per second.

$$R = \left| \frac{dN}{dt} \right| = R_0 e^{-\lambda t} = N_0 \lambda e^{-\lambda t} \qquad (44.8)$$

The half-life $T_{1/2}$ is the time required for half of a given number of radioactive nuclei to decay.

$$T_{1/2} = \frac{\ln 2}{\lambda} = \frac{0.693}{\lambda} \qquad (44.9)$$

When a nucleus decays by **alpha emission**, the parent nucleus loses two neutrons and two protons. For alpha emission to occur, the mass of the parent nucleus must be greater than the combined mass of the daughter nucleus and the emitted alpha particle. The mass difference is converted into energy and appears as kinetic energy, shared unequally by the alpha particle and the daughter nucleus.

$$^{A}_{Z}\text{X} \rightarrow ^{A-4}_{Z-2}\text{Y} + ^{4}_{2}\text{He} \qquad (44.10)$$

Two Example alpha decay equations are the decay of uranium into Thorium, and the decay of radium into radon.

$$^{238}_{92}\text{U} \rightarrow ^{234}_{90}\text{Th} + ^{4}_{2}\text{He} \qquad (44.11)$$

$$^{226}_{88}\text{Ra} \rightarrow ^{222}_{86}\text{Rn} + ^{4}_{2}\text{He} \qquad (44.12)$$

The disintegration energy Q can be calculated in MeV when the masses are expressed in u.

$$Q = \left(M_X - M_Y - M_\alpha \right)\left(931.494 \frac{\text{MeV}}{\text{u}} \right) \qquad (44.14)$$

When a radioactive nucleus undergoes **beta decay**, the daughter nucleus has the same mass number as the parent nucleus, but the charge number (or atomic number) increases or decreases by one. Two types of beta decay are shown, and corresponding examples are given.

$$_Z^A X \to _{Z+1}^A Y + e^- + \bar{\nu} \tag{44.15}$$

$$_Z^A X \to _{Z-1}^A Y + e^+ + \nu \tag{44.16}$$

$$_6^{14}C \to _7^{14}N + e^- + \bar{\nu} \tag{44.17}$$

$$_7^{12}N \to _6^{12}C + e^+ + \nu \tag{44.18}$$

The electron that is emitted is created within the parent nucleus by a process which can be represented by a neutron transformed into a proton and an electron. The total energy released in beta decay is greater than the combined kinetic energies of the electron and the daughter nucleus. This difference in energy is associated with a third particle called a neutrino.

$$n \to p + e^- + \bar{\nu} \tag{44.21}$$

Nuclei which undergo alpha or beta decay are often left in an excited energy state. The nucleus returns to the ground state by emission of one or more photons. Gamma decay results in no change in mass number or atomic number.

$$_Z^A X^* \to _Z^A X + \gamma \tag{44.23}$$

Nuclear reactions can occur when target nuclei are bombarded with energetic particles. In these reactions the structure, identity, or properties of the target nuclei are changed.

$$a + X \to Y + b$$

The quantity of energy required to balance the equation representing a nuclear reaction (e.g. Eq. 44.26) is called the Q value of the reaction. The Q value can be calculated in terms of the total mass of the reactants minus the total mass of the products or as the kinetic energy of the reactants. Q is positive in the case of exothermic reactions and negative for endothermic reactions.

$$Q = (M_a + M_X - M_Y - M_b)c^2 \qquad (44.27)$$

TABLE	Various Decay Pathways
Alpha Decay	$_{Z}^{A}X \rightarrow {}_{Z-2}^{A-4}Y + {}_{2}^{4}He$
Beta Decay (e⁻)	$_{Z}^{A}X \rightarrow {}_{Z+1}^{A}Y + e^- + \bar{\nu}$
Beta Decay (e⁺)	$_{Z}^{A}X \rightarrow {}_{Z-1}^{A}Y + e^+ + \nu$
Electron Capture	$_{Z}^{A}X + {}_{-1}^{0}e \rightarrow {}_{Z-1}^{A}Y + \nu$
Gamma Decay	$_{Z}^{A}X^* \rightarrow {}_{Z}^{A}X + \gamma$

SUGGESTIONS, SKILLS, AND STRATEGIES

The rest energy of a particle is given by $E_R = mc^2$. It is therefore often convenient to express the unified mass unit in terms of its equivalent energy, 1 u = 1.660540×10^{-27} kg or 1 u = 931.494 MeV/c^2. When masses are expressed in units of u, energy values are then $E_R = m(931.494 \text{ MeV/u})$.

Equation 44.7 can be solved for the particular time t after which the number of remaining nuclei will be some specified fraction of the original number N_0. This can be done by taking the natural log of each side of Equation 44.7 to find

$$t = \frac{1}{\lambda} \ln(N_0/N)$$

REVIEW CHECKLIST

▷ Use the appropriate nomenclature in describing the static properties of nuclei.

▷ Discuss nuclear stability in terms of the strong nuclear force and a plot of N vs. Z.

▷ Account for nuclear binding energy in terms of the Einstein mass-energy relationship. Describe the basis for energy released by fission and fusion in terms of the shape of the curve of binding energy per nucleon vs. mass number.

▷ Identify each of the components of radiation that are emitted by the nucleus through natural radioactive decay and describe the basic properties of each. Write out typical equations to illustrate the processes of transmutation by alpha and beta decay and explain why the neutrino must be considered in the analysis of beta decay.

▷ State and apply to the solution of related problems, the formula which expresses decay rate as a function of the decay constant and the number of radioactive nuclei. Describe the process of carbon dating as a means of determining the age of ancient objects.

▷ Calculate the Q value of given nuclear reactions and determine the threshold energy of endothermic reactions.

ANSWERS TO SELECTED CONCEPTUAL QUESTIONS

4. Why do nearly all the naturally occuring isotopes lie above the $N = Z$ line in Figure 44.3?

Answer As z increases, extra neutrons are required to overcome the increasing electrostatic repulsion of the protons.

☐   ☐   ☐   ☐

12. Two samples of the same radioactive nuclide are prepared. Sample A has twice the initial activity of sample B. How does the half-life of A compare with the half-life of B? After each has passed through five half-lives, what is the ratio of their activities?

Answer Since the two samples are of the same radioactive nuclide, they have the same half-life; the 2:1 difference in activity is due to a 2:1 difference in the mass of each sample. After 5 half lives, each will have decreased in mass by a power of $2^5 = 32$. However, since this simply means that the mass of each is 32 times smaller, the ratio of the masses will still be (2/32):(1/32), or 2:1. Therefore, the ratio of their activities will **always** be 2:1.

□   □   □   □

27. Suppose it could be shown that the cosmic ray intensity at the Earth's surface was much greater 10 000 years ago. How would this difference affect what we accept as valid carbon-dated values of the age of ancient samples of once-living matter?

Answer If the cosmic ray intensity at the Earth's surface was much greater 10 000 years ago, a greater fraction of the Earth's carbon dioxide would contain the heavy nuclide ^{14}C at that time. Thus, there would initially be a greater fraction of ^{14}C in the organic artifacts, and we would believe the artifact to be more recent than it actually is.

For example, suppose that the actual ratio of atmospheric ^{14}C to ^{12}C, two half-lives (11460 years) ago was 2.6×10^{-12}. The current ratio of isotopes would be

$$\left(2.6 \times 10^{-12}\right)\left(\tfrac{1}{2}\right)\left(\tfrac{1}{2}\right) = 0.65 \times 10^{-12}$$

We, believing the initial ratio to be 1.3×10^{-12}, would see the same current ratio, but would think that the artifact had died only one half-life (5730 years) ago.

$$\left(1.3 \times 10^{-12}\right)\left(\tfrac{1}{2}\right) = 0.65 \times 10^{-12}$$

□   □   □   □

Chapter 44

SOLUTIONS TO SELECTED END-OF-CHAPTER PROBLEMS

5. (a) Use energy methods to calculate the distance of closest approach for a head-on collision between an alpha particle having an initial energy of 0.500 MeV and a gold nucleus (^{197}Au) at rest. (Assume the gold nucleus remains at rest during the collision.) (b) What minimum initial speed must the alpha particle have in order to get as close as 300 fm?

Solution

G: The positively charged alpha particle ($q = +2e$) will be repelled by the positive gold nucleus ($Q = +79e$), so that the particles probably will not touch each other in this electrostatic "collision." Therefore, the closest the alpha particle can get to the gold nucleus would be if the two nuclei did touch, in which case the distance between their centers would be about 6 fm (using $r = r_0 A^{1/3}$ for the radius of each nucleus). To get this close, or even within 300 fm, the alpha particle must be traveling very fast, probably close to the speed of light (but of course v must be less than c).

O: At the distance of closest approach, r_{min}, the initial kinetic energy will equal the electrostatic potential energy between the alpha particle and gold nucleus.

A: (a) $K_\alpha = U = k_e \dfrac{qQ}{r_{min}}$

$$r_{min} = k_e \frac{qQ}{K_\alpha} = \frac{(8.99 \times 10^9 \text{ N} \cdot \text{m}^2/\text{C}^2)(2)(79)(1.60 \times 10^{-19} \text{ C})^2}{(0.500 \text{ MeV})(1.60 \times 10^{-13} \text{ J/MeV})} = 455 \text{ fm} \quad \Diamond$$

(b) Since $K_\alpha = \frac{1}{2}mv^2 = k_e \dfrac{qQ}{r_{min}}$, we find that $v = \sqrt{\dfrac{2k_e qQ}{mr_{min}}}$

$$v = \sqrt{\frac{2(8.99 \times 10^9 \text{ N} \cdot \text{m}^2/\text{C}^2)(2)(79)(1.60 \times 10^{-19} \text{ C})^2}{4(1.66 \times 10^{-27} \text{ kg})(3.00 \times 10^{-13} \text{ m})}} = 6.04 \times 10^6 \text{ m/s} \quad \Diamond$$

L: The minimum distance in part (a) is about 100 times greater than the combined radii of the particles. For part (b), the alpha particle must have more than 0.5 MeV of energy since it gets closer to the nucleus than the 455 fm found in part (a). Even so, the speed of the alpha particle in part (b) is only about 2% of the speed of light, so we are justified in not using a relativistic approach. In solving this problem, we ignored the effect of the electrons around the gold nucleus that tend to "screen" the nucleus so that the alpha particle sees a reduced positive charge. If this screening effect were considered, the potential energy would be slightly reduced and the alpha particle could get closer to the gold nucleus for the same initial energy.

---

9. A star ending its life with a mass of two times the mass of the Sun is expected to collapse, combining its protons and electrons to form a neutron star. Such a star could be thought of as a gigantic atomic nucleus. If a star of mass $2 \times 1.99 \times 10^{30}$ kg collapsed into neutrons $\left(m_n = 1.67 \times 10^{-27} \text{ kg}\right)$, what would the radius be? (Assume that $r = r_0 A^{1/3}$.)

Solution

The number of nucleons in a star of two solar masses is

$$A = \frac{2\left(1.99 \times 10^{30} \text{ kg}\right)}{1.67 \times 10^{-27} \text{ kg}} = 2.38 \times 10^{57}$$

Therefore, $\quad r = r_0 A^{1/3} = (1.20 \times 10^{-15} \text{ m})\sqrt[3]{2.38 \times 10^{57}} = 16.0$ km $\qquad \lozenge$

---

17. Nuclei having the same mass numbers are called **isobars**. The isotope $^{139}_{57}$La is stable. A radioactive isobar $^{139}_{59}$Pr is located below the line of stable nuclei in Figure 44.3 and decays by e^+ emission. Another radioactive isobar of ^{139}La, $^{139}_{55}$Cs, decays by e^- emission and is located above the line of stable nuclei in Figure 44.3. (a) Which of these three isobars has the highest neutron-to-proton ratio? (b) Which has the greatest binding energy per nucleon? (c) Which do you expect to be heavier, ^{139}Pr or ^{139}Cs?

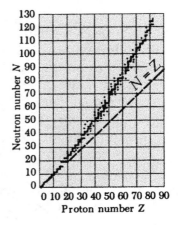

Figure 44.3

Solution

(a) For $^{139}_{59}$Pr the neutron number is $139 - 59 = 80$. For $^{139}_{55}$Cs the neutron number is 84, so the Cs isotope has the greatest neutron-to-proton ratio. ◊

(b) Binding energy per nucleon measures stability so it is greatest for the stable nucleus, the lanthanum isotope. Note also that it has a magic number of neutrons, 82. ◊

(c) Cs–139 has 55 protons and 84 neutrons. Pr–139 has 59 protons and 80 neutrons. When we plot both of them onto Figure 44.3, we see that Cesium is a little farther away from the center of the zone of stable nuclei. Being less stable goes with being able to lose more energy in decay, and with having more mass. We therefore expect Cesium to be heavier. ◊

19. A pair of nuclei for which $Z_1 = N_2$ and $Z_2 = N_1$ are called mirror isobars (the atomic and neutron numbers are interchanged). Binding energy measurements on these nuclei can be used to obtain evidence of the charge independence of nuclear forces (that is, proton-proton, proton-neutron, and neutron-neutron forces are approximately equal). Calculate the difference in binding energy for the two mirror isobars $^{15}_{8}$O and $^{15}_{7}$N.

Solution   For $^{15}_{8}O$ we have (using Equation 44.4)

$$E_b = [8(1.007825)\ u + 7(1.008665)\ u - (15.003065)\ u]\left(931.494\ \frac{MeV}{u}\right) = 111.96\ MeV$$

For $^{15}_{7}N$ we have

$$E_b = [7(1.007825)\ u + 8(1.008665)\ u - (15.000108)\ u]\left(931.5\ \frac{MeV}{u}\right) = 115.49\ MeV$$

Therefore, the difference in the two binding energies is $\Delta E_b = 3.54\ MeV$ ◊

---

23. Using the graph in Figure 44.8, estimate how much energy is released when a nucleus of mass number 200 is split into two nuclei each of mass number 100.

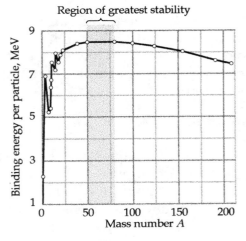

Figure 44.8

Solution

The curve of binding energy shows that a heavy nucleus of mass number A = 200 has total binding energy about

$$\left(7.4\ \frac{MeV}{nucleon}\right)(200\ nucleons) = 1.5\ GeV$$

Thus, it is less stable than its potential fission products, two middleweight nuclei of A = 100, having total binding energy

$$2(8.4\ MeV\ /\ nucleon)(100\ nucleons) = 1.7\ GeV$$

Fission the releases about $1.7\ GeV - 1.5\ GeV = 200\ MeV$ ◊

This is the energy source of uranium bombs and of nuclear electric-generating plants.

463

25. A sample of radioactive material contains 1.00×10^{15} atoms and has an activity of 6.00×10^{11} Bq. What is its half-life?

Solution $dN/dt = -\lambda N$

$$\lambda = \frac{1}{N}\left(-\frac{dN}{dt}\right) = \left(1.00 \times 10^{15} \text{ atoms}\right)^{-1}\left(6.00 \times 10^{11} \text{ s}^{-1}\right) = 6.00 \times 10^{-4} \text{ s}^{-1}$$

$$T_{1/2} = \frac{(\ln 2)}{\lambda} = 1160 \text{ s} \qquad \text{(This is also 19.3 minutes)} \qquad\qquad \lozenge$$

---

27. A freshly prepared sample of a certain radioactive isotope has an activity of 10.0 mCi. After 4.00 h, its activity is 8.00 mCi. (a) Find the decay constant and half-life. (b) How many atoms of the isotope were contained in the freshly prepared sample? (c) What is the sample's activity 30.0 h after it is prepared?

Solution

G: Over the course of 4 hours, this isotope lost 20% of its activity, so its half-life appears to be around 10 hours, which means that its activity after 30 hours (~3 half-lives) will be about 1 mCi. The decay constant and number of atoms are not so easy to estimate.

O: From the rate equation, $R = R_0 e^{-\lambda t}$, we can find the decay constant λ, which can then be used to find the half life, the original number of atoms, and the activity at any other time, t.

A: (a) $\lambda = \frac{1}{t}\ln\left(\frac{R_o}{R}\right) = \left(\frac{1}{(4.00 \text{ h})(60.0 \text{ s}/\text{h})}\right)\ln\left(\frac{10.0 \text{ mCi}}{8.00 \text{ mCi}}\right) = 1.55 \times 10^{-5} \text{ s}^{-1}$ $\lozenge$

$$T_{1/2} = \frac{\ln 2}{\lambda} = \frac{0.693}{0.0558 \text{ h}^{-1}} = 12.4 \text{ h} \qquad\qquad \lozenge$$

(b)   The number of original atoms can be found if we convert the initial activity from curies into becquerels (decays per second): $1\ Ci \equiv 3.7 \times 10^{10}\ Bq$

$$R_0 = 10.0\ mCi = \left(10.0 \times 10^{-3}\ Ci\right)\left(3.70 \times 10^{10}\ Bq\,/\,Ci\right) = 3.70 \times 10^8\ Bq$$

Since $R_0 = \lambda N_0$

$$N_0 = \frac{R_0}{\lambda} = \frac{3.70 \times 10^8\ decays\,/\,s}{1.55 \times 10^{-5}\ s} = 2.39 \times 10^{13}\ atoms \qquad \lozenge$$

(c) $R = R_0 e^{-\lambda t} = (10.0\ mCi)e^{-(5.58\times 10^{-2}\ h^{-1})(30.0\ h)} = 1.87\ mCi \qquad \lozenge$

L: Our estimate of the half life was about 20% short because we did not account for the non-linearity of the decay rate. Consequently, our estimate of the final activity also fell short, but both of these calculated results are close enough to be reasonable.

The number of atoms is much less than one mole, so this appears to be a very small sample. To get a sense of how small, we can assume that the molar mass is about 100 g/mol, so the sample has a mass of only

$$m \approx \left(2.4 \times 10^{13}\ atoms\right)(100\ g\,/\,mol)\Big/\left(6.02 \times 10^{23}\ atoms\,/\,mol\right) \approx 0.004\ \mu g$$

This sample is so small it cannot be measured by a commercial mass balance!

The problem states that this sample was "freshly prepared," from which we assumed that all the atoms within the sample are initially radioactive. Generally this is not the case, so that N_0 only accounts for the formerly radioactive atoms, and does not include additional atoms in the sample that were not radioactive. Realistically then, the sample mass should be significantly greater than our above estimate.

29. The radioactive isotope ^{198}Au has a half-life of 64.8 h. A sample containing this isotope has an initial activity ($t = 0$) of 40.0 μCi. Calculate the number of nuclei that decay in the time interval between $t_1 = 10.0$ h and $t_2 = 12.0$ h.

Solution   First, let us find λ and N_0 from the given information:

$$\lambda = \frac{\ln 2}{T_{1/2}} = \frac{0.693}{64.8 \text{ h}} = 0.0107 \text{ h}^{-1} = 2.97 \times 10^{-6} \text{ s}^{-1}$$

$$N_0 = \frac{R_0}{\lambda} = \left(\frac{40.0 \times 10^{-6} \text{ Ci}}{2.97 \times 10^{-6} \text{ s}^{-1}} \right) \left(\frac{3.70 \times 10^{10} \text{ decays}/\text{s}}{1 \text{ Ci}} \right) = 4.98 \times 10^{11} \text{ nuclei}$$

Since $N = N_0 e^{-\lambda t}$ the number of nuclei which decay between times t_1 and t_2 is

$$N_1 - N_2 = N_0(e^{-\lambda t_1} - e^{-\lambda t_2})$$

$$N_1 - N_2 = (4.98 \times 10^{11})\left[e^{-(0.0107 \text{ h}^{-1})(10.0 \text{ h})} - e^{-(0.0107 \text{ h}^{-1})(12.0 \text{ h})} \right]$$

$$N_1 - N_2 = 9.47 \times 10^9 \text{ nuclei} \qquad \lozenge$$

---

33.   Find the energy released in the alpha decay $^{238}_{92}\text{U} \rightarrow {}^{234}_{90}\text{Th} + {}^{4}_{2}\text{He}$

You will find the following mass values useful: $M\left({}^{238}_{92}\text{U}\right) = 238.050\ 784 \text{ u}$

$$M\left({}^{234}_{90}\text{Th}\right) = 234.043\ 593 \text{ u}$$

$$M\left({}^{4}_{2}\text{He}\right) = 4.002\ 602 \text{ u}$$

Solution $Q = (M_U - M_{Th} - M_{He})(931.494 \text{ MeV}/\text{u})$

$$Q = (238.050784 - 234.043593 - 4.002602)(931.494) = 4.27 \text{ MeV} \qquad \lozenge$$

37. The nucleus $^{15}_{8}$O decays by electron capture. Write (a) the basic nuclear process and (b) the decay process referring to neutral atoms. (c) Determine the energy of the neutrino. Disregard the daughter's recoil.

Solution

(a) $e^- + p \rightarrow n + v$ ◊

(b) Add 7 protons, 7 neutrons, and 7 electrons to each side to give
^{15}O atom $\rightarrow$ ^{15}N atom $+ v$ ◊

(c) From Table A.3, $m(^{15}\text{O}) = m(^{15}\text{N}) + Q / c^2$

$$\Delta m = 15.003065 - 15.000109 = 0.002956 \text{ u}$$

$$Q = (931.494 \text{ MeV/u})(0.002956 \text{ u}) = 2.75 \text{ MeV} \quad ◊$$

---

39. Enter the correct isotope symbol in each open square in Figure P44.39, which shows the sequences of decays starting with uranium-235 and ending with the stable isotope lead-207.

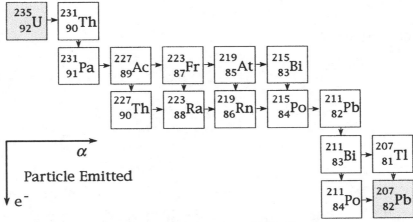

Figure P44.39 (modified)

Solution   Whenever an $\alpha = \frac{4}{2}$He is emitted, Z drops by 2 and A by 4. Whenever a $e^- = \frac{0}{-1}e^-$ is emitted, Z increases by 1 and A is unchanged. We find the chemical name by looking up Z in a periodic table. The values in the shaded boxes (^{235}U and ^{207}Pb) were given; all others have been filled in as part of the solution.

---

43.   Natural gold has only one isotope, $^{197}_{79}$Au. If natural gold is irradiated by a flux of slow neutrons, e^- particles are emitted. (a) Write the reaction equation. (b) Calculate the maximum energy of the emitted beta particles. The mass of $^{198}_{80}$Hg is 197.96673 u.

Solution   The $^{197}_{79}$Au will absorb a neutron to become $^{198}_{79}$Au, which emits an e^- to become $^{198}_{80}$Hg.

(a) For nuclei, the reaction is: $^{197}_{79}$Au $+ \frac{1}{0}$n $\rightarrow$ $^{198}_{80}$Hg $+ \frac{0}{-1}e^- + \bar{v}$ ◊

Adding 79 e^- to both sides: $^{197}_{79}$Au atom $+ \frac{1}{0}$n $\rightarrow$ $^{198}_{80}$Hg atom $+ \bar{v}$

(b) From Table A.3, $196.966543 + 1.008665 = 197.966743 + 0 + Q/c^2$

$Q = \Delta m c^2 = (0.008465 \text{ u})(931.494 \text{ MeV/u})$

$Q = 7.89 \text{ MeV}$ ◊

---

47.   Using the Q values of appropriate reactions and from Table 44.5, calculate the masses of ^{8}Be and ^{10}Be in atomic mass units to four decimal places.

Solution

G: The mass of each isotope in atomic mass units will be approximately the number of nucleons (8 or 10), also called the mass number. The electrons are much less massive and contribute only about 0.03% to the total mass.

O: In addition to summing the mass of the subatomic particles, the net mass of the isotopes must account for the binding energy that holds the atom together. Table 44.5 includes the energy released for each nuclear reaction. Precise atomic masses values are found in Table A.3.

A: The notation $\quad ^9\text{Be}\,(\gamma,\,\text{n})\,^8\text{Be} \quad$ with $\quad Q = -1.666\text{ MeV}$

means $\qquad ^9\text{Be} + \gamma \rightarrow\ ^8\text{Be} + \text{n} - 1.666\text{ MeV}$

Therefore $\qquad m\left(^8\text{Be}\right) = m\left(^9\text{Be}\right) - m_n + \dfrac{1.666\text{ MeV}}{931.5\text{ MeV}/\text{u}}$

$$m\left(^8\text{Be}\right) = 9.012174 - 1.008665 + 0.001789 = 8.0053\text{ u} \quad \lozenge$$

The notation $\quad ^9\text{Be}\,(\text{n},\,\gamma)\,^{10}\text{Be} \quad$ with $\quad Q = 6.810\text{ MeV}$

means $\qquad ^9\text{Be} + \text{n} \rightarrow\ ^{10}\text{Be} + \gamma + 6.810\text{ MeV}$

$$m\left(^{10}\text{Be}\right) = m\left(^9\text{Be}\right) + m_n + \dfrac{6.810\text{ MeV}}{931.5\text{ MeV}/\text{u}}$$

$$m\left(^{10}\text{Be}\right) = 9.012174 + 1.008665 - 0.001789 = 10.0135\text{ u} \quad \lozenge$$

L: As expected, both isotopes have masses slightly greater than their mass numbers. We were asked to calculate the masses to four decimal places, but with the available data, the results could be reported accurately to as many as six decimal places.

57. The decay of an unstable nucleus by alpha emission is represented by Equation 44.10. The disintegration energy Q given by Equation 44.13 must be shared by the alpha particle and the daughter nucleus in order to conserve both energy and momentum in the decay process. (a) Show that Q and K_α, the kinetic energy of the alpha particle, are related by the expression

$$Q = K_\alpha(1 + M_\alpha/M)$$

where M is the mass of the daughter nucleus. (b) Use the result of part (a) to find the energy of the alpha particle emitted in the decay of ^{226}Ra. (See Example 44.7 for the calculation of Q.)

Solution

(a) Let us assume that the parent nucleus (mass M_p) is initially at rest, and let us denote the masses of the daughter nucleus and alpha particle by M_d and M_α, respectively. Applying the equations of conservation of momentum and energy for the alpha decay process gives

$$M_d v_d = M_\alpha v_\alpha \tag{1}$$

$$M_p c^2 = M_d c^2 + M_\alpha c^2 + \tfrac{1}{2} M_\alpha v_\alpha^2 + \tfrac{1}{2} M_d v_d^2 \tag{2}$$

The disintegration energy Q is given by

$$Q = (M_p - M_d - M_\alpha)c^2 = \tfrac{1}{2} M_\alpha v_\alpha^2 + \tfrac{1}{2} M_d v_d^2 \tag{3}$$

Eliminating v_d from Equations (1) and (3) gives

$$Q = \tfrac{1}{2} M_\alpha v_\alpha^2 + \tfrac{1}{2} M_d \left(\frac{M_\alpha}{M_d} v_\alpha\right)^2 = \tfrac{1}{2} M_\alpha v_\alpha^2 + \tfrac{1}{2} \frac{M_\alpha^2}{M_d} v_\alpha^2$$

$$Q = \tfrac{1}{2} M_\alpha v_\alpha^2 \left(1 + \frac{M_\alpha}{M_d}\right) = K_\alpha \left(1 + \frac{M_\alpha}{M_d}\right)$$

(b)
$$K_\alpha = \frac{Q}{1 + M_\alpha/M_d} = \frac{4.87 \text{ MeV}}{1 + 4/222} = 4.78 \text{ MeV} \qquad \Diamond$$

69. "Free neutrons" have a characteristic half-life of 10.4 min. What fraction of a group of free neutrons at thermal energy (0.0400 eV) will decay before traveling a distance of 10.0 km?

Solution

The fraction that will remain is given by the ratio N/N_0, where $N/N_0 = e^{-\lambda t}$ and t is the time it takes the neutron to travel a distance of $d = 10.0$ km.

Since $K = \frac{1}{2}mv^2$, the time is given by

$$t = \frac{d}{v} = \frac{d}{\sqrt{\dfrac{2K}{m}}} = \frac{10.0 \times 10^3 \text{ m}}{\sqrt{\dfrac{2(0.0400 \text{ eV})(1.60 \times 10^{-19} \text{ J/eV})}{1.67 \times 10^{-27} \text{ kg}}}} = 3.61 \text{ s}$$

The decay constant is then,

$$\lambda = \frac{0.693}{T_{1/2}} = \frac{0.693}{(10.4 \text{ min})(60 \text{ s/min})} = 1.11 \times 10^{-3} \text{ s}^{-1}$$

Therefore, $\lambda t = (1.11 \times 10^{-3} \text{ s}^{-1})(3.61 \text{ s}) = 4.01 \times 10^{-3} = 0.00401$

so $\dfrac{N}{N_0} = e^{-\lambda t} = e^{-0.00401} = 0.9960$

Hence, the fraction that has decayed in this time is

$$1 - \frac{N}{N_0} = 0.00400 \quad \text{or} \quad 0.400\%$$

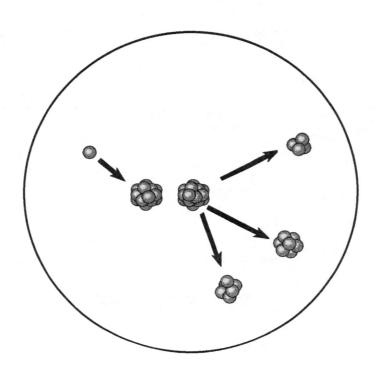

NUCLEAR FISSION
AND FUSION

NUCLEAR FISSION AND FUSION

INTRODUCTION

This chapter is concerned primarily with the two means by which energy can be derived from nuclear reactions: fission, in which a large nucleus splits (fissions) into two smaller nuclei, and fusion, in which two small nuclei fuse to form a larger one. In either case, there is a release of energy that can then be used either destructively (bombs) or constructively (production of electric power). We also examine the interaction of radiation with matter and several devices used to detect radiation. We conclude with a discussion of some industrial and biological applications of radiation.

EQUATIONS AND CONCEPTS

The fission of a uranium nucleus by bombardment with a low energy neutron results in the production of fission fragments and typically two or three neutrons. The energy released in the fission event appears in the form of kinetic energy of the fission fragments and the neutrons.

$$\begin{aligned}{}^{1}_{0}\text{n} + {}^{235}_{92}\text{U} &\to {}^{236}_{92}\text{U}\,^{*} \\ &\to \text{X} + \text{Y} + \text{neutrons}\end{aligned} \qquad (45.2)$$

These fusion reactions seem to be most likely to be used as the basis of the design and operation of a fusion power reactor. The Q values refer to the energy released from each reaction.

$$\begin{aligned}{}^{2}_{1}\text{H} + {}^{2}_{1}\text{H} &\to {}^{3}_{2}\text{He} + {}^{1}_{0}\text{n} \\ &(Q = 3.27 \text{ MeV})\end{aligned} \qquad (45.4)$$

$$\begin{aligned}{}^{2}_{1}\text{H} + {}^{2}_{1}\text{H} &\to {}^{3}_{1}\text{H} + {}^{1}_{1}\text{H} \\ &(Q = 4.03 \text{ MeV})\end{aligned}$$

$$\begin{aligned}{}^{2}_{1}\text{H} + {}^{3}_{1}\text{H} &\to {}^{4}_{2}\text{He} + {}^{1}_{0}\text{n} \\ &(Q = 17.59 \text{ MeV})\end{aligned}$$

473

Lawson's criterion states the conditions under which a net power output of a fusion reactor is possible. In these expressions, n is the plasma density (number of ions per cubic cm) and τ is the plasma confinement time (the time during which the interacting ions are maintained at a temperature equal to or greater than that required for the reaction to proceed).

$$n\tau \geq 10^{14} \text{ s} / \text{cm}^3 \qquad (45.5)$$

$$(\text{D} - \text{T interaction})$$

$$n\tau \geq 10^{16} \text{ s} / \text{cm}^3$$

$$(\text{D} - \text{D interaction})$$

The radiation dose in rem is the product of the dose in rad and the relative biological effectiveness factor.

$$\text{Dose in rem} \equiv \text{Dose in rad} \times \text{RBE} \qquad (45.6)$$

1 roentgen is the amount of ionizing radiation that deposits 0.0876 J of energy into 1 kg of air.

1 rad is the amount of radiation that deposits 0.01 J of energy into 1 kg of any absorbing material.

RBE is a factor defined as the number of rad of x-radiation or gamma-radiation that produces the same biological damage as 1 rad of the radiation being used.

REVIEW CHECKLIST

▷ Write an equation which represents a typical fission event and describe the sequence of events which occurs during the fission process. Use data obtained from the binding energy curve to estimate the disintegration energy of a typical fission event.

▷ Describe the basic design features and control mechanisms in a fission reactor including the functions of the moderator, control rods, and heat exchange system. Identify some major safety and environmental hazards in the operation of a fission reactor.

▷ Describe the basis of energy release in fusion and write out several nuclear reactions which might be used in a fusion powered reactor.

▷ Define the roentgen, rad, and rem as units of radiation exposure or dose.

▷ Describe the basic principle of operation of the Geiger counter, semiconductor diode detector, scintillation detector, photographic emulsion, cloud chamber, and bubble chamber.

ANSWERS TO SELECTED CONCEPTUAL QUESTIONS

2.   Why is water a better shield against neutrons than lead or steel?

Answer   The hydrogen nuclei in water molecules have mass similar to that of a neutron, so that they can efficiently rob a fast-moving neutron of kinetic energy as they scatter it. Once the neutron is slowed down, a hydrogen nucleus can absorb it in the reaction ${}_{0}^{1}\text{n} + {}_{1}^{1}\text{H} \rightarrow {}_{1}^{2}\text{H} + Q$.

□   □   □   □

8. Discuss the similarities and differences between fusion and fission.

Answer   Fusion of light nuclei to a heavier nucleus releases energy. Fission of a heavy nucleus to lighter nuclei releases energy. Both processes are steps towards greater stability on the curve of binding energy, Figure 44.8. The energy release per nucleon is typically greater for fusion, and this process is harder to control.

□   □   □   □

SOLUTIONS TO SELECTED END-OF-CHAPTER PROBLEMS

3. List the nuclear reactions required to produce ^{235}U from ^{232}Th under fast neutron bombardment.

Solution   First, the Thorium is bombarded: $^1_0n + ^{232}_{90}\text{Th} \rightarrow ^{233}_{90}\text{Th}$

Then, the Thorium decays by beta emission: $^{233}_{90}\text{Th} \rightarrow ^{233}_{91}\text{Pa} + e^- + \overline{\nu}$

Protactinium-233 has more neutrons than the more stable Protactinium-231, so it too decays by beta emission:

$$^{233}_{91}\text{Pa} \rightarrow ^{233}_{92}\text{U} + e^- + \overline{\nu} \quad \Diamond$$

7. Suppose enriched uranium containing 3.40% of the fissionable isotope $^{235}_{92}$U is used as fuel for a ship. The water exerts an average frictional drag of 1.00×10^5 N on the ship. How far can the ship travel per kilogram of fuel? Assume that the energy released per fission event is 208 MeV and that the ship's engine has an efficiency of 20.0%.

Solution

G: Nuclear fission is much more efficient for converting mass to energy than burning fossil fuels. However, without knowing the rate of diesel fuel consumption for a comparable ship, it is difficult to estimate the nuclear fuel rate. It seems plausible that a ship could cross the Atlantic ocean with only a few kilograms of nuclear fuel, so a reasonable range of uranium fuel consumption might be 10 km / kg to 10 000 km/kg.

O: The fuel consumption rate can be found from the energy released by the nuclear fuel and the work required to push the ship through the water.

A: One kg of enriched uranium contains 3.40% $^{235}_{92}\text{U}$, so

$$m_{235} = (1000 \text{ g})(0.0340) = 34.0 \text{ g}$$

In terms of number of nuclei, this is equivalent to

$$N_{235} = (34.0 \text{ g})\left(\frac{1}{235 \text{ g / mol}}\right)(6.02 \times 10^{23} \text{ atoms / mol}) = 8.71 \times 10^{22} \text{ nuclei}$$

If all these nuclei fission, the thermal energy released is equal to

$$(8.71 \times 10^{22} \text{ nuclei})\left(208 \frac{\text{MeV}}{\text{nucleus}}\right)(1.602 \times 10^{-19} \text{ J/eV}) = 2.90 \times 10^{12} \text{ J}$$

Now, for the engine, $\quad\text{efficiency} = \dfrac{\text{work output}}{\text{heat input}}\quad$ or $\quad e = \dfrac{fd \cos\theta}{Q_h}$

So the distance the ship can travel per kilogram of uranium fuel is

$$d = \frac{eQ_h}{f\cos(0)} = \frac{0.200(2.90 \times 10^{12} \text{ J})}{1.00 \times 10^5 \text{ N}} = 5.80 \times 10^6 \text{ m} \qquad\qquad \lozenge$$

L: The ship can travel 5 800 km/kg of uranium fuel, which is on the high end of our prediction range. The distance between New York and Paris is 5 851 km, so this ship could cross the Atlantic ocean on just one kilogram of uranium fuel.

---

9. It has been estimated that there is on the order of 10^9 tons of natural uranium available at concentrations exceeding 100 parts per million, of which 0.7% is ^{235}U. If all the world's energy use $(7 \times 10^{12}$ J/s) were to be supplied by ^{235}U fission, how long would this supply last? (This estimate of uranium supply was taken from K. S. Deffeyes and I. D. MacGregor, <u>Scientific American</u>, January 1980, p. 66.)

Solution

The mass of natural uranium reserves is $\quad m = 10^9 \times 10^3 \text{ kg} = 1 \times 10^{12} \text{ kg}$

The reserves of fissionable uranium are

$$N_{235} = \left(7 \times 10^{12} \text{ g}\right)\left(\frac{6.02 \times 10^{23} \text{ atoms}}{235 \text{ g}}\right) = 1.79 \times 10^{34} \text{ nuclei}$$

Following Example 45.1, we take the fission energy as 208 MeV/fission:

$$E = \left(1.79 \times 10^{34} \text{ nuclei}\right)\left(208 \frac{\text{MeV}}{\text{fission}}\right)\left(1.60 \times 10^{-19} \frac{\text{J}}{\text{eV}}\right) = 5.97 \times 10^{23} \text{ J}$$

Now, $\quad t = \dfrac{E}{\mathcal{P}} = \dfrac{5.97 \times 10^{23} \text{ J}}{7 \times 10^{12} \text{ J/s}} = 8.53 \times 10^{10} \text{ s} = 2700 \text{ yr}$ ◊

---

15. To understand why plasma containment is necessary, consider the rate at which an unconfined plasma would be lost. (a) Estimate the rms speed of deuterons in a plasma at 4.00×10^8 K. (b) Estimate the order of magnitude of the time such a plasma would remain in a 10-cm cube if no steps were taken to contain it.

Solution The average kinetic energy per particle $\left(\frac{1}{2}m\overline{v^2}\right)$ must equal the thermal energy $\frac{3}{2}k_BT$. Taking $m = 2m_p$ for deuterons,

(a) $\frac{1}{2}m\overline{v^2} = \frac{3}{2}k_BT$

$$v_{rms} = \sqrt{\frac{3k_BT}{2m_p}} = \sqrt{\frac{3\left(1.38\times10^{-23}\text{ J/K}\right)\left(4.00\times10^8\text{ K}\right)}{2\left(1.67\times10^{-27}\text{ kg}\right)}} = 2.23\times10^6\text{ m/s} \qquad \Diamond$$

(b) $t = \frac{x}{v} = \frac{0.100\text{ m}}{2.23\times10^6\text{ m/s}} \sim 10^{-7}\text{ s}$ $\qquad \Diamond$

21. A building has become accidentally contaminated with radioactivity. The longest-lived material in the building is strontium-90 ($^{90}_{38}$Sr has an atomic mass 89.9077, and its half-life is 29.1 yr.) If the building initially contained 5.00 kg of this substance uniformly distributed throughout the building (a very unlikely situation) and the safe level is less than 10.0 counts/min, how long will the building be unsafe?

Solution The number of nuclei in the original sample is

$$N_0 = \frac{\text{mass present}}{\text{mass of nucleus}} = \frac{5.00\text{ kg}}{(89.9077\text{ u})\left(1.66\times10^{-27}\text{ kg/u}\right)} = 3.35\times10^{25}\text{ nuclei}$$

$$\lambda = \frac{\ln 2}{T_{1/2}} = \frac{0.693}{29.1 \text{ yr}} = 2.38 \times 10^{-2} \text{ yr}^{-1} = 4.52 \times 10^{-8} \text{ min}^{-1}$$

$$R_0 = \lambda N_0 = \left(4.52 \times 10^{-8} \text{ min}^{-1}\right)\left(3.35 \times 10^{25} \text{ nuclei}\right) = 1.52 \times 10^{18} \frac{\text{counts}}{\text{min}}$$

$$R/R_0 = \frac{10.0 \text{ counts / min}}{1.52 \times 10^{18} \text{ counts / min}} = 6.599 \times 10^{-18} = e^{-\lambda t}$$

$$t = \frac{-\ln(R/R_0)}{\lambda} = \frac{-\ln\left(6.599 \times 10^{-18}\right)}{2.38 \times 10^{-2} \text{ yr}^{-1}} = 1660 \text{ yr} \qquad \lozenge$$

---

25. A "clever" technician decides to heat some water for his coffee with an x-ray machine. If the machine produces 10.0 rad/s, how long will it take to raise the temperature of a cup of water by 50.0 °C.

Solution

The energy required to heat the water is

$$\mathcal{P}t = Q = mc\Delta T$$

Noting that $1 \text{ rad} = 10^{-2} \text{ J / kg}$,

$$t = \frac{mc\Delta T}{\mathcal{P}} = \frac{m\left(4186 \text{ J / kg·°C}\right)\left(50.0 \text{ °C}\right)}{\left(10.0 \text{ rad / s}\right)\left(10^{-2} \text{ J / kg·rad}\right)m} = 2.09 \times 10^6 \text{ s} \cong 24 \text{ days} \quad \lozenge$$

(Note: the power $\mathcal{P}$ is the product of the dose rate and the mass).

---

29. In a Geiger tube, the voltage between the electrodes is typically 1.00 kV and the current pulse discharges a 5.00-pF capacitor. (a) What is the energy amplification of this device for a 0.500-MeV electron? (b) How many electrons are avalanched by the initial electron?

Solution

(a) $\dfrac{E}{E_0} = \dfrac{\frac{1}{2}CV^2}{0.500 \text{ MeV}} = \dfrac{\frac{1}{2}\left(5.00 \times 10^{-12} \text{ F}\right)\left(1.00 \times 10^3 \text{ V}\right)^2}{(0.500 \text{ MeV})\left(1.60 \times 10^{-13} \text{ J / MeV}\right)} = 3.12 \times 10^7$ ◊

(b) $N = \dfrac{Q}{e} = \dfrac{CV}{e} = \dfrac{\left(5.00 \times 10^{-12} \text{ F}\right)\left(1.00 \times 10^3 \text{ V}\right)}{1.60 \times 10^{-19} \text{ C}} = 3.12 \times 10^{10}$ electrons ◊

---

37. Carbon detonations are powerful nuclear reactions that temporarily tear apart the cores inside massive stars late in their lives. These blasts are produced by carbon fusion, which requires a temperature of about 6×10^8 K to overcome the strong Coulomb repulsion between carbon nuclei. (a) Estimate the repulsive energy barrier to fusion, using the required ignition temperature for carbon fusion. (In other words, what is the average kinetic energy of a carbon nucleus at 6×10^8 K?) (b) Calculate the energy (in MeV) released in each of these "carbon-burning" reactions:

$$^{12}\text{C} + {}^{12}\text{C} \rightarrow {}^{20}\text{Ne} + {}^4\text{He}$$

$$^{12}\text{C} + {}^{12}\text{C} \rightarrow {}^{24}\text{Mg} + \gamma$$

(c) Calculate the energy (in kWh) given off when 2.00 kg of carbon completely fuses according to the first reaction.

Solution

(a) At 6×10^8 K, each carbon nucleus has thermal energy of

$$\tfrac{3}{2}k_BT = (1.5)(8.62\times10^{-5} \text{ eV / K})(6\times10^8 \text{ K}) = 8\times10^4 \text{ eV} \qquad \Diamond$$

(b) The energy released is $\qquad E = \left[2m(C^{12}) - m(\text{Ne}) - m(\text{He}^4)\right]c^2$

$$E = (24.000000 - 19.992439 - 4.002603)(931.5) \text{ MeV} = 4.62 \text{ MeV} \qquad \Diamond$$

In the 2nd case, the energy released is $\quad E = \left[2m(C^{12}) - m(\text{Mg}^{24})\right](931.5) \text{ MeV / u}$

$$E = (24.000000 - 23.985042)(931.5) \text{ MeV} = 13.9 \text{ MeV} \qquad \Diamond$$

(c) Energy released = the energy of reaction of the number of carbon nuclei in a 2.00-kg sample, which corresponds to

$$\Delta E = \left(\frac{2000 \text{ g}}{12 \text{ g / mol C}}\right)\left(\frac{6.02\times10^{23} \text{ atom}}{1 \text{ mol C}}\right)\left(\frac{1 \text{ fusion}}{2 \text{ atom}}\right)\left(\frac{4.62 \text{ MeV}}{\text{fusion}}\right)\left(\frac{1 \text{ kWh}}{2.25\times10^{19} \text{ MeV}}\right)$$

$$\Delta E = 10.3\times10^6 \text{ kWh} \qquad \Diamond$$

---

41. The half-life of tritium is 12.3 yr. If the TFTR fusion reactor contained 50.0 m^3 of tritium at a density equal to 2.00×10^{14} ions / cm^3, how many curies of tritium were in the plasma? Compare this value with a fission inventory (the estimated supply of fissionable material) of 4×10^{10} Ci.

Solution

G: It is difficult to estimate the activity of the tritium in the fusion reactor without actually calculating it; however, we might expect it to be a small fraction of the fission (not fusion) inventory.

O: The decay rate (activity) can be found by multiplying the decay constant λ by the number of ^3_1H particles. The decay constant can be found from the half-life of tritium, and the number of particles from the density and volume of the plasma.

A: The number of Hydrogen-3 nuclei is

$$N = \left(50.0 \text{ m}^3\right)\left(2.00\times10^{14} \frac{\text{particles}}{\text{m}^3}\right)\left(100 \frac{\text{cm}}{\text{m}}\right)^3 = 1.00\times10^{22} \text{ particles}$$

$$\lambda = \frac{\ln 2}{T_{1/2}} = \frac{0.693}{12.3 \text{ yr}}\left(\frac{1 \text{ yr}}{3.16\times10^7 \text{ s}}\right) = 1.78\times10^{-9} \text{ s}^{-1}$$

The activity is then

$$R = \lambda N = \left(1.78\times10^{-9} \text{ s}^{-1}\right)\left(1.00\times10^{22} \text{ nuclei}\right) = 1.78\times10^{13} \text{ Bq}$$

$$R = \left(1.78\times10^{13} \text{ Bq}\right)\left(\frac{1 \text{ Ci}}{3.70\times10^{10} \text{ Bq}}\right) = 482 \text{ Ci} \qquad \lozenge$$

L: Even though 482 Ci is a large amount of radioactivity, it is smaller than $4.00\times10^{10} \text{ Ci}$ by about a hundred million. Therefore, loss of containment is a smaller hazard for a fusion power reactor than for a fission reactor. $\qquad \lozenge$

51. Assuming that a deuteron and a triton are at rest when they fuse according to $^2\text{H} + {}^3\text{H} \rightarrow {}^4\text{He} + \text{n} + 17.6\ \text{MeV}$, determine the kinetic energy acquired by the neutron.

Solution

G: The products of this nuclear reaction are an alpha particle and a neutron, with total kinetic energy of 17.6 MeV. In order to conserve momentum, the lighter neutron will have a larger velocity than the more massive alpha particle (which consists of two protons and two neutrons). Since the kinetic energy of the particles is proportional to the square of their velocities but only linearly proportional to their mass, the neutron should have the larger kinetic energy, somewhere between 8.8 and 17.6 MeV.

O: Conservation of linear momentum and energy can be applied to find the kinetic energy of the neutron. We first suppose the particles are moving nonrelativistically.

A: The momentum of the alpha particle and that of the neutron must add to zero, so their velocities must be in opposite directions with magnitudes related by

$$m_n \mathbf{v}_n + m_\alpha \mathbf{v}_\alpha = 0 \qquad \text{or} \qquad (1.0087\ \text{u})v_n = (4.0026\ \text{u})v_\alpha$$

At the same time, their kinetic energies must add to 17.6 MeV

$$E = \tfrac{1}{2}m_n v_n^2 + \tfrac{1}{2}m_\alpha v_\alpha^2 = \tfrac{1}{2}(1.0087\ \text{u})v_n^2 + \tfrac{1}{2}(4.0026\ \text{u})v_\alpha^2 = 17.6\ \text{MeV}$$

Substitute $v_\alpha = 0.2520 v_n$ to obtain

$$E = (0.50435\ \text{u})v_n^2 + (0.12710\ \text{u})v_n^2 = 17.6\ \text{MeV}\left(\frac{1\ \text{u}}{931.494\ \text{MeV}/c^2}\right)$$

$$v_n = \sqrt{\frac{0.0189c^2}{0.63145}} = 0.173c = 5.19 \times 10^7\ \text{m}/\text{s}$$

Since this speed is not too much greater than $0.1c$, we can get a reasonable estimate of the kinetic energy of the neutron from the classical equation,

$$K = \frac{1}{2}mv^2 = \frac{1}{2}(1.0087 \text{ u})(0.173c)\left(\frac{931.494 \text{ MeV}/c^2}{\text{u}}\right) = 14.1 \text{ MeV} \qquad \lozenge$$

L: The kinetic energy of the neutron is within the range we predicted. For a more accurate calculation of the kinetic energy, we should use relativistic expressions. Conservation of momentum gives

$$\gamma_n m_n \mathbf{v}_n + \gamma_\alpha m_\alpha \mathbf{v}_\alpha = 0$$

$$1.0087 \frac{v_n}{\sqrt{1-v_n^2/c^2}} = 4.0026 \frac{v_\alpha}{\sqrt{1-v_\alpha^2/c^2}}$$

yielding $\quad \dfrac{v_\alpha^2}{c^2} = \dfrac{v_n^2}{15.746c^2 - 14.746v_n^2}$

Then $\quad (\gamma_n - 1)m_n c^2 + (\gamma_\alpha - 1)m_\alpha c^2 = 17.6 \text{ MeV}$

and $\quad v_n = 0.171c, \quad$ implying that $\quad (\gamma_n - 1)m_n c^2 = 14.0 \text{ MeV} \qquad \lozenge$

---

53. (a) Calculate the energy (in kilowatt hours) released if 1.00 kg of ^{239}Pu undergoes complete fission and the energy released per fission event is 200 MeV. (b) Calculate the energy (in electron volts) released in the D-T fusion:

$$^2_1\text{H} + ^3_1\text{H} \rightarrow ^4_2\text{He} + ^1_0\text{n}$$

(c) Calculate the energy (in kilowatt hours) released if 1.00 kg of deuterium undergoes fusion according to this reaction. (d) Calculate the energy (in kilowatt hours) released by the combustion of 1.00 kg of coal if each $C + O_2 \rightarrow CO_2$ reaction yields 4.20 eV. (e) List the advantages and disadvantages of each of these methods of energy generation.

Solution We obtain the masses of the particles from Appendix A.3.

(a) $E = (1000 \text{ g}) \left(\dfrac{6.02 \times 10^{23} \text{ nuclei}}{239 \text{ g}} \right) \left(\dfrac{200 \times 10^6 \text{ eV}}{1 \text{ nucleus}} \right) \left(\dfrac{1.60 \times 10^{-19} \text{ J/eV}}{3.60 \times 10^6 \text{ J/kWh}} \right)$

$E = 2.24 \times 10^7 \text{ kWh}$ ◊

(b) $m_{before} = 2.014102 \text{ u} + 3.016049 \text{ u} = 5.030151 \text{ u}$

$m_{after} = 4.002603 \text{ u} + 1.008665 \text{ u} = 5.011268 \text{ u}$

$\Delta m = m_{before} - m_{after} = 0.018883 \text{ u}$

$\Delta E = \Delta m c^2 = (0.018883 \text{ u})(931.494 \text{ MeV/u}) = 17.6 \text{ MeV}$ ◊

(c) $E = \left(1000 \text{ g } {}_1^2\text{H} \right) \left(\dfrac{6.02 \times 10^{23} \text{ deuterons}}{2.014 \text{ g } {}_1^2\text{H}} \right) \left(\dfrac{17.6 \text{ MeV}}{{}_1^2\text{H fusion}} \right) \left(\dfrac{1.60 \times 10^{-13} \text{ J/MeV}}{3.60 \times 10^6 \text{ J/kWh}} \right)$

$E = 2.34 \times 10^8 \text{ kWh}$ ◊

(d) Coal is essentially pure carbon. Assuming complete combustion,

$(1000 \text{ g C}) \left(\dfrac{6.02 \times 10^{23} \text{ C atoms}}{12.0 \text{ g C}} \right) \left(\dfrac{4.20 \text{ eV}}{\text{C atom}} \right) \left(\dfrac{1.60 \times 10^{-19} \text{ J/eV}}{3.60 \times 10^6 \text{ J/kWh}} \right) = 9.36 \text{ kWh}$ ◊

(e) You likely pay the electric company to burn coal, because coal is cheap at this moment in human history. The limit on supply will inevitably drive up the price of fossil fuels. Worldwide, electric energy from fission is an immediate option. Fission may offer advantages of reduced overall pollution, and even a reduction in the radiation released into the atmosphere, compared to mining and burning coal. However, the Chernobyl explosion and the more recent accident in Japan demonstrated that there is a continuing significant risk of catastrophic disaster. For a fair comparison to a coal-fired plant, we should think of a nuclear generating station paying for long-term disposal of its waste and paying for its own insurance, without government subsidies. We hope that fusion reactors will become practical in the future, because fusion might be both safer and less expensive than either coal or fission.

55. Consider the two nuclear reactions

$$\text{(I)} \ A + B \rightarrow C + E \qquad \text{and} \qquad \text{(II)} \ C + D \rightarrow F + G$$

(a) Show that the net disintegration energy for these two reactions $(Q_{net} = Q_I + Q_{II})$ is identical to the disintegration energy for the net reaction $A + B + D \rightarrow E + F + G$. (b) One chain of reactions in the proton-proton cycle in the sun's interior is

$$^1_1\text{H} + ^1_1\text{H} \rightarrow ^2_1\text{H} + ^0_1\text{e} + \nu \qquad ^0_1\text{e} + ^0_{-1}\text{e} \rightarrow 2\gamma \qquad ^1_1\text{H} + ^2_1\text{H} \rightarrow ^3_2\text{He} + \gamma$$

$$^1_1\text{H} + ^3_2\text{He} \rightarrow ^4_2\text{He} + ^0_1\text{e} + \nu \qquad ^0_1\text{e} + ^0_{-1}\text{e} \rightarrow 2\gamma$$

Based on part (a), what is Q_{net} for this sequence?

Solution $Q_I = (m_A + m_B - m_C - m_E)c^2$

$$Q_{II} = (m_C + m_D - m_F - m_G)c^2$$

$$Q_{net} = (m_A + m_B - m_C - m_E + m_C + m_D - m_F - m_G)c^2$$

$$Q_{net} = (m_A + m_B + m_D - m_E - m_F - m_G)c^2$$

(a)   This value is identical to Q for the reaction $A+B+D \rightarrow E+F+G$. Thus, any product (e.g., "C") that is a reactant in a subsequent reaction disappears from the energy balance.   ◊

(b)   Adding all five reactions, we have $4\left({}_1^1H\right)+2\left({}_{-1}^0e\right) \rightarrow {}_2^4He+2v+Q_{net}/c^2$

Here the symbol ${}_1^1H$ represents a proton, the nucleus of a hydrogen-1 atom. So that we may use the tabulated masses of neutral atoms for calculation, we add two electrons to each side of the reaction. The four electrons on the initial side are enough to make four hydrogen atoms and the two electrons on the final side constitute, with the alpha particle, a neutral helium atom. The atomic-electronic binding energies are negligible compared to Q_{net}. We use the arbitrary symbol "${}_1^1H$ atom" to represent a neutral atom. Then we have

$$4\left({}_1^1H \text{ atom}\right) \rightarrow {}_2^4He \text{ atom} + 2v + Q_{net}/c^2$$

$$Q_{net} = \left[4(1.007825 \text{ u}) - 4.002602 \text{ u}\right](931.434 \text{ MeV / u}) = 26.7 \text{ MeV} ◊$$

57.   The carbon cycle, first proposed by Hans Bethe in 1939, is another cycle by which energy is released in stars as hydrogen is converted to helium. The carbon cycle requires higher temperatures than the proton-proton cycle. The series of reactions is

(1) ${}^{12}C+{}^1H \rightarrow {}^{13}N+\gamma$ (2) ${}^{13}N \rightarrow {}^{13}C+e^++v$

(3) $e^++e^- \rightarrow 2\gamma$ (4) ${}^{13}C+{}^1H \rightarrow {}^{14}N+\gamma$

(5) ${}^{14}N+{}^1H \rightarrow {}^{15}O+\gamma$ (6) ${}^{15}O \rightarrow {}^{15}N+e^++v$

(7) $e^++e^- \rightarrow 2\gamma$ (8) ${}^{15}N+{}^1H \rightarrow {}^{12}C+{}^4He$

(a) If the proton-proton cycle requires a temperature of 1.5×10^7 K, estimate by proportion the temperature required for the carbon cycle. (b) Calculate the Q value for each step in the carbon cycle and the overall energy released. (c) Do you think the energy carried off by the neutrinos is deposited in the star? Explain.

Solution   (a) The solar-core temperature 1.5×10^7 K gives particles a high enough kinetic energy to get past the Coulomb-repulsion barrier to $^1_1\text{H} + ^3_2\text{He} \rightarrow ^4_2\text{He} + e^+ + v$, estimated as $k_e e(2e)/r$. The Coulomb barrier to Bethe's fifth and eighth reactions is like $k_e e(7e)/r$, larger by $7/2$, so the temperature should be on the order of

$$\frac{7}{2}\left(1.5 \times 10^7 \text{ K}\right) \cong 5.3 \times 10^7 \text{ K} \qquad \Diamond$$

(b)   Remembering the conversion factor of $931.494 \dfrac{\text{MeV}/c^2}{\text{u}}$

$$Q_1 = (12 \text{ u} + 1.007825 \text{ u} - 13.005738 \text{ u})c^2 = 1.94 \text{ MeV} \qquad \Diamond$$

To calculate the energy released in the second step, add seven electrons to both sides to obtain the reaction for neutral atoms

$$^{13}\text{N atom} \rightarrow ^{13}\text{C atom} + e^- + e^+ + v + Q_2/c^2$$

$$Q_2 = (13.005738 \text{ u} - 13.003355 \text{ u} - 2(0.000549 \text{ u}))c^2 = 1.20 \text{ MeV} \qquad \Diamond$$

$$Q_3 = Q_7 = (2(0.000549 \text{ u}))c^2 = 1.02 \text{ MeV} \qquad \Diamond$$

$$Q_4 = (-14.003074 \text{ u} + 13.003355 \text{ u} + 1.007825 \text{ u})c^2 = 7.55 \text{ MeV} \qquad \Diamond$$

$$Q_5 = (1.007825 \text{ u} + 14.003074 \text{ u} - 15.003065 \text{ u})c^2 = 7.30 \text{ MeV} \qquad \Diamond$$

$$Q_6 = (15.003065 \text{ u} - 15.000108 \text{ u} - 2(0.000549 \text{ u}))c^2 = 1.73 \text{ MeV} \qquad \Diamond$$

$$Q_8 = (1.0007825 \text{ u} - 15.000108 \text{ u} - 4.002602 \text{ u})c^2 = 4.97 \text{ MeV} \qquad \Diamond$$

$$Q_{net} = (1.94 + 1.20 + 1.02 + 7.55 + 7.30 + 1.73 + 1.02 + 4.97) \text{ MeV} = 26.7 \text{ MeV} \quad \Diamond$$

This result is the same as for the proton-proton cycle, because the net reaction is the same as that in Problem 45.55, $4\left(^1_1\text{H atom}\right) \rightarrow ^4_2\text{He atom} + 2v$. The carbon is a catalyst.

(c)   Not all of the energy released heats the star. When a neutrino is created it will likely fly out of the star without interacting with any other particle.

Chapter 46

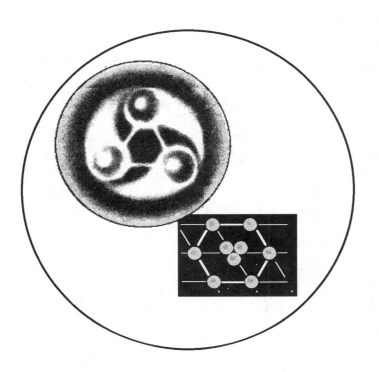

PARTICLE PHYSICS
AND COSMOLOGY

PARTICLE PHYSICS AND COSMOLOGY

INTRODUCTION

In this concluding chapter, we examine the properties and classifications of the various known subatomic particles and the fundamental interactions that govern their behavior. We also discuss the current theory of elementary particles, in which all matter is believed to be constructed from only two families of particles, quarks and leptons. Finally, we discuss how clarifications of such models might help scientists understand cosmology, which deals with the evolution of the Universe.

EQUATIONS AND CONCEPTS

Pions and muons are very unstable particles. A decay sequence is shown in Equation 46.1.

$$\pi^- \rightarrow \mu^- + \overline{\nu}_\mu \tag{46.1}$$

$$\mu^- \rightarrow e^- + \nu_\mu + \overline{\nu}_e$$

Hubble's law states a linear relationship between the velocity of a galaxy and its distance R from the Earth. The constant H is called the **Hubble parameter**.

$$v = HR \tag{46.7}$$

$$H \approx 17 \times 10^{-3} \text{ m / s} \cdot \text{ly}$$

The **critical mass density** of the Universe can be estimated based on energy considerations.

$$\rho_c = \frac{3H^2}{8\pi G}$$

REVIEW CHECKLIST

▷ Be aware of the four fundamental forces in nature and the corresponding field particles or quanta via which these forces are mediated.

▷ Understand the concepts of the antiparticle, pair production, and pair annihilation.

▷ Know the broad classification of particles and the characteristic properties of the several classes (relative mass value, spin, decay mode).

▷ Determine whether or not a suggested decay can occur based on the conservation of baryon number and the conservation of lepton number. Determine whether or not a predicted reaction/decay will occur based on the conservation of strangeness for the strong and electromagnetic interactions.

ANSWERS TO SELECTED CONCEPTUAL QUESTIONS

9. Identify the particle decays listed in Table 46.2 that occur by the electromagnetic interaction. Justify your answer.

Answer The decays of the neutral pion, eta, and neutral sigma occur by the electromagnetic interaction. These are the three shortest lifetimes in Table 46.2. All produce photons, which are the quanta of the electromagnetic force. All conserve strangeness.

□   □   □   □

14. How many quarks are there in (a) a baryon, (b) an antibaryon, (c) a meson, (d) an antimeson? How do you account for the fact that baryons have half-integral spins and mesons have spins of 0 or 1? (**Hint:** Quarks have spin 1/2.)

Answer   All baryons and antibaryons consist of three quarks. All mesons and antimesons consist of two quarks. Since quarks have spins of 1/2, it follows that all baryons (which consist of three quarks) must have half-integral spins, and all mesons (which consist of two quarks) must have spins of 0 or 1.

☐   ☐   ☐   ☐

SOLUTIONS TO SELECTED END-OF-CHAPTER PROBLEMS

3.   A photon with an energy $E_\gamma = 2.09$ GeV creates a proton-antiproton pair in which the proton has a kinetic energy of 95.0 MeV. What is the kinetic energy of the antiproton? ($m_p c^2 = 938.3$ MeV).

Solution

G: An antiproton has the same mass as a proton, so it seems reasonable to expect that both particles will have similar kinetic energies.

O: The total energy of each particle is the sum of its rest energy and its kinetic energy. Conservation of energy requires that the total energy before this pair production event equal the total energy after.

A: $E_\gamma = \left(E_{Rp} + K_p\right) + \left(E_{R\bar{p}} + K_{\bar{p}}\right)$

The energy of the photon is given as $E_\gamma = 2.09$ GeV $= 2.09 \times 10^3$ MeV. From Table 46.2, we see that the rest energy of both the proton and the antiproton is

$$E_{Rp} = E_{R\bar{p}} = m_p c^2 = 938.3 \text{ MeV}$$

If the kinetic energy of the proton is observed to be 95.0 MeV, the kinetic energy of the antiproton is

$$K_{\bar{p}} = E_\gamma - E_{R\bar{p}} - E_{Rp} - K_p = 2.09 \times 10^3 \text{ MeV} - 2(938.5 \text{ MeV}) - 95.0 \text{ MeV}$$

or $K_{\bar{p}} = 118 \text{ MeV}$ ◊

L: The kinetic energy of the antiproton is slightly (~20%) greater than the proton.   The two particles most likely have different shares in momentum of the gamma ray, and therefore will not have equal energies, either.

5.   One of the mediators of the weak interaction is the Z^o boson, with mass 93 GeV/c^2. Use this information to find the order of magnitude of the range of the weak interaction.

Solution

The rest energy of the Z^o boson is E_o = 93 GeV. The maximum time a virtual Z^o boson can exist is found from $\Delta E \, \Delta t \geq \hbar / 2$, or

$$\Delta t \approx \frac{\hbar}{2\Delta E} = \frac{1.055 \times 10^{-34} \text{ J} \cdot \text{s}}{2(93 \text{ GeV})(1.60 \times 10^{-10} \text{ J / GeV})} = 3.55 \times 10^{-27} \text{ s}$$

The maximum distance it can travel in this time is

$$d = c(\Delta t) = (3.00 \times 10^8 \text{ m / s})(3.55 \times 10^{-27} \text{ s}) \sim 10^{-18} \text{ m}$$ ◊

The distance d is an approximate value for the range of the weak interaction.

9. A neutral pion at rest decays into two photons according to

$$\pi^o \rightarrow \gamma + \gamma$$

Find the energy, momentum, and frequency of each photon.

Solution   Since the pion is at rest and momentum is conserved, the two gamma-rays must have equal momenta in opposite directions. So, they must share equally in the energy of the pion. By Table 46.2,

$$m_{\pi^o} = 135.0 \text{ MeV} / c^2$$

Therefore, $E_\gamma = 67.5 \text{ MeV} = 1.08 \times 10^{-11} \text{ J}$ ◊

$$p = \frac{E}{c} = \frac{67.5 \text{ MeV}}{3.00 \times 10^8 \text{ m/s}} = 3.61 \times 10^{-20} \text{ kg} \cdot \text{m/s}$$ ◊

$$f = \frac{E}{h} = 1.63 \times 10^{22} \text{ Hz}$$ ◊

---

11. Name one possible decay mode (see Table 46.2) for Ω^+, $\overline{K_S^0}$, $\overline{\Lambda}^0$, and $\overline{n}$.

Solution   The particles in this problem are the antiparticles of those listed in Table 46.2. Therefore, the decay modes include the antiparticles of those shown in the decay modes in Table 46.2:

$$\Omega^+ \rightarrow \overline{\Lambda}^0 + K^+$$

$$\overline{K_S}^0 \rightarrow \pi^+ + \pi^- \quad \text{or} \quad \pi^o + \pi^o$$

$$\overline{\Lambda}^0 \rightarrow \overline{p} + \pi^+$$

$$\overline{n} \rightarrow \overline{p} + e^+ + \nu_e$$ ◊

---

15. The following reactions or decays involve one or more neutrinos. In each case, supply the missing neutrino (v_e, v_μ, or v_τ).

(a) $\pi^- \rightarrow \mu^- + ?$

(b) $K^+ \rightarrow \mu^+ + ?$

(c) $? + p^+ \rightarrow n + e^+$

(d) $? + n \rightarrow p^+ + e^-$

(e) $? + n \rightarrow p^+ + \mu^-$

(f) $\mu^- \rightarrow e^- + ? + ?$

Solution

(a) $\pi^- \rightarrow \mu^- + \bar{v}_\mu$ $\qquad L_\mu: \ 0 \rightarrow 1-1$

(b) $K^+ \rightarrow \mu^+ + v_\mu$ $\qquad L_\mu: \ 0 \rightarrow -1+1$

(c) $\bar{v}_e + p^+ \rightarrow n + e^+$ $\qquad L_e: \ -1+0 \rightarrow 0-1$

(d) $v_e + n \rightarrow p^+ + e^-$ $\qquad L_e: \ 1+0 \rightarrow 0+1$

(e) $v_\mu + n \rightarrow p^+ + \mu^-$ $\qquad L_\mu: \ 1+0 \rightarrow 0+1$

(f) $\mu^- \rightarrow e^- + \bar{v}_e + v_\mu$ $\qquad L_\mu: \ 1 \rightarrow 0+0+1 \quad$ and $\quad L_e: \ 0 \rightarrow 1-1+0 \ \Diamond$

17. Determine which of the following reactions can occur. For those that cannot occur, determine the conservation law (or laws) violated.

(a) $p \rightarrow \pi^+ + \pi^0$

(b) $p + p \rightarrow p + p + \pi^0$

(c) $p + p \rightarrow p + \pi^+$

(d) $\pi^+ \rightarrow \mu^+ + v_\mu$

(e) $n \rightarrow p + e^- + \bar{v}_e$

(f) $\pi^+ \rightarrow \mu^+ + n$

Solution

(a) $p \rightarrow \pi^+ + \pi^0$ $\qquad$ Baryon number is violated: $1 \rightarrow 0 + 0$

(b) $p + p \rightarrow p + p + \pi^0$ $\qquad$ This reaction can occur.

(c) $p + p \rightarrow p + \pi^+$ $\qquad$ Baryon number is violated: $1 + 1 \rightarrow 1 + 0$

(d) $\pi^+ \rightarrow \mu^+ + v_\mu$ $\qquad$ This reaction can occur.

(e) $n \rightarrow p + e^- + \bar{v}_e$ $\qquad$ This reaction can occur.

(f) $\pi^+ \rightarrow \mu^+ + n$ $\qquad$ Violates baryon number: $0 \rightarrow 0 + 1 \quad$ and

$\qquad\qquad\qquad\qquad\qquad$ violates muon-lepton number: $0 \rightarrow -1 + 0 \quad \Diamond$

21. Determine whether strangeness is conserved in the following decays and reactions:

(a) $\Lambda^o \to p + \pi^-$ (d) $\pi^- + p \to \pi^- + \Sigma^+$

(b) $\pi^- + p \to \Lambda^o + K^o$ (e) $\Xi^- \to \Lambda^o + \pi^-$

(c) $\overline{p} + p \to \overline{\Lambda}^o + \Lambda^o$ (f) $\Xi^0 \to p + \pi^-$

Solution

We look up the strangeness quantum numbers in Table 46.2.

(a) $\Lambda^o \to p + \pi^-$ Strangeness: $-1 \to 0 + 0$
                               (-1 does not equal 0: so strangeness is not conserved) ◊

(b) $\pi^- + p \to \Lambda^o + K^o$ Strangeness: $0 + 0 \to -1 + 1$
                               (0 = 0: strangeness is conserved)   ◊

(c) $\overline{p} + p \to \overline{\Lambda}^o + \Lambda^o$ Strangeness: $0 + 0 \to +1 - 1$
                               (0 = 0: strangeness is conserved)   ◊

(d) $\pi^- + p \to \pi^- + \Sigma^+$ Strangeness: $0 + 0 \to 0 - 1$
                               (0 does not equal -1: strangeness is not conserved)   ◊

(e) $\Xi^- \to \Lambda^o + \pi^-$ Strangeness: $-2 \to -1 + 0$
                               (-2 does not equal -1: strangeness is not conserved) ◊

(f) $\Xi^0 \to p + \pi^-$ Strangeness: $-2 \to 0 + 0$
                               (-2 does not equal 0 : strangeness is not conserved)   ◊

31. Analyze each reaction in terms of constituent quarks:

(a) $\pi^- + p \rightarrow K^o + \Lambda^o$

(c) $K^- + p \rightarrow K^+ + K^o + \Omega^-$

(b) $\pi^+ + p \rightarrow K^+ + \Sigma^+$

(d) $p + p \rightarrow K^o + p + \pi^+ + ?$

In the last reaction, identify the mystery particle.

Solution   We look up the quark constituents of the particles in Table 46.4 and Table 46.5.

(a) $d\bar{u} + uud \rightarrow d\bar{s} + uds$

(b) $\bar{d}u + uud \rightarrow u\bar{s} + uus$

(c) $\bar{u}s + uud \rightarrow u\bar{s} + d\bar{s} + sss$

(d) $uud + uud \rightarrow d\bar{s} + uud + u\bar{d} + uds$ A uds is either a Λ^o or a Σ^o ◊

---

37. A distant quasar is moving away from Earth at such high speed that the blue 434-nm hydrogen line is observed at 650 nm, in the red portion of the spectrum. (a) How fast is the quasar receding? You may use the result of Problem 35. (b) Using Hubble's law, determine the distance from Earth to this quasar.

Solution

G: The problem states that the quasar is moving very fast, and since there is a significant red shift of the light, the quasar must be moving away from Earth at a relativistic speed ($v > 0.1c$). Quasars are very distant astronomical objects, and since our universe is estimated to be about 15 billion years old, we should expect this quasar to be ~10^9 light-years away.

O: As suggested, we can use the equation in Problem 35 to find the speed of the quasar from the Doppler red shift, and this speed can then be used to find the distance using Hubble's law.

A: (a) $\dfrac{\lambda'}{\lambda} = \dfrac{650 \text{ nm}}{434 \text{ nm}} = 1.498 = \sqrt{\dfrac{1 + v/c}{1 - v/c}}$ or squared, $\dfrac{1 + v/c}{1 - v/c} = 2.243$

Therefore, $v = 0.383c$ or 38.3% the speed of light ◊

(b) Hubble's law asserts that the universe is expanding at a constant rate so that the speeds of galaxies are proportional to their distance R from Earth, $v = HR$

so, $R = \dfrac{v}{H} = \dfrac{(0.383)\left(3.00 \times 10^8 \text{ m / s}\right)}{\left(1.70 \times 10^{-2} \text{ m / s} \cdot \text{ly}\right)} = 6.76 \times 10^9 \text{ ly}$ ◊

L: The speed and distance of this quasar are consistent with our predictions. It appears that this quasar is quite far from Earth but not the most distant object in the visible universe.

45. The energy flux carried by neutrinos from the Sun is estimated to be on the order of 0.4 W/m² at Earth's surface. Estimate the fractional mass loss of the Sun over 10^9 years due to the radiation of neutrinos. (The mass of the Sun is 2×10^{30} kg. The Earth-Sun distance is 1.5×10^{11} m.)

Solution

G: Our Sun is estimated to have a life span of about 10 billion years, so in this problem, we are examining the radiation of neutrinos over a considerable fraction of the Sun's life. However, the mass carried away by the neutrinos is a very small fraction of the total mass involved in the Sun's nuclear fusion process, so even over this long time, the mass of the Sun may not change significantly (probably less than 1%).

O: The change in mass of the Sun can be found from the energy flux received by the Earth and Einstein's famous equation, $E = mc^2$.

A: Since the neutrino flux from the Sun reaching the Earth is 0.4 W/m², the total energy emitted per second by the Sun in neutrinos in all directions is

$$\left(0.4 \text{ W / m}^2\right)\left(4\pi r^2\right) = \left(0.4 \text{ W / m}^2\right)(4\pi)\left(1.5 \times 10^{11} \text{ m}\right)^2 = 1.13 \times 10^{23} \text{ W}$$

In a period of 10^9 yr, the Sun emits a total energy of

$$\left(1.13 \times 10^{23} \text{ J / s}\right)\left(10^9 \text{ yr}\right)\left(3.156 \times 10^7 \text{ s / yr}\right) = 3.57 \times 10^{39} \text{ J}$$

in the form of neutrinos. This energy corresponds to an annihilated mass of

$$E = m_v c^2 = 3.57 \times 10^{39} \text{ J} \qquad \text{so} \qquad m_v = 3.97 \times 10^{22} \text{ kg}$$

Since the Sun has a mass of about 2×10^3 kg, this corresponds to a loss of only about 1 part in 50 000 000 of the Sun's mass over 10^9 yr in the form of neutrinos. ◊

L: It appears that the neutrino flux changes the mass of the Sun by so little that it would be difficult to measure the difference in mass, even over its lifetime!

49. Calculate the kinetic energies of the proton and pion resulting from the decay of a Λ^0 at rest:

$$\Lambda^o \rightarrow p + \pi^-$$

Solution

We first look up the energy of each particle: $m_\Lambda c^2 = 1115.6 \text{ MeV}$

$$m_p c^2 = 938.3 \text{ MeV}$$

$$m_\pi c^2 = 139.6 \text{ MeV}$$

The difference between starting mass-energy and final mass-energy is the kinetic energy of the products:

$$(K_p + K_\pi) = (1115.6 - 938.3 - 139.6) \text{ MeV} = 37.7 \text{ MeV}$$

In addition, since momentum is conserved,

$$\left| p_p \right| = \left| p_\pi \right| = p$$

Applying conservation of relativistic energy:

$$\left(\sqrt{(938.3)^2 + p^2 c^2} - 938.3 \right) + \left(\sqrt{(139.6)^2 + p^2 c^2} - 139.6 \right) = 37.7 \text{ MeV}$$

Solving the algebra yields $p_\pi c = p_p c = 100.4$ MeV.

Thus $K_p = \sqrt{(m_p c^2)^2 + (100.4)^2} - m_p c^2 = 5.35 \text{ MeV}$ ◊

and $K_\pi = \sqrt{(139.6)^2 + (100.4)^2} - 139.6 = 32.3 \text{ MeV}$ ◊

53. If a K_S^0 meson at rest decays in 0.900×10^{-10} s, how far will a K_S^0 meson travel if it is moving at $0.960c$ through a bubble chamber?

Solution The motion of the K_S^0 particle is relativistic. Just like the spaceman who leaves for a distant star, and returns to find his family long gone, the kaon appears to us to have a longer lifetime. That time-dilated lifetime is:

$$t = \gamma t_0 = \frac{0.900 \times 10^{-10} \text{ s}}{\sqrt{1 - v^2 / c^2}} = \frac{0.900 \times 10^{-10} \text{ s}}{\sqrt{1 - (0.960)^2}} = 3.21 \times 10^{-10} \text{ s}$$

During this time, we see the kaon travel at $0.960c$. It travels for a distance of

$$d = vt = \left[(0.960)\left(3.00 \times 10^8 \text{ m / s}\right)\right]\left(3.21 \times 10^{-10} \text{ s}\right)$$

$$d = 0.0926 \text{ m} = 9.26 \text{ cm} \qquad \lozenge$$

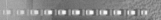

ILLUSTRATOR 8
f/x and design

Sherry London
T. Michael Clark

CORIOLIS

The Coriolis Group, LLC
14455 N. Hayden Road, Suite 220
Scottsdale, Arizona 85260

602/483-0192
FAX 602/483-0193
http://www.coriolis.com

Library of Congress Cataloging-In-Publication Data
London, Sherry.
    Illustrator 8 f/x and design / by Sherry London and T. Michael Clark.
        p. cm.
    Includes index.
    ISBN 1-57610-408-7
    1. Computer graphics. 2. Adobe Illustrator (Computer file) I. Clark, T. Michael, 1956–   . II. Title.
T385.L648 1999
006.6'869--dc21   99-11321
                        CIP

Printed in the United States of America
10 9 8 7 6 5 4 3 2

Publisher
Keith Weiskamp

Acquisitions Editor
Mariann Hansen Barsolo

Marketing Specialist
Gary Hull

Project Editor
Don Eamon

Technical Reviewer
David Xenakis

Production Coordinator
Jon Gabriel

Cover Design
Jody Winkler
additional art provided by Brandon Riza

Layout Design
April Nielsen

CD-ROM Developer
Robert Clarfield

OTHER TITLES FOR THE CREATIVE PROFESSIONAL

QuarkXPress 4 In Depth
by William Harrell & Elaine Betts

Adobe PageMill 3 f/x and design
by Daniel Gray

Looking Good In Presentations, Third Edition
by Molly Joss

To my husband Norm: You make my life worth living.
—Sherry London

To my lovely, and loving, wife Pamela.
—T. Michael Clark

❧

ABOUT THE AUTHORS

Sherry London is an artist, a writer, and a teacher, which is exactly what a going-into-college aptitude test predicted. Sherry is the author of a number of books, including *Photoshop 5.0: An Interactive Course*, *Photoshop Textures Magic*, *Photoshop 5 In Depth* (also with David Zenakis), *Painter 5 f/x* (with Rhoda Grossman and Sharon Evans), and *Photoshop 3 Special Effects How-To*. She has taught prepress and Photoshop in a number of schools, including Moore College of Art and Design in Philadelphia and Gloucester County College. Sherry has also spoken at a number of conferences, including the Thunder Lizard Photoshop Conference and the Professional Photographers of America convention.

Sherry has worked as a social studies teacher, an instructional systems designer, a programmer, a fiber artist, and a graphic designer. She is the principal of London Computing: PhotoFX, a full-service design studio. She has designed needlework for the Philadelphia Museum of Art gift store and the Horchow catalog. Her fiber art has been exhibited in many group shows, including shows at the Delaware Museum of Art, Ormond Memorial Art Museum, and the Brevard Art Center. She was a contributing editor for *Computer Artist* magazine and has written for *Pre*, *MacWeek*, *MacUser*, *Digital Vision*, and the combined *MacWorld/MacUser* magazine.

T. Michael Clark is the author of several prominent books on computer imaging software. Titles include *Teach Yourself Photoshop 5 in 21 Days* from SAMS, *Teach Yourself Paint Shop Pro 5 in 24 Hours,* also from SAMS, *Paint Shop Pro Web Techniques* from New Riders Publishing, and *Photoshop 5 Filters f/x and design* from The Coriolis Group. His clear, concise teaching style has won him acclaim from all corners of the globe, and Michael is very glad to be able to share his love of the Internet and digital graphics with his many readers and devoted fans.

With the advent of the Web, Michael is able to combine his many talents. An artist practically from birth, freelance photographer, computer consultant, certified programmer-analyst and full-time author, Michael also owns and operates GrafX Design (**www.grafx-design.com**), a Web site that offers online tutorials, product reviews, and a host of lessons on computer imaging topics. Michael also is an associate member of i/us, the site for visual professionals and is a regular contributor to Usenet graphics newsgroups.

ACKNOWLEDGMENTS

I would like to thank the vendors who have contributed to this project:

- Direct Imagination

- Extensis

- Ultimate Symbol

- Hot Door

- Vertigo

- Adobe

- RayFLECT

- ZaxWerks

I would also like to thank the artists whose work has so wonderfully enhanced the Color Gallery. You can learn so much about good design and about the creative use of Illustrator just by studying their images.

I also want to thank the staff at Coriolis, including: Stephanie Wall, who acquired this book, and Mariann Barsolo, who so ably continued as the shepherd of the project. I cannot possibly do an adequate enough job of thanking my project editor, Don Eamon, for his work on his book. He has been infinitely patient and always accessible and his caring attitude is intensely appreciated. Thanks also to Jon Gabriel, Production Coordinator, Jody Winkler, Cover Designer, April Nielsen, Layout Designer, and Robert Clarfield, CD-ROM specialist.

David Xenakis, the technical editor, has also done work "above and beyond…" David, your encouragement and your perfectionism have been valued beyond anything that I can possibly say.

I want to thank T. Michael Clark, my co-author, for the chapters that he has contributed to this book.

My husband, Norm, as always, has borne the brunt of my long hours, occasional temper tirades, and rotten cooking without complaint.

—*Sherry London*

I'd like to thank Sherry London for asking me to participate in this project.

Thanks to my agent, Margot.

A special thanks to the team at Coriolis for their help in turning out this incredibly cool book. A very special thanks to the Project Editor, Don Eamon. Thanks Don, for your patience, understanding, and guidance. You are a true professional, not to mention, a heck of a great guy.

A big thanks to the team at Adobe. Without every one of you, there would be no Illustrator.

To all of the artists who helped out with this project, and to one of my favorite models, Marianne Dodelet, thank you.

Finally, thanks to the following companies for their help and participation:

- Macromedia

- Hot Door

- Vertigo

—*T. Michael Clark*

Contents At A Glance

TABLE OF CONTENTS

I'm so glad that you're reading this page. Whether you've just purchased this volume or you're thinking about it, I want to tell you why Michael and I have written this book and what you can expect from it. There are many Illustrator books on the market, and each one does something slightly different—or does it in a different way. Almost all are good books, how do you begin to select which one or ones to buy?

Illustrator books seem to fall into several categories. There are the "manual replacements." These books are usually of mid-length and repeat the commands in the manual, restating them in case you couldn't understand Adobe's directions. These books are for total beginners or for infrequent users of Illustrator who need to be reminded of how to do a specific thing.

Then there are exhaustive reference volumes such as Ted Alspach's *Illustrator 8 Bible* and Deke McClelland's *Real World Illustrator*. These wonderful books belong on everyone's book shelf. You can learn something new from either of these books each time you look at them.

Finally, there are books that showcase the design tricks of a variety of artists. These books are like wonderful eye candy, and just looking at them can inspire you. They are sometimes a bit short on guidance, but they are fantastic sources of creative ideas.

About This Book

So what approach does this book take? It takes a bit of them all and a lot of none-of-the-above. In your hands, you hold a book written by two opinionated artists. We picked a wide (but by no means exhaustive) range of techniques and gave you a tutorial to follow. Although this book is really not for the rank beginner—we assume that you already know the basics of Illustrator—we go to extreme lengths to ensure that you can follow the instructions for the projects. Every step is clearly stated, so even if you're not an expert when you start this book, you'll be much closer to that status when you finish it. If you work your way through it at the computer (it isn't a novel—you really need a computer sitting in front of you as you read), you'll find that you learn things about Illustrator that you never knew. We've tried to give you a range of start-to-finish projects that go from basic idea to completed artwork.

We've tried very hard to find a balance in the amount of work needed to complete each project. Sometimes you start from a blank document; at other times, we give you the critical pieces. To make it easier to work through each chapter, we've included a CD-ROM with the book. For each Illustration (really an exercise or a project, but this is a book on Illustrator, so I came up with this catchy title for these elements), you'll find two figure areas on the CD-ROM. One area contains all the images that you need to start and build the project. The other folder contains the finished projects so that you can learn by looking at the completed work and see what objects were defined and how. Illustrator is much better than Photoshop in this regard. Many things that you do in Illustrator leave "tracks" so that it's easier to reconstruct the way that the project was worked. In Illustrator, for example, if I applied a gradient to an object on this book's CD-ROM, you will see by clicking on the object exactly what gradient was applied (and if you like it, you can reuse it in your own images). In Photoshop, there is no reasonable way to figure out the colors in a gradient. The raster format of Photoshop simply applies and moves on.

You also get the benefit of multiple authors in this book. Though I wrote 2/3 of the book, Michael's unique perspective adds value and is a good complement. It's always good to get multiple opinions, and in this book, you do. As you read each chapter, our bylines tell you which one of us is speaking. Because Michael and I are quite different, our approach to the program is different as well. You can never have too many teachers.

Teacher. That's what I've been for many years, and that's what I consider myself to be as I write this book. Teacher. Guide. Friend. The best part of this is seeing the "student" surpass me. Many of you will—and that makes this book worth writing (for me) and worth reading (for you). Enjoy! I had a great time writing this book, and I hope that you enjoy it.

Keep in touch. My email address is **slondon@earthlink.net**. Michael's is **tmc@grafx-design.com**.

—*Sherry London*

About Third-Party Filter Sets

Although this book is in the *f/x and design* series, we've chosen not to provide extensive coverage of most of the available third-party filter sets. Most third-party filters are Macintosh-only, and we really don't want to receive email flames from the Windows-only readers who may misunderstand our appearing to give preferential treatment to the Mac. The other (major) consideration is that Adobe changed the plug-in interface with Illustrator 8 so that many of the currently available third-party plug-ins are no longer 100-percent compatible (and few of them are even being sold).

Here is the status of Illustrator 6 and 7 filter sets that are considered incompatible with Illustrator 8. These sets are Macintosh-only.

Table I Illustrator filter set status.

Company	Filter Set Name	Retail Status
Alien Skin	Stylist	Not Being Sold
BeInfinite	InfinteF/X	Not Being Sold
Cytopia Software	CSI	Not Being Sold
Extensis	Vector Tools 2.0	Available
Letraset	Envelopes	Not Being Sold
MetaCreations	KPT Vector Effects	Not Being Sold

This leaves you with little choice. However, you can use some of these filters, after a fashion.

The Extensis Vector Tools still work if you remove everything to do with Vector Bars and Vector Tips. VectorTools also tries to install a menubar at the top of the screen to allow you to access the filters. This bar doesn't install properly, but seems to "find" itself after you access the menus a bit. It eventually pushes the Help menu over one spot to the right. The color and the shapes filters that remain all seem to work properly, and it is lovely to be able to randomize colors and adjust them.

The KPT Vector Effects filters all seem to work without problems, but this set was withdrawn from the marketplace. So, if you have them, try them on your system. I don't think that some of the previews work as well as they previously did, but the effects seem fully usable.

The CSI filters are also no longer for sale. They work if you remove anything that says Start-up or Plug-in from the set. The CSI filters gave you good control over 3D manipulations of vector objects (although you can do the same thing with the KPT Vector Effects filters).

The Letraset and BeInfinite filters have not been available for a number of years (although you can try them if you already own them).

Alien Skin Stylist will not work at all.

PART I

GETTING READY

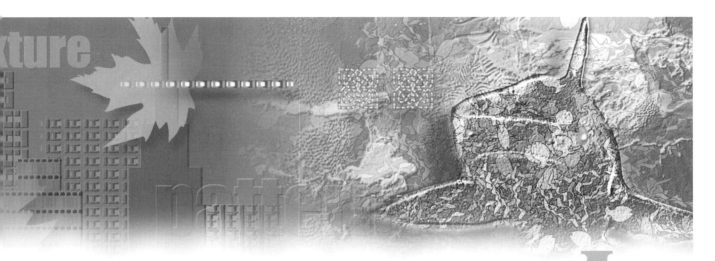

BASIC ISSUES AND WHAT'S NEW IN ILLUSTRATOR 8

T. MICHAEL CLARK

When you first run Illustrator 8, little may seem to have changed. Wrong. One hundred or so interface changes and a couple of tool changes/additions will amaze and mystify you.

You may be accustomed to using Illustrator to draw, but you've probably never thought that you could sketch or paint with it, right? Well, now you can! In this chapter, I'll give you an idea of how the new Illustrator 8 tools enable you to do things previously unheard of with most drawing programs. We'll go into greater depth in later chapters, and I'll also show you how to create some really cool effects.

I will not go into detail here about the smaller interface changes (such as the new keyboard shortcuts). These changes are outlined in the user guide and are available on the quick-reference card. Instead, I'll present and describe the new tools, palettes, and so on that are the bigger (and more exciting) changes. Along with the descriptions, I'll provide samples of some of the new changes so that you'll better understand what's possible with Illustrator 8 that wasn't possible with previous versions.

New Palettes

One of the significant interface changes is the addition of some new palettes. These palettes are tools that will make your work easier and will place in your hands far more power than what was available in previous versions of Illustrator. These new palettes are as follows:

- The Navigator palette

- The Links palette

- The Pathfinder palette

- The Actions palette

The following sections briefly describe each of these powerful new tools.

The Navigator Palette

The Navigator palette (see Figure 1.1) adds the same capability to Illustrator as it does to Photoshop.

You can easily locate and move to any area of your image by clicking and dragging the red rectangle (which appears in Figure 1.1 as a dark shaded gray rectangle).

Figure 1.1
The Navigator palette.

Along the bottom of the palette are four areas of interest. In the lower-left corner, you see the current zoomed percentage of the image you're currently working on. You can highlight the number there and enter any value you want, including decimal values. You can set the zoom, for example, to 102.55 percent.

Next to the current zoom value, you see a button that enables you to zoom out. This feature works in increments. With the zoom set to my crazy 102.55% number, one click brings the value to 100%, the next sets it to 66.67%, and a subsequent click sets it to 50%. This is a great way to quickly zoom out on an image.

To the right of the zoom-out button, you see a slider. Clicking and dragging the slider enables you to quickly set the zoom value from a ridiculously small size (3.13%) to an amazingly large size (6400%). This is the fastest way to zoom in and out on your image. The slider, coupled with the red rectangle that lets you zero in on any area, makes the tools in this new palette very powerful.

To the right of the slider is the zoom-in button. It works exactly like its counterpart, the zoom-out button (described in a previous paragraph).

The Links Palette

The Links palette (see Figure 1.2) enables you to manage your links. Rather than adding objects to the original image, you can link the objects from existing files. Doing so keeps the file size of the original image more manageable. Of course, you need to be able to control the links. This is where the Links palette comes in.

In Figure 1.2, you'll see that I currently have only one linked file. If more files were linked to the current image, they would also be shown within the palette.

Along the bottom of the palette are several buttons. These buttons and their descriptions are, from left to right, as follows:

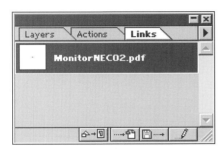

Figure 1.2
The Links palette.

- *Replace Link*—The Replace Link button is on the far left. Clicking on the Replace Link button brings up a dialog box that enables you to change the link to point to another file.

- *Go To Link*—The Go To Link button is the second button from the left. The Go To Link button brings you to the linked file in your current image and selects the object.

- *Update Link*—The Update Link button is the third button from the left. The Update Link button enables you to update the link to point to an updated version of the object.

- *Edit Original*—The Edit Original button is on the far right. The Edit Original button enables you to edit the artwork or object associated with the link. Clicking on the Edit Original button will actually run the program associated with the file. For example, clicking on the Edit Original button in Figure 1.2 brings up Adobe Exchange because the file that the link points to is a PDF file.

In addition to the buttons in the Link palette, there is more. You can get information about the file to which the link points by double-clicking on the entry for the linked file in the palette. Doing so brings up an information window that displays the file's name, location and size; the kind (type) of file; the dates on which the file was created and modified; and both whether transformations were done and the results of these transformations.

Additional information may also appear next to an entry. If, for example, the file you've linked to has changed or if the link is broken, you'll see different icons appearing to the right of the linked file's name.

If the icon shows a small question mark in a circle, the link has been broken. This means that Illustrator can no longer find the file at the location in which it was stored when the link was created.

If the icon you see has a small exclamation point in a triangle, the file that the link points to has been updated. Clicking on the Update Link button will update the link and display the updated file/object in place of the older, unmodified file/object.

You can change the way in which the information you see in the Links palette is presented. For example, you can sort the links. For more information on how to customize the palette, see the user guide.

The Pathfinder Palette

The Pathfinder palette (see Figure 1.3) is a cool addition to Illustrator 8. It enables you to quickly select and then apply any of the Pathfinder commands.

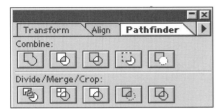

Figure 1.3
The Pathfinder palette.

The first row of buttons enables you to apply one of five Combine path commands. The five commands available are, from left to right, as follows:

- Unite

- Intersect

- Exclude

- Minus Front

- Minus Back

You'll see these commands applied at various times throughout the remainder of this book.

The second row of buttons enables you to apply the Divide/Merge/Crop path commands. The commands, from left to right, are as follows:

- Divide

- Trim

- Merge

- Crop

- Outline

Again, you'll see these options used at various times throughout this book.

Another row of buttons is normally hidden. To see this row, click the small, black triangle button to the immediate right of the Pathfinder tab and choose Show Options. Doing so will reveal the hidden row of three additional buttons (see Figure 1.4).

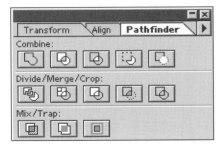

Figure 1.4
Additional (hidden) row of buttons in the Pathfinder palette.

This row contains three buttons that enable you to apply the Mix/Trap options, as follows:

- Hard Mix

- Soft Mix

- Trap

Using these options can provide you with some cool effects, such as faking transparency. We'll show you how these work later in the book.

The Actions Palette

In the Actions palette (see Figure 1.5), you can record and run actions. That's right; like Photoshop, Illustrator 8 now has the capability of actions! This change is really exciting. Actions can take the drudgery out of applying a similar set of commands over and over again. Now you can, instead, record the set of steps once and run them at any time as an action.

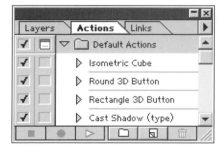

Figure 1.5
The Actions palette.

As well as designing your own actions, you can use any of the actions that ship with Illustrator 8. To access the default actions, simply click on the third icon from the left (below the palette tabs). Doing so drops down a list of the installed actions. These include the following:

- Isometric Cube

- Round 3D Button

- Rectangle 3D Button

- Cast Shadow (Type)

- Embossed (Type)

- Marble (Type)

- Multiple Outlines (Selection)

- Beveled Frame

- Train Tracks (Stroke)

- Rosewood

- Move Dialog (Selection)

- Swatch/Brush Cleanup

- Photo Crosshatch Filter

For some of these actions, you first need to make a selection; others will run only on type that you've entered. Play around with a couple of them to see how they work. We'll show you how to create your own actions in Chapter 6.

Brush Effects

The Paintbrush tool has been enhanced to enable you to draw by using different paintbrush strokes. These strokes, available under the Brushes palette, can give your work the appearance of being created with a paintbrush. Several new types of brushes are available. You can choose art brushes, scatter brushes, calligraphic brushes, and pattern brushes.

Illustrator ships with many brushes included on the CD-ROM. You can create your own brushes as well. Sherry covers brushes in more detail in Chapter 8.

Gradient Mesh

One of the most amazing new tools is the gradient mesh, which enables you to create effects that are impossible with any other vector program. Using the gradient mesh, you can add painterly effects that may remind you of watercolors. Because the gradient mesh enables you to add points and move them about as you would any other vector point, you have unbelievable control over the effects you create with this new tool.

The gradient mesh tools are covered in detail in Chapter 4.

Photo Crosshatch Filter

The Photo Crosshatch filter enables you to create illustrations that resemble pen-and-ink drawings. You can apply this filter to imported photographs. Figure 1.6 shows the results of applying the Photo Crosshatch filter to a photograph that first had the Photoshop Artistic Cutout filter applied to it.

You can see how the Photo Crosshatch filter has given the photo a pen-and-ink look. You can change several options in the filter's dialog box. We discuss this filter in greater depth in Chapter 5.

Figure 1.6
The Photo Crosshatch filter applied to a photograph that has the Photoshop Artistic Cutout filter applied.

Enhanced Blend Tool

The Blend tool has been upgraded. You can blend interactively by selecting a point on one object and a point on any other. You also can blend between open paths, closed paths, gradients, colors, and other blends.

When you create a blend between objects, a path is created. This path can be manipulated in the same way as any other path. You can add points, bend and twist the path, and so on. Figure 1.7 shows a blend between a circle filled with a gradient and a square filled with a solid color.

After blending the two shapes, I chose the Pen tool and added a point on the blend path. I then altered the point by Option+clicking (Mac) or Alt+clicking (Windows) on it. Using the Direct Selection tool, I then changed the path into the flowing curve that you see in the figure.

You can also edit any part of a blend, and the blend will update automatically.

Type Sampling

You can now use the Eyedropper and Paint Bucket tools to sample one type object and have the effects applied to any other type object. You can apply the fill and stroke and the font and size of any type object to any other selected or new type object.

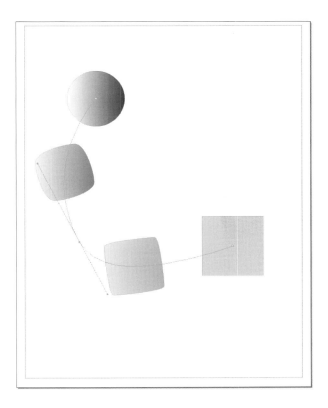

Figure 1.7
A blend between a gradient-filled circle and a solid-colored square.

This capability is quite powerful and is more useful than it may initially seem. You can simply select the Eyedropper tool and sample any existing type. The setting will then be applied to any new type that you enter or any type currently selected.

Enhanced Text Handling

Adobe enhanced the text handling so that you can now view nonprinting characters on screen. Support also was added for Japanese, wari-chu, and kinsoku shori in the Character and Paragraph palettes.

Enhanced Layers

Layers have a couple of enhancements in Illustrator 8.

One change allows you to create template layers, which enables you to base a new illustration on an existing piece of artwork. The template layer is locked, dimmed, and nonprinting. This makes it easy to trace existing artwork.

Other improvements are in showing, viewing, and printing layers in the Layers palette.

Finally, you can export layers to Photoshop by exporting the image as a Photoshop (PSD) file.

Color And Swatches Palette Enhancements

It's now easier to create and use process or spot colors. Spot color management, when merging colors from other files, has been improved.

Bounding Box Editing

The bounding box, which is displayed around object(s) selected, can now be used to scale, move, and duplicate objects. These processes can be accomplished by dragging one of the handles (hollow squares) that appear on the bounding box.

Free Transform Tool

The Free Transform tool enables you to reflect, rotate, shear, and distort selected objects in realtime without the need to enter values in the associated transform dialog box or by resorting to the Transform palette.

Pencil Tool

The Pencil tool now draws editable paths. You can easily draw a path and then change it by redrawing another path or portion of a path nearby. This comes pretty close to the feeling of sketching.

To change the line between any two points, simply place the Pencil tool where you want to start the change and draw toward the point where you want to finish the change. This process makes it extremely easy to shape and reshape lines.

Smooth Tool

The Smooth tool enables you to smooth out lines that you've drawn with the Pencil tool. Smooth removes points and smoothes the curves between the points left behind. To use the Smooth tool, simply click and drag between any two points on a line. The tool will remove points and smooth the curves between the start and end of your click-and-drag movement.

Erase Tool

The Erase tool enables you to erase portions of a line. You can erase portions of open or closed paths. To erase a portion of a line, simply click and drag over the area you want to erase.

The Erase tool makes it easy to accomplish all kinds of tasks. For example, copying and pasting the inner oval over the image and splitting the new oval into two paths with the Erase tool created the highlight and shadow on the stylized bowl in Figure 1.8. After splitting the new oval into two portions, the new portions were assigned different brush strokes and colors.

Figure 1.8
Stylized bow, created with ovals and lines by using the Erase tool.

Smart Guides

I've heard some say that the new Smart Guides feature alone is worth the cost of upgrading. Smart Guides are temporary guides that enable you to create, align, and edit objects relative to one another.

After turning on Smart Guides, you'll see the guides appear, with keywords that show you paths and anchor points that detail the movement and angle of the object(s) you're moving. Having Smart Guides turned on makes positioning objects a snap (no pun intended).

Registration Color

You can now mark any object with a color that will print on all the plates in a separation.

Getting Images Into Illustrator (Auto Trace, Place, And Streamline)

You can get your artwork into Illustrator in a number of ways. With the newer tools, for example, you may be tempted to just jump right in and start drawing and painting. If you prefer the more traditional approach of starting with a photo or a hand-drawn sketch, you're in luck; you can accomplish this in a number of relatively painless ways.

The first method I'll show uses Illustrator's built-in Auto Trace tool.

Auto Trace

The Auto Trace tool enables you to trace objects by simply clicking on them. To use Auto Trace, you'll need an existing image. The image you choose can be a line drawing that you've already scanned into Photoshop or some other TWAIN-compatible device.

You should scan the image you'll be using at fairly high resolution—300 dpi to 600 dpi should work just fine.

Note: TWAIN stands for Technology Without An Interesting Name. It's used to describe the interface between hardware, such as your desktop scanner, and your computer. Essentially, the technology is such that the program (such as Photoshop) doesn't really need to know the specifics of the hardware, such as your scanner; it simply lets the two communicate properly (most of the time).

After scanning the image into Photoshop, you may want to zoom in and clean up any areas that need it. You can remove spots, for example, introduced during the scanning process or from irregularities in the paper. You may also want to clean up any stray lines or that sort of thing.

For the following example, I sketched the character with pencil and traced over the pencil sketch with a marker. Doing so resulted in thick, bold lines that came out well in the scan. I scanned the image into Photoshop and cleaned it up a little. I also broke apart some of the lines to create objects that could be auto traced easily. By breaking up the lines, I could also manipulate the objects more easily in Illustrator.

After saving the image from Photoshop as a TIFF file, I simply opened the file in Illustrator (see Figure 1.9).

Figure 1.9

A scanned sketch, opened in Illustrator.

Before starting the auto trace process, you may want to adjust the settings for the Auto Trace tool. To do so, choose File|Preferences|Type & Auto Tracing. Doing so will bring up the Type & Auto Tracing dialog box (see Figure 1.10).

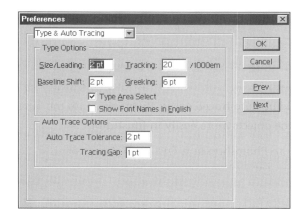

Figure 1.10
Illustrator's Type & Auto Tracing dialog box.

The setting that you'll be most concerned with is the Auto Trace Tolerance. The values you can enter range from 0 through 10 points. A value of 0 yields the most points, and a value of 10 yields the least. Figure 1.11 shows the leftmost strand of hair from the sketch traced twice. The object on the left was traced with the Auto Trace Tolerance set to 0 points, and the object on the right was traced with the tolerance set to 10 points.

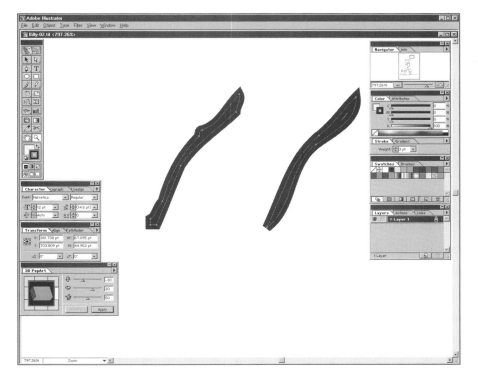

Figure 1.11
One object traced, using two different tolerance settings.

Notice that the object on the left has 9 points—just over twice as many as the 4 points that define the object on the right. On more complex objects, you may end up with a great number of points. Of course, more is not always better. To this end, you should choose a fairly low setting for the tolerance. Actually, the default of 2 works well most of the time.

The other setting, Tracing Gap, tells the Auto Trace tool to ignore gaps that are equal to or smaller than the value you set. Gaps can be introduced during drawing or scanning. Again, the default setting will be fine under most conditions. Figure 1.12 shows all the objects that make up the character's head, which was auto traced and moved to the side.

Figure 1.12
Several objects auto traced and moved to the side.

Note that these objects have been "vectorized" and can now be manipulated with any Illustrator tool, just as any other vector-graphic object might be.

Place

For a second method, placing an object into an Illustrator file enables you to create a template layer. By creating the layer as a template, you're telling Illustrator to lock and then dim the layer. This enables you to easily trace the image by using Illustrator's powerful drawing tools.

To place an image, simply choose File|Place. Doing so brings up the Place dialog box (see Figure 1.13).

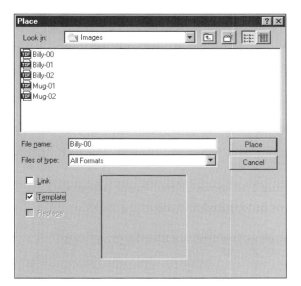

Figure 1.13
Illustrator's Place dialog box.

To place the file you're opening as a template, click on the Template option. Figure 1.14 shows the original scan placed into Illustrator as a template.

Notice how the image is dimmed to a gray. If you look at the layer in the Layers palette, you'll also notice that it's locked. This means you can't select or change anything on that particular layer. You can also trace over any object on that layer.

Figure 1.14
Original scanned artwork placed into Illustrator as a template layer.

You may notice that the layer that contains the dimmed and locked object is below another layer. You can select any drawing tools and draw on this layer, effectively tracing over the placed image.

Figure 1.15 shows the image, with most of it traced over with the Pencil tool.

Figure 1.15
The Placed image, mostly traced over manually with the Pencil tool.

You can also use this same method to create amazingly detailed illustrations of real-world objects. You might say, "If you can scan it, you can draw it." Figure 1.16 shows a scan of one of the calculators I own (I'm a bit of a nerd; I collect calculators).

With the scan in place, it was simple to trace the different objects, mostly lines, round-cornered squares or rectangles, and type, that make up the calculator. Figure 1.17 shows the final illustration of the calculator.

Re-creating the colors of the calculator was also simple. I only needed to sample the different colors from the actual scan by using the Eyedropper tool. If your real-world objects won't fit on your scanner, simply photograph them and use the scan of the photograph to capture an image that can be traced by using Illustrator.

Streamline

The third method requires another Adobe program. Streamline is a powerful program that enables you to trace a bitmap image. The results of

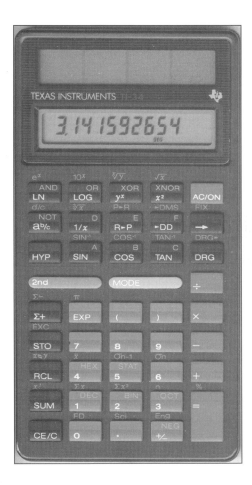

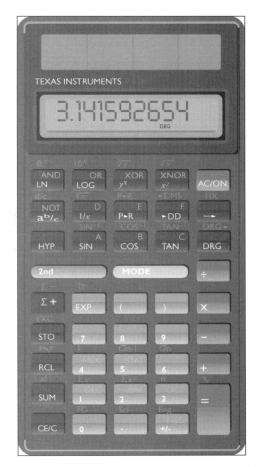

Figure 1.16 (left)

A "Placed" scan of a calculator.

Figure 1.17 (right)

The final illustration created by tracing the various shapes that make up the calculator.

the tracing can be saved as a vector image in Illustrator format (.AI). Streamline is quite powerful and has several useful options that enable you to trace all types of artwork.

Previously in this chapter, I showed an image (refer to Figure 1.6) that had been converted to a crosshatch pen-and-ink image. That image was originally a photograph scanned into Photoshop, where I lowered the number of colors by using the Artistic Cutout filter (see Figure 1.18).

I then opened the image in Streamline. Although it wasn't necessary to follow this process to create the pen-and-ink crosshatch image, having traced the photograph in Streamline meant that I had converted it from a bitmap into a vector image. Having done so meant that, once I opened the image in Illustrator, I could easily manipulate all the objects that make up the image.

In Streamline, you can control how an image will be traced. You can trace the outlines of different areas, you can trace through the center of lines, and so on.

Figure 1.18
A photograph opened in
Photoshop with the Artistic
Cutout filter applied.

To trace a drawing like the character sketch seen in previous figures, I set the following options:

1. Under Options|Settings, I chose Hand Drawn Sketch.

2. Under Options|Color/B&W Setup, I chose Black & White Only.

3. Finally, under Options|Conversion Setup, I chose the Outline and Centerline methods and set the Interactive Thinning to 15. This higher setting helped reduce the thicker lines drawn with the marker so that all (or at least most) of the lines would be the same width, and I would end up with lines rather than filled line-like shapes.

The final artwork, traced by Streamline, is shown in Figure 1.19. Figure 1.20 shows the artwork opened in Illustrator.

I've selected all the objects so that you can see how the conversion has created the lines.

Plenty of work remains to be done, though, depending on what the image is and what exactly you want from the finished image.

For example, many of the objects that I want to consist of closed paths are currently a collection of lines. I've selected the end points of each of the lines that make up the left side of the character's glasses and joined them all to create one object. In Figure 1.21, you can see the difference

Figure 1.19
Artwork traced in Adobe's Streamline.

Figure 1.20
Artwork scanned into Photoshop, converted to vectors in Streamline, and opened in Illustrator.

between the left side of the glasses (the one I've "fixed") and the right side. Figure 1.22 shows the work in progress.

I've closed some of the paths and colored in the face. I also changed the brushstroke to simulate a calligraphic look.

Of course, much more can be done. I can finish closing all the paths, and then I can choose and apply the colors I want. Using Streamline in

Figure 1.21
Lines joined to construct
a closed object.

Figure 1.22
The work in progress.

conjunction with Illustrator, however, has made it easy to get the original scanned sketch to the point where the drawing is currently.

A New Way To Draw (The New Pencil Tool)

Whether you're starting a drawing from scratch, tracing over a placed image, or simply editing an existing image, you'll be sure to appreciate the new Pencil tool. The Pencil tool is really three tools in one: the Pencil tool, the Smooth tool, and the Eraser tool.

The Pencil Tool

The Pencil tool enables you to draw freehand lines. To use the Pencil tool, employ the click-and-drag method. This process is the same used in previous versions of Illustrator. What's new, however, is that you can go back and reshape the line that you've previously drawn. Figure 1.23 shows two lines drawn with the Pencil tool.

The second line (the one on the right in Figure 1.23) was originally an exact copy of the first created via Edit|Copy and Edit|Paste. After pasting the line, however, I changed its path by clicking and dragging from one point on the line to another. Note how the line changes in the middle. I could keep going back and redrawing the line until I get the exact shape I'm after, and each time the line would be updated as I sketched.

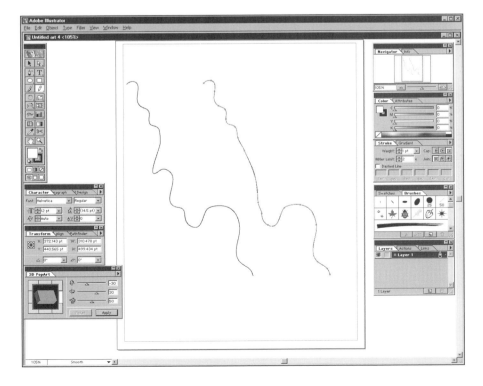

Figure 1.23

Two lines drawn with the Pencil tool.

Although this feature can be extremely powerful, it can also become quite annoying. The problem occurs when you draw one line and then add another line nearby. The second line, rather than being a new line, can change the first line. Argh!

This problem, however, is easily solved. To turn off the "sketching" feature, simply double-click on the Pencil tool's icon in the Toolbar to bring up the Pencil Tool Preferences dialog box (see Figure 1.24) and click on the Keep Selected option.

Doing so deselects any line you've just drawn, thus preventing it from being updated with the next stroke. To enable the feature again, simply click the option once more in the Pencil Tool Preferences dialog box.

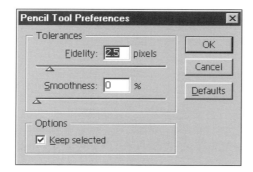

Figure 1.24
The Pencil Tool Preferences dialog box.

The Smooth Tool

The Smooth tool, available under the Pencil tool flyout menu, enables you to smooth the lines that you drew with the Pencil tool. To smooth a line after drawing it, simply click and drag the Smooth tool from one segment of the line to another. Doing so removes points and smoothes the curved segment between the two points where you start and finish the click-and-drag motion. Figures 1.25 and 1.26 show the result of using the Smooth tool on a freehand line.

I drew the nose of the character you see in Figure 1.25 by using the Pencil tool. I wasn't quite happy with the shape, however, so I used one stroke of the Smooth tool to get the shape that I wanted.

Before Illustrator 8, I had to remove points and reshape the line by using the Direct Selection tool. After using this new tool for a while yourself, I'm sure that you'll agree it's much easier than editing points and segments.

You can change the way in which the Smooth tool behaves. To do so, double-click on the tool's icon in the Toolbar to bring up the Smooth Tool Preferences dialog box (see Figure 1.27).

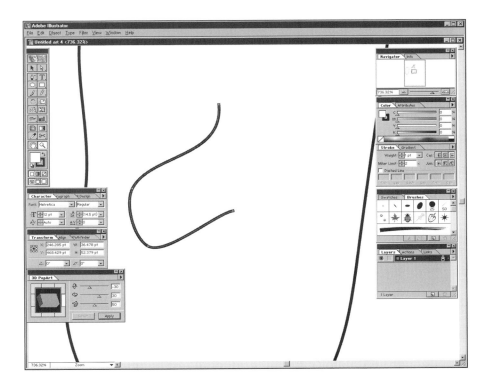

Figure 1.25
A freehand line drawn with the Pencil tool.

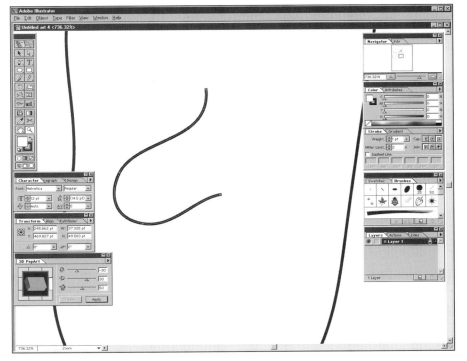

Figure 1.26
The same line as in Figure 1.25, but smoothed with the Smooth tool.

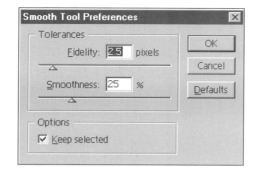

Figure 1.27
The Smooth Tool Preferences
dialog box.

You can adjust the fidelity and the smoothness. The Fidelity setting controls the distance that new curves will be drawn from the original curve. The lower the value you enter, the more angular the new curve becomes. The Smoothness setting controls, in percentages, how smooth or coarse the new curve will be. Higher values produce smoother curves.

The Erase Tool

The Erase tool (surprise, surprise) enables you to erase line segments. Using the Eraser tool is as easy as using the Pencil tool. Just click and drag over or near a segment, or segments, of a selected line, and the area you drag over will magically disappear.

Although the Erase tool is easy to use, you may find its operation a little peculiar. For example, you can't erase a portion of a line by crossing over a single segment. Instead, you must start and end on the same side of the line segment.

You also must draw (or erase, actually) quite near the segment. If you don't, and you just draw a big loop over the segment, you won't erase anything. It might take a little patience and practice, but this new tool is worth the effort.

Color Models And Illustrator

Color models are important in Illustrator, just as they are in any other digital-imaging program. You can choose from CMYK, grayscale, and RGB. The model you pick will most likely be decided on the basis of the final output you intend for your artwork. Artwork intended for print should be in CMYK or grayscale, and artwork intended for on-screen presentation should be in either RGB or grayscale.

Grayscale

The grayscale model simply means that your image will be represented by 256 or fewer levels of gray, ranging from black to white.

Note: When you erase a segment of a line (and if the line was open), the resulting line will be broken into separate pieces or objects. If the line was closed, the new line created by the Erase tool will be open.

If you want any color in your image, use one of the other models. However, you may choose the grayscale model and later add a spot color in Photoshop to simulate, for example, a sepia-tone image. Chances are, however, that you'll want to use one of the two color models for your artwork.

CMYK

CMYK (cyan, magenta, yellow, and percentage of black) is the model you choose if your artwork is intended for print. These four colors (actually, three colors and black) represent the colors of the inks used by printers and printing presses. Normally, if your work will be printed, you will separate the image into its color components and have color separations made for each color (one each of cyan, magenta, yellow, and black). These separations will then be used on the press to print your image.

All four colors in the CMYK model are expressed in percentages. These percentages relate to the percentages of ink of each color that will print when your artwork is processed through a press or color printer.

CMYK is the default color model for Illustrator because most illustrations created in Illustrator were traditionally intended to be printed. These numbers are now changing with the advent of file formats such as PDF (which you can view on screen) and with the growing popularity of graphics intended for the Web.

With these two areas growing daily, many illustrations are now intended solely for on-screen viewing and are better optimized in the native format of the CRT (cathode-ray tube, or computer screen), RGB.

RGB

With RGB (red, green, and blue), every color you see on your computer screen is made up of a combination of these three colors. When all three colors are displayed at full value (255), you see white. When all three colors are displayed at their lowest value (0), you see black. Each of the three colors in the RGB color mode represents the colors of the three electron guns that scan the coating of the back of your monitor's display. Because you can have 256 (0 through 255) values of each color, you have the possibility of using 256×256×256 (or 16.7 million) colors.

All computer images intended for on-screen viewing should be in RGB. If you plan to use your images on the Web, save them as either JPG or GIF files. Both of these file formats use the RGB model. I'll discuss these formats in greater detail in Chapter 12.

From One Model To The Other

You should be aware that converting back and forth from one color model to the other could result in degradation of your image (but not as much degradation when the image is in native Illustrator format). However, if you shuttle your image between programs such as Photoshop and Illustrator, you can lose some of the quality of the image. For example, some RGB colors don't map exactly to CMYK colors because the CMYK model is smaller than the RGB model. By smaller, I mean the CMYK model contains fewer colors than the RGB model. The number of colors available in any model is known as the *color space* (or more simply, *space*).

This can be important if you plan on using your images in both print and on screen. If you suddenly notice a change in colors between the printed version and the on-screen version of your artwork, chances are that (unless a major calibration problem exists between bits of your hardware—your scanner, printer, and monitor) you started with one model, and then somewhere along the line, you switched to the other model.

Deciding which model to choose may not be a simple matter. Today, people are designing brochures and then wanting to use the same artwork on the Web or in some other electronic format, such as PDF. With certain colors, such as the exact color of a corporate logo, this can pose quite a problem.

One helpful clue is that it's easier to go from CMYK to RGB than vice versa. If you've ever used a computer to design a logo in RGB and then switched the color model to CMYK to separate the image for printing, you know what I mean. Chances are good that some of your colors shifted noticeably.

Unfortunately, no simple solution is available. If you change the color mode of your image to CMYK in Illustrator and notice a big color shift, you will have to select the affected objects and choose new colors for them. Fortunately, this process is fairly easy in an object-oriented drawing program such as Illustrator.

Moving On

In this chapter, you read about the new additions to Illustrator 8, saw how powerful some of the new tools are, and saw a couple of the ways to get your artwork into Illustrator. Now, it's time to roll up your sleeves and begin the real work.

Sherry will walk you through the next few chapters, starting with Chapter 2, where she shows you how to use Illustrator in conjunction with Photoshop.

I'll be back in Chapter 6, where I'll show how Illustrator's new actions work. In that chapter, I'll show you how to use existing actions, how to create your own, and how to share them with others.

RASTER RAG

2

SHERRY LONDON

Adobe Illustrator is half of a perfect graphics program—Photoshop provides the other half.

What's This Raster Stuff?

Using Illustrator without Photoshop (or other pixel-based editing programs) is like buying basic cable without premium channels, buying a fast sports car to drive only in city traffic, or buying the introductory Lego construction kit without the Lego people. Illustrator alone can do wonderful things; so can Photoshop alone. Together, they exponentially expand your possibilities. You can place Illustrator images in Photoshop and Photoshop images in Illustrator and finally move both image types in PageMaker or QuarkXPress. In this chapter, you will look at a variety of techniques used to complete projects that require both raster and vector components. Because many of the projects later in this book need both Illustrator and Photoshop, you will get a chance to experiment with a number of different approaches here. (That way, I won't need to go into as much detail in the instructions afterward!)

Because this is an intermediate-level book, you should already know the difference between raster and vector images, but let's start here anyway. Adobe Illustrator is a vector-editing program. Therefore, the paths and shapes that you create are actually mathematical constructs. Rather than storing every point along a rectangle, for example, Illustrator stores only the coordinates of the upper-left and lower-right corners, along with the notation that the shape is a rectangle. Each path that you create is an individual object, and you can manipulate each path independently of the other. You can select any individual object from the place on screen where it lurks and know that Illustrator has retrieved every piece of it.

A raster program, however, doesn't know the difference between a drawing of the Mona Lisa and a picture of a garbage dump. A raster program understands only *pixels*—the little dots that form the smallest part of an image that a monitor can display. Adobe Photoshop, MetaCreations Painter, and Corel PhotoPaint are examples of raster image-editing programs. Although you can certainly make and move selections in a raster program, you need to identify to the program the group of pixels that it needs to move. The selection operation—far from being the point-and-click breeze it is in Illustrator—can represent hours of work and require an extraordinary amount of facility with the program.

You will sometimes hear Illustrator (and its relatives, Freehand and CorelDRAW) referred to as *object-oriented* programs. This is the same as saying that they are vector programs. Photoshop is a raster, or bit-mapped, program and uses pixels. Because Photoshop stores the color and location of every pixel in the image, it has no way of calculating which pixels belong together. In recent years, Adobe has tried to over-come that Photoshop limitation by allowing you to create layered images, where all the pixels on a layer can be treated as a group. Now you can place objects on their own layers, and, as long as they stay there, Photoshop knows how to select and change them for you.

All images are changed to raster images when they are printed. If you send images to a service bureau, you probably are familiar with the term *rip* (no, it doesn't mean "rest in peace"). This acronym stands for *raster image processor*—the machine that produces the film that is used to make printing plates. This machine reads the image files on your disk and changes any vector images to bitmapped (raster) images so that they can be printed. This process is similar to taking letter shapes and forcing them to fit into a regular grid. The rip decides where on the page to print each point from the processed shape. Because graphics professionals also have their jargon (why should they differ from academicians or government bureaucrats?), *rip* is also a verb (as in "I need to rip the file" or "There was an error when this image was ripped"). As a verb, rip is the process of sending an image to a raster image processor. This process is also known as *rasterizing.* It's what ATM (Adobe Type Manager) does continuously as it changes font outlines to smooth type on screen or allows you to print smooth type from a non-PostScript printer.

You can tell Illustrator to rasterize any vector shape that you have created and turn it into a pixel-based object. Illustrator then stores this collection of pixels as part of the file, along with the vector information. You can also rasterize entire Illustrator images to save them in Photoshop or an-other pixel-based format.

Illustration 2.1 shows the difference between a raster and a vector program.

PROJECT Illustration 2.1: Vector And Raster Graphics Differences

1. In Illustrator, open the image CELERATE.AI. Figure 2.1 shows the logo that I created in Illustrator for a fictional company.

2. Chose the Selection tool (V) and click on the logo to select it. (The logo is grouped so that one click selects the entire image. If your Area Select preference is set to Off, you need to click on the outline of the image to select it.)

3. With the logo still selected, double-click on the Scale tool to open its dialog box. Figure 2.2 shows the Scale dialog box.

Figure 2.1
The Celerate logo shown
at its original size.

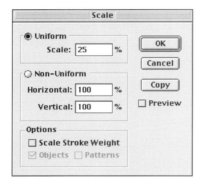

Figure 2.2
The Scale dialog box provides an
easy way to copy and scale an
object in one step.

4. Click on Uniform and enter "25" in the Scale box. Click on Copy to scale a copy rather than the original.

5. The copy is selected. Press Command (Mac) or Ctrl (Windows) to access the Selection tool and drag the copy away from the original. Figure 2.3 shows a screen capture of the original logo and its copy. Notice that the tiny logo is still smooth and intact.

6. Select the tiny logo again (if you deselected it). Double-click on the Scale tool. Change the Uniform Scale to 567% (yes, this is a deliberately wacky number). Click on Copy. Figure 2.4 shows the small copy in the top left of the figure and the enlargement copy below it. I hid the original logo when I captured the screen shot. Notice that the enlarged copy has suffered no image loss.

7. Close the image file. If you have enough RAM, launch Photoshop. If memory is tight, close Illustrator and then launch Photoshop.

8. Open the Photoshop Preferences (choose File|Preferences or Mac: Command+K; Windows: Ctrl+K). Figure 2.5 shows the Photoshop

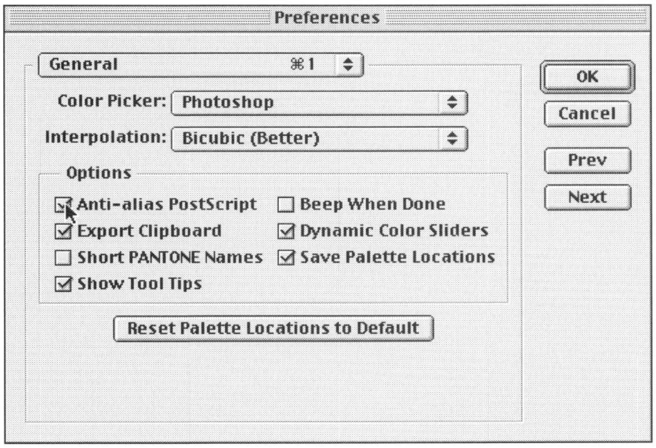

Figure 2.3 (top left)
Both the original logo and its tiny copy are nice and smooth.

Figure 2.4 (top right)
Enlarging a vector object doesn't degrade the quality of the image.

Figure 2.5
You can specify in Photoshop whether an imported Illustrator image should be anti-aliased.

General Preferences screen. Make sure that the Anti-alias PostScript checkbox is marked. Click on OK to exit the dialog box.

9. In Photoshop, create a new image (File|New) that is 500×500 pixels and CMYK color. Choose File|Place and select the CELERATE.AI image.

10. Double-click on the placed image to accept the size at which it appears. Figure 2.6 shows the image placed into Photoshop. Figure 2.7 shows an enlargement of the placed image. The lines are smooth and anti-aliased, but, as you can see in the enlargement, the anti-aliasing actually does not improve this design. Because the colors of the logo are black and pink, the anti-aliasing looks like a printer registration error where the paper color shows through.

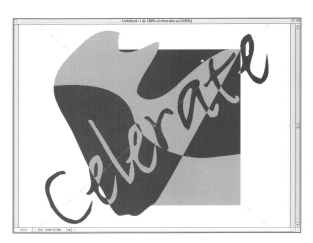

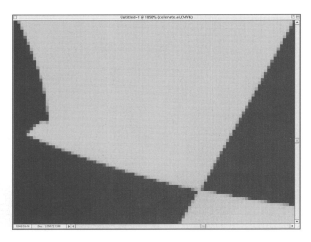

Figure 2.6 (left)

Photoshop places the Celerate logo into a box with an X across it so that you can resize or reposition the image before locking it down.

Figure 2.7 (right)

The placed image is anti-aliased, but this leaves a subtle and unintended line in the image.

11. Reduce the image size to 25% (Image|Image Size), as shown in Figure 2.8. Make sure that the Constrain Proportions and the Resample Image checkboxes are marked. Click on OK. When the image is resized, Photoshop throws away the pixels that it no longer needs. The logo is smooth but tiny—too small to print reasonably here. The type begins to get a bit too blurred for comfortable reading. Remember, at this size, the Illustrator version was very crisp and clear.

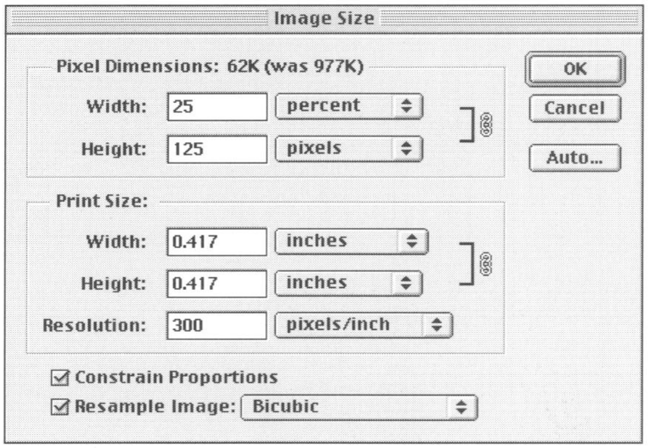

Figure 2.8

The Image Size dialog box allows you to resize an image.

12. Open the Image Size dialog box again and increase the image size by 567%. Figure 2.9 shows an enlargement of the result, which is not at all attractive. The anti-aliased edge has wandered all over the outlines of the image and is no longer smooth. It is now like an out-of-control snake. What happened? Photoshop didn't have enough information in the tiny file to "invent" the correct pixels needed to keep the image smooth.

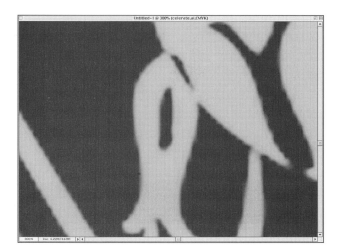

Figure 2.9
When you increase the image size in Photoshop, often there isn't enough detail to keep the image smooth.

13. If you think that the resized image looks bad, wait until you see the results if you do not have Photoshop anti-alias the placed image! Press Command+K (Mac) or Ctrl+K (Windows) to reopen the Preferences. Click to remove the checkmark next to the Anti-alias PostScript checkbox.

14. Repeat Steps 9 through 12. The only changes that need to be made in the instructions are in Steps 11 and 12. In the Image Size dialog box, select Nearest Neighbor as the Interpolation Method so that no anti-aliasing occurs at any point. Figure 2.10 shows a close-up of the result. Figure 2.11 shows a comparison of the original image (as reduced to 25% and enlarged to 567% in Illustrator) and the aliased and anti-aliased results of placing it into a Photoshop file. Although

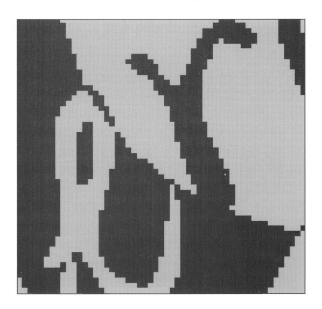

Figure 2.10
With Anti-alias turned off, the image is exceedingly jagged.

Figure 2.11
a) **(top left)** The anti-aliased image is too soft. b) **(top right)** The aliased image is too sharp. c) **(bottom)** The vector image is just right.

all three results might look acceptable here—at a size of three inches—you have seen the close-ups. We will discuss image resolution a bit later in this chapter.

So, what have you proved? One major reason for using a vector format is to create images that can be resized to any size (within reason—obviously, if you reduce an image to a small enough size, you lose the detail regardless of the file format). Vector images will also print at the resolution of the printer; raster files must specify the resolution in advance. Therefore, you get sharp print quality without needing to worry about the printing process itself. However, you lose the ability to work with continuous-tone (photographic) images. Although you can put a photograph into an Illustrator file, it still handles like a resolution-dependent pixel-based image.

Using Illustrator and Photoshop together allows you to mix the best of both types of graphics and to use the best format for the task at hand.

Illustrator-To-Photoshop Dynamics

Many of the Illustrations in this book have Photoshop components. You can move your Illustrator files into Photoshop in a number of ways. Before you consider why you might want to move files to Photoshop, let's spend some time exploring your various options. If you find that many Photoshop techniques are shown in this chapter, well, that's what this chapter is about. Don't worry. Enough Illustrator techniques appear in subsequent chapters to more than justify your purchase of this book.

Vector In: How?

There are five basic ways to get Illustrator images into Photoshop:

- You can use Photoshop's File|Open command.

- You can use Photoshop's File|Place command.

- You can use the Copy and Paste commands.

- You can drag images from Illustrator directly into a Photoshop file.

- You can save the image in Illustrator as a layered Photoshop file.

Open Sesame

Photoshop allows you to use either the File|Open or the File|Place command to import an Illustrator file. Opening the file (File|Open) brings in the Illustrator image just as if it were a normal Photoshop file. It is the most straightforward way to use an Illustrator file, and it does not require that any file be already open in Photoshop. Opening an Illustrator file allows you to specify the size of the image, the resolution, and whether it is to be anti-aliased. However, you cannot rotate the image or change its placement. In the next Illustration, you will open an image from the Racine's Costume Historique collection from Direct Imagination. You will then use this Japanese kimono print to fill a kimono shape.

Illustration 2.2: An Illustrator Image Needs Room To Rotate And Move

1. In Photoshop, select File|Open. Choose the image J-03.AI on the CD-ROM for this book. Figure 2.12 shows the dialog box that appears.

2. Make sure that the Constrain Proportions checkbox is marked. The Resolution should be set to 300 ppi and the Mode to CMYK Color. Set the Height to 900 pixels. The width is automatically calculated, as shown in Figure 2.13. Figure 2.14 shows the image that has been imported.

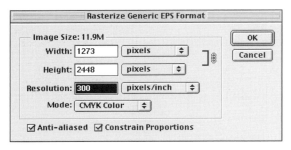

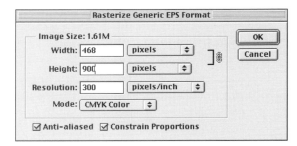

Figure 2.12 (left)

When you open an Illustrator file in Photoshop via the Open command, you have a number of options about how you want the image to appear.

Figure 2.13 (right)

Photoshop calculates the opposite dimension when you enter a width or height if the Constrain Proportions checkbox is marked.

Figure 2.14

J-03.AI is a lovely kimono pattern from Direct Imagination's Racine's Costume Historique.

3. Open the image KIMONOMASK.PSD. Choose the Move tool (V) and drag the kimono shape into the J-03.AI file. Center it as well as you can—it's too large to fit into the image.

4. Select the Free Transform command (Mac: Command+T; Windows: Ctrl+T). Press and hold Shift+Option (Mac) or Shift+Alt (Windows) and drag the upper-left bounding box corner toward the center. Coax the kimono to fit into the image (it might take a few tries). Figure 2.15 shows the kimono shape on top of the original image.

5. In the Layers palette, drag the kimono layer below that of the original image. Make the top layer (now Layer 1) active by clicking on it in the Layers palette. Group the layers (Mac: Command+G; Windows: Ctrl+G). Figure 2.16 shows the Layers palette.

6. Choose the Rectangular Marquee tool. Drag a marquee around the kimono so that it encloses the entire image and leaves a bit of extra room. Crop the image (Image|Crop). Figure 2.17 shows the final effect.

Because you used the Open command to bring the Illustrator image into Photoshop, you had no flexibility to move the fabric around. Because the fabric was rectangular, that is probably not a major flaw. However, let's continue Illustration 2.2 and build a much more interesting kimono (maneuvering room lets you try so many more things). You will again use the Open command, but this time you will composite the kimono inside a second image. This shows you another way to use an imported Illustrator image.

7. Close all the open files. Open the Preferences (Mac: Command+K, Command+5; Windows: Ctrl+K, Ctrl+5) and change the Units to Pixels.

8. Open the file KIMONO.PSD. It looks as if it is empty—the image is totally blank. There are three channels in the image, however, that enable you to reconstruct a kimono.

9. Click on the Body channel to make it active (Mac: Command+6; Windows: Ctrl+6).

10. Using the new Measure tool in Photoshop 5, drag it across the widest part of the kimono body to read the measurement in pixels (look at the Info palette to see the amount). Now you know the needed width for the fabric file (approximately 580 pixels).

11. Load the Body channel (Mac: Command+click on the channel name or press Command+Option+6; Windows: Ctrl+click on the channel name or press Ctrl+Alt+6). Make the Composite channel active (Mac: Command+~; Windows: Ctrl+~).

12. Make a new layer (click on the New Layer icon at the bottom of the Layers palette). Press D to set the colors back to the default of black and white. Fill the selection (Mac: Option+Delete; Windows: Alt+Delete). Deselect (Mac: Command+D; Windows: Ctrl+D).

13. Open the file J-03.AI. In the dialog box, enter your measurement amount (580 pixels) for the width. Make sure that the Constrain Proportions checkbox is marked.

14. Select the Move tool (or press the Command key on the Mac or the Ctrl key in Windows). Drag the fabric from the J-03.AI file into the KIMONO.PSD file. Place it over the body of the kimono. (Let it stay in its own layer.) Group the two layers (Mac: Command+G; Windows: Ctrl+G). Move the fabric so that it covers the entire body shape and so that there are flowers near the top right of the form. Figure 2.18 shows the result.

Figure 2.15
The kimono shape can be resized to fit on top of the rasterized fabric.

Figure 2.16
The Layers palette shows that the fabric (Layer 1) is on top and that it is grouped with the kimono shape (Layer 2).

Figure 2.17
The kimono fabric fills the shape when the images are grouped properly.

Figure 2.18

The fabric easily covers the kimono body.

The fabric for the right arm needs to make a perfect 90-degree miter. This is a bit tricky to accomplish. Here's how:

15. First, make a base for the clipping group. Make a new layer (click on the New Layer icon at the bottom of the Layers palette). Load the Right Sleeve channel (Mac: Command+click on the channel name or press Command+Option+7; Windows: Ctrl+click on the channel name or press Ctrl+Alt+7). Fill the selection with black (Mac: Option+Delete; Windows: Alt+Delete). Deselect (Mac: Command+D; Windows: Ctrl+D).

16. Drag the kimono fabric layer (Layer 2) to the New Layer icon at the bottom of the Layers palette to copy it. Move the thumbnail for the copy to the top of the Layers palette. Figure 2.19 shows the Layers palette at this point.

17. To miter something, you need to flip and then rotate it. Choose Edit|Transform|Flip Horizontal. Using the Move tool with the Shift key pressed to constrain movement to the horizontal, move the fabric in the Layer 2 copy to the right. Press the numeric 5 key to set the transparency to 50% so that you can see the fabric on the kimono body. Keep the Shift key pressed and move the fabric so that the flower on the sleeve layer is just to the right of the matching flower on the kimono body. Figure 2.20 shows this view of the image.

18. Choose the Free Transform command (Edit|Free Transform or Mac: Command+T; Windows: Ctrl+T). Drag the center of the rotation marker to the underarm seam (where the body and right sleeve meet). Figure 2.21 shows the repositioned center of rotation marker (but all the layers have been made lighter for the screen capture so that you can see the marker more clearly).

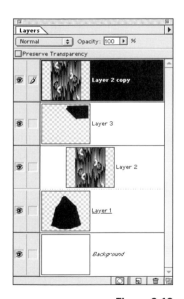

Figure 2.19

The Layers palette shows the grouped kimono body and the right sleeve ready to be mitered and grouped.

19. Place your cursor outside the bounding box. You will see a double-headed curved arrow cursor; this is the Rotation cursor. Press Shift and drag downward to rotate the image counterclockwise. It should take six "tugs" on the image to rotate the sleeve 90 degrees. (Shift constrains the rotation to 15-degree increments). When the fabric is in the right spot, press Return/Enter to set the transformation.

20. Group the two layers (Mac: Command+G; Windows: Ctrl+G). Press the numeric 0 key to return full opacity to the fabric. Move the fabric as needed (with the Move tool) so that you create a perfect matching miter. Figure 2.22 shows the kimono with the miter completed.

Figure 2.20 (left)
You can line up the flipped flower if you change the layer opacity to 50%.

Figure 2.21 (right)
You can change the center of the rotation or transformation by dragging its marker to a new location.

Figure 2.22
The kimono with a perfectly mitered sleeve.

21. Make a new layer (click on the New Layer icon at the bottom of the Layers palette). Load the Left Sleeve channel (Mac: Command+click on the channel name or press Command+Option+5; Windows: Ctrl+click on the channel name or press Ctrl+Alt+5). Fill the selection with black (Mac: Option+Delete; Windows: Alt+Delete). Deselect (Mac: Command+D; Windows: Ctrl+D).

22. Drag the fabric from the J-03.AI image into the KIMONO.PSD image again. Flip it horizontally (Edit|Transform|Flip Horizontal). Rotate the fabric –45° (Edit|Transform|Numeric; don't forget to enter the minus sign in front of the 45°). Group the two layers (Mac: Command+G; Windows: Ctrl+G). Move the fabric until the double flower pattern sits near the wrist on the left sleeve as shown in Figure 2.23.

Figure 2.23
The completed basic kimono shows both sleeves with a slightly different fabric treatment.

We could consider the kimono image complete (after all, you have now seen one reason why you might want to be able to move an Illustrator image around—which was the purpose of this section). However, you can make the effect more artistic, so let's play with this a bit more.

23. Let's add a trim for the garment. Click on the Foreground Color Swatch in the Toolbox and choose a true red (CMYK: 0, 100, 100, 0). Make a new layer (click on the New Layer icon at the bottom of the Layers palette). Load the Left Sleeve channel (Mac: Command+ Option+5; Windows: Ctrl+Alt+5). Stroke the selection (Edit|Stroke|8 pixels, Center). Deselect (Mac: Command+D; Windows: Ctrl+D).

24. Make the Channels palette active. Load the Body channel (Mac: Command+Option+6; Windows: Ctrl+Alt+6). Add the Right Sleeve channel to the selection (Mac: Shift+Command+click on channel 7; Windows: Shift+Ctrl+click on channel 7). The Shift key adds the

channel to the current selection. Because the two selections overlap, there is no line where the right sleeve and the body meet—which is precisely what we want. Stroke the selection (Edit|Stroke|8 pixels, Center). Deselect (Mac: Command+D; Windows: Ctrl+D). Figure 2.24 shows the stroked trim on the kimono (I have hidden the kimono itself for this screen shot).

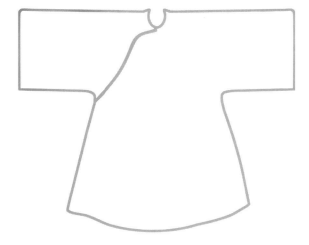

Figure 2.24

You can create trim by stroking the edges of the kimono components.

25. Let's add a bit of pattern to the trim; choose the Magic Wand tool (W). Press Enter/Return to show the Magic Wand options. Set the Tolerance to 15 and Anti-aliased to On. Do not mark the Use All Layers checkbox.

26. Make the Layer 2 copy layer (the fabric from the right sleeve) active. Click inside the upside-down yellow flower to select it. Press the Shift key and click inside the blue triple arches above it (this adds the blue thing to the selection). Copy it to the Clipboard (Mac: Command+C; Windows: Ctrl+C).

27. Create a new document (File|New). Accept the defaults—they will be the size of the copied selection. Paste in the selection from the Clipboard (Mac: Command+V; Windows: Ctrl+V). Turn off the Eye on the Background layer. Turning off the Background layer allows the final pattern to contain transparency so that it doesn't obliterate all the red outline trim.

28. Reduce the flower image to 35% of its original size (Image|Image Size, 35 Percent, Bicubic Interpolation). Select the entire image (Mac: Command+A; Windows: Ctrl+A). Define this as a pattern (Edit| Define Pattern).

29. Fill the trim layer with the pattern (Shift+Delete|Use: Pattern, 100 Percent Opacity, Normal, Preserve Transparency: On). When you fill with the Preserve Transparency checkbox marked, only the outline of the trim will "catch" any of the pattern. Figure 2.25 shows the trim that has had the pattern added to it.

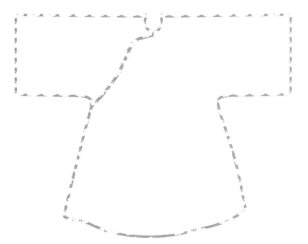

Figure 2.25

You can apply a pattern to only the trim by checking the Preserve Transparency box in the Fill dialog box.

Much more can be done to dress up this image, but there is much more to cover in this chapter, and it is time to move on. Figure 2.26 shows the final version (look at the color version as well). You can find the completed image on the CD-ROM for this book.

Figure 2.26

The finished kimono image has a background pattern and an added final border.

Here is a brief description of what I did. I added a drop shadow for both the garment and the upper-right neck edge. I created a mosaic (four-way) pattern from the double flower in the fabric and filled a layer with it. I used a Hue/Saturation Adjustment layer to turn it a light blue (and reduced the opacity of the layer as well). I increased the Canvas size to give me room to stroke the edge of the background pattern (on a new layer) and apply the same pattern to the stroke that I used as the garment trim. You will have an opportunity to try all these techniques in the course of the various Illustrations in this book (so do not worry if you are not sure how any of these were done).

When you open an Illustrator image in Photoshop, you also risk cutting off some of the image detail if you anti-alias the image. Unfortunately, this problem exists whether you open or place the file, so you need to be aware of the consequences of your decisions. This is an easy problem to resolve, once you recognize what is happening. Illustration 2.3 will make it clear.

PROJECT Illustration 2.3: An Illustrator Image Needs Room To Grow

1. In Photoshop, open the image BUTTERFLY.AI (File|Open). This is an image from the Ultimate Symbol's Nature Icons series that I have colored. Open the file at 600 pixels wide. Make sure that the Constrain Proportions checkbox is selected. Do not mark the Anti-aliased checkbox.

2. Use the Canvas Size command (Image|Image Size) to add 100 pixels to each dimension as shown in Figure 2.27.

3. Magnify the image until you can see the edge of the butterfly's right wing, where the arrow appears in Figure 2.28. Notice that this portion of the wing is quite straight and almost looks cut off. Because the

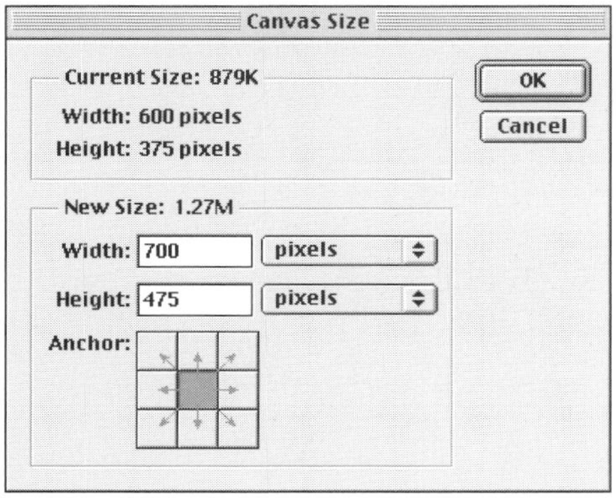

Figure 2.27

The Canvas Size command allows you to add more working space to your image.

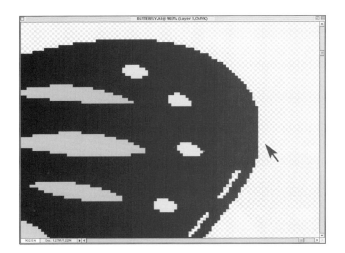

Figure 2.28
The arrow shows a flat area where the butterfly's wing butted the original image border.

image is not anti-aliased, it does not look that out of place. This was the area where the wing touched the edge of the original image.

4. Close the BUTTERFLY.AI image. Open it again (File|Open). This time, open the image with the Anti-aliased checkbox marked. Use the same dimensions as before. Repeat Steps 2 and 3. Figure 2.29 shows the close-up of the butterfly's wing. Notice that the same area on the wing is still straight, but that the rest of the image is anti-aliased.

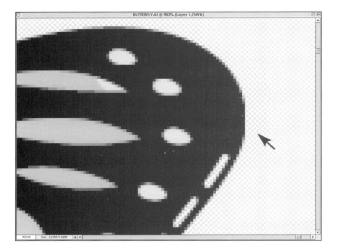

Figure 2.29
Close-up of butterfly wing, with anti-aliasing turned on; same area of wing is straight.

5. Repeat Steps 1 through 4, using the image BUTTERFLY2.AI. Figure 2.30 shows the jagged version, and Figure 2.31 shows the anti-aliased version. The jagged version still contains the same straight edge, but the anti-aliased version has a smooth edge on the wing.

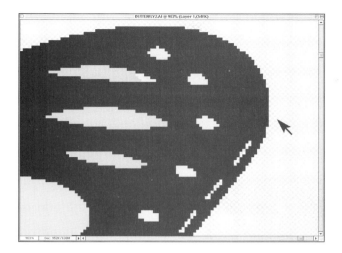

Figure 2.30

The aliased version of BUTTERFLY2.AI shows the same straight edge as BUTTERFLY.AI.

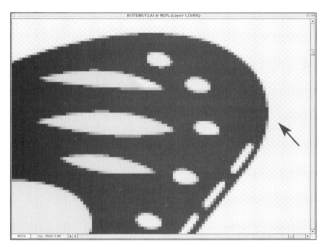

Figure 2.31

The anti-aliased version of BUTTERFLY2.AI sports a smooth wing.

BUTTERFLY.AI and BUTTERFLY2.AI seem to be the same image. What has changed? You might have noticed that the two images do not contain the same number of suggested pixels, even though you rasterized both at a width of 600 pixels. Switch to Adobe Illustrator, and we'll look at the two images.

6. In Illustrator, open the file BUTTERFLY.AI. Choose the Selection tool (V). Click on the butterfly. Notice that the box that surrounds the grouped objects hugs the boundary of the butterfly. Figure 2.32 shows BUTTERFLY.AI with all objects selected.

7. Still in Illustrator, open the file BUTTERFLY2.AI. Click on the butterfly with the Selection tool. As you can see in Figure 2.33, the rectangular, outside box is well away from the image itself.

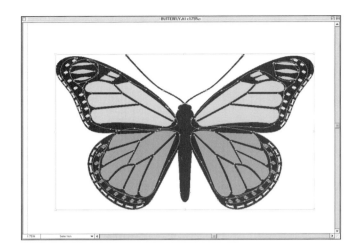

Figure 2.32
BUTTERFLY.AI's bounding box is directly up against the butterfly.

Figure 2.33
The bounding box of BUTTERFLY2.AI does not touch the butterfly itself.

The box that defines the perimeter of a shape is known as the *bounding box*. This is a PostScript notation that defines the height and width of the object. The bounding box of an object is the smallest rectangle that can contain the shape. A problem occurs when you either move the object into Photoshop or rasterize it in Illustrator. If you rasterize and anti-alias an object with a bounding box that touches it, the object parts that touch the bounding box cannot be anti-aliased because no room is left to do so—the bounding box is in the way. It's like trying to produce a Gaussian Blur in Photoshop with a selection marquee in the way—the selection cuts off part of the blurred edge.

If you group objects, the bounding box is the one that is farthest out. When the image is Placed into Photoshop, it is given the dimensions of the smallest rectangle that can hold all the objects in the file.

To restate the problem, if you bring an Illustrator file into Photoshop, and the bounding box of the image touches one or more of the objects, those objects will not be anti-aliased where they touch the bounding box. The obvious solution is to make sure that the bounding box of the file itself does not touch any of the objects. But how?

You only need to draw a new rectangular object that is the size of the area that you want to import. Give this object the attributes of no stroke and no fill; therefore, it is a transparent object. However, because PostScript images all objects without regard to their fill or stroke, this "dummy" rectangle is used as the real bounding box of the image and gives the image its new border. Try it with the very simple example shown in Illustration 2.4.

PROJECT Illustration 2.4: A New Bounding Box

1. Create a new file in Illustrator (File|New).

2. Select the Ellipse tool (L). Draw a circle with the tool anywhere in the image.

3. Click on the Fill icon in the Toolbox to make it active. With the circle selected, choose a Fill color from the Swatches palette.

4. Save the file as in Illustrator format (File|Save).

5. Open the file in Photoshop as an anti-aliased image using the Open command. Using the Move tool, move the circle and zoom in until you verify that the edges of the circle that touched the image borders are not anti-aliased. Close the image and do not save it.

6. Back in Illustrator, open the circle file (if you closed it). Click on the Rectangle tool (R). Draw a rectangle on top of the circle so that it more than covers it. Set the Stroke and Fill for the rectangle to None. Save the file again.

7. Open the file in Photoshop as you did before. This time, you should see an anti-aliased edge all the way around.

Placing Projects

The Place command in Photoshop is very similar to the Open command in that it displays the Illustrator image for you to edit. However, it has two significant differences: You must already have a file open when you invoke the Place command, and you can interactively resize, rotate, skew, or distort the image to be rasterized. The disadvantage is that you cannot specify on the spot whether to anti-alias the Illustrator artwork, nor can you specify the precise measurements to use.

So, which command should you use? It really depends on what you want to accomplish—which is why you are given the option to either open or place the file. Try this short example in Illustration 2.5 to decorate a tree branch with a Japanese maple leaf and a butterfly from Ultimate Symbol's Nature Icons collection.

PROJECT Illustration 2.5: Placing Images

1. In Photoshop, open the image BRANCH.PSD.

2. Open the Preferences file (Mac: Command+K; Windows: Ctrl+K). Deselect the Anti-alias PostScript checkbox. Close the Preferences dialog box.

3. Choose File|Place. Select the MAPLE.AI file. Figure 2.34 shows the image as it is being placed.

Photoshop automatically shows the placed image at the largest size that will fit into the host image. You see a bounding box with an X through it. The box has eight control handles—one at each corner and one at the center of each side. You use these handles to resize the image, just as you would in any other graphics program. If you move a corner handle, you can resize in both directions. Dragging a center handle resizes either horizontally or vertically. If you place the cursor outside the bounding box, it changes to a double-pointed curved arrow—the Rotate cursor. You can modify this behavior with several different keystrokes.

Figure 2.34

Placing the maple leaf image into the Branch file.

During a normal Photoshop transformation, you can also press the Shift, Command/Ctrl, or Option/Alt keys together to distort your image. However, neither the Distort nor the Perspective commands are available while you are placing an image.

4. Place your cursor on the lower-right handle of the bounding box. Keep the Shift key pressed and drag the cursor up and to the left to make the maple leaf smaller. When you are satisfied with its size, place your cursor inside the leaf (although not on the center point marker) and drag the leaf into the approximate place where you want it to go. Place your cursor outside the bounding box and rotate the leaf to the desired angle. Move it into position where you want it to actually appear in the image and press Enter/Return. Figure 2.35 shows the leaf just before the transformation is accepted.

5. Open the Preferences file (Mac: Command+K; Windows: Ctrl+K). Select the Anti-alias PostScript checkbox.

6. Place two more copies of MAPLE.AI (one at a time). Experiment with the Skew command by keeping the Command (Mac) or Ctrl (Windows) key pressed as you move the control handles. Also, see

Table 2.1 Some helpful keystrokes that you can use in transformations.

Keystroke	Result
Shift	Constrains the angle of rotations and skews; keeps resizing proportional
Command (Mac), Ctrl (Windows)	Allows you to skew images
Option (Mac), Alt (Windows)	Makes transformation happen from center point

what happens when you drag from the handle at the bottom center of the bounding box. When you are finished, place a copy of BUTTERFLY2.PSD into the image. Figure 2.36 shows my version of this exercise. (I also applied a yellow glow using Layer|Effects| Outside Glow to each of the placed objects. An easy way to apply the Layer Effect most efficiently is to link all the object layers by clicking on the second column in the Layers palette next to each affected layer, create the glow on one layer, and choose Layer|Effects|Copy Effect. Then select the command Layer|Effects|Paste Effects To Linked. All four object layers instantly sport the same glow.)

Why should you bother to correctly size and position the placed image when you could do it after you have placed the image? The simple answer is that it gives you the best possible results. Every time you resize, rotate, or otherwise mangle a raster image, you lose data, and the anti-aliasing that

Figure 2.35
The maple leaf is in position and ready to be rasterized into the image.

Figure 2.36
Using the Place command, three maple leaves and one butterfly are included in the BRANCH.PSD image.

HOW TO SAY "I ACCEPT"

You can finalize the Place command in several ways. Adobe prefers that you use the "newer" version of just pressing Enter/Return, but you can also double-click inside the bounding box as you did in early (preversion 4) incarnations of the program. In addition, if you press the right mouse button on Windows or the Control key on the Mac, you bring up a context-sensitive Help menu. One of the options on this menu is Place, which will also rasterize the placed image into your file.

occurs causes your image to lose sharpness. If you know the size and location that you want, you are much better off fiddling *before* you finalize the Place command. Try the brief (I promise) example in Illustration 2.6.

PROJECT Illustration 2.6: On The Benefits Of Scaling And Rotating When You Place

1. Launch Photoshop. In Photoshop, create a new document (File|New, Width: 500 pixels, Height: 500 pixels, Resolution: 300, Mode: CMYK Color, Contents: White).

2. Place the image LADYBUG.AI (File|Place). Do not finalize the placement yet.

3. Pull down the Edit menu (one of the few menus that aren't grayed out during this operation). Choose Transform|Numeric Transform. Deselect the Position and Skew checkboxes. Mark the Constrain Proportions checkbox under the Scale section and scale the image to 25%. Change the Angle Of Rotation to 62°. Click on OK. Figure 2.37 shows you this dialog box. Figure 2.38 shows a close-up of one of the ladybugs.

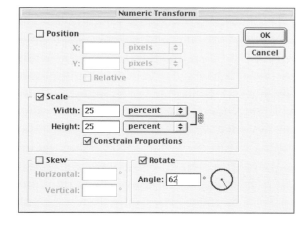

Figure 2.37
You can use the Numeric Transform dialog box when you Place an image into Photoshop.

4. Create a new document the same size and resolution as the one you created in Step 1.

5. Place the file LADYBUG.AI. This time, when you see the bounding box, simply press Enter/Return to set the Place command.

6. Choose Edit|Transform|Numeric Transform. It should contain the last transformation that you did. Change the Scale percentage to 74 (don't ask me why this gets you approximately the same number of pixels as you have in the first image—I had to ask my husband for this calculation—I just know that it works). Leave the Rotate angle

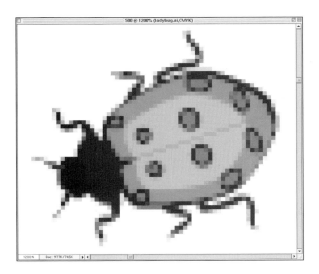

Figure 2.38
Notice that the ladybug is crisp and clear.

at 62%. Press Enter/Return to execute the transformation. Figure 2.39 shows a close-up of the same ladybug, which looks a bit worse for wear. Check the line across her back if you want to find one of the most clearly defined areas of damage.

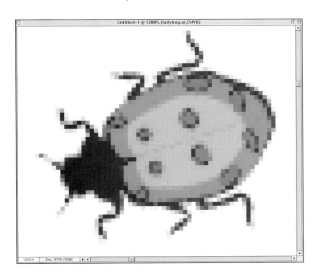

Figure 2.39
A too-blurry ladybug shows that it was placed first and rotated later.

The moral of this tale? Rotate your image to its final resting place when you first place it. Should you not discover what this number is until later, delete the layer and place the image again at the correct angle.

Be A Copy-And-Paste Cat

You can also copy and paste images between Illustrator and Photoshop. This is a fairly simply process, although it uses Clipboard memory and overwrites whatever was previously on the Clipboard. One advantage

CREATING A FILE THE SAME SIZE AS AN OPEN ONE

If you try to create a new image and the dialog box does not show the same size as the one that you last created, it means that you have data on the Clipboard. You can simply pull down the Window menu and select the name of the file whose attributes you want to duplicate. Photoshop creates a new image of the same size, resolution, and color mode as the image that you select.

Figure 2.40
You are given the choice of pasting pixels or paths when you transfer an object from Illustrator to Photoshop via the Clipboard.

that this method has over Open or Place is that you can copy a single path or element without having to bring over the entire file. You are given the option to paste the path itself into Photoshop or to rasterize the object that the path defines. You may also decide whether to anti-alias the pasted path if you opt to rasterize it. The object pastes at the resolution of the Photoshop file that you have created but at the size that was defined in Illustrator. You have no chance to scale or transform it.

If you want to try this method, follow along with the brief Illustration 2.7.

PROJECT Illustration 2.7: Copy And Paste

1. In Illustrator, create a new document.

2. Use the Polygon tool to create a filled hexagon (or any other shape that you prefer). Fill the polygon with a solid color.

3. Select the polygon with the Selection tool and copy it to the Clipboard (Mac: Command+C; Windows: Ctrl+C).

4. In Photoshop, create a new document (File|New, Width: 500 pixels, Height: 500 pixels, Resolution: 300, Mode: CMYK Color, Contents: White).

5. Paste in the selection from the Clipboard (Mac: Command+V; Windows: Ctrl+V). The dialog box shown in Figure 2.40 appears. Choose Paste As Pixels and Anti-Alias. Click on OK.

6. Select the Paste command again. This time, paste the object as Paths. We'll finish off this Illustration by stroking the path with the Paintbrush.

7. In the Paths palette, double-click the work path and save it (just accept the suggested name).

8. Choose a bright color as your foreground color. Click on the Paintbrush tool in the Toolbox and select a 65-pixel brush in the Brushes palette.

9. Click on the Stroke path icon at the bottom of the Paths palette (the second icon from the left). You see a bright, soft line around your image.

It's A Drag

You can drag a selection from Illustrator into Photoshop as long as you have a file open in Photoshop and you can see both the Illustrator and the Photoshop files at the same time. The drag-and-drop operation does not use Clipboard memory and does not replace the image on the Clipboard.

Figure 2.41 shows an object being dragged from Illustrator to Photoshop. Notice how a black line forms along the border of the Photoshop image when the image is ready to receive the object.

When you drag and drop, the dropped object will always be anti-aliased, regardless of the status of the Anti-alias PostScript toggle in the Preferences. If you want to paste paths, press Command (Mac) or Ctrl (Windows) as you drag and drop.

The Layered Look

Another method of bringing Illustrator images into Photoshop exists and it's new to version 8. Finally, you can save Illustrator layers as separate layers in a Photoshop document. This is an incredible time-saver and convenience, especially if you take the time to organize your Illustrator document into layers. This gives you the ability to move things around in Photoshop or to apply filters and effects on an object-by-object basis.

> **WARNING: POTENTIAL PASTING PROBLEMS**
>
> If you do not update Photoshop's Parser plug-ins with Illustrator 8 compatible ones, you might get a PostScript error if you try to copy and paste an object with a gradient fill. If you try to paste (or Open, Place, or anything else) an object that uses a brushstroke or a gradient mesh, you will get very odd results.

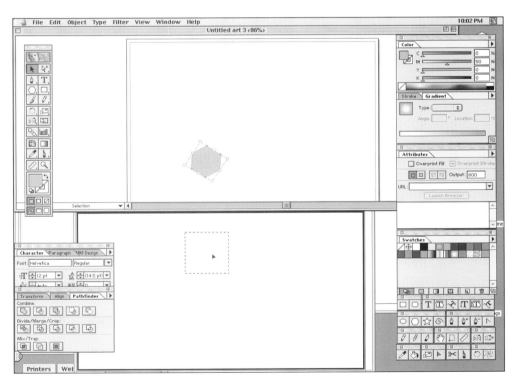

Figure 2.41
You can drag an object from Illustrator into Photoshop.

Vector In: Why?

We've spent a lot of time just getting to the point of discussing why you might want to take Illustrator images into Photoshop. The advantage of Illustrator (or other vector) images is that they are sharp and crisp. They also have the advantage of a small file size. Both of these advantages are lost when you move the vector images into Photoshop. However, Illustrator is very good at doing some things that are difficult or impossible in Photoshop:

- Illustrator has many more options for creating and designing text.

- You can much more easily manipulate paths in Illustrator than in Photoshop.

- If you need to create an image with a lot of pieces engineered from basic shapes, it is much faster to do so in Illustrator.

Many artists create all their basic shapes in Illustrator and then move the composition into Photoshop. Because we know that they lose the resolution independence and small file size when they bring images into Photoshop, what benefits are gained from the transfer?

- You can apply transparency and blend modes to the images.

- You can more efficiently apply filters to the imported images (we'll tackle this point again later in the chapter).

- You can use Layer Effects to add glows, bevels, embosses, and drop shadows that are blurred and transparent (although later in the book, you will learn how to produce Illustrator versions of these effects).

Ultimately, you need to decide on a case-by-case basis which program is best for the job at hand. You might want to take the image back into Illustrator when you are done to either link or embed it, or you might want to leave some vector shapes in Illustrator and composite the pieces from the two programs in PageMaker or QuarkXPress.

I strongly urge you to create your text in Illustrator, save it as outlines, and bring it into Photoshop when you need to work with text in Photoshop. Although the text handling in Photoshop 5 is much better than in the past, Illustrator allows you to have text on a path, which you cannot do in Photoshop, and in general, you will probably find the experience of creating text to be more pleasant in Illustrator. If you need to have text at a small point size (10 to 14 points) in a Photoshop document, don't use Photoshop for the text. In this situation, I strongly advise you either to use Illustrator or to enter the text directly in your page-layout program on top of the placed Photoshop image. When the edges of black text anti-alias to a white background, the resulting text can look soft and sloppy. If you need black text on a white background, create it in the page-layout program. If you need decorative type or are going to fill the type with a photo (which you can also do in Illustrator), it is reasonable to move the text into Photoshop.

The following Illustration 2.8, which is fairly long, creates a piece of advertising art that shows many of the uses of Photoshop for images begun in Illustrator. Although you are not starting this image from the beginning (as I did when I created it), you will get a chance to field test many reasons to exploit the Photoshop/Illustrator combination.

Illustration 2.8: The Hard-Wrap Candy Case

Creating The Basic Image

To create the basic image, take these steps:

1. In Illustrator, open the image CANDY.AI. Figure 2.42 shows the image in Illustrator format.

Figure 2.42
The original CANDY.AI image.

2. Your first task is to prepare this file to be saved in a layered Photoshop format. If you look at the Layers palette in Illustrator, you will see that, currently, the file contains only one layer. Each group or element—and there are nine of them—needs to be on its own layer. With the Selection tool, click on the solid text. Both the solid

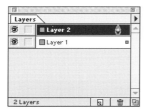

Figure 2.43

The Layers palette shows the status of the layers that are present in your document.

and the open text are selected (that's because I already created the appropriate groups for you).

3. Click on the New Layer icon at the bottom of the Layers palette. Figure 2.43 shows the Layers palette after the new layer is created. Notice that the Pen icon on the top layer and the highlighting on the top layer show that this layer is currently active.

4. You need to move your selection from the bottom layer to the top layer. Notice the tiny dot to the right of the layer entry on the bottom layer. This indicates an active selection. Drag the dot to the top layer. The bounding box around the selection changes color to match the color swatch shown on the left of the text entry of the top layer. (It was originally blue and then turns red, at least on my Mac. On a PC, it may appear slightly different.)

5. Double-click on the layer name and change it by entering the word "TEXT", as shown in Figure 2.44.

Figure 2.44

You can rename layers to make it easier to remember what is on each one.

6. Repeat Steps 2 through 5 for each piece of candy and for the scribble. When you have finished, double-click on Layer 1 and change its name to White Rectangle (yes, there is another shape there, and it will show up in Photoshop eventually). A bounding box "control" object (unfilled and unstroked) also lurks on that layer. Figure 2.45 shows my Layers palette with all the layers named.

7. Choose File|Export and locate the directory into which you want to save the new file. Name it CANDY.PSD and select the Photoshop 5 file format from the drop-down list. Figure 2.46 shows the Export dialog box.

8. When you click on OK, you will see another dialog box, shown in Figure 2.47. The Photoshop Options dialog box allows you to select the resolution and color mode (choose CMYK). You can also determine whether you want to save the layers (you do) and whether to anti-alias (set Anti-alias to On). Select the Other resolution and enter "125" dpi into the box as shown. The image is

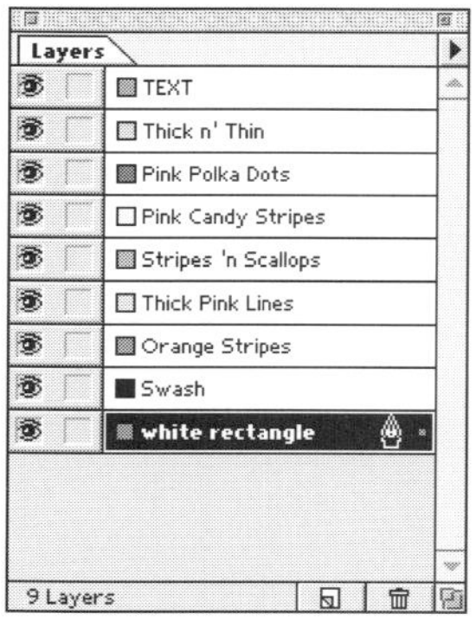

Figure 2.45

CANDY.AI now contains nine layers.

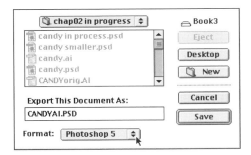
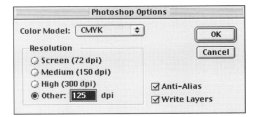

Figure 2.46 (left)
The Export dialog box allows you to choose the raster file format to use.

Figure 2.47 (right)
The Photoshop Options dialog box lets you choose the characteristics of the Photoshop file to be saved.

quite large. When I designed it at 300 dpi, I ended up with a 128MB file in Photoshop (28MB on disk) after all the effects were completed. If you are low on RAM, you might want to export the image at 72 dpi.

9. Close Illustrator now if you want. You will not need it for a while. It's time to see how you use Photoshop's special features to enhance this design (and if you decide later that you prefer "before" to "after," that's all right, too).

10. Open Photoshop. In Photoshop, open the CANDY.PSD file that you just created.

11. To begin, let's get the TEXT and Swash layers out of our faces. Click on the Eye icons next to the layers TEXT and Swash to hide them from view. We will deal with these later.

12. Press D to set the colors back to the default of black and white. Choose the Eyedropper tool and change the Foreground color to the same orange as the bottom candy (the Orange Stripes layer). Make the Background layer active. Fill the layer (Mac: Option+Delete; Windows: Alt+Delete). Surprise! Only the border area turns orange because of the practically invisible rectangle on the White Rectangle layer.

13. Some of the candies look a bit too close together. Double-click on the Move tool to open the Move Options palette. Mark the Auto-Select Layer checkbox. This toggle, when turned on, sets the active layer to the layer that has the highest opacity percentage in the spot where you clicked. (If more than one layer exists with 100% opacity, Auto-Select chooses the one closest to the top of the layer stack.) Usually, I like to keep it off because it puts me in the wrong place if I'm not paying attention, but for rearranging objects, it is wonderfully convenient. Move the candies as you want them to be. You can rearrange the stacking order of the candies by dragging their thumbnails in the Layers palette. Figure 2.48 shows the spacing that I chose.

WHY WORK IN CMYK?

You don't have to export your Illustrator image in CMYK mode. However, if you have worked in CMYK color space inside of Illustrator and you don't need to apply any RGB-only filters to the CMYK image, there's no reason to convert to RGB only to convert back to CMYK for printing. In this image, you aren't using any photographic image (another reason why you might want to stay in RGB color space during the creation process). If you have an ICC profile that you usually use with CMYK images, you can choose to embed it by choosing that option in the Color Settings dialog box. (If you don't know what I'm talking about, you might want to invest in either, or both, of the following books: *Photoshop 5 In Depth*, by David Xenakis and Sherry London, The Coriolis Group, or *Real World Photoshop 5*, by Bruce Fraser and David Blatner, Peachpit Press.)

Figure 2.48
You can move each candy within Photoshop to change the spacing between them.

AN ALTERNATIVE TO AUTO-SELECT

If choosing Auto-Select makes you crazy every time that you land in the "wrong" place, but you like the convenience of not having to use the Layers palette to choose a layer, try this: When the Move tool is selected, hold down the Command/Ctrl key and click to select a layer.

14. Let's add some Layer Effects to the candy. Make one of the candies active (it doesn't matter which one). Choose Layer|Effects|Drop Shadow. The Drop Shadow defaults to black in Multiply mode. Change the Opacity to 67% and the Angle to 120° and mark the Use Global Angle checkbox. Set the Distance to 24 pixels and the Blur to 19 pixels. Leave the Intensity set to 0. Do not click on OK yet. Figure 2.49 shows these settings.

15. Let's also add an Inner Shadow. This is one of my favorite new effects. I like the play of shadow that it creates. From the Effects drop-down menu on the Layer Effects dialog box, select Inner Shadow. Place a checkmark next to the name of the effect to turn it on. Set the Opacity to 20%, the Angle to 120° (leave Global Angle selected), the Distance to 53 pixels, the Blur to 12 pixels, and the Intensity to 0. Figure 2.50 shows this dialog box. Click on OK.

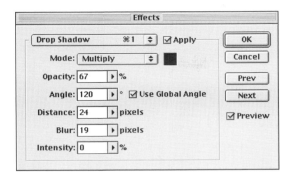

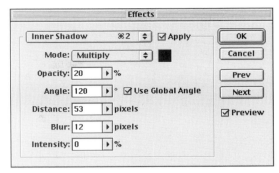

Figure 2.49 (left)
Setting the options for a Layer Effects drop shadow.

Figure 2.50 (right)
You can also set up additional effects, such as an Inner Shadow, at the same time you create a drop shadow.

16. It's very easy to give these same effects to the other pieces of candy. With your current candy layer still active, click on the empty column next to the Eye column in the Layers palette for all the candy layers (except the active one—it already has a Paintbrush icon in that column). This links the layers together so that they move and transform as one. Figure 2.51 shows the Layers palette with all the candy layers linked.

17. Now that the layers are linked, they can all be given the same Layer Effects. Choose Layer|Effects|Copy Effects. Then select Layer|Effects| Paste Effects To Linked. Figure 2.52 shows the image in progress with the Layer Effects applied.

18. We will now create a pattern for the background. Figure 2.53 shows the area of the image that I marqueed. If you want to be obsessive about it, I used a rectangular marquee that was 42×576 pixels. You can be on any layer and drag the marquee over any area that looks interesting. Then choose Edit|Define Pattern. It is as if Photoshop takes a picture of whatever is present in the image and makes that area inside the rectangle into a pattern tile. Therefore, you will pick up the shadows and Layer Effects in the area as well as the objects themselves. (If you want to see where you are going with this example before you choose your pattern, look in the Color Studio section of the book or open the finished example on the CD-ROM.)

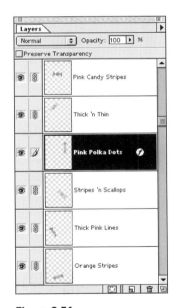

Figure 2.51
Linking layers in Photoshop allows them to move and transform together.

19. Deselect the pattern tile after you have defined the pattern and make the White Rectangle layer active. Drag the White Rectangle layer to the New Layer icon at the bottom of the Layers palette to copy it. Double-click on the thumbnail for the copied layer (which is above the original) and change its name to Pattern. Click on OK to close the dialog box. Fill the image with the pattern (Shift+Delete, Using: Pattern, 100% opacity, Mode: Normal, Preserve Transparency: On). Figure 2.54 shows the Fill dialog box. Figure 2.55 shows the result.

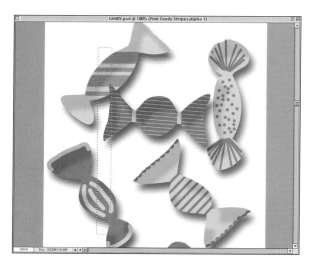

Figure 2.52 (left)

CANDY.PSD now sports some Layer Effects and a border.

Figure 2.53 (right)

The rectangular marquee shows the area that I selected to use as a background pattern.

20. The result is a bit garish (to my eye, at least). Let's tame it down. Reduce the Opacity of the layer to 90%. The Color Studio section shows several possibilities for the final version, but let's add an Adjustment layer for now—to make a monotone of the pattern. That is one way to help it fade into the background and not fight with the candies. Select the nontransparent Layer pixels of the Pattern layer (Mac: Command+click; Windows: Ctrl+click on the Layer name in the Layers palette). Create a new Adjustment layer (Mac: Command+click on New Layer icon; Windows: Ctrl+click on New Layer icon). Select the Hue/Saturation Adjustment layer. Click on Colorize *before* you start to adjust the settings. I set the Hue to 131 (a green), the Saturation to 50, and the Lightness to +69. Figure 2.56 shows the Hue/Saturation dialog box and the settings that I used.

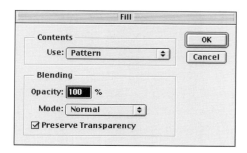

Figure 2.54

You can preserve the transparency of the layer as you fill it with a pattern.

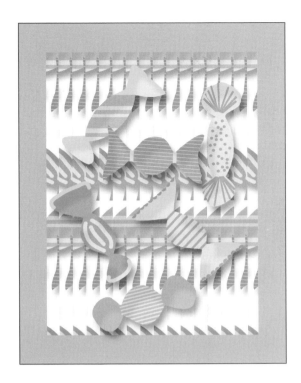

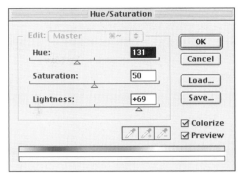

Figure 2.55 (left)
You can create a pattern fill based on a slice of the image.

Figure 2.56 (right)
By colorizing the pattern, you can make it less of a distraction in the image.

21. Let's add a bit of texture to some of the candy. Click on the Orange Stripe candy to select it. Press D to set the colors back to the default of black and white. Use the Eyedropper tool (I) to set your Foreground color to the orange of the candy wrapper end. Press X to exchange the Foreground and Background colors. (Or you could press the Option/Alt key as you click to set the Background color to begin with.) When you apply a Pointillize filter, the color of your Background swatch determines the background color behind the color dots that the filter produces.

22. Choose Filter|Pixelate|Pointillize. Set the Cell Size to 5. Figure 2.57 shows the filter dialog box.

23. Change your Background color (Option/Alt click with the Eyedropper tool) to the green used in the bands on the Pink Candy Stripe layer. Make that layer active. Choose Filter|Pixelate|Pointillize again or Mac: Option+Command+F; Windows: Alt+Ctrl+F to reopen the Filter dialog box. Set the Cell Size to 3. Click on OK. Now choose Filter|Fade Pointillize. Change the Opacity to 60%. Click on OK. Figure 2.58 shows the Fade Filter dialog box.

24. Save your work as CANDY.PSD.

TRANSPARENCY AND THE PATTERN

I left the White Rectangle layer in the image so that I could reduce the opacity of the pattern as low as I wanted it to go without revealing the orange Background layer underneath. If you are low on memory, you can merge the White Rectangle layer into the Background layer if you want. The white will still be there as needed behind the pattern layer.

Figure 2.57 (left)
The Pointillize filter adds wonderful dots of color to the object.

Figure 2.58 (right)
You can reduce the effect of any filter after you have applied it by choosing the Filter|Fade Filter command.

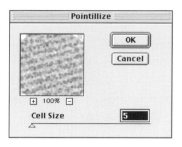

You are now going to create a layer of pattern that uses three of the candies in a repeat that is on a 45-degree angle. Figure 2.59 shows what the pattern layer looks like when it is done; however, the background of the pattern layer will be transparent (not white), and you will be able to blend it into the Hard Wrap candy image using an Apply mode. Because creating the pattern takes a number of steps, I am going to start the step-by-step directions back at Step 1.

When I first designed this project, I used Photoshop to create the pattern. To get Photoshop to slant the pattern diagonally, I had to fiddle with the size of the image. Basically, to get enough room to repeat the pattern at a

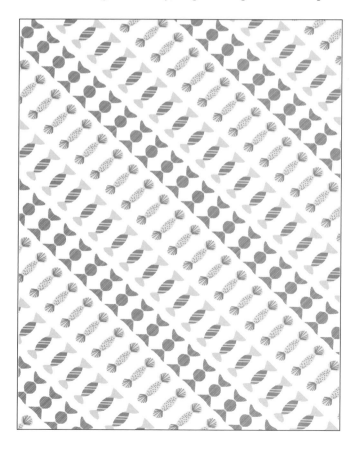

Figure 2.59
This is the pattern layer that you will create.

45-degree angle, I had to rotate the canvas (Image|Rotate Canvas|Arbitrary) 45 degrees clockwise. This actually made the document larger so that the rotated old canvas fit inside the new document border. I then filled a layer with the pattern that I had created horizontally. Now, I rotated the canvas back 45 degrees counterclockwise (which made the image even bigger) and then cropped the canvas back to its original size. This process brought my computer to its knees. My original file for this project was over twice the size at which I recommend that you work right now, and, although I had over 200MB of RAM as my Photoshop partition (on a 275MHz G3 chip), it took a long time to complete this action. I want you to know how to do this in Photoshop, but I also want you to know—especially because this *is* an Illustrator book—that it's more efficient and more accurate to build this pattern in Illustrator and then import the pattern as a layer. So, the instructions for making the pattern send you back to Illustrator.

By the way, if you want to skip over the entire pattern-making process, PATLAYER.AI is on the CD-ROM.

Creating The Pattern Layer

Now, to take things further, follow along with these steps:

1. Launch Illustrator (you may close Photoshop if you do not have enough RAM for both of them to remain open).

2. Open the file CANDY.AI (this is the original on the CD-ROM). Use the Selection tool (V) to select the pink candy with the thin green stripes, the green candy with the pink polka dots, and the yellow-green candy with the thick and thin orange lines.

3. Create a new document (File|New). Drag the selected candies from the CANDY.AI file into the new image. You may close CANDY.AI if you want.

4. Deselect (Mac: Shift+Command+A; Windows: Shift+Ctrl+A). Select the orange and yellow-green candy. Choose the Rotate tool (R). Press the Shift key and drag to constrain rotation to 45°.

5. Select the polka dot candy. With the rotate tool, keep the Shift key pressed and rotate the candy 90 degrees clockwise.

6. Select the thin-striped candy. Double-click on the Rotate tool. In the dialog box, change the angle of rotation to 4.2° (yes, it took some experimentation to figure out that this exact amount would make the stripes lie straight, horizontally across the candy). Figure 2.60 shows the three candies properly rotated.

Figure 2.60

The pattern-to-be candies are rotated to make them look horizontal.

7. Select the entire image (Mac: Command+A; Windows: Ctrl+A). Double-click on the Scale tool. In the dialog box, change the Uniform Scale amount to 27% as shown in Figure 2.61.

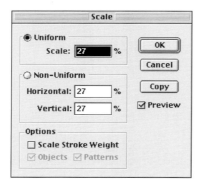

Figure 2.61

You need to scale the candies to 27%.

8. Drag the candies into a straight line so that the pink candy is on the middle and the polka dot candy on the left. Select the orange and yellow-green candy as shown in Figure 2.62. Notice that its bounding box is the only one that really looks angled. Choose Object| Transform|Reset Bounding Box. This makes the bounding box hug the new orientation of the object and can help avoid problems (It isn't strictly necessary to do this. However, because Adobe included the feature, it seems safer to use it.)

9. Select the Rectangle tool (M). Click one in the image to open the Rectangle Options dialog box. Change the width to 2.556 inches and the height to 0.403 inches. Click on OK. Change the color of the box to gray (or any color that isn't in the candies—you are using this box only as a temporary spacer). Send the box to the back (Mac:

Figure 2.62

You need to reset the bounding box around the candy on the right.

Shift+Command+[; Windows: Shift+Ctrl+[). Drag the candies (in their current order) into the box.

10. Deselect (Mac: Shift+Command+A; Windows: Shift+Ctrl+A). Select just the gray box and the polka dot candy on the left. Click on the Align Left icon on the Align palette. It is the leftmost icon on the top row shown in Figure 2.63. This ensures that the box and the left-most candy share the same left edge.

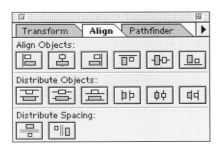

Figure 2.63
The Align palette lets you easily space your objects.

11. Drag the candy on the right so that it is inside the box. Select it and the box. Click on the Align Right icon in the Align palette (third icon on the top row).

12. Now that the left and right objects are in the correct place, deselect the gray box and select all three candies. Click on the Vertical Align Center icon (next-to-the-last one of the top row). If you do not see three rows of icons on the Align palette, choose Show Options from the sidebar menu on the palette. This reveals the Distribute Spacing commands. Click on the Horizontal Distribute Spacing icon to evenly space the three candies. Figure 2.64 shows the candies now correctly lined up inside the gray box.

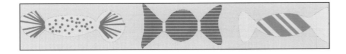

Figure 2.64
The candies are evenly spaced and are horizontally aligned with one another.

13. Click on the gray box and delete it because it's no longer needed. Select the three objects (Mac: Command+A; Windows: Ctrl+A) and then group them (Mac: Command+G; Windows: Ctrl+G).

14. Select the Rectangle tool (R). Click in the image to open the Rectangle Options dialog box. This time, create a filled, unstroked rectangle that is 2.681 inches wide and 0.458 inches high. This is the final size of the pattern.

15. Send the candies to the front (Mac: Shift+Command+]; Windows: Shift+Ctrl+]). Drag the candies on top of the new box. It is not critical that they be centered (although I centered them by eye) as long as they are entirely within the box. Change the Fill and Stroke for the box to None. Group the candies and the box and drag the group to the Swatches palette. Your pattern is now complete. (For more on creating patterns, see Chapter 5.)

16. Let's actually create the pattern layer now. For safety's sake, choose File|Save A Copy and save the image in the Swatches Library folder of your Illustrator installation. That way, you can either edit the original pattern again if you want or import the Swatch containing the pattern into another document.

17. Select the entire image (Mac: Command+A; Windows: Ctrl+A). Delete. This leaves the pattern still in the Swatches palette.

18. You need to create a new box that is exactly the same size (in inches or points) as your Photoshop file. I'll save you the trouble of looking. Choose the Rectangle tool (R) and click in the image to open the Rectangle Options dialog box. The box needs to be 6.672 inches high by 8.864 inches wide.

19. The box should have no stroke. Click on the new candy pattern to make the pattern into the box fill. Figure 2.65 shows the filled box.

20. Next, you need to rotate the pattern fill. Click on the sidebar menu in the Transform palette as shown in Figure 2.66 and select the Transform Fill option. Then click on the Rotation drop-down menu at the bottom left of the Transform palette and select 45° as the angle of rotation. The pattern rotates to look almost as it did in Figure 2.59.

21. You may, if you want to, be very obsessive (as I am) and force the pink candy to appear in the top-left corner of the box. I can't give you an exact setting here because it depends on where in the box you placed the candies and where on the page the box is located.

WAS ALL THIS MEASURING REALLY NECESSARY?

No, it wasn't really necessary to fool around with the gray spacer box. If you had been designing this image, you could have made the pattern any size that you wanted. Because you are reproducing my results, I needed to find a way for you to exactly duplicate my measurements. This was the easiest (short of my just giving you the pattern).

The second box was needed. The precise measurements are mine, but you do need an invisible box to set the boundary of your pattern, or else the candies would touch one another when the pattern is repeated. The purpose of the outer box is to define the amount of white space around the repeat. There will be much more about this in Chapter 5.

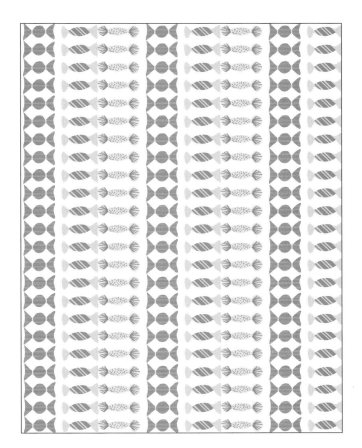

Figure 2.65 (left)
The candy pattern is used to fill a box that is the same size in inches as your Photoshop document.

Figure 2.66 (right)
You can elect to transform only the pattern fill from the Transform palette.

Turn on the Rulers (Mac: Command+R; Windows: Ctrl+R). If you centered the candies as I did, drag the ruler origin from the upper-left corner of the document window so that the horizontal line is parallel to (and touches) the top of the box, and the vertical line is at about the minus-one-half-inch mark on the vertical ruler. Undo (Mac: Command+Z; Windows: Ctrl+Z) the original change if you do not get it right and try again. It may take several tries. Figure 2.67 attempts to show the correct location for centering the candy pattern.

RETURNING TO PHOTOSHOP

How do you get the pattern layer back into Photoshop? You have just tried a number of different methods. The easiest method, which we will use in a few minutes, is simply to place the file into the Hard Wrap image. You can also drag and drop. The drag-and-drop feature works perfectly in this instance. Copy and Paste is not a wonderful idea because the image is really too large for the Clipboard. You could, instead, export the pattern layer within Illustrator as a Photoshop file and open it in Photoshop, but this approach is much trickier that you may expect. Because you rotated the pattern, the rasterizer in Illustrator cannot properly anti-alias it. If you leave the pattern fill alone and Export the image, the candies are distressingly jagged, although some anti-aliasing is done. Should you need to export a rotated or skewed pattern fill, you need to expand (Object|Expand) it first. This works, although it produces a Photoshop file with white space around the edges that needs to be cropped to size before placing it into the image. In Chapter 5, you will learn more about the Expand command.

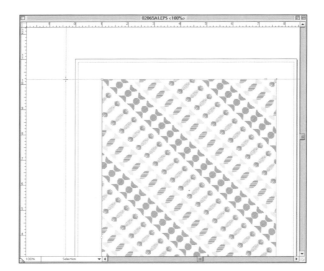

Figure 2.67

In my image, dragging the origin point as shown centered the pink candy in the upper-left corner of the object.

22. Save the document as PATLAYER.AI. You may quit Illustrator now if you cannot run both Illustrator and Photoshop at the same time. The image that you created looks exactly like Figure 2.59.

Finishing Up

To wrap up this Illustration, take the following steps:

1. Launch Photoshop.

2. Open your saved copy of CANDY.PSD.

3. Make the TEXT layer active.

4. Choose File|Place and locate the PATLAYER.AI image. Choose it and click on OK. The preview box should fit exactly and doesn't need to be adjusted. Press Enter/Return to execute the Place command.

5. Change the Apply mode to Multiply and the Opacity for the pattern layer to 40%. Figure 2.68 shows the Hard Wrap image at this point.

6. Change the Foreground color to CMYK: 69, 39, 64, 77 (a dark charcoal) and the Background color to CMYK: 31, 0, 31, 0 (a light green).

7. Make the TEXT layer active. Fill the text with the Foreground color with Preserve Transparency On (Mac: Shift+Option+Delete; Windows: Shift+Alt+Backspace). Change the Apply mode to Color Burn at 100% opacity. The effect needs more punch.

8. Drag the TEXT layer to the New Layer icon at the bottom of the Layers palette to duplicate it. Do this again so that you have a third identical layer.

MOVING A PATTERN WITHIN AN OBJECT

You can move the starting location of a pattern in an object by changing the 0,0 ruler origin point of the document. Just drag the ruler origin from the upper-left corner until you like the new location of the pattern in the selected object. This is not one of Illustrator's most shining moments. You cannot preview the result, and it takes a frustratingly long time to get the pattern exactly where you want it.

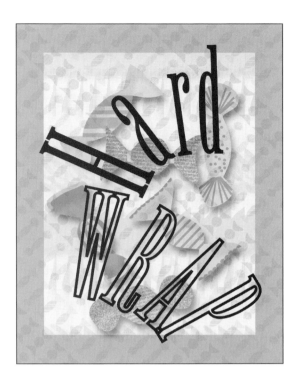

Figure 2.68
Changing the pattern layer Apply mode to Multiply and lowering its opacity softly mixes it (blends it in) with the rest of the image.

9. Make another copy of the TEXT layer the same way (this is called TEXT copy 3).

10. Make the TEXT layer active (yes, the original one at the bottom of the collection). Fill the layer with the Background color with Preserve Transparency On (Mac: Shift+Command+Delete; Windows: Shift+Ctrl+Backspace). Change the Apply mode for this layer back to Normal and lower the Opacity to 60%. The text is now nicely shaded and stands out quite well. It just needs a little more dimensionality.

11. Let's create a white drop shadow for the text. Although Photoshop has Layer Effects drop shadows, we will create this the old way because we need to mask it.

12. Drag the TEXT layer to the New Layer icon at the bottom of the Layers palette to duplicate it (again). Move this copy below the TEXT layer in the Layers palette. Double-click on the layer thumbnail and change the layer name to Text Glow. Change the layer opacity to 100% (now you know why we will need to mask it—the wonderful text shading disappears).

13. Press D to set the colors back to the default of black and white. Fill the layer with the Background color with Preserve Transparency On (Mac: Shift+Command+Delete; Windows: Shift+Ctrl+Backspace).

14. Select the Move tool (V). Press the Right arrow key three times and the Up arrow key three times. This offsets the white drop shadow.

15. With the White Glow layer still active, select the nontransparent Layer pixels on the TEXT layer (Mac: Command+click; Windows: Ctrl+click on the TEXT Layer thumbnail in the Layers palette). You do not need to have the layer active to be able to load the layer transparency.

16. Choose Layer|Add Layer Mask|Hide Selection (or Option/Alt click on the Add Layer Mask icon at the bottom of the Layers palette). Figure 2.69 shows the Hard Wrap image after all the text has been styled.

Figure 2.69
The text is finally in place for the Hard Wrap image.

17. At this point in the creation process, I began to feel as if the candy was fighting a little bit with the text. Therefore, your next step is to lighten just the candies in the image (but in such a way that you can easily change your mind). Make the Pink Striped Candy layer active (the highest of the candy layers). Select the nontransparent Layer pixels (Mac: Command+click; Windows: Ctrl+click on the Layer name in the Layers palette). Press and hold Shift and load the nontransparent pixels on all the other candy layers (Shift allows you to add to the selection). When you are done, only the six candies should be selected.

18. Create a new Adjustment layer (Mac: Command+click on New Layer icon; Windows: Ctrl+click on New Layer icon). The selection is automatically changed into a mask for the Adjustment layer. Choose Hue/Saturation as the type of layer to create. Increase the Saturation to +23 and the Lightness to +25 as shown in Figure 2.70. Click on OK.

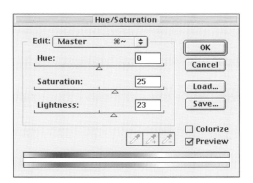

Figure 2.70
You can change the Saturation and Lightness of just the candies in the image all at the same time.

Only one more effect needs to be created. There is still the Swash layer to fix. The design seems to need a celebration, so I added a ribbon swirl (that's the layer that is still hidden). I feel that the original black is much too strong for the packaging, although you are certainly free to disagree with me. Let's decorate this layer and soften it.

19. Make the Swash layer active. Change your Foreground color to CMYK: 31, 0, 31, 0. This is the same light green as the base TEXT layer. Fill the text with the Foreground color with Preserve Transparency On (Mac: Shift+Option+Delete; Windows: Shift+Alt+Backspace).

20. Use the Eyedropper tool with the Option/Alt key pressed to make the orange in the border the Background color. Apply the Pointillize filter (Filter|Pixelate|Pointillize) with a Cell Size of 11.

21. Choose Filter|Distort|Wave and use the settings shown in Figure 2.71.

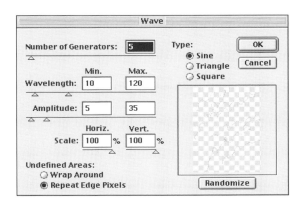

Figure 2.71
Use these Wave filter settings to make the Swash layer flow.

22. Your Foreground color should still be the light green. Fill the Swash with the Foreground color with Preserve Transparency On (Mac: Shift+Option+Delete; Windows: Shift+Alt+Backspace). This, of course, covers the result of the Pixellate filter, so choose Filter|Fade Fill and change the Opacity to 40%. Save your work and take a rest. You've earned it! Figure 2.72 shows the final image, but it looks much better in color.

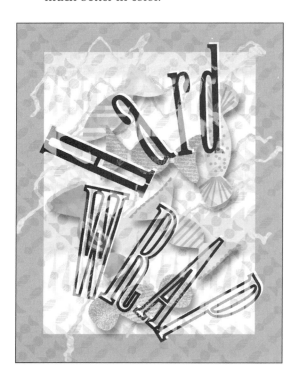

Figure 2.72
The Hard Wrap candy box is done.

What have you done in this image that you could not do in Illustrator? The drop shadow on the candies wouldn't have been as smooth or convincing. The inner shadows are not really possible in Illustrator because they require Apply mode changes. You used Apply modes on the text and the pattern layer. You also created Adjustment layers that allow you to easily change your mind about the specifics.

Could you have built the entire image in Photoshop? Probably, except that the candy would not have been as crisp, the candy pattern would have seriously anti-aliased as you reduced and rotated it, and you could not have set type on a path at all.

Moving On

This chapter has looked at how Photoshop treats Illustrator images. You have learned how to open, place, cut and paste, and drag and drop images that are in Illustrator format. Most of these techniques work with Freehand images as well.

You have also learned how to export layered images from Illustrator. We discussed the advantages and disadvantages of each method of rasterizing an Illustrator file. We briefly touched on resolution and some of the Illustrator-to-Photoshop "gotcha's" that exist.

In the next chapter, you look at the same raster-vector issue from the other perspective: How do you bring Photoshop images into Illustrator, and why would you want to?

RASTER IMAGES
IN ILLUSTRATOR

3

SHERRY LONDON

Photoshop can serve as the best source for Illustrator imagery. It's easy to get scanned line art from Photoshop and change it to vector format. You can also place raster images directly into Illustrator and leave them there.

Photoshop To Illustrator: Going My Way

In the olden days—oh, about ten years ago—the only way to get an Illustrator drawing started was to draw it yourself. You could get a primitive scan. I remember that I had a tiny hand scanner for the PC by 1990. Illustrator 88 (Mac version—the first Windows incarnation of the program was terrible) had an autotrace tool. If you clicked near a line on the imported bitmapped image (your template art), you sometimes ended up with a usable shape.

In the 10 years since Illustrator 88, so much has changed in the world of digital graphics that it's like moving from the Middle Ages into the space age in about three months. One of the nicer changes is the ability to place raster and vector images in the same program and to leave them where they make the most sense to output. In Chapter 2, we looked at how you can take Illustrator images and "finish them off" (not by doing them in, I hope) in Photoshop. In this chapter, you take the voyage the other way and explore ways in which Photoshop-originated imagery can be used in Illustrator. As before, you look first at how you can move the images back and forth and then consider why you would want to do this. Finally, you will look at some examples of images that wander from program to program before coming to their ultimate resting place.

Photo In: How?

Just as you have a number of ways to get a Photoshop image into Illustrator, you have an equally extensive list of options for bringing raster or originally raster images into Illustrator:

- You can open a Photoshop file directly into Illustrator.

- You can place a Photoshop image into an open document.

- You can copy and paste Photoshop images into Illustrator.

- You can use the drag-and-drop method to move images from Photoshop into Illustrator.

- You can pass Photoshop paths to Illustrator using any of the first three methods listed here.

Because you practiced using the Open, Place, drag-and-drop, and copy-and-paste features from Illustrator to Photoshop in Chapter 2, there is no reason to make you do it in reverse. The process is the same. However, a few gotchas exist for the Photoshop-to-Illustrator route that make it either different or more complex than the Illustrator-to-Photoshop route.

It's A Real Drag

Image resolution seems to be a tricky and complex topic for most people. I usually urge folks to work in pixels inside Photoshop because pixels are fixed: A 900-pixel-wide image contains 900 pixels regardless of its ppi. If the ppi is set to 300, the image prints at 3 inches wide; if the ppi is set to 100, the image prints at 9 inches wide. However, the pixels that make up the image do not change (as long as the Resample Image checkbox is unmarked in the Image Size dialog box in Photoshop).

Things become more complex when you move your image into Illustrator. Three of the transfer methods are no problem; however, when you drag and drop an image, the image resolution changes to 72 ppi. Unlike the previous Photoshop example (the unchanging 900-pixel image), when you drag a 900-pixel-wide image from Photoshop to Illustrator and the Photoshop resolution is 300 ppi (a 3-inch width), you get an image in Illustrator that is effectively only 216 pixels wide (3 inches × 72 ppi). Obviously, your image quality consequently suffers. Figure 3.1 shows a Photoshop image (at 400 percent magnification) that was placed into an Illustrator file. Figure 3.2 shows the same image (still at 400 percent magnification), but this time it was dragged and dropped. In the screen capture, if you look carefully, you can see the pixelization. The placed image has enough resolution that you don't see the individual pixels.

Does this render the drag-and-drop feature useless? Not if you really want to use it. Drag and drop is faster and more convenient, so it's fine for spur-of-the-moment "how does this look?" adjustments. If, however, you want to keep the full resolution while dragging and dropping, here's what you need to do.

Figure 3.1 (left)
An image placed into Illustrator keeps its original resolution.

Figure 3.2 (right)
The image resolution reverts to 72 ppi when you drag and drop from Photoshop into Illustrator.

PROJECT Illustration 3.1: Full-Resolution Drag-And-Drop

1. Make sure that Illustrator is running and that a document is open.

2. Open the image MACAW.PSD in Photoshop.

3. Change its resolution to 72 ppi (Image|Image Size, 72 ppi, Do Not Resample). Figure 3.3 shows the Image Size dialog box in Photoshop.

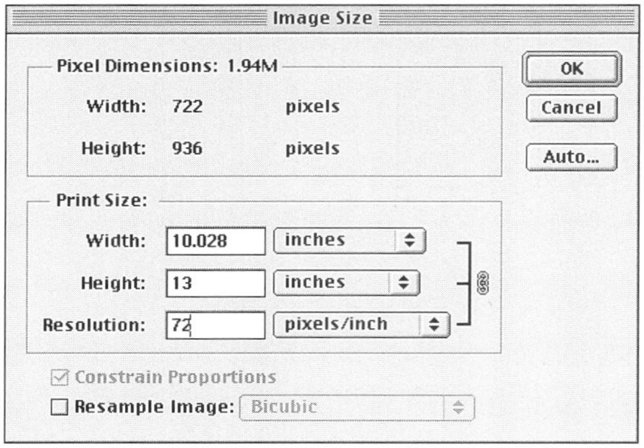

Figure 3.3

You can change only the resolution of the image by unmarking the Resample Image checkbox.

4. Select the Move tool (M). Drag the macaw image into the open image in Illustrator (the Illustrator document window border changes color briefly when you drag the image into it). Release the mouse button and wait a short while. The application doesn't transfer to Illustrator, but you should see the image appear in the Illustrator window. Figure 3.4 shows the comparison between the size of the macaw when placed at 300 ppi and the dragged-and-dropped macaw at its new 72 ppi Photoshop resolution. Notice how much bigger it appears.

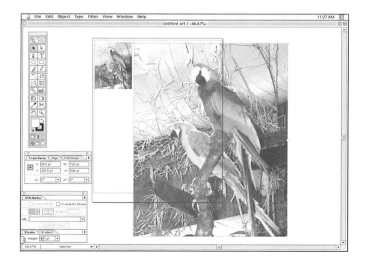

Figure 3.4

When you drag and drop a 72-ppi image from Photoshop into Illustrator, it's much larger than when placed at 300 ppi.

5. Make the Illustrator document active. Click on the dragged image to select it. To verify that the image was dragged at 72 ppi, choose File|Selection Info|Embedded Images. Figure 3.5 shows the information window. It describes the macaw image and shows its resolution as 72×72 ppi.

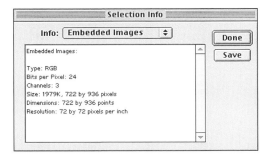

Figure 3.5

The Selection Info window shows that the resolution of the dragged macaw is 72 ppi.

You need to change the image resolution back to 300 ppi in Illustrator if you want the image to print well. If you scale the image inside Illustrator, you will increase the resolution as you decrease the physical size of the image. Why? Because you are packing the same number of pixels into a smaller space, which causes the resolution to go up. Several ways are available to calculate the amount to which to scale your image, but I prefer the following because it requires no math.

6. In Photoshop, change the image resolution back to 300 ppi with Resample Image Off. In the Image Size dialog box (shown in Figure 3.6), you can see the size (in inches) of your image at 300 ppi.

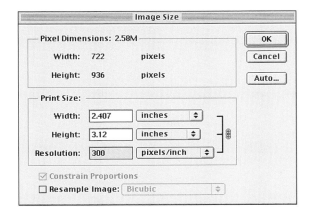

Figure 3.6

In Photoshop, you can see that the size of the macaw should be 2.407 inches wide by 3.12 inches high.

7. In Illustrator, change the General Units Preference (File|Preferences| Units & Undo, General) to Inches, as shown in Figure 3.7.

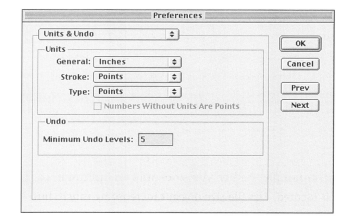

Figure 3.7

If you change the General Units Preference to Inches, it is easy to transform images to the exact size.

8. Enter the correct height and width (as determined in Photoshop) in the Transform palette, as shown in Figure 3.8. Remember to press Enter/Return after each entry.

9. Choose File|Selection Info|Embedded Images again. This time, as you can see in Figure 3.9, the resolution is much closer to the desired 300 ppi. Although a rounding error in the math shows that the resolution is not quite 300 ppi, you do not need to do anything more.

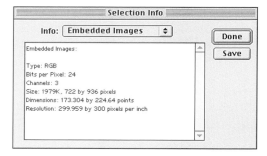

Figure 3.8 (left)

The Transform palette.

Figure 3.9 (right)

The Selection Info window shows the increase in resolution that comes from making the macaw physically smaller in Illustrator.

The only issue remaining with drag-and-drop is that the dragged image is embedded inside Illustrator rather than linked. We tackle this topic next.

To Place Or Parse (Or To Link Or Embed)

In versions prior to 8, Illustrator allowed you to *place* or *parse* raster images. Although euphonious, this language has never allowed me to remember which is which. I like the new terms *link* or *embed* much better. For whatever reason, embed is far more descriptive in my mind than parse and sounds much less like computer jargon. An artist would not have coined the term parse for the process of including the original image directly in the Illustrator file structure.

When you embed a raster file, you actually include the image in the Illustrator file. All its data are converted to PostScript format (although you still get a single raster object rather than editable paths), and no external files are needed when you go to print. If you embed a large file, the file size increases dramatically. However, if you want to apply filters to a raster image in Illustrator, you must embed the file. Images that are either drag and dropped or pasted into Illustrator are automatically embedded.

When you link a file (using the Place command), you don't store the image inside the Illustrator file. Rather, you store only an indication of where the original is located. The file size doesn't increase by much, but you lose the capability to apply filters to the image. You need to remember to include the originals when you take the image to a service bureau. However, you gain flexibility. If you decide to change the original image, you only need to do so in one place (the original image), and Illustrator can follow the link to retrieve the updated image. If you use the Place command, Link is the default method of placing the file.

You can tell whether a file is linked or embedded in either of the following two ways:

- A linked file is shown with an X through it when the object is selected, as shown in Figure 3.10.

- The file name shows in the Links palette, which is new to Illustrator 8. If the file is embedded, an object icon appears after it in the Links palette, as you can see in the bottom two entries in Figure 3.11.

You can take other actions with the Links palette to help you manage the images that you place into Illustrator. You can see information about the scale, size, and orientation of the placed images. You can also quickly embed a linked image from the Links palette.

Figure 3.10 (left)
Linked files are shown with an X across the box.

Figure 3.11 (right)
The Links palette also helps you spot the linked and the embedded files.

The Passing Of The Path

Photoshop allows you to create vector paths by using the Pen tool. Chapter 2 showed ways to take an object from Illustrator and use its path in Photoshop. You can also take a path created in Photoshop and send it to Illustrator. Paths can be sent individually (through copy-and-paste) or exported by using the Paths To Illustrator option in the File|Export menu. Additionally, paths are saved with any file that you store in Photoshop format and are available for use if you open the file in Illustrator.

Although the Illustrator manual seems to hint that you can drag and drop paths from Photoshop to Illustrator, you cannot. However, enough other ways exist to accomplish the same goal. You will try some of these ways a bit later in this chapter.

Photo In: Why?

Although it's more common to move Illustrator images into Photoshop, many reasons exist to take them in the other direction. You can scan a drawing or photograph in Photoshop and then move it to Illustrator to prepare it further. You can use Adobe Illustrator to create a clipping path that you want to later use in Photoshop. You might need to place a Photoshop image into Illustrator so that you can use it as a template for type to be composited in a page-layout program. You might also want to create a texture in Photoshop and use the texture in Illustrator to add surface interest. In the sections that follow, we explore these topics a bit more.

The Best AutoTrace Tool In Town

I find it very frustrating to scan an image and then have to spend hours getting the image into a usable vector format. If *I* can see the "edges" of the shapes that I want, why can't the computer? Of course, the computer can't "see" anything—and that is just the problem. Adobe provides a variety of ways for you to tell the computer how to create outlines from line art or continuous-tone images.

You can use the Illustrator AutoTrace tool—although normally this is a bad idea. It doesn't work well, and it gives you only one line at a time. You can use Adobe Streamline. This slick autotrace program can work with line art or full-color images and can be tweaked to decently automate the entire process. A number of folks have had a lot of success with this method, although I am not one of them. I lack the patience or whatever to sufficiently tweak my images. Actually, if I were only to scan, Streamline, and print, I too would probably be satisfied. However, I usually manipulate an image much more than that, and I am typically not happy with the individual shapes that come out of Streamline (although again this is a personal preference).

There's another way to get editable shapes from a raster image. Place the image in Illustrator into a template layer in the layer's palette (new to Illustrator 8) and then use the Pencil or Pen tool to trace over the shapes by hand. This method gives the maximum possible control over your result. However, I rarely have enough time to use this approach.

My favorite method of getting usable shapes is to select the areas inside Photoshop, create paths from the selections, and send the paths into Illustrator. This gives you a marvelously easy autotrace facility that can handle multiple shapes at one time. The result can be either simple or complex, depending on your starting image. Let's work though two examples—one easy and the other one more intricate.

For the first project, we'll revisit the candy that you used in Chapter 2. The candy shapes came from a scan of some glass candies that I have displayed on a knick-knack shelf at home. I simply placed each piece of glass on my scanner and acquired the scan. The image that you will use as a starting place was colored for a long-ago project. All that you want from it are the shapes of the candy.

PROJECT Illustration 3.2: Candy Raster To Candy Vector

1. Open the image RAWCANDY.PSD in Photoshop.

2. Press Command (Mac) or Ctrl (Windows) and click on the thumbnail of Layer 1 in the Layers palette. This loads the nontransparent pixels on the layer as a selection. (In practical terms, it selects the entire object in the layer.)

3. Press and hold Shift and then press Command (Mac) or Ctrl (Windows) and click on the thumbnail of Layer 2 in the Layers palette to add Layer 2 to the current selection.

4. Repeat Step 3 to add Layers 3 through 6 to the selection. Figure 3.12 shows the final selection.

5. Select Make Work Path from the side menu in the Paths palette, as shown in Figure 3.13.

6. Accept the default Tolerance of 1.0 in the dialog box that appears (see Figure 3.14). *Tolerance* refers to the fidelity with which the path hugs the original. A tolerance of 1.0 is quite faithful to the original (0.5 is the tightest tolerance). A lower tolerance also means that more points are created to define the shape. Although this has the potential to cause printing problems, the candy shapes are not complex enough to cause difficulties.

> **WARNING: WATCH OUT FOR EMBEDDED GRADIENT MESHES!**
>
> In an easy-to-miss portion of the Adobe Illustrator 8 user manual, Adobe states that you must save gradient mesh images as PostScript Level 1 EPS files if you want to embed them into other Illustrator documents.

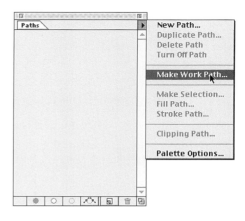

Figure 3.12 (left)

All six layers of candy are selected.

Figure 3.13 (right)

Make Work Path changes the selection into a path.

7. Double-click on the work path in the Paths palette. Accept the default path name. This saves your paths.

8. Choose File|Export|Paths to Illustrator. Figure 3.15 shows the dialog box. Save the document as RAWCANDY.AI. Notice that a box allows you to select which path to save. If you have multiple paths, you can save them all at one time. It was just as easy here to create all the paths as one. If the paths were more complex, this might not have been a good idea.

What have you created? Open Illustrator and find out.

9. In Adobe Illustrator, open the RAWCANDY.AI file that you just saved. Figure 3.16 shows a screen capture of the result that you will see. *Don't panic!* I know that it looks basically empty.

10. Why *does* the image look empty? Photoshop creates paths that have no stroke and no fill. Therefore, when you open the image in Illustrator, everything is invisible. The easiest way to make things show up is to select the entire image (Mac: Command+A; Windows: Ctrl+A). As you can see in Figure 3.17, everything is still there.

Figure 3.14 (left)

The Make Work Path dialog box allows you to determine how faithfully the path should reproduce the original selection.

Figure 3.15 (right)

Saving the paths as an Illustrator document is as easy as saving any other file.

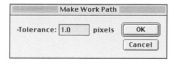

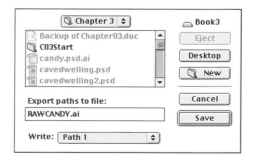

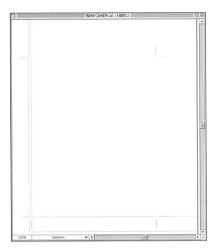

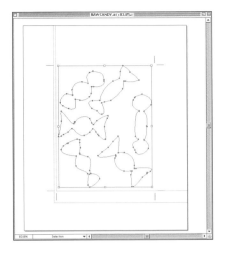

11. Make the Stroke Color selector active and choose black as the stroke color. This allows you to find the paths more easily. With the paths stroked in black, you can edit them as you wish and change the colors when you are finished. This is how I prepared the candies for you to use in Chapter 2.

12. When you import paths from Photoshop, you will often want to adjust them to make them smoother or to remove extra points. Now that you have the shapes in Illustrator, it's easy to edit them as you want. Figure 3.18 shows one candy in a "before" (on the left) and "after" (on the right) view. I changed "before" into "after" by removing all the points on the right side of the candy body and by using the Pencil tool to draw a new top wrapper for the candy.

You may play with the shapes if you want or close this file and go on to the next exercise. You now know how I started the Hard Wrap project so that you can design your own version or work on a similar project.

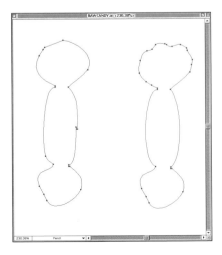

Figure 3.18
(Left) The candy shape as placed into Illustrator. (Right) The edited candy shape.

ABOUT IMPORTED CROP MARKS

When you export paths from Photoshop, Photoshop writes the boundaries of the document into the file. These boundaries show up as crop marks. If you want to keep them—so that you can export the same-size document back to Photoshop—just make sure that your drawing falls completely inside of them. Should you prefer to delete them, choose Object|Crop Marks|Release, and they will turn into an unstroked, unfilled rectangle around the candies. You can then delete this rectangle.

WARNING: LARGE FILE AHEAD

DAVID.PSD is a large file that will use over 29MB of RAM inside Photoshop. If you are working on a low-RAM machine, you can reduce the image by 50 to 66 percent before you continue with the instructions.

I used the candy project to show how to use the layer transparency of an object as a basis for a selection and how to make exportable paths from that selection. The candies were easy because they were solid objects. They would have been fairly easy even if they had been flat scanned objects against the Background layer. If you had scanned the glass candies, you would have needed to make a selection around each candy and place it into its own layer (or color each candy a solid black and use the Magic Wand to select all the candies).

PROJECT Illustration 3.3: Exporting Complex Paths

This next project is more complex. It is an extension of my Photoshop woodcut technique that I tend to overuse because I like it so well. Although this project starts with a photographic scan, the photo will not gain paths. I have already created the selection paths and channels that you will need, so that you can concentrate on the part of the exercise that is most instructive. Follow along with Illustration 3.3.

Creating Complex Paths In Photoshop

Meet my nephew David (who was growing new teeth when this photo was taken several years ago). Luckily, I needed only my sister's permission to use this image—not David's—who might not forgive me for publishing it!

1. Open the image DAVID.PSD in Photoshop. Figure 3.19 shows the original image.

2. Load the Alpha 1 channel (Mac: Option+Command+4; Windows: Alt+Ctrl+4). This step selects David's shirt.

3. Create a New Layer Via Copy from the selection (Mac: Command+J; Windows: Ctrl+J).

4. Choose Filter|Other|High Pass, Amount: 1.6. This step creates an image that looks as though it has lost all detail and all contrast. It has lost contrast and isn't usable as is, but the High Pass filter has left highly detailed edges embedded in the image, and the next step will bring them out.

5. Select Image|Adjust|Threshold. Set the slider to 129 as shown in Figure 3.20. Figure 3.21 shows the result. Notice how much detail appears in the stripes of the shirt.

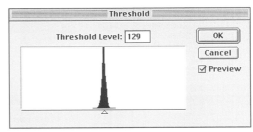

Figure 3.19 (left)
This is David, in the original Photoshop version.

Figure 3.20 (right)
You can force an image to black and white by adjusting the image threshold.

Figure 3.21
After applying the threshold command to David's shirt, the stripe detail appears crisp and sharp.

The Blur-and-Levels technique shown in Steps 6 and 7 comes from a tip posted by Kai Krause (of MetaCreations) on AOL many years ago. It is a marvelous technique for making scanned line art smooth. The basic premise is that you take your starting black-and-white (bitmap) scan and slightly blur it (1 or 2 pixels), which creates gray tones that can then be forced to white or black again by using the Levels command. You can manipulate these gray tones by moving the White and Black Input sliders in the Levels dialog box closer together. If you move the group of three sliders leftward, the image becomes lighter; as you move them rightward, it becomes darker.

Next, you need to smooth out the lines that were created and to get a more solid look to the black and white.

6. Choose Filter|Blur|Gaussian Blur, Amount: 3. This step adds a large number of gray tones to the image. Usually, you should not blur the image this much, but I wanted to create more grays than needed to force this detail to a fairly solid black and white.

7. Choose Image|Adjust|Levels (Mac: Command+L; Windows: Ctrl+L). Figure 3.22 shows the Levels setting that I used. I brought the Black Input slider to 118 and the White Input slider to 152. Click on OK to exit the Levels dialog box. Figure 3.23 shows the result.

8. Make the Background layer active. Load the Alpha 2 channel (Mac: Option+Command+5; Windows: Alt+Ctrl+5).

9. Create a New Layer Via Copy from the selection (Mac: Command+J; Windows: Ctrl+J). Now David's head is on its own layer.

10. Choose Filter|Other|High Pass, Amount: 1.6.

11. Select Image|Adjust|Threshold. This time, use a setting of 128. (The head was processed separately from the body because we needed to be able to select a different Threshold setting. Also, if we had used all of David's face instead of just the features that I selected, he would have looked diseased.)

12. Apply a Gaussian Blur of 2 (Filter|Blur|Gaussian Blur, Amount: 2.0).

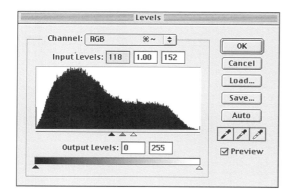

Figure 3.22 (left)
The Levels command is used to smooth the edges of a blurred selection.

Figure 3.23 (right)
David's shirt now has more solid areas of black and white, but the edges are quite smooth.

13. Choose Image|Adjust|Levels, as shown in Figure 3.24. I used a Black Input level of 131 and a White Input level of 168. Figure 3.25 shows the result.

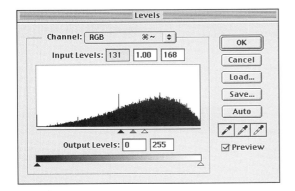

Figure 3.24 (left)

The Levels command settings used for David's head.

Figure 3.25 (right)

David's head and body are now ready to convert to paths.

14. Now you need to prepare the line art for tracing. Make the Background layer active. Create a new layer (click on the Create New Layer icon at the bottom of the Layers palette). Fill the layer with white. You now have a choice: To preserve the body and head layers for some future purpose and if you don't care whether the image grows to 29MB of RAM, go to Step 15. If you want a smaller working file, follow Step 16. In either case, continue to Step 17.

15. Make the top layer active. Make a new layer (click on the New Layer icon at the bottom of the Layers palette). Create a merged image in the new layer (Mac: Shift+Option+Command+E; Windows: Shift+Alt+Ctrl+E). Figure 3.26 shows the completed line art.

16. Make the top layer active. Merge Down (Mac: Command+E; Windows: Ctrl+E). Merge Down again. Now you have only two layers. Figure 3.26 shows the completed line art.

17. Make sure that the line art layer is the active layer in the Layers palette. Load the Composite channel of the image (Mac: Command+Option+~; Windows: Alt+Ctrl+~), which selects all the white pixels in

Figure 3.26
David has now become all line art.

the image. You need the *black* pixels to be selected, so reverse the selection (Select|Inverse, Mac: Shift+Command+I; Windows: Shift+Ctrl+I).

18. In the Paths palette, choose Make Work Path from the side menu. Leave the Tolerance at 1.0. Save the path (double-click on the path name in the Paths palette and accept the default name offered). This process can take a while. Figure 3.27 shows the paths "posed" against a white background (for the purpose of this screen shot).

19. Choose File|Export|Paths to Illustrator. The default name of DAVID.PSD.AI is fine. Write only the line art path (it should be

Figure 3.27
The line art version of David has been selected and converted to paths.

named Path 1). Export, too, is a bit slow, so do not be concerned if it seems to take a long time to write the new file.

20. Save your work in Photoshop, as you are not yet done with the Photoshop part of this image. If RAM limitations require it, close Photoshop.

21. In Illustrator, open the DAVID.PSD.AI file that you just saved. (Remember, it will open and look empty.)

22. Select the entire image (Mac: Command+A; Windows: Ctrl+A). Set the fill to black. Figure 3.28 shows this image. Notice that it seems as if you have lost some shirt detail and that David's teeth are black. We'll fix that next.

Figure 3.28
The paths that you saved in Photoshop are filled with black in Illustrator.

23. Let's fix David's mouth first because this is the most distracting part of the image. Open the Preferences (Mac: Command+K; Windows: Ctrl+K) and mark both the Bounding Box and the Area Select checkboxes. Close the General Preferences dialog box. Magnify David's mouth area and select the outside shape of his lips (click near the outside border of his lips to select them). You should have one large shape, as shown in Figure 3.29.

24. Choose Object|Hide Selection. The mouth temporarily disappears, making it much easier to find the teeth. Figure 3.30 shows three separate tooth shapes selected. Change their fill to white. Choose Object|Show All. Figure 3.31 shows David with his teeth filled with white. You can fix David's smile more by poking around to find more black shapes that look better filled with white. The area between the two left teeth can be filled with white.

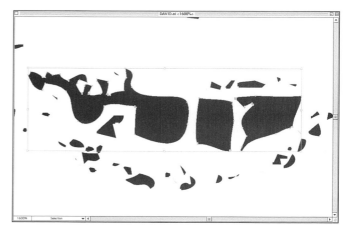

Figure 3.29 (left)
David's mouth is selected.

Figure 3.30 (right)
David's teeth are selected so that they can be filled with white.

25. Some black shapes can still be seen on top of David's teeth. They add a woodcut feel to the image. If you don't like them, select these spots and delete them. If you want to make up your mind later, create a new layer and onto it, move the "tooth spots." This way, you can view the image "with" and "without" as many times as you want until you decide.

Notice how much less detail shows in the stripes in David's shirt when compared to the image that you saved in Photoshop. The reason for this is the same as the reason for the black teeth. Photoshop creates what actually are compound paths but does not write compound paths to the exported file. Therefore, when you import the Photoshop paths and fill them with black, every path is filled with black, and nothing is subtracted, as it would be in a true compound path. If you like the way the stripes look, you can leave them. You can recover either all the lost detail (a bit tedious) or most of the lost detail (much easier). Follow along.

26. If you want to recover most of the woodcut-like detail, drag a marquee with the Selection tool across the bottom half of the image, as shown in Figure 3.32. This selects all of David's shirt and doesn't select the short line that defines his chin. Choose Object|Compound Paths|Make

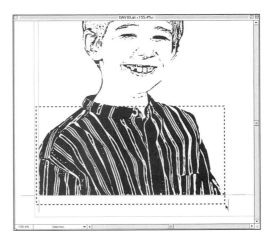

(Mac: Command+8; Windows: Ctrl+8). Figure 3.33 shows the image with the compound path. Although I have included a tip for recovering all the image detail, I recommend that you try this on a copy of the working image. For the steps that you will use to finish the Illustration, you really need the compound path because you actually need the areas that *look* white to be transparent.

Figure 3.31 (left)
When you show all the objects, David's mouth looks much better.

Figure 3.32 (right)
Carefully drag a marquee across David's shirt to recover the woodcut look.

Style Is The Key

After the line art is colored as you need it to be in Illustrator, turn your attention to making the image into a stylized finished project. Because we have chosen to make this a vector drawing, let's place a background and create an only-possible-in-Illustrator effect on it. The Illustration continues, but we'll start the numbering back to Step 1 because this is a new phase to the example. You can peek at the final image in color in the Color Studio section of this book, if you want to see where we are heading with this image.

1. In Photoshop, open the original DAVID.PSD image (the one from the CD-ROM that you did not edit). Choose Filter|Blur|Gaussian Blur, 30. This applies a significant blur to the image. Change the mode to CMYK (Image|Mode|CMYK). Save the file as DAVIDBLUR.PSD. If necessary, close Photoshop. Figure 3.35 shows the blurred image.

HOW TO GET THE MAXIMUM DETAIL FROM THE IMAGE

To get as much detail as possible in the shirt stripes, you need to do more work—and *not* create a compound path. First, select the large black shapes in the shirt (there are only three or four). As you find each major piece, hide it from view. Finally, select the entire shirt area (with the large pieces hidden) and move the remaining pieces to a new layer. Set the fill for these objects to white. Show all of the image. The major pieces are on the bottom layer, and you can protect them from change by locking the layer (click in the Lock Layer column on the Layers palette). Some of the shapes that you just filled with white should really be filled with black. You need to find and fill them individually—basically, by clicking where you think they might be. No shortcut exists that I can recommend for this. Figure 3.34 shows the image with as much detail restored (and it might be more detail than you actually want).

Figure 3.33 (top)
Creating a compound path is an easy way to restore detail to the stripes in David's shirt.

Figure 3.34 (bottom)
You can recover even more detail in the stripes if you seek and select the shapes by hand.

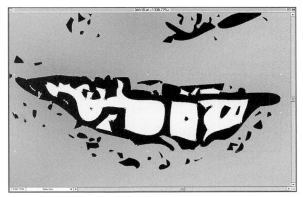

Figure 3.35
A 30-pixel blur applied to the entire image makes it into a suitable background.

Figure 3.36
All the objects in David's mouth area are selected so that a compound path can be created.

2. In Illustrator, first create a new layer and drag it to the bottom of the layer stack. Leave it as the active layer and choose File|Place. Select the DAVIDBLUR.PSD image that you just saved (yes, it's also in the folder on the CD-ROM, just in case you are feeling lazy). Drag the image into position behind the line art layer.

3. It looks good—expect for the glaringly white teeth! We need to do something about that. I warned you that a compound path was needed rather than white paths so that David's shirt will look right. Figure 3.36 shows an enlargement of David's teeth with all the objects inside his mouth selected. You need to do the same thing on your image (select all the pieces of his mouth). Now, choose Object|Compound Paths|Make. Like magic, the white areas of the teeth drop out, and the problem disappears (as you can see in Figure 3.37).

4. Lock all the layers in the Layers palette so that nothing in the image can be changed. Then create a new layer (click on the New Layer icon at the bottom of the Layers palette). Drag the empty layer so that it rests just above the bottom layer of the image.

5. Create a rectangle that's the same size and in the same location as the imported background image. You can do this by eye, as the exact measurements are not critical.

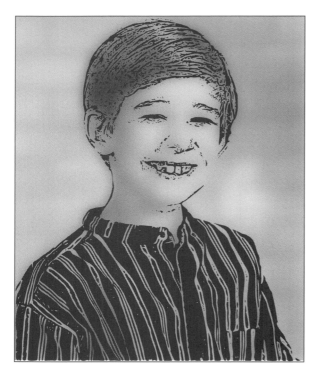 

Figure 3.37 (left)

After you make a compound path, the white areas of David's teeth become transparent.

Figure 3.38 (right)

A gradient fill has been placed over the background in preparation for the next step.

6. Change the fill for the rectangle to a black-to-white gradient with the black area on the right, as shown in Figure 3.38.

7. Use the Hatch Effects feature in Illustrator to change the gradient rectangle into a series of gradated objects that form a texture over the background of the image. Choose Filter|Pen and Ink|Hatch Effects. A complex dialog box, shown in Figure 3.39, appears. For now, just enter the settings I request. Choose these settings:

- Hatch Effect: Angled Lines Gradient

- Hatch: Swatch

- Match Object's Color: Off

- Keep Object's Fill Color: Off

- Density: 40%

- Dispersion: 30%–200%, Random

- Thickness: N/A

- Scale: 200%, Constant

- Rotation: –70%, Constant

- Fade: Use Gradient, 0%

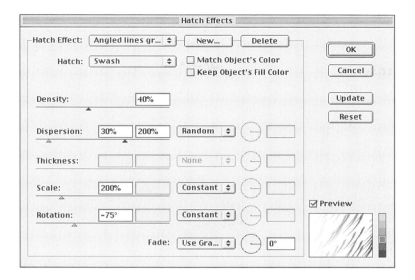

Figure 3.39
The Hatch Effects settings for the gradation that cover the background image imported from Photoshop.

Figure 3.40 shows the result—just what's needed to perk up the background image, except that it also covers David's face, hair, and shirt. We'll fix that next.

You need to create a mask so that the hatch effects show up only in the background of the image—not inside David, where it looks odd. The mask that you need already exists in Photoshop; let's retrieve it.

Figure 3.40
The hatch pattern adds an interesting texture but is out of place on David's hair, face, and shirt.

8. In Photoshop, open the original DAVID.PSD image (or, if you already have it open, you can use one of the other copies). The David Outline path is the first path in the Paths palette. Press Command (Mac) or Ctrl (Windows) and click on the David Outline Path palette entry to load it as a selection. This selection is the *opposite* of what we need. An Illustrator mask shows only what is inside the masked area (which is the exact opposite of how a Photoshop layer mask works). Reverse the selection (Select|Inverse, Mac: Shift+Command+I; Windows: Shift+Ctrl+I). In the Paths palette, choose Make Work Path from the side menu. Use a Tolerance of 1. Double-click on the path name to save it.

9. Select the Arrow tool from the Pen tool pop-out menu in the Toolbox. Position your image so that the window that contains it is larger than the image itself. With the Arrow tool, drag a marquee around the entire image. This selects all the points on the path. Copy it to the clipboard (Mac: Command+C; Windows: Ctrl+C). Close Photoshop, if necessary.

10. In Illustrator, add a new layer (click on the New Layer icon at the bottom of the Layers palette). This layer should be directly above the Hatch Effects layer. All the layers (except the new one) should be locked. With the new layer active, paste in the path from the clipboard (Mac: Command+V; Windows: Ctrl+V). Press D to set the colors back to the default of black stroke and white fill. Drag the path into place so that it lines up with the background and the hatch effects.

11. Remove the lock on the Hatch Effects layer (click on the second column to remove the pen-with-a-line-through-it icon). Select the entire unlocked image (Mac: Command+A; Windows: Ctrl+A). Choose Object|Masks|Make (Mac: Command+7; Windows: Ctrl+7). Although the hatched objects are already masked by the rectangle that originally contained the gradient, they accept the new mask. Finally, the vision of a woodcut image with texture and some color is realized, and it took both Photoshop and Illustrator to do it. Figure 3.41 shows the finished image.

If you check the Links palette, it shows that you need DAVIDBLUR.PSD available when you print.

This was a long example, but it shows a good use of features from both Photoshop and Illustrator and, because of the Hatch Effects, also shows a technique that couldn't be done in Photoshop alone. The other benefit of placing the line art layer in Illustrator rather than leaving it in Photoshop is that you have gained the ability to print the image at varying sizes,

Figure 3.41
David—a stylized version of a photograph.

with no loss of resolution. Because of the blur in the background image DAVIDBLUR.PSD, you can scale up the image much more than would be possible with a finely detail photographic image, and the vector objects are infinitely scalable.

Clipping Paths

You can save Photoshop images that have clipping paths in them and place them into Illustrator. Actually, this topic had trouble finding a home for itself. So many ways are available to use and create clipping paths or masks that I had trouble trying to decide whether it was an Illustrator-to-Photoshop thing or a Photoshop-to-Illustrator thing. In reality, it is both or either.

You can do the following:

- Create a shape in Illustrator, pass the path to Photoshop, and create a clipping path in Photoshop.

- Create the entire clipping path and image in Photoshop and send it to Illustrator.

- Place a Photoshop image into Illustrator and create a shape in Illustrator to use as a mask.

What's the difference between a clipping path and an Illustrator mask? Only the name and the way in which it's used differ. A clipping path is a vector object you use to cut away part of a raster image. It gives a smooth, hard edge to a raster image when the EPS file that holds the clipping path is printed from a PostScript program. A masked object in Illustrator is one in which a vector shape is used to reveal only part of another vector or raster shape in the program.

You might want to mask out part of a raster image for a number of reasons. Typically, you do this if you need to change the background of the image—for example, if you have a female model and want to place her against a solid background. You can also use a clipping path to trim a texture and make it conform to a shape. The most successful clipping path is one in which the image is a bit larger than the path that contains it. Because a clipping path is a PostScript vector object, it prints at the resolution of the printer. Therefore, the exact borders of the path are not set until the image goes through the rip. At that time, the outline is matched to the printing grid. The clipping path then permits any part of the raster image that falls inside of it to print. Until that moment, it isn't clear exactly which pixels near the edge of the clipping path will actually print. If there is extra raster image to cut off, the edge will always be smooth and crisp, with no accidental white space or background peeking through.

The other important thing to remember about a clipping path is that you should have the boundary of the rectangular image as close to the borders of the clipping path as possible. When the file is imaged by the rip, the rip must draw the entire raster image in memory before it decides what to clip off. Therefore, to have a small clipping path on a large image is a waste of valuable and expensive rip time.

In the section "Out, Around, And About" later in this chapter, you will have a chance to work on a greeting card that contains clipping paths and masks.

Critical Locations

You might want to place Photoshop images into Adobe Illustrator to check for placement so that you can combine the image at printing or so that you can link images that need Illustrator elements. Illustration 3.4 uses an example of an image that started in Illustrator, was built in Photoshop, and was sent back to Illustrator to retain a sharp outline. Throughout the process, the Photoshop and Illustrator documents need to be kept in register.

WARNING: POSSIBLE SWATCHES PALETTE BUG ALERT!

I have had trouble with seeing all my swatches since one of the early betas for Illustrator 8. When the default Swatches palette appears, much of it is shown in black on black. However, all the swatches are present. If you cannot find anything in your Swatches palette that looks like a rainbow gradient, use the side menu on the palette to view the entries by name. Now, you can easily choose the rainbow gradient.

To fix this bug, make sure that you have the latest drivers for your graphics card.

In this following project, you will play in a free manner with the new gradient mesh, add some calligraphic pen flowers, and export the image to Photoshop. In Photoshop, you add some effects to the flowers, save the image, and place it back into Illustrator, where the project will finish. This is an easy, lighthearted project to complete, but it shows you another facet of the Photoshop/Illustrator connection.

PROJECT Illustration 3.4: Rainbow Flowers (Are Falling On My Head)

1. In Illustrator, create a new document (File|New).

2. In the Toolbox, set your colors to black fill with no stroke. Choose the Rectangle tool. Drag your cursor into the center of the image (by eye) and Option (Mac) or Ctrl (Windows) as you click. This opens the New Rectangle box. Enter inch values "6" for the width and "8" for the height.

3. Click on the Rainbow gradient in the Swatches palette to set the fill for the rectangle.

4. Show the Gradient palette (see Figure 3.42). Click the first color slider to the left. This makes that color visible in the Color palette. Drag the color from the Color palette to your Swatches palette. Repeat this for every color in the gradient slider (there are six colors). Why? This gives you easy access to the correct colors when you start mucking around with the gradient mesh in the next few steps.

5. Choose Object|Expand. You see the dialog box shown in Figure 3.43. Choose Expand Fill and Gradient To Gradient Mesh to turn the rainbow gradient into a gradient mesh object. A line appears down the center of each color band (which makes six vertical lines, if you count the outside edges). You learn about the Gradient Mesh tool in Chapter 4, but for now just follow the instructions without a lot of explanation.

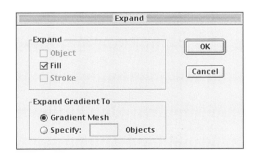

Figure 3.42 (left)
The Gradient palette defines the colors in the gradient.

Figure 3.43 (right)
The Expand dialog box allows you to change an existing gradient into a gradient mesh object.

6. Select the Gradient Mesh tool from the Toolbox. This tool can be tricky if you don't know what it is (or what you are) doing. Figure 3.44 might allow you to see the difference between a selected and an unselected point along the gradient mesh. A selected point is solid, whereas an unselected point is hollow. With the Gradient Mesh tool, carefully click on the center of the vertical line that runs through the yellow color on the gradient. If you click directly on the line, you leave a horizontal line where you click. (If you miss the line, you leave both a horizontal and a vertical line at the click point—quickly Undo if this happens and try again.) Click on the same vertical line two more times and try to space the clicks fairly evenly between the center click and the top or bottom of the image. You have now added three horizontal control lines to the gradient mesh.

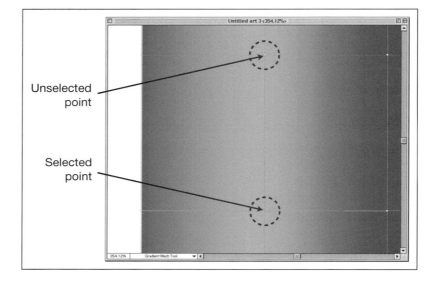

Figure 3.44
Selected points are solid and unselected points are hollow in a gradient mesh.

7. If you move the cursor on top of a control point, the cursor icon for the Gradient Mesh tool loses its plus sign (+); if you click on one of the control points (and hit it exactly), the point becomes solid. Click on some of the control points and then move them slightly to distort the color mix in the gradient. Figure 3.45 shows (I hope) some of the moved gradient mesh points (because the image is in grayscale, examine the shape of the gradient mesh lines—the color shading is meaningless).

8. When you feel comfortable moving the mesh points, click on a control point and then click on one of the gradient colors that you moved into the Swatches palette at the start of this exercise. The area around the control point changes to the clicked-on color, which blends into the gradient. Place a few "out-of-order" areas into the gradient mesh (keep the colors of the gradient, however).

9. Lock the layer and create a new one on top of the gradient mesh layer. Choose both the Paintbrush tool and the 10-pt. Oval brush from the Brushes palette. Even if you feel that you can't draw, you can draw a five-petal daisy. Draw one similar to the one shown in Figure 3.46. The nice thing about a vector program is that, if you aren't happy with your initial effort, you can fine-tune the daisy after drawing it.

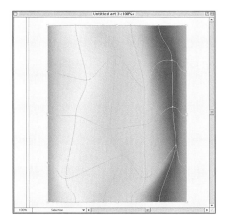 

10. Use the Selection tool to select the daisy. Press Option (Mac) or Alt (Windows) to drag a copy of the daisy to a new location. With the copy still selected, make a new layer (click on the New Layer icon at the bottom of the Layers palette). Drag the dot on the previous layer to the new layer to move the selected object into the new layer. Repeat this procedure to create a third daisy (also on its own layer). Your image should now contain four layers. Arrange the daisies in a way that pleases you. Figure 3.47 shows my arrangement, in which two of the daisies are out of the gradient mesh boundary.

11. Create a new layer and drag it to the top (if it isn't created there). Turn on Smart Guides (Mac: Command+U; Windows: Ctrl+U). Select the Rectangle tool. As you drag the cursor around the image, look for the prompt from the Smart Guides that you are over the center of the gradient mesh object. When you find the center, press Option (Mac) or Alt (Windows) and click. Create a rectangle 6 inches wide by 8 inches high. This should align exactly with the gradient mesh.

Figure 3.45 (left)
You can move the gradient mesh control points to alter the mix of the colors.

Figure 3.46 (right)
Anyone can draw a five-petal daisy.

Figure 3.47
One possible arrangement of
three daisies.

12. Turn off Smart Guides (Mac: Command+U; Windows: Ctrl+U). Select
    the entire image (Mac: Command+A; Windows: Ctrl+A). Make a
    mask (Object|Masks|Make, Mac: Command+7; Windows: Ctrl+7).
    The rectangle on top of the objects hides the parts of the daisy that
    are outside the gradient mesh boundary.

13. Export the image (File|Export). Choose the Photoshop 5 format and
    name the file FLOWERLAYERS.PSD. I used a medium resolution (as
    you can see in Figure 3.48), but you may save it at 72 dpi if that is
    easier for you. Use CMYK. Unmark the Anti-Alias checkbox and
    select the Write Layers checkbox as shown. Click on OK.

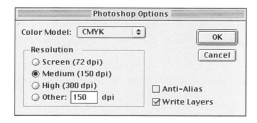

Figure 3.48
Here are the export options that I
used to save the flower to a
Photoshop file.

14. Save the file that you are using in Illustrator as FLOWER.AI and close
    Illustrator (if you do not have enough RAM to open Photoshop as well).

15. In Photoshop, open the image FLOWERLAYERS.PSD. As you can see
    in Figure 3.49, all five layers appear. Turn off the Eye next to the
    Background layer and Layer 1. Make Layer 5 active.

16. Create a merged image in the active layer (Mac: Shift+Option+ Command+E; Windows: Shift+Alt+Ctrl+E). You can now drag Layers 2 and 3 to the Layer palette trashcan. You will not need them again.

17. Select the Move tool (M). Turn the Eye on to reveal Layer 1, and turn the Eye off on Layer 5 to hide it for the moment. Move the flower in Layer 3 around until you like the part of the gradient that shows through it.

18. Press D to set the colors back to the default of black and white. Using the Paintbucket tool (K), fill the flower with black. Because you did not anti-alias the image, it fills perfectly. Select the nontransparent Layer pixels (Mac: Command+click; Windows: Ctrl+click on the layer name in the Layers palette). Drag Layer 3 (the flower layer) to the Layer palette trashcan and make Layer 1 the active layer.

19. Create a New Layer Via Copy from the selection (Mac: Command+J; Windows: Ctrl+J). Turn the Eye back on to reveal the outlines in Layer 5. Let Layer 6 (the new flower layer) remain active. Drag the flower underneath the outline for the lower-left flower. Do not register exactly. I left it somewhat to the right and above the outline, as if it were a watercolor that was not outlined exactly over the image. Figure 3.50 shows this in an image that was artificially constructed (in terms of color) to show the offset, as gray on gray doesn't print very well.

20. While still using the Move tool, press Option (Mac) or Alt (Windows) and drag a copy of the flower so that it's offset under the

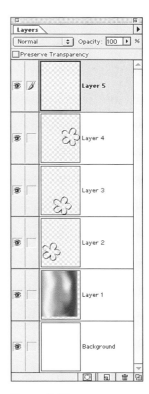

Figure 3.49

The exported Illustrator file places each Illustrator layer into its own Photoshop layer.

Figure 3.50

This simulation of the image shows the offset of the gray flower behind the lower-left flower outline.

next outline. The Option/Alt key copies as you drag and makes a new layer in the process.

21. Drag the icon for Layer 1 to the New Layer icon (the center icon at the bottom of the Layers palette). This creates a Layer 1 copy. Drag this layer to below the top layer (Layer 5) in the Layers palette. Change the Apply mode to Dissolve and lower the layer opacity to 20%. This creates a type of noise that you cannot duplicate with the Noise filter or with any facility in Illustrator.

22. Make Layer 6 active. Choose Layer|Effects|Drop Shadow. Figure 3.51 shows the settings that I chose. Figures 3.52 and 3.53 show the settings that I used on the other two flower layers (the angle and the blur differ for each daisy).

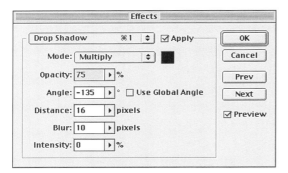

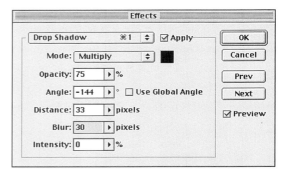

Figure 3.51 (top left)
The Layer Effects dialog box settings used to create a drop shadow for the first daisy.

Figure 3.52 (top right)
The Layer Effects dialog box settings used to create a drop shadow for the second daisy.

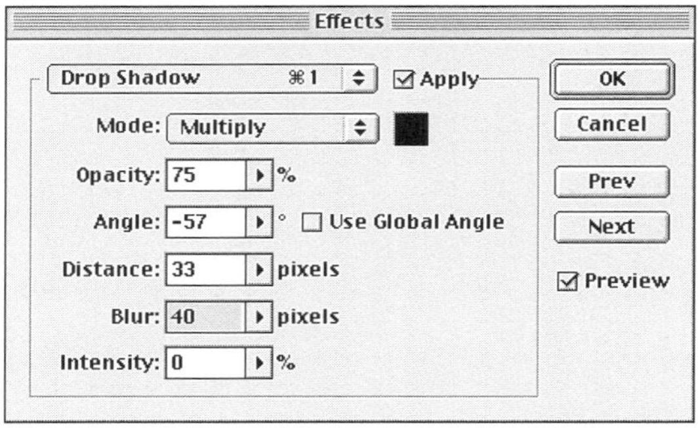

Figure 3.53 (bottom)
The Layer Effects dialog box settings used to create a drop shadow for the third daisy.

23. If the daisy edges look too hard, you can apply a Gaussian Blur of 1 to 2 pixels to each daisy layer (Filter|Blur|Gaussian Blur).

24. Make Layer 1 active. Select the nontransparent layer pixels (Mac: Command+click; Windows: Ctrl+click on the layer name in the Layers palette). Choose Image|Crop. Figure 3.54 shows the result.

25. Drag Layer 5 to the Layer palette trashcan. Choose Layer|Flatten Image. Save the flattened image as BACKGROUND.PSD. Close Photoshop.

Figure 3.54
The flower image is cropped to
the boundaries of the gradient
mesh layer.

26. Back in Illustrator, open the Illustrator image-in-progress if it isn't
    already open. Select the gradient mesh object on Layer 1 and delete
    it. Leave Layer 1 as the active layer.

27. Choose File|Place and place the BACKGROUND.PSD layer as a
    linked file. You can easily position it by lining up the bottom and left
    edges against the cut flower outlines caused by the mask. The image
    should look much like it did in Photoshop when the outline layer
    was visible. However, the outlines here will print smoothly, which
    they would not have done in Photoshop.

28. We have a bit more to clean up. First, you need to remove the part of
    the daisy that overlaps another (if you have any of these areas). Figure
    3.55 shows the area on my image that needs to be fixed. Use the Direct
    Selection tool to select the one lower daisy and then the Scissors tool to
    cut away the area of overlap. Notice on the fixed image in Figure 3.56
    that I cut away excess outline from flower to flower, rather than outline
    to outline. After you make the cut, select and delete the cut area.

29. As a final touch, make a center for the daisies. Choose the Ellipse
    tool with a black fill and no stroke. Press Shift+Option (Mac) or
    Shift+Alt (Windows) and click in the center of each daisy to drag out
    a perfect circle. The circles should be similar in size, but they do not
    need to be exact.

30. Use the Selection tool to select one of the circles. Choose Filter|Distort|
    Roughen. Figure 3.57 shows the filter dialog box and the settings

Figure 3.55 (left)

This area has overlapping outlines that need to be cleaned up.

Figure 3.56 (right)

The Scissors tool cuts away the excess outline.

that I used. Save and print your image. Figure 3.58 shows the finished image in grayscale.

You might wonder why we flattened the entire image in Photoshop and didn't just save the flowers to place on top of the gradient mesh object. If we had done that, the placed image would have had a white background, and you would not have been able to see the gradient mesh object once the Photoshop file was placed on top of it. You would have needed to create a mask, and the soft edge of the drop shadow that we created in Photoshop is almost impossible to properly mask. Therefore, the image needed to be saved as an entire background in Photoshop and placed into Illustrator to take advantage of the smooth vector flower shapes and centers.

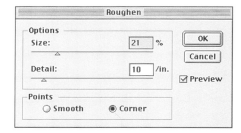

Figure 3.57 (left)

The Roughen filter adds some variety to the center of the daisies.

Figure 3.58 (right)

The finished flower image.

That Touchable Texture

You can use Photoshop to create bitmap textures that add interest to an Illustrator image. In Illustration 3.5, you develop a texture from a photograph that I took of a rhinoceros in the zoo and use this texture to enhance the image in Illustrator. Because a bitmap EPS file allows the white to drop out, no mask or clipping path is needed.

Illustration 3.5: Rhino Texture

1. In Illustrator, create a new document.

2. Choose File|Place, then select the image RHINOTEMP.PSD. In the dialog box, mark the Template checkbox, as shown in Figure 3.59. Figure 3.60 shows the Layers palette with the placed image as a write-protected template.

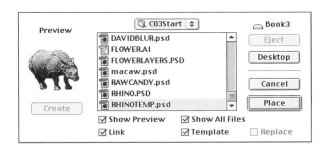

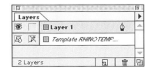

Figure 3.59 (left)
You can place a Photoshop file into a template layer that is dimmed and write-protected.

Figure 3.60 (right)
The Layers palette shows the write-protected template layer.

3. Using the Paintbrush tool and the 6-point flat calligraphic brush (the first one in the default Brushes palette), stroke over the most significant lines in the image in a manner similar to the one shown in Figure 3.61. Figure 3.62 shows just the lines (without the template).

Figure 3.61
You can create a sketch of the rhino by stroking his most significant lines based on the template.

Figure 3.62

When you remove the template, the strokes should be able to stand up on their own.

4. In Photoshop, open RHINO.PSD. This is the original image, and I added a path so that you can easily create a rhino shape. In the Paths palette, click on the Rhino Outline path. Choose the arrow tool from the Pen tool flyaway in the Toolbox. Drag a marquee with the Arrow tool around the entire path to select it and copy it to the Clipboard (Mac: Command+C; Windows: Ctrl+C).

5. In Illustrator, paste in the path from the Clipboard (Mac: Command+V; Windows: Ctrl+V). Create a new layer and transfer the selected path to the new layer. Drag the layer in the Layers palette into position above the template layer. Drag the path on top of the rhino so that it falls into place.

6. With the Eyedropper tool, pick up a rhino color from the template layer. Use this color as the fill for the rhino path. The stroke should be set to None.

7. Now that the path is colored, offset it a bit from the calligraphic strokes in a manner similar to Figure 3.63.

8. Save the image as RHINO.AI. You now need to develop a texture for the rhino in Photoshop.

9. Back in Photoshop, open the RHINO.PSD image (if it isn't already open).

You can develop a texture in many ways. The result needs to be a bitmap (black and white) EPS file, with transparent whites. To get to that point, you could convert your image to grayscale and then into a bitmap by using the 50% Threshold, Diffusion Dither, Pattern Dither, or Halftone conversions built into the grayscale-to-bitmap conversion process. Based

Figure 3.63
The stroked and colored rhino.

on your original image, you might (or you might not) like what happens. The trick exerts some control over the resulting bitmap. You can do this by controlling the values in your image and by controlling the way in which the bitmap pattern is created. As a first try, use the Reticulation filter to create an interesting pattern that lends itself to being bitmapped.

10. Choose Image|Duplicate, OK to make another copy of the RHINO.PSD image. Drag the Background layer to the Layers palette trashcan. This leaves only the rhino itself. Select the nontransparent layer pixels (Mac: Command+click; Windows: Ctrl+click on the layer name in the Layers palette). Select Layer|Flatten. The rhino stays selected.

11. Choose Image|Mode|Grayscale, then click on OK to discard color. The rhino is still selected.

12. Open the Levels dialog box (Mac: Command+L; Windows: Ctrl+L). Drag the White Input slider to the left until it reaches the start of the data in the image. Drag the Black Output slider to the right until it reads 128, as shown in Figure 3.64. This cuts out half the values in

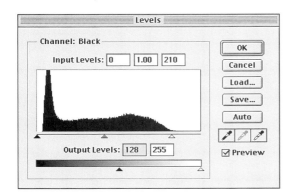

Figure 3.64
The Levels command is used to remove all the dark values from the image.

the image and significantly lightens it, which makes a small but critical difference to the filter that you'll apply next.

13. Choose Filter|Sketch|Reticulation and use the settings shown in Figure 3.65. Use a Density of 40, a Black Level of 16, and a White Level of 5. Figure 3.66 shows the filtered rhino.

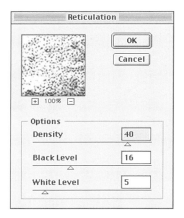

Figure 3.65 (left)
The Reticulation filter brings out image detail as texture.

Figure 3.66 (right)
The rhino has been filtered with the Reticulation filter.

14. Invoke the Levels command again, but this time press Option+Command+L (Mac) or Alt+Ctrl+L (Windows) to reuse the last settings (the ones that you used in Step 12). Click on OK.

15. Choose Image|Mode|Bitmap and use the Diffusion Dither, as shown in Figure 3.67. Figure 3.68 shows the result. Because we used the Levels command to remove the dark values, the result is fairly light and unobtrusive.

16. Save the image as RHINOBMAP.EPS. Make sure that you mark the Transparent Whites checkbox in the dialog box, as shown in Figure 3.69.

17. In Illustrator, open the RHINO.AI image if it's closed. Create a new layer (click on the New Layer icon at the bottom of the Layers palette). Lock the original layers.

18. Choose File|Place and select the RHINOBMAP.EPS file. Unmark the Link checkbox in the Place dialog box. You need to embed the EPS image, or the image preview is truly rotten. The file is placed on the top layer.

19. Drag the layer to the middle of the layer stack (between the fill and the outlines).

20. Drag Layer 3 into position. Figure 3.70 shows the finished image.

TRY THE SKETCH FILTERS

The Sketch filters were part of the original Adobe Gallery Effects series of filters (formerly Aldus Gallery Effects and Silicon Software Gallery Effects). These filters generally create two-color versions of the image that look similar to sketches (therefore, the filter category). They range from mezzotint-type effects to posterization effects. Try them all when you have the time to experiment.

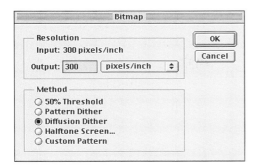

You can force a bit more contrast into the final bitmap by using the Threshold command to separately process the darkest areas of the image. RHINOADJ.PSD shows how I did this. I copied the reticulation-filtered image to a new file and duplicated the starting image (the top layer in the RHINOADJ.PSD image, and it has the Eye turned off). I used the

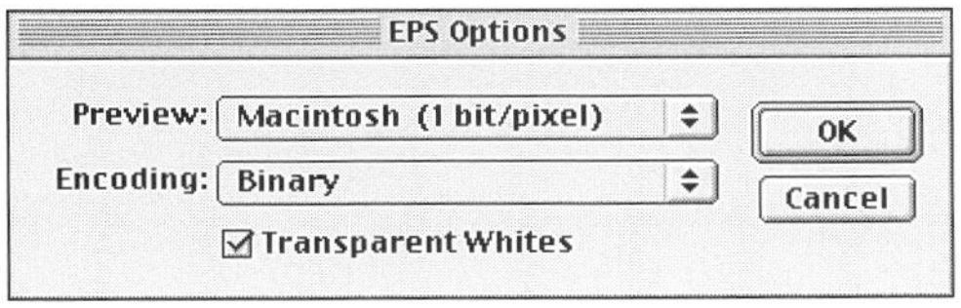

Figure 3.67 (left)
You can convert an image to Bitmap mode in a variety of ways. Here, we use the Diffusion Dither.

Figure 3.68 (right)
The rhino is now in bitmap form.

Threshold command on this layer to make a solid area of black. I then touched up the black area to make it even more solid and painted on the layer in white to separate the black areas so that I could easily select them with the Magic Wand. I used this to build a selection mask.

I duplicated the Background layer again and, with the selection still active, created a Threshold Adjustment layer with a Threshold of 69. This confined the Threshold operation to only the area that had been selected. I then used black with a soft Paintbrush to create a blend area between

Figure 3.69 (left)
The EPS Options dialog box.

Figure 3.70 (right)
The Diffusion-dithered EPS image adds an interesting texture to the rhino.

the area that was thresholded and the area that was not. At that point, I saved RHINOADJ.PSD. Finally, I duplicated the image as merged layers to a new file and converted it to a bitmap. Figure 3.71 shows this version (which has more detail but required more effort).

Figure 3.71
A more detailed texture can be created with a bit more work.

21. Before you leave this topic, let's see what happens to the bitmap if it's created by using a custom pattern. In Photoshop, if you cannot use the History palette to get back to the last Levels command from Step 15, open the image RHINORETIC.PSD. This is my image from that step.

22. Open the image STARPAT.PSD. Select the entire image (Mac: Command+A; Windows: Ctrl+A). Define this as a pattern (Edit| Define Pattern).

23. Make the RHINORETIC.PSD image active. Choose Image|Mode|Bitmap and use the Custom Pattern option at 300 ppi. Save the image as RHINOSTAR.EPS with Transparent Whites in Photoshop EPS format.

24. Hide Layer 3 in the RHINO.AI image. Create a new layer and place the EPS file as an embedded file. Repeat Steps 19 and 20. Figure 3.72 shows the result.

Rasterizing

You don't always need Photoshop to change vector images into raster images. Illustrator allows you to rasterize images within Illustrator and then use them as embedded objects. This cuts out a step if you only want to create some objects, manipulate them as pixels, and then leave them in Illustrator.

DO YOU WANT A LARGER AND LESS DENSE TEXTURE?

If you look at the bitmapped image and would prefer a texture that has larger dots that were not as closely packed, you can create that effect in two ways. You can create the bitmap at 72 ppi and the same physical size in inches, or you can reduce the image before it is bitmapped to about 33 percent of its size, change it to a bitmap, and enlarge by 300 percent (keep track of the number of pixels so that the file goes back to its original size). Either way, you get a texture that is much more coarse.

Figure 3.72
Starry, starry rhino is complete.

Filtered Pixels

The main reason that you might want to rasterize areas of an Illustrator image is so that you can apply Photoshop filters to them. You can apply filters to RGB raster objects and sometimes to CMYK or grayscale raster objects. You might also rasterize an object when you want to use the Create Object Mosaic or the Photo Crosshatch filters, which work only on raster objects.

Be aware, however, that Photoshop still gives you more control because you can fade filter effects and change their Apply modes. Also, because you cannot enlarge the size of the bounding box that surrounds the raster object, any filter that visually seems to increase the number of pixels in the image (such as the Gaussian Blur filter) might look as if it is cutting off part of the filter effect. Therefore, use the Rasterize command for a fast filter, but depend on Photoshop for the rest.

Rasterize For The Nice Rip

The "nice rip" has a nasty habit of both taking a long time to rasterize a complex Illustrator image and then, on occasion, ending up with a PostScript limit-check error for its efforts. A limit-check error occurs when too many points exist on an object for the rip to handle. Too many points occur when a clipping path or mask contains an overabundance of points or when the object has simply become too complex. While steps exist that you can take in Illustrator to help avoid the problem (increasing Flatness settings, breaking up long paths, or keeping compound paths to the minimum), you also can rasterize the problem. After it is changed to a raster image, the rip no longer needs to do anything other than write it out—all the processing is already done.

NOT CUTTING IT OFF

What do you do when you need to apply a filter, such as a Gaussian Blur, and you really do not or cannot use Photoshop? If you add an unstroked, unfilled rectangle around the object that you want to blur and select it as well as the object when you choose the Rasterize command, the rectangle (rather than the smaller object) will define the raster object's bounding box. This is an easy way to give yourself room for a filter when you rasterize.

I once had a service bureau tell me that it was taking a long time to rip my files because I had used gradient. I pointed out to them the fact that the gradient happened to be located in a *Photoshop* file, but they couldn't see why that mattered (I used their services only once). Believe me, it matters. A gradient in a Photoshop file has no need to be rasterized—it already is. It cannot possibly choke the rip. If you find that an *Illustrator* gradient is too complex, rasterize it. If you find that the gradient is banding, rasterize it and place it in Photoshop, where you can add a small but different amount of noise to three of the four channels. The banding will disappear.

Out, Around, And About

Many images spend a lot of time in transit between Photoshop and Illustrator. They start out in one, visit the other, and end up back where they started (or go through the cycle again). All my images end their journey in a page-layout program (usually QuarkXPress—although I'm not endorsing it over PageMaker). I don't need to place an image into a page-layout program if it's a single page. However, I know that most service bureaus are more comfortable producing output from a Quark file, so I send them one. Process-color output is too expensive for me to experiment happily (I would rather spend my extra income on clothes, books, software, yarn, and Broadway shows). Therefore, you don't have to use a page-layout program, but you will probably do so anyway.

Before we take a whack at compositing in page-layout programs, look at two situations in which you will almost always go back and forth a few times between applications.

Crossing Clipping Paths

As we discussed previously, the shapes for clipping paths or masks can easily be created in either Illustrator or Photoshop. They can then be traded where they will do the most good. In Illustration 3.6, we will try out a variety of ways as you create a holiday greeting card. I like the technique that we will use so well that, for the moment, I am going to dodge the issue of "Will it print?" Let's just look at it on screen!

PROJECT Illustration 3.6: Season's Greetings

1. In Illustrator, open the image XMASTEXT.AI. Figure 3.73 shows the two small snippets of text that the file contains.

2. Using the Selection tool, select the Merry Christmas object. This text was turned into outlines. The original font is ErikRightHand (from

MAKING A PATTERN BRUSH FROM TEXT

When you create a Pattern brush from text, you need to create outlines from the text before you can create the pattern.

the very talented type designers at FontFont). The object also contains a diamond from the Zapf Dingbats font. Copy the selection to the clipboard (Mac: Command+C; Windows: Ctrl+C).

3. Create a new document (Mac: Command+N; Windows: Ctrl+N). Paste in the selection from the clipboard (Mac: Command+V; Windows: Ctrl+V).

4. Drag the selection to the Brushes palette. The dialog box shown in Figure 3.74 appears. Click on Pattern Brush.

5. After you decide to create a new Pattern brush, the dialog box shown in Figure 3.75 asks for more details. Name the brush and give it a 200% Scale with 5% Spacing. Let it approximate the path. You do not need to flip it or colorize it in any way.

Figure 3.73
"Merry Christmas" and "TO YOU" will be used in the Season's Greetings project to construct a greeting card.

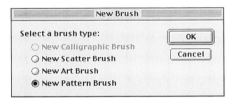

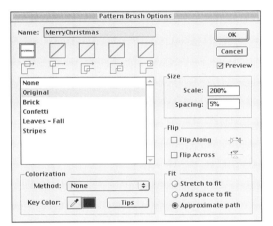

Figure 3.74 (left)
The New Brush dialog box allows you to specify the type of brush that you want to create.

Figure 3.75 (right)
The Pattern Brush Options dialog box allows you to name and size the brush.

6. Delete the text that you used to create the Pattern brush. Because the text is saved as a brush, you won't need it again.

7. Select the Paintbrush tool and draw a curvy path that goes back and forth to describe a stylized version of a Christmas tree, as shown in Figure 3.76 (this image was altered to show you the path). As soon as you draw your path, it should look like Figure 3.77 (unless you have accidentally set a fill). If your fill is not set to None, change it now.

8. To change the shape of the path, you can use the Direct Selection tool to edit it as you would any other Illustrator path. Press Shift and click on a point along the path to let the tool know where to begin. If you want to make changes in the scale of the pattern or in its spacing, double-click on the Brush entry in the Brushes palette. If you cannot produce the shape that you want and get frustrated (although it isn't difficult to draw or edit the path), you can open my

Figure 3.76 (left)
Draw this path with the Paint-brush tool.

Figure 3.77 (right)
This is how your brush pattern should look.

ready-made TREEPATH.AI image, select the path, and choose the Christmas Tree brush with the Paintbrush tool.

You will paste the tree path into Photoshop as a path. The next few steps are in preparation.

9. Select the tree-text-on-a-path object. Choose Object|Expand and expand the Stroke and the Object. Select None. Drag the Selection tool close to one of the curves on the tree and select only the original, curvy path that you drew. Do not select any text. Delete the original path. If you don't delete the starting path, the rest of this Illustration won't work properly. Figure 3.78 shows a comparison of the tree when it is a selected Pattern brush object (on the left) and when it is a selected expanded object with the original path removed (on the right).

Figure 3.78
(left) The tree-text-on-a-path object. (right) The expanded Pattern brush.

10. Select the entire image (Mac: Command+A; Windows: Ctrl+A). Group the selection (Mac: Command+G; Windows: Ctrl+G). Copy it to the clipboard (Mac: Command+C; Windows: Ctrl+C). Save and close the image (and Illustrator, if you need to reclaim the RAM).

11. In Photoshop, open the image TEXTURE.PSD. This is a mixed Noise and Pixellate texture in a basic Christmas green.

12. Paste in the selection from the clipboard (Mac: Command+V; Windows: Ctrl+V). In the dialog box that appears, choose Paste As Paths.

13. If the paths are larger than the image, zoom out (using the Navigator palette) until you can see both the entire image and the all the paths. Choose the Arrow tool and drag a marquee so that it encloses (selects) all the paths. Use the Select The Free Transform Points command (Mac: Command+T; Windows: Ctrl+T). Drag a corner handle while pressing Shift+Option (Mac) or Shift+Alt (Windows) to transform the paths from the center as you constrain the aspect ratio. When the paths fit into the image, release the mouse button. Press Return/Enter to complete the transformation. (If your paths are much smaller than the image, perform the same set of steps, but enlarge rather than reduce the paths.) Figure 3.79 shows the selected, reduced paths against a background that was significantly lightened so that you can see the selected paths in print.

14. Save the path by double-clicking on the entry in the Paths palette. From the side menu in the Paths palette, choose Clipping Path and identify the path that you want to use (it should be called Path 1). Do not place a number in the Flatness field. When you leave this field blank, the Flatness is calculated from your selected printer.

EXPANDING A PATH PATTERN

When you expand a Pattern brush, you get both the pattern on the path and the original path. Depending on what you need to do with the expanded pattern, you will most likely need to delete the original path.

Figure 3.79
The paths are selected and have been proportionally reduced to fit within the Photoshop image boundaries.

WARNING: LARGE FILE ALERT

This exercise produces a huge file (about 52MB when finished and saved on disk). If this size is not comfortable for you, resize the TEXTURE.PSD image to 72 ppi (by choosing Image|Image Size) and set Resolution to 72 ppi. Mark the Resample Image and Constrain Proportions checkboxes.

15. Save the file as TREECLIP1.EPS as a Photoshop EPS file. Pick an 8-bit preview. The clipping path is automatically saved.

16. With the new TREECLIP1.EPS image still open, create a new Hue/Saturation Adjustment Layer (Mac: Command+click on New Layer icon; Windows: Ctrl+click on New Layer icon). Drag the Hue slider to –121, which turns the texture a nice Christmas red (at least it does on my monitor—fiddle with the suggested setting if your monitor doesn't match mine). Merge Down (Mac: Command+E; Windows: Ctrl+E).

17. Choose File|Save As and name it TREECLIP2.EPS. Close Photoshop if you need to reclaim the RAM.

18. In Illustrator, create a new document (Mac: Command+N; Windows: Ctrl+N). Choose File|Place, then select TREECLIP1.EPS. Unmark the Link checkbox in the Place dialog box. You need to embed this image. Click on OK. Figure 3.80 shows the image at this point.

Figure 3.80

TREECLIP1.EPS embedded in a new Illustrator image.

19. Make a new layer (click on the New Layer icon at the bottom of the Layers palette). Place TREECLIP2.EPS into the new layer. Do not mark the Link checkbox in the Place dialog box.

20. Move the TREECLIP2 image so that it is to the right and down and fits inside the empty spaces in the green image. (This shows you another method of working with the paths.) Figure 3.81 shows both EPS images placed.

21. Make a new layer (click on the New Layer icon at the bottom of the Layers palette). Open the file XMASTEXT.AI if it isn't already open. From the file, copy the words "TO YOU" and paste them into your

Figure 3.81

The red and green layers are offset from each other and look quite festive in color.

working image of the tree. This text extract uses Copal Solid from the Adobe Type Library. Copal has the advantage of being an extremely heavy typeface, which makes it useful when you want to place a design inside the letters.

22. Paste "TO YOU" in position under the tree so that it looks like a short tree trunk. With the object selected, press Option (Mac) or Alt (Windows) and then drag a copy below the first one. Press Shift after you begin to drag-copy the image so that it constrains the direction of the copy. Make another two copies (which gives you four copies of "TO YOU"). Select all four copies and click on Vertical Distribute Top in the Align palette. Group the selection (Mac: Command+G; Windows: Ctrl+G). Figure 3.82 shows the image with the tree trunk in place.

23. Choose Object|Rasterize. In the dialog box that appears, select a high-resolution image (unless you are running short on RAM), anti-alias and make a mask, and make an RGB object. Figure 3.83 shows these setting selected in the dialog box.

24. Choose Filter|Pixelate|Pointillize with a size of 30. Then choose Filter|Brush Strokes|Crosshatch with Stroke Length: 50, Sharpness: 10, and Strength: 3. The large Pointillize values give the Crosshatch filter a good start. Figure 3.84 shows the Crosshatch dialog box, and Figure 3.85 shows a close-up of the filtered raster object. Because you created a mask when you rasterized, only the text shows the effects of the filter.

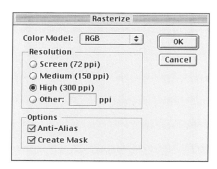

Figure 3.82 (left)

Multiple copies of the words "TO YOU" form the trunk of the tree.

Figure 3.83 (right)

This dialog box allows you to change a vector object into a raster object.

25. When you rasterized the text, you selected an RGB colorspace. The Crosshatch filter doesn't work in CMYK color. Now that you've applied the filter, you need to convert the raster object back to CMYK colorspace. Choose Filter|Colors|Convert to CMYK.

Every Christmas tree needs an angel. This time, let's build an object to rasterize that we will mask with a piece of clip art (the angel from Ultimate Symbol's Design Elements collection).

26. Select the Rectangle tool. Click near the top-left corner of the image to create a rectangle 1.5 inches wide by 1.2 inches high. Choose Window|Swatch Libraries|Other, and select the GOLDGRAD.AI image from the start folder for this chapter on the CD-ROM. The gold

Figure 3.84 (left)

The Crosshatch filter dialog box.

Figure 3.85 (right)

A close-up of the tree trunk after the Crosshatch filter has been applied.

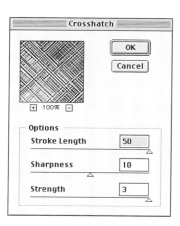

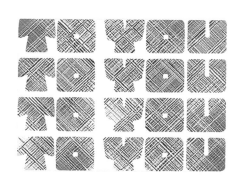

metallic gradient is now available. Drag the swatch into the rectangle that you created. Also, drag it into the Swatches palette for this image.

27. Chose Object|Rasterize. Make it a high-resolution, RGB object. Anti-alias, but *do not create a mask*. Click on OK. Choose Filter|Distort|Ocean Ripple, with a Ripple Size of 5 and a Ripple Magnitude of 20. Figure 3.86 shows a close-up of the result of this filter.

28. Open the image ANGEL.AI. Copy it to the clipboard (Mac: Command+C; Windows: Ctrl+C). Return to your working tree image. Paste in the selection from the clipboard (Mac: Command+V; Windows: Ctrl+V). You could also drag and drop the angel into the working image. Drag the angel on top of the filtered rectangle and scale the angel proportionally until it fits. (Yes, it's a cupid angel, but that's okay.) Position the angel inside the rectangle until you like the place in the gradient that it covers (it's easier to see, although harder to move, if you remove the fill). When you are satisfied, select both the angel and the rectangle and create a mask (Object|Masks|Make, Mac: Command+7; Windows: Ctrl+7). Then group the two objects (Mac: Command+G; Windows: Ctrl+G). Figure 3.87 shows a close-up of the angel.

PLACING ILLUSTRATOR FILES

I ask you to place the Cupid file instead of opening, copying, and pasting because if you placed and linked the file, you couldn't use it as a mask. If you embedded it in the image, it contains an invisible mask around it that needs to be released, located, and deleted to be able to use the angel as a mask.

29. Drag the angel into position on top of the tree. Choose Filter|Colors|Convert to CMYK. Figure 3.88 shows the result.

Let's try one more thing—create an ornament shape using a filter that is not available in Illustrator. We will then make a path from the ornament and copy it into Illustrator to use it as a mask for an Illustrator gradient.

30. Save your work in Illustrator. Close the program if you need to and open Photoshop. In Photoshop, create a new document (Mac: Command+N; Windows: Ctrl+N). Make the document 491×550 pixels, 300 ppi, Grayscale.

Figure 3.86 (left)
The Ocean Ripple filter is applied to a rectangle with a gold metallic gradient.

Figure 3.87 (right)
The angel creates a mask so that the ocean ripples texture you created now takes the form of Cupid.

Figure 3.88

The angel now sits atop the tree.

31. Select the Elliptical Marquee tool. Press Return/Enter to bring up the Options palette. Select Anti-Aliased with a Feather of 0. Choose Fixed Size for the Style and enter a size of 450×450 pixels. Click near the center of the image to place the circle. Press D to set the colors back to the default of black and white. Fill the area (Mac: Option+Delete; Windows: Alt+Delete). Unselect (Mac: Command+D, Windows: Ctrl+D).

32. Choose Filter|Distort|Polar Coordinates, Rectangular to Polar. Repeat this filter (Mac: Option+Command+F; Windows: Alt+Ctrl+F) seven more times. Figure 3.89 shows the unusual shape that appears.

33. Select the Move tool (V) and drag the shape up toward the top of the image. This step automatically changes the Background layer into Layer 0. Choose Layer|Flatten Image to get it back to the Background layer.

34. Apply a Gaussian Blur of 2.0 to the image (Filter|Blur|Gaussian Blur, 2.0). Open the Levels dialog box (Mac: Command+L; Windows: Ctrl+L). Drag the Black Input slider to the right until all three sliders touch. The Black Input level will be 230. Click on OK. The object is now much smoother and a bit thicker.

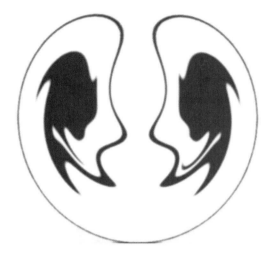

Figure 3.89
The Polar Coordinates filter creates an unusual shape when used multiple times on an object.

35. Load the values of the layer (Mac: Command+Option+~; Windows: Alt+Ctrl+~). Reverse the selection (Select|Inverse, Mac: Shift+Command+I, Windows: Shift+Ctrl+I). Now the black areas are selected.

36. From the Paths palette side menu, choose Make Work Path. Give it a Tolerance of 2 (you don't need to make it hug this shape exactly). Double-click on the Work Path entry in the Paths palette to save the path as Path 1.

37. Use the Arrow tool from the Pen palette pop-out to select the entire path. Copy it to the clipboard (Mac: Command+C; Windows: Ctrl+C). Save your shape and close Photoshop.

38. Back in Illustrator, open the tree image if it isn't still open.

39. Make a new layer (click on the New Layer icon at the bottom of the Layers palette). Drag the layer to the bottom of the Layers palette. Select the Rectangle tool. Click in an unused spot in the image to create a rectangle .79 inches wide by .74 inches wide. Click on the Rainbow gradient in the Swatches palette to use this as the fill.

40. In the Gradient palette, change the Angle to 90˚.

41. Make a new layer (click on the New Layer icon at the bottom of the Layers palette). Paste in the selection from the clipboard (Mac: Command+V; Windows: Ctrl+V). Change its fill to black so that you can clearly see it. Drag it on top of the rainbow gradient. Use the bounding box control handles on the ornament shape to scale it so that it fits inside the rainbow gradient rectangle. Double-check to ensure that the ornament is *on top* of the gradient.

42. Select both the rainbow gradient and the ornament and make a mask (Object|Masks|Make, Mac: Command+7; Windows: Ctrl+7). Select both objects and group them (Mac: Command+G; Windows: Ctrl+G) Figure 3.90 shows the completed ornament (which is an all-vector shape).

Figure 3.90
The masked ornament is composed totally of vector shapes.

43. Lock the layers that contain the two tree clipping paths, the tree trunk, and the angel. Drag the ornament over to the tree and "hang" it. Because it's on a bottom layer, it should appear behind the other layers so that you can see it through the tree. Use Option (Mac) or Alt (Windows) to duplicate the ornament as you drag it. Place about five ornaments. Figure 3.91 shows the finished greeting card.

This exercise was somewhat long, but you have learned how to use a variety of clipping path types:

- An Illustrator path placed into Photoshop serves as a clipping path for the object that is then placed back into Illustrator.

- A vector object in Illustrator is rasterized so that its outlines become its mask. The filter is visible only in the unmasked areas.

- A vector object is rasterized and filtered and then masked with a piece of vector clip art.

- Photoshop is used to create a path, and the path (when pasted into Illustrator) becomes the mask for another vector shape.

You should now be able to mix and match techniques to get the best use of the clipping path strengths of each program.

Figure 3.91
The greeting card design is now complete.

Me And My Shadow

I wish that drop shadows weren't such popular and useful design devices because they can be major headaches! A shadow can be perpetual problem, regardless of the program in which it is created. A hard shadow (a stylized shadow with a sharp edge) causes little difficulty. However, a natural-looking shadow is rarely hard. Soft shadows are a blend of the foreground shadow color and the background. They bleed gradually into nothing.

Why should fading away cause a problem? The problem is twofold: First, how do you make a color fade into nothing (i.e., the background)? Second, what if you want to place the object and its shadow on another background? The "how" depends on the program that you are using. QuarkXPress cannot create a drop shadow without either faking it or using an XTension that actually burns the shadow into an underlying TIF file. Illustrator can create a soft shadow by using the Blend tool, but no transparency is really involved, so if you need a soft shadow over a patterned background, you can't get it in Illustrator. Photoshop creates a true soft edge that can blend over any background, but only if it's originally created over that background.

This brings us to the second part of the dilemma. If you need to place just an object and its shadow from Photoshop into either XPress or PageMaker or Illustrator and think that a clipping path will do the trick, you probably won't be happy with the result. However, again, different scenarios exist. If you have an apple with a soft shadow over a white background in Photoshop and you want to use a clipping path to place it into an Illustrator document where the background will again be white, have at it. Just ensure that your clipping path is well outside the image data. Placing white on white causes no problem if you don't try to clip too close to the shadow.

If you want the apple to cast a shadow onto a checkered tablecloth, you need the tablecloth under the apple when you create the shadow. If you don't want any more of the tablecloth than necessary to be transferred to the Illustrator document, you need to ensure that the CMYK percents defined in Photoshop are the same as the CMYK definition of the tablecloth colors in Illustrator. You also need to align the image in Illustrator so that it appears over the exact same portion of the checkerboard. Your shadow transports best when you take the entire background along with it.

Consequently, another one of the most common reasons for bouncing from Photoshop to Illustrator and vice versa is to create shadows. You just need to be careful of what is under the shadow when you decide what to rasterize—or whether you are going to place the shadow back into Illustrator or composite it in your page-layout application.

Page Layout Composites

One reasonable question that you might have at this point is, "Where do I do my image compositing?" Is it better to leave rasterize and place Photoshop TIF or EPS files into a page-layout program or to place raster images into Illustrator, or do you get the best results by using the programs individually and then layering the resulting images in XPress or PageMaker?

I spoke to the folks at Adobe tech support, and their "official" response was that it should make no difference. This, of course, is somewhat simplistic. It probably doesn't make enough of a difference that you need to avoid placing Photoshop images into Illustrator. However, use common sense if "official" wisdom is lacking. If all I need from Illustrator is to add some text to a Photoshop image, I will create the text in Illustrator and place it over the Photoshop image in XPress. I see no reason to burden my Illustrator file with a large raster object (or even a linked raster object).

However, the "gotcha" for the text-over-a-photo example is trapping. If my text is black, I can select the Overprint checkbox in Illustrator, and (assuming that I don't exceed my ink limit) all is well. However, when you place an EPS file into a page-layout program, it cannot be trapped by the application. The PostScript commands in it are "encapsulated" (which is what EPS means), and nothing else can touch the instructions that the file carries. XPress will report the EPS file as "knocking out" its background, and there is no way to change this. In brief, any trapping that is needed must be done in Illustrator or by your service bureau by using a high-end trapping application (if you are lucky enough to deal with a service bureau that can afford this kind of equipment).

Check the appendix in this book for a brief discussion of Illustrator and trapping, written by my technical reviewer—and dear friend—David Xenakis, one of the most knowledgeable folks on this earth about all things prepress.

You should have no problems placing an EPS or TIF file into an Illustrator document and printing it from QuarkXPress. There should also be no problem if you place an Illustrator file with an embedded raster object into XPress. However, if you saved the raster image as a DCS file (DCS 1.0 or 2.0), beware. You will not see an accurate rendering of the image if you print from a composite printer (such as an Epson 3000 or 5000—or almost any other composite printer), and if you try to rip the image, you are likely to get no film. This isn't a problem with Illustrator, however.

As long as you confine yourself to EPS or TIF files, you can let convenience and comfort be your guide as to where you want to do your image compositing.

Moving On

You should now be an expert in moving images into and out of Photoshop and Illustrator. In this chapter, you learned how to adjust the resolution of placed raster images, create a woodcut image in Illustrator from a black-and-white Photoshop image, use Photoshop to create a texture bitmap to use on top of an Illustrator image, and use clipping paths to move any which way.

In Chapter 4, you learn how to take advantage of the strengths of other vector programs to add features to Illustrator that it lacks.

PART II

SETTING THE STAGE (BASIC TECHNIQUES)

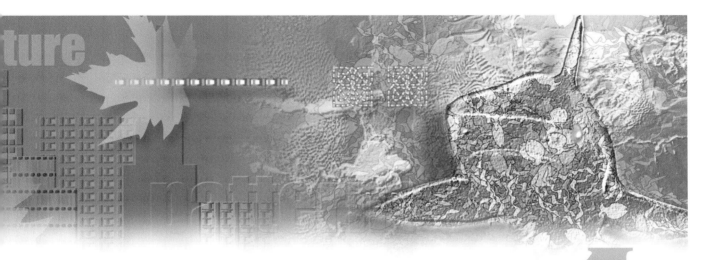

GRADIENTS, BLENDS, AND MESHES

SHERRY LONDON

Although I typically think of flat colors when the words "vector art" are used, Illustrator contains such a wealth of color gradation tools that you never need to settle for flat just because you are using an illustration program. From the original masked blends, Illustrator has flowered in features to include a gradient editor, live blends, and a new gradient mesh object that lets you create free-form gradients to follow a shape.

A Short Introduction

Blends, gradients, and Gradient Mesh objects allow you to add glowing color and depth to your Illustrator projects. I can say a huge amount about each of these topics in this chapter, but space limitations prevent this (we cannot publish the *Encyclopedia Illustratannica* here). To cover everything would require another book! Therefore, I need to keep this chapter brief (you do, after all, want to learn more about Illustrator than how to make blends and gradients). In most of my chapters, I tried to give you complete projects. In this chapter, however, you will simply work with a lot of pieces (and with a few full projects interspersed). In this way, I should be able to cover more effects using blends, gradients, and gradient mesh objects.

Blends

In the beginning, there were blends. A blend allowed you to change colors from shape A through shape B. Illustrator had no gradients in its earliest incarnations. To simulate gradients, you had to blend from one rectangle to another and then mask the result to get it into the shape that you wanted. Times have changed.

The blends have also changed. New to version 8 is the live blend. Now you can change a blend after it is created by editing one of the shapes involved in the blend.

Basic Blends

Because this is an intermediate- to advanced-level F/X book, I really don't want to cover very much that you can just as easily find in the Illustrator manual or the online help. However, I'll quickly review some of the most basic operations here.

The Fundamentals

The Blend tool allows you to create intermediate objects from two objects. The objects are blended by color and by shape so that you can change a black square into a gray circle, for example. You can control the number of steps in the blend so that you can blend smoothly (as in Figure 4.1) or do it in discrete steps (as in Figure 4.2).

You can also blend multiple objects together smoothly (as in Figure 4.3) or in discrete steps (Figure 4.4).

Figure 4.1
You can create a smooth blend between two differently shaped and colored objects.

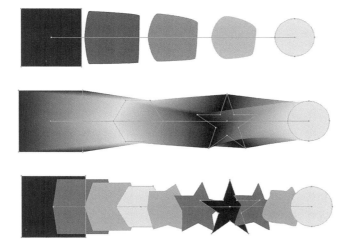

Figure 4.2 (top)
You can create a stepped blend between two differently shaped and colored objects.

Figure 4.3 (center)
You can create a smooth blend between multiple differently shaped and colored objects.

Figure 4.4 (bottom)
You can create a stepped blend between multiple differently shaped and colored objects.

You can edit the starting or ending shapes in the blend and select the point on each object from which to blend. The starting point for the blend is significant because it affects the shape of the finished blend. For example, two circles blended from their top points look like Figure 4.5, whereas the same two circles blended from the top point on circle 1 to the bottom point on circle 2 look like Figure 4.6, where the blend shrinks to nothing and then thickens again.

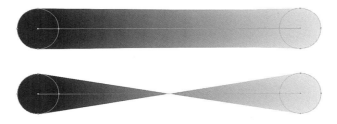

Figure 4.5 (top)
You need to carefully select the points to blend. These two circles are blended from their top points, resulting in a smooth blend.

Figure 4.6 (bottom)
This blend moves from the top point on circle 1 to the bottom point on circle 2.

Because the blends are now "live" (you can see the line between the starting and ending shapes on any of the figures shown so far), you can change the blended objects without having to re-create the blend. You can use the Direct Selection tool or the Group Selection tool to choose one of the blend participants. You can then change the object's color or shape.

If you blend between two stroked objects, the Stroke color, as you can see in Figure 4.7, is blended as well. If you try to blend between two patterned objects, Illustrator gets confused and is not able to produce the blend, as you can see in Figure 4.8. However, if you blend between a solid object and a gradient, Illustrator does a much better job. You can see that blend in the Color Studio because it would not show up in grayscale.

Figure 4.7
Blends between two stroked objects are successful.

Figure 4.8
Blends between two patterned objects are not very successful.

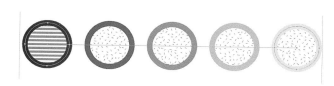

Of course, most blends don't occur in a straight line. Before the Gradient Mesh objects debuted in Illustrator 8, a blend was the only way to create chiaroscuro (realistic highlights and shadows) on an object. Using a blend, you can put the shine on an apple—a very primitive example of which shows up in Figure 4.9.

Figure 4.9
By creating a blend whose ending shape is on top of the starting shape, you can create a highlight in an object.

The Spine Switch

One of the more interesting (if possibly less-than-useful) things that you can do with a blend is to change the shape of its spine. This is the line that connects the shapes in the blend. Try this short Illustration just for fun.

PROJECT Illustration 4.1: New Vertebrae

1. Create a new document in Illustrator.

2. Change your Fill to None and your Stroke color to yellow. Select the Ellipse tool. Click once near the upper-left corner of the image. Enter 1.5 inches in the height and width fields in the dialog box.

3. Press Option (Mac) or Alt (Windows) and drag a duplicate of the circle to the top-right edge of the document directly across from the original as shown in Figure 4.10. Press Shift after you begin to drag to constrain movement to the horizontal. Change the Stroke color to purple.

Figure 4.10
Two circles, waiting to be blended.

4. Select the entire image (Mac: Command+A; Windows: Ctrl+A).

5. Choose the Blend tool. Click on the control point at the top of the yellow circle, then click on the control point at the top of the purple circle. A straight, smooth blend appears.

6. Choose Object|Blends|Blend Options and select Specified Steps in the Spacing pop-down. Enter "45" as the number of desired steps, as shown in Figure 4.11. Now the blend consists of interlocked circles.

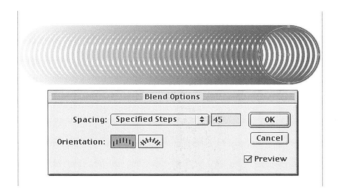

Figure 4.11
You can create a specific number of objects in a blend.

7. After you have the blend created, you can change the shape of the spine. One way to do this is to edit the spine itself. Deselect the blend. Choose the Direct Selection tool. Click approximately on the center of the blend to select only the spine.

8. Choose the Pen tool. Use the Pen tool to add control points to the spine and then use the Direct Selection tool (or the Pen tool with Command/Ctrl pressed) to move the control points. You can use the Pen tool (with Alt/Option pressed) to change the corner points that you added into smooth points. Figure 4.12 shows the slinky that results.

9. You can also create an object in isolation and replace the original blend spine with it. Select the Spiral tool. Click once inside the drawing area. In the dialog box, enter a Radius of 1 inch, a Decay of 80%, and 10 segments. Then select the top Style radio button as shown in Figure 4.13.

WAYS TO GET SMOOTH POINTS

You can also use the Convert Point tool and drag handles out of the end nodes of the spine. If you do, all points that you add to the line are automatically smooth points (because you made the spine into a curved path).

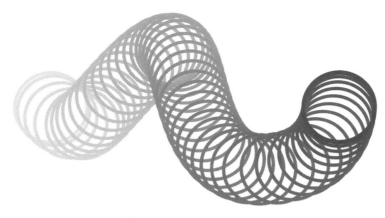

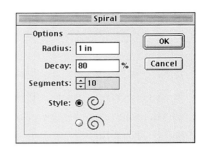

Figure 4.12 (left)

You can change the tube into a slinky by adding and manipulating the control points along the spine of the blend.

Figure 4.13 (right)

The Spiral Tool Options dialog box.

10. Select the entire image, which is the blend and the spiral (Mac: Command+A; Windows: Ctrl+A). Choose Object|Blends|Replace Spine. Figure 4.14 shows the result.

11. You can coax this blend into a snail-shell shape. Deselect the blend. Select just the purple circle with the Group Selection tool. Double-click on the Scale tool and enter a Uniform value of 30% in the dialog box. Click on OK. Figure 4.15 shows what happens.

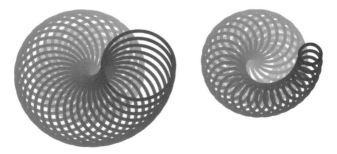

Figure 4.14 (left)

When you replace a linear spine with a spiral, the entire blend curls up on itself.

Figure 4.15 (right)

Now you have a curl.

12. With the blend selected, choose Object|Blends|Reverse Front To Back. As you can see in Figure 4.16, you can almost picture the little snail inside.

13. You can also reverse the direction of the blend. Choose Object|Blends|Reverse Spine. Now you have a seashell. (Figure 4.17).

Figure 4.16 (left)

Switching the Front-To-Back orientation of the blend makes the spiral blend resemble a snail in its shell.

Figure 4.17 (right)

You can create a conch shell by reversing the blend's direction.

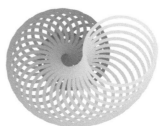

14. If you have the need to "set" the blend so that it is no longer live or if you want to manipulate any of the objects that the blend has created, you can select the Object|Blends|Expand command. Try it.

Linear Blends

Although one normally associates the Blend command with filled shapes, as you saw in Illustration 4.1, one can blend strokes. In the first Illustration, you used stroked, closed objects. In the next several "mini-Illustrations," you will use open paths.

Weaves

You can use linear blends to mimic grids and woven goods. Here are two fast examples—and I'm not even going to attempt to suggest a use for these.

 Illustration 4.2: Waffle Weave

1. Open a new document in Illustrator.

2. Turn on the Rulers (Mac: Command+R; Windows: Ctrl+R). Drag a guide to the 4-inch mark and another guide to the 8-inch mark along the side ruler. Create vertical guides at the 2- and the 6-inch marks on the top ruler.

3. Select the Pen tool. Click at the top-left intersection of the guides, press Shift, and click at the top-right intersection of the guides. This leaves a perfectly straight line. Set the Fill to None and the Stroke color to green, and then set the Stroke weight to 6 points.

4. Press Option (Mac) or Alt (Windows) and drag a duplicate of the stroke to the lower horizontal guide. Press Shift after you begin to drag to constrain movement to the vertical. Change the Stroke color for the duplicated line to yellow.

5. Select both lines and choose the Blend tool. Click in the leftmost point of both lines to set the blend points.

6. Choose Object|Blends|Blend Options and set the Specified Steps Spacing option to 20. Click on OK. Figure 4.18 shows the venetian blind effect that results.

7. With the blend selected, double-click on the Rotate tool. Enter 90 degrees as the Angle of rotation and click on Copy. Figure 4.19 shows the resulting waffle weave.

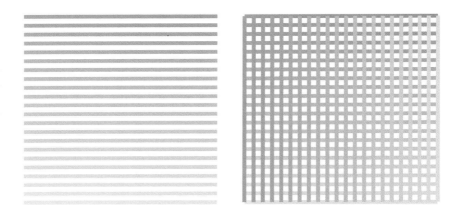

Figure 4.18 (left)

When you blend one straight line into another in specified steps, you can create venetian blinds that show a gradation of color.

Figure 4.19 (right)

When you rotate a copy of the open blend, you can produce a fake weave.

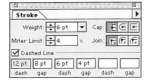

Figure 4.20

The Stroke palette.

8. You can do the same thing with wavy dashed lines if you mask the result. Open a new document in Illustrator.

9. Use the Pencil tool to create a wavy green stroke that is 6 points wide. In the Stroke palette, select the Dashed Line checkbox and enter 12 points, 8 points, 6 points, and 4 points, respectively, in the first four boxes, as shown in Figure 4.20.

10. Press Option (Mac) or Alt (Windows) and drag a duplicate of stroke (press Shift after you begin to drag to constrain movement to the vertical) down about four inches. Change the color of the copied stroke to gold.

11. Select both lines and click on the leftmost point of each line with the Blend tool. Figure 4.21 shows the blend. The white spaces in the blend are caused by the dashed line.

12. Double-click on the Rotate tool and enter 90 degrees in the Angle field. Click on Copy. Deselect all objects.

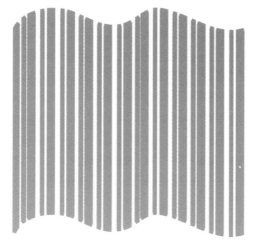

Figure 4.21

Blending two dashed lines yields stripes.

13. Select the Rectangle tool. Put your mouse cursor in the center of the blends, press Shift+Option (Mac) or Shift+Alt (Windows), and drag out a square from the center.

14. Select the entire image (Mac: Command+A; Windows: Ctrl+A). Make a mask (Object|Masks|Make; Mac: Command+7; Windows: Ctrl+7). Figure 4.22 shows the result.

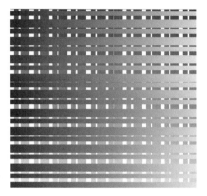

Figure 4.22

You can also create a weave by using a dashed line as the blend element.

This technique has some intriguing possibilities. Try a blended dashed line with a spiral spine. Try overlapping the objects. Try placing various colored or gradated backgrounds behind the weaves.

The Great Ribbon Fantasy

One of my most favorite tricks with linear blends is "The Great Ribbon Fantasy." This simple technique is the first technique that I ever learned for using blends and fascinated me when I saw it—probably in a Mac magazine—so many years ago in the days of Freehand 2 or 3.

As you will see throughout this book, anything that involves a lot of colors or looks tie-dyed appeals to the latent hippie in me. In any case, it is an interesting technique and, with some additional new twists that I've not seen written about before, bears repeating.

PROJECT Illustration 4.3: Irregular Line Blends

1. Create a new document in Illustrator.

2. Using the Pencil tool with no Fill and a color that you like as the Stroke (I chose cyan), draw a very wavy line either straight across the page or at a diagonal. The line that I used is shown in Figure 4.23.

3. Pick a different Stroke color and draw another wavy line, but make it very different from the first line. Figure 4.24 shows my second line.

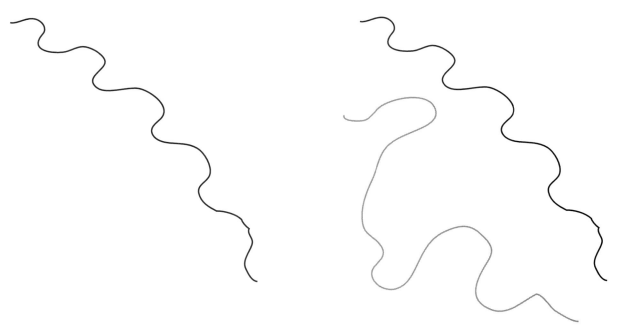

Figure 4.23 (left)
This is the starting line for my ribbon fantasy.

Figure 4.24 (right)
This figure shows both lines before the blend is created.

4. Select both lines. Select Object|Blends|Blend Options and enter a Specified Steps amount of 30. Click on OK. Choose the Blend tool and click on the left points on both lines. Figure 4.25 shows the result in grayscale and you can see the color in the Color Studio. Also in the Color Studio is a second version of the image. This one is twisted out of shape by replacing the spine of the blend with a short, wavy line.

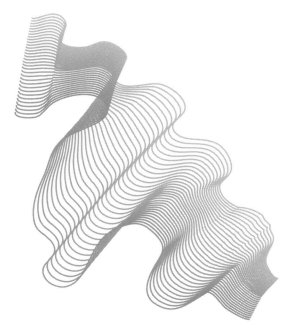

Figure 4.25
The Great Ribbon Fantasy.

I could go on about the different effects you can get by changing the lines and changing the shape of the spine. However, as I prepared this image for publication, it occurred to me that this could be an interesting technique to use with clip art—to cut apart a simple shape and then blend the top line against the bottom one. I prepared two more short examples, using clip art from Ultimate Symbol's Nature Icons collection. The stylized shapes of the clip art in this collection work beautifully in the treatment that I have in mind.

PROJECT Illustration 4.4: Clip Art Blends

1. Open the file ANGELFISH.AI from the companion CD-ROM. Figure 4.26 shows the fish, which has already been split apart into the layers shown in Figure 4.27.

2. Hide all the layers except for the Fish layer. Swap the Fill and the Stroke so that the fish now has no Fill and a black Stroke.

3. Choose the Scissors tool. Click on the two points that are circled in Figure 4.28 to cut the fish apart.

4. Deselect. Select the top half of the fish with the Selection tool. Change the Stroke color to orange. Press the Up arrow key several times to move the top half a bit away from the bottom half.

5. Select the bottom half of the fish and change its Stroke color to yellow.

6. Choose Object|Blends|Blend Options. Set the Specified Steps amount to 30. Click on OK.

7. Select both lines that make up the fish. With the Blend tool, click on the leftmost endpoints of the two lines. Figure 4.29 shows the result.

8. Turn on all the layers. You'll notice that the objects on the Gill and Mouth Gills layers are white, which doesn't show up too well. (Forgive me, by the way, if I am verbally mutilating fish anatomy. Biology was never my thing.) Select the objects on both those layers and change the Fill to a more appropriate color. Black is fine; orange or yellow should work as well.

9. Select the fin on the Upper Fin layer and lock all the other layers.

10. Swap the stroke and fill colors on the fin so that it has no Fill and a black Stroke.

11. Select the Scissors tool. Click on the leftmost of the two circled points on Figure 4.30 with the Scissors tool. Then click on the circled point on the right (the two arrows make these points easier to find). Delete the selected line at the bottom of the fin. You'll need to press Delete twice.

Figure 4.26
The Angel Fish from Ultimate Symbol's Nature Icons collection.

Figure 4.27
The Layers palette shows how I have divided the fish into layers.

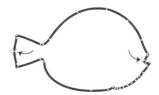

Figure 4.28
The arrows show the two circled points on which to click to sever the fish into two parts.

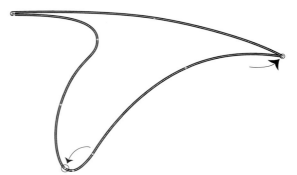

Figure 4.29 (left)

The blended lines re-form the body of the fish.

Figure 4.30 (right)

Using the Scissors tool to fin-tune the fish.

12. With the Scissors tool, click on the point that is at the tip of the fin (where the two "arms" come together).

13. Deselect. With the Selection tool, select the upper line and press the Up arrow key twice to move the top fin far enough away to be able to select both points.

14. Select both lines. With the Blend tool, click on the two points at what was originally the sharp tip of the fin.

Figure 4.31

The finished Angel Fish.

15. Deselect the blend. With the Group Selection tool, click on the top line (the one that you moved). Press the Down arrow key twice to move it back into its original place. Figure 4.31 shows the fish with all its layers turned on. You can also see it in color in the Color Studio. The Color Studio also shows a variation of the fish created in the same way. The difference is that I did not separate the objects on the Mouth Gills and Flippers layers from the original fish body, so I got a dramatically different blend.

16. You can blend more than two lines at a time. Open the SEAHORSE.AI image. This is another image in Ultimate Symbol's Nature Icons collection. As I did in the Angel Fish image, I cut the original shape into component parts and put each part (there are only three here) on its own layer.

17. Keep only the Seahorse layer visible. Swap the Stroke and the Fill on the seahorse so that it has a black Stroke and a Fill of None.

18. With the Scissors tool, cut the seahorse at the tip of his tail and the center of his snout (in both cases, no control points exist—you need to click on the stroke itself).

19. Choose Object|Blends|Blend Options and set the Specified Steps number to 4. For this technique, you need six paths in total. The Specified Steps number counts the *extra* paths to add. Click on OK.

20. Move one of the paths a little bit away from the other one so that you can find the endpoint on which to click. Select both paths. With the Blend tool, click on the two endpoints that are on the seahorse's snout. You should end up with six lines, as shown in Figure 4.32.

21. Choose Object|Blends|Expand. Ungroup (Mac: Shift+Command+G; Windows: Shift+Ctrl+G). Deselect.

22. Start with the leftmost path. Select the path. Change its stroke to red-orange. Change the remaining paths (in left-to-right order) to yellow, green, cyan, purple, and hot pink. You have made a color wheel of the strokes.

23. Select the entire image (Mac: Command+A; Windows: Ctrl+A). With the Blend tool, carefully click the endpoint of each path near the seahorse's snout—from the red-orange to the yellow, green, cyan, purple, and hot-pink path. The Blend tool turns black and shows a plus sign (+) when it is over a path. You will need to wait for the blend to be calculated as you click on each path.

24. If you have more than four paths between each of your six colors, choose Object|Blends|Blend Options and fix the number of Specified Steps.

25. Choose Object|Blends|Reverse Front To Back. Figure 4.33 shows the rainbow result in a grayscale rainbow—you can see a color version in the Color Studio.

Gonna Build A Mountain

You can use linear blends to create scenery. I can picture this technique being used with the stylized clouds, mountains, and seas of traditional Chinese art, but in Illustration 4.5, I will demonstrate it in a form more suited to science fiction or fantasy illustration. You can build a mountain from a single line.

PROJECT Illustration 4.5: Linear Landscapes

1. Create a new document in Illustrator.

2. You need to draw a line to define the top of the mountain first. Use the Pencil tool to create an irregular line. Give it a Fill of None and a Stroke of at least 5. For now, the color doesn't matter (as long as it isn't white). Figure 4.34 shows my starting line.

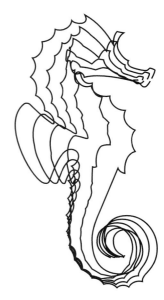

Figure 4.32
Blending the two main lines on the seahorse with a Specified Steps number of 4 yields a total of six lines.

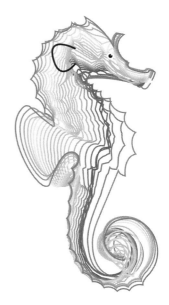

Figure 4.33
The rainbow seahorse in grayscale—with four blends between each color line.

Figure 4.34

This line will become the top of a mountain.

3. Choose the Selection tool. Press Option (Mac) or Alt (Windows) and drag a duplicate of stroke toward the bottom of the image, as far down on the page as you want your mountain to be high.

4. Press Option (Mac) or Alt (Windows) and use the center control handles on the top and sides of the bounding box to flatten and widen the stroke to form the base of the mountain. Figure 4.35 shows the two lines.

5. Choose Object|Blends|Blend Options and set the Specified Steps number to between 3 and 7 (your choice). Select the entire image (Mac: Command+A; Windows: Ctrl+A); both lines will be selected. With the Blend tool, click on the leftmost endpoints of top and bottom lines to create a stepped blend. Figure 4.36 shows the five intermediate steps that I created.

Figure 4.35 (left)

The bottom line is a flattened and widened version of the first line.

Figure 4.36 (right)

This blend has five intermediate steps.

6. Choose Object|Blends|Expand to free the seven lines from one another. Ungroup (Mac: Shift+Command+G; Windows: Shift+Ctrl+G). Deselect.

7. You can now rearrange the lines as you wish—to get a less even blend than would otherwise occur. Your next step will be to color the lines and reblend them smoothly to form the mountain. Now is your chance to redecorate. You can remove some of the lines, widen or shorten the space between them, or stretch or shorten some of the lines. The changes that you make here determine the shape of your mountain. I tossed away the bottom two intermediate lines, stretched the remaining lines, moved them, and duplicated the lowest line. Figure 4.37 shows the new arrangement before coloring and blending.

Figure 4.37
The altered structure lines for the mountain are ready to be recolored and blended again.

8. Change the stroke color on the lines to create the coloring for the mountain. My choices are (starting at the top) a snowy gray-white, a pebbly gray dirt, golden pebbles and dirt, rich green trees, dark-green leaves, to golden dirt again.

9. Choose Object|Blends|Blend Options and set the Spacing to Smooth Color. Click on OK.

10. Select the entire image (Mac: Command+A; Windows: Ctrl+A). With the Blend tool, click on the leftmost endpoint of each line in turn to include it in the blend. Instant mountain! Figure 4.38 shows the result.

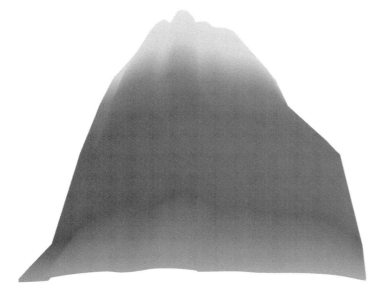

Figure 4.38
You have created a single mountain, and you can, with a bit of imagination, expand on that to create an entire landscape.

Water, Water, Everywhere

Illustration 4.6 is a short project that shows how to create a stylized Japanese-type wavy-water scene using identical wavy lines and a Specified Step blend. Because you already know how to do the "technical" aspects of the blend, I created most of the pieces for you. After you work through this exercise, you can change the shape of the line, the gradient, or any other aspect of the image to see how that affects the final result. This is a mostly blend-and-go Illustration. Repeating multiple copies of the blended objects, masking the result, and placing it over a shaded background creates the water effect.

PROJECT Illustration 4.6: Creating Turtle Water

1. Open the image WATER.AI from the companion CD-ROM. Figure 4.39 shows the two lines that are visible when you open the file.

Figure 4.39
Wavy water lines are all that's
visible when you open the file
WATER.AI.

2. Choose Object|Blends|Blend Options and set the Specified Steps number to 9.

3. Select the entire image (Mac: Command+A; Windows: Ctrl+A).

4. Choose the Blend tool. Click on the leftmost endpoint of the top line to set the start of the blend. Click on the leftmost endpoint of the bottom line to end the blend. Figure 4.40 shows the blended lines.

5. With the Selection tool, drag the blend up to the top of the page.

6. Press Option (Mac) or Alt (Windows) and drag a copy of the blend down directly under the first blend (press Shift after you begin to drag to constrain movement to the vertical). Leave as much space between the objects as there is space between the lines in the blend. Figure 4.41 shows the second blend in place.

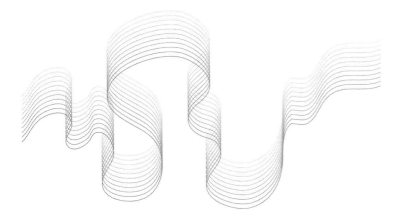

Figure 4.40
The wavy line makes a beautiful ribbon.

Figure 4.41
Drag a copy of the blend down so that it looks like a continuation of the original ribbon.

7. Transform Again (Mac: Command+D; Windows: Ctrl+D) until the blend reaches the bottom of the image.

8. Turn on the Background layer by clicking in the first column in the Layers palette next to the entry for the Background layer. Do not remove the lock on the layer. The Background layer contains a gradient.

9. Delete any of the blends that fall totally outside the areas occupied by the background rectangle. I'm asking you to do this to optimize print speed. There is no point in having the rip rasterize a blend, only to toss it away. Figure 4.42 shows the rectangle with the blends that fall on top of it.

10. Select the Rectangle tool. Choose a Fill of None. The Stroke color (or whether there is a stroke) doesn't matter. Place your mouse cursor at the top-left corner of the Background layer gradient (though the Water Line layer is the active layer). Drag out a rectangle from the upper-left corner of the Background layer gradient to the lower-right

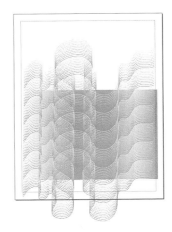

Figure 4.42
All the remaining blends overlap the gradient on the Background layer.

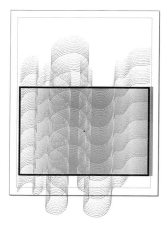

Figure 4.43

Draw a rectangle that will become a mask.

corner, where the blends end. Figure 4.43 shows this rectangle stroked in a very wide black stroke so that you can see it.

11. Unlock and make visible all the layers. Select the entire image (Mac: Command+A; Windows: Ctrl+A). Make a mask (Object|Masks|Make; Mac: Command+7; Windows: Ctrl+7).

12. Lock the Water Line layer and select two or three of the turtles. Unlock the Water Line layer (don't deselect the turtles) and move the selected rectangle from the Turtle layer to the Water Line layer. This puts the turtles in front of the water. Figure 4.44 shows the final effect in grayscale. You can see it, and a variation that contains a different line, in the Color Studio.

Figure 4.44

The finished water scene—turtles and all.

What makes this effect work? The shape of the line itself plays a big part in the success of the water scene. I tried a number of different lines before I found one that I liked. The spacing of the lines (i.e., the number of intermediate steps) also matters. One of the biggest factors, however, is the closeness of the colors that I chose for the starting and ending lines. I had to strike the finest balance to find shades that were obviously different but yet close enough that they didn't cause the water to form an obvious pattern.

We're going to move on now to talk about gradients. You'll see more about blends in Chapter 6, when Michael shows you how to create an action that produces a soft shadow.

Gradients

One of the fervent cries of Illustrator users was for real gradients—ones that worked like the gradients in Macromedia Freehand—where you didn't have to create a blend and mask it. CorelDRAW users already had that gradient (or Fountain) tool. Several versions ago (frankly, I forget which one), Adobe obliged. Gradients made it easier to control the direction and angle of the color ramp. Because a gradient was a type of fill, no mask was necessary—which also simplified things quite a bit. In this section, once we cover the basic steps needed to create a gradient, I'll move on to some gradient "tricks." Because this book is really at an intermediate level, I really think that most of you have already used gradients.

Gradient Basics

When you apply a gradient to an object, you can control several attributes, such as the color ramp, the angle of the ramp, and the whether the ramp is linear or circular. At its most basic, a gradient is applied by simply clicking on a square in the Swatches palette that contains a gradient. You can modify this gradient as you wish.

You can also create your own. Figure 4.45 shows the default gradient that you see if you haven't used any other gradient within Illustrator's memory. This is the basic black-to-white linear gradient.

Here's a fast recap of how to create a gradient:

- *To change a color*—Click on a Color Stop to select it. Use the Colors palette to change the Stop to another color.

- *To pick up a color*—Click on a Color Stop to select it or add a Color Stop. Choose the Eyedropper tool. Press Shift and click on the desired color from anywhere in the image.

Figure 4.45
The Gradient palette is used to create and edit gradients. This example shows a basic black-to-white linear gradient.

- *To add a color*—Click under the Color Slider where there is not currently a Color Stop tab. A new Color Stop will appear.

- *To remove a color*—Drag the Color Stop off the Gradient palette by dragging it downward.

- *To copy a Color Stop*—Press Option (Mac) or Alt (Windows) and drag the Color Stop to a new location.

- *To change the spacing*—Drag the Color Stop closer to or farther away from its neighboring stops to change the distance over which Color A changes to Color B.

- *To set the Gradient Type*—Change the pop-down menu from Linear to Radial and vice versa.

- *To set a new global angle*—Enter a number into the Angle field.

Using the facilities on the Gradient palette, you can create gradients that are quite complex and exciting and that move to many colors. You can also create sharp stripes or color changes by setting the location of two colors directly next to one another as shown in Figure 4.46.

You can change Linear gradients into Radial gradients and vice versa by changing the gradient Type in the pop-down menu. You can also change the angle of the gradient either by entering into the Angle field or by interactively using the Gradient tool—which works exactly like the Gradient tool in Photoshop. The next Illustration shows how to control the Gradient tool using multiple objects.

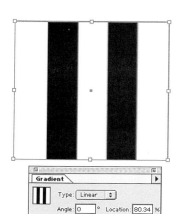

Figure 4.46
You can create stripes by placing two Color Stops next to one another and placing two stops of the same color a distance apart.

Angled Moves

Illustration 4.7 takes a look at a curiosity that I discovered when I was fiddling with gradients. Although you can get some of the same results just by carefully using the Gradient tool, my technique is easily repeatable. Stated simply, if you change a radial gradient that has been individually placed into multiple objects into a linear gradient (with all of the objects selected at the same time), the angle at which the radial gradient was applied remains the same. This can create interesting angled text. Here's how it works.

PROJECT Illustration 4.7: Working The Angles

1. Open the image WELCOME.AI on the companion CD-ROM. The image simply says "WELCOME" in 72-point Adobe Copal Solid type that has been converted to outlines. I like the Copal typeface

for lettering effects because the letterforms are so wide that there is a lot of room for decoration. You should also see guides in the image. If you don't, choose View|Show Guides. Figure 4.47 shows a screen capture of the image so that you can see the placements of the guides.

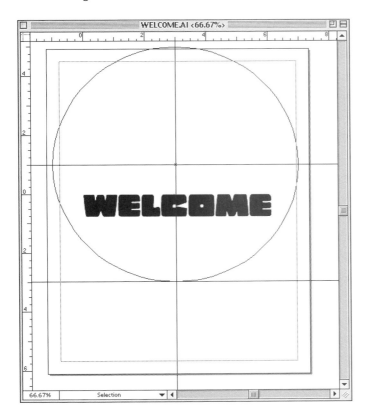

Figure 4.47
"WELCOME" text shows up, surrounded by guides.

A GUIDED EXPLANATION

You're probably wondering why the guides are set up in the WELCOME image as they are and how I got that circle in place. I first created the text and changed it to outlines. I then turned on the rulers. I selected the text objects and dragged the origin point of the rulers to the top-left corner of the bounding box. The text was just a bit wider than 6 inches (6.102, to be precise). The halfway point was a bit beyond 3 inches, where I placed the one vertical guide. That marks the vertical center of the text.

I changed my Fill to the MODI RAINBOW radial gradient (which you will use soon) and, with all the text selected, used the Gradient tool to find the starting and ending points where I liked the gradient fill. I made note of those points along the side ruler and dragged guides to those points. They are the two horizontal guides in the image.

I then selected the Ellipse tool and placed the cursor at the intersection of the vertical guide and the top horizontal guide. This intersection becomes the center of a circle drawn from the intersection to the bottom guide with Shift and Alt/Option pressed to constrain the Ellipse tool to draw a circle from the center. I change the selected circle to a guide by selecting View|Make Guides. I now have a circle that allows me to control the angle of the Gradient tool as I apply a gradient individually to several objects.

2. Choose Window|Swatch Libraries|Other Libraries and locate the KAIGRAD.AI file on the companion CD-ROM. This loads a palette that contains two swatches. Click each swatch in the palette (MODI RAINBOW and KAI GRADIENT) to transfer these swatches to the Swatches palette attached to the WELCOME.AI image. You may close the KAIGRAD.AI swatch library after you've transferred the swatches.

3. Select the entire image (Mac: Command+A; Windows: Ctrl+A). Click on the MODI RAINBOW swatch in the Swatches palette. Figure 4.48 shows the text with the radial gradient applied.

Figure 4.48
Text with radial gradient applied to all objects.

4. Deselect everything. Select only the letter W.

5. Choose the Gradient tool. Place your mouse cursor on the intersection of the vertical and horizontal guides (which from now on I'll call "center circle"). Press and hold the mouse button. Drag the Gradient tool line from the center circle through the letter W until it reaches the circular guide. Release the mouse button. Although you have a number of lines that you could draw with those directions, aim for the lower-left corner of the W.

6. Select the letter E—the second letter in "WELCOME." Again, drag the Gradient tool line from the center circle to the circular guide. This time, drag it through the E. As you drag the Gradient tool line, try to keep the line going though only one letter on its passage to the circular guide. If possible, in each letter, aim for the lowest point on the letter that both misses another letter and is closest to the left side of letter (for "WEL") or the right side of the letter (for "OME"). Drag the Gradient tool directly down for the letter C that is in the center of the word.

7. Select each of the remaining letters in turn and apply the radial gradient with the Gradient tool. For each letter, drag from the center circle through the selected letter to the circular guide. Figure 4.49 shows the text after all the letters have been filled. Even though each letter was filled individually, the gradient looks even because you have used a consistent starting point and drag length.

Figure 4.49

Each letter is filled with the same radial gradient applied at a different angle.

8. Select the entire image (Mac: Command+A; Windows: Ctrl+A). Change the Gradient Type in the Gradient palette from Radial to Linear. Figure 4.50 shows the result.

Figure 4.50

Changing a radial gradient to a linear gradient applies the gradient at the same angle that the radial gradient had used.

9. If you like the way the linear gradient falls, you're done. However, you can play this out a bit more. If you select a different linear gradient, your letters will keep the same gradient angles (if you apply a gradient defined as radial, all bets are off). With the letters still selected, click on the KAI GRADIENT swatch that you transferred into your Swatches palette (and the gradient name is explain in the sidebar, "What's A 'Kai Gradient?'" on page 160). Figure 4.51 shows the new look for the text.

![WELCOME text with angled linear gradient fill]

Figure 4.51

Change the Fill for the text to the KAI GRADIENT swatch.

10. Repeat Steps 4 through 7 using the new linear gradient. Your result, shown in Figure 4.52, looks like a radial gradient, but it's not. However, to expand the gradient for some reason, it could be easier to handle than a true radial gradient. Figure 4.53 shows a "before" and "after" image of the text if you applied the MODI RAINBOW radial gradient to all the text at once and then changed it to a linear gradient. As you can see, you lose the angled application of the linear gradient—everything gets the same angle. If you want to play some more, import the Swatches palette from the YELLORGRN.AI image on the companion CD-ROM. The custom radial gradient that it contains also looks good with this technique.

![WELCOME text with radial-looking linear gradient fill]

Figure 4.52

Applying a linear gradient to multiple objects from a consitent starting point with a consistent line length makes the linear gradient look radial.

Figure 4.53
The top text object shows the MODI RAINBOW gradient applied to the entire word at one time. The bottom text object shows what happens when you change the gradient type from radial to linear.

Metallic Looks

Metallic looking images are always very popular—especially on text. Here's a fast technique to create metallic letters from a linear gradient.

WHAT'S A "KAI GRADIENT?"

If you own Kai's Power Tools from MetaCreations, you can use the wonderfully complex gradient presets that come with the product. Of course, KPT allows you to create gradients in shapes that are impossible in Illustrator, but you can also find very interesting linear or radial gradient presets—or you can modify a color ramp that you like so that it is either linear or radial. You can then re-create the gradient's color ramp inside Illustrator. Here's how.

Fill a rectangle with your chosen gradient by using Kai's Power Tools from within Photoshop. Save the document that contains the original gradient-filled rectangle. Make another copy of the image and change it to Indexed Color mode in Photoshop. I used 20 colors as my cutoff. Save this image too. It's the "work plan" for the gradient. The bands of discrete color make it easier for you to find the defining colors in the gradient.

Open the Indexed Color image in Illustrator. Place the original gradient in the same file. Draw a rectangle between the two bitmapped images that spans the width of the document. Use the Selection tool to stretch the bitmaps until they are the same size as your vector rectangle. You can see this layout if you open KAIGRAD.AI in Illustrator.

Select the rectangle. Choose the Eyedropper tool. Click on the first color band in the Indexed Color image. This sets the rectangle to that color. Then drag the Fill swatch from the Color palette onto the left edge of the Gradient Color slider in the Gradient palette. You now have a ramp from the new color to whatever was already in the Color slider. You can remove the other Color Stops from the Color slider. Click at the end of the Color slider to create an ending Color Stop. Use the Eyedropper with Shift pressed to pick up the last color in the gradient from the Indexed Color image.

Keep adding colors to the Gradient palette by selecting them from either the Indexed Color image or the original bitmapped gradient (with Shift pressed). When you add a new stop to the Color Bar, leave room between the last and the new Color Stops. You can move the Color Stop into position as soon as you're sure it's colored properly. Use the original gradient as your guide for spacing (which is why you made all three rectangles the same size). You can add colors from the original gradient that didn't make it into the Indexed Color version (like the orange in the KAI GRADIENT). When you have a group of colors defined in the Color slider, you can press Option (Mac) or Alt (Windows) and drag the color to create a duplicate Color Stop (if you need to use the color more than once). Make sure that you release the mouse button before you release the modifier key.

A word about color modes: The two images from Photoshop will come over as RGB images unless you have saved them in CMYK mode (which is impossible with Indexed Color if you keep it indexed). However, after the bitmaps are inside Illustrator, you can use the Filter|Colors|Convert To CMYK filter to change the object to CMYK. This makes it easier to define a CMYK gradient, and you'll probably get a warning from Illustrator if you don't have them as CMYK gradients when you go to print.

PROJECT Illustration 4.8: All That Glitters...

1. Open the document METAL.AI on the companion CD-ROM. It contains the word "METAL" in Adobe Copal Solid, which was changed to outlines and scaled to fit the page width. The tracking is opened a bit. I set the type originally at 72 points with 60 thousandths of an em tracking.

2. Select the entire image (Mac: Command+A; Windows: Ctrl+A).

3. Make sure that the Fill swatch is active. Click on the Metal Gradient swatch in the Swatches palette (this is the last swatch in the palette).

4. With all the letters selected, choose Object|Transform|Transform Each. Set both the Horizontal and Vertical Scale amounts to 80%, as shown in Figure 4.54. Click on Copy. Do not deselect.

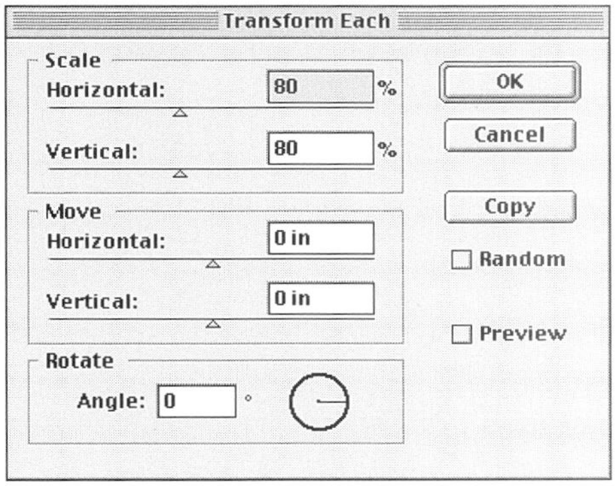

Figure 4.54
The Transform Each command allows you to scale the entire group of letters and copy them in one command.

5. Create a new layer. Move the selected object rectangle from Layer 1 to Layer 2 in the Layers palette (moving the scaled text to Layer 2).

6. With the letters in Layer 2 still selected, enter "31" into the Angle box on the Gradient palette. Deselect. Figure 4.55 shows the result (you can also see it in color in the Color Studio).

The metallic gradient technique in Illustration 4.8 works well on Web objects, such as buttons, and need not be confined to use on text. The secret of the technique is to apply the same gradient to the inner and outer copies of the object but at different angles.

Figure 4.55
"Metal" now looks metallic.

Of course, the gradient that you apply needs to look metallic before it is applied to anything. A metallic gradient is a color ramp that moves unevenly from the base color (silver, gold, bronze, or whatever) to a highlight of white or near white and then cycles back and forth to the base color.

Sometimes, the realism of a metallic gradient is improved by adding a few stranger colors into it. A *stranger* color is my name for a color close to the primary color in value but slightly different in hue. It's a color that you would tend to say "clashes" with the base color if you were to try to wear both colors together. You can also form an additional color by adding black to the base color. This method doesn't clash as much and isn't of the same value as the base color. It's also an effective way to add tone to the gradient.

Illustration 4.8 makes use of two copies of the text for a very basic reason: Illustrator doesn't allow you to add a gradient to a stroke. Therefore, to create a wider lip for the rim of the letters, you need to use the text and create a copy to serve as the inner form. You must deal with two filled objects.

You can also apply the effect by using the gradient tool and stretching the gradient across the entire word. However, because the character of metal is to be reflective, I felt that it looked better to apply the gradient so that each letter contained the full range of colors. Of course, you are free to try it the other way. Also, nothing is sacred about the gradient angles that I used. They can be anything that you want, as long as they are different for both the inner and the outer shapes. Don't forget to try the technique using the transferred KPT gradient from Illustration 4.7 (this is also shown in the Color Studio).

David Xenakis, artist and technical reviewer for this book, contributed MOLDED.AI for the companion CD-ROM. The file shows three stunning examples of achieving dimensionality by using gradients. In the self-teaching example, David shows the amazing elevations changes possible simply by rotating one of the shapes. Check it out.

Gradients For Realism

Of course, one of the major reasons to use gradients is to add realism to your image. In the hands of a master illustrator, the image would appear to be a photograph. That takes a special talent (of which I possess a limited amount). Perhaps the most critical skill needed—in addition to an ability to draw—is the ability to organize and segment. You need to analyze every inch of your subject to see how it reacts to light. You need to know the front-to-back layering needed to make the object look real.

The Pen

In a departure from the norm in this book, I invite you to take apart the image PEN.AI in the area of the companion CD-ROM where we store the finished images for this chapter. The pen is the creation of David Xenakis (also my coauthor on *Photoshop 5 In Depth,* by The Coriolis Group). David is also a tremendously skilled illustrator. The pen is a marvel of realism. By clicking on each object, you can see exactly how it was constructed. The shadow is a rasterized copy of the image on which the Unite filter was run. Figure 4.56 shows the pen in grayscale. You can see it in color in the Color Studio (or on the companion CD-ROM).

Figure 4.56

This pen, by David Xenakis, shows how realistic an object can look when created using gradients.

Spray Those Germs Away

The spray can that you'll re-create in Illustration 4.9 is my attempt at realism. Although it is simple drawing, there are enough shapes in the image that I have given you the starting points rather than directed you to create your own.

When I started this chapter, I knew that I wanted to have one Illustration that showed how to apply a metallic-looking gradient to a real object so that I could demonstrate how to look at the light and try to capture the reflections of metal. While at the dentist's office, I became so fascinated by the play of light on a spray can that the dental assistant finally insisted that I take it home. The result is Illustration 4.9.

Illustration 4.9: Realistic Metallic Gradients

1. Open the document CAN.AI from the companion CD-ROM. Figure 4.57 shows the starting image. It is flat and lifeless.

2. Poke around the image before you actually start editing. Figure 4.58 shows the image in pseudo-line mode. I filled most of the shapes with None and stroked them in black so that you can see the actual objects. You can easily match the very long Layers palette (shown in Figure 4.59) to the objects on each layer by using Command+Option+ Click (Mac) or Alt+Ctrl+Click (Windows) on a layer in the Layers palette to select the object(s) on that layer. My naming isn't scientific (or accurate), so you might need to guess what I mean when I say, "Select the Squirt Stand." Lock all the layers after you finish exploring.

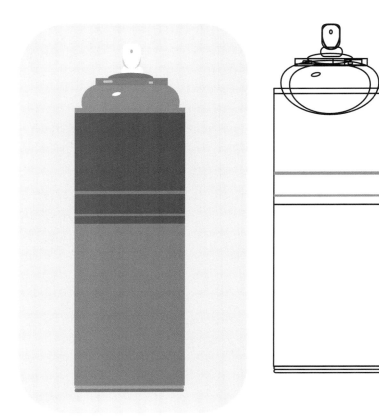

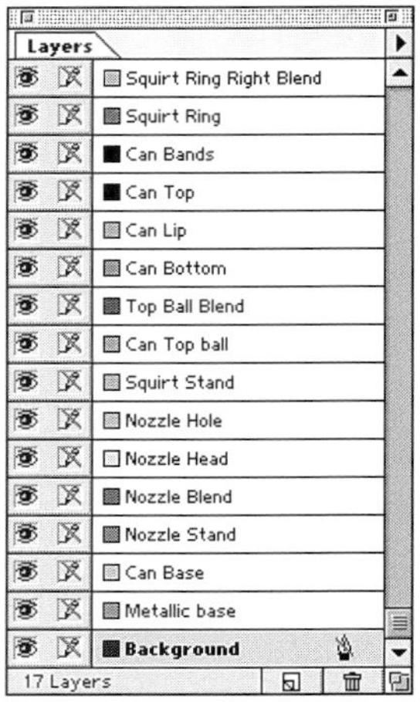

Figure 4.57 (left)

This is the starting point for a spray can that is about to gain depth and dimension.

Figure 4.58 (center)

The can looks much like it would in Artwork mode.

Figure 4.59 (right)

The Layers palette has many entries that you need to match to the objects they contain.

Figure 4.60

The New Pink Gradient swatch as defined in the Gradient palette.

3. Let's create the main body of the can first—the areas where the pink bottom of the can meets the blue top of the can. My premise is that if you want the spray can to look realistic, you need to match the gradients as they change color. The play of light needs to be identical even if the hue set changes. Remove the lock on the Can Bottom layer. Select the entire image (Mac: Command+A; Windows: Ctrl+A). Only the pink bottom of the can is selected.

4. Click on the New Pink Gradient swatch in the Swatches palette to apply it to the object (make certain that the Fill is the active Color palette component) to apply a soft, satiny pink metallic gradient to the object. Figure 4.60 shows the Gradient palette so that you can see the gradient Color Stops. Lock the layer again.

5. Unlock the Can Top layer. This is the blue top of the can—remember that you can use the Command+Option+Click (Mac) or Alt+Ctrl+Click (Windows) command on any layer name in the Layers palette to select the object(s) on that layer. If you are experienced at designing gradients, you can simply click on the New Gradient Swatch 6 in the Swatches palette and go to Step 7. If you want to practice turning pink into blue, go to Step 6.

6. Select the blue can top object on the Can Top layer. Click on the New Pink Gradient to apply it. Hmmmm…. This object is supposed to be blue (but it has to match). Click on the first Color Stop on the left of the Color slider in the Gradient palette. Set the Color to CMYK: 97, 88, 0, 0. Click on the second Color Stop. Here is a list of values for the Color Stops in the gradient. Figure 4.61 shows the Gradient palette for the blue gradient. Notice that its stops are in the identical locations as the pink gradient:

- Color Stop 1—CMYK: 97, 88, 0, 0
- Color Stop 2—CMYK: 96, 54, 0, 0
- Color Stop 3—CMYK: 97, 77, 0, 0
- Color Stop 4—CMYK: 96, 54, 0, 0
- Color Stop 5—CMYK: 97, 78, 0, 0
- Color Stop 6—CMYK: 96, 54, 0, 0
- Color Stop 7—CMYK: 97, 88, 0, 0
- Color Stop 8—CMYK: 97, 88, 0, 27

7. Unlock the Metallic Base layer. Select the metallic base (the gray shape at the very bottom of the can). Click on the Gray Gradient in the Layers palette to apply it. Lock the layer again. As you apply my preset gradients, look to see how each one is defined. The Can Base layer remains a solid pink, so there is no need to touch that layer. Lock all the layers.

8. Unlock the Can Lip layer. Click on the Gray Gradient to apply it. Change the Angle to 90 degrees. That's better, but the shading isn't quite right. Choose the Gradient tool. Drag the Gradient tool line from the bottom of the Can Lip object vertically up until it is out of the object by a distance equal to the height of the object. Figure 4.62

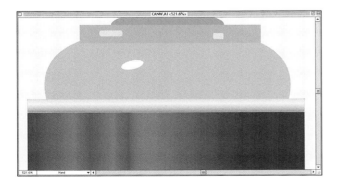

Figure 4.61

The blue version of the metallic gradient uses the same Color Stops as the pink gradient in Figure 4.60.

Figure 4.62

The lip of the can needs to be delicately shaded.

shows a close-up of the Can Lip object with the gradient correctly applied. Lock the layer.

9. The only other layer that gets a "simple" gradient is the Squirt Stand layer (the gray, rounded shape beneath the nozzle assembly). Unlock the layer and select the object. Apply the Squirt Stand gradient to it (this is actually the same as the Gray Gradient applied at an angle of 1.64 degrees). Lock the layer again.

10. The Can Top Ball layer and the Squirt Ring, Nozzle Hole, Nozzle Stand, and Nozzle Head layers are solid and already correctly colored. You do not need to touch them. Five more layers remain to color, and each of these contains a blend rather than a gradient. Let's start with the easiest one. Unlock the Can Band layer. Fill both bands with the New Pink gradient. Choose Object|Blends|Blend Options and change the Specified Steps number to 2. Choose the Blend tool and click on the top-left corner of both shapes to create the blend. You can blend shapes that contain gradients, and when both the starting and ending shapes use the same gradient, it is easy for Illustrator to calculate the correct result. Because the bands use the same gradient as the bottom shape, the lighting is automatically in the correct location. Lock the layer.

11. Unlock the Top Ball Blend layer. Figure 4.63 shows the layer before the blend is applied. The objects on the layer are a gray shape (the same color gray as the Can Top Ball layer that is the background for this blend) and a small highlight shape. Select the entire image (Mac: Command+A; Windows: Ctrl+A), which selects just the gray and the highlight shapes. Choose Object|Blends|Blend Options and set the Spacing to Smooth Color. Choose the Blend tool. Click on the center point inside the highlight shape to set your first point. Then click on the center point of the gray shape to finish the blend. Figure 4.64 shows the finished blend. Because the figure is from a screen capture, you can also see the spine of the blend. Lock the layer.

12. Unlock the Nozzle Blend layer. The two shapes on this layer are a white front piece and a gray back piece. Select both shapes. With the Blend tool, click on the control points (see Figure 4.65) to establish the smooth blend. Lock the layer.

13. The remaining two blends form the dual highlights on the rim of the can (which I have called the "squirt ring" because I was running low on part names). These are the shapes on the Squirt Ring Left Blend and Squirt Ring Right Blend layers. Unlock the Squirt Ring Right

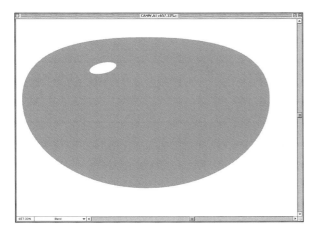

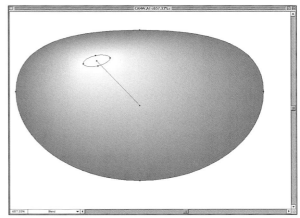

Blend layer. Again, I have used the trick of making a smaller "stop blend" shape that is the same color as its base object. This smaller shape keeps the highlight from spreading farther than it logically should. Select both shapes on the layer. With the Blend tool, click on the center marker inside the highlight shape and then click on the center marker inside the gray shape. Lock the layer again.

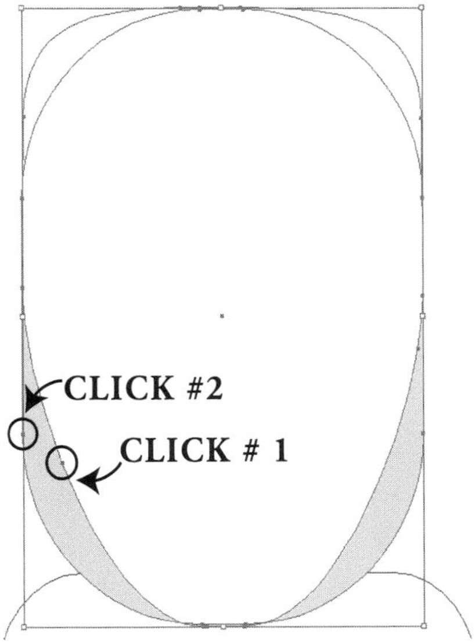

Figure 4.63 (left)
The Top Ball Blend layer before being blended. There are just two shapes.

Figure 4.64 (right)
The Top Ball Blend layer after the blend has been applied shows the two shapes and the spine of the blend.

CLICK #2

CLICK # 1

Figure 4.65
Click on the control points as marked to create this blend.

14. Unlock the Squirt Ring Left Blend layer, select the objects, and blend from the center point of the highlight to the center point of the gray stop-blend shape as you did in Step 13. If you have hidden any layers or changed them to Artwork mode, make everything visible now and admire your handiwork. Figure 4.66 shows the finished spray can in grayscale.

Figure 4.66

The finished spray can is much closer to reality than the flat object that you first opened.

The most difficult part of creating Illustration 4.9 was creating the metallic blend. The Swatches palette still contains some of my early attempts. In the light of my studio, the metallic surface of the can gleamed with many more light-to-dark vertical lines than I put in the blend. I discovered that you need to be very cautious about balanced blends (A-B-C-D-C-B-A) that are used to simulate metallics. The eye sees depth so easily that all too often my blend made it look as if the object itself was rolling in and out. Because I wasn't trying to make the spray can look like a theater curtain, I needed to very carefully edit the gradient and watch that the colors didn't vary too much along the Color slider. The worst dimensionality occurred when there were sharp differences between the light and the dark tones. I had to work at the gradients until I saw a round can rather than a tin ribbon.

Gradient Mesh Objects

In Illustrator 8, Adobe has given the artist a marvelous new tool for capturing light and dark and subtleties of shading. The Gradient Mesh tool is one of the most exciting features that I've found in any Illustrator upgrade (though it is nearly equaled by the wonderful new brush features). Along with exciting possibilities, however, comes a fairly steep learning curve. This is a complex tool. The mechanics of it, as you'll discover in this section, are not that difficult. The challenge comes from trying to make it do what you want it to do. Bluntly, you can train an ape to sit at the typewriter and make letters appear on a piece of paper, but only Shakespeare can write Romeo and Juliet. I make no claims to be Shakespeare, and I'm certainly not Picasso, but I can give you some guidance in the basics of using this tool. You can then take it as far as your own capabilities allow (the more experience you have in painting and in creating chiaroscuro, the easier you'll find it to be). I'll freely confess that this isn't the strongest part of my artistic ability.

Now that you are warned, a Gradient Mesh is a new type of gradient that allows you to create a shaded object by filling points or patches on a grid with color that blends softly into the surrounding color. You can control the mesh by adding or deleting points along its surface, editing the points as if they were control points on an object, and selecting the colors and locations at which to place the colors. If you worked through this book in order, you used the Gradient Mesh in Chapter 3.

Mesh Basics

The Gradient Mesh tool is complex enough that I don't want to take the time or space to do an in-depth review of the tool. Do take the time to read the manual or to review the online Help for the Gradient Mesh tool. The "nutshell" review of the tool is as follows:

- *To create a Gradient Mesh*—Click on an object with the gradient Mesh tool or choose Object|Create Gradient Mesh.

- *To remove the entire Gradient Mesh*—Sorry, you lose! It cannot be done except by deleting the entire object.

- *To color a Gradient Mesh point*—Click on the point with the Gradient Mesh tool, drag and drop a color swatch onto the point, or use the Paint Bucket tool.

- *To color a Gradient Mesh patch*—Click in the patch with the Gradient Mesh tool, drag-and-drop a color swatch onto the patch, or use the Paint Bucket tool.

- *To edit the shape of the Gradient Mesh*—Use the Gradient Mesh tool or the Direct Selection tool and move the anchor points and direction lines as if there were regular Path anchor points and direction lines.

- *To slide a point along a mesh line*—Press Shift as you move it.

- *To add a new mesh line*—Click on the object with the Gradient Mesh tool where no mesh line currently exists (the cursor arrow will show a plus sign).

- *To add a mesh without a color change*—Press Shift and click on the object with the Gradient Mesh tool where no mesh line currently exists (the cursor arrow will show a plus sign).

- *To delete a mesh line*—Press Option/Alt and click on a mesh line (the cursor arrow will show a minus sign).

In the next several Illustrations, you'll get plenty of practice using these techniques. Figure 4.67 shows an anatomy of a Gradient Mesh object.

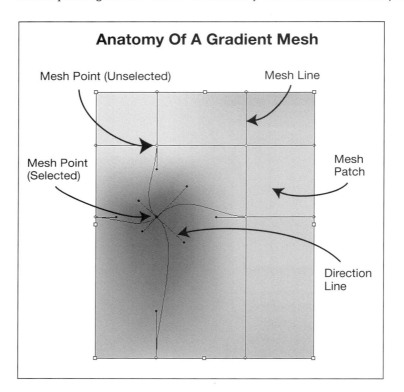

Figure 4.67
The Gradient Mesh object guide map.

Gradient Mesh Objects: Inside And Out

This section provides a hands-on introduction to Gradient Meshes through two simple Illustrations. In the first, you'll see how you can use a Gradient Mesh object as the background for scratch-it-off artwork, and in the second, you'll see how you can adapt a popular style of Illustrator art to use Gradient Mesh objects.

Scratch Art

You can buy scratchboard now and easily create scratch art, though I remember doing this in grammar school with crayons and poster paint. If you keep your scratch design simple, you can even use Illustrator to simu-late scratch art. There are many ways to do this, and the way that I've selected is not necessarily the most efficient, but I'll explain my choices as we go along. I have chosen to create a compound path on the top shape so that a unified Gradient Mesh shows up underneath it.

PROJECT Illustration 4.10: Scratch-It-Off (Or Make A Compound Path Of It)

1. Open the image SWIRL.AI. This is another piece of clip art from Ultimate Symbol's Design Elements collection (and one of their demo images). Figure 4.68 shows the original image.

2. Choose Object|Compound Paths|Release. You'll see what looks like black flower petals on a black shape.

3. Deselect. Click in the center of the image to select the largest black object. Delete. This leaves just the petal shapes.

4. Lock the layer. Create a new layer by clicking on the New Layer icon at the bottom of the Layers palette. Drag the new layer below Layer 1.

5. With nothing selected, change the Stroke to None and the Fill to yellow.

6. Choose the Rectangle tool. Press Option (Mac) or Alt (Windows) and click once in the center of the petal form. This opens the Rectangle Tool Options dialog box. Enter 4 inches for the width and 3.2 inches for the height.

7. Move the rectangle to the left so that the flower form is closer to the right edge. Figure 4.69 shows this positioning.

Figure 4.68
SWIRL.AI: clip art from Ultimate Symbol's Design Elements collec-tion.

Figure 4.69
Move the rectangle to the left so that the flower is closer to the edge.

8. Deselect everything. Change your Fill color to green. Select the Gra-dient Mesh tool. As you drag the cursor over the rectangle, the cursor changes to show a plus sign (+), signifying that you can add a Gra-dient Mesh there. Click on one of the petals (remember, the petal

layer is locked and you're actually painting *underneath* it). You've created one Gradient Mesh line.

9. Deselect. Change your Fill to hot pink. Click on another petal to leave a second Gradient Mesh line.

10. Continue to deselect. Change the Fill color and create another two or three points. They don't all have to be on the petals (though only those will show when you are done).

11. Use the Gradient Mesh tool or the Direct Selection tool to move mesh points around. Figure 4.70 shows a screen shot of my Gradient Mesh.

12. Lock Layer 2. Create a new layer directly above Layer 2.

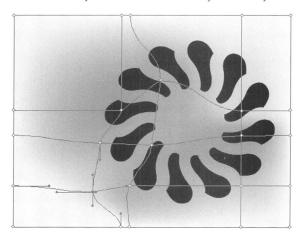

Figure 4.70
Gradient Mesh showing mesh points that have been moved.

13. Change your Fill color to black with no Stroke. Choose the Rectangle tool and draw a rectangle just a tiny bit larger than the Gradient Mesh object.

14. Unlock Layer 1. Select the entire image (Mac: Command+A; Windows: Ctrl+A). Click on Exclude in the Pathfinder palette. Figure 4.71 shows the result.

Figure 4.71
By excluding the petal shapes from a solid rectangle, you create holes through which the Gradient Mesh object can show.

15. Lock all the layers and create a new layer that is at the top of the layer stack.

16. Set the Fill to None and the Stroke to white. Choose the Paintbrush tool with the second Calligraphic brush in the Brushes palette (the 10-point oval).

17. Draw approximately five lines with the Paintbrush. Keep the lines short and simple and make all of them start outside the left edge of the rectangle. They need to be similar to those shown in Figure 4.72.

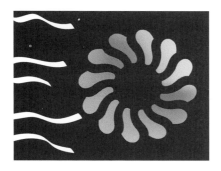

Figure 4.72
Add some simple calligraphic strokes to the left edge of the image.

18. Select the entire image (Mac: Command+A; Windows: Ctrl+A). Choose Object|Expand. Expand the object and deselect the Stroke and Fill options. Ungroup (Mac: Shift+Command+G; Windows: Shift+Ctrl+G). Deselect.

19. Use the Selection tool (and Shift) to select each of the lines individually until they are all selected. Choose Object|Hide Selection. Select the entire image (Mac: Command+A; Windows: Ctrl+A). Now you should see a bunch of paths that have no stroke or fill. Delete. Choose Object|Show All. Deselect.

20. Unlock Layer 1. Select one white line and the black rectangle. Click on Minus Front on the Pathfinder palette. Press Shift and click on the next white line. Click on Minus Front on the Pathfinder palette. Continue to press Shift and select the next line, then use the Minus Front command until you either run out of lines or Illustrator tells you that the path is too complex. (For now, because you're probably not going to print this exercise, you can ignore the warning.) Figure 4.73 shows the finished effect in grayscale.

What can you do if you want to create a complicated scratch art image? There are several possibilities. You could create a large number of small Gradient Mesh objects and fill each one with a different set of colors. However, this would lose the continuity of the supposed underlying art board.

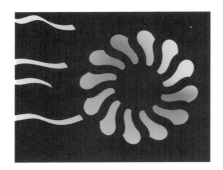

Figure 4.73
Scratch art, done.

Another possibility is to segment the black rectangle. Instead of drawing it all in one piece, create it out of a large number of smaller rectangles so that you reduce the complexity of the paths.

Stroke Pictures

If you flip through the pages of almost any clip art collection, you'll see images that use a delightfully breezy style of a simple outline over a quick calligraphic stroke. Figure 4.74 shows an example of this style artwork using an image from volume 1 of Ultimate Symbol's Design Elements collection.

Figure 4.74
Coloring not quite within the lines is a very good way to personalize clip art. This outline is from Ultimate Symbol's Design Elements collection.

The Gradient Mesh in Illustrator 8 adds a new dimension to this style. You can use a Gradient Mesh as the scribble—though not quite as easily as it would seem. Let's try it. You'll work two short versions of this exercise. In the first part, you'll use the Gradient Mesh as a soft background. In the second part, you'll use the Gradient Mesh as the sketch stroke itself.

PROJECT Illustration 4.11: Sketch Art

1. Open the image STARFISH.AI. Again, I have to thank the folks at Ultimate Symbol for allowing me to use their demo images. This one is from the Nature Icons collection. I have already modified it a bit here to make the Illustration easier (and faster) to work. Figure 4.75 shows the starting image.

Figure 4.75
Starfish, from Ultimate Symbol's Nature Icons collection (brushstroke and background by the author).

2. Lock the Starfish and Brush Stroke layers. Select the dark gold rectangle on the Background layer.

3. Select the Eyedropper tool and click inside the starfish. The rectangle changes to the same color as the starfish.

4. Choose Object|Create Gradient Mesh. Figure 4.76 shows the dialog box. Enter 4 rows and 4 columns and set the Appearance: To Edge with a Highlight of 100%. Setting the Appearance To Edge places a highlight at the outer edges of the object. You can control the amount of the highlight by entering a number lower than 100%. Click on OK.

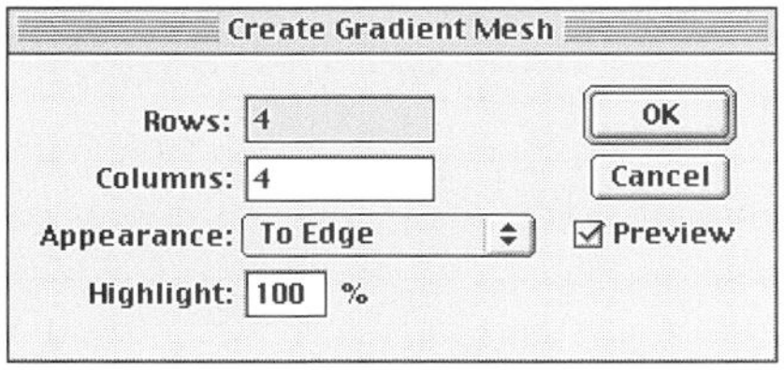

Figure 4.76
The Create Gradient Mesh dialog box.

5. Unlock the Starfish layer and drag its entry in the Layers palette to the New Layer icon at the bottom of the Layers palette to create a copy of the layer. Drag the copy below the original in the Layers palette.

6. Change the Stroke color to black or 75% gray. Drag the object just slightly to the right and down to act as a shadow. Figure 4.77 shows the finished example.

Figure 4.77
You can finish off a sketched object with a soft Gradient Mesh background.

7. Save a *copy* of the image under a different name and revert the STARFISH.AI image back to its original.

8. Hide the Background layer and the Starfish layer. You can change the Brush Stroke layer into a Gradient Mesh object. Select the Gradient Mesh tool (do not select any objects yet). Set your Stroke to None and the Fill to white. With the Gradient Mesh tool, click on the Brush Stroke object to set a white highlight somewhere in the brush stroke. Figure 4.78 shows the rather messy result in a screen shot of the selected object. When you create a Gradient Mesh by clicking on the object with the Gradient Mesh tool (or even when you use the Create Gradient Mesh dialog box on a nonrectangular shape), the mesh lines follow the shape of the object. This could be very good, but here it's just confusing and leaves a jumbled highlight as well.

9. Undo the Gradient Mesh. You'll tackle this a different way.

10. Drag the Background layer entry to the New layer icon at the bottom of the Layers palette. Drag the copy below the original and rename the layer "Gradient Mesh". Lock the layer.

Figure 4.78
A less-than-successful attempt to create a Gradient Mesh.

11. Make sure that both the Starfish and the Gradient Mesh layers are locked. Select the entire image (Mac: Command+A; Windows: Ctrl+A). Create a Compound path (Mac: Command+8; Windows: Ctrl+8). Set the Fill for the compound path to white. Now only the stroke and the starfish seem visible.

12. There is no longer anything on the Background layer. You can toss it into the Layers palette trashcan.

13. Lock the Brush Stroke layer and unlock the Gradient Mesh layer.

14. Change the Fill color on the Gradient Mesh layer object to something a bit lighter.

15. Deselect the object, choose a color, select the Gradient Mesh tool, and add a mesh point. Keep adding mesh points and editing them as you wish. Make a shadow as you did in Steps 5 and 6. Figure 4.79 shows my version in grayscale.

At the start of this topic, I stated that creating a stroke sketch would not be as easy as it seemed it should be. As you've seen, a complex shape is not the easiest to control when used as a Gradient Mesh object. The simpler the shape with a Gradient Mesh, the better time you'll have adding colors to it.

Another problem with this technique (that I kept from happening in this Illustration) is that your stroke can become so complex that you cannot fuse it together to make a single shape. How did I get the stroke? Exactly the same way that you used in Illustration 4.10 to add lines to the swirl image. I used a calligraphic brush stroke and then expanded it. I needed to find the skeleton no-fill-no-stroke strokes and delete them, and then I used the Unite command on the Pathfinder palette to glue the strokes together into a single object. I discovered that it is much easier to unite

---

WHAT IF THE MAKE COMPOUND COMMAND DOESN'T WORK?

Sometimes, if your path is a bit complex, the Object I Compound Paths I Make command either does nothing or gives you an error message. If that happens, try the Exclude command in the Pathfinder palette. It also creates a compound path but doesn't seem to be as fussy about it.

If the compound doesn't work, it's usually because some inner shapes have paths going in the wrong direction. To solve this mystery, use the Direct Select tool to select the shapes and click on one of the two buttons almost in the center of the Attributes palette. Click on the one that isn't slightly darkened. That should fix the problem.

Figure 4.79
The Gradient Mesh object becomes the background behind a compound path made from the original brush stroke.

the expanded strokes if you create only single strokes that go in one direction and do not double back on themselves. Scribbling a stroke is a recipe for hours of extra work. You need to lay down each stroke carefully.

Gradient Mesh Object From A Bitmap

One of my "gee whiz" discoveries with the Gradient Mesh feature is that it's possible to turn a bitmapped image into a Gradient Mesh object. Of course, processing slows almost to a crawl if the image is large and the Gradient Mesh object that you create is complex, but what fun is it if you don't push the edge sometimes?

This final Illustration is a perfect finish for the chapter on gradients, blends, and meshes and provides a summing up of knowledge gained from Chapter 1. It brings together elements of everything that you've done so far. It also gives you the freedom to spend five minutes seeing the basic effect possible or several hours to reconstruct all my work and take it farther than I did for this book.

I've constructed an abstract image of a photograph that I took several years ago, on Halloween, at a local schoolyard. Work through the Illustration, just to see what a bitmap looks like when converted to a Gradient Mesh object, and then I'll share with you the steps that I used to create the starting image.

PROJECT Illustration 4.12: Halloween Horror

1. Open the image HALLOWEEN.AI on the companion CD-ROM. When you open the image, you'll notice that almost all the layers are locked and that two of them are hidden, as you can see in Figure 4.80.

2. Select the entire image (Mac: Command+A; Windows: Ctrl+A). Because all but the Bitmap layer are locked, only the bitmapped object embedded in the image is selected. This image is a cropped version of the original with a clipping path on it. The original image is shown in grayscale in Figure 4.81.

3. Zoom in to the image so that it fills the screen. Choose Object|Create Gradient Mesh. Try a variety of values for rows and columns from four rows, four columns to as high as you can go without killing your machine or your patience. See what each setting does to the image and select the one that you like best (remember to select the Preview checkbox so that you don't need to click on OK until you are actually ready to perform the conversion). Figure 4.82 shows the settings that I eventually used. Click on OK when you are finished experimenting.

4. Make the bottom two layers visible.

5. Change your page layout so that it is in Landscape mode.

6. Unlock all the layers. Select the entire image (Mac: Command+A; Windows: Ctrl+A). Move the image to the upper-left corner. Press Shift and drag the lower-right corner of the bounding box as far down and

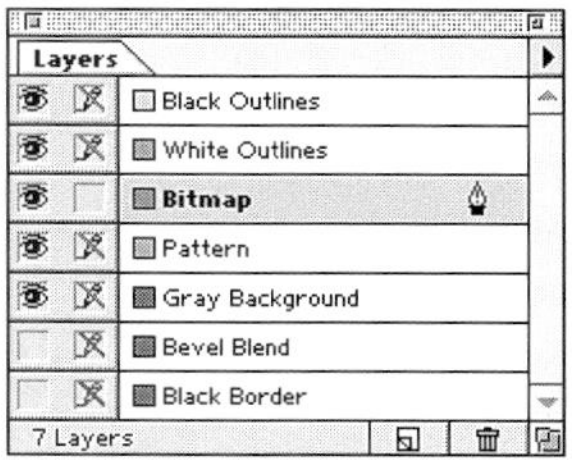

Figure 4.80
The Layers palette from HALLOWEEN.AI shows that two layers are hidden and that all but one are locked.

Figure 4.81
The original image of children dressed in Halloween costumes at school.

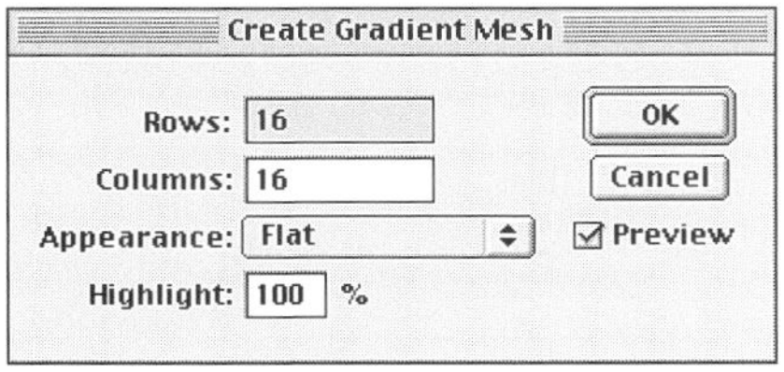

Figure 4.82
The Create Gradient Mesh dialog box, showing the author's final settings.

Figure 4.83

The bitmap layer has become a Gradient Mesh object, and a wide border with a bevel gives the finishing touch to the image.

to the right as it will go while still remaining inside the page boundaries. Save your work. Figure 4.83 shows the finished example.

That very short Illustration required quite a lot of work to produce the starting image—and you're not necessarily finished with the final image, either. In brief, here's what I did:

- I opened the KIDS.PSD image (you have it on the companion CD-ROM) in Photoshop.

- I loaded the Cropping Channel alpha and chose Image|Crop.

- I used the Shape channel to select just the children and move them to a new layer.

- I used the process described in the David example in Chapter 3 to develop a set of black lines for the image (your shortcut would be to use the Black Outlines channel in the KIDS.PSD image to re-create the layer or the selection). I used the selection to save a path named "Black Outlines". (Another shortcut: In the ASSEMBLE folder on this book's CD-ROM is BLACKPATH.AI, all ready for you to use).

- I made a duplicate copy of the layer containing the children and applied the Smart Blur filter to it in Photoshop. Figure 4.84 shows the settings that I used.

- With the magic Wand Tolerance set to 0 and the Anti-Alias checkbox deselected, I clicked on a white outline in the image. I then chose Select|Similar. I saved the selection to a channel, blurred the channel, and used the tricks shown in Chapter 3 to thicken the line. (The Filter|Other|Maximum filter can also help thicken the line.) If you're feeling lazy, you can use the White Outlines channel to make the selection to create the White Outlines path. (If you're feeling even lazier, you can use the WHITEPATH.AI image in the ASSEMBLE folder.)

- I used the Shape channel—or the selection of the nontransparent Layer pixels (Mac: Command+click; Windows: Ctrl+click on the Layer name in the Layers palette)—to create a Shape path which I then turned into a Clipping Path. I tossed out all the layers except for the image of the children that used the Shape channel and saved the image as an EPS file so that the Clipping Path was saved.

- I exported the Black Outline path and White Outline path each in its own file as Paths to Illustrator. (You need to export them individually, or the paths become intertwined and you can't get them colored or untangled.)

- I assembled the whole thing in Illustrator. I opened the EPS file first. Then I opened the BLACKPATH.AI file and filled the shapes with black and opened the WHITEPATH.AI file and filled the shapes with white (if you use mine, they are already filled for you). I grouped each of the path images. I dragged both images (each on a separate layer) into the EPS file, selected everything, and used the Align palette to align them.

- I created the Background layer, made a small pattern with a transparent background to place over it, created the large black border, and made the shapes for the bevel blend. You can see how each of these steps was done by pulling apart the starting image. The final image is shown in Figure 4.85.

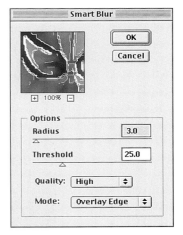

Figure 4.84

I used these Smart Blur settings to create white outlines on the image.

Where can you go from here? I gave you the "simple" route to creating the Gradient Mesh object. You can go back into the object and edit it to add colors left out from the original or to change the coloring completely. You can make that as realistic or abstract as you wish. You could also revert to the original starting point and, instead of converting the bitmap layer into a Gradient Mesh object, use it as a basis for creating your own Gradient Mesh object. This Gradient Mesh could be as large as the bitmapped image or could be done as several smaller Gradient Mesh objects. The choice is yours. Enjoy!

Figure 4.85
The completed Halloween image.

Moving On

Although I could say much more about any topic in this chapter, it is regretfully time to move on. You have learned how to create blends and to use them either to mimic shaded objects or to create linear blends, which can be opened or closed.

You've discovered many ways to create and edit gradients and ways to use them, in conjunction with blends, to simulate reality. There was neither time nor room to show you how to create an entire image made up of blends and to segment the image into sections that could be adjusted with the Filter|Colors|Adjust Colors filter. However, if you remember this suggestion, you'll have all the knowledge that you need to create this effect when you have finished the book.

Finally, you had the chance to create Gradient Mesh objects and to learn how they react. They are a wonderful tool when you learn how to deal with them.

In the next chapter, you learn some techniques to create patterns and Pattern brushes and to use the new Photo Crosshatch filter to give a *Wall Street Journal* look to your images.

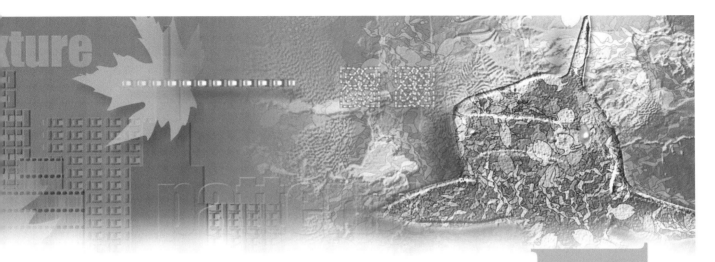

PATTERNS AND TEXTURES 5

SHERRY LONDON

Patterns are all around us: on the clothes we wear, on floors and walls, and on furnishings and household items. So often, Illustrator is used to create patterns with only rectangular repeats because this is all that most artists know how to create. The possibilities for using patterns are endless, and some of the best uses of patterns are almost subliminally subtle. In this chapter, you'll look at different patterning systems and some ways that you can use patterns. You'll also try out and tame the new Photo Crosshatch filter.

I have a confession—if you haven't already guessed, I'm a pattern junkie. For many years, I designed knitwear and needlework. I still do in my "spare" time. My first introduction to personal computers was to use them to automate the pattern-design process. So, this is more than "just another chapter" to me. In fact, I hope to share with you a more complete version of my knowledge about creating patterns in a book devoted totally to this subject. This chapter, then, is an appetizer.

The Language Of Pattern

People created patterns long before there were computers. Patterned fabric and pottery remains predate the written word by hundreds of centuries. The word *pattern* has several meanings. Although it's commonly used to refer to the template used to cut fabric for clothing or to instructions to knit or stitch something, it also refers to design elements that repeat across a surface. It's that context in which I'll use the word *pattern* in this chapter.

Language can certainly be imprecise, and I can make the words mean whatever I say they mean, so I want to define terms before I take you into this world of largely unknown terminology.

- *Motif*—The subject of the pattern, the design to be repeated (and is a more popular term that means much the same thing as *repeat unit*).

- *Pattern*—A recognizable motif that repeats at regular and predictable intervals along an invisible grid or lattice.

- *Texture*—A series of nonrepresentational motifs that repeat at regular and predictable intervals along an invisible grid but are arranged in such a way as to look completely random. You cannot easily distinguish where each repeat begins or ends in a properly done texture. In addition, a true texture has no distinguishing feature that would allow you to locate the motif from which it's constructed.

- *Symmetry*—The ways in which an arbitrary motif can be manipulated to form a repeat. This is a mathematical concept. There are only 17 ways in which it's possible to repeat a motif to cover a surface (and seven ways to repeat a pattern on a band or border).

- *Generating Unit*—The smallest area of a motif from which an entire repeat unit can be constructed.

- *Repeat Unit*—The smallest area that can be used to create the repeat pattern (it can contain multiple copies of the pattern-generating unit). I'm also inclined to call this a "tile." Because Illustrator can create only rectangular block repeats, the repeat unit (as used here) will always be the smallest rectangle that can be defined as a pattern to re-create the selected symmetry and grid structure.

- *Pattern Skeleton*—This is my own term (see, I told you that I make them up). This is the no-stroke, no-fill rectangle that sits behind a pattern repeat unit in Illustrator and defines its boundaries.

- *Grid*—The invisible geometric construct on which the pattern is based. A block repeat, for example, uses a rectangular grid. A grid can also be a parallelogram, a triangle, a diamond, or a hexagon. The terms *lattice* and *network* are also used to refer to the grid and are actually somewhat more accurate, if not as commonly understood.

This list might already be more than you wanted to know about patterns, but after you really get into working with patterns, they become entrancing and addictive. Don't worry about the dry terminology used in this section; I'm just setting the stage.

Dropping Patterns

Most patterns use a rectangular grid network as the repeat structure. Figure 5.1 shows a simple motif with a visible pattern skeleton behind it, repeated along a rectangular grid. The rows and columns are lined up like soldiers. It doesn't have to be that way. Figure 5.2 shows the same motif using a brick repeat (look at the pattern skeleton—it resembles the pattern formed by bricks). The rows are all even, but the columns don't line up. Figure 5.3 shows the pattern varied in another way. This is the classic *half-drop* repeat beloved of wallpaper designers around the world. The columns line up, but the rows don't.

Figure 5.1
A standard rectangular grid makes this pattern comfortingly familiar.

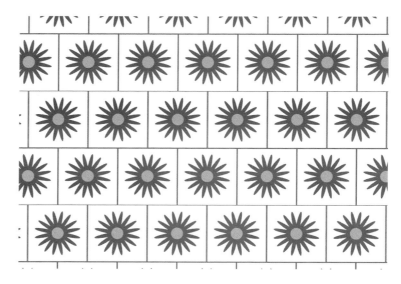

Figure 5.2
This example uses a brick repeat, in which the rows are even but the columns aren't.

Figure 5.3
The half-drop repeat creates the impression of vertical lines and is a common wallpaper repeat structure.

Figure 5.4 shows a repeat that uses a *diaper network*. I'll bet you thought that a diaper network was a support group for wet, young bottoms! It's a traditional pattern design term that refers to a structure that repeats on the diagonal. Many other repeat networks are possible—circles and hexagons are behind many Celtic, Chinese, and Arabic patterning systems. Although these structures are beyond the scope of this book, let's look at Illustration 5.1 to see how the basic pattern structures are created.

Illustration 5.1: Rectangular Repeats

1. Open the image GUITAR.AI. It doesn't look much like a guitar, but you get the idea. It's small, and it tiles nicely.

Figure 5.4

A pattern that repeats along a diagonal is called a diaper pattern.

2. Select the guitar and drag it into the Swatches palette. This is the easy way to create a pattern (though, as you'll see, not the best way). Double-click on the new pattern tile in the Swatches palette and name it "Too Close To Play".

3. Choose the Rectangle tool. Click near the bottom of the image and enter a width of 7 inches and height of 4 inches. Click on OK. Make the Fill color swatch active and click on the Too Close To Play pattern. Guess what? Those guitars really are too close together, as you can see in Figure 5.5. Of course, if you like it this way, there's nothing wrong with this method of defining a pattern.

4. You can control the distance between the guitars by the use of a pattern skeleton. This invisible rectangle sits behind your motifs and defines the boundary of the repeat. Deselect the test pattern rectangle. Select the Rectangle tool. Set the Fill and Stroke to None. Click near the small guitar motif and enter 1 inch for both the height and the width.

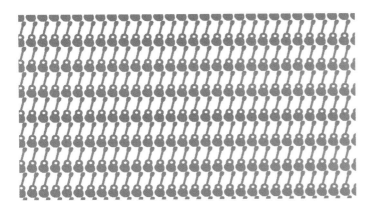

Figure 5.5

Because the guitar had no space around it when you dragged it to the Swatches palette, the motifs are close together within the pattern.

5. Choose the Selection tool. Drag the rectangle so that it surrounds the guitar motif. Choose Object|Arrange|Send To Back. Press Shift and add the guitar to your selection. Drag both the guitar and the pattern skeleton to the Swatches palette. Deselect. Double-click on the new pattern swatch and name it "One Inch". (If you name it with the repeat unit selected, you'll fill the objects with the new pattern.)

6. Select the test pattern rectangle. Click on the One Inch pattern to fill the rectangle. Figure 5.6 shows this image. Now it's too far apart.

Figure 5.6

One inch is probably too far apart for this guitar motif to repeat; however, a pattern skeleton rectangle of one inch sets the distance.

7. You'll probably have an easier time setting your desired spacing by eye. Draw a no-Stroke, no-Fill rectangle around the guitar as I did in Figure 5.7. You can easily center the motif by clicking on Horizontal Align Center and Vertical Align Center in the Pathfinder palette, but it isn't really necessary—the pattern will repeat just fine as long as the entire repeat unit is inside the pattern skeleton. You should send the pattern skeleton to the back, however (though as long as the repeat unit is totally inside it, the stacking order of the pattern skeleton doesn't matter either). Select the guitar and the pattern skeleton and drag them to the Swatches palette. Deselect. Name the swatch "Right Size". You can drag the swatch directly from the Swatches palette onto the test pattern rectangle to change the fill. Figure 5.8 shows the Right Size pattern.

8. You might find it easier to visualize your spacing (though the construction is trickier) if you attach the motif to the top-left corner of the pattern skeleton. Zoom in as closely as possible to the guitar motif and the pattern skeleton. Toggle Artwork mode (Mac: Command+Y; Windows: Ctrl+Y). Bear with me. Although this process might seem Byzantine, it becomes a timesaver when you try to create the other pattern network types. You'll also get a better understanding of what happens when you tile a motif, as you follow along.

Figure 5.7 (left)

Draw a rectangle around your motif to provide the needed spacing.

Figure 5.8 (right)

The pattern looks much better when you control the spacing of the repeat.

9. Deselect everything. Turn on Smart Guides (Mac: Command+U; Windows: Ctrl+U). Select the guitar motif. Place your cursor over the center point of the guitar (the spot right in the center of the "hole") and drag it to the top-left corner of the pattern skeleton (you'll see the word "anchor" when you're at the right spot).

10. Select both the guitar and the pattern skeleton. Press Option (Mac) or Alt (Windows) and drag the objects to the right (from the center of the guitar's hole) until the copy snaps to the original (press Shift after you begin to drag to constrain movement to the horizontal). You need to be closely zoomed in for the tolerances to be tight enough to work. Figure 5.9 shows the two guitars.

11. Zoom out so that you can see both repeats with room to spare. Select both guitars and their invisible boxes. Again, drag from the center of the guitar's hole (whichever one you prefer) to move a copy of the motifs. This time, drag downward (with Alt/Option and, after you start to drag, Shift pressed) until the copied objects snap to the bottom of the originals. Figure 5.10 shows this step.

12. Deselect everything. Select and delete the three copied rectangles (leave the original and leave all the guitars). Figure 5.11 shows the remaining pattern skeleton and the four guitars. Zoom in to about 4800% on each corner of the pattern skeleton where it meets a guitar. Double-check to make sure that all the guitars are exactly

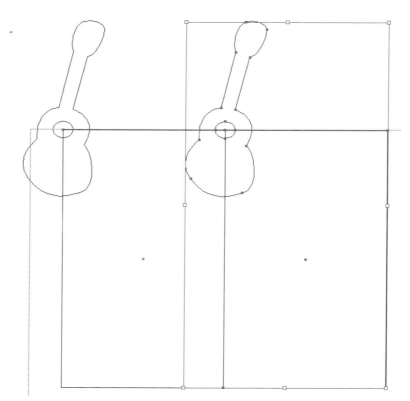

Figure 5.9
Notice how the selected guitar overlaps the pattern skeleton rectangle of the first motif.

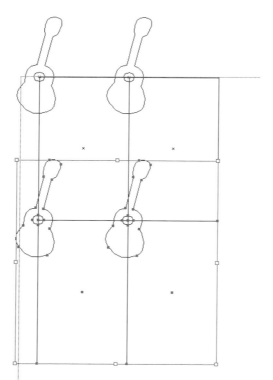

Figure 5.10
Drag a copy of both guitars and their invisible rectangles downward until it snaps to the bottom of the originals.

snapped to the points of the rectangle. Turn off Smart Guides (Mac: Command+U; Windows: Ctrl+U). Toggle Artwork mode (Mac: Command+Y; Windows: Ctrl+Y).

13. Select the entire repeat unit (the four guitars and the pattern skeleton). Drag it to the Swatches palette. Deselect.

14. Name the pattern swatch "Four Guitars". Zoom out to Fit In Window (Mac: Command+; Windows: Ctrl+0). Select the test pattern rectangle and apply the Four Guitars pattern. It should look no different than it did before (although the start of the pattern will probably shift). The four corner elements should meet seamlessly. If you feel the need to check this out, zoom into the test pattern and examine it. Don't close your practice image; you'll use it in the other Illustrations in this section.

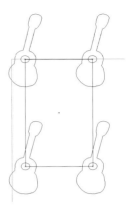

Figure 5.11
The repeat unit consists of four guitars and a pattern skeleton.

The screen display of the pattern might look more ragged than it did when you simply surrounded the single motif with a rectangle. Not to worry: That is only screen artifacting and will not show up in print (unless, of course, you really misaligned the tiles).

These two questions might come to mind as you finish this first exercise:

• *Should I use the Crop command to make the repeat unit the same size as the pattern skeleton?* No. You could crop, and it wouldn't change things—in theory. In practice, I've seen cropping cause the pattern to meet improperly. The "sure bet" is to drag the pattern skeleton and all the stuff hanging onto it and let Illustrator determine the pattern boundaries. This also makes it easier to edit the pattern again, if needed.

• *What happens if the pattern skeleton isn't the bottom object in the repeat unit?* The short answer: Your tile won't tile properly. Figure 5.12 shows this best.

After you go through the agony of figuring out the complex way to make a single motif repeat, it's simple to make a brick or half-drop repeat out of it, which is your next task (see Illustration 5.2).

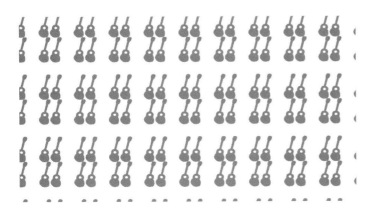

Figure 5.12
This is what the pattern looks like if you forget to make the pattern skeleton the base object in your repeat unit.

PROJECT Illustration 5.2: Brick It Up

1. Select the four-guitar repeat and drag a copy of it to another location in your image. Zoom in to the copied pattern repeat unit. To make a brick repeat, you need two guitars on top and one on the bottom—dead center between the two top guitars.

2. The next step seems odd, but it really works. It's a technique recommended many years ago by Luanne Seymour Cohen in her landmark work on Illustrator patterns (the manual for the original Adobe Collector Series patterns, long out of print). How do you create an object in between two others? Use the Blend tool. Select the two bottom guitars. Choose the Blend tool. In the Blend Options dialog box (Object| Blends|Blend Options), set the Specified Steps to 1. Click on the center point of the leftmost guitar on the bottom and then click on the center point of the rightmost guitar on the bottom. Figure 5.13 shows you the five guitars (four and a blend).

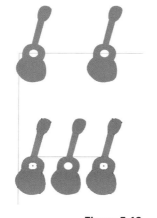

Figure 5.13

To place an object dead-center between two others that are identical, use the Blend tool with a step of 1.

3. Because Illustrator 8's blends are "live," you need to expand the blend (Object|Blends|Expand) so that the blended object can live on its own. After you've expanded the blend, deselect it, select it again, and ungroup it (Mac: Shift+Command+G; Windows: Shift+Ctrl+G). Deselect.

4. Select the two end guitars (the start and end points for the blend that you just expanded) and delete them. Look carefully at Figure 5.14. It shows the three remaining guitars and the pattern skeleton (all selected so that you can see the pattern skeleton rectangle). If you look inside the pattern skeleton, you'll notice that the pieces of the guitars are not enough to make a repeating pattern. The rectangle itself doesn't contain even one complete guitar. Therefore, you need to enlarge the pattern skeleton.

5. Toggle Artwork mode (Mac: Command+Y; Windows: Ctrl+Y). With the Selection tool, select the pattern skeleton. Double-click on the Scale tool and select Non-Uniform. Enter a Horizontal scale amount of 100% and a Vertical scale amount of 200%. Click on OK. Figure 5.15 shows the Artwork view. Now the pattern skeleton encompasses one complete guitar and two halves (which can join together to form another whole).

6. Toggle Artwork mode (Mac: Command+Y; Windows: Ctrl+Y). Zoom out to Fit In Window (Mac: Command+; Windows: Ctrl+0). Select the

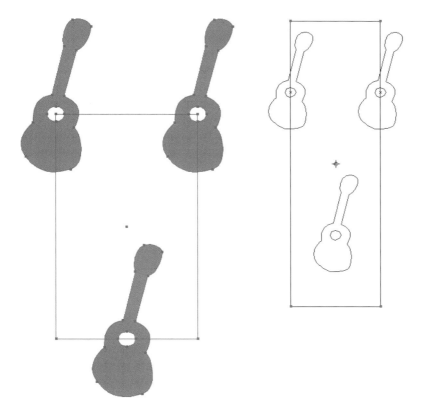

entire repeat unit (if necessary, first send the pattern skeleton to the back) and drag it into the Swatches window. Deselect the repeat unit and name the new pattern "Bricked Guitars". Select the test pattern rectangle and fill it with the Bricked Guitars pattern. Figure 5.16 shows the pattern applied to the rectangle.

Let's create a half-drop repeat in Illustration 5.3 so that you can see the differences.

Figure 5.14 (left)
The pieces of guitar that fall within the pattern skeleton are not enough to reconstitute an entire guitar.

Figure 5.15 (right)
When you double the vertical measurement of the pattern skeleton, it becomes large enough to hold a full brick repeat.

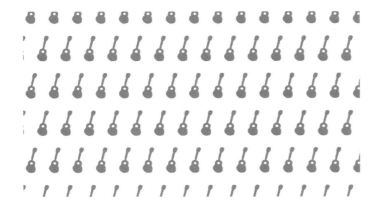

Figure 5.16
The guitars are now in a brick repeat.

PROJECT Illustration 5.3: The Half-Drop Repeat

1. Zoom in to your image on the original four-guitar repeat. Select the four guitars and the invisible pattern skeleton. Drag a copy of the selection to another location.

2. Select the top-right and bottom-right guitars. With the Blend tool, click on the center points of the top and then the bottom selected guitar. The Blend Options should be set to a specified step of 1.

3. Expand the blend (Object|Blends|Expand), deselect, select again, and ungroup.

4. Select and delete the top-right and bottom-right guitars. Figure 5.17 shows the almost-ready repeat unit. Again, you'll notice that the pattern skeleton is not large enough to hold a whole repeat unit.

5. Select the pattern skeleton rectangle only. Double-click on the Scale tool. Select Non-Uniform. Enter a Horizontal scale amount of 200% and a Vertical scale amount of 100% (this is the exact opposite of the procedure to create a bricked repeat). Click on OK. Figure 5.18 shows the finished repeat unit.

6. Zoom out to Fit In Window (Mac: Command+; Windows: Ctrl+0). Select the entire repeat unit (if needed, send the pattern skeleton

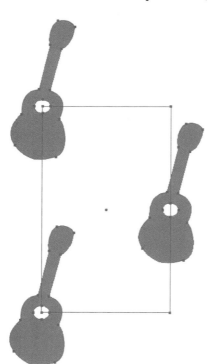

Figure 5.17

After you create a blended top-to-bottom guitar, remove the two originals on the right of the repeat unit.

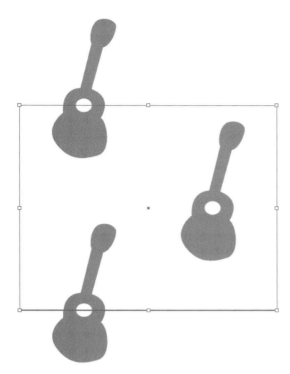

Figure 5.18
The finished half-drop repeat unit is wider than it's long (whereas the bricked repeat unit was longer that it was wide).

to the back first) and drag it into the Swatches window. Deselect the repeat unit and name the new pattern "Half-Dropped Guitars". Select the test pattern rectangle and fill it with the Half-Dropped Guitars pattern. Figure 5.19 shows the pattern applied to the rectangle.

Now that you've made brick and half-drop repeats, the diaper pattern you'll create in Illustration 5.4 is easy. Because this is the third time through on similar instructions, I made these instructions brief. Review Illustrations 5.1, 5.2, and 5.3 if you have trouble remembering a command or shortcut.

Figure 5.19
The Half-Dropped Guitars show straight columns and alternating rows.

PROJECT Illustration 5.4: Diapering Patterns

1. Make another copy of the original four-guitar pattern repeat unit.

2. Switch to Artwork mode. Turn on Smart Guides. Drag a copy of one of the guitars from its center point to the center point of the pattern skeleton.

3. Turn off Artwork mode and turn off the Smart Guides.

4. Select the center guitar. Choose the Rotate tool and rotate the center guitar until it's leaning slightly in the opposite direction.

5. Select the entire pattern repeat and drag it to the Swatches palette. Deselect. Name the new pattern "Diapered Guitars". Select the test pattern rectangle and fill it with the diaper pattern. Figure 5.20 shows the pattern repeat unit, and Figure 5.21 shows the tiled pattern.

Before you leave this topic, try one of the slightly more complex patterns—one of the 17 symmetries (see Illustration 5.5). If you have the Terrazzo filter from Xaos Tools for Photoshop (one of my favorite plug-ins), you can see that this is the symmetry they call "Honey Bees." It involves two rotations and a reflection but is easy to do if you follow along.

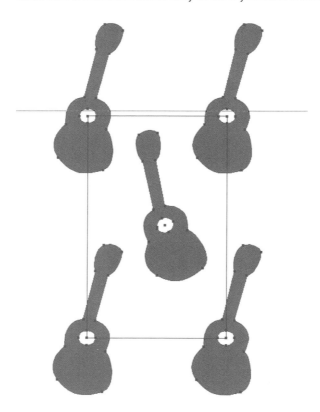

Figure 5.20

This is the repeat unit for the diapered guitars.

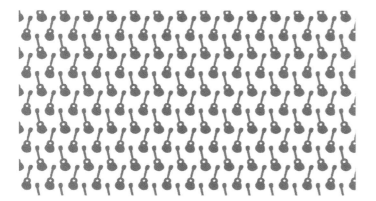

Figure 5.21
Guitars, arranged in a diaper pattern (a pattern structure that repeats diagonally in both directions).

PROJECT Illustration 5.5: Honey Bees

1. Make a copy of one of the guitars and the original pattern skeleton. Drag it to a new location in the image. With both objects selected, click on Horizontal Align Center and Vertical Align Center in the Pathfinder palette to center the guitar inside the pattern skeleton rectangle. Toggle Artwork mode (Mac: Command+Y; Windows: Ctrl+Y). Select just the pattern skeleton. Resize the pattern skeleton (press Shift and Alt or Option to constrain aspect ratio and scale from the center) so that it hugs the guitar more closely as in Figure 5.22.

2. With both objects selected, choose the Rotate tool. Press Option (Mac) or Alt (Windows) and click on the lower-right corner of the

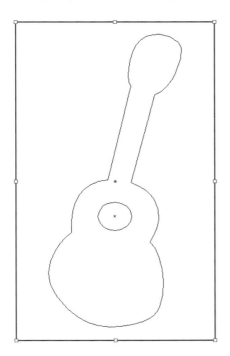

Figure 5.22
Resize the pattern skeleton to fit more tightly around the guitar.

pattern skeleton. Enter 180 degrees in the Amount field. Click on Copy. Figure 5.23 shows the result.

3. Select both guitars and pattern skeleton rectangles. Double-click on the Reflect tool. Select the Vertical radio button and click on Copy. Figure 5.24 shows the pattern in progress.

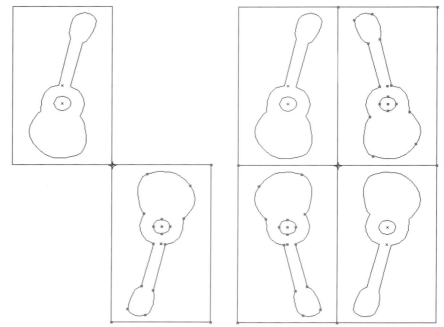

Figure 5.23 (left)
Rotate a copy of the motif and the pattern skeleton 180 degrees from the bottom-right corner.

Figure 5.24 (right)
Reflect the two guitars and the two pattern skeletons vertically from the default center point of the selected objects.

4. The next step is the only tricky one. Leave only the two reflected guitars selected (they were selected at the end of Step 3). Choose Object|Transform|Transform Each. Set the Horizontal and Vertical scale to 100%, move the objects vertically and horizontally 0 points (in other words, *don't* move them), and change the Angle of Rotation to 180 degrees. Deselect the Random checkbox. Click on OK (not Copy). Figure 5.25 shows the correctly filled-in dialog box, and Figure 5.26 shows the pattern repeat unit.

5. Toggle Artwork mode off (Mac: Command+Y; Windows: Ctrl+Y). Because there's white space around the guitars, you can simply select all four pattern-generating units and drag the entire repeat unit to the Swatches palette. Name the pattern and fill your test rectangle as you've done before. Figure 5.27 shows the tiled pattern. (If the pattern were more complex and some of the elements crossed over pattern skeleton rectangles, you would need to unite the four pattern skeleton rectangles that actually form the skeleton of the pattern.)

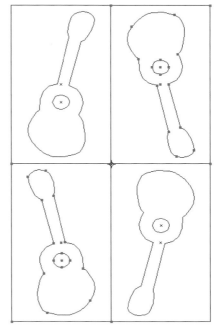

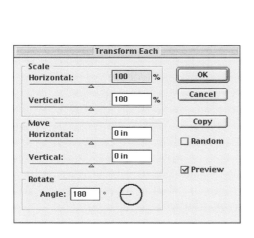

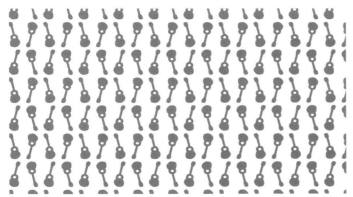

Figure 5.25 (left)
The Transform Each dialog box lets you rotate each guitar individually.

Figure 5.26 (right)
The final pattern repeat unit.

Figure 5.27
The Honey Bees repeat shows another version of the guitars.

The Pattern Brush

In Chapter 3, you discovered how easy it is to take a piece of text and define it as a Pattern brush. You're going to use the guitar again to create an interlocking Pattern brush—a bit more challenging but still simple. The same procedures that work for creating a pattern also work when creating a Pattern brush. The only difference is that you need to exercise a bit of care when defining a Pattern brush pattern because it needs to fit along a path. Therefore, a wide pattern is better for a Pattern brush than a tall one. A side benefit of Illustration 5.6 is that it will segue nicely into our discussion of creating interlocking patterns.

▚ Illustration 5.6: A Long Line Of Guitars

1. Drag a copy of one of the guitars in the original pattern repeat unit to a new location in the image. Toggle Artwork mode on (Mac: Command+Y; Windows: Ctrl+Y). Do not copy the pattern skeleton, as you need to make a new one.

2. Choose the Rotate tool. Press Option (Mac) or Alt (Windows) and click on the top-right anchor point of the guitar (it's actually on the right side of the fret, but it's the highest point on the right side). Enter 180 degrees as the angle of rotation and click on Copy. Group the two guitars (Mac: Command+G; Windows: Ctrl+G).

3. Draw a pattern skeleton around the double-guitar object, giving it a bit of white space. With the pattern skeleton and the two guitars selected, click on Horizontal Align Center and Vertical Align Center in the Align palette to center the guitars in the rectangle. (Of course, if you draw the pattern skeleton from the center—the point where the two guitars join—the rectangle will automatically be centered.) Figure 5.28 shows the pattern-generating unit.

4. Select the guitars and the pattern skeleton. Press Option (Mac) or Alt (Windows) and move the selection to the right (press Shift after you begin to drag to constrain movement to the horizontal) until the left edge of the pattern skeleton rectangle copy clears the guitar as shown in Figure 5.29.

5. Transform Again (Mac: Command+D; Windows: Ctrl+D) two more times. Figure 5.30 shows the four pattern-generating units in a row.

6. Select the second pattern skeleton rectangle from the left. Hide it (Object|Hide Selection). Select and delete the remaining pattern

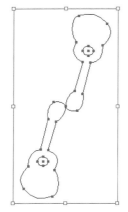

Figure 5.28

The rotated guitars and the new pattern skeleton form the pattern-generating unit from which the new repeat unit is created.

Figure 5.29 (left)

Move a copy of the guitars and the pattern skeleton to the right.

Figure 5.30 (right)

Create two more copies of the guitars and pattern skeleton with the same spacing by using the Transform Again command.

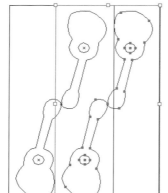

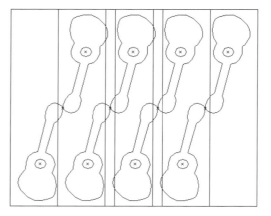

skeleton rectangles. Show All (Object|Show All). Zoom in closely to the selected pattern skeleton rectangle and drag each side of the rectangle toward the center until the sides are precisely between two connected guitars as shown in the circled side control handles in Figure 5.31.

7.   Select the pattern skeleton and send it to the back.

8.   Select and drag the pattern repeat unit into the Brushes palette. Figure 5.32 shows the dialog box that pops up. Select New Pattern Brush.

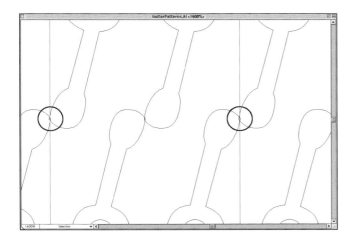

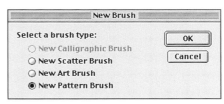

Figure 5.31 (left)
Move the control handles of the bounding box of the pattern skeleton toward the center until each side lies between two connected guitars.

Figure 5.32 (right)
The New Brush dialog box allows you to create a Pattern brush.

9.   Figure 5.33 shows the next dialog box. Name the Pattern brush "Rotated Guitars" and click on the Inner Corner icon. Select the Honey Bees pattern that is in the list as the corner element. Click on the Outer Corner icon and select the Honey Bees pattern again. Click OK.

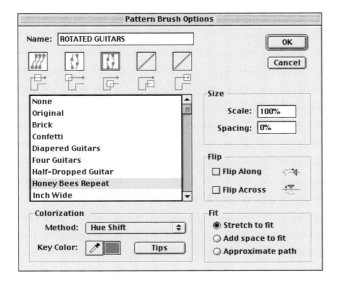

Figure 5.33
The New Pattern Brush dialog box.

10. Select your test pattern rectangle. Change its Fill to None. Choose the Paintbrush tool. Click on the new Pattern brush to use the brush as the stroke for the rectangle. Figure 5.34 shows the rectangle and some painted strokes.

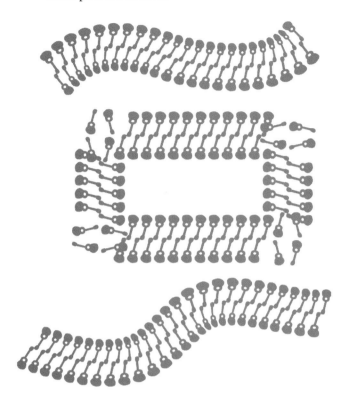

Figure 5.34

The rotated guitars are applied as a stroke using the Pattern brush.

Having learned how to define patterns, you now have the basics of the Pattern brush as well. In Chapter 7, you'll learn how to miter a corner for the Pattern brush, and you'll learn some additional tricks in Chapter 9. Oh, you can close the guitar image now. It has served us well.

Escher Patterns

M. C. Escher created some of the most exciting patterns that the world has ever seen. He was a master at making hands look like they were drawing each other and at making two-dimensional lizards pop out of the canvas and seem to walk away. He created interlocking patterns in which geese faced in alternate directions. The incredible thing about many of his patterns was they way in which they interlocked to form animals and plants yet tiled seamlessly.

Escher was a master at the mathematics of symmetry. I can't even begin to image how he worked out most of his effects (and I certainly couldn't duplicate them), but I do have a few interlocking tricks up my sleeve.

The Three-Corner Trick

If you draw a line of any shape that hits three corners of a rectangle, when you repeat the rectangle, the shapes will seamlessly interlock. This can be a bit of a fiddly project, but its worst offense is that once you master the technique, it can easily become addictive.

One of the things that I love about Illustrator is the large number of ways in which you can construct an object. It allows me to exercise my propensity for getting (as my tenth-grade geometry teacher once put it) "to New York by way of Chicago"—and I grew up in Philadelphia, so New York is only 90 miles away. If there's a short, easy way to do it, my natural inclination (and first tries) will be anything but that. However, I promise you that I tried (and rejected) at least 12 other ways of constructing this effect until I developed the procedure in Illustration 5.7. This is as easy as I can find to explain it.

PROJECT Illustration 5.7: Creating Interlocking Patterns

1. Create a new document in Illustrator.

2. Select the Rectangle tool. Click on the intersection of the top and left margins in the document. In the Rectangle Options dialog box, enter a width of 2 inches and a height of 1 inch. Click on OK. Choose View|Make Guides.

3. Turn on Smart Guides (Mac: Command+U; Windows: Ctrl+U). Select the Pencil tool. Toggle Artwork mode on (Mac: Command+Y; Windows: Ctrl+Y).

4. Begin in the lower-left corner of the rectangle guide and draw any type of squiggly line that you want. It must do three things:

   - It must start in the lower-left corner.

   - It must touch the upper-left corner.

   - It must end at the upper-right corner.

   The line may go outside the rectangle a bit as long as it meets these three criteria. As you gain more experience, you'll notice that it's even possible to break the "it must touch the upper-left corner" rule if you obey the other rules. Figure 5.35 shows the line that I drew.

5. If you zoom in on the corners of the rectangle, you'll probably discover that you didn't quite get the path to exactly touch the corners. Use the Direct Selection tool to attach both endpoints to

WARNING!

Be careful when you select a pattern fill. It's possible to fill a stroke with a pattern, but it's not bright. You might not see that you've applied a pattern to your stroke on screen (unless it has a wide stroke width), and you'll probably get a long wait and, perhaps, an error, when you go to print (and trapping the stroke is not an amusing day's task). Always make sure that you are actually changing the Fill swatch when you click on a pattern, because it's much too easy to get this wrong.

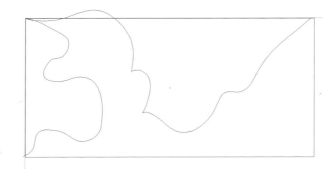

Figure 5.35

Draw a squiggly line that touches three corners of the rectangle.

their respective corners of the guides. Unfortunately, for this technique to work correctly, you must be exact.

6. Once you fix the endpoints of the path, take a closer look at how the path hits the upper-left corner of the guides. It has a nasty tendency to form a loop rather than a point, which is what happened in Figure 5.36. As you can see in the selected corner tip, the two parts of the path cross over each other. You need to use the Direct Selection tool and move the direction lines so that the path no longer overlaps and loops. Figure 5.37 shows the problem area fixed. Now you're ready to play.

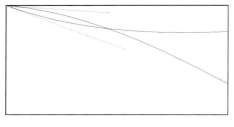

Figure 5.36 (left)

Leaving a loop in a corner is a good way to make a mess.

Figure 5.37 (right)

The loop is untangled and looks much better now.

7. Choose the Scissors tool. Click on the top-left corner point of the path (the point that you just fixed). This breaks your line into two separate paths.

8. If it is not already selected, select the lower-left-to-upper-left corner path. Choose Object|Lock.

9. Zoom in to the image as closely as you can while still seeing the entire path. Choose the Selection tool and deselect everything. Position your cursor at the top-left corner of the top path. The Smart Guides should tell you that it is an anchor point. Press Shift+Option (Mac) or Shift+Alt (Windows) and keep them pressed. Click on the anchor point (still holding down the modifier keys) while dragging

the path downward. Drag the copy down until the word "intersect" appears along the left margin. Release the mouse button and then the modifier keys.

10. Choose Object|Unlock All and then Edit|Select Inverse. Then choose Object|Lock Selected. This prevents you from accidentally selecting the top or bottom paths—an easy thing to do in this instance.

11. Position your cursor at the top-left corner of the side path. The Smart Guides should tell you that it is an anchor point. Press Shift+Option (Mac) or Shift+Alt (Windows) and keep them pressed. Then click on the anchor point (while still holding down the modifier keys) while dragging the path to the right. Drag the copy to the right until the word "intersect" appears along the top margin. Release the mouse button and then the modifier keys. Figure 5.38 shows the puzzle piece that I created. Yours should be somewhat similar.

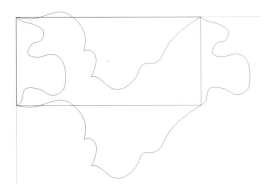

12. Choose Object|Unlock All. Deselect. Turn off Smart Guides (Mac: Command+U; Windows: Ctrl+U). Although you can work with them on, I find them unpleasant and in my face when I don't really need them, so I leave them active for as short a time as possible.

13. You now need to join all four corners so that you form a closed, filled shape. To do this, use the Direct Selection tool to select the two points that intersect to form the corner of the shape. Average the points (Mac: Command+Option+J; Windows: Alt+Ctrl+J) both vertically and horizontally. Join the points (Mac: Command+J; Windows: Ctrl+J) with a corner point (unless, as in the lower-right corner of my example, you feel that a smooth point would work better). Repeat this for each of the four "corners" (the places where the cut paths meet).

14. Select a Fill color for the shape. Toggle Artwork mode off (Mac: Command+Y; Windows: Ctrl+Y). Turn on Smart Guides (Mac: Command+U; Windows: Ctrl+U).

Figure 5.38
By copying the top and side segments of the path, you have created a completely interlocking shape.

15. Choose the Selection tool. Position your cursor at the top-left corner of the shape. The Smart Guides should tell you that it is an anchor point. Press Shift+Option (Mac) or Shift+Alt (Windows) and keep them pressed. Then click on the anchor point (while still holding down the modifier keys) while dragging the path downward. Drag the copy down until the word "intersect" appears along the left margin. Release the mouse button and then the modifier keys.

16. Transform Again (Mac: Command+D; Windows: Ctrl+D) two more times. You have four of the shapes in your document. Select shape 2 and change its Fill color. Select shape 4. Use the Eyedropper tool to give it the same Fill color as shape 2.

17. Select the entire image (Mac: Command+A; Windows: Ctrl+A). Choose the Selection tool. Position your cursor at the top-left corner of the top shape (this is not the same as the bounding box control handle, so you are less likely to have problems). The Smart Guides should tell you that it is an anchor point. Press Option (Mac) or Alt (Windows) and click on the anchor point (while still holding down the modifier key). After you start to drag the shapes to the right, press Shift to constrain the movement to the horizontal. Drag the copy to the right until the word "intersect" appears along the top margin. Release the mouse button and then the modifier keys.

18. Transform Again (Mac: Command+D; Windows: Ctrl+D). You have a grid of shapes four rows by three columns deep. You need to flip the colors of the center column of shapes. The easiest way is to select the Eyedropper tool. Use Command or Ctrl to access the Selection tool to select the object and then release the key to pick the color for the shape. Figure 5.39 shows the finished group of interlocking shapes.

Figure 5.39

You've created an interlocking grid of shapes using counterchange coloring (i.e., alternating colors).

19. Select the Rectangle tool and choose a Fill and a Stroke of None. Press Option (Mac) or Alt (Windows) and click near the center of the group of shapes. Enter a width of 4 inches and a height of 2 inches into the Rectangle Tool Options dialog box. Click on OK. Choose Object|Arrange|Send To Back. You don't have to try to find the exact center of the image. As long as the rectangle is over the solidly filled shapes, your pattern will tile properly.

20. Select the entire image (Mac: Command+A; Windows: Ctrl+A). Press Shift and drag the lower-right-corner control handle on the bounding box up and to the left to make the pattern tile smaller. Don't decrease its size too much, as you can always scale the pattern after it's applied. However, we created the pattern at a size that's probably too large for comfort.

21. Select the entire object again and drag it to the Swatches palette. Make a test rectangle and try your new pattern. Figure 5.40 shows my pattern applied to a clothing shape in a very stylized image.

Figure 5.40

The young lady in the image is wearing a jumper that sports the latest fashion in interlocking patterns.

How does this effect work? Why does it work? The "secret" is that you are really creating an odd-shaped rectangle with identical parallel sides. When you take a rectangle and make changes to a side, as long as you make the same changes to the opposing side, the rectangle will always tile. You merely created two very strange-looking sides for a rectangle and repeated them.

The other trick in this exercise is that you must know your starting tile size. Even though you reduced the pattern repeat before making it into a pattern, up until the last moment you knew the original repeat size (it was the 2-by-1 rectangle that started the procedure). Your final tile must be an even multiple of both directions of that first rectangle. If you had wanted only to repeat your single-colored tile (though there isn't much point in doing that), the tile size would have been the same as the original. Because you used two colors, the full repeat was twice the original rectangle size.

You can also use this technique to create wonderful Pattern brushes. If you wanted to define your interlocking pattern as a Pattern brush, you would use the irregularly shaped top and bottom edges as they are and, for a two-color pattern, make a pattern skeleton that was 4 inches wide (twice your original rectangle).

The Ogee

You can also create interlocking patterns that aren't based on rectangles. Hexagons tiles seamlessly just as they are. Diamonds also tile by nature. The ogee form is another shape that always tiles if constructed properly. Never heard of an ogee? You've seen it even if you didn't know it had a name. It's a pattern long used in Arabic, Islamic, and Persian design. It also looks like an old-fashioned Christmas tree ornament—the ones that were pointed on top and bottom and round in the middle. That's an ogee. As a motif, it frequently appears in wallpaper design. (Too frequently—we lived for much too long with a green velvet flocked ogee pattern on a metallic gold background that a previous owner of our house put up on the bedroom walls. Yes, it was as awful as it sounds.) However, the ogee has a respectable lineage and can be quite lovely as a pattern device when used in good taste. In Illustration 5.8, you'll create and decorate one.

PROJECT Illustration 5.8: O, Gee What A Beautiful Morning

1. Create a new document in Illustrator.

2. Choose a Fill color (I used a soft green) and set the Stroke to None. Select the Ellipse tool. Click near the center of the image and enter 1.5 into the Width and Height fields of the Ellipse Tool Options dialog box. Click on OK.

3. Turn on the Rulers (Mac: Command+R; Windows: Ctrl+R). Turn on Smart Guides (Mac: Command+U; Windows: Ctrl+U). Drag a horizontal and vertical guide so that they intersect in the exact center of

the filled circle. (Does your circle lack a center point? Click on the square-with-a-period-in-it icon on the Attributes palette.)

4. You only need one quarter of the circle. Choose the Rectangle tool. Draw a rectangle to the right of the vertical guide so that it completely covers the right side of the circle as shown in Figure 5.41.

5. Select the entire image (Mac: Command+A; Windows: Ctrl+A). In the Pathfinder palette, click on Minus Front. Now you have a half-circle.

6. Draw another rectangle. This time let it cover the bottom half of the half-circle as shown in Figure 5.42.

Figure 5.41 (left)
Cover the right side of the circle with a rectangle.

Figure 5.42 (right)
When you subtract the rectangle from the half-circle, you'll leave only a quarter of the original circle.

7. Swap the Stroke and the Fill for the quarter-circle so that it's stroked but not filled. Toggle Artwork mode on (Mac: Command+Y; Windows: Ctrl+Y). Choose the Scissors tool. Click on the two points of the circle where the arc intersects the guides. Figure 5.43 shows the quarter-circle cut with direction lines coming from the cut anchor points. For the screen capture, I also changed stroke colors so that you could see the two pieces that come from the scissors cut.

8. Select the right-angle portion of the quarter-circle and delete it. Only the arc remains.

9. Select the arc. Choose the Rotate tool. Press Option (Mac) or Alt (Windows) and click on the top end point of the arc (the one that touches the vertical guide). This sets the top point as the center of the transformation. Watch the Smart Guides to make sure that you are clicking in the right place. Enter 180 degrees into the Rotate Tool Options dialog box. Click on Copy. Figure 5.44 shows the path that becomes the template for all the sides of the ogee.

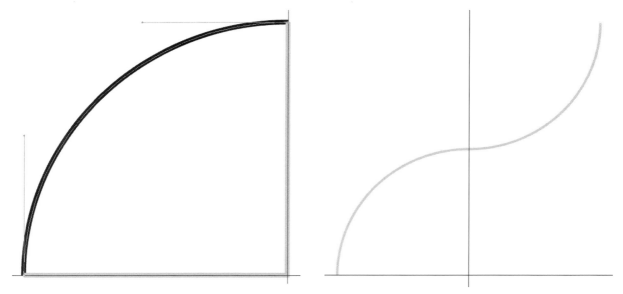

Figure 5.43 (left)

Cutting the arc from the bottom
and the side of the quarter-circle
leaves only the arc.

Figure 5.44 (right)

The ogee will tile because its side
is made from a path that is rotated to form the second half, and
all its sides are flipped or rotated
copies of the orginal path.

10. Choose the Direct Selection tool. Drag a marquee around the two points on the arc where the copy joins the original (on the vertical guide). Average the points (Mac: Command+Option+J; Windows: Alt+Ctrl+J) both vertically and horizontally. Join the points (Mac: Command+J; Windows: Ctrl+J) with a smooth point. Figure 5.45 shows the almost lyrical picture of the curved arcs with perfectly straight direction lines. Deselect.

11. Choose the Selection tool and select the curved line. Choose the Reflect tool. Press Option (Mac) or Alt (Windows) and click on the top

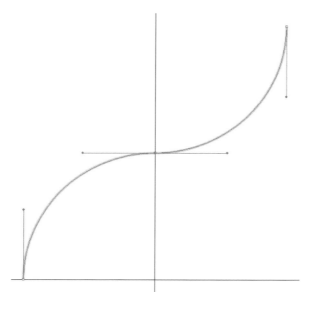

Figure 5.45

The curve of the joined ogee
with its direction lines is almost
magical.

endpoint on the curved line. Select the Vertical radio button in the Reflect Tool Options dialog box. Click on Copy. Use the Direct Selection tool to select the top points on the ogee-to-be. Figure 5.46 shows this point as a screen capture of the line after the Reflect tool was used. Average the points (Mac: Command+Option+J; Windows: Alt+Ctrl+J) both vertically and horizontally. Join the points (Mac: Command+J; Windows: Ctrl+J) with a corner point.

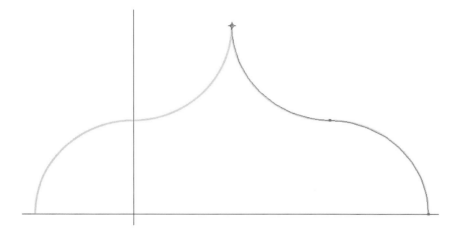

Figure 5.46
Select the top points of the two curves and average and join them.

12. Select the entire image (Mac: Command+A; Windows: Ctrl+A). With the Reflect tool, press Option (Mac) or Alt (Windows) and click on either of the endpoints of the path that lies along the horizontal guide. Select the Horizontal radio button in the Reflect Tool Options dialog box. Click on Copy. Figure 5.47 shows the finished ogee outline.

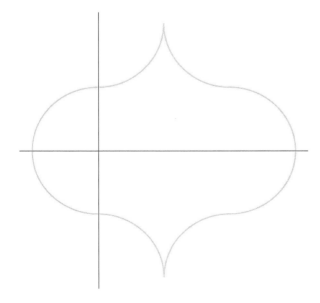

Figure 5.47
The finished ogee outline.

13. With the Direct Selection tool, select the two points that form the left side of the ogee and touch the horizontal guide. Average the points (Mac: Command+Option+J; Windows: Alt+Ctrl+J) both vertically and horizontally. Join the points (Mac: Command+J; Windows: Ctrl+J) with a smooth point.

14. With the Direct Selection tool, select the two points that form the right side of the ogee and touch the horizontal guide. Average the points (Mac: Command+Option+J; Windows: Alt+Ctrl+J) both vertically and horizontally. Join the points (Mac: Command+J; Windows: Ctrl+J) with a smooth point. Select the entire image (Mac: Command+A; Windows: Ctrl+A). Toggle Artwork mode off (Mac: Command+Y; Windows: Ctrl+Y). Swap the Fill and the Stroke so that the ogee is filled. Move the ogee so that its bottom point snaps to the intersection of the guides as shown in Figure 5.48. (Deselect everything and move your cursor over the bottom point until the Smart Guides indicate the anchor point. Press the mouse and drag. This selects the object at the same time).

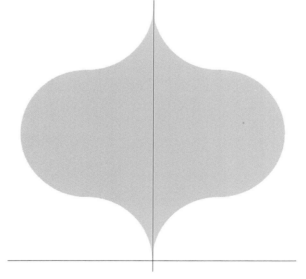

Figure 5.48

Snap the bottom point of the filled ogee to the intersection of the guides.

15. Hide the guides. Decorate your ogee with designs or squiggles as you wish. (You might want to turn off Smart Guides if they annoy you.) Figure 5.49 shows my very abstract decoration. (You're going to be cutting the ogee and its decoration apart, so don't get too carried away.) Don't use gradients, patterns, meshes, or the Paintbrush tool (or masks or compound paths).

16. Select the entire image (Mac: Command+A; Windows: Ctrl+A). Drag a copy of the ogee into the pasteboard area for safekeeping. Create a new layer and move the copied ogee's selection rectangle onto the new layer. Lock and hide the layer. Now it's not in your way, but it's there if you need it. Make Layer 1 active.

17. Select the entire image (Mac: Command+A; Windows: Ctrl+A). Click on Merge in the Pathfinder palette.

18. Show the guides (View|Show Guides). Turn on Smart Guides (Mac: Command+U; Windows: Ctrl+U). Select the entire image (Mac: Command+A; Windows: Ctrl+A). Drag a horizontal guide to the first set of anchor points below the top point of the ogee. Drag a horizontal guide to the last set of anchor points before the bottom point of the ogee as shown in Figure 5.50.

Figure 5.49
Decorate the ogee using whatever and however many colors you want.

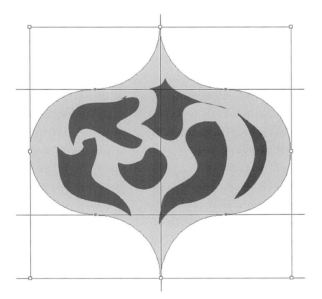

Figure 5.50
Create two more horizontal guides at the second-from-the-top and the second-from-the-bottom points on the ogee.

19. Deselect. Choose the Pen tool. The color doesn't matter, but you need to select a Fill of None and a stroke of whatever. Draw a straight line across the ogee using the top horizontal guide as your ruler. The line must be wider than the ogee. Press Option (Mac) or Alt (Windows) and drag a copy of the line (press Shift after you begin to drag to constrain movement to the vertical) down to the next horizontal guide as shown in Figure 5.51.

20. Select both lines. Select Object|Blends|Blend Options and set the Specified Steps to 1. Choose the Blend tool and click on the leftmost end point of each line. Select the blend. Choose Object|Blends|Expand (see Figure 5.52).

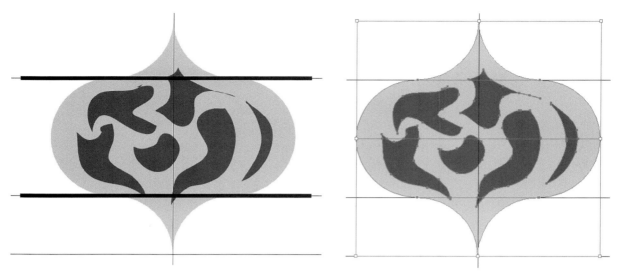

Figure 5.51 (left)
You need to have a line that runs across the width of the ogee on both guides that you added to the image. I've deliberately thickened the lines in this figure.

Figure 5.52 (right)
Three lines are used to cut the lines apart.

Figure 5.53
The color bands show the objects to be grouped into horizontal strips.

21. Select the entire image (Mac: Command+A; Windows: Ctrl+A). Click on Divide in the Pathfinder palette. The lines disappear, leaving a cut-up ogee behind. If I'm lucky, you can see the cut lines in the selected image shown in Figure 5.52.

22. Now you need to put the ogee back together again across horizontal strips. Select all the pieces in a strip and group them. Figure 5.53 shows the reconstructed ogee. I've changed the color for each strip so that you can see it. Don't change the colors in your own ogee. You might need to zoom in close to the tiny cuts if any of the really small pieces fall near a horizontal cut. Select the small pieces on a strip first. That makes it easier to see which ones have been selected. You also might find it easier to hide the guides until you have rebuilt the ogee.

23. Show the guides if you have hidden them and turn on Smart Guides if they are off. Drag a horizontal guide to about 1 inch above the first horizontal guide in the image. With nothing selected, place your cursor over the top of point of the ogee. Press the mouse button and drag the ogee up to the new guide. Press Shift after you begin to drag to constrain movement to the vertical.

24. Select the entire image (Mac: Command+A; Windows: Ctrl+A). Click on Vertical Distribute Center in the Align palette. (If the command doesn't work, you've got a tiny ungrouped cut somewhere that's gumming up the works.) Figure 5.54 shows the cut and parted ogee.

25. Select the entire image (Mac: Command+A; Windows: Ctrl+A). Group the shapes (Mac: Command+G; Windows: Ctrl+G). Press Shift after you begin to drag to constrain movement to the vertical and drag the ogee up slightly along the vertical guide. Move it about the same distance as the space between the strips of the ogee.

26. Drag a small square from the intersection of the horizontal and vertical guides to act as a spacer. It should be about twice the size as the space between the strips of ogee. There is no good automated way that I've found to do this without measuring precisely and driving everyone bonkers. Just work by eye and drag the bottom point of the ogee to the top of the spacer square. Then drag a copy of the ogee directly down until the top of the copy touches the bottom of the spacer square. Drag another copy to the left of the original so that it's centered on the left side of the spacer square. Drag a copy of the left side to the right side of the spacer square. Your image should look like Figure 5.55.

27. Delete the spacer square. Clear the guides. It's time to create new ones. Zoom in closely. You need to frame the ogee with four guides— from the top and bottom points of the side ogees, you need horizontal guides. From the same top and bottom points on the side ogees, you need vertical guides. The smart guides help you place them accurately by reporting on "anchor point" and/or "intersect." Figure 5.56 shows the ogee with its four new guides.

28. Select the Rectangle tool with no Fill and no Stroke. Drag a rectangle from the top-left intersection of the guides to the bottom-right intersection of the guides. You might want to turn off Smart Guides before you do this. I find them more trouble than they're worth when I already have guides in the image. Send the new rectangle to the back so that it can act as a pattern skeleton.

29. Select the entire image (Mac: Command+A; Windows: Ctrl+A). Drag it to the Swatches palette. Drag your pattern repeat unit to the pasteboard for now. Create a test rectangle and fill it with the new pattern. Double-click on the Scale tool and deselect the Scale Objects checkbox, leaving the Patterns checkbox selected. Set a Uniform Scale amount of 40%. Click on OK. Figure 5.57 shows the tiled pattern.

Figure 5.54
The ogee has been cut, and the pieces have moved apart.

HOW DO I KNOW WHETHER I'VE GOT THEM ALL?

The easiest way to make sure that you have selected all the little cuts in the horizontal strip is to hide the selected objects after you've selected a group of them. You can keep selecting and hiding until none of them are left in the strip. When you choose Object|Show All, all the hidden objects appear selected. You can now easily group them.

Figure 5.55 (left)
Four copies of the ogee frame a spacer square in the center.

Figure 5.56 (right)
You'll use the four new guides to make a pattern skeleton.

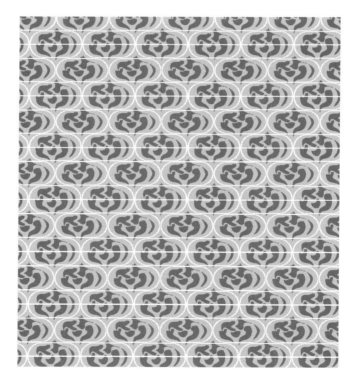

Figure 5.57
The ogee pattern makes an interesting tile.

Manipulating Patterns

Illustrator gives you a large number of manipulations and functions that you can use on patterns. Although the online help and the manual list all of them, I wanted to give you a mini-list as well.

- You can recover the objects from which the pattern is created by dragging the pattern from the Swatches palette directly into your document.

- You can change the starting point of your pattern in the object by either changing the zero point on the rulers or using the Move command from the Transform menu and selecting the Pattern Only option.

- You can perform any transformation on your pattern once it is applied to an object by using any and all of the transformation commands on the Transform menu with the Patterns Only option selected. Don't go too crazy with this ability, as it will haunt you when you print the image. Patterns, in general, can be difficult to image because they can be so complex. Therefore, a word of caution is in order.

- You can change the color of a pattern by dragging it out of the Swatches palette, editing the objects, and then dragging it in again under a new name.

Embossed Patterns

When you define a pattern with a no-stroke, no-fill pattern skeleton, the background of the pattern is transparent. That allows you to create effects such as layering one pattern on top of another and making the pattern look embossed. Illustration 5.9 presents a fast poster that you can create from one of the guitar patterns from earlier in this chapter.

PROJECT Illustration 5.9: Jazz Band Is Coming

1. Create a new document in Illustrator.

2. Choose Window|Swatch Libraries|Other Library and select the GUITARPATTERNS.AI file from the enclosed CD-ROM.

3. Drag the Half-Drop Guitar pattern from the Swatches palette to the document.

4. Make three copies of the pattern tiles.

5. Decide what color you want the background to be. I placed a white rectangle on top of a rectangle filled with the original guitar color and used the Soft Mix command at 50% to create a new color. Select the guitars in one copy of the pattern (use the Group Select tool) and change their Fill to the background color (name it "BGK One"). Make another group of guitars a light tint of the background color (named "Light One") and make the third copy a shade of gray (to keep the pattern subtle, I used a light gray; name it "Dark One").

6. Drag each newly colored pattern group back into the Swatches palette and name it appropriately.

7. Create a poster-sized rectangle and fill it with the background color.

8. Create another rectangle the same size on top of the first one. Set its Fill to the darkest pattern.

9. Create another rectangle of the same size on top of the group. Set its fill to the white or the lightest pattern. With the Selection tool, move the rectangle up and to the left just enough to offset the pattern.

10. Make a final rectangle and fill it with the background color pattern. Position this rectangle so that the dark and light patterns below look embossed.

11. Open the file JAZZBAND.AI from the enclosed CD-ROM. Drag the contents into the poster file and position as you want. Figure 5.58 shows the final effect.

You can also take your patterns into Photoshop, though you'll need to crop them before you save the tile for transfer (unless you apply the pattern in Illustrator and then bring the filled shape into Photoshop). Photoshop requires a rectangular tile and doesn't recognize Illustrator's convention of

Figure 5.58
You can create an embossed look to patterns through layering.

the transparent pattern skeleton. However, once you have the pattern in Photoshop, you can easily recolor it, use it with Apply modes, control the transparency, or filter it. You can get an interesting embossed look by applying the Emboss filter to the pattern and then using Hard Light mode on the embossed layer to add texture to the layer beneath.

Shadows

The Jazz Band poster is a bit anemic and could use a bit of spice. Let's use some patterns in a slightly different way in Illustration 5.10 to jazz it up a bit.

PROJECT Illustration 5.10: Pattern Play—Adding Some Jazz

1. Drag the Light One pattern from your Swatches palette to the pasteboard area of the image. Ungroup (Mac: Shift+Command+G; Windows: Shift+Ctrl+G).

2. Create a new layer. Drag Layer 3 between Layers 1 and 2 in the Layers palette list. Lock Layers 1 and 2.

3. Drag one tiny light guitar to the center of your image. Press Shift+Option (Mac) or Shift+Alt (Windows) to resize from the center and maintain the aspect ratio. Make the guitar quite large and rotate it to a pleasing angle as shown in Figure 5.59.

4. You'll now create a patterned shadow for the guitar. Drag a copy of the Confetti pattern (one of the default patterns) from the Swatches palette onto the pasteboard of your image. Figure 5.60 shows the original pattern. It's very light. It's even lighter than the image shown in the figure because I felt that it wouldn't show up at its original weight.

5. Ungroup the pattern. The pattern consists of a number of straight lines and a pattern skeleton rectangle that's smaller than the combined shapes. Deselect. Select just the pattern skeleton. Look at the Fill and Stroke swatches on the Toolbox. The pattern skeleton rectangle has no Stroke, but it has a Fill of white. This means that it will also have an opaque white background so that it cannot be used, as it is currently, to texture an object. Change the Fill to None. Now the pattern is transparent and can be applied over another color.

6. Select the entire pattern again and deselect the pattern skeleton. Only the short strokes in the pattern should be selected. Change the weight of the strokes to 3 points. Change the Fill color to a shade of

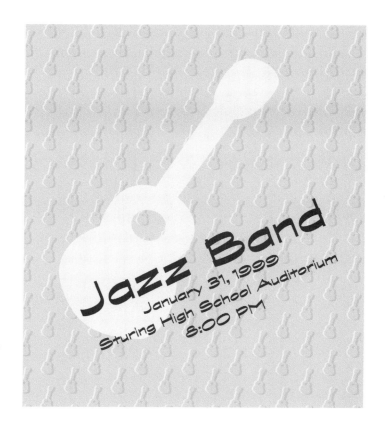

Figure 5.59
Enlarge the tiny guitar from the pattern to use it as a soft background element and rotate it to work better in the design.

gray slightly darker than the one that you used on the Dark One pattern. Figure 5.61 shows the altered pattern.

7. Select the new pattern repeat and drag it onto the Swatches palette. Deselect. Name the pattern "Transparent Confetti".

8. Press Option (Mac) or Alt (Windows) and drag a copy of the large guitar to the left and slightly toward the top of the image. It will be your drop shadow. Change the Fill to the Transparent Confetti pattern that you just created. Figure 5.62 shows the image.

9. The image still needs more pizzazz. Select the Rectangle tool and drag a wider-than-it-is-high rectangle across the center of the image.

Figure 5.60 (left)
The original Confetti pattern.

Figure 5.61 (right)
The "new" version of Confetti has thicker lines and a medium gray Fill and is transparent (as you can see from the shape that I placed underneath for purposes of this figure—it's not in your image).

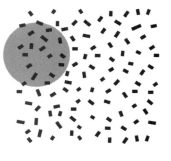

Figure 5.62
You can use a transparent pattern as a shadow element.

Use the Rotate tool to interactively rotate it so that the right end is higher than the left. Figure 5.63 shows my rotated rectangle.

10. You should have two tiny guitars "left over" from the pattern that you dragged onto the pasteboard. Change the Fill color of one of them to the same color that you used in the Transparent Confetti pattern. Do not enlarge the size of the guitar. Drag the re-colored guitar to the Brushes palette. Define it as a Pattern brush. Click on OK.

11. Select the rotated rectangle and, with the Stroke as the active toolbox swatch, click on the guitar Pattern brush to apply it to the rectangle. I think that the guitars need to be smaller and spaced farther apart. Double-click on the guitar Pattern brush in the Brushes palette to reopen the Brush Options dialog box. Change the Scale to 70% and Spacing to 10% (to leave a bit more room between elements). Click on OK. When asked, click on Apply To Strokes.

12. As a final task, you need to mask out the corners of the guitar-stroked rectangle. The rectangle is likely to extend over the edge of your image. Although the pieces near the page boundary might not print, and the areas outside the boundary definitely won't print, it's a good idea to control things so that only what you really want to

Figure 5.63

Add a rotated rectangle to the image.

print shows up on the page. Anything that falls within the page margin will definitely print. Unlock the layers. If necessary, size all the elements so they are approximately the same proportions as the work you did in Step 11. Make Layer 2 (the top layer) active.

13. Draw a no-stroke, no-fill rectangle that is the same size as the page margin.

14. Select all. Make a mask (Object|Masks|Make; Mac: Command+7; Windows: Ctrl+7). Figure 5.64 shows the finished poster.

Textures: Splotches, Spatters, And Spots

Creating Random Textures

The confetti pattern that you used in Illustration 5.10 is a good example of a texture. It certainly fits the definition that you were given at the start of this chapter.

Figure 5.64
The jazzed-up Jazz Band poster.

You can add wonderful subtlety to your images by apply either a pattern or a texture with a transparent background over an object—especially if you keep the color of the texture close to the value of the object (the closer it is in value, the more subtle the effect). You can add flair to your image by controlling where the shapes contain texture. Figure 5.65 shows an example using simple shapes.

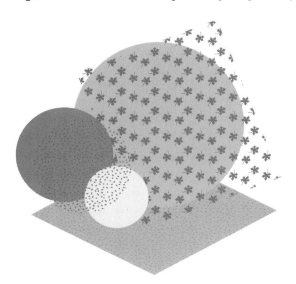

Figure 5.65
A simple composition is enhanced by the use of texture that doesn't quite conform to the confines of the main shapes.

Illustration 5.11 shows how to create a randomly spaced texture that tiles seamlessly.

PROJECT Illustration 5.11: A Random Stroll

1. Create a new document in Illustrator.

2. Select a Stroke of light blue (or whatever color you want), and a Fill of None.

3. Choose the Spiral tool. Click in your image to open the Spiral Tool Options dialog box. Enter a radius of .25 inches and click on OK.

4. Select the spiral and drag it to the Brushes palette. Click on New Scatter Brush. Complete the dialog box as shown in Figure 5.66. (You'll create more Scatter brushes in Chapter 9.) Click on OK. Deselect.

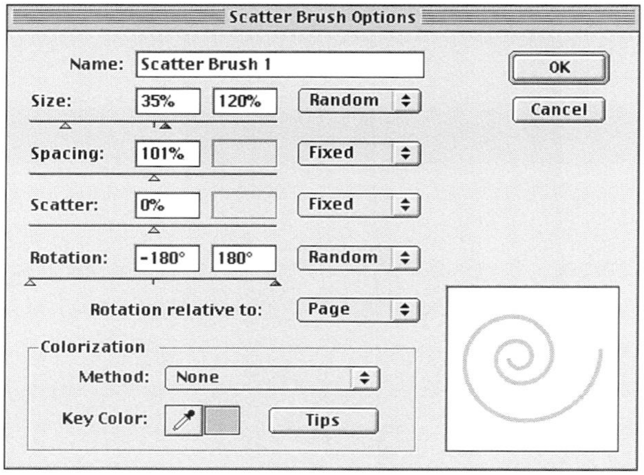

Figure 5.66
Use these settings to create a new Scatter brush.

5. Select the Paintbrush tool. Click on the spiral Scatter brush to select it. Draw a curvy path as shown in Figure 5.67.

Figure 5.67
Because of the random settings, you have many different sizes and angles of spirals when you paint a curvy line.

6. With the brush stroke selected, choose Object|Expand and expand the object. Ungroup (Mac: Shift+Command+G; Windows: Shift+Ctrl+G).

7. Select just the skeleton stroke and delete it. You are left with a wide choice of sizes and angles for the spiral motif.

8. Choose the Rectangle tool. Select a Fill and Stroke of None. Click in your image and enter a width and height of 2 inches in the Rectangle Tool Options dialog box. Click on OK.

9. Toggle Artwork mode on (Mac: Command+Y; Windows: Ctrl+Y).

10. Drag some of the spirals onto the rectangle so that they intersect the left and top edges only. Do not cover any corners. Figure 5.68 shows my developing texture.

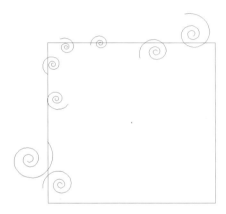

Figure 5.68
Place spiral motifs over the top and left sides of the rectangle so that they intersect the rectangle but don't cover any corners.

11. Turn on Smart Guides (Mac: Command+U; Windows: Ctrl+U). Select the entire texture-in-progress. Press Option (Mac) or Alt (Windows) and drag the objects to the right from the left side of the rectangle (press Shift after you begin to drag to constrain movement to the horizontal) until the word "intersect" appears as you drag over the right side of the original rectangle.

12. Select both rectangles and their spirals. Repeat Step 11, dragging a copy of the assembly down from the top of the rectangle until the copy intersects the bottom of the original rectangle. Figure 5.69 shows the four rectangles and their spirals.

13. Turn off Smart Guides (Mac: Command+U; Windows: Ctrl+U). Delete the three copied rectangles. Move the extra spirals that don't cover the edges of the original rectangle into the center of the rectangle. Scatter them randomly but try to space them so that they don't bunch up. *Do not move any spiral that touches an edge, and don't place*

any of the new spirals over an edge. This is the key rule that will allow your pattern to tile seamlessly. You can place anything else anywhere within the rectangle, but the edges are sacred. If you have trouble, select the pattern skeleton and the edge spirals and lock them. That way, you cannot move them accidentally. Figure 5.70 shows my texture. Notice that the bottom-right corner shows two edge spirals that overlap.

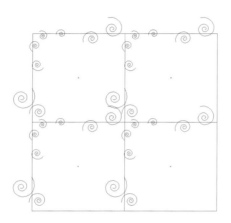 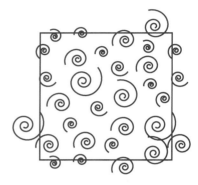

Figure 5.69 (left)
Copy the texture-in-progress to the right and then to the bottom of the orignal unit.

Figure 5.70 (right)
The nearly complete texture has a problem of two overlapping edge spirals.

14. You need to fix the spirals because no other pairs of spirals overlap, and allowing it to remain would create a break in the texture. However, the top and bottom edges of the rectangle are matched pairs. They need to remain identical. The solution is to select the *matching* spiral along opposite edges. Figure 5.71 shows the improved spacing achieved by moving *pairs* of edge spirals.

Figure 5.71
Fix the spacing of your spirals if necessary by moving pairs of edge spirals. Adjust the inner spirals as you want to compensate for the edge changes.

15. Select the pattern skeleton. Send it to the back.

16. Select the spirals and the pattern skeleton. Drag the texture into the Swatches palette. Deselect. Select and drag the original repeat unit to the pasteboard. Create a test pattern rectangle and fill it with your texture. Figure 5.72 shows the texture tiled.

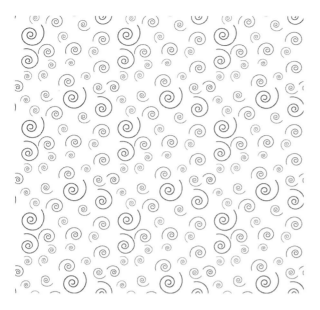

Figure 5.72
The spirals form a random texture.

The Photo Crosshatch Filter

Adobe's marketing literature for Illustrator 8 touts the new Photo Cross-hatch filter as an easy way to create *Wall Street Journal*–type graphics. Certainly, the *Wall Street Journal* is known for its wonderful textured line drawings. The crosshatched patterns show the image detail and value through the weight and spacing of the lines. That's what the Photo Cross-hatch filter is also supposed to do.

I'm sure you just caught the word "supposed." Adobe PR folks, don't choke, please. I still love you. I just have my doubts about the Photo Crosshatch filter. The theory: The filter takes a continuous-tone photo and uses the values in the image as a basis for creating a textured line drawing. It does precisely what is claimed. I'm just not too happy with the result. The filter is easy to use but diabolically difficult to make look good. The default settings on the filter produce marginal to unsatisfactory results.

I'm not saying that you can't find other ways to make the filter yield decent results. I have. Illustration 5.12 presents the results of my exploration.

The problem with the filter, oddly enough, is that it almost gives you too many choices. It is powerful but slow to apply, and complex images take a long time to redraw (the filter is also prone to the same printing ghoulies as the Pen And Ink filter). Because the best way to discover optimal settings is to experiment, you need a lot of time and patience to get good results. The Photo Crosshatch filter actually changes the bitmapped image into a vector drawing (much like the way you were able to use the colors from a bitmap in the Gradient Mesh object Illustration in Chapter 4). Each line created by the Photo Crosshatch filter is a separate object. There are so many objects that you cannot see the image with all of them selected.

Before you start the Illustration, I want to show how some simple cleanup in Photoshop can vastly improve even the default settings. Figure 5.73 shows a photograph taken by a friend several years ago in Frankfurt, Germany. Figure 5.74 shows the same scene with the Photo Crosshatch filter applied in Illustrator using the default settings (which you'll see shortly) and five layers. Notice that the sky posterizes badly (and the electric wires don't add anything to the scene, either). Figure 5.75 shows the same settings in Illustrator after I cleaned up the image in Photoshop to fill the sky with solid color.

Figure 5.73
A street scene in Frankfurt.

Figure 5.74
The default settings in the Photo Crosshatch filter are applied in five layers to the Frankfurt street scene.

You can get more abstract looks with the Photo Crosshatch filter, but the longer the lines that you use, the less detail you see. Figure 5.76 shows the Frankfurt scene heavily edited and posterized in Photoshop (actually, I practically redrew the scene in order to control the values in the buildings). Figure 5.77 shows a finished image from Illustrator using longer line settings than the default and changing the angles of the lines.

Figure 5.75
I reused the default settings to the Frankfurt street scene but made the sky a solid color in Photoshop before I reapplied the filter.

Figure 5.76
A posterized and redrawn Frankfurt street scene.

Figure 5.77
The Photo Crosshatch filter with a border added.

To use the Photo Crosshatch filter, place a raster image into Illustrator and select it. Then choose Filter|Pen and Ink|Photo Crosshatch. The dialog box shown in Figure 5.78 appears. It shows the default values, unless you've changed them during a session.

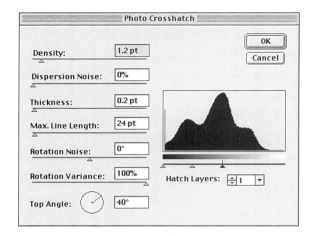

Figure 5.78
The default values for the Photo Crosshatch filter.

Here's a brief explanation:

- According to Illustrator's manual, the Density slider adjusts the number of hatch elements applied to the image, from .5 points to 10 points. A more useful explanation of density is that this setting controls the amount of coverage that you get from the filter. A low density setting (.5) results in a heavy coverage that nearly obscures the shape. At a setting of 10 points, the hatch elements are far apart.

- The Illustrator manual states that the Dispersion Noise setting controls the spacing of hatch elements (0% to 300%). It does, but not in the same way that the density does. The Dispersion Noise setting controls the variance in the spaces between the hatch elements. The higher the Dispersion Noise setting, the more random looking the lines become. Although they keep the same angle (if you don't change the angle setting), they are scattered about the object.

- Thickness controls the stroke weight of the hatch elements (minimum: .1 point, maximum: 10 points).

- You can set the Max. Line Length of the hatch elements (5 points to 999 points). You'll see more detail with a shorter line length.

- Rotation Noise sets the amount of random rotation of objects within the hatch layers, from −360 degrees to 360 degrees.

- Rotation Variance sets the amount that each layer is rotated from the previous layer, from 0% to 100%.

- Top Angle sets the angle of rotation for the topmost hatch layer, from −360 degrees to 360 degrees.

It's time for you to try out the filter yourself. The image that you'll use is an old photograph of the mother of a friend. This is a good choice of image because it has simple lines and a reasonable range of values.

PROJECT Illustration 5.12: Crosshatching Photos

1. Create a new document in Illustrator. Place the image RITASMUM.PSD from the enclosed CD-ROM. Figure 5.79 shows the original photograph.

Figure 5.79
A photograph taken in the 1930s is the basis for Illustration 5.12.

2. Choose Filter|Pen and Ink|Photo Crosshatch. Accept the defaults and click on OK. Except for the lightest areas of the image, the entire image is covered in a uniform mass (or mess) of lines. Undo. Try it again with five layers—that's better. Undo. Figure 5.80 shows the image with one layer and the default settings, and Figure 5.81 shows it with five layers and the default settings.

3. Try it again. This time, enter the settings shown in Figure 5.82. All the settings, except for Rotation Variance and Top Angle, have been modified. This is a much more successful image, as you can see in Figure 5.83.

4. Undo. Apply the filter again. This time, choose a line length of 100 points. Figure 5.84 shows the enormous change that this causes in the image. The line length reduces the image detail, but for a different image it could produce an appropriate and exciting effect.

Figure 5.80 (left)
Default settings, one layer.

Figure 5.81 (right)
Default settings, five layers.

5. You can also help the Photo Crosshatch filter find the correct values in the locations where you want them to be by posterizing the photograph in Photoshop before you place it in Illustrator. Delete the placed file RITASMUM. Instead, place your Illustrator image 5TONE.PSD. This is the same image, but it's posterized by Photoshop

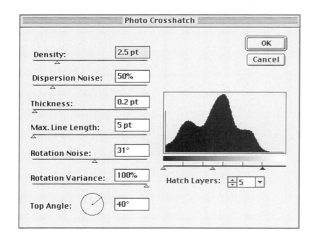

Figure 5.82 (left)
New settings for the Photo Crosshatch filter.

Figure 5.83 (right)
These settings produce a nicely shaded image that resembles a mezzotint because of the short line length and reduced density.

into five values and then reworked a bit. Experiment with a variety of settings on this image as well. When you posterize an image, it loses detail. Because of the loss of detail, it becomes a perfect candidate for abstract and stylized treatments.

Figure 5.85 shows the finished image that I created using the 5TONE.PSD version of the photograph. I used the settings shown in Figure 5.86 to filter

the posterized photo. I then locked the layer, created a new one underneath it, and placed the 5TONE photo again. I adjusted the colors in the photo with the Filter|Colors|Adjust Colors filter to remove 50% of the black. I then locked that layer and created a another new one below it. I filled the layer with a dark brown and turned it into a three-by-three Gradient Mesh object with 70% Highlight To Edge. I added a no-stroke, no-fill rectangle around it to give it some growing room, selected both rectangles, and rasterized them. I unlocked Layer 2 (the photo) and applied the Auto F/X Photo/Graphic Edges filter to both raster objects at the same time. The edge treatments in the Photo/Graphic Edges filter are wonderful and add so much to an image. The filter is among my favorites.

Figure 5.84 (left)

Changing the line length to 100 points reduces the detail in the image.

Figure 5.85 (right)

The 5TONE photo is used as a basis for this image, which is finished off with an edge treatment from Auto F/X applied to a rasterized Gradient Mesh object.

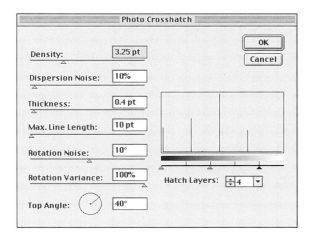

Figure 5.86

The Photo Crosshatch filter settings used to create the final image.

Moving On

You learned a lot of different techniques in this chapter for creating and using patterns and textures. You can create rectangular, half-drop, and brick repeats and make diaper patterns. You have learned how to create patterns that interlock and how to apply them to a path using a Pattern brush. You can create interlocking geometric patterns, such as ogee forms.

You also learned how to create seamless, seemingly random textures, and you learned to use the new Photo Crosshatch filter. You also learned some tricks about that filter.

We've played only with simple, one-motif patterns, but you can combine multiple motifs to create ever more complex and interesting patterns. You can, for example, combine multiple motifs in the diaper pattern that you created or colored each spiral in the texture differently (as long as the edge spirals matched). I hope that I also managed to communicate my love for pattern and the excitement that it generates.

In Chapter 6, Michael will tell you about the new Actions feature in Illustrator 8.

PART III

TRICKS OF THE TRADE

ILLUSTRATOR 8 STUDIO

This studio's purpose is twofold. The first part showcases images of selected Illustrations (projects) found in various chapters within this book. The second part presents the beauty and variety of Illustrator art from some of the world's best digital artists.

Using a TWAIN compliant scanner and software, such as Adobe Photoshop, you can scan in real-world objects. After saving the file in a bitmap format, the resulting file can be opened as a template in Illustrator. Then you can use Illustrator's tools, and the techniques from this book, to turn the scan into an editable, printable illustration.

This series of images shows the advantages Illustrator/Photoshop combined usage. This piece of advertising art (top left) shows many of the uses of Photoshop for images begun in Illustrator. The Hard Wrap candy box is done. A monochromatic color scheme covers the original pattern in the background. For complete details, see Chapter 2.

You might like the box with the original bright pattern showing in the background (top-right).

You can also soften the original colors in the pattern showing in the background (bottom).

Using the Place command, you can place three maple leaves and one butterfly in the BRANCH.PSD image. This image shows the result of a project that you can duplicate in Chapter 2.

This illustration shows another example of using Photoshop and Illustrator in concert. The author opened an image from the Racine Costume Historique collection from Direct Imagination and used the Japanese kimono print to fill a kimono shape. The finished Kimono image has a background pattern and a final border added to it.

This finished flower image uses an embedded image started in Illustrator, modified in Photoshop, and linked to the Illustrator file for printing.

This greeting card design uses Photoshop-created textures applied to Illustrator Pattern brushes and objects. It is printed from an Illustrator file.

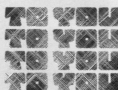

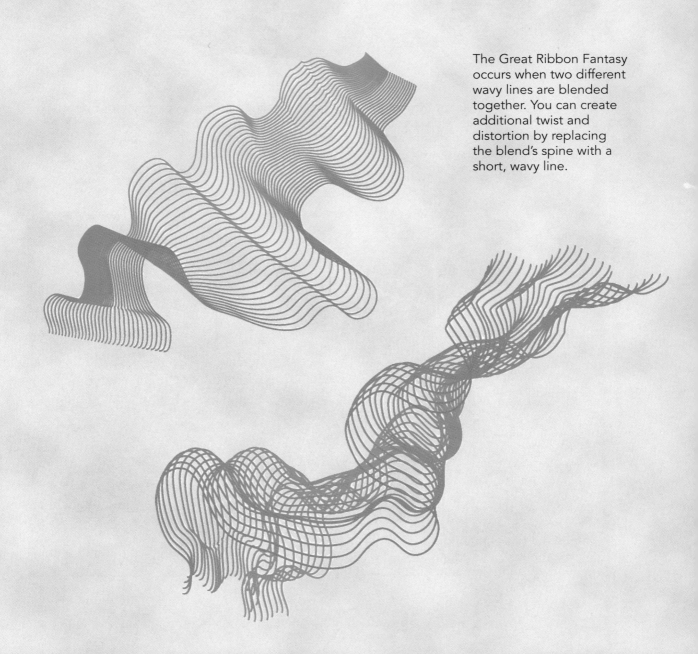

When you blend between a solid object and a gradient, Illustrator tries to find intermediate colors.

The Great Ribbon Fantasy occurs when two different wavy lines are blended together. You can create additional twist and distortion by replacing the blend's spine with a short, wavy line.

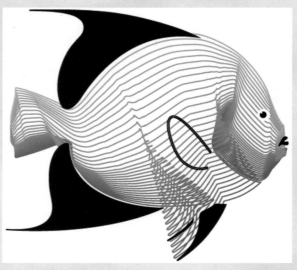

This angel fish (from the Ultimate Symbol Nature's Icons collection) has been cut apart, and some of its major lines were used as the basis to form blends (see Chapter 4).

This is a slightly different treatment for the angel fish. I did not separate the body in the same way, causing the lines to create a totally different-looking blend.

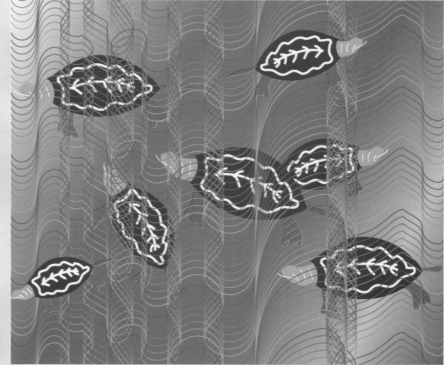

The effect of a wavy linear blend over a gradated background depends upon the shape of the original line, the angle and color of the gradient, and the objects (in this case, turtles) placed in the image. Changing any of these components changes the look and feel of the image.

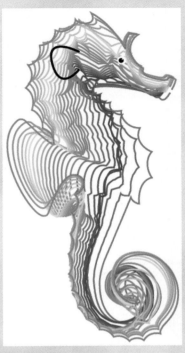

A Gradient Mesh object becomes the background behind a compound path made from the original brush stroke.

This seahorse also started out as a piece of clip art in the Ultimate Symbol Nature's Icons collection. It shows how you can create a multi-hued blend by blending to intermediate colors along the way.

You can create metallic-looking text by applying the same gradient but at different angles to layered copies of the same text.

If the gradient has more twists and bends in it, you might want to use the Gradient tool to stretch the gradient across the entire set of text (see Chapter 4).

This image of children dressed in Halloween costumes at school is the basis for the Gradient Mesh image that you create at the end of Chapter 4.

The children at the school yard now look properly scary in this cropped and abstracted image that also has the advantage of being completely flexible in size because it is all vector.

The young lady in this image is wearing a jumper that sports an interlocking pattern (see Chapter 5).

Notice the use of pattern as texture and pattern as shadow in this simple poster for a high school jazz band.

A 5-tone posterized version of a photograph is used as a basis for this image, which is first filtered with the Photo Crosshatch filter and then finished off with an edge treatment from Auto F/X applied to a rasterize Gradient Mesh object.

After scanning an image into a TWAIN compliant program such as Photoshop, the scan can be saved in a bitmap format such as TIFF. The file can then be opened in Illustrator, where it's possible to use drawing tools to create outlines of the shapes (see Chapter 6).

After drawing over a shape, the resulting vector path can be manipulated in an infinite number of ways. Here, the vector representation of a leaf (created from the scanned photograph) was colored and copied in random ways by using an action, a new feature in Illustrator 8. With just a few mouse clicks, running the action can result in many randomly placed, sized, and colored copies of the original vector path.

This advertisement for a pizza parlor is created using a variety of transformation commands (see Chapter 7).

You can stroke the string art paths with the History brush in Photoshop to apply a previously created gradient. The image on the right was created by using string art stars that were composited with other Illustrator elements in Photoshop.

This mandala (from Chapter 7) was cut into five segments, filled with a rainbow gradient, and then rotated. All of the various cuts were then put in order to make the rotation seamless.

This ogee design was built from an expanded radial gradient and assorted shapes, and cut into sections using the Divide command.

This final mandala rendering is a composite of multiple manipulated layers. Subtle design differences may be created by changing the order of the layers.

This finished coupon makes use of dashed strokes and contains soft brush strokes of color.

This cityscape image was finished in Photoshop, which allows you to add special effects (see Chapter 8).

This is an example that I created by using flourishes and a circle design from Ultimate Symbol's Design Elements collection. I stroked all the flourishes with a Pattern brush made from a dashed stroke.

This is a "promo tag" for a company called "Put On Your Sundae Clothes." It uses the Calligraphic brushes that are new to Illustrator 8 and was finished in Photoshop.

This image is all knotted up, with dashed patterns and with copies of itself used as corner elements in a Pattern brush (see Chapter 8).

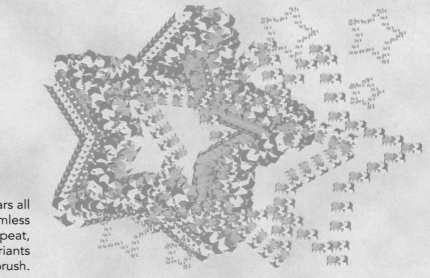

These stars all use the same seamless overlapping star repeat, but they are variants of a Scatter brush.

A bare tree (from the Ultimate Symbol Nature, Icons collection) stroked with a leaf Art brush also created from clip art in the same collection.

These sample fish are created from an Art brush built from scratch in Illustrator.

The top image is the starting point of a project that allows you to colorize the image and add snow. The bottom image—the snowflake image with no whitecaps—has been manipulated in Photoshop to control the impact of the snowflakes on the village.

Using vertical type can add a dynamic and fun quality to your images (see Chapter 10).

Using Illustrator's Paragraph Type tool, you can create images completely from type.

The fish image is embedded in a photographic collage.

Around and Around

You can get blended colors in text without using a gradient by applying the Blend Horizontal color filter, as shown in this example.

© 1999 Robert Burger

Artist
Robert Burger

Email
burgerbobz@aol.com

URL
www.voicenet.com/~rbbb

Robert starts all of his work with a pencil sketch. He then scans and uses the sketch as an Illustrator template. When everything is drawn, he assigns each element a Fill of white and a Stroke of black. He then assigns colors, gradients, patterns, and textures.

This image was created as an editorial illustration for *Internet World* to accompany an article about the opportunities available to women within Internet-related businesses.

© 1999 Robert Burger

Magnetic Attraction was created as an editorial illustration for *Information Week Magazine* to accompany an article about companies keeping high-tech employees on the payroll.

Artist

K. Daniel Clark

URL

www.artdude.com

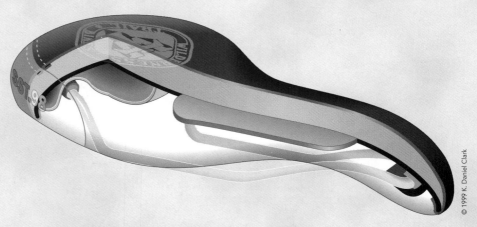

© 1999 K. Daniel Clark

Artist

Joe Jones
Art Works Studio

Email

dujave@aol.com

URL

www.artworksstudio.com

Joe created the basic shape of the structures and objects in Port Merillia by creating the profile shape, and then the shapes as you would see from the top view. He then pulled these two-line elements into RayDream 5 to create the 3D modeled buildings. He also used Illustrator's texture mapping feature for the buildings. The building panels and dome maps were originally created on separate layers in Illustrator, then each layer was imported into Photoshop to create all the beveling, edge treatments, and weathered look.

© 1998 Art Works Studio

© 1998 Art Works Studio

The High Prairie Logo is an example of high-end logo work, incorporating dynamic visual elements and effective type.

Artist
Jose Cruz

Email
members.aol.com/cruz13

I try to keep all my designs to this philosophy: "Less is more, more or less." The simplest thing can be the most striking. All of Jose's illustrations incorporate humor, color, and—above all—good design.

Artist
Pamela Drury Wattenmaker

Email
pamela@wattenmaker.com

URL
www.wattenmaker.com

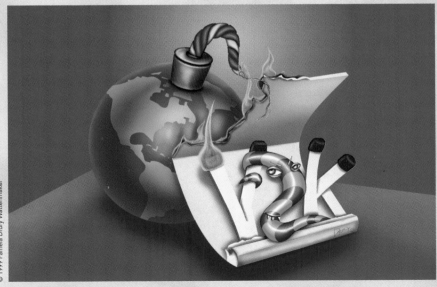

These whimsical illustrations were created by using illustrator and Photoshop. Usually Pamela starts with a traditional sketch, then scans the sketch into the computer. She then redraws in Illustrator, makes changes, colorblocks (her custom color palette is in Illustrator), and to finalize, takes the art to its "highest level" in Photoshop.

Artist
Anne Darrah
Darrah Design+Marketing

Email
AnneDarrah@aol.com

© 1999 Anne Darrah

Inspired by African-American krewe's style of costumes and masks, Anne sketched this image used for a Mardi Gras poster. The orange ribbons above the mask and the beads around the neck seem to glow because of the graduated colors, which Illustrator can make, and also because of the warm-over-cool color choice.

This cover image for an annual report for a rural utility cooperative was designed to convey the company's vision. When working color into an illustration, Anne finds it helpful to create a color palette with several shades of each tone to use for shadows and highlights. The landscape inside the binoculars was drawn separately, then masked to fit the shape. When drawing complex illustrations, she draws each element on its own layer. Then she can hide or lock layers as needed.

© 1999 Anne Darrah

Angel was created in Illustrator 8 for a holiday card. Anne sketched on paper until she acheived the desired personality and design, then scanned it as a template.

©1999 Louis Fishauf

Artist

Louis Fishauf
Louis Fishauf Design
Limited and Reactor
Art+Design Limited

Email

fishauf@reactor.ca

URL

www.total.net/~fishauf/

Louis created the background spider-web motif by using the Punk filter on a circle, blending between large and small copies of the resulting shape, then applying brushtrokes to the path segments. A Pattern brush was used for the spider's "hairy" legs. This particular image is a spot illustration for *Fortune* magazine.

©1999 Ian Giblin

©1999 Ian Giblin

©1999 Ian Giblin

GIBLIN

Artist

Ian Giblin
Fine Design

Email

giblins@ix.netcom.com

Ian creates portraits produced primarily using Illustrator's Paintbrush tool and a Wacom tablet. He scanned the subject's photograph and placed it on Layer 1. On Layer 2, Ian roughed out the basic compositional blocks that defined the face by using the Pencil tool; these were filled with random colors. On Layer 3, Ian used various stroke widths and a variety of different Art brushes to give the portrait more texture.

Artist
Jason Fabbri

Email
jfabbri@stickfigures.org

URL
www.stickfigures.org

©1999 Jason Fabbri

©1999 Jason Fabbri

Fabbri

Jason primarily uses Illustrator's Brush tool so that he can create a look that appears hand-drawn, yet preserves the flexibility of working digitally. First, the outlines of each image are created in black. By using the Brush tool, he creates rough shapes that outline various sections of the image. Next, he selects a solid color with which to fill each section (using the Brush tool). Bringing the outlines to the top gives the fine edge to outline the image, but still allows filled areas to display beyond the black, which gives the rough freehand look.

Artist
Zach Trenholm

Email
zachtrenholm@earthlink.net

©1999 Zach Trenholm

©1999 Zach Trenholm

Caricatures of
Courtney Love
and Julia Roberts.

Artist
Doug Ross

Email
doug@dougross.com

URL
www.dougross.com

ILLUSTRATOR 8 STUDIO

Artist
Marty Smith

© 1999 Marty Smith

Marty Smith

Marty is conventionally trained in airbrush and pen-and-ink and began using the computer in 1988. Marty works exclusively in Adobe Illustrator, transforming blueprints, photos, or actual parts into digital renderings. He specializes in line art, exploded views, and photorealistic cutaways.

Artist
Tim Webb
Tim Webb Illustration

Email
spyder@southwind.net

URL
www.timwebb.com

Tim takes a high-tech approach to his conceptual art, using Adobe Illustrator while drawing on the "not-so-high-tech" suggestions of his wife and five sons. He keeps his finger on the pulse of the big cities via the Internet and proves by his innovative work that for a small-town boy, "there's no place like home."

Artist
Doug Musto

Email
dmusto@adobe.com
Musto3@ix.netcom.com

Doug has a fine arts background, and says he's not always as concerned about design elements as he is about the total impact of the illustration. His approach to an illustration is to build from the background to the foreground, paying close attention to light and dark. Doug learned to use Illustrator by drawing with the mouse—he still does today. Many lines in his illustrations are drawn with the Freehand tool, not the Pen tool. His favorite parts of the program are layers, "which make it easy to build from back to front, and gradients, which allow me to have some shapes blend into others."

Artist
Dava Mayumi
Dava Production

Email
dava@t3.rim.or.jp

Dava's designs show a delicate approach to using Illustrator as an artistic canvas.

Yume-hime

Yume-hime

ACTIONS 6

T. MICHAEL CLARK

Lights... Camera... Action! Okay, seriously, though, one of the nicest new features in Illustrator 8 is the addition of actions. If actions become half the hit they are with Photoshop, you'll see new Web sites springing up overnight that offer more actions than you can shake a stick at.

What Are Actions?

Actions are a series of steps that have been recorded in Illustrator. You can then play back these steps later with the click of a mouse. If you have worked with actions in Photoshop or with macro features found in many other applications (such as Microsoft Word 6 or later), you have an accurate image of what actions are and how they work.

If you have Illustrator 8, you can create actions. These recorded steps can then be used by anyone who has Illustrator. You can copy your actions and give them to friends, set up a Web site and give them away, or even package them all together on a CD-ROM and sell them (believe me, it will be done).

The Actions Palette

Adobe has included a few actions that are installed by default along with the program. You can see these actions listed within the Actions palette (see Figure 6.1).

If you don't see the Actions palette, choose Window|Show Actions.

In Figure 6.1, you can see the first entry, Default Actions. This entry isn't one of the actions; but rather, it is the Default Action set. Notice the small folder icon to the left of the title. This icon denotes the fact that this entry is a set. When you first open the Actions palette, you might not see any actions listed below the Default Actions entry. To see the actions included in this set, simply click on the small triangular icon to the left of the folder icon. Doing so displays all the actions included in the set.

Along the bottom of the Actions palette, you'll find several small controls, in this order:

• Stop Playing/Recording

• Begin Recording

• Play Current Selection

• Create New Set

• Create New Action

• Delete Selection

These controls allow you to create, record, play back, and delete actions. You'll look at these controls in greater depth as you progress through the rest of this chapter. You can display the Actions palette in either of two modes. The default (List) mode is shown in Figure 6.1, and Button mode is shown in Figure 6.2.

Figure 6.1

Illustrator's Actions palette.

In Button mode, the entries are color-coded. Some of the information present in List mode, such as the controls along the bottom of the palette, isn't shown.

When you create a new action, as you'll do later, you can give the action a color (as well as a function key). The color you select, if any, will be the color used for the entry for that action in Button mode.

With the palette displaying the actions in List mode, you can get all types of information about the actions. When you open a set by clicking on the small triangular icon, you get a list of actions contained in that set. You can break down each action further into its set of commands by clicking on the small icon next to the action. Finally, you can get a listing of the recorded value used for each command by breaking down the command by clicking on its triangular icon (see Figure 6.3).

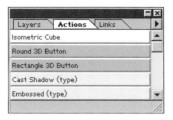

Figure 6.2

The Actions palette in Button mode.

Figure 6.3

The Actions palette showing the recorded values for the different command values used in the various steps in an action.

In Figure 6.3, note that the action is named Isometric Cube, that its first step is called Layer, and that the command's values are as follows:

- New Layer

- Name: Isometric Cube

- Template: No

- Show: Yes

- Lock: No

- Preview: Yes

- Print: Yes

- Dim Image: No

All these values correspond to the values you would enter into the New Layer dialog box. This shows that Illustrator has recorded all the values from the dialog box as the action itself was recorded. Reading through an action created by someone else in this way can be helpful when you're learning how certain techniques are accomplished in Illustrator.

The Actions Menu

Along with the Actions palette is an associated sidebar menu. To access this menu (see Figure 6.4), click on the small right-pointing triangular icon to the far right of the palette tabs.

Figure 6.4
The Actions menu.

Note in Figure 6.4 that some of the choices are also available along the bottom of the Actions palette. From this menu, you can create a new action set or a new action. You can also duplicate, delete, or play an existing action. You can start or continue a recording, and you can insert a menu item, a stop, a select path command, or an object. (You'll explore some of these options as I guide you in the creation of an action later in this chapter.) You can also select action or playback options and clear, reset, load, replace, and save your actions. Finally, you can toggle Button mode on and off.

Installed Actions

Illustrator 8 installs several actions during the installation process. You can use these actions by selecting Default Actions in the Actions palette. In addition, you can explore these actions by breaking down each one into its components, as explained earlier. This will give you some insight into how actions work.

The actions included in the Default Actions set are fairly basic and include, for example, an isometric cube, 3D round and rectangular buttons, cast shadows and embossing for type, and even an action that creates a rosewood texture. Although these actions are fairly simple, they are worth exploring if only for learning the ins and outs of both how actions work and what they can and can't do.

Extras Actions

Several collections of actions in Illustrator 8 are included, but they aren't installed by default. You can find these on the Illustrator 8 Application CD-ROM.

These actions are stored in the Illustrator|Illustrator Extras|Action Sets folder. You'll find sets for the following:

- Buttons

- Clip Art

- Commands

- Line Effects

- Shape Effects

- Shapes

- Text Effects—Raster

- Text Effects—Vector

- Textures

You can make these actions available by loading them into the Actions palette. To do so, click on the small triangular arrow button in the upper-right corner of the Actions palette (to bring up the sidebar menu) and choose Load Actions. In the Load Actions dialog box, browse to the folder on the CD-ROM that contains the extras actions sets and select the set or sets you want to load. To make the sets more readily available in the future (especially if you find a set or two that you like and intend to use often), copy the file or files to the folder on your hard drive that Illustrator uses to store action sets. The default is the Action Sets folder under the Illustrator 8 hierarchy.

You might want to explore the extras actions sets, which contain some cool actions. After playing around with the default and extras actions, you might start itching to create your own actions. The next section deals with this topic.

Creating Actions

Creating your own actions is a process you'll want to learn. Although creating an action, especially a complex one, can be time consuming, it saves you time in the long run.

After an action has been created, tweaked (if necessary), and saved, it can be applied with a few simple mouse clicks. Illustration 6.1 shows how to create an action that generates a drop shadow.

PROJECT Illustration 6.1: Creating A Drop-Shadow Action

To save time when creating an action, you might want to outline the details on paper. If the action is simple, you might want to simply run through the processes and commands once or twice to see how it all ties together before you begin. Doing this might highlight any snags that you would rather not run into when creating the actual action.

To show you how to create a working action, I'll walk you through the process of creating a drop shadow.

The Dry Run

I wrote the following set of steps to create a drop-shadow effect:

1. Create type, set it in place on the screen, and select a color for the fill (this is not included in the action).

2. Show Actions palette.

3. Choose New Set and name it.

4. Choose New Action and name it.

5. Click on Record on the Actions palette.

6. Choose Edit|Copy (the type you entered should already be selected).

7. Choose Object|Lock.

8. Choose Edit|Paste In Back.

9. Choose Object|Group.

10. Set fill color to Black.

11. Nudge down and right 5 pixels.

12. Choose Edit|Copy.

13. Choose Edit|Paste In Back.

14. Set fill color to White (this assumes that the type will be placed on a white background).

15. Nudge down and right 5 pixels.

16. Scale the grouped objects by clicking and dragging the lower-right corner handle (this is a bit subjective, and you might change it to use the more precise menu command).

17. Choose Edit|Select All.

18. Double-click on the Blend tool icon in the Toolbox and enter a value of 20 in the Specified Steps option.

19. Click on a handle on the black type.

20. Click the corresponding handle on the white type.

21. Stop recording.

These steps were written after I created a drop shadow; they might need a little refining. Additionally, you'll need—or at least want—to add a stop that identifies the action and its author.

The Real Thing

After writing the steps for your new action, you might want to do a dry run to see whether you've forgotten anything. If everything works the way you planned, it's time to create the actual action.

For now, you can simply follow along as I create the drop-shadow action. To do so, follow these steps, which are similar to the previously outlined steps:

1. Create some type. To do so, select the Type tool, select a font and a size, click somewhere in the work area, and enter the type. I entered "Shadow" in Lithograph Bold at 100 points.

2. Set the fill color of the type and leave the type selected (this is all you'll need to do in the future to create a drop shadow).

Follow these steps to create the action:

1. If the Actions palette is not visible, Choose Window|Show Actions.

2. Select New Set from the sidebar menu of the Actions palette. Doing so brings up the New Set dialog box (see Figure 6.5), which allows you to name the new set. As you can see in Figure 6.5, I named the set "Illustrator f/x". You can choose a name that depicts the type of actions you'll be saving in the set you create.

WINDOWS USERS, TAKE NOTE

Actually, I ended up using the wonderfully descriptive name "My Actions". I originally tried to call the Action Set "Illustrator f/x". However, this caused a serious problem; I couldn't save the file, and I went through all kinds of contortions trying to get the name to stick. This resulted in the loss (several times) of the action I had laboriously created. After a lot of sweat and creating the action for the third or fourth time, I realized that the system probably didn't appreciate the slash ("/") character in the action set name. I went with the simpler name "My Actions" and solved the problem. The lesson: You should stick to regular characters for the set and action names.

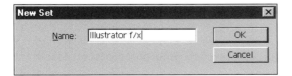

Figure 6.5
The New Set dialog box.

Note: *You must create a new set for your actions because this is the only way that you can save the actions you create. Doing so is also a good way to keep separate actions for different types of jobs filed. You might have, for example, a collection of actions that work on type and another that you use for prepress.*

3. Select New Action from the sidebar menu of the Actions palette. Doing so brings up the New Action dialog box (see Figure 6.6), which allows you to name the action you're about to create. As you can see in Figure 6.6, I named the action "Drop Shadow (type)." I added the parenthetical "type" as a reminder to myself (or as a hint to others who may use the action) that the action expects to be run against a type object.

4. Click on Record to close the dialog box and start recording your new action.

5. Choose Edit|Copy.

6. Choose Object|Lock to lock the original type. This will allow you to select and deselect the objects you create during the course of recording the action without worrying that the original type will be changed or selected.

7. Choose Edit|Paste In Back.

8. Set the fill color to Black.

9. Choose Type|Create Outlines (this is one step I forgot during the creation of my original notes). This step is needed so that the Blend tool will work.

10. Choose Object|Group to group all the letters together. This is needed because changing the type to outlines creates separate objects from each letter.

11. Nudge the new type objects down and to the right 5 pixels. You do so by tapping both the down arrow and the right arrow keys five times each.

12. Choose Edit|Copy.

13. Choose Edit|Paste In Back.

14. Set the fill color to White.

15. Nudge the new objects down and to the right 5 pixels.

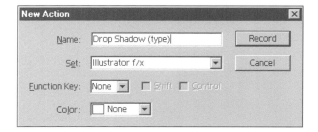

Figure 6.6
The New Action dialog box.

16. Select the Selection tool from the Toolbox and scale the new objects by clicking and dragging the corner handle at the lower right of the bounding box down and to the right (see Figure 6.7).

Figure 6.7
Click and drag the lower right handle down and to the right to scale the second group of objects.

17. Choose Edit|Select All.

18. Double-click on the Blend tool to bring up the Blend Options dialog box (see Figure 6.8).

Figure 6.8
The Blend Options dialog box.

19. Choose the Specified Steps option and enter "20" for the number of steps.

20. Choose Object|Blends|Make. This will blend the two objects, creating a soft drop-shadow effect.

21. Choose Edit|Unlock All.

22. Choose Object|Deselect All.

23. Click on Stop Recording on the Actions palette.

That's it! You've just created your first new action. You can test it by clearing the type you used to create the action, entering some new type, selecting the action from the Actions palette, and clicking on Play.

You might notice that the Blend Options dialog box pops up during the test run. You can prevent this from happening by clicking on the Toggle dialog box's on/off button in the Actions palette (see Figure 6.9).

You might also notice that I didn't follow the plan exactly as I had laid it out in the first list. One of the changes occurred because I discovered that I had left a step out. Another change occurred when, during testing, I discovered that the Blend tool wouldn't work when I clicked on the two

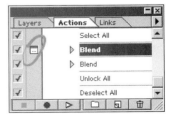

Figure 6.9
The Toggle dialog box's on/off button in the Actions palette.

points. I made the change to choosing Object|Blends|Make and found that that method worked fine.

However, having the plan in place before I began helped me keep track of where I was as I entered the action, and it allowed me to smoothly make the corrections required to get the action to run properly.

Signing Actions

With your spiffy new action created, tested, and ready to run, only one thing remains to be done—you need to sign it! To do so, take the following steps:

1. Click on the title of your action in the Actions palette to highlight it.

2. Expand the action list to show the commands if they're not visible.

3. Highlight the first command (this is necessary because leaving the title highlighted will add the stop to the bottom of the list, requiring you to move it back to the top, where you want it to be).

4. Click on the small triangular icon in the upper-right corner and select Insert Stop. Doing so will bring up the Record Stop dialog box (see Figure 6.10).

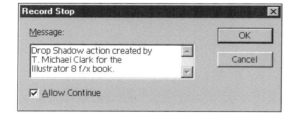

Figure 6.10
The Record Stop dialog box.

5. You can enter a message in the space provided. You can use Stops to inform a user to make changes in, for example, upcoming dialog boxes. Also, adding a stop at the beginning of an action is a great way to sign the action and add a copyright notice.

6. You can see from Figure 6.10 that I've added my name and the name of the book. I've also selected the Allow Continue checkbox. This allows the user to continue to run the action after reading the message.

7. Because the stop was added below the first command, it needs to be dragged into place. To do so, simply click and drag the stop from its position into place above the first command.

When the action is run, the user will first see the message and then be able to continue running the action by clicking on Continue. After running the action, the user will see a soft drop shadow added to the type (see Figure 6.11).

Using Stops As A Debugging Tool

Stops can also be used as a helpful debugging tool. Placing stops throughout an action allows you, as the creator of the action, to track down and (it is hoped) fix any problems that your action is experiencing.

Saving Actions

To save your actions so that they can be loaded later and/or shared with others, take these steps:

1. Choose Save Actions from the sidebar menu of the Actions palette.

2. In the Save Actions dialog box, name your actions file and save it. The actions are saved by default to the Action Sets folder under the Illustrator 8 hierarchy, with the .AIA file extension.

To load your actions in the future, choose Load Actions from the sidebar menu of the Actions palette. This brings up the Load Actions dialog box, which allows you to choose from the action sets you've saved.

To share your actions with others, simply give them a copy of the AIA file that you saved.

Adding Interactivity

During the creation of the previous action, I alluded to the fact that you can add interactivity to the actions you create. This is true, and I'll show you how to do so by creating another action. Illustration 6.2 shows how to extrude your type to give it some dimensionality—a bit of a 3D effect.

PROJECT Illustration 6.2: An Interactive Extrusion Action

If you've run the action created in the previous example, you saw how the dialog boxes pop up. You can control, or toggle, whether they do so with one of the small icons that appear next to the step in the Actions palette (see Figure 6.12).

When the icon appears as a very small dialog box (see Figure 6.12), the dialog box will pop up as each action runs, which allows the user to enter or change the available values in the dialog box. When the icon appears empty, the dialog box will use the values entered at the time that the action was created, and the dialog box will not pop up.

You can add interactivity to your actions by toggling on the dialog boxes that you want to allow the user to enter values for and by adding a stop command with appropriate instructions. To show you how this is done, I'll

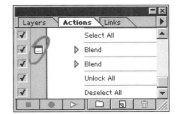

Figure 6.12

The Toggle dialog box's on/off icon in the Actions palette.

walk you through the creation of another action. After the action has been created, you'll see how and when to add the appropriate stops to allow the users of the action to have some say in how the action affects their image.

Creating A To-Do List

Again, I'll start by laying out a small plan of attack just to keep my thoughts in order and to give me a starting point for creating the action's instructions. Note that the action will assume that type has been entered and colored and is now selected. The list I constructed is as follows:

1. Create outlines from the type.

2. Group the new objects.

3. Run the Transform|Move command and make a copy of the grouped objects.

4. Run the Filter|Color command to darken the new copy of the type.

5. Move the group to the back.

6. Blend the two groups with the same number of steps used in the move from Step 3.

Creating The Action

With the general steps laid out, it's time to start creating the new action. To get started, add some type to a new image and change the color of the type (I've used a pale blue). I used the Helvetica font at 120 points. With the type entered and colored, it's time to create the actual action. Here are the steps involved in doing so:

1. Assuming that you still have the new action group you created previously, highlight that group and click on Create at the bottom of the Actions palette.

2. In the New Action dialog box, name the action (call this one "Extrude (type)") and click on Record.

3. Choose Type|Create Outlines.

4. Choose Object|Group.

5. To name the grouped object, choose Window|Show Attributes. Choose Show Note from the sidebar menu of the Attributes palette (see Figure 6.13).

Figure 6.13 shows the Note section of the Attributes palette. Notice that I've entered "Type" in the note section. This allows me to refer to and select the object later in the Actions palette sidebar menu. Note that

SELECTING OBJECTS IN AN ACTION

You might have noticed that if you run the previous action with other objects present, the action causes all sorts of weirdness as it selects all objects (Edit|Select All). This happens because, during the recording of an action, you cannot select an object by using the selection tools.

A quick-and-dirty option (as programmers like to say) was to simply select all the objects present and assume that the user had just created the type and run the action with no other artwork present on the screen. This works well enough under that assumption but fails miserably if other objects are present. However, there is a solution.

With any object selected, you can give that particular object a name. Copying that object, as you're doing during the creation of this action, gives the copy the same name as the original object. With the object named, you can select it by using a menu choice available from the Actions palette sidebar menu. The next step shows you how to name the selected object, and an upcoming step will show you how to select the named object so that you can run a command against it.

"Type" is just the name I've given the selection and that it has nothing to do with the type of object. I could just as easily have named the object "Elephant" if I had chosen to do so.

6. Choose Object|Transform|Move.

7. Enter "–15" and "15" for the Horizontal and Vertical values, respectively. This step moves the type up and to the left. Later, I'll show how to enable the user to change the values here.

8. Click on Copy to make a copy of the type.

9. Choose Object|Arrange|Send To Back to move the copied and moved type behind the original type.

10. Choose Filter|Colors|Adjust Colors to bring up the Adjust Colors dialog box (see Figure 6.14).

11. In the Adjust Colors dialog box, drag the Black slider to the right until it shows 33%, then click on OK. This will darken the new type.

12. From the sidebar menu of the Actions palette, choose Select Object to bring up the Set Selection dialog box (see Figure 6.15).

13. Enter the same name that you gave the object in the Attributes palette (Step 5). I entered "Type" because that's the name I used.

14. You'll note that all the type is now selected but that nothing else is (that is, if any other objects are present).

15. Double-click on the Blend tool in the Toolbox to bring up the Blend Options dialog box.

16. Choose Specified Steps for the Spacing and enter "15", the same value you used for the move in Step 7.

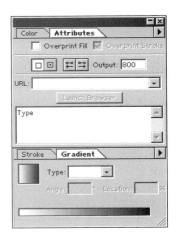

Figure 6.13
The Attributes palette showing the Note section.

> **Note:** In Step 12, you'll select the named objects, as mentioned in the previous tip.

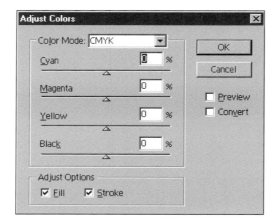

Figure 6.14
The Adjust Colors dialog box.

Figure 6.15
The Set Selection dialog box.

17. Choose Object|Blends|Make to apply the blend to the two type groups.

18. Choose Edit|Deselect All.

19. To stop recording the action, click on Stop Playing/Recording in the lower-left corner of the Actions palette.

You should now have extruded type that resembles the type shown in Figure 6.16.

Figure 6.16
Extruded type created with a recorded action.

Wow! Amazing 3D type has been created in minutes in Illustrator. Even more amazing is that the simple click of a mouse will re-create this effect. Still more amazing is that you'll now add some stops to the action and encourage the user to change the action to suit his or her needs.

Before you begin modifying the action, however, it might be prudent to save the action. To do so, highlight the action set by clicking on its name in the Actions palette. With the action set highlighted, choose Save Actions from the sidebar menu of the Actions palette. From the Save Set To dialog box (where you can either select an existing action, enter a new name, or browse your system for an existing name), save it with the name "My Actions". When I do so, Illustrator warns me that the set exists and asks whether I want to replace it. If you're following along and have created this new action in the same set as the previous action, click on Yes to replace the existing action set. Doing so resaves the set with the newest addition.

Now it's time to add the interactivity. You should add stops at two points (not including the vanity screen, demonstrated in the previous action) to inform the user what's expected. The first point is just before the move, and the second is just before the blend.

Adding Interactivity

1. To add the first stop, highlight the command before the point at which you want the stop placed. In other words, highlight the command that precedes the command you want to insert instructions for. In the case of the first stop in this particular action, I'll highlight the Attribute Setting command (see Figure 6.17). Adding the stop at this point (just before the Move command) will make the action stop just prior to running the next command, which is the Move command.

2. Highlight the Attribute Setting command; choose the sidebar menu of the Actions palette to bring up the Actions menu.

3. Choose Insert Stop to bring up the Record Stop dialog box.

4. In the dialog box, enter the following instructions to the user: "Enter the same value for both the Horizontal and the Vertical options. Use negative values to move left or down, then click on Copy."

5. After entering the instructions to the user, click on Allow Continue. The Stop dialog box will now appear when either you or anyone else runs the action.

6. The second stop needs to go just before the first Blend command. To insert the stop command there, highlight the Set Selection command in the Actions palette and choose Insert Stop from the Actions sidebar menu.

7. Enter instructions to the user: "Enter the same value for the steps that you entered for the Horizontal and Vertical values in the Move dialog box."

8. I also clicked on Allow Continue again.

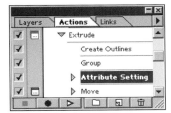

Figure 6.17
Highlighting the Attribute Setting command in the Actions palette.

Note: *All the type you enter must fit into the dialog box without scrolling. Although you can enter plenty of type here and let the dialog box scroll, the type that falls past the visible window will not be displayed to the user when the stop shows up during the running of the action.*

That should just about do it. You might want to run the action a few times using different values to see how it works. Oh, and be sure to save it again. Before you do so, however, you might want to add the vanity stop at the beginning to inform the user what the action does and who programmed it.

Figure 6.18 shows two of the unlimited possibilities of type effects produced with a couple of clicks of the mouse (after constructing the action, that is).

Figure 6.18
Two examples of extruded type created with an action.

Seeing The Forest Through The Leaves—Controlled Randomness

After working through the two structured actions in the previous sections, you might have the impression that creating actions is more mechanical than artistic. To some degree, this is true. Creating an action is somewhat similar to programming. However, it is possible to add a little artistic flair to the process while removing most of the structure. This following final action shows how it's possible to build an action on the fly. I'll also show you a couple of tricks. I'll first show how to include a path in the action, and then I'll show how to rearrange and edit the commands contained within any of your actions.

The following action shows you how to fill the artboard with a predefined shape. After the action has been recorded and tweaked, a single click of the mouse will fill the artboard with the path that you've created. The action I created for this section will randomly place leaves of varying shades of the current fill color and random sizes at different angles throughout the artboard.

First Things First

Although here we're dispensing with the overly structured process of the previous actions, something must come first.

Illustrator's actions won't allow you to record the process of creating a path. You can, however, record an existing path. Confused? Don't be. During the recording of an action, you cannot draw a path and have the drawing process recorded. However, you can draw a path or copy and paste one from another image and record the actual path in the action.

All this means is that the path can be recorded, not the process of drawing (creating) the path. If you're still confused, perhaps a real-world example will help.

Because I wanted a leaf shape, I started with a photograph of some fall leaves. If you want to follow along, you'll find the leaves photograph on this book's CD-ROM.

With the bitmap image that was previously scanned into Photoshop placed into the current artwork, it is a simple matter to draw over one of the leaves using the Pencil tool (see Figure 6.19).

After tracing the leaf, you can safely delete the bitmap file. Delete it from the drawing, that is, not from your hard drive.

With the path drawn, it's time to start creating the action. Make sure that the path is selected. You can also set the fill and stroke colors to something appropriate. I chose a shade of green for the fill and none for the stroke. However, this step isn't necessary because you'll be able to select a new fill and stroke before each running of the action.

As in the previous sections, open a set that you want this action stored in or create a new set. I opened the My Actions set, which you'll find included on this book's accompanying CD-ROM, and started the new action, which I called "Leaves (random)".

To create a new set, click on the small folder icon along the bottom of the Actions palette. Remember, if the Actions palette is not visible, you can make it so by choosing Window|Show Actions.

To create a new action, simply click on the small icon that resembles a small sheet of paper with the lower-left corner turned up.

In the New Action dialog box, name the action and press Enter to start the recording process. To store the path that you created by drawing over the leaf from the bitmap image, you can click on the small black triangular icon in the upper-right corner of the Actions palette.

To make it a little easier to follow along, I'll number the steps from here.

Figure 6.19
A tracing of one of the leaves
from a placed bitmap file.

Illustration 6.3: The Randomness Action

1. Select the path that you want to use.

2. Choose Insert Select Path from the sidebar menu of the Actions palette. This will record the actual path that you created previously. Note that you can record any path that you've created.

Okay, you've created a path—in this case, the outline of a leaf. What's next? Remember that you're working from the seat of your pants here.

Because the object is to create many copies of the leaf and have them change their size (randomly) and where they're placed on the screen, the next step is to copy the leaf into a new position, with a new size and placed at a random angle.

Sounds like a lot, but it's easily done:

3. Choose Object|Transform|Transform Each. Doing so will bring up the Transform Each dialog box (see Figure 6.20).

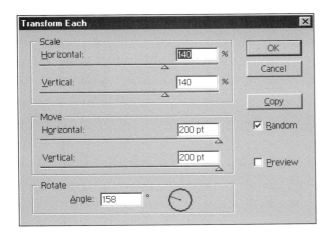

Figure 6.20
The Transform Each dialog box.

In the Transform Each dialog box, you can set the horizontal and vertical scale, the horizontal and vertical move, and the angle of rotation. Better still, you can have the whole process randomized. Notice that I marked the Random checkbox. You should do the same.

4. Instead of clicking on OK, click on Copy. This step copies the selected path, and you'll end up with two leaves (or two of whatever path you've chosen to create the action).

Before continuing with building the random action, I'll tell you a bit about how the Random option works in conjunction with the various settings available in the Transform Each dialog box.

With the Random option selected, the transform process will transform the selected path by using the settings as limits. This means the copy of the leaf will be scaled between 100% (the same size as the original) and 140% (as I set it for this particular transform). The copy will also be placed anywhere within 200 points, both horizontally and vertically, of the original. Additionally, it will be rotated up to 158 degrees from the angle of the selected or original path.

As you'll see in upcoming steps, I chose different values for the Scale, Move, and Rotate settings. These differing values result in a fairly good random collection of leaves in various sizes scattered about the artboard, in different places and at different angles.

For the first step, I chose 140 for the horizontal and vertical scale, 200 for the horizontal and vertical move, and 158 for the angle of rotation (you can't use all discrete numbers if you're going for that random look).

After clicking on Copy, you should see two leaves on the artboard (see Figure 6.21).

Figure 6.21
Original leaf copied, moved, scaled, and rotated with the Transform Each command.

It would be nice if the leaf could be given a different shade as well as being transformed. This is entirely possible.

5. Choose Filter|Colors|Adjust Colors. This will bring up the Adjust Colors dialog box (see Figure 6.22).

6. Adjust the Black slider to the left to lighten the color or to the right to darken the color. As you move the slider, if you've turned on the Preview option, you'll see the effect it has on the leaf. The trick will be to darken some and lighten others. For now, move the slider to the right until the value is about 20. This will darken the new leaf a little.

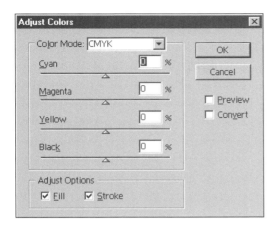

Figure 6.22
The Adjust Colors dialog box.

You now have three steps in your new action. Believe it or not, you're nearly done with creating new steps. From here on, you'll duplicate and edit the existing steps to create new ones.

7. Click on Stop Playing/Recording in the lower-left corner of the Actions palette to stop recording.

The next step requires that you duplicate the last two steps you recorded. To do so, highlight the two steps. The last step should already be highlighted, and you can add a highlight to the previous step by holding down the Shift key and clicking on the entry in the Actions palette.

With both steps selected, choose Duplicate from the Actions sidebar menu. This will add another copy of the two steps you already recorded. This is good, but to keep the whole process as random as possible without making subsequent leaves bigger and bigger and darker and darker, these new steps need adjusting.

8. Click on one of the steps that are not highlighted to deselect the new steps and then double-click on the Transform Each step that was created by the Duplicate command.

9. In the Transform Each dialog box, enter "60" for each of the scale values, "–200" for each of the move values, and "332" for the angle of rotation. Click on Copy so that the command will produce a new leaf with the new settings. If, for some reason, the dialog box doesn't appear, make sure that one of the leaves is still selected.

10. It's time to color the newest leaf. Double-click on the new Adjust Colors entry and enter "–40" for the Black. This lightens the newest leaf. You should have three leaves in the artboard: the original; a larger, darker one; and a smaller, lighter one.

You can stop here if you want, or you can continue. To go on, duplicate the last four instructions and reset the values in each subsequent instruction.

The final action you'll find on this book's CD-ROM has eight iterations. It creates a total of nine leaves in varying shades of whatever the current fill color was when you started the action. Each leaf is different in size and is placed in a random angle and position within the artboard. Actually, some of the leaves might wander slightly off the artboard, but these way-ward (or wind blown) leaves can easily be selected and moved into a new position. The whole idea is that a simple click of the mouse will produce a variety of random leaves. You can find an example of this in this book's Color Studio.

To have some real fun with this action, run it, set a new fill color, run it again, and so on. Figure 6.23 shows the result of running the action five or six times, using green, red, orange, yellow, and brown.

Figure 6.23

A collection of leaves in various colors, shades, and sizes created by running the Leaves action.

You need to do a couple of things to finalize your new action:

11. Add a step at the end that deselects the final leaf. To add this step, make sure that you're recording and that one of the leaves is selected. Also, make sure that you've highlighted the last step so that the Deselect command will be in the correct place.

12. To turn off the selection, choose Edit|Deselect All.

If you want to, you can also add a vanity message, as was done in the previous actions.

Moving On

At this point, I'll hand you back over to Sherry, who will start the next chapter by showing you some cool tricks you can accomplish with trans-formations and distortions.

I'll be back in Chapter 10, where I'll show you some incredible type tricks. Sit up... Lie down... Extrude. Good type.

MACINTOSH CAVEAT

If the action finishes and does not fill the shapes, here's how to fix it: Choose a fill color before you run the action.

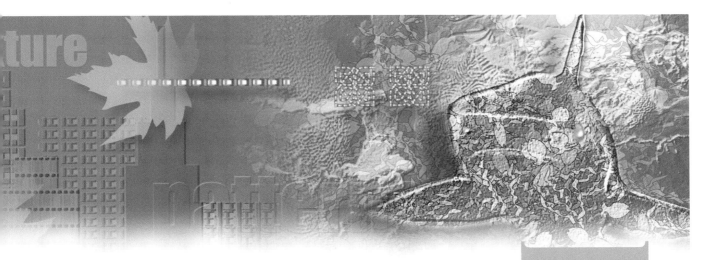

TRANSFORMATIONS 7

SHERRY LONDON

Transformations are what happen when you take simple shapes and, using only Illustrator commands, turn them into something else. This chapter looks at some of my favorite ways of turning simple things into complicated ones (although not the instructions, I hope).

Illustrator is bursting at the seams with commands to change one shape into another. You can use the Transform palette, the Pathfinder palette, the Transform tools (Free Transform, Rotate, Reflect, Skew, and Scale), and a host of other Path commands and filters. You have already used some of these tools in previous Illustrations, yet there are too many of them to try to use everything here. Instead, I have somewhat arbitrarily picked out techniques that I like and developed three projects for this chapter.

Extra Cheese

It's difficult to believe that I never tasted a "real" pizza until I was in high school. Pizza shops did exist, but I was a fussy eater, and I really hated the pizzas served by the Philadelphia school system (a slice of stale bread topped with a spoonful of tomato sauce and a slice of cheese, then cooked until the cheese was nicely browned). If that was a pizza, I wanted no part of it!

How Do You Like Your Pizza?

The pizza that you are going to create in this Illustration 7.1 is more appetizing (if not totally realistic). While you are building its ingredients, you will also learn about the Free Transform tool, perspective text, the Add Points command, and the Transform Each command. Because the project is quite long, I divided it into a number of Illustrations.

PROJECT Illustration 7.1: Building The Perfect Pizza

1. Create a new document in Illustrator.

2. Choose the Ellipse tool. Create an oval or circular shape in your image. It doesn't matter where or how large—we'll fix that in the next step.

3. Although I could have had you create the circle at the correct size, we'll use the Transform palette, as this is the Transformations chapter. Leave the circle selected. Enter 4.6 inches into both the width and the height fields in the Transform palette. Press Enter/Return to execute the transformation, as shown in Figure 7.1.

4. Change the Stroke to None and the Fill to a warm, crusty golden-brown (I used CMYK: 20, 42, 70, 7). Double-click on the Layer entry in the Layers palette and change the Layer name to "Pizza Dough". You'll use a lot of layers in this example.

5. Lock the Pizza Dough layer. Change your Fill color to CMYK: 20, 75, 70, 7 (or your favorite tomato sauce color). Drag the Pizza Dough layer entry in the Layers palette to the New Layer icon at the bottom of the Layers palette (this is a painless way to make an exact copy in place). The layer is named "Pizza Dough Copy". Double-click on the layer entry and change the name to "Sauce".

6. Unlock the Sauce layer. Drag the Fill color swatch from the Toolbox (or the Color palette) onto the unselected circle. The circle immediately changes to the new color. You can drag any color swatch onto an unselected object—a useful technique when you have the desired color in the Toolbox but know that it will disappear as soon as you click to select another object.

7. Select the circle. Double-click on the Scale tool and enter a Uniform Scale amount of 80%. Click on OK (not on Copy).

8. Choose Object|Path|Add Anchor Points. The circle had four points; now it has eight because the Add Anchor Points command places one point midway between every original pair of points in the object. Choose Object|Path|Add Anchor Points again. Now the circle has 16 points as shown in Figure 7.2, and your object is much more interesting if you want to apply filters that tweak the points.

9. Choose Filter|Distort|Scribble And Tweak. Select the Preview checkbox (annoyingly, Illustrator turns it back off after each use). All three options checkboxes should be selected by default as shown in Figure 7.3. Change the Horizontal and Vertical sliders as you want, but don't let the sauce spill out of the pizza dough. The filter produces random effects, and moving the sliders back to a setting that you've looked at previously will give you different results. When you like what you see, click on OK. Figure 7.4 shows the sauce—now spread out nicely on the dough.

> ## EFFICIENCY
>
> Copying the Pizza Dough layer was probably not the most efficient way to create a scaled copy of an object. A more direct route to this end is to simply double-click on the Scale tool and create an 80-percent copy. Then fill it with sauce color. However, Illustrator has a variety of paths to the same end. Efficiency isn't everything. You should also know the side roads.

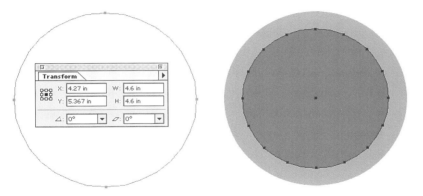

Figure 7.1 (left)
You can use the Transform palette to make changes in the size and position of your objects.

Figure 7.2 (right)
The Add Anchor Points command doubles the number of points in an object.

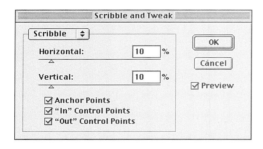

Figure 7.3 (left)
The Scribble And Tweak dialog box lets you control how random changes are applied to the control points in the selected object.

Figure 7.4 (right)
Instant tomato sauce (after adding a dash of Scribble And Tweak).

10. Do you like meatballs? They're tasty on a pizza, so you'll add some now. Create a new layer and name it "Meatballs". Lock the other two layers. Select the Ellipse tool and click somewhere on top of the sauce. In the dialog box, enter both width and height dimensions of 0.6. Leave the meatball selected.

11. Set the Stroke color to None. Set your Fill color to CMYK: 20, 59, 96, 6 (a light meat-brown). Drag the Fill swatch to the left end of the Gradient slider in the Gradient palette. If your Gradient palette contains any other color stops, drag them off the palette to get rid of them (except for the one at the far right). Click on the color stop at the far right of the Gradient slider and change its color to CMYK: 20, 70, 75, 42 (a darker meatball color). Change the gradient Type to Radial.

12. Press Option (Mac) or Alt (Windows) and use the Selection tool to drag copies of the meatball and scatter them over the sauce. Figure 7.5 shows the pizza in progress.

13. You are now going to create the crusty rim on the pizza. You have four choices as to how you like your pizza:

    • Neat sauce, light crust

    • Neat sauce, dark crust

    • Messy sauce, light crust

    • Messy sauce, dark crust

The general steps for all options are the same. You first need to create the crust itself. Make the Pizza Dough layer active and lock all the layers (including the Pizza Dough layer). Duplicate the layer as you did in Step 5 and name it "Crust".

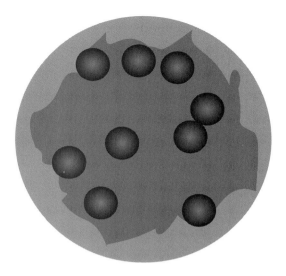

Figure 7.5
Meatballs add a nice flavor to a pizza.

14. Remove the lock from the Crust layer. Select the circle. Use the Transform palette to change the width and height of the circle to 4.45 inches. Press Enter/Return to execute the transformation. Visually re-center the crust object.

15. Double-click on the Scale tool and enter a Uniform Scale amount of 80%. Click on Copy. Select the unlocked shapes in the image (Mac: Command+A; Windows: Ctrl+A). Create a Compound path (Mac: Command+8; Windows: Ctrl+8). You now have a donut shape with a huge hole.

16. Choose the Eyedropper tool and click on a meatball. This makes the gradient that you used for the meatball the current Fill color. Press Option (Mac) or Alt (Windows) and click inside the compound path that you created in Step 15. This transfers the gradient to the crust.

17. The crust has a slight problem with this gradient—it isn't really showing up at all. First, you need to define a highlight color for the crust. Click just under the 50% location on the gradient slider to create a color stop. Click on the new color stop and give it a value of CMYK: 3, 21, 45, 0. The gradient still doesn't appear. The problem is that the crust is in the default part of the gradient, where no large color change occurs.

18. The easiest solution is to create a pizza with a dark crust. To do this, drag the highlight color stop (now at the 50% location) toward the right until it nears the last color stop. The highlight is now visible in the crust. Position the color stop so that the highlight is in approximately the center of the crust edge. On my image, its location is

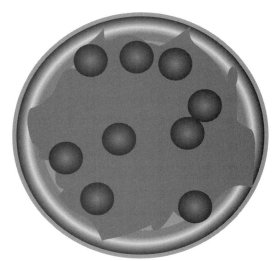

Figure 7.6 (left)
The Gradient palette shows the locations of the three color stops needed to create a gradient for the crust.

Figure 7.7 (right)
The pizza now has a dark, messy crust.

91.01%. Then drag the leftmost color stop until it also nears the right edge (85.39%). Figure 7.6 shows the Gradient palette, and Figure 7.7 shows the pizza.

19. A light-crust pizza is a bit trickier. One way is to simply move the two color stops on the left toward the left until both the highlight and the leftmost brown are farther from the right edge. This lightens the crust. However, it still leaves a browned outer crust. If you like your pizza less well done, you need to move all three color stops toward the center of the Gradient slider as shown in Figure 7.8. Choose the Gradient tool. To get the center highlight to appear in the middle of your crust, you need to first position the Gradient tool in the center of the crust shape (i.e., in the middle of the "hole") and press and hold the mouse button. Press Shift and drag the Gradient line past the right edge of the document. This makes the gradient think that its center is in the center of your crust and gives you the lightly toasted look shown in Figure 7.9.

20. If you prefer a neat pizza with no messy sauce on the crust, move the Crust layer up until it is just under the Meatballs layer. If you also want your meatballs to look as if they are cut off at the crust, move the Crust layer until it is above the Meatballs layer as shown in Figure 7.10.

21. Let's create the cheese. Lock all the layers. Create a new layer and name it "American Cheese". Set your Stroke color to None and the Fill color to CMYK: 2, 6, 38, 0—a medium-light yellow.

22. Choose the Rectangle tool and click near the top-left corner of the bounding box for the pizza (or where it would be if the pizza were selected). Create a rectangle that is 0.1 inches wide and 0.7 inches

BAKING A DARK CRUST

If you prefer the dark-crust pizza, you might want to stroke the crust in the same color as the outer edge of the crust. A 5- or 6-point stroke works well to blend the dark color into the outer lip of the pizza that you created in Step 14, when you reduced the size of the original pizza dough circle.

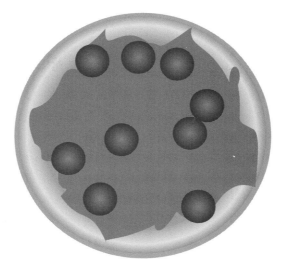

Figure 7.8 (left)
Use these Gradient settings for a light crust.

Figure 7.9 (right)
By employing the Gradient tool, you can get a lightly toasted pizza crust.

high. Choose the Selection tool. Press Option (Mac) or Alt (Windows) and drag a copy of the cheese shred to the right about one cheese shred width away. Press Shift after you begin to drag to constrain movement to the horizontal. Transform Again (Mac: Command+D; Windows: Ctrl+D) until you have created an entire horizontal row that covers the pizza as shown in Figure 7.11.

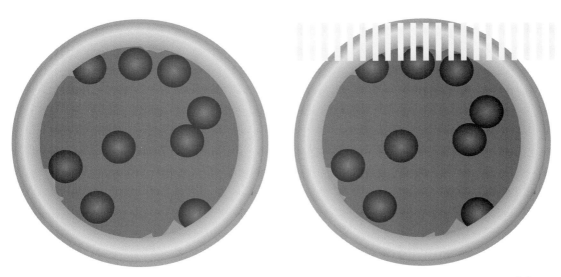

23. Select the unlocked shapes (Mac: Command+A; Windows: Ctrl+A). Press Option (Mac) or Alt (Windows) and drag a copy of the cheese shred down about one cheese shred width away. Press Shift after you begin to drag to constrain movement to the vertical. Transform Again (Mac: Command+D; Windows: Ctrl+D) until you have created an entire vertical row that covers the pizza as shown in Figure 7.12.

Figure 7.10 (left)
You can achieve a neat pizza by moving the Crust layer above the Meatballs layer.

Figure 7.11 (right)
Create a horizontal row of shredded cheese (note horizontal lines at the top of the pizza).

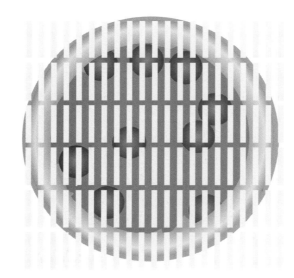

Figure 7.12
Create vertical rows of shredded cheese across the pizza.

24. The arrangement of the cheese shreds is unnatural—you need to fix that now. Select the unlocked cheese shreds (Mac: Command+A; Windows: Ctrl+A). Choose Object|Transform|Transform Each. Leave the Scale at 100%. Allow movement between –0.7 inches horizontally and 0.7 inches vertically. Set the Rotation to 359 degrees and select the Random checkbox as shown in Figure 7.13. Preview the effect to make sure that you like it—what you see is what you get. When you are satisfied, click on OK. Figure 7.14 shows the layer of scattered cheese. Do not deselect.

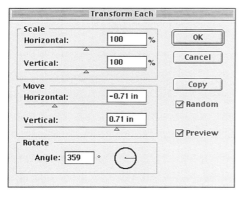

Figure 7.13 (left)
Use these settings in the Transform Each dialog box.

Figure 7.14 (right)
This is a more natural arrangement of cheese.

25. Next, you need to create a layer of cheddar cheese (I think that I forgot to mention that you are creating a three-cheese pizza). Choose Object|Transform|Transform Each. The previous settings should still be there. Accept these settings and click on Copy. Do not deselect. Create a new layer named "Cheddar Cheese". Move the selected object rectangle from the American Cheese to the Cheddar Cheese layer (thus moving the newly dispersed layer of cheese onto the new layer).

26. Select a more orange color for the Fill. I used CMYK: 2, 30, 38, 0. Now the cheese looks more like cheddar (or Colby).

27. Let's add a bit of curve to the cheese. Unlock the American Cheese layer. Now you can select both cheese layers. Choose Filter|Distort| Twirl and enter an amount of 30 degrees. Because this filter works from the center of the selection to the edges, it applies just a gentle curve. Figure 9.15 shows the pizza with two layers of cheese.

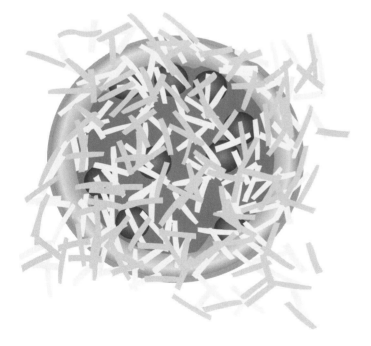

Figure 7.15
The pizza now has two layers of cheese scattered about.

28. The cheese is really a mess. It's all over the table. You need to clean it up. Lock all the layers. Drag the Layer entry for the Pizza Dough layer to the New Layer icon at the bottom of the Layers palette. Change its name to "Cheese Mask". Drag the layer to the top of the Layer stack. Unlock the American Cheese and Cheddar Cheese layers. Select all the unlocked objects (Mac: Command+A; Windows: Ctrl+A). Make a

mask (Object|Masks|Make; Mac: Command+7; Windows: Ctrl+7). Lock the Cheddar Cheese and American Cheese layers again.

29. You need to decide how much cheese you want on your crust. If you do not scale the cheese mask, you have cheese all over the crust. If that's way you like your pizza, that's fine with me. If you want the crust to be somewhat free of cheese, you can double-click on the Scale tool and enter a Uniform amount of 85%. If you do not like cheese on the crust at all, you can scale the Cheese mask uniformly at 80%. If you want some scattered cheese on the crust, you can apply the Add Anchor Points command to the mask and use the Roughen filter as you did when you made the crust (but you need to be sure that the filter doesn't remove any of the cheese from on top of the sauce). I scaled the mask to 85% as you can see in Figure 7.16.

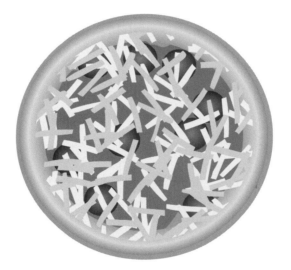

Figure 7.16
The pizza's cheese is now mostly on top of the tomato sauce.

30. The top layer of cheese looks a bit flat. Drag the icon for the Cheddar Cheese layer to the New Layer icon at the bottom of the Layers palette to make a copy. Name the copy "Cheddar Cheese Shadow". Move the layer under the Cheddar Cheese layer. Lock all the layers except for the Cheddar Cheese Shadow layer. Select the entire image (Mac: Command+A; Windows: Ctrl+A). Set the Fill to CMYK: 11, 43, 92, 2 (a sort of Velveeta color). Press the Down Arrow key two times and then press the Right Arrow key two times to offset the shadow.

31. Next, you need to create a layer of Parmesan cheese. This grated cheese tends to stick in clumps (or else can be very coarsely added in small chunks). Create a new layer named (surprise) "Parmesan Cheese." Drag it below the Cheddar Cheese shadow layer. Lock all of

the other layers, and hide all of the cheese and mask layers (except for the Parmesan cheese layer). Set your Stroke to None and your Fill color to a very pale, light yellow (almost white). I used CMYK: 5, 8, 23, 0. Create an ellipse (with the Ellipse tool) that is approximately 0.5 inches wide by 0.3 inches high somewhere on top of the sauce.

32. Alt/Option drag-copy 20 to 30 of these little clumps and scatter them about in the sauce. Select all of the cheese clumps (Mac: Command+A; Windows: Ctrl+A). Choose Object|Transform|Transform Each. Allow the shapes to scale horizontally by 85% and vertically by 115% and rotate up to 356 degrees. If you like your arrangement, set the horizontal and vertical movement to zero. Select the Random checkbox and click on OK.

33. Use the Roughen filter (Filter|Distort|Roughen) to add some irregularity to the shapes. Use Smooth Points. Figure 7.17 shows the settings that I used. If the blobs are too thick, use the Transform Each command to scale only the vertical direction about 75%. Figure 7.18 shows the Parmesan cheese on top of the tomato sauce and meatballs. The other cheese layers are hidden.

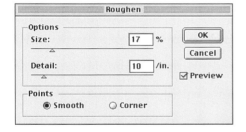

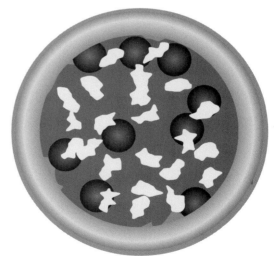

Figure 7.17 (left)
The Roughen filter allows you to add randomness to the control points in a shape.

Figure 7.18 (right)
The Parmesan cheese blobs.

34. The pizza needs a few more meatballs. Lock the Parmesan Cheese layer and unlock the Meatballs layer. Select one meatball. Create a new layer named "More Meatballs" and move it to the top of the Layer stack. Move the selected object rectangle from the Meatballs layer to the More Meatballs layer. Lock the Meatballs layer.

35. Press Option (Mac) or Alt (Windows) and drag a duplicate of the meatball until you have three meatballs on this layer. Lock the More Meatballs layer.

36. Now you need to add some extra cheese. Drag the icon for the American Cheese layer to the New Layer icon at the bottom of the Layers palette. Name the layer "Extra Cheese" and drag it to the top of the Layer stack. This pops the cheese out of the mask—which is fine for now.

37. Select the unlocked cheese shreds (Mac: Command+A; Windows: Ctrl+A). Choose the Rotate tool. Rotate the layer about 45 degrees from its center point (which means that all you need to do is drag the mouse until the image rotates as much as you want it to). Deselect everything.

38. Use the Selection tool to select rectangular areas of cheese that go outside the crust. If you also select some of the cheese inside the crust at the same time, that's fine. The cheese layer is too thick, and this helps to thin it out. Feel free to leave a few odd strands of cheese on the crust. Anything too perfect screams that it was drawn by computer. You are also welcome to add a few more meatballs if you see the need. Figure 7.19 shows the pizza that is now fully assembled.

39. Save your work—it would be a shame to lose it now.

Figure 7.19
One meatball pizza with extra cheese, hold the pepperoni.

Keep Your Perspective

Now that you have built a pizza, you are going to use it in an ad for a pizza parlor called Pizza Pizza. In Illustration 7.2, you will change the angle from which you view the pizza.

PROJECT Illustration 7.2: From A New Perspective

1. You are going to make the pizza look as if it is on a table that is being viewed more from in front than top down. To keep the mask around the cheese layers, you need to group the mask and the cheese layers first (if you don't group the layers, the mask won't scale when you scale the rest of the image). Unlock the Cheese Mask, Cheddar Cheese, Cheddar Cheese Shadow, Parmesan Cheese, and American Cheese layers. Select all the unlocked objects (Mac: Command+A; Windows: Ctrl+A). Group them (Mac: Command+G; Windows: Ctrl+G). All the objects move to the Cheese Mask layer, so you can delete all the other unlocked layers because they are now empty. Figure 7.20 shows the Layers palette after the cheese layers are deleted.

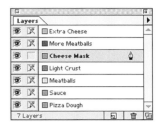

Figure 7.20
After you group the cheeses with the mask, these are the layers that still contain objects.

2. Unlock all the layers. Double-click on the Scale tool. Select Non-Uniform and choose a Horizontal Scale of 100% (no change) and a Vertical Scale of 46%. Click on OK. Lock all the layers.

3. The perspective of the pizza has dramatically changed. Some of it looks a bit flat—as if everything has melted. Let's add a bit of punch back into it. Unlock the More Meatballs layer. Click on a meatball, press Option (Mac) or Alt (Windows) and drag the top-center handle on the bounding box to make the meatball a bit more round. It should not flatten out as much as the rest of the pizza because the meatballs are a bit higher. Then select the Gradient tool. Move the highlight of the radial gradient back to where the new top of the meatball would be. Fix as many meatballs as you want.

4. Lock the More Meatballs layer. Unlock the Cheese Mask layer. Using the Group Selection tool, choose about 10 slices of cheese. With these slices still selected, add a new layer (name it "Top Cheese") and transfer the selected objects to the new layer by moving the selected object rectangle from the Cheese Mask layer to the Top Cheese layer. Lock the Cheese Mask layer again.

5. Drag the icon for the Top Cheese layer to the New Layer icon at the bottom of the Layers palette. Place the duplicate layer under the original and name it "Top Cheese Shadow". Lock the Top Cheese layer.

6. Select everything on the Top Cheese Shadow layer and change the fill to an orangish color similar to the Velveeta cheese shadow color that you used in Illustration 7.1. Press the Down Arrow key two times and the Right Arrow key two times to offset the shadow.

7. Unlock the Top Cheese layer. Select the two unlocked layers (Mac: Command+A; Windows: Ctrl+A). Choose Filter|Distort|Punk And Bloat and move the slider to 50%. Click on OK. Now you have a bit more depth in the image as you can see in Figure 7.21. Lock all the layers.

Figure 7.21

The pizza with a change in perspective now looks a bit less flat.

Turn A Neat Square Corner

You are ready to construct the border for the pizza advertisement. You will use the Pathfinder commands and learn how to manipulate a pattern tile to build a mitered corner unit.

PROJECT Illustration 7.3: Border Crossings

1. You need to build a border around the pizza for the ad. Because I didn't give a specific location in the image as your starting point, you need to create the border from wherever your pizza is currently located. Create a new layer named "Border". Turn on the Rulers (Mac: Command+R; Windows: Ctrl+R). Drag a guide to the top of the pizza. Drag guides from the side ruler until they just touch the right and left sides of the pizza. Place the mouse on the zero-point where the two rulers meet at the top of the image. Drag the Zero-point until it hugs the upper-left edge of the virtual rectangle that you have made with the guides.

2. Drag a vertical guide from the side ruler until it is halfway between the left and right guides. Drag a horizontal guide from the top ruler approximately one-quarter of the height of the pizza lower than the top guide. Select the Rectangle tool and use a Fill of None with a Stroke of any color that you want for now. Press Option (Mac) or Alt (Windows) and click with the Rectangle tool at the intersection of the center guide and the one-quarter-distance guide. In the dialog box that pops up, enter a rectangle width of 6 inches and height of 4 inches. Click on OK.

3. Drag another guide from the top ruler until it lies on top of the top stroke of the new rectangle. Drag down another guide and place it 1 inch higher than the top of the rectangle.

4. Choose the Ellipse tool. Move your cursor to the intersection of the top-of-the-rectangle guide and the center vertical guide. Press Option (Mac) or Alt (Windows) and drag an ellipse until its top reaches the highest guide in the image and its sides reach the guides that mark each side of the pizza. Release the mouse button. Figure 7.22 shows a screen shot of the image with the guides placed as requested and both the rectangle and the ellipse in place.

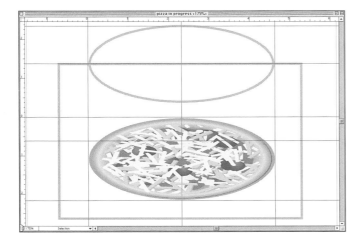

Figure 7.22
The border elements are now drawn.

5. Select both the rectangle and the ellipse. Click on the Unite command in the Pathfinder palette. You now have a shaped panel that acts as a border. I have already created the border Pattern brush for you. You will create the corner motif by mitering the pattern unit.

6. Open the file PATTERN.AI on the enclosed CD-ROM. It contains one pattern tile and the same border structure that you're using in the pizza example. Figure 7.23 shows a many-times enlargement of the pattern tile for the border.

7. Select the pattern tile, press Option (Mac) or Alt (Windows), and drag a duplicate of the shape. Make another, pressing Shift after you begin to drag to constrain movement to the horizontal.

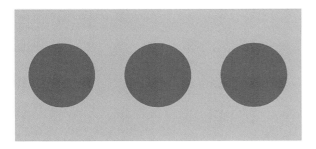

Figure 7.23
The pattern tile is an orange rectangle that contains three green circles.

8. Select the last copy and Ungroup (Mac: Shift+Command+G; Windows: Shift+Ctrl+G). Delete the three circles, leaving only the rectangle. Change the Fill color to any other dark contrast color. Figure 7.24 shows the arrangement. Leave the rectangle selected.

Figure 7.24
Ungroup the center tile and delete the circles.

9. Turn on Smart Guides (Mac: Command+U; Windows: Ctrl+U). Select the Pen tool. Drag your mouse over the upper-left corner of the rectangle. You will see a line appear that says "on" and "align 135 degrees" as shown in Figure 7.25. Click the Pen to leave a point on that guideline above the rectangle. Follow the Align On 135 Degrees guideline below the rectangle and click to set another point. You now have a stroke that cuts the top-left corner of the rectangle at a 45-degree angle. This is your miter line to create a corner for a border.

10. Turn off Smart Guides (Mac: Command+U; Windows: Ctrl+U). Select *only* the miter line stroke. Choose Object|Path|Slice. You now have a triangle and an object that looks like a knife blade. Select the "knife blade" object and delete it.

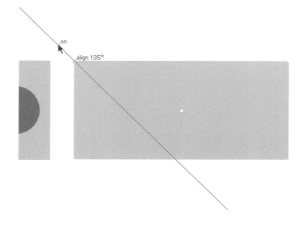

Figure 7.25
The Smart Guides easily let you find the proper angle needed to miter a corner.

11. Select the triangle and the remaining pattern tile copy. Click on Horizontal Align Left in the Align palette, then click on Vertical Align Bottom. The triangle needs to be on top of the pattern tile. With both shapes selected, click on Crop in the Pathfinder palette. You have a 45-degree corner cut from the original pattern tile as shown in Figure 7.26.

12. To make a mitered corner, you need to create a mirror image of the corner. The entire miter takes two transformations: a flip and a rotate. Neither rotating nor flipping alone can produce the miter. Therefore, the first step is to flip the shape horizontally. You can do this in two ways. You can duplicate the shape, open the Transform palette sidebar menu, and select the Flip Horizontal command, or you can do both a flip and copy in one step by using the Reflect tool. If you use the Reflect tool, Alt/Option-click on the bottom-right point of the triangular shape. Select Vertical Axis and click on Copy. Figure 7.27 shows this process.

Figure 7.26
You have created one-half of the corner tile.

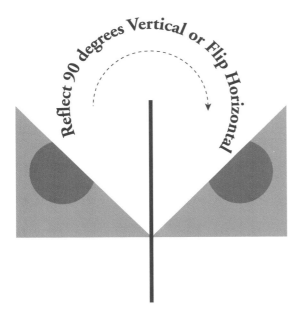

Figure 7.27
To make a mitered corner, you must first flip or reflect the corner tile.

13. The next step in building the mitered corner is to rotate the flipped unit by 90 degrees. Select the flipped copy. Alt/Option-click on the lower-left corner of the flipped triangle. Enter 90 degrees into the dialog box. Click on OK. To make certain that the copies are exactly aligned, select both triangles, click on Horizontal Align Left in the Align palette, then click on Vertical Align Bottom. Group the two shapes (Mac: Command+G; Windows: Ctrl+G). Figure 7.28 shows the process and the final corner.

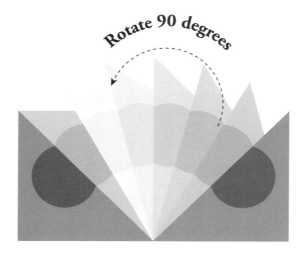

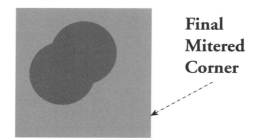

Figure 7.28
To make the mitered corner tile, you need to rotate the flipped copy 90 degrees.

14. If you look carefully at the Brushes palette in Figure 7.29, you see that the Brush icon occupies the second and third squares of a six-square row. You will learn much more about this in Chapter 9, but for now you just need to know that the two squares show the brush to be applied to the straight and curved portions of a stroke. Squares 1 and 4 are reserved for the two types of corner tiles. Therefore, drag the grouped corner tile onto the first square in the brush entry in the Brushes palette. Important: Press Option (Mac) or Alt (Windows) *after* you start to drag the object. Click on OK. Drag the shape again (with Option/Alt, as you did before) into the fourth square on the brush. Click on OK in the dialog box.

15. Select the border shape with the Selection tool and then click on the Pattern brush to apply it to the stroke. Figure 7.30 shows the border, and it looks as if the corner tiles are flipped so that they don't merge properly with the straight pattern.

16. The secret of designing a corner tile is to set it up the way the upper-left corner of a Pattern-brushed rectangle should look. Your instructions didn't bother asking you to do that before you defined

Figure 7.29
The Pattern brush "lives" in a six-square entry in the Brushes palette.

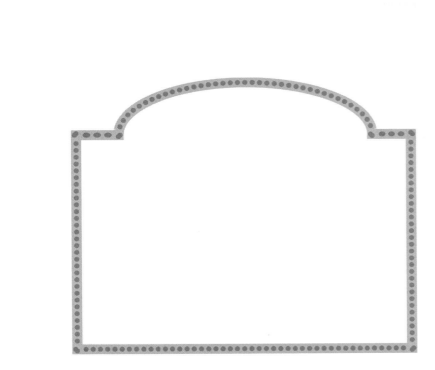

Figure 7.30
Applying the pattern to the border shows a problem with the way that the corner elements are oriented.

the corner. Let's do it now. Deselect all image elements. Click on the brush entry and drag the entry into the image. The pattern tiles appear as objects (this is an excellent way to obtain the pattern elements if all you have kept is the brush). Ungroup (Mac: Shift+Command+G; Windows: Shift+Ctrl+G).

17. Position the straight edge to the right of a corner. Duplicate the straight edge, rotate it 90 degrees, and place it under the corner as shown in Figure 7.31.

18. It is even more obvious from Figure 7.31 that something is amiss. The solution is simple. Flip the corner tile until it is where you prefer it to be. Select the corner tile. From the Transform palette sidebar menu, choose Flip Vertical and choose Flip Horizontal. The tile lands in a much happier place. Figure 7.32 shows its new position.

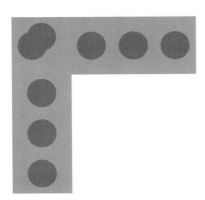
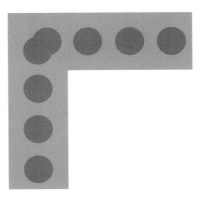

Figure 7.31 (left)
Simulating a border that contains a corner tile.

Figure 7.32 (right)
Flipping the corner tile both horizontally and vertically points it in a much better direction.

19. Repeat Step 14 to drag a copy of the corner into squares 1 and 4 of the Pattern brush entry. Click on Apply To Strokes each time that you are asked. Figure 7.33 shows the border correctly cornered.

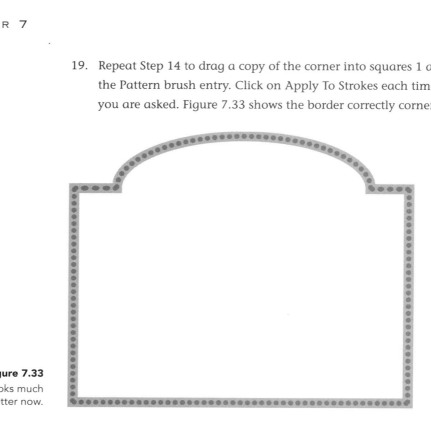

Figure 7.33
The border looks much
better now.

20. Save the PATTERN.AI file to your hard drive as BORDER.AI and close the document. Return to the document that contains the pizza.

21. In the pizza image, choose Window|Brush Libraries|Other Libraries. Select BORDER.AI as the library to open. With nothing selected, click on the brush in the BORDER.AI library. From the sidebar menu on the BORDER.AI Brush library, select Add To Brushes. Close the library. The brush is now added to the Brushes palette for your document.

22. Select the border panel and click on the new Pattern brush to apply it.

Putting Type In Perspective

An ad has to have some text to let you know whose product to buy. In Illustration 7.4, you'll use imported text and create a distorted shadow for it. I set the word "Pizza" using ICGChoc. I used a copy of the pizza dough ellipse as the path. I changed the text to Outlines. I then used the Free Transform Perspective tool (which you will also use later) to make the word larger at the left than at the right. I reflected a copy of the word across a 90-degree vertical axis and then saved the file.

PROJECT Illustration 7.4: Distortions Of Meaning

1. Open the file PIZZATEXT.AI from the enclosed CD-ROM. This contains the saved pizza text.

2. Use the Selection tool to drag the text (which is grouped) from the PIZZATEXT.AI file into your working pizza file.

3. Turn on the Rulers (Mac: Command+R; Windows: Ctrl+R) if they aren't already visible. If any guides currently show in your image, choose View|Clear Guides. Drag a horizontal guide from the top ruler to just underneath the top horizontal edge of the border (formed from the original rectangle in Step 9 of Illustration 7.3). Drag the vertical guides from the left-side ruler so that they are at the inner corners of the border where the rectangle meets the ellipse.

4. Select the Rectangle tool with a black Stroke and a Fill of None. Draw a rectangle that begins at the left intersection of the vertical and horizontal guides, extends to the intersection of the horizontal guide and the rightmost vertical guide, and continues down not quite as far as the start of the pizza.

5. Turn on Smart Guides (Mac: Command+U; Windows: Ctrl+U). Click on the Pizza Pizza text at the upper-left point of the letter "P" and drag this point until it intersects with the upper-left corner of the rectangle that you drew in Step 3. Turn on Smart Guides (Mac: Command+U; Windows: Ctrl+U). Select both the text and the rectangle and click on Align Horizontal Center in the Align palette. Deselect. Select the rectangle and delete it. Clear the guides. Figure 7.34 shows the text before clearing the guides so that you can see their position.

Figure 7.34

Position the text by centering it on a "dummy rectangle" created along the guidelines shown.

6. Duplicate the text layer by dragging the Pizza Pizza layer entry to the New Layer icon at the bottom of the Layers palette. Lock the Pizza Pizza layer. Rename the copied layer "Pizza Pizza Shadow". Change its Fill color to 30% black and drag the layer below the Pizza Pizza layer in the Layers palette.

7. Grab the top-center handle of the text's bounding box and drag it down until the bounding box is just under the small letter "a" in pizza as shown in Figure 7.35.

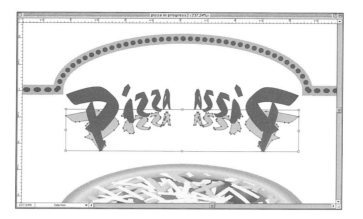

Figure 7.35
Resize the text shadow as shown.

8. Choose the Free Transform tool. Place your cursor over the control handle at the top-left corner of the shadow text and start to drag the handle to the left. Just after you begin to drag the control handle, press Command+Option+Shift (Mac) or Ctrl+Alt+Shift (Windows) in the same order as I've listed them. Watch what happens to your shape as you add each additional key. The Command/Ctrl key enables you to distort the image by moving a single corner, adding Option/Alt allows you to skew the shape, and adding Shift allows you to apply perspective.

Drag the handle (with the three keys still pressed) until it reaches the center of the border. Release the mouse button before you release the keys. Lock the Pizza Pizza Shadow layer. One more task remains.

A Pizza Plate

You need to create a tray to hold the pizza. The Punk And Bloat filter makes short work of this task, as is covered in Illustration 7.5.

PROJECT Illustration 7.5: Green Plate Special

1. Lock all the layers. Drag the entry for the Pizza Dough layer to the New Layer icon at the bottom of the Layers palette. Double-click on the copied layer and name it "Plate". Drag the Plate layer to the very bottom of the Layer stack.

2. Unlock the Plate layer and select the elliptical object. Choose Object|Path|Add Anchor Points. The ellipse receives four new points.

3. Double-click on the Scale tool and enter a Uniform amount of 120%. Click on OK (do not make a copy).

4. With the ellipse selected, choose Filter|Distort|Punk And Bloat and move the slider to 12%, as shown in Figure 7.36. This creates a flower shape for the tray beneath the pizza.

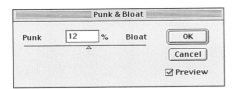

Figure 7.36

The Punk And Bloat filter controls the shape of curved path segments so that you can create flowers or a wide variety of unusual forms.

5. Finally, click on the Stripes patter swatch (the last one on row 2 of the Default Swatches palette) to set the Fill for the tray. Figure 7.37 shows the finished pizza ad in grayscale here and in color in the Color Studio.

Figure 7.37

You have created an advertisement for a pizza parlor using a variety of transformation commands.

String Art

When I was little, I used to love drawing with the Spirograph toy. I still enjoy it enough that I bought one for all my nieces and nephews. A tiny—almost a throwaway—note in Deke McClelland's book, *Real World Illustrator 7*, inspired this next project. He just happened to mention that pressing the tilde (~) key as you use any of the shape tools in the Toolbox causes you to draw multiple copies of the shape. That gave me an idea, and you now have the opportunity to profit from it.

In Illustration 7.6, you will create a series of string-art stars and use them to build an illustration in Photoshop. I will also discuss a variety of things that you could do with the shapes while keeping them in Illustrator.

The Shape Tools

The Ellipse pop-out in the Illustrator Toolbox contains four tools: the Ellipse, the Polygon, the Star, and the Spiral. The Rectangle pop-out has two tools: the Rectangle tool and the Round-Corner Rectangle tool. All the shape tools have certain similarities in the way in which they are applied. Before we create a finished project, let's just play a bit with these tools.

Illustration 7.6: Get In Shape

1. Create a new document in Illustrator.

2. Select the Ellipse tool. Choose a Fill of None and a black Stroke of 1 point. Click the mouse button and draw an ellipse. As you draw, press and hold the tilde key. Figure 7.38 shows one of my doodles with the Ellipse tool. You can produce amazing sculptural forms.

3. If you press Shift in addition to the tilde key as you draw, you can create interlocking circles. However, the effect is much harder to control because Shift also acts as a constrain key and the copies tend to go in lines at a 45-degree angle as you can see in Figure 7.39. The free-form ellipses occurred when I released Shift.

4. The Rectangle tool and the Round-Corner Rectangle tool work the same way. You can interactively control the radius of the rounded corner by pressing the Up and Down Arrow keys as you draw the shape. I don't think I could coordinate that with using the tilde key (at least not without three hands). Try drawing some shapes and changing the roundness of the corners. They try using the tilde key as you draw.

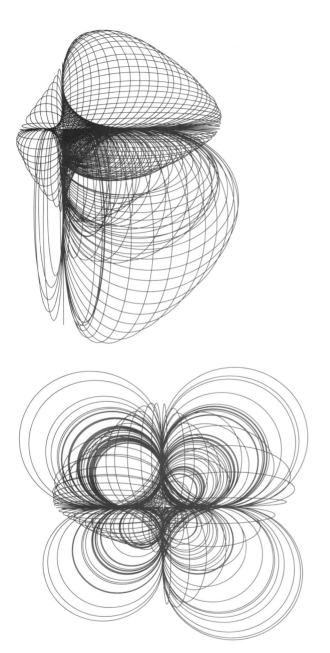

Figure 7.38
This form almost resembles a seashell.

Figure 7.39
The Shift key makes it harder to draw circles where you want them to be.

5. The Polygon tool also lets you use the Up and Down Arrows as you draw. The arrows increase and decrease the number of sides to the shape. The Polygon makes wonderful string art. Figure 7.40 shows a form that resembles a rose.

6. The Spiral tool has even more options. Like the polygon tool, you can rotate it as you draw. You can also change the number of seg-

Figure 7.40
Rose is a rose is a series of inter-locked polygons.

ments by pressing the Up and Down Arrow keys. If you use the Spiral tool Options dialog box to set the size, you can change the tightness of the wraps on the spiral (decay) and the direction in which the form spirals. Figure 7.41 shows the options available, and Figure 7.42 shows a variety of different spirals. Try drawing some spirals. Before you change the settings in the dialog box, make sure you know what they were. It is very easy to create a spiral that won't spiral, and once you have changed the options, they remain in effect until you use the dialog box to change them again.

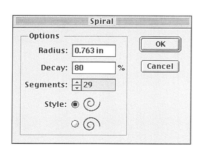

Figure 7.41 (left)
The Spiral tool Options dialog box lets you control some things that you cannot set interactively.

Figure 7.42 (right)
A variety of spirals.

7. When you use the Star tool, you can change the number of points on the star by pressing the Up and Down Arrow keys. You can move the location of the star (and the other shapes as well) as you draw by pressing the spacebar. You can also change the sharpness of the points by setting the inner and outer star radius, but you need to click in the image to open the Star Options dialog box to make changes to the radius. Figure 7.43 shows the Star Options dialog box, and Figure 7.44 shows some stars. You will create stars using the tilde key in Illustration 7.7, but practice making a variety of stars right now.

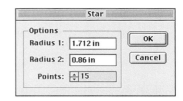

Figure 7.43
The Star Options dialog box allows you to set the inner and outer radii of the star as well as change the number of sides.

The Astronomer's Eye

When you create string art with the help of the tilde key, you can create lovely, complex forms. They are stunning as black-and-white compositions. If you want to add color to these creations, you will find it to be a bit more difficult. Because so many shapes are available, Illustrator can take a long time to redraw the screen—even with a fast machine and a lot of RAM. If you add to that complexity, the redraw times get worse.

You might also have an issue with patience. I don't have enough to individually select each shape and recolor it. You can make a global change, but you cannot, for example, add a gradient to the strokes—which would be a wonderful effect—because Illustrator doesn't do gradients on strokes. You can laboriously hand-select some stars and change their stroke weight or color. That is what I did in Figure 7.45, but it is really quite tedious.

Once you have finished Chapters 9 and 10, you will be able to re-create the image in Figure 7.46. I used a the Front To Back color filter to color the stars and created a gradient that I expanded into a Pattern brush so that I could stroke the stars with multiple colors.

Figure 7.44
You can draw an infinite variety of star forms.

Figure 7.45
You can hand-select certain stars and change their stroke color or weight.

Figure 7.46
You can apply a gradient to a stroke if you make it into a pattern brush first. This complex image was much easier to create than it looks.

Because it can be so difficult to colorize a string art image inside of Illustrator, I decided to take the paths into Photoshop and to stroke them or color in that program. Therefore, although Illustration 7.7 starts out in Illustrator, it ends up in Photoshop.

PROJECT Illustration 7.7: String Art Stars

SLOW SYSTEM ALERT!

If you are short on RAM or machine speed, your string art creation should resemble Figure 7.47b because this one contains the fewest paths.

1. Create a new document in Illustrator.

2. Select the Star tool. Use a traditional 5-point star (with Radius 1 at about 2 inches and Radius 2 at about 1 inch).

3. Set the current Stroke to None and the current Fill to black.

4. Start near the center of the image area and draw a star. Press tilde (~) as you draw to draw out multiple copies. As you are drawing the multiple copies, drag the mouse (with the mouse button still down) in a large circle around your original star. You might need to either try several drawings or undo and try again a few times until you like the results. Try to leave a distinct, visible star in the center of the image. Figure 7.47 shows three of my "creations."

Figure 7.47
Three string art star designs.

5. This project is easier if you are able to open both Photoshop and Illustrator at the same time. In Illustrator, select the entire image (Mac: Command+A; Windows: Ctrl+A). Copy it to the clipboard (Mac: Command+C; Windows: Ctrl+C). Save your image as STRINGART.AI.

6. Launch Photoshop.

7. Create a new document (File|New, Width: 1000 pixels, Height: 1007 pixels, Resolution: 300, Mode: CMYK Color, Contents: White).

Figure 7.48
You can paste paths into Photoshop directly from Illustrator.

8. Paste in the paths from the clipboard (Mac: Command+V; Windows: Ctrl+V). As shown in Figure 7.48, select Paste As Paths.

9. In the Paths palette, double-click on the Work Path to save it as "Path 1". Notice that you see much less of your design than you did in Illustrator. The paths extend beyond the confines of the visible image (you could scale them using the Transform Paths command if you really wanted to, but that is not part of this exercise).

10. Leave the paths active and switch to the Layers palette. Make a new layer (click on the New Layer icon at the bottom of the Layers palette).

11. Choose the Angle Gradient tool from the Gradient pop-out menu on the Toolbox. In the Gradient Tool Options palette, select the Dither checkbox and choose the Spectrum gradient. Place your mouse cursor in the center of the star that you left in the paths and click to set the starting point for the gradient. Drag the rubber-band gradient line to the upper-right corner of the image. Release the mouse button. You have placed an angled gradient in the layer.

12. In the History palette, create a new snapshot.

13. Undo (Mac: Command+Z; Windows: Ctrl+Z) the gradient.

14. Back in the History palette, move the History Brush icon into the column to the left of the gradient snapshot (so that the snapshot becomes the active History document).

15. Choose the History brush in the Toolbox (Y). In the Brushes palette, select the second brush from the left on the top row (the two-pixel hard brush).

16. Select the Stroke Path option from the Paths palette sidebar menu. Figure 7.49 shows you the result in glorious grayscale (it looks much better in the Color Studio).

17. Make a new layer (click on the New Layer icon at the bottom of the Layers palette).

18. Select the Paintbrush tool. Press D to set the colors back to the default of black and white. In the Brushes palette, select the second brush from the left on the second row. Make sure that Path 1 is still active and click on the Stroke Path icon at the bottom of the Paths palette. Depending on the density of your paths, you might need to select a different brush. You should get more coverage on the stroked paths than you did in Step 16, but some white space should still be there. Change the brush size as necessary until you get the coverage that you need. Figure 7.50 shows the desired amount of coverage.

 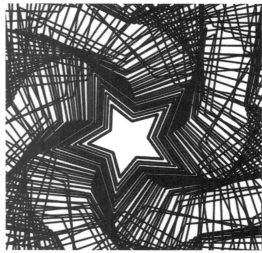

19. Click underneath Path 1 in the Paths palette to turn off the path.

20. Select the Linear Gradient tool. Click on Edit in the Gradient Options palette. From the sidebar menu, select Load. Load the CH07GRAD.GRD gradient on the companion CD-ROM. This document adds a gradient named AIFX-Red to Yellow to your gradient list. Select this gradient and click on OK to close the Edit gradient dialog box.

21. Make sure that Layer 2 is active. Select the Preserve Transparency checkbox on the Layers palette.

22. Drag the gradient cursor from the lower-left corner of the image to the upper-right corner.

23. Deselect the Preserve Transparency checkbox in the Layers palette.

24. Drag the thumbnail for Layer 2 beneath that of Layer 1 in the Layers palette.

Figure 7.49 (left)
You can stroke the string art paths with the History brush in Photoshop to apply a previously created gradient.

Figure 7.50 (right)
Use a larger brush to cover more of the paths.

25. Drag the icon for Layer 2 to the New Layer icon (the center icon at the bottom of the Layers palette). Drag the Layer 2 copy beneath that of Layer 2. With this bottom copy active, choose Filter|Blur| Gaussian Blur. I used an amount of 5 pixels, but you can judge what looks best on your image. I simply wanted to spread the color a bit.

26. Make Layer 1 the active layer.

27. Open the image ASTRONOMY.PSD from the enclosed CD-ROM. Figure 7.51 shows this image. The image contains four layers: clouds created in Photoshop using the Clouds filter, an embossed eye from the Ultimate Symbol Design Elements collection, text on a spiral created in Illustrator, and a telescope (also created in Illustrator) that you should have been able to create for yourself easily on the basis of the material in Chapter 4. All the elements except for the clouds are available in the ASTRONOMY.AI image on the enclosed CD-ROM if you want to explore them in their native Illustrator forms.

Figure 7.51
You will add a telescope, text, and an icon to your string art image.

28. Choose the Move tool (V). Press Shift and drag the image in the ASTRONOMY.PSD file into your working image. Shift causes the dragged file to appear centered in the working image. Because I linked all the layers in the ASTRONOMY.PSD file, when you drag directly from the image, all the linked layers come along for the ride.

29. Drag the thumbnail for Layer 3 (the Clouds layer) so that it is just above the Background layer in the Layers palette.

30. Make Layer 1 the active layer.

31. Choose File|Place and select the STRINGART.AI image that you previously saved. Because you are placing the file, you have a choice of sizes. Press Shift to constrain the aspect ratio and drag the upper-left corner of the bounding box to reduce the size of the placed image to about three-quarters the size of the document. This places the fill string art star in the lower-right corner of the image. Press Enter/ Return to execute the Place command.

32. Select the Preserve Transparency checkbox on the Layers palette. Choose the Eyedropper tool and change the foreground color to the green of the Eye icon in Layer 5. Fill the layer (Mac: Option+Delete; Windows: Alt+Delete).

33. Deselect the Preserve Transparency checkbox on the Layers palette. If the fill looks a bit anemic, select the nontransparent Layer pixels (Mac: Command+click; Windows: Ctrl+click) on the Layer name in the Layers palette. Choose Select|Modify|Expand, 1 pixel. Fill the selection (Mac: Option+Delete; Windows: Alt+Delete). Deselect (Mac: Command+D; Windows: Ctrl+D).

34. Change the Apply mode to Multiply. Use the Move tool to position the layer as desired. Reduce the Opacity of the layer to between 30% and 65%. Figure 7.52 shows the final image in grayscale. You can see one version in color in the Color Studio.

Figure 7.52
The string art stars have been combined with other Illustrator elements in Photoshop.

Mandalas

A mandala is a complex circular image that can help you to restore peace and inner harmony by contemplating its center. Because I love kaleidoscopes, I am also attracted to mandalas and other circular design forms (I'm also a child of the 1960s who forgot to grow up—so long as I have tie-dyed in my world, I'm happy). Illustration 7.8, which you'll work on next, can bring peace and inner harmony contemplating it—but only after it's done. It is a bit of a nasty project—it was a bear to create the first time, until I got all the kinks out, but I have tamed it so that you at least should not throw this book through your computer screen.

If this is a nasty project, why put it in the book? I think it's pretty. I also think that once you work through it, you'll have a better understanding of how some of the Pathfinder commands work, and you'll learn ways to search for a solution when the command doesn't work—but you think that it should.

Next, you will create two levels of revolving objects, and you will also learn how to transform the straight edge of an image into a ragged border.

A Tale Of Over And Under

In the first part this Illustration, you will rotate a decorated ogee around its bottom edge to create the start of a mandala. The challenge of the project is to make part of the last object rotated (the one on top) tuck under the first object so that the pattern is maintained throughout the entire rotation. The only way to do that is to cut the ogee into the necessary segments before you rotate it. To make sure that the cut ogee stays in its proper position, all the original cutting must be done "in place" with objects locked and/or hidden to allow you to select the correct ones. Although the process might seem confusing, work each step carefully in order, and it will come out right.

BUILDING AN OGEE

I have already created the ogee for you because I don't want this project to occupy the entire book, but I want to explain how it was created and show you the files that you can use to build your own version.

In Chapter 5, you created an ogee from scratch to use as a pattern. An ogee is a geometric object that can interlock because all its sides use rotated copies of the same stroke. You already know how to create the shape. However, OGEEPLAIN.AI (shown in Figure 7.53) gives you an already-built shape to use if you don't like my design (or simply want to make one of your own). The lines on top of the ogee are the ones that I used to divide the ogee into sections.

Figure 7.54 shows a close-up of the single ogee design. You can also see it in color because the gradient in the center doesn't show up well in grayscale. This design is the major element that you'll rotate. The solid ogee was filled with a radial gradient that was expanded into 25 steps. I then placed a variety of shapes on top of the ogee. I used the lines in the PLAINOGEE.AI file along with the Divide command on the Pathfinder palette to cut the ogee into strips. I then moved each section apart to create the white spaces.

PROJECT Illustration 7.8: Slice And Dice

1. Open the file OGEEDESIGN.AI. It contains a number of layers, but only one (the one shown in Figure 7.54) is visible. Figure 7.55 shows the Layers palette.

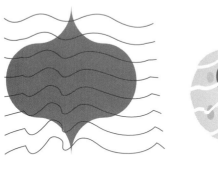

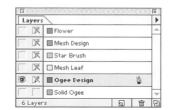

Figure 7.53 (left)
This solid ogee can be decorated as you wish and then sliced up.

Figure 7.54 (center)
The decorated ogee will be used in Illustration 7.8.

Figure 7.55 (right)
The Layers palette for OGEEDESIGN.AI contains a number of layers that are not visible.

2. You will need only the Solid Ogee and Ogee Design layers for the first part of this Illustration. Turn on the Rulers (Mac: Command+R; Windows: Ctrl+R).

3. Select the Ogee Design. Drag a horizontal and a vertical guide so that they intersect at the bottom-center control handle on the object's bounding box as shown in Figure 7.56.

4. Lock the Ogee Design layer. Reveal the Solid Ogee layer. Select the entire image (Mac: Command+A; Windows: Ctrl+A). Only the solid ogee is selected.

5. Zoom into your image until you can very clearly see the intersection of the guides. Choose the Rotate tool. Press Option (Mac) or Alt (Windows) and click on the point where the guides intersect. Enter -30 degrees in the Rotate Tool Options dialog box and click on Copy.

6. Change the fill of the copy to dark blue.

7. When objects cross one another, the Pathfinder commands do not always work as expected. Select the Pen tool. Place the Pen over the bottom point on the dark blue ogee (you know that you are over a point when you see a minus sign next to the Pen icon). Click to remove the bottom point from the shape. Figure 7.57 shows a close-up of the lower points of the objects. Figure 7.58 shows the objects once you have deleted the bottom point. Deselect.

WARNING!

When you create a shape that you plan to cut apart, you should always save a copy of the shape *before* you slice and dice it. If you need it again (and in this technique, you will), it's easier to have a whole shape than to attempt to put Humpty Dumpty back together again. Trust me—I've discovered this the hard way.

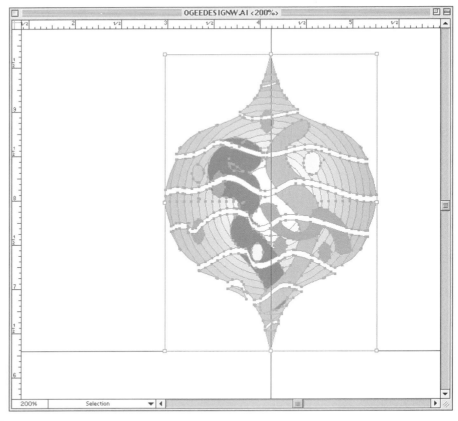

Figure 7.56

Drag guides so that they intersect at the bottom point of the
ogee (which is also the bottom-
center control handle on the
bounding box).

8. Hide the Ogee Design layer. Select the entire image (Mac:
   Command+A; Windows: Ctrl+A). Click on Divide in the Pathfinder palette.

9. Ungroup (Mac: Shift+Command+G; Windows: Shift+Ctrl+G). Deselect.

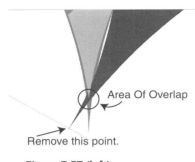

Figure 7.57 (left)

Delete the bottom point on the
dark ogee to remove the area
where the two points overlap.

Figure 7.58 (right)

The shapes no longer cross one
another.

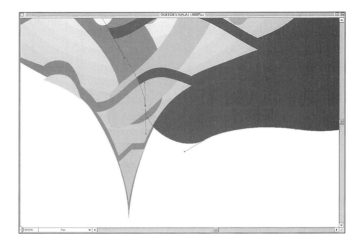

10. Select the right portion of the dark blue ogee. Delete. This leaves only the two sections that form the original ogee (one magenta and one dark blue as shown in Figure 7.59).

11. Turn on the Ogee Design layer again (don't unlock it) and zoom in to your image until you can very clearly see the intersection of the guides. Select the dark blue segment of the solid ogee.

12. Choose the Rotate tool. Press Option (Mac) or Alt (Windows) and click on the point where the guides intersect. Enter –30 degrees in the Rotate Tool Options dialog box and click on Copy.

13. Change the Fill color of the copy to yellow. Hide the Ogee Design layer.

14. Select both the yellow copy and the dark blue ogee segment. Click on Divide in the Pathfinder palette.

15. Ungroup (Mac: Shift+Command+G; Windows: Shift+Ctrl+G). Deselect. Select the right segment of the yellow ogee and delete it. Figure 7.60 shows the finished template.

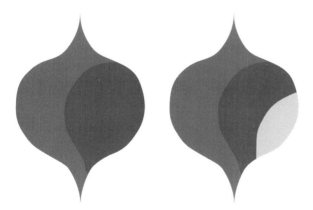

Figure 7.59 (left)
The original soild ogee is now divided into two segments.

Figure 7.60 (right)
You have divided the original solid ogee into three pieces.

16. Drag the icon for the unlocked Ogee Design layer to the New Layer icon (the center icon at the bottom of the Layers palette). Drag the new layer to the bottom of the layer stack.

17. Lock and hide the Ogee Design layer.

18. Choose the Selection tool. Select the dark blue and the yellow segments of the solid ogee. Choose Edit|Select|Select Inverse. Now the magenta segment and the ogee design on the bottom layer are selected.

HOW MANY PIECES DO WE NEED?

If this was a project of your own and you were using a different shape, how would you know the number of pieces that you need in the final cutting template? You need to cut your original shape into as many pieces as it takes for your rotation to rotate out of the original image. When you rotated the shape the first time, the copy overlapped. When you rotated the copy, it still overlapped the original. If you were to rotate that third piece, it would completely clear the original—showing that you need only three pieces.

Examine Figure 7.61. This is a differently shaped object that you are planning to rotate at 15 degrees. That would move it in the opposite direction from the ogee and take 24 copies to complete a circle (360 degrees/15 degrees = 24 rotations). You can see in the figure that it takes four additional shapes before the rotation clears the starting object. Figure 7.62 shows the cutting template that you would need. To make something this complex, you would just repeat Steps 11 to 15 with the newest copy until you have divided your original shape into the needed number of elements.

As you will see a bit later, you need the cutting template only because the rotated design is a grouped shape. If it were solid, you could work with the original design and not need the template at all (but it would be much less interesting). Figure 7.63 shows the finished rotation with the cut shape left in the template colors (if it were solid, there would really be no reason to cut it up because you would not be able to see the design anyway). In the Color Studio, you can see a version of this image that was filled with the Rainbow gradient and then rotated. All the cuts were needed to make the shape come out right.

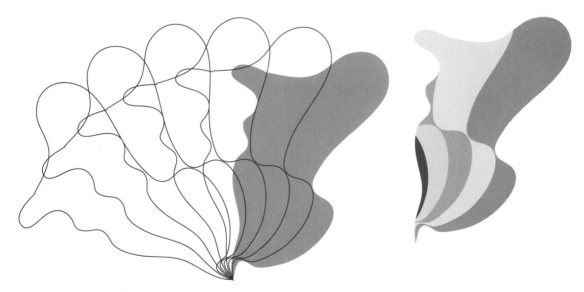

Figure 7.61 (left)
This object needs five cuts for the rotation to clear the original shape.

Figure 7.62 (right)
The cutting template for the complex shape.

19. Click on Crop in the Pathfinder palette. Figure 7.64 shows the result.

20. Only the Solid Ogee layer contains an active selection. Drag the selected object rectangle from the Solid Ogee layer to the Ogee Design copy layer. Lock the Ogee Design copy layer.

21. Unlock the Ogee Design layer. Drag the icon for the unlocked Ogee Design layer to the New Layer icon (the center icon at the bottom of the Layers palette). Drag the new layer just under the Solid Ogee layer.

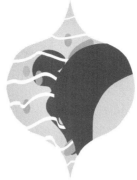

Figure 7.63 (left)
After the pieces are put in order, the rotated cutting shape shows the complexity of the design.

Figure 7.64 (right)
Cropping the magenta segment and the designed ogee leaves the designed ogee in the area originally occupied by the magenta cutting template.

22. Lock and hide the Ogee Design layer.

23. Choose the Selection tool. Select the yellow segment of the solid ogee. Choose Edit|Select|Select Inverse. Now the dark blue segment and the ogee design on the Ogee Design Copy 2 layer are selected.

24. Click on Crop in the Pathfinder palette. Figure 7.65 shows the result.

25. Only the Solid Ogee layer contains an active selection. Drag the selected object rectangle from the Solid Ogee layer to the Ogee Design Copy 2 layer. Lock the Ogee Design Copy 2 layer.

26. Reveal and unlock the Ogee Design layer. Drag its thumbnail in the Layers palette underneath the Solid Ogee layer. (Because this is the last cut, you don't need to make a copy.)

27. Select the entire image (Mac: Command+A; Windows: Ctrl+A). The ogee design and the yellow solid segment are selected.

28. Click on Crop in the Pathfinder palette. Now the ogee design looks exactly as it did when you first opened the file—except that the ogee is in three parts. Move the selected object rectangle from the Solid Ogee layer to the Ogee Design layer. Drag the Solid Ogee layer (which has nothing on it) to the trashcan on the Layers palette.

29. Unlock the two other ogee design layers. You are ready to rotate the shape.

Figure 7.65
Cropping the dark blue segment and the designed ogee leaves the designed ogee in the area originally occupied by the dark blue cutting template.

Let's recap what you did before we move on. You took the template shape and created enough rotations to see that you needed to slice the shape into three parts. You then created the three cutting templates and used these templates to crop copies of the ogee design into the same three segments. Why not just use the originals, rotate them, and use the Divide command as you did when making the templates? If the shapes were solid, you could. However, when you use the Divide command and then Ungroup to delete the extra areas, all the shapes in the segment are also ungrouped. This becomes very ugly if you need to decide, on a piece-by-piece basis, which shapes to keep and which to toss. I selected the easiest method that I could manage in the hope that my logic makes sense to you.

In theory, you could also have used the Minus Front command and a different set of templates. However, the Minus Front and Minus Back commands in the Pathfinder palette work in a much different way than you think they might. If you have a group of objects and a single shape, and the single shape is on top, I feel that it is logical to expect that the Minus Front command would remove the top shape from the grouped object below it. Not so. Rather, the command removes everything but the very bottom object in the group from the very bottom object in the grouping. This makes the command less than useful when you try to use it with a grouped object. Because the Minus Back command works the same way, it too is worth using only on solid objects.

In Illustration 7.9, you'll create the mandala form and learn why you had to make mincemeat out of the shape in the first place.

PROJECT Illustration 7.9: As The Ogee Turns

1. Rearrange the layer order so that the Ogee Design layer is on the bottom of the layer stack, the Ogee Design Copy 2 layer is on top of that, and the Ogee Design layer copy is the top of the three layers. With this reordering of layers, your mandala will be perfect when you rotate it.

2. Select the entire image (Mac: Command+A; Windows: Ctrl+A). This selects all three parts of the ogee.

3. Choose the Rotate tool. Press Option (Mac) or Alt (Windows) and click on the intersection of the two guides at the bottom point of the ogee. Enter –30 degrees into the Rotate Tool Options dialog box. Click on Copy.

4. Transform Again (Mac: Command+D; Windows: Ctrl+D) 10 times. This completes the circle. Figure 7.66 shows the perfectly seamless mandala. Because you kept each segment of the shape on its own

layer, when you rotated the shape, the stacking order was correctly maintained. Compare this to Figure 7.67, where the shapes were not sliced up.

5. Just for fun, switch the order of the Ogee Design layers and see the different patterns that emerge (you can see some of them in this book's Color Studio).

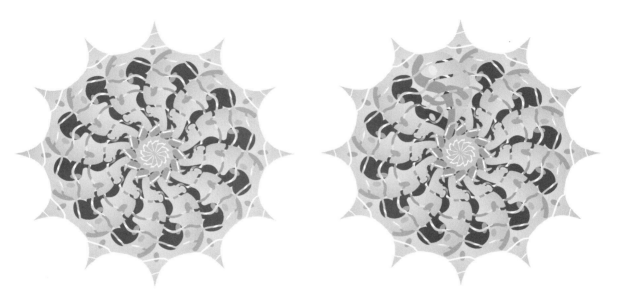

Around The Mandala Again

You still have four layers in the mandala that you haven't seen. These four layers contain additional elements to build a second layer of the mandala. No trick is involved here. You'll only rotate the shapes. However, one of these shapes, the element on the Mesh Leaf layer, is a gradient mesh. I've created it so that it doesn't overlap. If it did, all you could do is learn to live with it. You cannot use any Pathfinder filters on a Mesh object, and you can't expand a Mesh object. If you had to make it overlap and then make it layer properly, your only recourse (other than slicing the object before you make it into a Gradient Mesh) would be to put it into Photoshop and force it to work there.

The other layers are complex as well, but they don't touch one another. The Flower layer is built from several radial gradients, the Mesh Design layer contains another gradient mesh, and the Star brush layer is a Scatter brush.

As it is, though, this particular design will be easy. However, it does slow Illustrator down quite a bit. Because of that, you'll hide the mandala that is already drawn until the end. In Illustration 7.10, you'll learn how to make these changes.

Figure 7.66 (left)
When you create a mandala from a sliced and layered shape, you get a perfectly seamless rotation.

Figure 7.67 (right)
When you create a mandala from a whole shape, the last rotation sits on top, and there is no way to tuck part of it in to make it look like every segment is on top of every other one.

PROJECT Illustration 7.10: Revolving Door

1. Choose View|Clear Guides to remove the original guides. They're on the Ogee Design layer that you need to hide (but not yet).

2. Reveal the four hidden layers and unlock them.

3. Make the Mesh Leaf layer active by clicking on its name in the Layers palette.

4. Drag a horizontal and vertical guide into the image so that they intersect at the center of the mandala.

5. Hide the Ogee Design layers (all three of them).

6. Select the entire image (Mac: Command+A; Windows: Ctrl+A).

7. Select the Rotate tool. Press Option (Mac) or Alt (Windows) and click on the intersection of the guides. Enter 30 degrees in the Rotate Tool Options dialog box. This sends the new elements in the opposite direction.

8. Transform Again (Mac: Command+D; Windows: Ctrl+D) 10 times. You can either wait for the screen redraw or press the command keys 10 times as you count to make sure you've got it right. This is a very slow screen redraw. Figure 7.68 shows the new mandala.

Figure 7.68

Create a secondary mandala in your image.

9. Reveal the Ogee Design layers to see what you've drawn.

10. Unlock all the layers. Select the entire image (Mac: Command+A; Windows: Ctrl+A). Double-click on the Scale tool and enter 80% in the Uniform Scale box. Click on OK. This reduces the size of all the elements and gives you enough room to create a starburst and a background. Lock all the layers.

On The Edge

In the final part (Illustration 7.11), you will create several starbursts behind the mandala and create an interesting edge treatment for the image.

PROJECT Illustration 7.11: A Bit Edgy

1. Create a new layer. Drag it to the bottom of the Layer stack. Name it "Starburst1".

2. Select yellow as your Fill color and set the stroke to None.

3. Choose the Ellipse Tool. Place your mouse cursor at the intersection of the two guides, press Shift+Option (Mac) or Shift+Alt (Windows) and drag a perfect circle from the center until the outline reaches the center of the mesh designs in the mandala. Release the mouse button before you release the modifier keys.

4. Although you cannot see the circle, you can still see the selection's bounding box. Choose Object|Path|Add Anchor Points. Choose the Add Anchor Points command two more times. There are now 32 points on the circle.

5. Select Filter|Distort|Punk and Bloat. Use a setting of –59. This drags the points into a nice starburst. Check to make sure that you can see the starburst and adjust the settings if you cannot. Click on OK.

6. To keep your sanity as the screen redraws, hide all the layers except the Starburst1 layer.

7. Select the entire image (Mac: Command+A; Windows: Ctrl+A). (Only the starburst is selected.) Double-click on the Scale tool and enter 90% in the Uniform Scale box. Click on Copy.

8. Choose the Rotate tool. By eye, rotate the copy so that its points fall between those of the first starburst.

9. Select the larger starburst. Fill it with the Rainbow gradient. Figure 7.69 shows the two starbursts. Lock the Starburst1 layer.

Figure 7.69

Two starbursts.

10. Make the Mesh Leaf layer visible (though locked) if that is where your guides are located (if not, view the layer where they are).

11. Create a new layer. Name it "Background". Drag it to the bottom of the Layer stack.

12. Choose the Rectangle tool. Set your Fill color to black with a Stroke of None. Place your mouse cursor at the intersection of the two guides, press Shift+Option (Mac) or Shift+Alt (Windows), and drag a perfect circle from the center until the outline reaches a bit past the ends of the starbursts. Release the mouse button before you release the modifier keys. Figure 7.70 shows the visible image with the background in place.

13. With the black rectangle selected, choose Object|Path|Add Anchor Points three times.

14. Choose Filter|Distort|Roughen. Select a Size of about 4, an Amount of about 10 per inch, and Corner points as shown in Figure 7.71.

15. Double-click on the Scale tool and enter a Uniform amount of 110% in the dialog box. Click on Copy. Change the Fill of the copy to magenta or hot pink. Send the copy to the back.

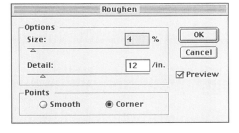

Figure 7.70
Create a black square to act as
the background of the image.

Figure 7.71
The Roughen filter dialog box
allows you to create a rough
edge to the background panel.

16. Select the Rotate tool and drag the magenta copy to rotate it a little bit.

17. Select Filter|Distort|Punk And Bloat and drag the slider to about –6.
    Click on OK. Deselect.

18. Turn on all the layers. Your mandala is finished. Figure 7.72 shows
    the finished image in grayscale.

You can build up very complex edge effects by the combination of filter
and layers of background that you use. Although you copied the inner
edge to make the outer edge, you certainly could have used a totally different filter or filter setting on the outer edge.

Figure 7.72
The finished mandala.

Moving On

In this chapter, you learned to use nearly every transformation command that Illustrator contains. You worked between Illustrator and Photoshop and saw how to use the string art paths as the basis for a Photoshop design. You learned how to miter the corner of a border on a Pattern brush. You also learned how to cut apart a shape so that it will seamlessly rotate.

In Chapter 8, you'll explore the wonderful effects that you can obtain using dashed lines in the Stroke palette.

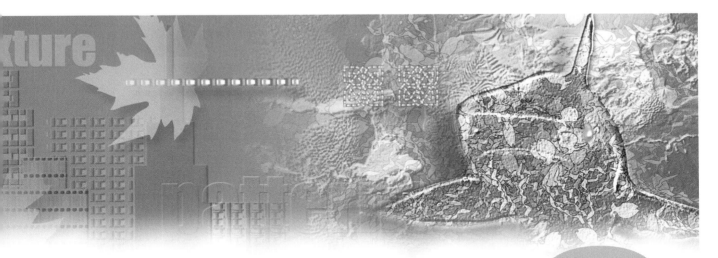

A BIT OF DASH

8

SHERRY LONDON

Dots and dashes can really spice up a stroke. This chapter looks at some ways to make the most of this feature.

I really enjoy playing with the dashes and gaps that change the look of strokes. I am fascinated by the changes that dotted lines can make in a design. In this chapter, you work four illustrations: one simple, one practical, and two fanciful. I hope that you will enjoy the never-ending variety of dashes and gaps.

Workin' On The Railroad

You can create a variety of strokes by using dashed lines. In Illustration 8.1, you will create a circle of railroad tracks. Of course, it isn't the tracks that are important, as you probably don't have much call for them on a regular basis. However, once you get the idea, you can take your train in many new directions.

 Illustration 8.1: Basic Dashes And Gaps

1. Create a new document in Illustrator.

2. Change the Fill to None and the Stroke to black. Figure 8.1 shows the Stroke palette before I created any objects.

3. Select the Ellipse tool. Click once in the center of your image and enter a width of 6 inches and a height of 4 inches, as shown in Figure 8.2. Move the resulting ellipse back into the center of your image if necessary.

Figure 8.1 (left)
These are typical default settings for strokes in Illustrator. However, the Stroke palette usually shows the last settings you used.

Figure 8.2 (right)
Illustrator makes it easy to create an ellipse at exactly the size you want.

4. Change the stroke width to 40 in the Stroke palette, as shown in Figure 8.3.

5. Select the Dashed Line checkbox in the Stroke palette. Enter a dash of 2 points and a gap of 6 points, as shown in Figure 8.4 (which also shows the result). Instant pickup sticks!

6. Leave the dashed ellipse selected and double-click on the Scale tool. With the Uniform Scale radio button selected and the scale amount set to 100%, click on Copy. Figure 8.5 shows the dialog box. Change the stroke weight to 2 points and deselect the Dashed Line checkbox.

7. Double-click on the Scale tool again, then click on Copy with the same settings used in Step 6. This makes another copy of the ellipse.

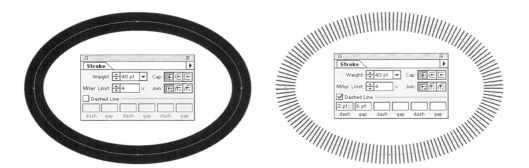

8. Choose the Selection tool. Press the Option key (Mac) or Alt key (Windows) and drag the bounding box from a corner. You need to scale the ellipse so that it's almost the same size as the outer edge of the "tracks." Pressing the Option/Alt key scales the object from the center. You could constrain the aspect ratio as well by pressing the Shift key as you scale, but don't do it. For some reason, it doesn't give you as even an enlargement as working by eye. Figure 8.6 shows the outer edge for the tracks scaled (while the inner edge is still in the center of the tracks).

Figure 8.3 (left)
A stroke width of 40 points makes a very large outlined edge on the ellipse.

Figure 8.4 (right)
You can create parallel straight lines by using a wide stroke on a dashed line.

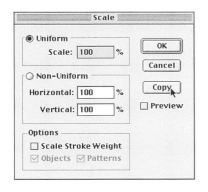

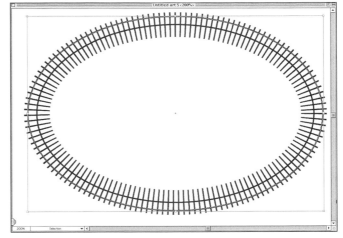

9. Select the unscaled ellipse. With the Option key (Mac) or Alt key (Windows) pressed, make the ellipse small enough to almost cover the inner edge of the tracks. Figure 8.7 shows the finished train tracks.

This was a simple technique. You can spin it out in many directions (although it isn't as easy to create the rails if you started with a wiggly line because you could not simply scale the line and expect it to fit). However, here's what you can do with a wiggly line, courtesy of David Xenakis.

1. Copy the first path.

Figure 8.5 (left)
You can make a copy in place by simply clicking on Copy with the Scale tool set to 100%.

Figure 8.6 (right)
The top ellipse is scaled to almost reach the outer edge of the dashed line.

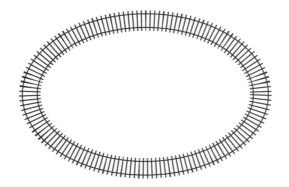

Figure 8.7

The train can now go around the tracks.

2. Paste in Front.

3. Change the stroke weight to about 8 points less than the original, and remove the dash.

4. Outline the Path.

5. Release the compound path (which leaves two separate paths).

6. Apply track strokes to the two paths.

The following two examples take the technique a bit further.

La Vie En Danse

A dashed line has long been the symbol for "cut here" on a coupon. However, no rule states that a coupon has to be rectangular. In Illustration 8.2, you will create a coupon that incorporates its topic (a ballet dancer) into the coupon itself. The dotted line both adds to the design and shows a more playful treatment for the traditional coupon.

Illustration 8.2: Cut Here: A Dotted-Line Coupon

1. Open the image DANCER.AI from the companion CD-ROM. It looks like a blank image (because it was saved as paths from Photoshop), but it is not. Select the entire image (Mac: Command+A; Windows: Ctrl+A). Change the stroke to black. Make sure that the Dashed Line checkbox is not selected in the Stroke palette. Set the Stroke Width to 4 points. Figure 8.8 shows the original dancer stroked in black.

2. Choose Object|Crop Marks|Release. This removes the crop marks placed in the image when it was created by the Photoshop Paths To Illustrator command. Delete the invisible object that appears in the bounding box (the crop marks, when released, turn into a rectangle with no stroke and no fill).

3. Double-click on the Scale tool, select the Uniform radio button, and enter a scale percentage of 70. Click on OK (do not make a copy). Move the dancer toward the left of the document so that you have room to create the coupon. Figure 8.9 shows the dancer, reduced in size.

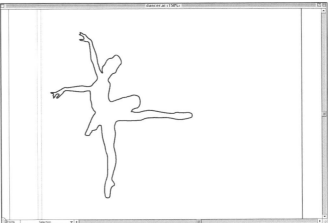

Figure 8.8 (left)
The image of a dancer, stroked with black to make it visible, is the starting point of this Illustration.

Figure 8.9 (right)
Scale the dancer and move her to the left of the document.

4. Double-click on the Scale tool. Select the Uniform radio button and enter a scale percentage of 100. Click on Copy. With the new copy selected, create a new layer (click on the Create New Layer icon at the bottom of the Layers palette). Move the selected object to the newly created layer by dragging the tiny rectangle at the right of Layer 1 in the Layers palette from Layer 1 to Layer 2. Click on the Eye icon in the Layers palette to hide Layer 2.

5. Create a new layer (Layer 3). Drag it below Layer 1 in the Layers palette. Protect Layer 1 from changes by clicking on its Lock toggle. Figure 8.10 shows the Layers palette, as it should now look. Layer 3 is the active layer.

6. Select the Rectangle tool. Make sure that the Dashed Line checkbox is not selected in the Stroke palette. Set the Stroke Width to 4 points. Click in the image and set the rectangle's width to 4.26 inches and height to 2.461 inches, as shown in Figure 8.11.

7. Now that the coupon rectangle is created, you need to get it into position and size it properly. This is a bit tricky. Figure 8.12 shows an enlargement of the dancer with circles placed to show the exact intersection points for the coupon and the dancer. I have also

Figure 8.10
The Layers palette with three layers created, one hidden, and one locked.

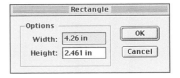

Figure 8.11
Set the width and height of the rectangle that will become the body of the coupon.

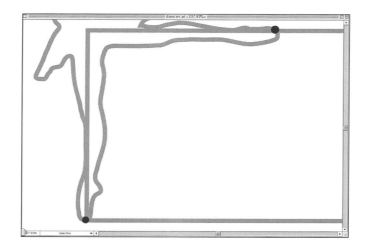

Figure 8.12

The black circles show the points at which the coupon needs to intersect with the dancer's legs.

changed the strokes to light gray so that you can clearly see the black circles. Use the Selection tool to move the rectangle into position at the dancer's leg that forms the left edge of the coupon. Then pull up on the control point at the top center of the bounding box to correctly size the height of the coupon until it intersects with the dancer's outstretched leg. The hardest point to set is where the lower-left corner of the rectangle meets the dancer's toe. You need to ensure that the rectangle doesn't overlap the outside edge of her weight-bearing leg. To make it easier to see where you're placing the rectangle, toggle artwork mode on (Mac: Command+Y; Windows: Ctrl+Y) and turn on Smart Guides (Mac: Command+U; Windows: Ctrl+U). You can then move the lower-left corner of the rectangle to the dancer's toe and easily snap it into place. Figure 8.13 shows a close-up of the left side of the dancer's legs and location of the rectangle.

8. You will fuse the dancer and the rectangle a bit later to form the body of the coupon. In the process, you'll remove the inside portion of the dancer's legs. However, you need to save this area for reuse. Here's the easiest way: Lock Layer 3 (the rectangle layer) and unlock Layer 1 (the dancer). Leave Smart Guides and Artwork mode turned on. Choose the Direct Selection tool. Drag the tool from the lower-right corner of the locked rectangle to the upper-left corner of the locked rectangle to select the area of the dancer inside the rectangle (use the Smart Guides to let you know you're in the right place). Figure 8.14 shows the selected points as filled squares.

9. Turn off Smart Guides (Mac: Command+U; Windows: Ctrl+U). Toggle artwork mode off (Mac: Command+Y; Windows: Ctrl+Y). Copy the

Figure 8.13 (left)
You need to watch the position of the rectangle along the left edge of the dancer's leg very carefully to make sure that it doesn't overlap (which would give the dancer an unnaturally straight edge to the front of her leg).

Figure 8.14 (right)
Select the area of the dancer's legs that are within the rectangle.

selected area to the Clipboard. Lock Layer 1. Create a new Layer (Layer 4). Choose Edit|Paste in Front (yes, I use the Copy and Paste commands when it's really necessary). Hide Layer 1. Figure 8.15 shows the area inside the rectangle pasted into Layer 4 with the dancer on Layer 1 hidden. Use the Direct Selection tool to clean up stray points outside the rectangle. Lock and hide Layer 4 for now. Make Layer 1 visible again.

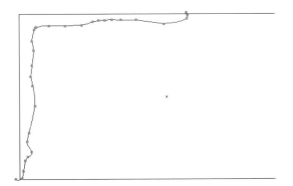

Figure 8.15
Check the pasted area for points that lie outside of the rectangle, and remove them.

10. Remove the locks from Layers 1 and 3. Select both the dancer and the coupon. The easiest way to join them into one shape is to click on Unite in the Pathfinder palette. If the shapes are properly aligned, the Unite command forms one seamless shape, as shown in Figure 8.16.

11. Coupons typically use dotted or dashed lines to show where to cut. With the dancer/coupon selected, select the Dashed Line checkbox on the Stroke palette. Set both the dash and the gap to 6 points. Figure 8.17 shows the result.

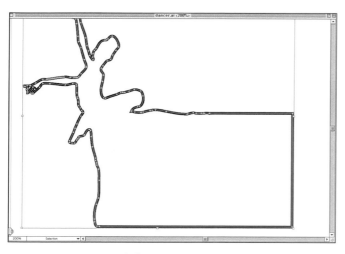

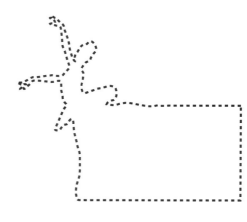

Figure 8.16 (left)
The dancer and the coupon form one seamless shape, using the Unite command on the Path-finder palette.

Figure 8.17 (right)
The dancer/coupon with a 6-point dash and gap.

12. Experiment with a variety of dash and gap settings. Experiment with the stroke width as well. When you are finished playing, use these: Stroke Weight: 4 points, Dash: 2 points, Gap: 6 points, Cap: Round, Join: Round. Figure 8.18 shows these settings, and Figure 8.19 shows the result.

13. Make Layer 4 active and make its Eye icon visible so that you can see what is on the layer. Lock Layers 1 and 3 (although Layer 3 should contain nothing). Figure 8.20 shows that you have managed to salvage the "cut" portion of the dancer's legs so that she is whole again.

14. Select the Pencil tool. With Layer 4 still active (but with no objects selected), draw a wavy line that finishes the bottom of the dancer's tutu (her skirt for those of you unfamiliar with ballet dress). This too is a 4-point black solid stroke. Figure 8.21 shows the dancer with her skirt completed.

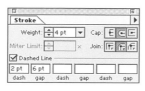

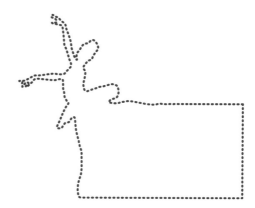

Figure 8.18 (left)
These stroke settings create an elliptical dotted line around the coupon.

Figure 8.19 (right)
The coupon with the settings from Figure 8.18.

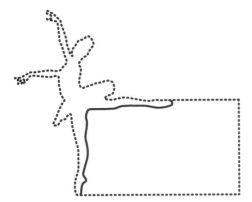

Figure 8.20

You need to replace the missing portions of the dancer's legs, but these should not be part of the dotted "cut here" lines.

15. Now it is time to import the text of the coupon. You may create your own if you prefer. I liked the way that the Image Club (now Eye Wire) Fajita Picante font looked. I have changed the text to outlines so that you can use it. Make Layer 3 active, visible, and unlocked. There should currently be nothing on this layer (the coupon, which was originally on this layer, was moved to Layer 1 when you combined the coupon and the dancer). Choose File|Place. Select the Show All Files checkbox so that the BALLET.AI file on this book's companion CD-ROM for this book is selectable. Click on Place to bring the text into your working image. Use the Selection tool to center the text in the coupon. Figure 8.22 shows the result. (By the way, if this were a real ad, an address and phone number would be a lovely gesture for the prospective customer.)

16. Remember that copy of the dancer that you have hidden in Layer 2? You are going to fill it with a text pattern that you need to create now. Use Adobe Garamond bold for the font if you have it; if not, Times Roman will suffice. Set the point size to 12. Enter "Dance Dance Dance". Leave a space between each word. Press Enter (Mac) or Return (Windows) and type "Dance" four times on the next line. Press Enter/Return again and repeat line 1. Select all the text. Click on Align Center in the Paragraph palette. Reduce the leading between the lines until there is almost no space between the lines. On my version of Adobe Garamond, 10-point leading looks best. Figure 8.23 shows the way the text should look.

17. Select the text object with the Selection tool. Choose Text|Create Outlines. Group the text (Mac: Command+G; Windows: Ctrl+G).

18. Select the Rectangle tool with no fill and a black stroke with a width of 1 point. Place your cursor between the first "a" and "n" in the second line of text. Draw a rectangle that extends up over the first

TROUBLESHOOTING THE UNITE COMMAND

What happens if the shapes don't intersect properly? You might get an unwanted line segment when you use the Unite command. If this happens, you can select the unwanted area with the Direct Selection tool and delete it, or you can unite the two shapes manually.

To unite the shapes manually, use the Direct Selection tool and select the last point on the dancer's weight-bearing leg and the closest point on the coupon. Choose Object|Path|Average (Mac: Command+Option+J; Windows: Alt+Ctrl+J), select the Both radio button. Click on OK. Then select Object|Path|Join (Mac: Command+J; Windows: Ctrl+J). Select a Smooth join. Repeat the same process for the points on the dancer's outstretched leg and the closest coupon point. This makes the same seamless shape as the Unite command, but it is not as easy.

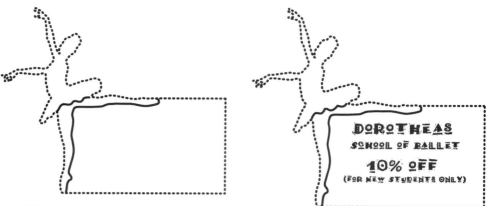

Figure 8.21 (left)
Use the Pencil tool to finish the dancer's tutu.

Figure 8.22 (right)
The Place command is used to add the text to the coupon.

line of text and ends between the last "a" and "n" on the second line. Figure 8.24 shows the correct placement of the rectangle.

19. Move the control handle in the center of the bottom side of the rectangle to resize the rectangle so that it ends just above the lowercase letters in the third line. Move the center-top control handle so that the rectangle starts just above the lowercase letters in the first line. Figure 8.25 shows this location.

Figure 8.23
This is the start of the text pattern fill.

Dance Dance Dance
Dance Dance Dance Dance
Dance Dance Dance

Figure 8.24
This rectangle will set the final size for the pattern.

Dance Dance Dance
Dance Dance Dance Dance
Dance Dance Dance

Figure 8.25
The rectangle now encloses exactly one pattern repeat.

Dance Dance Dance
Dance Dance Dance Dance
Dance Dance Dance

20. Because it is difficult to be sure that the rectangle is positioned so perfectly that the letter "D" will repeat seamlessly, you can take advantage of the room for error that the white space between the lines provides. Use the Selection tool to move the rectangle up, until the horizontal edges are not touching any text. Press Shift after you start to move the rectangle to constrain movement to the vertical direction only. Figure 8.26 shows a good position for the rectangle.

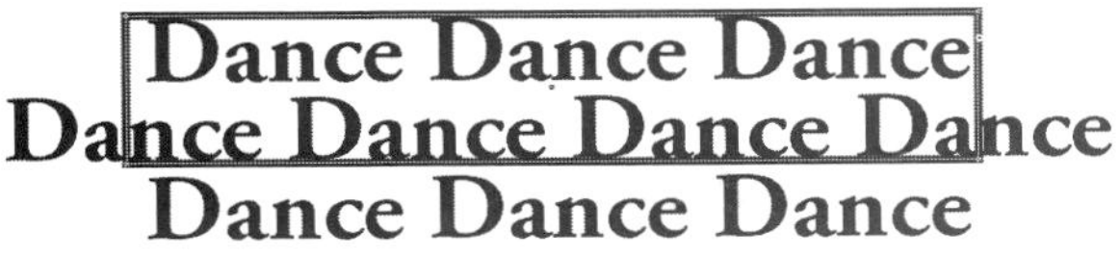

Dance Dance Dance Dance Dance Dance Dance Dance Dance Dance

Figure 8.26
The rectangle is moved to ensure that the letters will not show a seam.

21. The ideal next step would be to create a mask from the rectangle. Unfortunately, a pattern tile cannot contain a mask. Therefore, you need to ungroup (Mac: Shift+Command+G; Windows: Shift+Ctrl+G) the text and, using the Selection tool, select each letter that is outside the rectangle and delete it. Figure 8.27 shows what's left after this step is done. Change the stroke on the rectangle to None (remember, it also has no fill). Figure 8.28 shows the finished pattern.

Dance Dance Dance nce Dance Dance Da

Figure 8.27
Delete all the letters outside the rectangle.

Dance Dance Dance nce Dance Dance Da

Figure 8.28
The finished pattern tile.

22. Select the invisible rectangle and the text. You may group it all together for easier transport if you want. Drag the pattern into the Swatches palette.

23. Make Layer 2 visible. Drag the layer to the bottom of the Layers palette. Select the dancer. Change her stroke to None and click on the new pattern in the Swatches palette to make it the fill. Figure 8.29 shows the almost-finished coupon.

24. If you were creating this coupon for a newspaper, it would now be adequate for an inexpensive run. In basic black and white, not

Figure 8.29
The coupon is almost complete.

Figure 8.30

Use the Watercolor-Wet brush to add some excitement and color to the coupon.

much can be done to mess up the printing. However, you might want to add some color to it. The new brush strokes provide the perfect background. Add a new layer (Layer 5) and drag it to the bottom of the Layers palette. Write-protect Layers 1 to 4 by turning on their lock icons.

25. Choose a pastel yellow for your stroke color. Choose Windows|Brush Library|Artistic Sample. This is a library provided with Illustrator 8. If, for some reason, it does not show up in your library list, you should have it on the Illustrator 8 installation CD-ROM. Select the Paintbrush tool and click on the last brush in the library (it is called Watercolor-Wet) shown in Figure 8.30. Paint a squiggle behind the dancer.

26. Change the stroke color to a pastel green and paint another squiggle. Make one stroke each in pastel blue, light lavender, and pastel pink. Each stroke should be a bit larger than the previous one and should finally cover the entire area occupied by the coupon. Because each stroke is higher in the layering order than the previous one, your final stroke in light pink might tend to dominate the image. In my original, I decided that I preferred the strokes in reverse order. If you would too, select your first stroke (the yellow one) and send it to the back (Object|Arrange|Send To Back). Then select the green stroke and send it to the back. Select and send to the back the blue stroke, then the purple stroke, and finally the pink stroke. You have rearranged the layering order, and now the smallest of the strokes is the one that is fully visible. Figure 8.31 shows the finished coupon, in grayscale. It looks better in this book's chapter's Practice Files on the CD-ROM.

The pastel brush strokes are an excellent introduction to the new Art brushes included with Illustrator 8. I'll talk a bit more about these new brushes in Chapter 9.

Figure 8.31

The finished coupon has soft brush strokes of color in it.

Cityscape

In Illustration 8.3, you will learn how to create textures using dotted and dashed lines on open paths. The textures that combinations of dashes and gaps can create are truly spectacular. You will practice a few different methods of working with these stroked patterns and learn the various trade-offs that are necessary when deciding which method to use.

Architecture 101—Design And Construction

An open path doesn't behave in the same manner as a closed object. If you apply a dash pattern to an open path, either the Pathfinder filters do not work or they remove all stroke from the path. Therefore, although it's easy to create pattern fill swatches that contain dashed-line patterns, it can be tricky to manipulate these patterns if you want to make a pattern tile from only *part* of them. I know that this sounds confusing but try it, and you'll see what I mean.

PROJECT Illustration 8.3: A Dash Of This And A Dot Of That

1. Open the image CITYSCAPE.AI from the companion CD-ROM. Figure 8.32 shows this image. It's an abstract image of city buildings.

2. I created this image as separate shapes all on a single layer. You should fix this image before you start to work on it (yes, I'm deliberately making more work for you). Select each building and move it onto its own layer. Name each layer so that it is meaningful. Where the building is composed of more than one shape, select all the component shapes and use the Pathfinder palette's Unite command to make them into one shape. Figure 8.33 shows the Layers palette after this has been done.

EASY LAYERING

It is easiest to keep the stacking order of the objects intact as you move them onto new layers. Start by selecting the object closest to the front (the short cyan building). Create a new layer and then move the "selected object" rectangle from Layer 1 to the new layer on the Layers palette. Double-click on Layer 2 and change its name to "Aqua Building". Select the next closest-to-the-front object (i.e., the short green building). This makes Layer 1 the active layer again. Now, when you create a new layer, it is placed automatically above Layer 1 but below all the other layers. Move the "selected object" rectangle onto the new layer and change its name. For all the remaining buildings, select the building first, *then* create the new layer for it. If the building contains more than one shape, unite the shapes before you move them to their new layer.

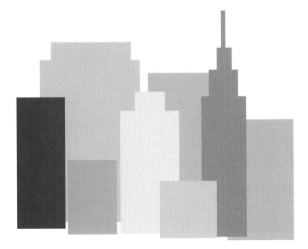

Figure 8.32

An abstract series of building shapes is the starting point for a texture exploration.

3. Select the building on the left (I named it "Purple Building"). Choose View|Make Guides. Toggle the View|Lock Guides menu item so that a checkmark appears in front of it on the menu. The purple object loses its fill, but you can now snap points to it. Toggle the Snap To Point option to select it as well. Figure 8.34 shows the View menu with the correct options selected.

4. Select the Pen tool. Click to place a point on the bottom-left corner of the building guides. Press the Shift key to constrain the next point to the horizontal and place it on the bottom-right point of the building guides. (Zoom in so that you can see the building well enough.) Click on the Swap Fill And Stroke arrow in the Toolbox to give the new path no fill and the purple stroke from the original building. Change the stroke Weight to 8 points.

5. In the Stroke palette, select the Dashed Line checkbox and enter the values Dash: 3 points, Gap: 7 points, Dash: 8 points, Gap: 5 points, as shown in Figure 8.35. Select a Butt Cap and a Miter Join.

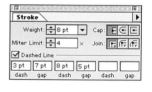

Figure 8.33 (left)
It is helpful to organize the buildings into individual layers before you begin the project.

Figure 8.34 (center)
Toggle Snap To Point and Lock Guides to select both options.

Figure 8.35 (right)
Creating the first dash-and-gap pattern.

6. With the line selected, double-click on the Scale tool. Select the Uniform radio button and change the scale to 100%. Click on Copy. This copies the stroke in place. Change the stroke color to white. In the Stroke palette, change the Stroke Weight to 2. Enter the values Dash: 1 point, Gap: 9 points, Dash: 6 points, Gap: 7 points for the dashed line. Press the right Arrow key once to move the line 1 point

Figure 8.36
The first stroke pattern.

to the right to center the new stroke over the original one. Figure 8.36 shows an enlargement of the combined stroke.

7. Select both the purple and the white line and group them (Mac: Command+G; Windows: Ctrl+G). Leave the group selected.

8. Choose Object|Transform|Move. Use the settings in Figure 8.37 (Horizontal: 0 inches, Vertical: 0.13 inches, Distance: 0.13 inches, Angle: 90 degrees). Click on Copy. You now have two strokes in your building (as shown in Figure 8.38).

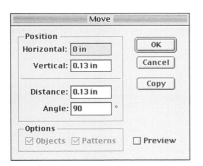

Figure 8.37 (left)
Use these settings to create a copy of your grouped stroke.

Figure 8.38 (right)
This is the start of the building pattern.

9. Use the Transform Again command (Mac: Command+D; Windows: Ctrl+D) to copy and move the stroke to cover the rest of the building guides. Figure 8.39 shows the cityscape image with one building in pattern.

10. Select all the component strokes in the building and group them (an easy way to do this is to lock all the other layers and then Select All).

11. The stroke pattern that you created is partially transparent—the gaps allow whatever is behind them to show through. Although a transparent building is okay, let's put a solid fill behind this one. With no objects selected, change the current Fill color to an orange-brown and set the Stroke to None. Choose the Rectangle tool and draw a rectangle that snaps to the building guides. Select Object| Arrange|Send Backward to put the fill behind the stroke pattern.

12. Lock the Purple Building layer. Unlock the Lavender Building layer. Select the lavender building. To make it easier to see this building, hide the Purple Building, Short Green Building, and Yellow Building layers. Repeat Steps 3 and 4 to make guides and draw the foundation line stroke.

SELECTING A STROKE COLOR

If you aware of what is happening, you can easily preserve the color of the building to use it for your strokes. When you select a building, the color of that object becomes the current fill color. When you change the building into guides, the current fill color doesn't change. After you create your foundation stroke, you can swap the current fill color for the current stroke color (which is None because the buildings are not stroked).

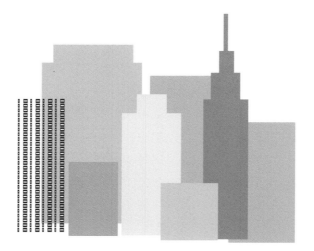

Figure 8.39

Your cityscape now contains one patterned building.

13. Set the Stroke weight to 20 points and create a dashed line with Dash: 10 points, Gap: 2 points, Dash: 6 points, Gap: 2 points. Use the lavender of the original building as your stroke color.

14. Repeat Step 6. Use white as the stroke color for the copied line and give it a Stroke Weight of 7 points. Use the same dash-and-gap pattern as before. Using the Selection tool, shorten the white stroke so that it starts at the beginning of the second dash on the left and ends at the end of the next-to-last dash on the right. Figure 8.40 shows a close-up of this stroke.

Figure 8.40

By changing the starting and ending position of the superimposed stroke, you can create a more interesting pattern.

15. Repeat Steps 7 and 8. However, use the value 0.32 as the amount by which to move the stroke vertically. Repeat the transformation until the stroke covers the height of the entire building. Figure 8.41 shows the building at this point.

16. Select the two stroke groups that cover the top of the building (where it diminishes in size). Drag the control handles on both sides of the selection so that the stroke covers only the shape of the building. Figure 8.42 shows the finished building.

17. The small green building is your next "victim." You have completed two buildings so far, and the only difference between them is that one gets smaller at the top. On both patterns, you have left white space between the strokes. It would be faster to fill the building shape if you could simply define the stroke as a pattern swatch and

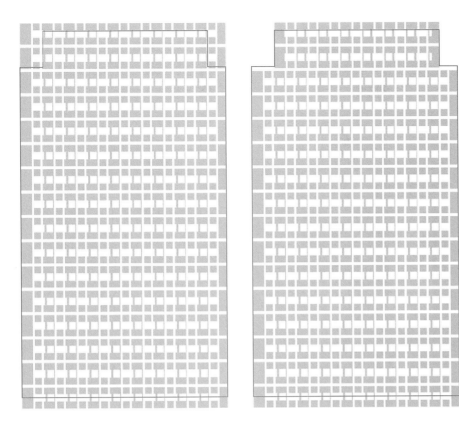

be done with it. As it happens, you can—and you will. Work in the
pasteboard area of your image. Remove the lock from the Short
Green Building layer (and lock all the other layers). Select the green
building (an easy way to make the correct green your current fill
color). Deselect the green building and choose the Pen tool. Swap the
Fill and Stroke colors. Working in the pasteboard area, click to create
a point, press the Shift key, and create another point about 2 inches
to the right of the first point (you don't need to measure this).

18. Select the Dashed Line checkbox in the Stroke palette. Create a
    Dash: 6 points, Gap: 2 points, Dash: 8 points, Gap: 2 points se-
    quence. Set the stroke width to 6 points. Make another copy of the
    line as you did in Step 6.

19. Change the stroke color for the copy to white. Change the stroke
    width to 2 points. Create a dashed line pattern of Dash: 4 points,
    Gap: 4 points, Dash: 10 points, Gap: 3 points. Group both strokes
    together and drag the group into the Swatches palette. After it is
    safely in the Swatches palette, you may delete the original from
    the document.

20. Select the short green building. Change the fill color to the pattern that you just created (click on the new pattern in the Swatches palette). Figure 8.43 shows the cityscape with three buildings textured.

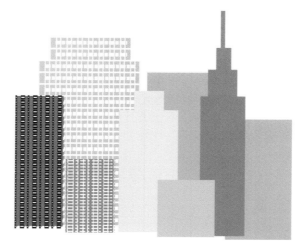

Figure 8.43
Three buildings are now complete.

21. The yellow building is next on your list. For this building, you will create a more complex series of strokes to use as a pattern fill. Make the Yellow Building layer the only unlocked layer. Select the yellow building to set the current fill color. Deselect the yellow building and choose the Pen tool. Swap the Fill and Stroke colors. Working in the pasteboard area, click to create a point, press the Shift key, and create another point about 2 inches to the right of the first point.

22. Set the stroke width to 40 points. Mark the Dashed Line checkbox in the Stroke palette. Create a Dash: 2 points, Gap: 2 points, Dash: 8 points sequence. Notice that you are using only three slots on the dash-and-gap area. Figure 8.44 shows the stroke that results. See whether you can figure out what is strange about it.

Figure 8.44
Using only three spots on the dash-and-gap list makes the sequence of dashes and gaps reverse: You have a dash of 2, gap of 2, and dash of 8 and then a gap of 2, dash of 2, and gap of 8 before the sequence repeats.

23. Make another copy of the line as you did in Step 6. Change the stroke color for the copy to light green (about the same value as the yellow). Press the right Arrow key to move the copied line to the right until the green stripes appear in the gaps in the yellow line. Shorten

the right edge of the green line so that it ends where the yellow line ends. Figure 8.45 shows this effect (with the contrast darkened so that it shows up in grayscale).

Figure 8.45
When moved to the right, the dashes in the green line appear in the gaps in the yellow line.

24. A white gap remains at the start of the line (after the first yellow dash). There is no efficient way to get rid of it. You cannot define a dashed line so that it begins with a gap—which would be the easiest solution here. However, you can "fake" it. Make another copy of the line. Change it to a solid stroke and send it behind the yellow line (choose Object|Arrange|Send Backward twice). This action "plugs" up the gap but creates a third line that needs to be included in the grouping.

25. Select all three overlapping lines. Choose Object|Transform|Move and move the line 40 points vertically and 90 degrees. Click on Copy. Deselect the shapes. Figure 8.46 shows the Move dialog box.

```
┌─────────── Move ───────────┐
│ ┌─Position──────────┐      │
│ │ Horizontal: [0 in ] │ ┌────OK────┐ │
│ │                    │ └──────────┘ │
│ │ Vertical: [0.556 in]│ ┌──Cancel──┐ │
│ │                    │ └──────────┘ │
│ │ Distance: [40 points]│ ┌───Copy───┐ │
│ │ Angle: [90    ] °  │ └──────────┘ │
│ └──────────────────┘      │
│ ┌─Options───────────┐      │
│ │ ☑ Objects ☐ Patterns │ ☑ Preview │
│ └──────────────────┘      │
└────────────────────────────┘
```

Figure 8.46
To move an object by a specific number of points, you need to enter "points" in the dialog box.

26. You now need to reverse the colors in the copied line. There is no automatic way to do this, not even an automatic way to "hold" one color while selecting the "other" object. The least fiddly way that I can find is to first select the green line and drag the green stroke color into the Swatches palette and then select the yellow part of the line and drag the yellow stroke color into the Swatches palette. Now you at least have both colors safely tucked away.

27. Select the yellow part of the copied line. Click on the green swatch in the Swatches palette to change it to a green stroke. Hide the selection

(Mac: Command+3; Windows: Ctrl+3). You are left with what looks to be a solid green line. Remember, it's both the green dashed line and the green solid line beneath. Fix both at once; select both, and then select yellow as the stroke color. Show the hidden selection (Mac: Command+Option+3; Windows: Alt+Ctrl+3). Your line now consists of stripes that counterchange in the middle. Let's add a bit more dash (actually, more dot).

28. Copy any of the lines in the pattern so far (all you need is an easy way to get a straight line). Set the stroke width for the copy to 4 points and make the stroke a deep green-brown. Select both this stroke and the yellow stroke in the bottom set of stripes (you must select the brown stroke first and then the yellow stroke). Click on Vertical Align Bottom on the Align palette to make the brown stroke line up with the base of the bottom strokes. Select only the brown stroke. In the Stroke palette, change the dash pattern to Dash: 1 point, Gap: 6 points. Select a Round Cap. You may leave the Miter Join setting as is. Stretch or contract the line so that the dots start and end within the larger stripe. Figure 8.47 shows the current stroke pattern.

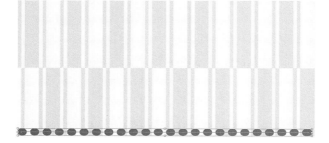

Figure 8.47
Adding a dotted line to the dashed pattern provides some spice.

29. Select the dotted brown line. You need to make three more copies across the pattern. Choose Object|Transform|Move and move the line vertically 20 points (one-half of the 40-point width of each stripe). Click on Copy. Use the Transform Again command two more times (Mac: Command+D; Windows: Ctrl+D).

30. Select the bottom dotted brown line. Copy it the way that you did in Step 6. Set the stroke color to yellow and the stroke width to 2 points. Deselect the Dashed Line checkbox (you now have a solid yellow stripe). Change the Cap to Butt. Lengthen or shorten the line so that it is the same length as the main stripe.

31. Choose Object|Transform|Move. Set the Horizontal Position to 0 and the Vertical Position to 3 points. Click on OK (do not make a copy this time). This moves the yellow line just above the dotted line. Leave the line selected.

32. Choose Object|Transform|Move. Set the Horizontal Position to 0 and the Vertical Position to 20 points. Click on Copy. The yellow line is copied just above the second-lowest dotted line. Use Transform Again two more times (Mac: Command+D; Windows: Ctrl+D).

33. Select all four solid yellow lines. Choose Object|Transform|Move. Set the Horizontal Position to 0 and the Vertical Position to .2 inches. Click on Copy. The yellow solid lines move up to frame the dotted lines, and the highest one butts up against the top of the wide stripe (where it will frame the lowest dotted line as a repeat pattern). Figure 8.48 shows this finished pattern.

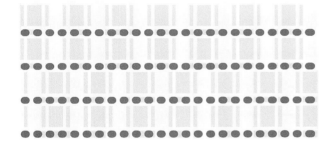

Figure 8.48
The complex stripe pattern is complete.

34. Select the entire stripe and drag it into the Swatches palette. Change the fill of the yellow building to the new pattern. You may delete the pattern, but I recommend instead that you cut it from this image and paste and save it to a new file. There is no good way to reconstruct the starting pattern by using only the pattern swatch.

Architecture 102—Windows And Paint

You have now applied two stroke sequences and two complete patterns to buildings. It was undoubtedly easier to make the pattern fill conform to the uneven shape of the yellow building than it was to make the stripes conform to the shape of the lavender building. You can create a pattern from a stripe so that no space remains between strokes. You can also create a pattern that allows for white space by simply adding an unstroked, unfilled rectangle to the selection as a spacer. Try that next (on an easy pattern), in Illustration 8.4.

PROJECT Illustration 8.4: Spaced Out (And Pulled In) Strokes

1. Make the Orange Building layer the only unlocked layer. Select the orange building to make its fill the active Fill color. Deselect the building and use the Pen tool to make a straight line. Swap the Fill and Stroke colors. Set the stroke width to 10 points, with a Butt Cap. Select the Dashed Line checkbox and create a pattern of Dash: 10, Gap: 6, Dash 4, Gap: 2. (This is a reverse Fibonacci progression with a base of 2 points.)

2. Copy the line and change its stroke to white and the stroke width to 6 points. Reverse the stroke pattern (Dash: 2 points, Gap: 4 points, Dash: 6 points: Gap 10 points). Select the Rectangle tool. Set both the Fill and the Stroke color to None. Create an invisible rectangle around the stroke that is a bit taller than the stroke. Select the two strokes and the rectangle and drag them into the Swatches palette. Select the orange building and set the fill for the building to the newly defined pattern. Figure 8.49 shows the image with five buildings completed. Do not delete the original pattern stroke.

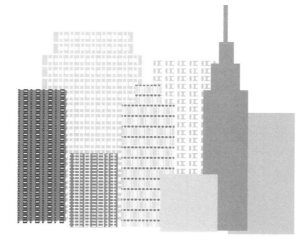

Figure 8.49
The fifth building created shows how you can leave white space in a pattern.

3. You are going to see what happens when you try to use less than one full stroke height as your repeat. To begin, use the same pattern as the one you saved in Step 34 of Illustration 8.3. Select the white stroke on top of the orange stroke. Hide the selection (Mac: Command+3; Windows: Ctrl+3).

4. Unlock the Mid-Size Green Building layer (don't lock the Orange Building layer, however, or you will also lock your pattern-to-be). Select the midsize green building to make its fill the Current fill color.

Deselect it. Drag the Fill color into the Swatches palette. Select the orange stroke. Change its stroke to the green in the Swatches palette. Reveal the white stroke (Mac: Command+Option+3; Windows: Alt+Ctrl+3).

5. Select both strokes and Group them (Mac: Command+G; Windows: Ctrl+G). Press Alt (Windows) or Option (Mac to constrain movement to the vertical only. Figure 8.50 shows both stripes.

Figure 8.50
Copy the stripe about twice the stroke width distance above the original.

6. Select both stripes. Alt/Option drag a copy of the two strokes up so that they fall between the existing strokes, as shown in Figure 8.51.

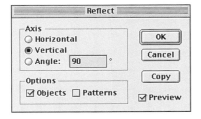

Figure 8.51
A second copy of the two strokes closes the gaps.

7. Leave the two strokes selected. Choose Object|Transform|Reflect and select the Vertical axis radio button, as shown in Figure 8.52. Click on OK.

```
┌──────────────── Reflect ────────────────┐
│ ┌─Axis ──────────────┐                    │
│ │ ○ Horizontal       │   ┌────────────┐  │
│ │ ● Vertical         │   │     OK     │  │
│ │ ○ Angle:  [ 90 ] ° │   └────────────┘  │
│ └────────────────────┘   ┌────────────┐  │
│                          │   Cancel   │  │
│                          └────────────┘  │
│ ┌─Options ───────────┐   ┌────────────┐  │
│ │ ☑ Objects ☐ Patterns│  │    Copy    │  │
│ └────────────────────┘   └────────────┘  │
│                          ☑ Preview        │
└──────────────────────────────────────────┘
```

Figure 8.52
Flip the alternate strokes so that they face in the opposite direction.

8. Select all four stroke sets and group them (Mac: Command+G; Windows: Ctrl+G).

9. You can easily define this four-stroke combination as a pattern. However, let's see what happens if you want to repeat this "set" about three-quarters of the way up rather than a whole repeat. Select the grouping. Make sure that you have enough room to repeat the pattern a number of times (you need at least as much room above the group as the height of the midsize green building). Choose Object| Transform|Move and set the vertical position to .417 inches (I have already done the math for you). Use the Transform Again command (Mac: Command+D; Windows: Ctrl+D) until the "stack" seems larger than the height of the building. Figure 8.53 shows this stroke stack.

Figure 8.53

Strokes are repeated less than their full width so that they overlap each other.

Figure 8.54

Compare the regularity of the pattern in the rectangle on the right (made from individual strokes) with the pattern breaks in the outlined rectangle on the left (made from a pattern tile that did not meet properly).

10. Select the top stroke group and then select the next one down (so that two groups are selected). Drag the two groups onto the Swatches palette to make a pattern tile. If you have room, create a rectangle next to your repeated stroke that is the same size as the entire stroke grouping. Fill the rectangle with the newly created pattern. Figure 8.54 shows the pattern rectangle on the left (with a black stroke around it) and the individual strokes on the right. You will see a subtle but definite difference between them. The pattern that you made by hand is regular throughout the length of the shape. The pattern tile shows the breaks where it could not overlap the strokes. Every two strokes, the pattern seems to change. You may decide which one you prefer to use.

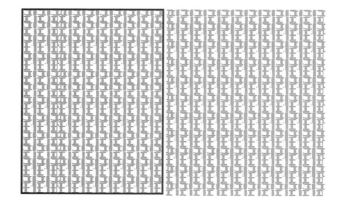

11. If you decide to use the individual strokes, you can get them inside the shape in another way. When you used the repeat-the-stroke method in buildings 1 and 2, you started by making guides of the original buildings. Now, select the entire stroke group. Transfer the selection rectangle from the Orange Building layer to the Mid-Size Green Building layer. Lock all the layers except for the Mid-Size Green Building layer. Drag the selection on top of the midsize green building. Group it, send it to the back, and then select the entire layer (Mac: Command+A; Windows: Ctrl+A). Make a mask with Object|Masks|Make (Mac: Command+7; Windows: Ctrl+7).

12. You have two remaining shapes. I chose to leave the pink building solid and to create only a small blue and white pattern to the aqua building in the foreground. Finish the two shapes with any stroke pattern you want. This is an excellent opportunity for you to apply the lessons in this chapter. Figure 8.55 shows my final Illustrator image.

I also took this image and saved it as a layered Photoshop file. I added to it a scan of turtles swimming in a sea (taken from *Japanese Stencil Designs*,

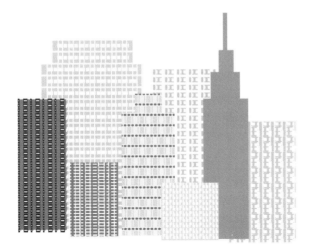

Figure 8.55
This is my finished Illustrator cityscape.

by Andrew W. Tuer, Dover Books, 1967). I embossed both the sea design and the buildings and added some gradients and layer masks. You may inspect the Photoshop layered file as it is on the companion CD-ROM (CITYSCAPE2.PSD). Figure 8.56 shows my finished Photoshop image (and, of course, it looks better in the Color Studio).

Figure 8.56
You can move the cityscape image to Photoshop and add special effects.

Knotting Up Loose Ends

You've seen several of the possibilities and uses for dashed lines in this chapter. Some limitations exist as well. You cannot slice (Object|Path|Slice) an open path to sever it, although the Slice command works as expected on a dashed, stroked object that is closed. The Expand command (which I use repeatedly throughout this book) does not work well on dashed strokes. When you expand a stroke with a dashed line in it, you get a solid line (not the little boxes that I hoped would magically appear).

However, you've seen that you can create patterns from dashed stroked paths. You can also create Pattern brushes, which expand into shapes. In the final Illustration, you will create a Pattern brush out of a dashed path. This technique is excellent for the combined dash patterns that you have made (redundant for a simple dash sequence because you can directly stroke its path with dashes in the Stroke palette). Combined dash patterns look quite well when placed on printer's ornaments and curlicues (you have several examples in this book's CD-ROM clip art sample from Ultimate Symbol). Figure 8.57 shows an example that I created using flourishes and a circle design from Ultimate Symbol's Design Elements collection. I stroked all the flourishes with a Pattern brush made from a dashed stroke.

Figure 8.57
Printer's flourish and ornament design, using clip art from Ultimate Symbol's Design Elements collection.

In Illustration 8.5, you too can play with a Pattern brush. To save you the hours it took me to create it (and I didn't do nearly the job the monks in Ireland did by hand in the Middle Ages), you'll use a Celtic knotwork image as your starting point, along with a dash pattern you created for the previous cityscape image.

![PROJECT] Illustration 8.5: Tying Knots

1. Open the image CELTIC.AI from the CD-ROM that accompanies this book. Figure 8.58 shows the starting image. The image contains two layers—a green continuous knot and a black knot that has been cut apart to show the weaving of the continuous line. (This allows those of you who want to try it the opportunity to see whether you can do a better job of weaving the ribbon than I did—and I'm sure that you can because I had limited time in which to create this image!)

Figure 8.58
The original Celtic knot image.

2. Open the DASHPATTERN.AI image from the CD-ROM. Select the entire image (Mac: Command+A; Windows: Ctrl+A). Copy it to the clipboard (Mac: Command+C; Windows: Ctrl+C). Close the DASHPATTERN.AI image and make the CELTIC.AI image active. Paste in the selection from the clipboard (Mac: Command-V; Windows: Ctrl+V). The dash pattern shown in Figure 8.59 is the I-beam pattern that you have used before.

Figure 8.59
The I-beam dash pattern, enlarged.

3. Drag the dash pattern onto the Brushes palette and select the New Pattern Brush radio button in the resulting dialog box. In the New Pattern Brush dialog box that appears, name the pattern "I-Beam Dash" and change the Colorization method to Tints And Shades, as shown in Figure 8.60.

4. Hide the layer that contains the green continuous knot. Select the broken knot. Click on the I-Beam Dash brush in the Brushes palette. Change the stroke color to dark green. Figure 8.61 shows the result.

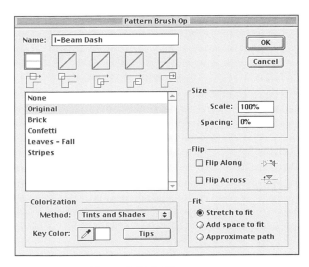

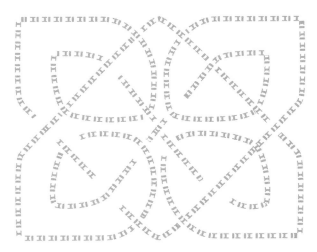

Figure 8.60 (left)

Setting the options for the Pattern brush.

Figure 8.61 (right)

Stroke the broken knot paths with the Pattern brush.

5. Click on the Eye icon next to the Continuous Knot layer to turn it back on. Notice that you can see the green of that layer right through the brushed path. Therefore, the Pattern brush has areas that are transparent because of the gaps in the original dashed line. Transparent areas are fine, unless they annoy you. They annoyed me as I was doing this example, so you are going to add an opaque background to the stroke and resave the Pattern brush.

6. Deselect everything in the image. Hide the Continuous Knot layer again. Choose a stroke of None and a fill of dark blue. Select the Rectangle tool. Draw a rectangle over the dashed line that you pasted onto the pasteboard area so that it's as close as possible in size to the dashed line. Choose Object|Arrange|Send To Back. Figure 8.62 shows the new stroke pattern.

Figure 8.62

When you place an opaque object behind a dashed line, the entire line becomes an opaque pattern.

7. Select the entire stroke pattern and drag it onto the Brushes palette. Make it a Pattern brush and give it the same options as you did in Step 3. Name the brush "Dash Pattern 2". Select the Broken Knot and click on Dash Pattern 2 to apply it. Figure 8.63 shows the result. Were you to reveal the Continuous Knot layer now, it would show through much less than it did before (you don't need to do this, however).

8. If you look carefully at the outer corners of the knot, you notice gaps in the line. The Pattern brush allows you to create special tiles for the corners of a line. Let's do that now. Select the entire dash pattern

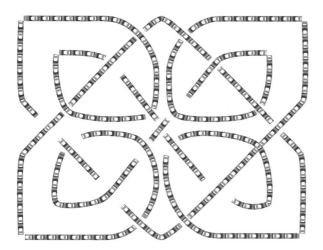

Figure 8.63
The broken knot with an opaque Pattern brush.

that is still on the pasteboard. Press Option (Mac) or Alt (Windows) and drag it to make a duplicate. Work in the duplicate. Ungroup it (Mac: Shift+Command+G; Windows: Shift+Crtl+G), then deselect.

9. With the Selection tool, click on the dashed line in the center of the line pattern. You will know that you have selected the desired piece if the fill color in the Toolbox is set to None and the stroke to white. You should be able to see the dashed-line pattern in the Stroke palette. Choose the Scissors tool and click on the path line about the same distance from the left edge as the line is high (the idea is to end up with a mostly square chunk that you can use as a corner). Delete the selected chunk of line. Figure 8.64 shows the line pattern after the first piece has been removed.

Figure 8.64
The small chunk to the left of the line is all that will be left when the corner tile is complete.

10. Select the medium-blue dashed line by clicking at the center of the line where there is no white. Use the Scissors tool and click in the same spot as you did in Step 9. Delete the selected portion to the right. Only the deep-blue rectangle is left at the original length.

11. Select the deep-blue rectangle and drag its bounding box handle on the right side toward the left until it is a tiny bit larger than the blue and white stripe. Select all three pieces and group them.

12. Press the Option key (Mac) or Alt key (Windows) after you start to drag the square tile to the Brushes palette. Drag the tile and drop it into the first division on the Brushes palette on the entry that contains the Dash Pattern 2 brush. The arrow in Figure 8.65 shows the

Figure 8.65
Drag the corner tile into the first slot on the Brushes palette in the Dash Pattern 2 brush.

square on which to drop the tile. The square is outlined in black if you drag it with the Option/Alt key pressed. If you drop it in the wrong place, you will not get the expected results.

13. Figure 8.66 shows the Pattern Brush Options dialog box with the new tile placed in slot 2 in the dialog box (this is the outside corner slot, and the confusing thing is that it is at a different position than the slot onto which you dragged the tile). Click on OK. Click on Apply To Strokes when that dialog box appears.

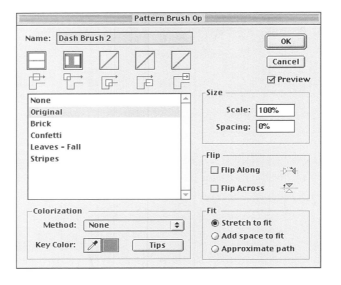

Figure 8.66

The Pattern Brush Options dialog box shows you whether you dragged the tile to the right spot.

Figure 8.67

Drag the same corner tile to the Inner Corner tile location in slot 4.

14. Repeats Steps 12 and 13. This time, however, Option/Alt drag the corner tile to the fourth division (out of six) on the Brushes palette. You need to watch this carefully. The pattern itself occupies two slots. Figure 8.67 shows the correct place to drop the tile. Again, select Apply To Strokes. All your corners now have tiles in them.

15. Now you can have some fun with this. Select the pattern-brushed knot, press Option or Alt, and drag a copy of the knot to the pasteboard. Work on the copy on the pasteboard.

16. Select the pattern-brushed knot on the pasteboard and make the Transform palette active. Type ".773" for both the height and the width of the knot. Press Enter (Windows) or Return (Mac) to accept the values. The knot becomes tiny, and the pattern is all mixed up. Figure 8.68 shows an enlargement of this tiny image.

17. With the tiny knot still selected, click on the second icon from the left at the bottom of the Brushes palette. This allows you to customize the options for a selected item. Change the scale to 35%, as shown in

Figure 8.68
The knot has a pattern that is too big for it and is all tangled up.

Figure 8.69, then click on OK. Figure 8.70 shows the knot with the scaled pattern.

18. Select the tiny knot and Option/Alt drag to make a duplicate of it. Choose Object|Expand and expand the object, its stroke, and fill. Group it. (You cannot create a Pattern brush from something that contains a Pattern brush. Expanding the object makes all its attributes "real" and removes the brush pattern by re-creating it in permanent objects.)

19. Select the tiny expanded knot group and drag it onto the Brushes palette to create a new Pattern brush. Name it "Knot Pattern". Set

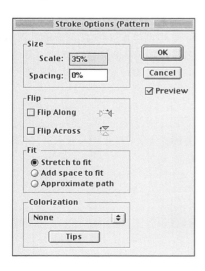

Figure 8.69 (left)
You can set specific options for each item to which you apply a custom brush.

Figure 8.70 (right)
The pattern fits much better now on the tiny knot.

the Colorization to Tints And Shades. Click on OK. Press the Alt/Option key and drag the tiny expanded knot group back on the Knot Pattern entry in the Brushes palette in slot 1 (to make an outside corner). Click on OK. Press the Option/Alt key and drag the tiny expanded knot group back on the Knot Pattern entry in the Brushes palette in slot 4 (to make an inside corner).

20. Select the large knot in the center of your image. Click on the Knot Pattern brush to apply it to the large knot. Click on the second icon on the left at the bottom of the Brushes palette to open the options for the selected object. Set the Scale to 75%. You might need to do some cleanup. If there are stray points or lines in the image that show up as blobs of dark color, you can use the Direct Selection tool to find and delete them. You might also want to space the breaks in the knot differently. Edit the shape as you see fit. Figure 8.71 shows the image at this point (after some cleanup has been done).

Figure 8.71
The knot now contains many tiny knots as the pattern.

21. Unfortunately, although you made the original brush pattern fully opaque, in this knotted version, the areas where the knot is not are not opaque (sorry about that). To rectify that, select the large center knot (your image area). Double-click on the Scale tool and select a uniform scale of 100%. Click on Copy. This places a copy of the knot directly on top of the first one.

22. Change the stroke to None to remove the brush (or click on the first icon on the left at the bottom of the Brushes palette). Select white as the stroke color and change the stroke width to 45 points. Choose Object|Arrange|Send To Back.

23. With the white knot still selected, repeat Step 21. Change the stroke color to black. Choose Object|Arrange|Send To Back. Do not deselect. Press the Right arrow key twice times and the Down arrow key three times to move the black shadow. Deselect.

24. Select the top-most large center knot in the image area. Choose a medium green for the stroke. If the image doesn't change, open the selected item options and set the Colorization to Tints And Shades.

25. Double-click on the Scale tool and select a uniform scale of 100%. Click on Copy. This places a copy of the knot directly on top of the first one. Click on the Dash Pattern 2 brush to change the brush stroke back to the thinner brush. Click on the second icon from the left at the bottom of the Brushes palette and set the scale for the pattern to 100%. Change the Colorization to Tints And Shades and click on OK. Choose a gold for the stroke.

26. Double-click on the Scale tool and select a uniform scale of 60%. Click on Copy. Click on the second icon from the left at the bottom of the Brushes palette and set the scale for the pattern to 90%. Change the Colorization to None and click on OK. Figure 8.72 shows the finished image (which looks better in the Color Studio).

Figure 8.72
The finished image is now all knotted up.

Moving On

You should now be an expert in making patterns from dashed and dotted lines. You have seen how to use dash-and-gap patterns to create decorative lines such as train tracks. You have also used a variety of dashed lines to make a coupon and a group of textures. You have learned how to create patterns and Pattern brushes from dashed lines and learned about some of the commands (such as Expand) that might not work as expected on dashed open strokes.

In Chapter 9, you will learn about some of the other stroke brushes: the Art and Scatter brushes and the Calligraphic brush. You will have another opportunity to use recursion (making something from itself the way that you used the knot to create a pattern tile for the knot).

DIFFERENT STROKES 9

SHERRY LONDON

Illustrator 8 sports some unique brush tricks that allow you to create calligraphic strokes, paint with an object or scatter it along a path, make pattern brushes, and use dots and dashes along the stroked edge of an object.

Brush Up Your Brush Skills

Adobe packed a lot of surprises in the Illustrator 8 upgrade. Among the more lighthearted new features are a bevy of brush types that add new meaning to the "Different Strokes" of this chapter title. Although Meta-Creations might be a bit peeved that Adobe added some of the brush features that make Expression unique, copying is still one of the most sincere forms of appreciation. The Art brush, modeled after the one in Expression (though not containing as many features), allows you to draw a stroke with an object. The object follows the shape of the stroke, which allows you to create left- and right-facing leaves or shapes that undulate and curl.

You'll explore a variety of brush types in this chapter, which builds on many of the exercises in Chapter 8. Let's start by comparing the four types of brushes.

A Quartet Of Brush Types

When you select the Paintbrush tool in the toolbox, you can apply any of the four types of brushes to be found in the Brushes palette. Figure 9.1 shows all the Brush Libraries that are included with Illustrator 8. However, only the Default Brushes palette contains all four of the brush types. Figure 9.2 shows a closer look at the Default Brushes palette. As you can see, it is divided into four sections.

The Calligraphic brushes are at the top. They are easily identified because they all are some variation on a black circle—round, oval, squished, and/or rotated. You cannot create your own calligraphic shape from scratch, but you can modify a circle into the nib shape of choice. Calligraphic brushes allow you to achieve thick and thin strokes. This brush type is excellent for lettering or for oriental brushwork techniques. You can add more control to the calligraphic brush stroke through the use of a pressure-sensitive tablet (such as a Wacom tablet) as you draw. Figure 9.3 shows an example of an image that was stroked with a variety of Calligraphic brushes. Although the Calligraphic brush is new to Illustrator 8, it replaces the older Calligraphic pen, which created filled shapes. The brush adds thick-and-thin to the individual strokes—something that was not possible before.

The Scatter brushes are next. You can recognize them in the Brushes palette because each brush fits into a small square. Scatter brushes take a single design element (actually, whatever objects you drag to the Brushes palette) and repeat them along a path. Unlike the Pattern brushes, however, you

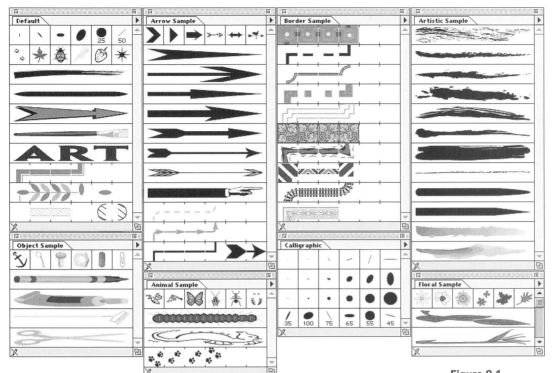

Figure 9.1

Illustrator 8 contains a variety of premade brushes that you can use.

can have these elements disperse themselves about in an orderly or random fashion. Figure 9.4 shows an example using Scatter brushes.

The Art brush stretches an object along a path. Each Art brush occupies one entire row of the Brushes palette. They are the third set of brushes in the palette. The Art brush looks best when it uses a shape that can logically be stretched, such as a combined, decorative stroke like the one you used behind the coupon example in Chapter 8, or an element, such as a

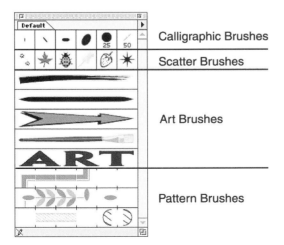

Figure 9.2

The Default Brushes palette contains four types of brushes: Calligraphic, Scatter, Art, and Pattern.

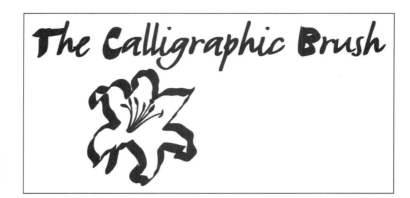

Figure 9.3
Calligraphic brushes are excellent for lettering or oriental brushwork.

Figure 9.4
Scatter brushes place image elements along a path in either a random or an orderly fashion.

dachshund that is long and stretchy anyway. You can stack individual instances of an Art brush in sequence to create animations. Figure 9.5 shows some examples using Art brushes.

You have already met the Pattern brush in Chapters 5 and 8. A Pattern brush repeats the element along a path. This is the Illustrator 8 replacement for the Illustrator 7 Path Pattern. The new implementation makes it even easier to use. As you have seen, you can create a pattern for the straight areas of stroke and special tiles for inner and outer corners and for the starting and ending tiles of a stroke. This allows you to create excellent borders as well as apply patterns along a path. Figure 9.6 shows some examples of Pattern brushes on hand-drawn text and a border.

Penmanship

Although the Brushes palette has four types of brushes, the Calligraphic brush is not quite in the same category as the others. It is the only brush type that doesn't permit you to drag an object in the palette to create a new brush. To make a new Calligraphic brush, you need to click on the

Figure 9.5
Art brushes stretch elements along a path.

Figure 9.6
The Pattern brush repeats the element along the path.

New Brush icon near the bottom right of the Brushes palette or choose New Brush from the sidebar menu on the Brushes palette.

In concept, the brush is very easy to use. (It's even easy to use in practice; it's just a bit harder to get decent results.) Figure 9.7 shows the process of designing a new Calligraphic brush. Once you click on New Brush and choose a Calligraphic brush, you simply fill in the parameters that you want your brush to have. You can set the angle interactively just by dragging the arrow on the picture of the brush. You can also set the roundness of the brush interactively, but it is not obvious how you do it. If you look carefully at the brush diagram, you will see two small dots at the perimeter of the brush across the center. By moving those dots closer together or farther apart, you can set the roundness of your nib. You can also set a diameter for the brush and specify whether it produces a fixed stroke, a random stroke, or a pressure-controlled stroke (if you have a pressure-sensitive tablet). Your best results with the brush are from pressure-controlled settings, as long as you intend to draw your strokes individually.

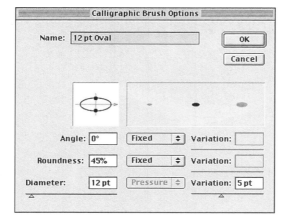

Figure 9.7
The Calligraphic Brush Options dialog box lets you set the performance of the brush.

Illustration 9.1 combines the uses of text and clip art with the Calligraphic brush. Along the way, you will learn some of the pitfalls and techniques for working with this new feature.

 Illustration 9.1: The Write Stuff

1. Open the image SUN.EPS. This is an image from Ultimate Symbol's Design Elements sampler (though it is in the Chapter 9 Start folder on your CD-ROM). Figure 9.8 shows the original image.

2. Turn on the Rulers (Mac: Command+R; Windows: Ctrl+R). Drag the zero point of the rulers to the upper-left corner of your image. Set the Units preference to "inches" if it isn't set up that way now. Drag a vertical guide from the left margin to the 4 1/8-inch tick mark on the horizontal ruler. Drag a horizontal guide from the top ruler to the 5 1/2-inch tick mark on the vertical ruler. This marks the center of the image. Lock the guides (View|Lock Guides).

3. Select the entire image (Mac: Command+A; Windows: Ctrl+A). In the Transform palette, enter both a width and a height of 3 inches for the sun. Press Enter/Return to execute the transform. Drag the

Figure 9.8
Sun image from Ultimate Symbol's Design Elements collection.

sun image so that it's centered on the intersection of the two guides. (The easiest way is to drag it by eye near the center of the image and then, with the bounding box showing, use the arrow keys to nudge it until the center control handles on the bounding box are on the guide lines.)

4. Ungroup (Mac: Shift+Command+G; Windows: Shift+Ctrl+G).

5. Carefully select the sun's face (but none of the rays). Create a new layer and transfer the selection rectangle on the Layers palette to the new layer. Turn off the Eye icon next to the new layer. The sun's features should disappear as shown in Figure 9.9.

6. You are going to use the Calligraphic brush to redraw the sun's rays. Currently, the sun's rays are filled shapes. If you change them into outlines, the Calligraphic brush strokes only the edges. You need to actually redraw each ray as a single stroke. To do this, you need to take some preparatory steps first. Change the fill color for the sun's rays to 50% black (use the Grayscale slider in the Color palette). Double-click on Layer 1 to open the Layer Options dialog box. Select the Template checkbox to make this a layer both nonwriteable and nonprintable.

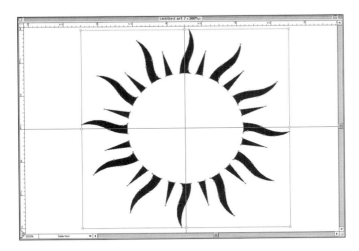

Figure 9.9
The sun's features disappear when you hide the layer to which you have moved them.

7. Create a new layer. Select a Fill of None and a Stroke of black.

8. In the Brushes palette, choose New Brush from the sidebar menu. Accept the Calligraphic brush default. Enter the settings shown in Figure 9.10 for the new brush.

9. Select the Paintbrush tool and the new brush. Carefully paint over all the sun's rays. Draw only one stroke per ray, from tip to just inside the

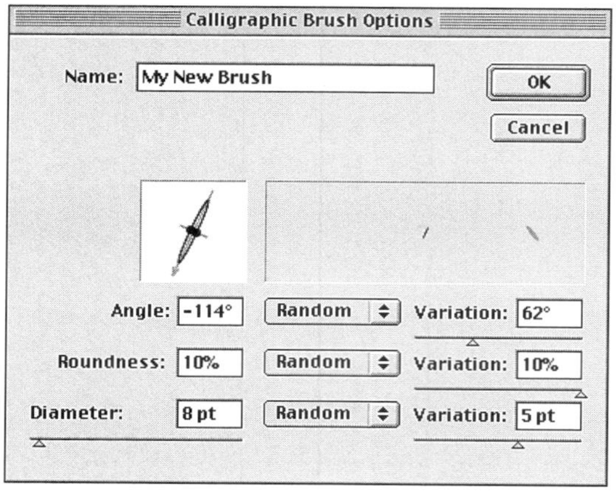

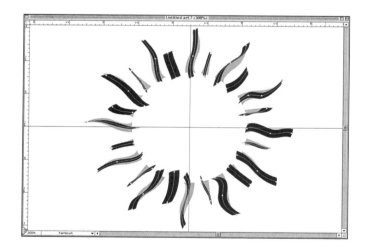

Figure 9.10
Settings for the Calligraphic brush that you will use to redraw the sun's rays.

empty circle (you may stroke from the circle out as well). A built-in randomness in the brush will give you results differ somewhat from mine, shown in Figure 9.11. The rays are selected in the image so that you can see their composition.

Figure 9.11
The sun's rays redrawn as single strokes.

10. Select the entire image (Mac: Command+A; Windows: Ctrl+A), although only the rays are selected. Try out several of the other Calligraphic brushes on the image (simply click on them to apply them to the strokes). If you like them better, leave them. If not, either click on the brush that you created (which will generate a completely different set of strokes) or undo back to your own brush strokes.

11. You now need to make a mask to get rid of the points of the strokes showing in the center of the sun's face. Make a new layer. Layer 4 should appear just above Layer 3.

12. Select the rectangle tool. Draw a rectangle that encloses the sun's rays. The actual size is not critical so long as it is larger than the sun.

13. Select the Ellipse tool. Place your cursor at the intersection of the guides (i.e., the center of the image). Press Option (Mac) or Alt (Windows) to draw the circle from the center. Press Shift as you start to draw. This makes a true circle. Draw a circle that is the same size as the circle around the sun's face (until the circle reaches the start of the gray sun's rays on the template layer).

14. Select both the circle and the rectangle. In the top row of the Pathfinder palette, click on Minus Front. This punches a hole out of the rectangle and creates the needed mask shape (you want the image to show the sun's rays).

15. Select the entire image (Mac: Command+A; Windows: Ctrl+A). Make a mask (Object|Masks|Make; Mac: Command+7; Windows: Ctrl+7). Figure 9.12 shows the ends removed from the new sun ray strokes.

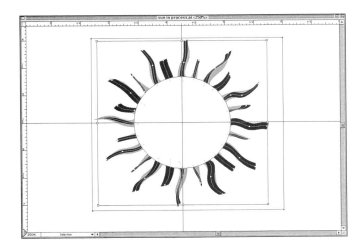

Figure 9.12
A mask makes the edges of the strokes look clean and sharp and shows the original circle.

16. Hide the Template layer and reveal Layer 2 (the sun's face). Figure 9.13 shows the sun image with calligraphic rays.

17. Make a new layer and drag it to the top of the Layer stack. Open the image FLORISH.EPS. This is a copy of a clip art image from Ultimate Symbol's Design Elements collection. Redraw it using the Pen tool so that instead of filled shapes it is a single stroke. Figure 9.14 shows the flourish.

18. Select the entire flourish and either copy it to the clipboard and paste it into the sun image or drag and drop it into the image. In either case, when you're finished with this step, the sun image should be active, and a copy of the flourish should be in the sun image.

DO YOU NEED PRACTICE USING THE PEN TOOL?

The image FLOURISH.EPS is a copy of the original image from Ultimate Symbol. If you want to spend the time, you can trace over the flourish using the Pen tool. I traced half the image and then duplicated and flipped the finished half. I averaged the points at the top of the flourish and at the bottom where the two halves meet (two different operations). I then joined both sets of points (bottom to bottom and top to top) to make one shape.

Figure 9.13 (left)
The sun now sports calligraphic rays.

Figure 9.14 (right)
The printer's flourish from Ultimate Symbol has been redrawn as a single stroked line.

19. Drag the flourish so that the center-top control handle on its bounding box is at the exact center of your image (you need to show the Template layer again because that is where your guides are located). Once the flourish is in position, choose Object|Transform|Move. Enter a vertical distance of 2.5 inches and a horizontal distance of 0. Click on OK. Figure 9.15 shows the result.

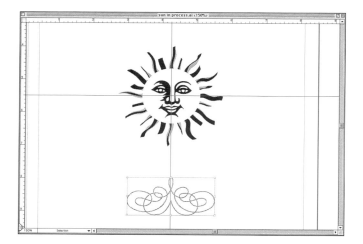

Figure 9.15
Move the flourish 2.5 inches down from the center of the image.

20. Choose the Rotate tool. Press Option (Mac) or Alt (Windows) and click at the exact center of the image. Enter a rotation amount of 30 degrees in the dialog box and click on Copy. Transform Again (Mac: Command+D; Windows: Ctrl+D) 10 times. This gives you 12 copies of the flourish as shown in Figure 9.16.

21. Drag the Flat 6 Pt brush from the upper left of the Default Brushes palette to the New Brush icon at the bottom of the Brushes palette to duplicate the brush. Double-click on the copy and change its angle to 78 degrees and the diameter to 2 points.

22. Lock Layers 2, 3, and 4. Select the entire image (Mac: Command+A; Windows: Ctrl+A). Click on the newly copied brush to apply it to the circle of flourishes. Figure 9.17 shows the brush's calligraphic effect.

Figure 9.16 (left)
Twelve copies of the printer's flourish frame the sun's face.

Figure 9.17 (right)
The tiny Calligraphic brush leaves thick and thin strokes.

23. Lock Layer 5. Create a new layer. Select the Ellipse tool. Place your cursor on the center of your image at the intersection of the guides. Press Option (Mac) or Alt (Windows) and click. Enter 3.5 inches as both the width and the height in the dialog box that appears. This circle is the basis for the text that you will create.

24. Select the Path Type tool. Click on the circle. Using Adobe Garamond Bold (if you have it or another serif font if you do not), enter "Put On Your" at 48 points. Choose the Selection tool and leave the type circle selected.

25. Double-click on the Scale tool. Choose a Uniform Scale of 100% and click on Copy. With the Path Type tool, edit this new text so that it reads "Sundae Clothes". Choose the Selection tool.

26. Grab the large I-beam in front of the words "Sundae Clothes" and drag the text to the bottom of the circle. As you drag, flip the words so that they read properly. If the Baseline setting on the Character palette is not visible, select Show Options from the sidebar menu in the Character palette. Set the Baseline Shift to –33 pt (or the amount needed to make your font move from the inside to the outside of the circle). Drag the text "Put On Your" and center it at the top of the circle.

27. Lock Layer 6 and unlock Layer 2 (the sun face layer). Select the entire image (Mac: Command+A; Windows: Ctrl+A). Choose the rotate tool and rotate the sun face slightly to the right (approximately 6.3 degrees). Figure 9.18 shows the image with the rotated sun face and the added text.

WHY ISN'T THIS AS EASY AS I THOUGHT IT WOULD BE?

Hand stroking text is really a pain—unless you love calligraphy anyway. Adobe has not made it any less painful by adding their wonderful new Pencil and Paintbrush editing behavior. When you draw near a selected shape, Illustrator assumes that you *want to change the shape of the selected object.* This actually is wonderful when you *do* want to edit an object. However, it is much less wonderful for freehand drawing. When you finish a stroke, the stroke stays selected. Therefore, if your next stroke is in the general vicinity, it is going to change your last stroke rather than create a new one (sigh). When you redrew the sun's rays, they were not close enough to one another to make Illustrator go into edit mode. However, unless you remember to press Command (Mac) or Ctrl (Windows) and deselect every object before you draw the next stroke, you are going to spend as much time cursing as you do drawing.

Drawing the strokes to re-create serif text is also time consuming. Your strokes will look better if you do not try to draw each letterform in one pass. Each serif and each down stroke should be done as a single unit. To draw the "P" in "Put," for example, uses three strokes: the serif at the foot of the "P", the stem of the "P", and bowl of the "P". Take your time on this.

28. Lock all the layers except Layer 6. Select the entire image (Mac: Command+A; Windows: Ctrl+A). Change the Fill color to 50% gray. Double-click on the Layer entry in the Layers palette and designate Layer 6 (the Text layer) as a Template layer. Create a new layer (Layer 7).

29. Set the Fill to None and the Stroke to black. Select the Paintbrush tool. You may use any calligraphic brush (or brushes) that you want. You might need to create one or more new ones. They should be fairly tiny because the text is small. Draw over each letter on the Template layer to re-create the text as combinations of single strokes. The original 6 pt Flat brush works quite well for this. Figure 9.19 shows the finished image.

What else can you do with this image when it is finished (besides use it in black and white)? In the Color Studio, you can see what I did with it. (Figure 9.20 shows this in less than glorious grayscale.) I opened the image in Photoshop. (If you have Photoshop 5.1, you should be able to open the image—even with its Calligraphic brushes; if you didn't bother to put in the minor upgrade to Photoshop 5 or are using an earlier version, you need to select everything in the finished sun face file and expand it. Then save it as an Illustrator file.)

SUNDAE.PSD, in the CH09ENDS folder, contains all the layers and channels that I used. I created an ice cream–colored background and used the KPT Pixel-Storm filter from the KPT 2.0 filter set to mix all the pixels together. I embossed a copy of the result and used Hard Light mode to make the texture in the paper. I then dragged the finished sun face image to the top and gave it a Gaussian Blur of about 2.5 (just to get some gray shades). I used the Levels command to make the area

for the sun face a bit larger. (If you move the Levels Input sliders one way, you expand your image area after you have applied a blur; in the other direction, you contract it. The direction depends on whether more dark or light is in your image.) I then embossed the result and very slightly (about 2.3 pixels) blurred the result. The final image appears to be softly stamped on watercolor paper. I used the Airbrush in Dissolve mode in a channel to make the selection to remove part of the hard edge on the image.

Figure 9.18 (left)
Text has been added and the sun face given a new direction.

Figure 9.19 (right)
All the text has been redrawn.

Figure 9.20
A "promo tag" for "Put On Your Sundae Clothes."

Before I leave the topic of Calligraphic brushes completely, I need to discuss the Expand command (again). This is an extremely useful command to use with Calligraphic brushes. When you created a calligraphic stroke in Illustrator 7, you had a filled shape with nodes that could be edited. With the Calligraphic brush, you do not. Although you can add points along the stroke, you cannot selectively alter the width of the calligraphic line. However, if you use the Expand command when you are done (and you *are sure* that you are done), you can tug on the shape of the line to make it anything you want. Figure 9.21 shows the Calligraphic brush-stroked letter on the left and the same letter expanded into filled shapes on the right. I edited the letter on the right to change the shape of the fill. You could not get the same letter shape using only strokes. Although I think that the stroked letter is more elegant, sometimes that is not the look that you are going for.

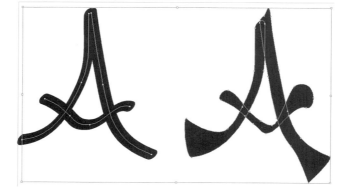

Figure 9.21
The stroked letterform is on the left, and the edited letter that has been expanded into a filled shape is on the right.

Seeing Stars

Now that we've thoroughly covered the Calligraphic brush, turn your attention to the three remaining brush types: the Pattern brush, the Art brush, and the Scatter brush. In Illustration 9.2, you will work a short example that shows you how the brushes differ from one another—even when defined with the same tile.

In this Illustration, you use a star as the pattern element and create a tile that can seamlessly repeat. For this reason, you will be given detailed instructions on dimensions, even though the specific dimensions are not critical to the success of the project (it is also easier for you to duplicate my work if I give you specific measurements). I am not enamoured of using guides when I draw, but they make communicating the expected results much more predictable—which is why I teach with them more often than I use them in practice.

PROJECT Illustration 9.2: One Tile—Three Brush Types

1. Create a new document in Illustrator.

2. Turn on the Rulers (Mac: Command+R; Windows: Ctrl+R). Drag a horizontal guide from the top ruler to the 8-inch mark on the vertical ruler. (I am assuming the default arrangement of your page where Illustrator has numbered the vertical ruler from zero at the bottom.) Drag another guide down to the 7-inch mark on the vertical ruler. Drag a vertical guide from the side ruler to the 2-inch mark on the top, horizontal ruler. Magnify your screen to 800% so that you can see the area where the three guides intersect. Unlock the guides if you need to move them now that you have zoomed in so closely (you'll usually notice that the guides were not placed exactly). Drag a third horizontal guide to the tick mark on the vertical ruler that is three small ticks above the 7.5-inch mark (technically, it is at 7 35/64). Figure 9.22 shows the layout of your document.

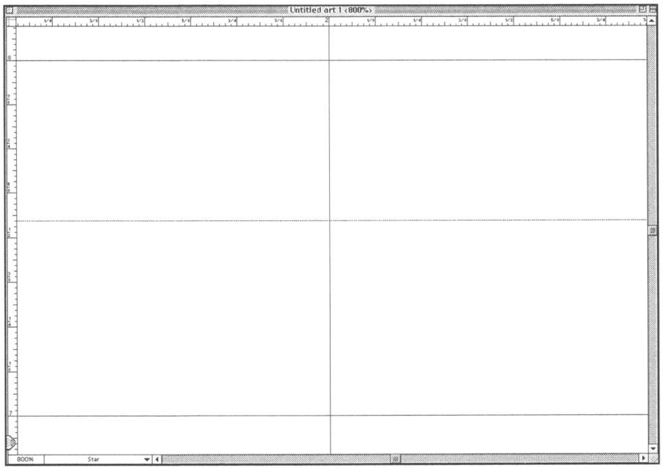

Figure 9.22

Preparing your image with three horizontal guides and one vertical guide.

3. Select the Star tool (it pops out of the same spot as the Ellipse tool). Choose a Stroke of None and a Fill of bright green. Place your cursor on the middle horizontal guide where it intersects the vertical guide. Press the mouse button and drag the cursor straight down to the bottom guide where it intersects the vertical guide. This creates a star about one-inch high that points directly down. Figure 9.23 shows an enlargement.

4. With the star selected, drag a vertical guide from the side ruler until it touches the control point marked with the arrow in Figure 9.24. (I

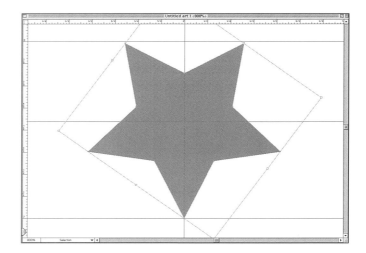

Figure 9.23
You have created a downward-pointing star.

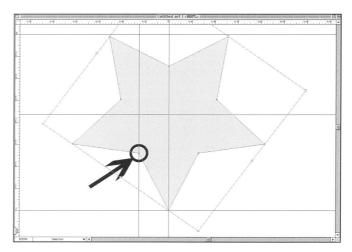

Figure 9.24
Drag a vertical guide to the control point marked by the arrow.

have lightened the color of the star so that you can see the guides and control points better.)

5. Choose the Selection tool. Press Option (Mac) or Alt (Windows) and drag a copy of the star to the right until its leftmost point touches the new guide that you placed in Step 4. Press Shift after you start to drag the star to constrain motion to the horizontal. Leave the new star selected. Fill the star with cyan. Figure 9.25 shows the result.

6. Zoom out to 300%. Transform Again (Mac: Command+D; Windows: Ctrl+D). A third star moves the same distance away. Fill it with light purple. Transform Again (Mac: Command+D; Windows: Ctrl+D). Fill the new star with pink. Transform Again (Mac: Command+D; Windows: Ctrl+D). Fill this star with yellow. Leave it selected. Figure 9.26 shows the five stars that you have created.

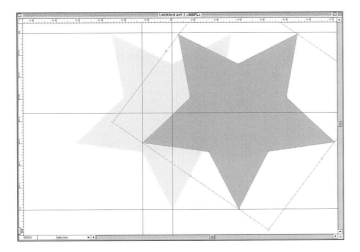

Figure 9.25
You now have two stars.

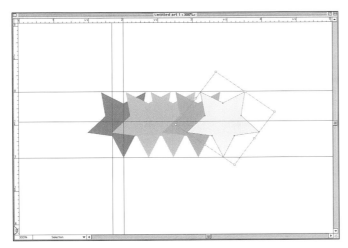

Figure 9.26
You have made a five-star lineup.

7. The tile that you are creating is to be seamless. Therefore, you will
   need to repeat the five-star sequence to obtain one complete repeat
   (because the design is overlapping, as you learned in Chapter 5).
   Therefore, Transform Again (Mac: Command+D; Windows:
   Ctrl+D). Use the Eyedropper to set the fill to the *same* green as you
   used for the first star. To be totally safe and sure of a good repeat,
   Transform Again (Mac: Command+D; Windows: Ctrl+D) and fill
   the final star with the same cyan used in the second star. You now
   have seven stars.

8. Select all the stars, press Option (Mac) or Alt (Windows), and drag a
   duplicate of these stars for safekeeping to the pasteboard area of
   your image. Lock the guides if they want to come along for the ride
   when you drag and copy the stars. Deselect the stars after you make
   the copy and return to the 300% view of your pattern-in-process.

9. Drag a vertical guide from the left ruler until it intersects the control point at the bottom of the rightmost *green* star (that's one in from the last star on the right). Figure 9.27 shows the position of the guide.

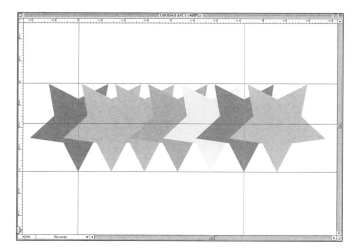

Figure 9.27
Place another vertical guide.

10. Unlock the guides (if they are locked). Drag the leftmost vertical guide off the image. (This leaves the two vertical guides that intersect the bottom points of the stars.)

11. Select the Rectangle tool. Make sure that Snap To Points is selected in the View menu. Draw a rectangle between the two vertical guides and the 7- and 8-inch marks (the top and bottom horizontal guides). Figure 9.28 shows the drawn rectangle.

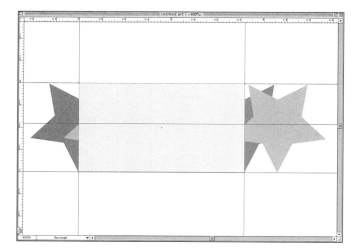

Figure 9.28
This rectangle sets the size for
the finished pattern tile.

12. Select the stars and the rectangle in the image area. Click on Crop in the Pathfinder palette. The stars are trimmed to an exact repeat size.

Group the stars (Mac: Command+G; Windows: Ctrl+G). Figure 9.29 shows the cropped stars.

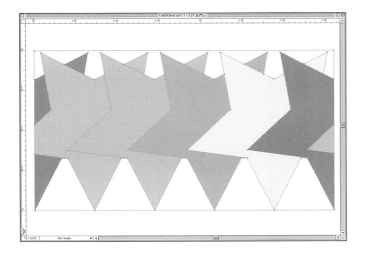

Figure 9.29
The Crop command uses the rectangle to trim the stars to an exact repeat.

13. Because you have already created Pattern brushes, let's start there. Drag the grouped stars into the Brushes palette. Define a Pattern brush (let's not worry about corner tiles right now). Accept all the default settings. Click on OK. Deselect the grouped stars (clicking on the stars a second time applies the new brush to *them*—which is not what you want to do).

14. Select and drag the grouped stars into the Brushes palette again. This time, select the Art Brush radio button. Again, accept the defaults as shown in Figure 9.30.

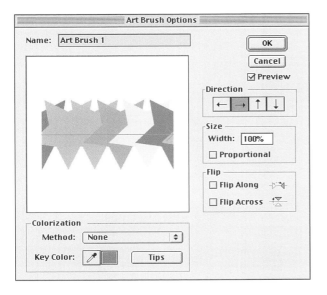

Figure 9.30
The New Art Brush Options dialog box.

15. Deselect the grouped stars. Select and drag the grouped stars into the Brushes palette. Select the Scatter brush. Accept the defaults as shown in Figure 9.31. Deselect.

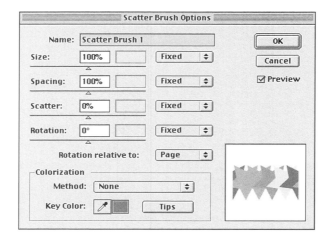

Figure 9.31
The New Scattter Brush Options dialog box.

16. Drag the original cropped tile to the pasteboard area of your image. Clear the guides (View|Clear Guides). In the image area, draw as large a star as will fit. Set the Fill to None. Click on the stars Pattern brush to apply it. Figure 9.32 shows the result. Because there isn't a corner tile, the points of the star are empty.

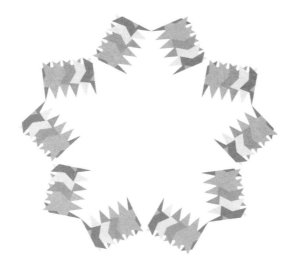

Figure 9.32
The star pattern decorates the large star.

17. Let's fix that. Select the yellow star in the seven-star pattern that you copied prior to cropping the stars. Drag it (with Option/Alt pressed *after* you start to drag) to the first box—the outer corner box— in the stars Pattern brush that you created. Accept the defaults in the Brushes dialog box and click on Apply To Strokes. Then select the

pink star and follow the same procedure, except that you need to drag it to the fourth box in the Brushes palette entry for the stars Pattern brush. Again click on Apply To Strokes. Figure 9.33 shows that the star now has a whole star at each point.

18. The pattern is still too big for the size of the star. Double-click on the star Pattern brush and change the Size to 60%. Figure 9.34 shows the result. If you want a smaller star for the corner, you can scale a copy of the individual yellow star (on the pasteboard) to 60% and replace the corner star by pressing Option (Mac) or Alt (Windows) and dragging the scaled yellow star into slot 1 of the star Pattern brush. Figure 9.35 shows this effect.

Figure 9.33 (left)
The single stars fill the corners of the large star.

Figure 9.34 (right)
Scale the star pattern to 60% to better fit the proportions of the object to which it is applied.

19. With the large star still selected, click on the star Scatter brush entry near the top of the Brushes palette. The image changes to look like Figure 9.36. This is not very attractive and does not show that it was applied to a star.

20. Double-click on the star Scatter brush entry to open the Brush Options dialog box. Set the Size to 30% (or 31% as it shows up in my screen capture in Figure 9.37). Set the Spacing to 30%. Click on OK and Apply To Strokes. Figure 9.38 shows the changed star. Now you can see that the image elements are applied to a star shape.

21. Open the Scatter Brush Options dialog box for the star again. This time play with all the controls until you are sure that you know how

Figure 9.35 (left)
The corner element is scaled to 60% in Illustrator and replaces the original corner star in the Pattern brush.

Figure 9.36 (right)
The Scatter brush scatters the stars in such a way as to completely obscure the star-shaped path.

each control affects the final result. Switch the Rotation Relative To setting from Page to Path. Figure 9.39 shows some changes in the settings from Figure 9.37, and Figure 9.40 shows the result. The star sets march around the larger star as if they were band members at a half-time maneuver. Explore some other settings. Figures 9.41 to 9.44 show two other sets of settings and their matching stars.

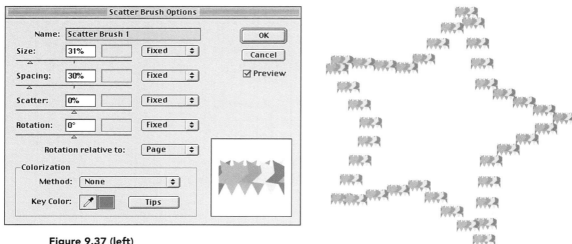

Figure 9.37 (left)
Change the Scale and Spacing in the Scatter Brush Options dialog box.

Figure 9.38 (right)
When you change the brush options, the result is much more attractive.

22. With the large star shape selected, click on the star Art brush that you created to apply the brush to the large star. Figure 9.45 shows the somewhat less than wonderful result. The single tile of five stars is stretched and elongated over the length of the star path.

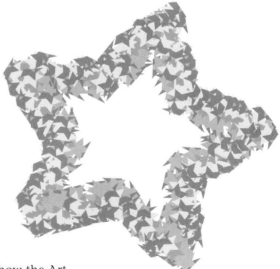

23. You can change the look of the Art brush by adjusting its Brush options. In Figure 9.46, I changed the direction in which the tile is stretched and applied. I also changed the Scale to 40%. If you click on Proportional, the image tile keeps its aspect ratio as it stretches around the shape. Except for short strokes, this is rarely practical. Figure 9.47 shows the result of the settings changes. Experiment with the settings until you find some that you like.

Figure 9.39 (left)
Change the angle of Rotation and the Rotation Relative To settings in the Scatter Brush Options dialog box.

Figure 9.40 (right)
The tiles march around the star.

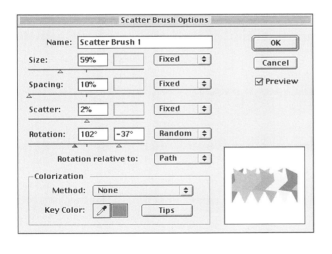

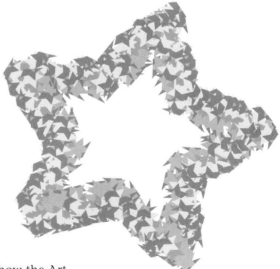

24. You can also split large strokes to take control of how the Art brush is applied. Select the Scissors tool. With the large star selected, click on every control point to split the star into 10 separate paths. With all the segments selected, set the Width in the Brush Options dialog box to 40% and the Direction to Stroke From Left To Right. (The tile also looks good Stroked From Bottom To Top.) Figure 9.48 shows the results.

Figure 9.41 (left)
Set some random parameters in the Scatter Brush Options dialog box.

Figure 9.42 (right)
With random rotation and a tight spacing, the star looks stamped with a linoleum block.

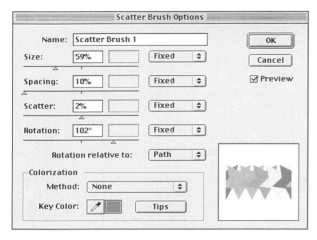

Figure 9.43 (left)
Set a fixed angle of Rotation in the previous settings in the Scatter Brush Options dialog box.

Figure 9.44 (right)
A regular pattern emerges that could not easily be achieved with the Pattern brush.

25. What else can you do with this brush and the star shape? So that you can get more wiggle for your stroke, you can apply the Roughen filter (Filter|Distort|Roughen). Try it. Select all the segments. Use the filter settings shown in Figure 9.49. By using relatively few new smooth

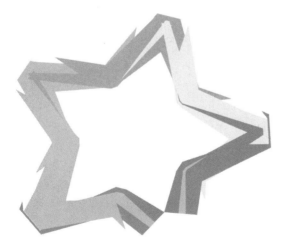

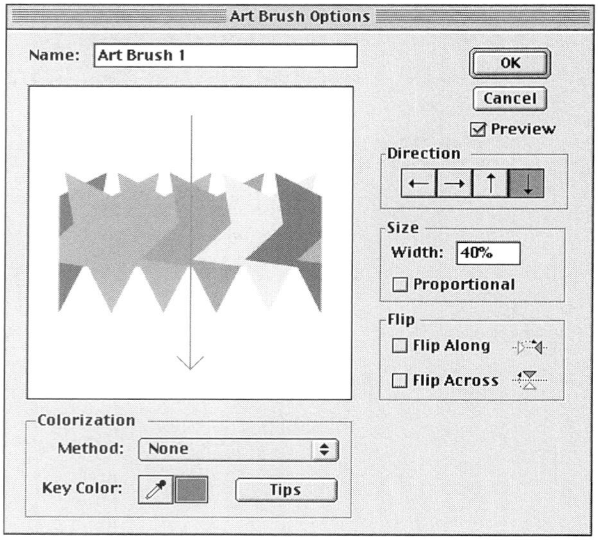

Figure 9.45 (left)
The default settings on the Art brush aren't the best look for the star shape.

Figure 9.46 (right)
You can change the direction in which the tile is applied.

points but a significant amount of change, you can create interesting waves in the star shape that are automatically followed by the Art brush. Figure 9.50 shows the result. (I also could not resist showing you the same image Stroked From Bottom To Top. See Figure 9.51.)

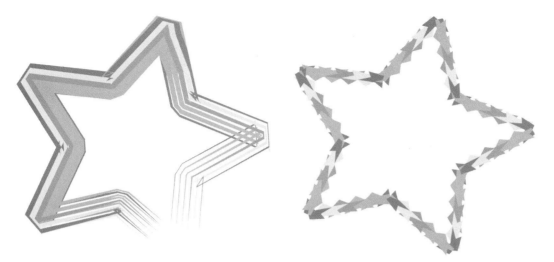

If you now had to decide when to use which brush type, you need a clearer idea of the strengths and weakness and the possibilities of each. Here are some very general guidelines:

- If you want to apply a uniform tile that repeats end to end without variation, use the Pattern brush. The length of the stroke is not a factor.

- If you want to randomize where the elements are applied or change the angle of their application, use the Scatter brush. The length of the stroke does not matter here, either.

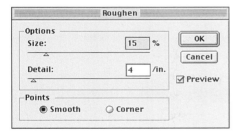

Figure 9.47 (left)
Changing the apply direction and the scale results in a totally different design.

Figure 9.48 (right)
Splitting the star into segments reduces the stretch on the image tile.

Figure 9.49
Use these settings for the Roughen filter.

- If you want to apply a single element along an entire stroke, use the Art brush. If the element is pictorial in nature, it will remain recognizable only if the path is fairly short. However, if the design element is abstract, the path length becomes much less significant.

- If your path contains many corner points that are a significant element in the design, a Pattern brush is suitable only if you also design corner elements for it.

- If the start and endpoints of a stroke need to be rounded or contain a special element, you can do that only with a Pattern brush. The Scatter

Figure 9.50 (left)

The Roughen filter adds interest to the strokes that use the Art brush.

Figure 9.51 (right)

This is the same image as Figure 9.50 but is Stroked From Bottom To Top.

brush makes no accommodation whatsoever, and the Art brush can change only the start and end of a stroke if you define the brush shape that way from the start (such as an arrow).

A Painterly Stroke

While I was writing this book, MetaCreations released Painter 5.5, an upgrade to the version of Painter that they placed on the market in late 1997. Among the new things in this upgrade is a series of wonderful new brushes that are designed to be used by Web artists. The brushes are optimized for Web use because they are not anti-aliased. In other words, they have hard edges that paint in a solid color. Although the edges produce a single color that does not fade to the background color, the shape of the edges differs with each brush and is quite interesting. I thought that it would be fun to use some of these new Painter brushes in Illustration 9.3 to create strokes that can be applied with Illustrator's Art brush. For those of you who do not have Painter (or this version of Painter), I have included a Photoshop document that contains sample brush strokes. (If you don't have Photoshop either, you can start with the Illustrator path documents created in Photoshop.)

PROJECT Illustration 9.3: Abstract Art Strokes

1. Open the image STROKES2.PSD in Photoshop. If you don't have Photoshop, skip to Step 7. Figure 9.52 shows the original image of several strokes created using some of Painter's new Web brushes.

2. Although several strokes are available for you to try, we will start with the thick stroke that is second from the bottom. Select the rectangular marquee tool. Draw a marquee around the stroke. Figure 9.53 shows the selection.

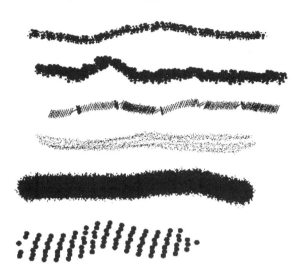

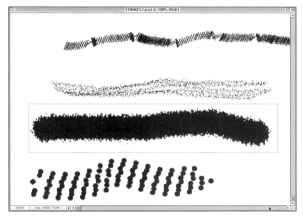

Figure 9.52 (left)
Web brush strokes created in MetaCreation's Painter.

Figure 9.53 (right)
Select the stroke that is second from the bottom.

3. Copy the selection to the clipboard (Mac: Command+C; Windows: Ctrl+C). Create a new document (File|New). Accept the defaults. Paste in the selection from the clipboard (Mac: Command+V; Windows: Ctrl+V).

4. Choose Image|Image Size. Select the Constrain Proportions checkbox. Select the Resample Image checkbox. Change the Size to 800%. Change the Interpolation Method to Nearest Neighbor. This allows you to keep the image aliased and sharp while you resize it large enough to capture an adequate path (which is why you copied the original stroke to a new image).

5. Double-click on the Magic Wand tool to open the Options palette. Set the Tolerance to 0 and deselect the Anti-Alias checkbox. Click on a black pixel in the new image. Choose Select|Similar. All the black pixels in the image are selected.

6. Make the Paths palette active. Choose Make Work Path from the sidebar menu on the Paths palette. Select a Tolerance of 1, as shown in Figure 9.54. That draws a tight path around the selection. Double-click on the Work Path entry in the Paths palette to save it. You may use the default name of Path 1.

7. Choose File|Export|Paths To Illustrator. Save the file as BIGSTROKE.AI.

8. In Illustrator, open the file BIGSTROKE.AI (it is also on your CD-ROM). Because it is a Photoshop export file, it looks empty. Select the entire

Figure 9.54

Enter a Tolerance of 1 to draw a tight path around a selection.

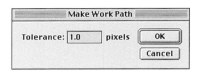

image (Mac: Command+A; Windows: Ctrl+A). Change the Fill to black. Choose Object|Crop Marks|Release and delete the empty rectangle around the large filled stroke object.

9. The stroke is gigantic. Select the entire image (Mac: Command+A; Windows: Ctrl+A). Double-click on the Scale tool and enter a Uniform Scale of 20%. Click on OK.

10. Choose the Selection tool. Group all the objects in the image (Mac: Command+G; Windows: Ctrl+G).

Figure 9.55

The stroke is colored gray and reduced to 20%.

11. Change the Fill to 50% gray. Figure 9.55 shows the reduced-in-size Stroke in gray.

12. Press Option (Mac) or Alt (Windows) and drag a duplicate of stroke. Fill the duplicate with 20% gray and overlap the lighter gray stroke near the bottom of the original stroke.

13. Press Option (Mac) or Alt (Windows) and drag a duplicate of the light gray stroke. Set its Fill to 80% black. Let this stroke overlap the top of the medium gray stroke but send it to the back (Object|Arrange|Send To Back). Figure 9.56 shows the three overlapping strokes.

Figure 9.56

Three copies of the original stroke overlap one another.

14. Group the three strokes (Mac: Command+G; Windows: Ctrl+G). Drag them onto the Brushes palette and define a new Art brush as shown in Figure 9.57. Remember to change the Colorization method to Tints And Shades.

15. Drag the original stroke into the pasteboard area for safekeeping. With nothing selected, change the Fill color to None and the Stroke color to a soft pink.

16. Choose the Paintbrush tool and click on the BigStroke Art brush. Draw. Figure 9.58 shows a fast sketch that I did just to let you see how the brush looks.

Obviously, we have just started to scratch the surface of possibilities with this technique, but we need to move on. Try applying the strokes top to bottom and bottom to top as well as horizontally.

Something's Fishy

As you can see, it's easy to create strokes that look like watercolor using the Art brush. The Art brush also is a wonderful tool when you want to create

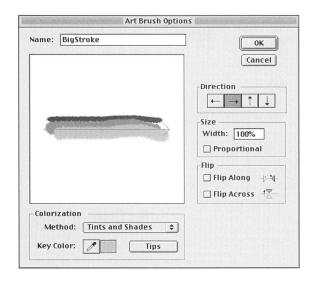

Figure 9.57 (left)
The Art Brush settings for the BigStroke brush that you have just created.

Figure 9.58 (right)
A fast sketch made with the BigStroke Art brush.

an object that can be enhanced by bending and stretching along its length. Leaves come to mind as an almost perfect application. Figure 9.59 shows a bare tree (from Ultimate Symbol's Nature Icons collection) to which I have

Figure 9.59
A bare tree (from Ultimate Symbol's Nature Icons collection) stroked with leaf Art brush.

added leaves (from the same clip art set). I defined a single leaf as an Art brush to be Stroked From Bottom To Top. Every leaf in the image looks different because the shape of the leaf as an Art brush follows the shape of the stroke used to paint it. The time savings is tremendous—if I had to draw each individual leaf, it would have taken much longer than the five minutes that I spent putting the image together.

Although I changed the colors of the individual leaves, most natural leaves have gradations of color in them. An Art brush that used a gradient would be lovely. Unfortunately, it is also impossible—at least if you try to define a shape containing a gradient as an Art brush. None of the object-defined brushes can contain masks, compound paths, gradients, patterns, or gradient meshes. They need to be simple objects. In Illustration 9.4, you see how you can get around this obstacle and define any brush that you want.

Illustration 9.4: Expand Your Horizons

1. In Illustrator, open the image FISH.AI or draw your own simple fish outline. Figure 9.60 shows the starting image.

2. Select the fish. Press Option (Mac) or Alt (Windows) and drag a duplicate to the pasteboard area for safekeeping.

3. Change the Units preferences to points. Create a new layer. Select the Rectangle tool and click inside the image area. Create a rectangle that is 250 points wide by 10 points high (this gives you approximately nine stripes in the fish). Set the stroke of the rectangle to None. Set the Fill of the rectangle to the Rainbow Linear gradient. Figure 9.61 shows the Gradient palette with the Rainbow gradient selected.

4. In the Gradient palette, drag the last four colors on the right (purple, pink, blue, and green) off the palette to remove them. This leaves only the orange and the yellow as shown in Figure 9.62. Drag the yellow pointer to the right edge of the color slider as shown in Figure 9.63.

5. Drag the gradient rectangle to a position just above the top of the fish. Lock the Fish layer (Layer 1). Choose Object|Transform|Move. Move the rectangle horizontally 0 points and vertically –9 points (one height distance). Click on Copy. Transform Again (Mac: Command+D; Windows: Ctrl+D) until you have covered the fish. You should have about 12 gradient rectangles as shown selected in Figure 9.64.

Figure 9.60
A simple fish outline provides the start of a new Art brush.

Figure 9.61
The Rainbow gradient is selected.

Figure 9.62
Remove the last four colors in the Gradient palette.

6. Select the bottom rectangle and every other rectangle (select one, skip one until you select six). In the Gradient palette, exchange the placement of the yellow and orange points by sliding them toward the center. Then put the yellow pointer at the left end and the orange pointer at the right end. Drag the Position slider to 70% to lengthen the yellow transition area. Figure 9.65 shows the result.

7. Select the fourth rectangle from the bottom and the fifth rectangle from the top (they are both orange-on-the-left gradients). Drag the yellow pointer from its position at the right edge of the slider to the 50% mark to keep the yellow in the gradient solid for a longer time. Figure 9.66 shows the new transition areas. You are making this change to stagger the position of the transition area so that it is less regular.

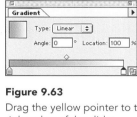

Figure 9.63
Drag the yellow pointer to the right edge of the slider.

Figure 9.64
The 12 rectangles are stacked on top of one another.

Figure 9.65
Swap the position of the yellow and orange pointers in the Gradient slider.

Figure 9.66
Stagger the position of the gradient transitions.

8. Select the entire image (Mac: Command+A; Windows: Ctrl+A). Because you have Layer 1 locked, only the gradient bands should be selected. Choose Object|Expand. Expand the fill to 12 objects. This

will posterize the gradient, but it will also prevent any printing problems that might occur and add a bit of sparkle to this particular image. In the fish, I feel that the banding can actually enhance the design—which would not be the case in other situations. You might not have any printing problems even if you used the full 256 objects to describe the gradient, but you don't need to put that theory to the test right now. Figure 9.67 shows the Expand dialog box.

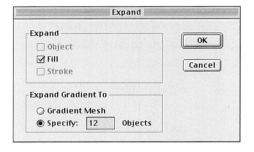

Figure 9.67
The Expand dialog box.

9. Group the objects (Mac: Command+G; Windows: Ctrl+G) and deselect them.

10. Remove the lock on Layer 1 and drag it above Layer 2. You can now see the outline of the fish on top of the gradient squares as shown in Figure 9.68.

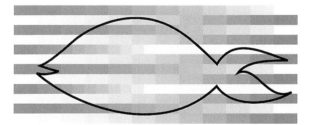

Figure 9.68
Drag the fish layer on top of the gradients.

11. Select the fish outline. Double-click on the Scale tool and enter a Uniform Scale of 100%. Click on Copy. Hide the selection (Mac: Command+3; Windows: Ctrl+3). You'll need this extra copy in a little while.

12. Select the fish and all gradient squares beneath it. In the Pathfinder palette, click on Crop. Figure 9.69 shows the cropped fish. You need to use the Crop command if you want to make the rectangle conform to the shape of the fish. The only other way to get the fish shape is to use the fish as a mask, but if you did that, you wouldn't be able to define the result as an Art brush because masks aren't allowed.

13. Show the hidden selection (Mac: Command+Option+3; Windows: Alt+Ctrl+3). Now the fish has a stroke around it. Change the Stroke weight to 1 point.

14. Lock Layer 1. Drag Layer 2 above it in the Layers palette (Layer 2 is now empty because cropping the fish moved everything onto Layer 1).

15. Select the Circle tool. Set the Fill color to white and the Stroke color to None. Press Shift+Option (Mac) or Shift+Alt (Windows) and draw a circle as large as you want for the white of the fish's eye. Select the fish outline. Double-click on the Scale tool and enter a Uniform Scale of 35%. Click on Copy. Change the Fill of the copy to black. Figure 9.70 shows the finished fish.

16. Select all the pieces of the fish and drag the fish into the Brushes palette. Create a new Art brush using the settings shown in Figure 9.71. This is the first time that I have instructed you to choose the Hue Shift method of Colorization. When you use this method (which works like the Hue slider in Photoshop's Hue/Saturation command to move all the colors around the color wheel), you need to select a key color from the Art brush. This is the color that is used as the controlling color to calculate the distance of the shift around the color circle. Use the Eyedropper tool inside the Art Brush dialog box and click the solid yellow of the stripe that you see third from the bottom of the fish. Click on OK to exit the dialog box.

17. Deselect the fish and then select and drag it to the pasteboard area for safekeeping.

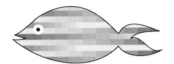

Figure 9.69
Crop the rectangle to conform to the shape of the fish.

Figure 9.70
The fish is ready to become an Art brush.

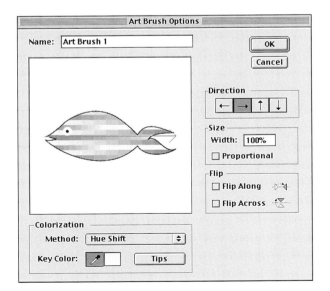

Figure 9.71
Setting the fish Art brush options.

ON USING STROKES AND FILLS

You could also make the fish's eye by using a black fill with a very large white stroke around it. That would seem to be more efficient. However, when you make an Art brush from the fish and apply it to a short stroke, the black center of the eye disappears because no room exists for it any longer in the image. When you define the eye as two filled shapes, both shapes appear, regardless of the length of the stroke that you make with the Art brush.

18. Deselect everything. Pick a strong yellow for the Stroke and None for the Fill. Choose the Paintbrush tool and the Fish brush and draw a school of fish. Select some of the fish and change their stroke colors. Select some of the fish and use the Selected Object icon on the bottom of the Brushes palette to change the Colorization method to Tints And Shades only for the selected strokes. The two fish that are swimming against the crowd in Figure 9.72 were able to do so by selecting the Flip Along checkbox in the Selected Object Art Brush Options dialog box. The two hot pink and orange fish (visible in the Color Studio) were Stroked From Top To Bottom in a copy of the Fish brush.

Figure 9.72
Sample fish.

Let It Snow

The Expand command might be one of the best tools for creating interesting brushes that seem to make use of "forbidden" elements such as gradients, but you can play some additional tricks with brushes. The Expand command also allows you to create brushes *from* brushes in much the same way as you did in Chapter 8, where you used the Celtic knot brush as the corner element in the Celtic knot.

The computer term for this process is *recursion*. Recursion is something that calls or refers to itself. A good example in "common" language is a company named MAGIC, where MAGIC is the acronym for MAGIC Auto Glass Installation Company and the MAGIC in MAGIC Auto Glass Installation Company stands for MAGIC Auto Glass Installation Company and—you have the idea. It could go on forever. I am fascinated by the possibilities of recursion in artwork. One of the most entrancing first tricks I learned in Photoshop from Kai Krause was how to take an image and turn it into a custom bitmap to use it as the halftone for the same image (so when you looked closely at the printed image it was printed with smaller copies of itself).

This attraction that I feel toward recursive images was sparked by the book *Gödel, Escher, Bach,* written in 1979 by Douglas Hofstadter. In this incredible book of mind games and wordplay, Hofstadter frequently creates words that use embedded letters to spell out other words (a "yes" made up of tiny copies of the word "no," or the word "holism" sort of written twice but in such a way that reading the letters that make up each major letter spells "reductionism").

Illustration 9.5 is my playful tribute to recursion and to the infinite ways in which recursion can recurse.

PROJECT Illustration 9.5: Recursive Play: Scatter Brushes From Scatter Brushes And Images From Pattern Brushes

1. Create a new document in Illustrator. You are going to draft a large snowflake.

2. Neither the stroke nor the fill color nor the exact size of the shapes matter in this step, but for consistency, set the Fill to None and the Stroke to black. Select the Rectangle tool. Press Option (Mac) or Alt (Windows) and click in the image window in center about a quarter the way down from the top of the page to open the Rectangle Options dialog box. Enter a width of 1.5 inches and a height of 2.0 inches.

3. Turn on Smart Guides (Mac: Command+U; Windows: Ctrl+U). Select the Ellipse tool. Move the cursor over the rectangle until you can see the Smart Guide that says Align 90°. Move your cursor up until it is still on the Align 90° line and the word "intersect" appears as you reach the top of the rectangle as shown in Figure 9.73. Press Option (Mac) or Alt (Windows) and the mouse button and drag the circle toward the right side of the rectangle. Press Shift after you begin to drag and release the mouse button when the word "path" appears by the right edge of the rectangle. Figure 9.74 shows the circle on top of the rectangle.

4. Select both the rectangle and the circle and click on Combine in the Pathfinder palette. You have a shape that looks unfortunately just

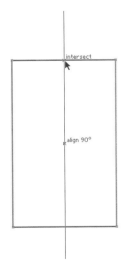

Figure 9.73
Start your circle at the center of the top edge of the rectangle.

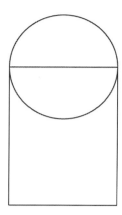

Figure 9.74
The circle and the rectangle.

Figure 9.75
Split the shape in half.

like a tombstone (but it won't for long). Your next step is to divide the shape in half vertically.

5. Select the tombstone and choose the Pen tool. You need to place another point in the exact center of the bottom edge of the object. Move your mouse until the Smart Guide reads both Align 90° and Intersect when you are over the bottom line, then click to add a point. Now you know where to divide the shape.

6. Turn off Smart Guides (Mac: Command+U; Windows: Ctrl+U), or you might not be able to accurately use the Scissors tool. Select the Scissors tool. Click on the control point at the center top of the object. Click on the control point that you added at the bottom center of the object. Press Delete/Backspace until half the tombstone has disappeared. If you prefer, you could instead sweep select half the object with the Direct Selection tool and then press Delete. That way, you would not need to use the Scissors tool. Figure 9.75 shows the new form—now looking somewhat like half a feather (which is a much happier description).

7. Why did you need to split the shape? Most snowflakes are formed from six identical spokes, and each spoke exhibits bilateral symmetry (i.e., one half is the mirror image of the other). You are now in the process of creating the "master" spoke (or half spoke). Let's add some interest to this half spoke. Select the shape. Choose Filter|Distort|Zig-Zag. Be conservative in your settings (don't use too much of the filter). Figure 9.76 shows the settings that I used.

8. Choose Filter|Distort|Roughen. Figure 9.77 shows the settings that I used. The Roughen filter is completely random. If you do not like the preview, keep moving the sliders until you do. If you come back to a

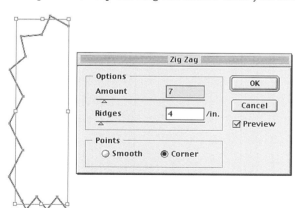

Figure 9.76
The Zig-Zag filter adds some new points to the snowflake spoke.

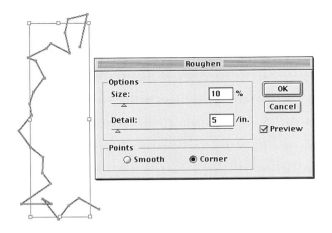

Figure 9.77
Use the Roughen filter to make the shape more irregular.

previous setting, the result will differ from what you saw before. Do not let the line cross itself if you can avoid it. In Figure 9.77, you will see a small area of overlap at the bottom of the shape. This is okay at the bottom edge but not anywhere else in the shape.

9. Leave the shape selected. Choose the Reflect tool. Place your cursor on the bottom end point of the path, press Option (Mac) or Alt (Windows), and click to open the Reflect Tool Options dialog box. Select Vertical Axis and click on Copy to make a mirror image of the half-spoke. Figure 9.78 shows the result of the Reflect command.

10. You now need to join the two halves of the shape. Choose the Direct Selection tool. Draw a small marquee around the two bottom points where the Reflect tool placed the right half of the shape. Average the points (Mac: Command+Option+J; Windows: Alt+Ctrl+J) both vertically and horizontally and then join the points (Mac: Command+J; Windows: Ctrl+J) with a corner point.

11. Select the two endpoints at the tops of the shapes. They are probably not touching. Average the points (Mac: Command+Option+J; Windows: Alt+Ctrl+J) both vertically and horizontally. Join the points (Mac: Command+J; Windows: Ctrl+J) with a corner point. You now have one shape as shown in Figure 9.79.

12. Select the entire shape using the Selection tool and then choose the Rotate tool. Press Option (Mac) or Alt (Windows) and click on the point at the bottom center of the shape (where the two halves meet). Enter an Angle of 60 degrees in the Rotate Tool Options dialog box and click on Copy. Transform Again (Mac: Command+D; Windows: Ctrl+D) four more times. You now have a snowflake.

Figure 9.78
The Reflect command places a mirror image of the half-spoke at your "click point."

Figure 9.79
One spoke of the snowflake is complete.

DO YOU NEED SMOOTH POINTS OR CORNER POINTS?

I chose corner points for the snowflake in both the Zig-Zag filter and the Roughen filter. Smooth points actually look prettier and create a really beautiful snowflake. However, every time I have used them and try to combine the six snowflake spokes, Illustrator complains about the complexity of the shape and threatens that it might not print. Although I might be willing to take that chance and deal with it in my own work, I prefer not to give you techniques that might not print. So try it if you want, but be forewarned.

13. Even though the snowflake form is endlessly fascinating and each one is different, and the snowflake probably looks best as an outline, you need to put these spokes together into one shape for this Illustration. Select the entire image (Mac: Command+A; Windows: Ctrl+A). Drag a copy of the combined form to the Pasteboard area of the document for safekeeping. Press Option (Mac) or Alt (Windows) to make a copy as you drag. Create a new layer and move the extra snowflake's "selected rectangle" from Layer 1 to Layer 2 in the Layers palette. Lock Layer 2.

14. Select the entire image (Mac: Command+A; Windows: Ctrl+A). Because you locked Layer 2, only the snowflake in Layer 1 is selected. Click on Combine in the Pathfinder palette. Figure 9.81 shows the single outline that results.

15. My snowflake now looks solid; yours probably will, too. However, even if you can't see it, the snowflake is part of a compound path. The next step is to release the compound path and get rid of the extra bits and points. With the snowflake selected, choose Object|Compound Paths|Release. The object is not grouped, so you don't need to ungroup it. Do not deselect.

16. Press Shift and click on the outer edge of the snowflake outline. This deselects the main shape. Delete whatever is left.

17. The output of this portion of the Illustration is a Brush Library that you will use a bit later. You need to empty the current brushes out of the Brushes palette so that you are not saving duplicates. In the Brushes palette sidebar menu, choose Select All Unused. Because you have not used any brushes in this image, all the brushes are selected. From the same sidebar menu, select Delete Brush as shown in Figure 9.82. Click on OK to the Delete Selected Brushes? query. The Brushes palette is now empty.

18. Select the snowflake and give it a Fill of 50% gray and a Stroke of None. Drag the snowflake into the Brushes palette and define it as a Scatter brush. Leave the Size and Spacing at 100% and the Scatter at 0%. Change the Rotation to Random and set the sliders from –180 degrees to 180 degrees to allow for maximum rotation. Set the Rotation Relative To option to Path. Click on OK to accept the settings that are shown in Figure 9.83. Deselect the snowflake.

19. You are going to create some variations in the snowflake to make a variety of brushes. Here's an easy way to reuse the settings that you have just made. Drag the thumbnail for the Simple Snowflake brush

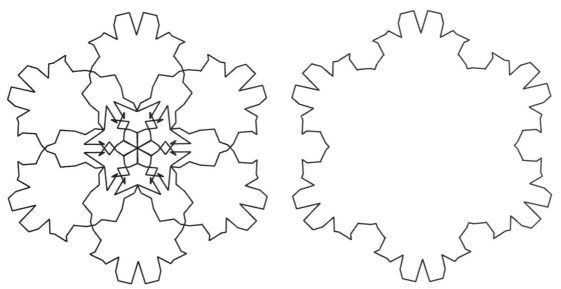

Figure 9.80 (left)
From a single line, a snowflake is born.

Figure 9.81 (right)
The Combine command merges all six sides of the snowflake.

to the New Brush icon at the bottom of the Brushes palette. This makes a copy of the brush.

20. With the Selection tool, select the original large snowflake that you used to create the first brush (hereafter I will just call it the "snowflake shape"). Change the Fill to 30% gray and give it a Stroke of 60% gray with a Stroke weight of 4 points. Drag the snowflake shape into the Brushes palette—pressing Option (Mac) or Alt (Windows) after you start to drag the shape—and drop onto the thumbnail of the copy of the Simple Snowflake brush that you just created. This reopens the Scatter Brush Options dialog box and changes the brush itself to the new version. Because you want to use the same settings, change the name to "30 Percent Outline" and click on OK to close the dialog box.

New Brush...
Duplicate Brush
Delete Brush

Remove Brush Stroke
Brush Options...
Options of Selected Object...
Select All Unused

View By Name
✓ Show Calligraphic Brushes
✓ Show Scatter Brushes
✓ Show Art Brushes
✓ Show Pattern Brushes

Figure 9.82
The sidebar menu in the Brushes palette allows you to perform a variety of different tasks.

Scatter Brush Options

Name: Flat Snowflake   OK

Size: 100%   Fixed   Cancel

Spacing: 100%   Fixed

Scatter: 0%   Fixed

Rotation: -180° 180° Random

Rotation relative to: Path

Colorization
Method: Tints and Shades
Key Color: [eyedropper]   Tips

Figure 9.83
Setting options for the new Scatter brush.

21. You can make other variations using other combinations of gray values for fill and stroke (or for only fill or stroke). Make a snowflake brush that uses a radial gradient in shades of gray from 10% to 75%. Build it exactly like you did in the Fish exercise (using Expand and Crop). Expand the gradient to about 50 shades. Unlock the "safety" copy of the snowflake on the pasteboard and change the Stroke of the uncombined shapes to white with a Stroke weight of 9 points and Miter Limit of 5 points. Option/Alt drag this into another copy of the Simple Snowflake brush. You may even create some other snowflakes if you want. If you choose to create ones with smooth rather than corner points, do not use the Combine command on them. Instead, just select all six spokes and drag them to the Brushes palette. These can be either filled or stroked (but not both—stroking the uncombined spokes will not create a symmetrical snowflake).

22. Re-create several of the variants—a plain filled snowflake, a stroked and filled snowflake, a stroked snowflake, and an uncombined stroked snowflake as Pattern brushes. Remember to save corner snowflakes for these Pattern brushes.

23. You will need another element that I have already created for you. It is a brush made out the word "Snow" from FFEricRightHand (another of my favorite FontFont fonts). You might just as well save it into the library that you are building. Choose Window|Brush Libraries|Other and negotiate to the Chapter 9 Starts directory. Choose the file SNOW.AI. To transfer the brush to your document's own Brushes palette, just use the brush in the image. Even if you toss away the brush stroke, the brush remains in your Brushes palette. The SNOW.AI document has two brushes; if you like the snowflake, you may transfer it as well.

24. Save the document as SNOWFLAKES.AI and then close it. You will use it (or its Brushes palette) shortly.

So what happens now? Where do we go from here? (Yes, you can ask.) You created a variety of snowflake brushes and placed them into a document that can be used as a Brush library. In the next part of this Illustration, you will modify a Japanese snow scene that I created for this book (from diverse elements redrawn but taken from *Japanese Border Designs*, Theodore Menten, Dover Books, 1975. This volume is part of the Dover Pictorial Archive of royalty-free images).

In Illustration 9.6, you will turn the snow scene into a brush so that you can select a monochromatic color scheme for it. You will then use the snowflake brushes that you created earlier to decorate the image. I have saved two versions of the snow scene. SNOWSCENE.AI contains mostly white elements (the sails and houses are all white). In SNOWSCENE2.AI, these white elements have been colored. I think the SNOWSCENE.AI is the more attractive design, but the colors in SNOWSCENE2.AI allow themselves to be colorized. It is your choice which version to use.

PROJECT Illustration 9.6: The Big Brush Switch

1. Open the image SNOWSCENE.AI (or SNOWSCENE2.AI) on the enclosed CD-ROM. Figure 9.84 shows the original SNOWSCENE.AI.

Figure 9.84

This image was drawn from elements in *Japanese Border Designs*, Theodore Menten, Dover Books, 1975.

2. To prepare the image to be used as a brush, you need to rid it of any illegal elements, such as gradients, masks, and dynamic brushes. The image, by design, has no masks or compound paths. However, the mountains (in the Houses layer) and the flat waves (in the Flat Waves layer) contain Calligraphic brushes. The Whitecaps are a Pattern brush. You need to expand these elements.

   Lock all the layers *except* for the Houses layer. Select the entire image (Mac: Command+A; Windows: Ctrl+A). Choose Object|Expand. Expand the Object, the Fill, and the Stroke. Lock the layer.

3. Unlock the Whitecap layer. Select the entire image (Mac: Command+A; Windows: Ctrl+A). Choose Object|Expand. Expand the Object, the Fill, and the Stroke. Lock the layer again.

4. Unlock the Flat Waves layer. Select the entire image (Mac: Command+A; Windows: Ctrl+A). Choose Object|Expand. Expand the Object, the Fill, and the Stroke. Lock the layer again.

5. Unlock the Boats layer. Calligraphic strokes were applied here as well. Select the entire image (Mac: Command+A; Windows: Ctrl+A). Choose Object|Expand. Expand the Object, the Fill, and the Stroke. Lock the layer again. The image looks no different, but Illustrator will now permit you to create a Pattern brush from it.

6. In the sidebar menu for the Brushes palette, choose Select All Unused. Then, from the same menu, choose Delete Brush. Click on OK to the warning message.

7. Unlock all the layers. Select the entire image (Mac: Command+A; Windows: Ctrl+A). Drag the entire image into the Brushes palette. Select the Pattern Brush radio button. Name the brush "Image Brush". Leave the Scale at 100% and change the Colorization method to Tints And Shades.

8. Save the document as SSBRUSH.AI and close it.

9. Create a new Illustrator document. Choose File|Page Setup to change the orientation of your document to Landscape mode (wider than long).

10. Turn on the Rulers (Mac: Command+R; Windows: Ctrl+R). Drag a guide from the top ruler to the 5 1/4-inch mark on the left vertical ruler. Set the Fill color to None and the Stroke color to 50% black. Select the Pen tool. Click on the left side of the guide where the guide meets the image margin. Press Shift and click again on the guide where it meets the right edge of the document (you are going past the margin on the right side). Turn off the Rulers (Mac: Command+R; Windows: Ctrl+R). Choose View|Clear Guides.

11. Choose Window|Brush Libraries|Other Library. Select the SSBRUSH.AI that you created in Step 8. Select the stroke that you created. Click on the Image Brush brush. Use the arrow keys to move the image into position top to bottom (it should be close). If necessary, stretch the path a bit to make the image fill the document area within the margin guides. Figure 9.85 shows the SNOWSCENE2.AI applied back to the image as a brush. The image has changed very little.

12. Experiment with changing the Stroke color to change the coloring of the entire image. As you drag the Stroke selection slider, you see a

Figure 9.85
The SNOWSCENE2.AI image is applied as a Pattern brush using a single tile.

variety of monochromatic color schemes appear. I eventually set my stroke color to CMYK: 0, 50,70, 21 (a light, grayed peach). You can see this image in the Color Studio.

13. Lock Layer 1 (with the image in it). Create a new layer.

14. Choose Window|Brush Libraries|Other Library and open the SNOWFLAKELIB.AI that you created (or mine, if you did not complete Illustration 9.5).

15. Select the Paintbrush tool and scatter one or two large simple snowflakes about the page. Choose Object|Expand. Expand the Object, the Fill, and the Stroke. Ungroup (Mac: Shift+Command+G; Windows: Shift+Ctrl+G) two times. With Shift, click on the snowflakes to deselect them and then delete the skeleton stroke. Figure 9.86 shows the image—somewhat overwhelmed by two overly large flakes.

16. These large snowflakes can now be used as the path for more snowflakes. They are now just regular paths. Apply the SNOW text brush as a Scatter brush to one of the snowflakes. Change the Stroke to white and then change the brush settings to Accept Tints And Shades. You may also change the other brush settings. If the snowflake is too large for your image, resize it by pressing Shift as you move a corner control handle on the bounding box. The brush that has been applied will automatically conform to the new object size.

17. Press Option (Mac) or Alt (Windows) and drag a duplicate of snowflake object. You can resize it as you want, but this is an easy way to grab the stroke. You can then apply any brush that you choose. Try

Figure 9.86
These large snowflakes are currently out of proportion to
anything suitable for this image.

the various brushes. Keep the snowflakes either white or light blue or light pink-purple. Figure 9.87 shows my solution in grayscale here and on the CD-ROM as FLAKESDONE.AI (so you can poke around and see my settings).

When you have finished your image in Illustrator, you might want to export it to Photoshop. If you move each large snowflake stroke brush to its own layer and export layers, you can get very fine control in Photoshop over your results. I took the image into Photoshop (after first creating a few

Figure 9.87
Snowflake Scatter brushes applied to the snow scene.

more snowflake groupings that were much too bold to apply in Illustrator) and took all the layers into Photoshop. I embossed most of the snowflake layers and then changed their Apply mode to Hard Light. I decided that the whitecaps in the image were too fussy, so I went back to the original layered-in-Illustrator version that was saved when I created the image brush. I simply turned off the Whitecap layer (ah, the benefits of using layers) and defined a new brush, which I then substituted for the image layer. Figure 9.88 shows the grayscale result, but you can see the color version in the Color Studio. The benefits of using Photoshop for this are that you can both change the Apply modes, selectively recolor portions of the image (or snowflakes), and change the opacity at which the element appears. None of these actions can be done in Illustrator (but Photoshop would not have allowed you to easily create the image at all).

Figure 9.88
The snowflake image with no whitecaps has been manipulated in Photoshop to control the impact of the snowflakes on the village.

Moving On

You have covered a lot of information in this very long chapter and completed some very challenging projects. You should have both an excellent idea of how to create Calligraphic, Pattern, Scatter, and Art brushes and a good appreciation for the pitfalls and strengths of each brush type. You have learned how the Combine and Expand commands can be used to help create brushes that would otherwise not be possible.

You have also gotten a "preview" project from Chapter 11. The ability to define an image as a brush and to use Tints And Shades to recolor it is a powerful color effect. I used it here because it fit into the exercise so well, but it could also have gone into Chapter 10. With this technique, you can turn an image into sepia tones (or shades of gray). If you define Hue Shift as your Colorization method, you could move the colors in the original around the color wheel. Once you have your color scheme set in the brush, you could render it by expanding the brush back into individual shapes. The colors are then "set." They could still be altered globally (or individually) by the Adjust Colors filter or by the Negative or Complement option on the Color palette's sidebar menu. We didn't have the time to try those tricks in this chapter.

Chapter 11 discusses a variety of other color tricks that you can play in Illustrator. Now I'll to pass you on to Mike to learn some Type techniques in Chapter 10.

TYPE EFFECTS 10

T. MICHAEL CLARK

A main strength of Illustrator is in type handling. You can choose from many fonts, set the baseline, automatically and manually kern, set the leading, and more. Although strictly speaking, Illustrator isn't a layout program, you can use it for many of your type layout needs. In this chapter, I'll show you some cool tricks to make your text stand out.

Some effects in this chapter use type in such a manner that the type can be easily reentered or edited. That is, the type remains type and isn't converted to outlines. Other tricks require you to convert the type to outlines, and others require that you take the drastic action of converting your type to bitmap or exporting/importing it into Photoshop for the final effect.

Circular Type (Or Type On A Path)

One strength of using an illustration program to enter your type is that you can, quite literally, bend it to your will (or at least bend it to fit on a path of your choosing). Surfing the Internet newsgroups (where people discuss, comment on, and ask questions about all sorts of topics) and especially the graphics newsgroups often turns up the question, "How do I create text in a circle?" Using Illustrator, this is easy.

To follow along with this exercise, open a new file in Illustrator and do the following:

1. Select the Ellipse tool and draw a large circle (holding down Shift while drawing the ellipse will ensure that you draw a circle rather than an ellipse).

2. Set the default white fill and black outline. This can be done quickly by clicking on the smaller white-over-black (default Fill/Stroke) icon just below and to the left of the larger Fill/Stroke icon near the bottom of the Toolbox or by pressing D.

3. Click and hold your mouse over the Type tool. When the flyout menu appears, select the Path Type tool (see Figure 10.1).

4. Click anywhere on the circle that you created in Step 1.

5. Enter some text using the keyboard.

It's as easy as that. Figure 10.2 shows the text that I entered using Steps 1 through 5.

One of the really cool things is that you can change the text even after it has been created and drawn along the path. Simply click and drag the Path Type tool over the existing type, and it will be selected.

Once the type is selected (it will be quite obvious that the text is selected, as the selected text will be shown as negative; i.e., the black-on-white type will be shown as white on black), you can change any of its characteristics. The next example shows how this is done.

You'll also see, in Illustration 10.1, how easy it is to draw text along any existing path. To follow along, first create a new file.

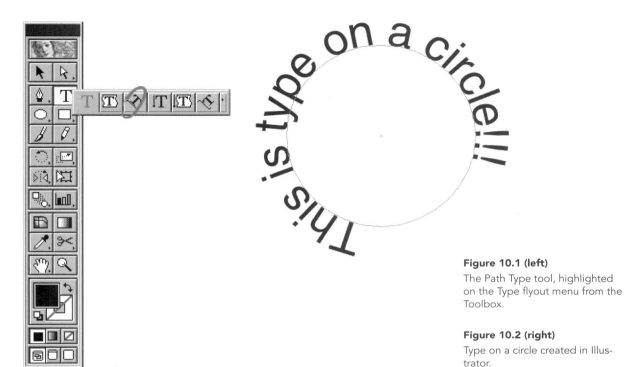

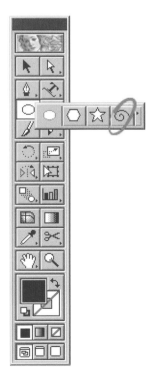

Figure 10.1 (left)
The Path Type tool, highlighted on the Type flyout menu from the Toolbox.

Figure 10.2 (right)
Type on a circle created in Illustrator.

PROJECT Illustration 10.1: Circular Locution

1. Press and hold the mouse button over the Ellipse tool. When the flyout menu appears, select the Spiral tool (see Figure 10.3).

2. Click and drag to create a spiral shape. You can hold down Shift to keep the spiral's width and height aspects equal. The spiral I created is shown in Figure 10.4.

3. With the spiral drawn, select the Path Type tool and click near the top of the spiral to enter your text. I entered "Down, down into the churning depths we sank" as shown in Figure 10.5.

4. With the Path Type tool still active, click and drag it along the first word. The word that you selected should appear highlighted as shown in Figure 10.6.

5. I had set the font characteristics to Helvetica Regular at 72 points. You may want to take a moment to set your font likewise to make it easier to follow along. After highlighting the first word, I set its size to 75 points. My reasoning will become clear in a moment.

The point size of a font can be set in a number ways. You can simply highlight the font size in the Character palette (see Figure 10.7), or you

Figure 10.3
The Spiral tool.

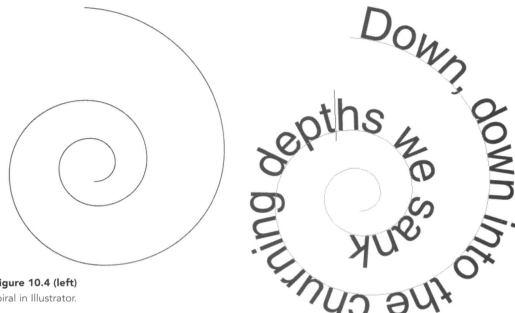

Figure 10.4 (left)
Drawing a spiral in Illustrator.

Figure 10.5 (right)
Entering type along a spiral path.

can choose Type|Size from the main menu; this highlights the font size in the Character palette. Either selection allows you to change the size of the font. You can also select a preset size from the menu selection or from the palette. Finally, you can scroll up and down through the sizes, point by point, using the spin controls next to the size entry in the Character palette. I prefer to simply highlight the control and enter a new value. You can also use keyboard shortcuts to increase/decrease the font size. Using Ctrl+Shift (Command+Shift on a Mac) and pressing either the > or the < key increases or decreases the font by one point size. Adding Alt (Option on a Mac) to the shortcut increases or decrease the point size by using Illustrator's pre-figured sizes.

6. Using the same method as in Step 4, select the second word and change its point size to 70.

7. Continue in this manner, selecting each subsequent word and decreasing its point size by 5 until you've selected and changed each word that you entered along the spiral path.

8. Clicking away from the text with the Selection Tool will deselect the type and the path. Holding down Ctrl (Command on a Mac) changes the type cursor to the selection tool.

You should end up with something that resembles Figure 10.8.

Note: I selected the comma and the first trailing space when I selected the first word. As I select each subsequent word, I'll do the same to keep the spacing consistent with the rest of the text. You should also note that double-clicking within a word selects the word, the trailng comma, and the following space.

Figure 10.6 (left)
Highlighting text to change its
characteristics.

Figure 10.7 (right)
The Character palette with the
font size highlighted.

Using the previous two methods, you can draw text not only along a
circle or a spiral but also along any path, both closed and open. The type
in Figure 10.9 was drawn very quickly.

To do so, I first drew the square shape by using the Rectangle tool. I then
entered "This is a box…" (note the trailing space after the ellipses) once,

*Note: At any time, you can
add new text simply by clicking at the end of the current
text with the Path Type tool
and typing. I added type to
the end of the current text as
room became available. You
can also use this method to
edit the text that already
exists. Click anywhere within a
sentence and use the arrow
keys, the backspace and
Delete keys, and any alpha-
numeric keys to edit your text.*

Figure 10.8
Increasingly smaller text drawn
along a spiral path in Illustrator.

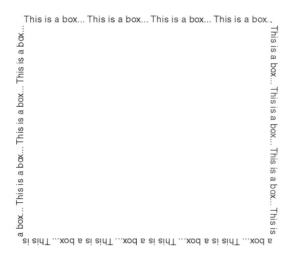

Figure 10.9
Text along a square entered
using copy and paste.

using the Path Type tool. With the Path Type tool, I highlighted the type and chose Edit|Copy (Mac: Command+C; Windows: Ctrl+C) and then clicked away from the type to deselect it. I then used Command+V (Mac) or Ctrl+V (Windows) several times until the type was quickly and easily added along the square path.

Kidnapper's Type

In addition to selecting each word separately, as we did to create the decreasing type size on the spiral, you can select each character. Doing so can yield some unusual results.

The "kidnapper's" type seen in Figure 10.10 was created by first entering the text and then selecting and randomly changing the characters sizes and fonts, separately and in groups.

Using different sizes and fonts gives the impression that the type was created by cutting out the words and letters from a newspaper or magazine.

To take this concept further, you can change each character's color, font, and size. You can even change the stroke size and color. And all these

Figure 10.10
Type held hostage.

If **you** ever want **To** see **your** type al**ive** **a**gain, Send the **MONe**y to the **follow**ing A**D**dress changes can be done while the type is still type (as opposed to having been changed to outlines). This means that all this seemingly haphazard type is still fully editable, and you can even spellcheck it.

Changing Where The Type Lies On A Path

Along with changing the font characteristics, you can change where on the path your type will lie. For example, text can be outside or inside a path. In addition, you can change where the beginning of the type will be. You can make all these changes happen even after you build the path and enter the text.

Outside Or Inside?

Type can lie along either the outside or the inside of a path. The results are more dramatic with a closed path. To see the effects of which side of the path your text lies on, try the following steps:

1. Open a new document and draw a circle by using the Ellipse tool.

2. Select the Path Type tool and enter some text (this is done exactly the same way as in the first exercise in this chapter). Your text should resemble that shown in Figure 10.11.

If your type doesn't fill the circle as mine does, simply increase the font size as I demonstrated earlier. To move the existing type to the inside of the path, do the following:

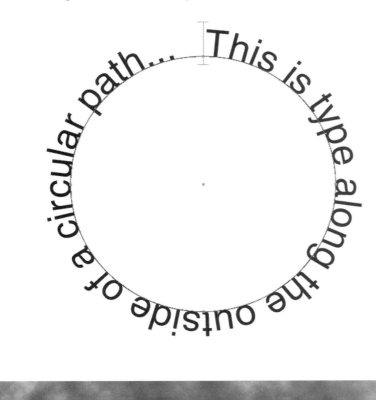

Figure 10.11
An example of type around the outside of a circle.

3. Select the Direct Selection tool.

4. Click on the I-beam cursor that indicates the text. It should be at the very beginning of your text, and it will appear as soon as you select the Direct Selection tool. You can see the I-beam in Figure 10.11 as I'm just getting ready to transform the text.

5. While holding the mouse button down, drag the I-beam toward the center of the circular path. Doing so will cause the text to go from the outside of the path to the inside. You can see the change that this makes in Figure 10.12.

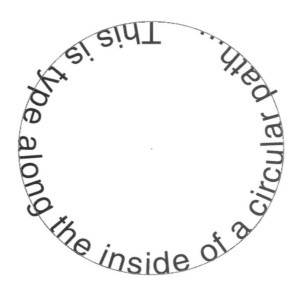

Figure 10.12
Type dragged into the inside of the circular path.

I've changed the font size to something a few points lower to accommodate the smaller surroundings and brought the type from the outside to inside the circular path.

Pushing Your Type Around

Beyond moving the type from one side of the path to the other, you can change where on the path the text begins. (You can use the type that's been moved to the inside of the path for this part of the exercise.) To move the beginning of the type to another location on the path and, subsequently, the rest of the type along with it, try the following:

1. Select the Direct Selection tool (unless it's still active from the previous exercise, in which case it won't need to be selected).

2. Click and drag the I-beam cursor around the inside of the circle.

3. As you do this, note that the type is being dragged about with the I-beam.

Figure 10.13 shows my type after I've moved the I-beam from the top of the circle to the bottom.

A word of caution might be helpful here. With a closed path, such as with the type inside the circle, the type simply moves around within the path. However, this is not the case when the type lies along an open path. Figure 10.14 shows some type along an open path.

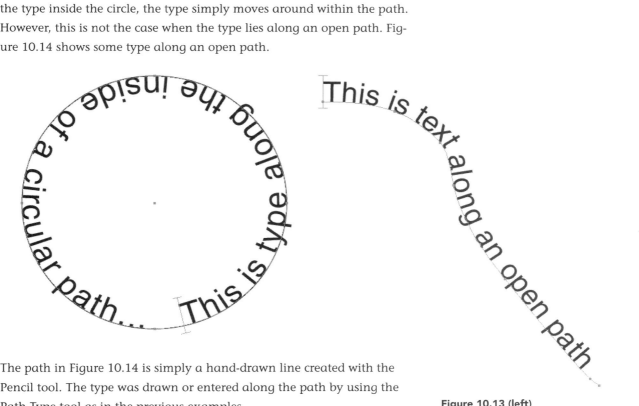

The path in Figure 10.14 is simply a hand-drawn line created with the Pencil tool. The type was drawn or entered along the path by using the Path Type tool as in the previous examples.

Before following along with Illustration 10.2, draw a path by using the Pencil tool and enter some text on it with the Path Type tool.

Figure 10.13 (left)
Text moved around within the circular path.

Figure 10.14 (right)
Type along an open path.

PROJECT Illustration 10.2: Type On A Path

1. Select all the text with the Path Type tool and change the font size so that the text fills the path from beginning to end as in Figure 10.14.

2. Select the Direct Selection tool and drag the I-beam so that the beginning of your text is about one-third to one-half the way down the path (see Figure 10.15).

Notice how the text falls off the end of the path? Where does it go? Well, to be honest, the caution I mentioned earlier is not all that serious, as you'll see in the final step of this exercise.

3. In the Character palette, reduce the size of the font; you'll see that all your text is still there.

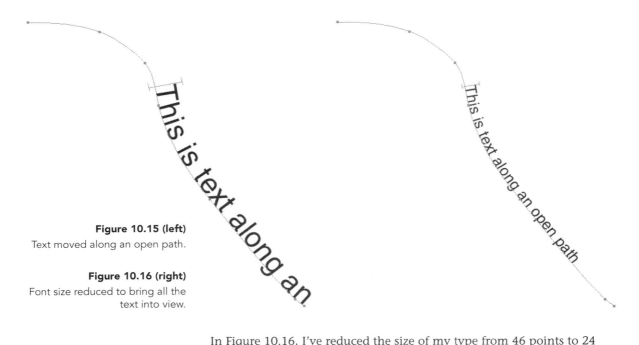

Figure 10.15 (left)
Text moved along an open path.

Figure 10.16 (right)
Font size reduced to bring all the text into view.

In Figure 10.16, I've reduced the size of my type from 46 points to 24 points, and all my text has magically reappeared.

That's about as far as I want to go with type on a path. I encourage you to try the examples, if you haven't already, and to see what you can come up with on your own as well.

Vertical Type

Each of the Type tools in Illustrator has a vertical counterpart. These tools enable you to enter type that flows vertically instead of in the usual horizontal axis. This is not to say that the text has been rotated 90 degrees but rather that each individual character is entered along a vertical axis. If the distinction has you scratching your head, take a look at Figure 10.17.

The type on the left was entered normally and then rotated 90 degrees. It was rotated by selecting it with the Selection tool and then selecting Object|Transform|Rotate. The type on the right was entered with the Vertical Type tool.

As I mentioned earlier, each Type tool has a vertical counterpart. All six of the Type tools, three regular and three vertical, can be found under the Type tool flyout menu in the Toolbox (see Figure 10.18).

In Figure 10.18, starting from the left, you'll see the Type tool, Area Type tool, Path Type tool, Vertical Type tool, Vertical Area Type tool, and Vertical Path Type tool. Shift+T will cycle through each of the Type tools in turn.

Figure 10.17
Horizontal type rotated 90 degrees and vertical type.

Vertical type can be used to add drama or flair to an image. Figure 10.19 shows an example of a travel poster or brochure that uses vertical text to add to the image's look.

You might not want to overdo it, but the Vertical Type tools can sometimes come in handy when you need to give a boost to an image.

Almost any of the type in all the techniques I demonstrate in this and in other chapters (and any other type you enter with Illustrator) can be entered both vertically and horizontally by using the Vertical Type tools.

Shaped Type

Using the Area Type tool, it's possible to create artwork by filling shapes with type. You can enter the type right into the shape by using the Area Type tool, or you can copy and paste from a word processor or text editor.

As you'll see later, the results of this technique can be amazing. I'll walk you through creating a simple piece and follow up by showing you a piece created for a brochure by an artist who works with type.

The design I'll be creating is from one of the first things that many artists draw in an art class: a still life. The sketch I used is from one of the first exercises I did when I first tried MetaCreation's Painter.

To start, I used a technique from Chapter 1. I loaded the sketch into Photoshop and traced around the body of the apple. I then saved that tracing as a TIFF file that I opened in Adobe's Streamline. Using the centerline method, I converted the sketched outline to vectors. The resulting converted file was saved in Illustrator format (see Figure 10.20).

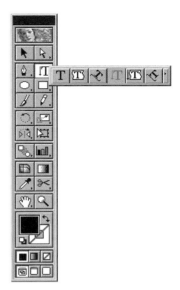

Figure 10.18
The Type tool flyout menu.

Figure 10.19
Vertical type gives this travel poster some added flair.

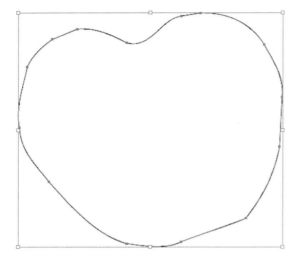

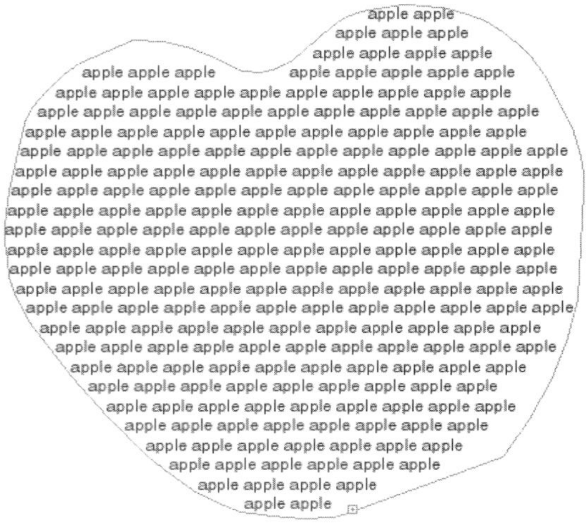

Figure 10.20 (left)
Traced sketch of an apple opened in Illustrator.

Figure 10.21 (right)
Apple shape filled with type.

To fill the apple with the word "apple" (see Figure 10.21), I selected the Area Type tool, entered "apple... " (note the added space at the end) once and highlighted it. I then copied the word and pasted it into the shape by holding down Command+V (Mac) or Ctrl+V (Windows). I continued holding down the key until the entire shape was filled. You can speed the process by using the Ctrl+V (Windows) or Command+V (Mac) shortcut a few times to paste several copies, re-select these copies and copy and paste all of them.

I set the type size and font by selecting the text, using the Area Type tool, and changing the settings in the Character palette (see Figure 10.22).

You might notice in Figure 10.22 that the type doesn't quite conform to the shape. Many ways are available to correct this, and some are simpler than others.

Figure 10.22
Type characteristics changed.

Depending on the final effect you're after, you can change the font of separate words, change the style from normal to bold, or even change the size of some of the words or characters. I chose to judiciously add spaces between some of the words. The result can be seen in Figure 10.23.

If you need to add spaces to your own artwork, simply click within the text by using the Area Type tool and press the spacebar. For a final touch, I selected all the text and changed its color to a deep red—but there's more.

A nice, fresh apple would probably have a stem and maybe a small leaf attached to it. To create the leaf that you see in Figure 10.24, I used exactly the same method as that used to create the apple. However, rather than sketching the simple leaf shape in another program, I simply drew it with the Pencil tool.

The stem was created by using a technique that I'll cover later in this chapter in the section "The Path Is The Type." Now, as promised, I'll show what a real type artist can do when using area type. Figure 10.25 is part of a promotional brochure created for Evansday Design by Rob Day.

The dog in the brochure was created by using MyriadMM, an Adobe multiple master font. A multiple master font enables the artist to manipulate the width and/or weight of the type without distorting the letterforms. By subtly adjusting the weight and width of individual letters, the artist can achieve the illusion of tonality.

Figure 10.23 (left)
Spaces added to help the type fill the shape.

Figure 10.24 (right)
Final apple created completely with type.

Figure 10.25

Image from promotional brochure created for Evansday Design by Rob Day. (Used with permission of the artist.)

Mr. Day started out by scanning in a photograph of a dog, saving it as a grayscale PICT file, and opening it in Illustrator. The file was placed on the bottom layer and dimmed to 50 percent (this is one of the methods described in Chapter 1). He then created a range of about 30 instances of the Myriad type by using the heaviest and most extended type for the deepest shadows and the lightest and most condensed type for the highlights.

An outline of the dog was created and the text flowed into the outline. The weight and width of each letter was then adjusted on the basis of a grayscale template created for this purpose. To help with the process, Mr. Day used Adobe's Type Reunion, a tool that allows the artist to assign names to each instance of the multiple master font.

Rough Type

This next technique is one that I see being asked about often on the Internet. Text of this type (no pun intended) goes by many names, and a couple of fonts have even been created to simulate this effect.

The effect I'm talking about is stencil, or rubber stamp, type. Such type was probably made famous in its use in the credits for the television series *M*A*S*H*. The U.S. Army, in need of a way to mark many items with the same text and codes (such as serial numbers) simply created stencils and painted the letters onto the various objects by rolling or spraying paint through the stencil.

Because of the nature of vector objects, this effect is easily reproduced in Illustrator. Start by selecting the Type tool and choosing a fairly large serif font. I chose Times New Roman MT Extra Bold set at 100 points (see Figure 10.26).

Figure 10.26
A heavy serif font to be used for rough type.

You can see from the dialog box in Figure 10.26 that I also increased the tracking to 10 to add a little space between the letters. With the text entered, it's time to break it apart so that it will resemble stencil type. To do so, follow these steps:

1. Click on the Selection tool to select the type that you entered.

2. Choose Type|Create Outlines to change the type into outlines.

3. Going from letter to letter, create a shape (usually a rectangle) that you can use to break apart each letter as needed. Figure 10.27 shows how this is done, using the letter "F" in my example.

You don't have to be really precise, but be sure to use a fairly narrow rectangle.

Figure 10.27
Use a small rectangle shape to cut pieces from each letter.

4. With the rectangle in place, use the Selection tool to select the letter and the rectangle.

5. Click on the Minus Front icon in the Pathfinder palette (see Figure 10.28).

Figure 10.28
Cutting the letter up by using the Minus Front command.

Figure 10.29 shows the letter after using the Minus Front command.

Figure 10.29
The letter after using the Minus Front command to remove the rectangular object from the letter.

With some letters, you might need to rotate the rectangle into place. To do so, simply move the rectangle near to where you want to make the cut, select the Rotate tool, and rotate the rectangle into place (see Figure 10.30).

Some letters, such as "A", require more cuts. If necessary, repeat the process (see Figure 10.31).

If you're not sure where a letter should be cut, you might be able to find a stencil at your local stationery store. It's been some time since I played around with such a stencil, but I'm sure you can still purchase them.

Figure 10.30
Rotating a rectangle into place on the letter "A".

Figure 10.31
Cutting a second piece from the letter "A".

Figure 10.32 shows the finished letters (as far as cutting the pieces out, anyway).

FRAGILE

Figure 10.32
All the letters with the pieces removed.

The final step in creating the rough type is to rough up the type a little. To do so, choose Filter|Distort|Roughen. In the Roughen dialog box, set the size very low, set the detail quite low, and set the Points to Smooth (see Figure 10.33).

The small Size setting (2 in my example) means that the effect isn't overdone. The fairly small Detail setting (5 in my example) gives a smoothness to the roughen effect. A smaller setting makes the effect too lumpy, whereas a larger setting makes the effect too spiky.

I set the Points to Smooth because this setting yields a better effect than the Corner setting. What I'm shooting for is the effect of having a full paint roller rolled over a stencil. Figure 10.34 shows my final type.

Figure 10.33
Roughing up the type.

Figure 10.34
Final rough type image.

After creating the type in Illustrator, I opened it in Photoshop and pasted it over a wood texture that was created in Photoshop.

To get the effect of a paint-slopped stencil over a wood texture, I duplicated the type layer and set both the type layers to Soft Light. This allowed the wood texture to show through the type just enough to make the whole image believable (see Figure 10.35).

By the way, the wooden texture in Figure 10.35 was created by using a technique that I wrote about in *Photoshop 5 Filters f/x and design,* also available from The Coriolis Group.

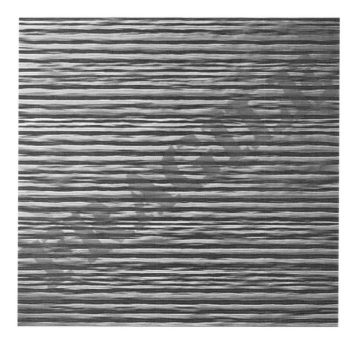

Figure 10.35
Rough type placed over a
wooden texture in Photoshop.

The Path Is The Type

Using the brushes in Illustrator 8, it's possible to make the path become the type. This is a new (and really cool) feature of Illustrator 8. To use this technique, you must first create a type brush. If you've already glanced through the manual or read a review or two, you might wonder what you've missed. "There is no Type brush," you're saying to yourself, right? Okay you're right—partly right, that is.

A new brush type, called Art, can be used for type. To create a Type (Art) brush, take these steps:

1. Use the Type tool to enter some type. Select a fairly large font. I selected Caricature set at 100 points.

2. Select the Selection tool to highlight the type.

3. Choose Type|Create Outlines.

4. Click on the small bent-cornered icon (the New Brush icon) at the bottom of the Brushes palette to create a new brush.

5. This brings up the New Brush dialog box, enabling you to select the type of brush you want to create. The choices are New Calligraphic Brush, New Scatter Brush, New Art Brush, and New Pattern Brush.

6. Choose New Art Brush. Doing so will bring up the Art Brush Options dialog box (see Figure 10.36).

You can name the brush and set various options. For the Type brush, I left the defaults and named the brush "Waves". After creating your Type brush, you can convert any line to the type, which flows along the path. The type in Figure 10.37 flows along a circle created with the Ellipse tool.

You can see that this is a completely different effect from that of using the Path Type tool as seen earlier in this chapter. I'll use the Art Brush type effect in the next example as well.

Simulating Envelopes

Some drawing programs have a method of stuffing type into different shapes. For example, in CorelDRAW! you can use envelopes (ready-to-use shapes) to which your type will conform. Although this feature isn't available in the same way in Illustrator, it's a fairly simple albeit time-consuming process to stuff your text.

For this example, I use an existing default Art brush to construct the initial type, which I then bend and twist until I get it into the shape I want.

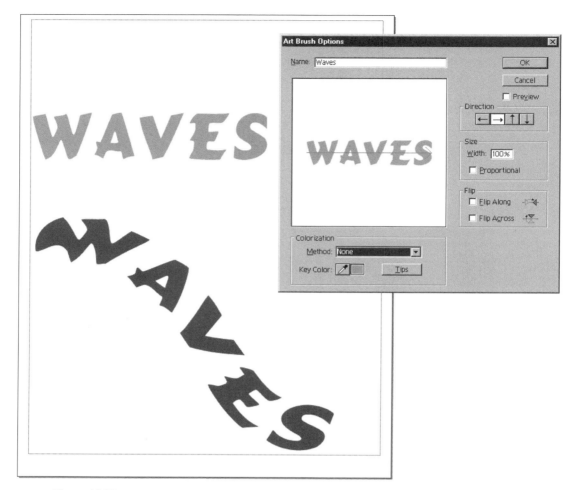

Figure 10.36

The type used to create a new Art brush (top left), the Art Brush Options dialog box, and a path painted with the new brush (bottom left).

The reason I decided to use the Art Brush type was that I wanted the type to be shaped in a semicircle. Using the Art brush meant that half the battle would be won simply by creating an elliptical path and painting the type over it.

To follow along, do the following:

1. Draw an ellipse with the Ellipse tool. It should be almost as wide as the work area and quite narrow.

2. Use the Scissors tool to cut the ellipse in half along the horizontal axis. Doing so makes type follow along a half ellipse (see Figure 10.38) rather than along a complete ellipse.

3. Click on the default Art Brush—named, appropriately enough, Art.

You should have a semicircular word, as shown in Figure 10.38.

Figure 10.37 (left)
Type created with an Art brush constructed from type outlines flows along an elliptical path.

Figure 10.38 (right)
Art bent to a semicircular shape.

If you're not completely happy with the size or placement, you can resize or move the ellipse, thereby resizing or moving the type. I stretched mine a little higher and moved it more toward the center of the work area.

The problem with creating the type as a brush is that you can't really work with it. You can't change the shape of the individual letters, which is necessary for the envelope shape effect. Naturally, a solution exists. To change the type from the Art brush into outlines that can be manipulated, choose Object|Expand. Doing so brings up the Expand dialog box (see Figure 10.39).

Leave the default settings and click on OK. Doing so will magically turn the path, which has become an Art brush, into editable outlines.

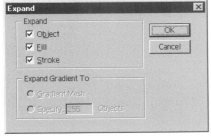

Figure 10.39
The Expand dialog box.

The type outlines will still be connected to, or grouped with, the original ellipse. To remove the ellipse and leave the type intact, choose Object Ungroup, then, using the Selection tool, click on the ellipse (it might be hidden, but it won't be hard to find). A simple method is to select the type and then choose Edit|Select|Inverse. Of course, this assumes that the only objects present are the ellipse and the type.

Now that the type is drawn (at least partly) in the shape that we're after, it's time to proceed with the rest of the type shaping.

4. Select the Pen tool and click below and to the left of the type.

5. While holding down Shift, click below and to the right of the type; this ensures that you get a perfectly horizontal line. This line will be our guide in creating the bottom of the envelope (see Figure 10.40).

Figure 10.40
A line placed along the bottom of the type just prior to being converted into a guideline.

6. To make the line a guide, choose View|Make Guides. The line should dim a little and become deselected. It's now a guide that you can use to line up the bottom of the type.

7. Using the Direct Selection tool and working with one letter at a time, line up the bottom points of the letter with the new guideline.

You might find that you can delete the points between the corners as you work. This will give you a nice straight line.

Figure 10.41 shows the work in progress.

Figure 10.41
Moving the points of the bottoms of the letters into place along the guideline.

This might seem a long, drawn-out process, but it really takes only a few minutes. My final type, bent into shape, can be seen in Figure 10.42.

To get rid of the guide after you're done with it, choose View|Clear Guides.

Figure 10.42
Final type bent into shape.

For the final piece, seen in Figure 10.43, I mirrored the type by using Object|Transform|Reflect and set the Axis to Horizontal and the Angle to 0 degrees. I then applied a perspective transform on the reflected text (Mac: Option+Command+Shift; Windows: Alt+Ctrl+Shift while using the Free Transform tool).

Figure 10.43
The final piece of artwork created by using type that was curved and bent into shape with a couple of different techniques.

The background is simply two rectangles filled with different gradients; a yellow-to-red radial gradient on the top and a white-to-black linear gradient on the bottom. The perspective type is also filled with a black-to-white linear gradient.

3D Type

Last but not least, I'll show you a technique that you can use to create 3D, extruded type. I created the same effect in my book *Photoshop 5 Filters f/x and design* by using Photoshop. However, it's much easier to create with Illustrator. Just don't tell that to the folks you know who love Photoshop.

There's a lot of talk and fuss going on over 3D these days. Everywhere you turn, in print and on television, you see 3D images, and you can't buy a video game that doesn't say that it's 3D.

All this fussing leads me to ponder what exactly 3D is, at least in the context of graphics or, more important, computer graphics. The 3D that all the hype is about is actually rendered images. Images of this type are created with 3D-rendering software. The software uses mathematics to calculate the (virtual) rays of light and how they affect the surfaces they are lighting. Of course, none of this is possible in Illustrator.

I don't think it matters that much, unless you're going for that 3D-rendered computer graphics look. After all, the image is still really going to be represented in two dimensions. When you look at a screen or at a piece of paper with an image on it, you really only see width and height. Any depth that you see is a trick of light and shadows. It's all in how the light and shadows are used to fool the eye into believing that some depth exists in the image or in the objects in the image.

With a little work on your part, you can construct some text, extrude it, and add highlights and shadows to the point where the text will appear as though it has depth. Illustration 10.3 shows one way to do this.

PROJECT Illustration 10.3: Illuminations

1. Open a new image.

2. Add a new layer by clicking on the Create New Layer icon at the bottom of the Layers palette. This will be Layer 2 (we won't need to rename it for the purposes of this exercise).

3. On the new layer, add some type by using the Type tool. I entered "GrafX" in black, using 160-point Helvetica.

4. Select the Selection tool and move the type into the middle of the work area.

5. Choose Edit|Copy and make the first layer (Layer 1) active by clicking its title in the Layers palette.

6. Choose Edit|Paste to paste the type.

Normally, I would use Edit|Paste Behind, but this just pastes the type back into Layer 2 as does Paste In Front. You just have to bite the bullet and move the new type into place by hand.

You should now have type on both layers and the first layer set as active.

7. Change the fill to about 50% black.

8. Move the type into place behind the type on Layer 2. It should completely disappear behind the type on Layer 2.

9. Lock Layer 2 for now and turn off its visibility so that you can access the type on Layer 1 and not have to worry about the type on Layer 2.

10. Use the Selection tool to select the type.

11. Choose Type|Create Outlines.

12. Choose Object|Group. This will ensure that the Blend operation will work correctly a few steps from now.

13. Choose Edit, Copy, and then Edit|Paste In Back to paste another copy of the gray text behind the first.

14. Use the arrow keys to nudge the new type up and to the left 10 times each (i.e., press the Up Arrow key 10 times, followed by the Left Arrow key 10 times). This begins the extrusion. However, we need to do more, so let's move on.

Note that you can set the amount of movement for the cursor keys under File|Preferences|General.

15. Select the Selection tool and select both copies of the type.

16. Choose Object|Blend|Blend Options.

17. In the Blend Options dialog box, set the Spacing to Specified Steps and the number to 50.

At this point, you might ask yourself, Why use 50 blending steps if the two copies of the objects are only 10 units apart? Good question. Through trial and error, I came up with this number. Using too low a number, such as the same number as the distance between the objects, will result in jagged diagonal lines in the final image as seen on the left in Figure 10.44.

Figure 10.44
A blend of 10 steps versus a blend of 50 steps between two objects that are spaced 10 units apart.

Okay, you're now asking, Why not use a much large number, say, 250, to *really* smooth the diagonals out? Another good question. If you try this, though, you'll likely find out that Illustrator will run out of memory when you later try to unite the blends (a few steps from now). I tried this on a machine with 256MB of RAM and ran out of memory. This was part of the trial-and-error method I mentioned earlier. I think 50 steps is a good compromise between memory needs, speed, and smoothness of the finished image. However, if want, you can raise or lower the number as necessary, depending on what you want in the final image. (You may find that raising the value takes more processing time as well as memory. On my dual Pentium II 266MHz machine, it takes a couple of seconds to run the Unite process.) For example, you might want a deeper or a shallower extrusion.

18. Select the Blend tool and click somewhere on the foremost type, then click on the corresponding point on the other type.

You should get a nice blend between the two type objects. However, I find (and I'm not sure why or how this happens) that any holes in the letters get filled in. For example, in the lowercase "a", you'll see that it's filled in (see Figure 10.45).

I've found that by choosing Edit|Undo and rerunning the blend, you can correct this problem (see Figure 10.46).

Figure 10.45
A small glitch in the blend process.

Figure 10.46
The glitch corrected.

19. Choose Object|Expand. Leave the default settings and click on OK.

20. Choose Object|Ungroup.

21. Select the Selection tool and click away from the objects to deselect them.

22. Draw a rectangle over the first letter. If this selects all the letters, choose Object|Ungroup again (another small glitch?), and you should be able to deselect all the type and then select the first letter.

23. With the first letter selected, click on the Unite icon on the Pathfinder palette (see Figure 10.47).

Figure 10.47
Uniting the blended objects into a single object.

24. Select the second combination (the letter "r") and unite it.

25. Continue until all the objects have been united.

The whole point of these steps was to construct separate objects that appear to be extrusions of each letter. By blending the letters, we give them the appearance of thickness, and, now that each combination of blended objects has been united, we can use gradients to give the appearance of highlights and shadows.

26. Turn on the visibility of Layer 2 but leave it locked. You should now have something that resembles Figure 10.48.

27. Use the Selection tool to select the gray portion of the letter "G".

28. Click on Gradient (below the current Fill/Stroke icons in the Toolbox) to fill the letter with a gradient instead of a solid color.

Figure 10.48
Type with extrusion behind it.

29. Change the gradient in the Gradient palette so that three colors appear. The color should be black at both ends and white in the middle (you may want to use different colors, depending on the image you're creating). See Figure 10.49 for a look at the way I set up the gradient for this example.

Figure 10.49
Using a gradient to add highlights and shadows to the extruded letters.

Notice in Figure 10.49 that I've also set the gradient to Linear and the angle to 45 degrees. These settings will work for all the letters. You'll need to change only where the white area lies in relation to the ends of the gradient. For now, select each letter and apply the same gradient (see Figure 10.50).

Note that you don't need to apply the gradient to any letters that don't have a curved section. For example, I haven't applied the gradient to the "X".

You may notice that the gradient seems a little off on some of the letters. Ignoring for the moment that even the straight areas are filled with the

Figure 10.50
Gradient highlights and shadows applied to all curved letters.

gradient, you'll see that the highlights aren't quite right, but that's easy to fix. Simply select each letter in turn and move the white slider back and forth until the highlight looks right. Figure 10.51 shows my example with the corrected highlights on top and the previous image below.

Figure 10.51
Corrected highlights versus original highlights.

I've simply moved the white slider to the right a little for both the "r" and the "f".

We're almost done. You'll notice that the highlights and shadows spill over onto the straight areas of the type, and this isn't correct. To correct this situation, the best method is to simply add new objects over the straight-edged areas of the letters. You'll want to zoom in to help with the placement of the new objects.

30. After zooming in, use the Pen tool to construct a new straight-edged object over any areas that need it.

31. To construct the new object, simply click in each corner of the area. Make sure that you finish each area by clicking on the original corner; that is, count to five as you click the corners of the areas that need to be covered. The first click goes in one corner, the next click goes in the next corner, and so on until you click on the starting corner to close the object (see Figure 10.52).

The areas that lie toward the top of the letters should get a 50% black fill, and areas toward the left of the letters should be somewhat lighter. A 40% black fill should work. The whole point is to give the illusion that some areas are closer to, or more directed toward, the light source. In this example, I'm assuming a light coming from above, to the left, and slightly behind the letters.

32. Keep adding objects until you've covered all the straight edges. Don't worry if you come to an area where the object isn't a trapezoid but

Note: *If you find that you're having trouble lining up the new objects, zoom way in. I zoomed in as much as 2,600% to help me line up the corners of the objects. After zooming in, use the Direct Selection tool to select a corner and move it into place.*

Figure 10.52
Constructing straight-edged objects to cover up the noncurved areas of the extruded letters.

rather a triangle. Just add a triangle or complete the trapezoid; it will be hidden behind the type on Layer 2 anyway.

The "X" is a special case because it has a surface that lies perpendicular to the light in this example. The areas that lie perpendicular to the light should be much brighter than any other areas. These planes are, in effect, highlights. However, I find it a little unnatural to make them too bright, so I assigned a value of 15% black. Because these areas are the most dominant on the "X", I assigned the 15% black fill to the whole area and then added new objects where needed. Figure 10.53 shows my finished 3D type.

Figure 10.53
Final 3D type created completely in Illustrator.

You can apply these same techniques to any object that you want to give depth to.

After all this work (more than 30 steps), you may wonder whether a better way is available to do all this. The answer is yes.

If you plan to construct many 3D objects in Illustrator 8, you might want to invest in the excellent 3D Pop Art plug-in from Vertigo (on the Web at **www.vertigo3d.com**). This plug-in allows you to quickly create 3D objects from your two-dimensional drawings. You can set the depth and angle and choose from eight different light directions.

Figure 10.54 is an example of the type from the previous exercise created in seconds with a couple of mouse clicks.

Figure 10.54

3D type created in Illustrator with the help of Vertigo's 3D Pop Art plug-in.

Compare the image in Figure 10.54 to the image in Figure 10.53. They're both quite similar, but the latter took seconds, whereas the former took much longer.

Moving On

I hope that you'll find these type techniques useful and that, as always, they will be a starting point for explorations of your own. Every hour that you spend with Illustrator is worth its weight in gold, even if only for the learning experience.

I'm going to pass you back to Sherry now, but I'll be back in Chapter 12, where I'll show you a few tricks to get your art ready for the Web.

COLOR EFFECTS

II

SHERRY LONDON

Illustrator 8 provides some interesting capabilities for changing colors. Although the choices are now fewer than before, in this chapter I'll discuss some exciting possibilities.

This chapter was originally planned to be larger, longer, and more exciting than it is. The blame, I guess, rests partially with Adobe and partially with the audience for Illustrator. For the story, read on.

Ode To The Lost Filters

The market for third-party filters for Illustrator is not large; it seems that many fewer Illustrator users than Photoshop users want to add plug-ins to the program. Additionally, Adobe made some changes to the filter interface in Illustrator 8 so that the earlier filter sets (e.g., KPT Vector Effects, Extensis Vector Tools, and the discontinued ones from BeInfinite and Cytopia Software) are no longer guaranteed to work inside Illustrator. Many of them still do work, however, and most of them also work in Freehand. (See my note in the Introduction for more specifics.)

Neither Extensis nor MetaCreations has plans to revise their filter sets for Illustrator 8. If you already own the filter sets, you'll want to keep using the filters that still work. If you don't currently own these sets, try out the demos of the products before spending money on them. Windows users, however, don't need to mourn the loss—these filters were never available for Windows.

So what facilities has Adobe given you for color manipulation in Illustrator? You have some color filters, and you have the Mix Hard or Mix Soft commands on the Pathfinder palette. Also fully compatible and cross-platform is the third-party plug-in, Transparency (from HotDoor), which allows Illustrator to simulate the transparency lens effects found in Freehand and CorelDRAW. Let's look at what these various commands and filters can do.

The Color Filters

Figure 11.1 shows the Color Filters menu options. You can adjust the components of the CMYK colors in an object; you can blend colors front to back, horizontally, or vertically; you can change colors to CMYK, RGB, or grayscale; and you can invert and saturate colors. However, you can't randomize colors, fade them, or globally alter their hue as you can in Photoshop or as you could with the now-obsolete Extensis and MetaCreations filter sets.

The blending options are somewhat intriguing, however, and merit a closer look.

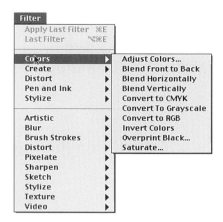

Figure 11.1
The Color filters available in Illustrator 8.

Blend Horizontal

You can use the Blend Horizontal color filter to simulate a quick two-color gradient if your shapes don't contain any compound paths. Although I have not found anything that I can't also do with a gradient, the Blend Horizontal filter keeps each shape in a single color—which isn't always the case with the Gradient tool. Try Illustration 11.1.

PROJECT Illustration 11.1: Around And About

1. Create a new Illustrator document.

2. Choose the Ellipse tool. With no stroke and no fill, draw a circle approximately 4 inches in diameter. This will be the path for circular text.

3. Select the Text-On-A-Path tool. I used Hiroshige Black, 60 points, but you may use any font that is installed on your system. The technique shows up best if you use a fairly wide font (Copal, Bees Knees, and most black weights are good choices).

4. Enter "Around and Around". After you enter the text, switch to the Selection tool. Figure 11.2 shows the position of the text as I first entered it.

5. Grab the I-beam and drag the text to the left so that it moves around the circle. Center the text (see Figure 11.3).

6. Choose Text|Create Outlines. While all the text is selected, choose Object|Compound Paths|Release. Figure 11.4 shows the text with the compound paths released (all the counters in the letters fill in).

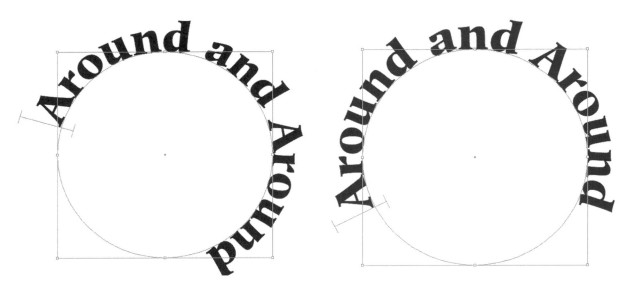

Figure 11.2 (left)
The text is not centered around the circle.

Figure 11.3 (right)
Moving the I-beam with the Selection tool changes the starting position for text on a path.

7. Drag a marquee around the "Around" text on the left. Change the fill to orange-red.

8. Drag a marquee around the "Around" text on the right. Change the fill to cyan.

9. Select the entire image (Mac: Command+A; Windows: Ctrl+A).

10. Choose Filter|Colors|Blend Horizontally. The colors blend across the letters from orange to cyan.

11. Deselect all the text. Now you need to put back the compound paths. Unfortunately, you cannot create the compound path in one step the way you released it. If you try, you will discover that you create one large compound path of all the letters and that only one color is in it. Instead, use the Selection tool to drag a tiny marquee through the first letter that needs to be fixed (the "A" in "Around"). Press Command+8 (Mac) or Ctrl+8 (Windows). The compound path appears good as new. Repeat for any letter that is missing its counter. Figure 11.5 shows the finished effect.

What does the Blend Horizontal color filter do? It takes the color in the shapes that is farthest to the left and blends it into the color fill that is farthest to the right. The filter only blends the fill color; it doesn't affect the stroke color.

The filter also gives you a fast way to generate a palette of blended colors if you don't want to bother creating a gradient and then expanding it in steps. You could create a small rectangle and duplicate it a number of times, dragging each new copy to the right (let's say you wanted 10 steps

Figure 11.4 (left)
Many letterforms require compound paths and look odd when they are released.

Figure 11.5 (right)
Blended colors in text.

between red and green). Fill the leftmost box with red and the rightmost box with green. Blend Horizontal. Click on each box in turn and drag the color swatch onto the Swatches palette for permanent storage. If you save the document to the Swatches folder of your Illustrator install, you can open this Swatches document any time you choose.

The Blend Top To Bottom filter is similar to the Blend Horizontal filter. It blends from the object closest to the top of the document to the object closest to the bottom.

Blend Front To Back

The Blend Front To Back filter is similar to the Blend Horizontal filter. In this one, the color blends from the object that is the highest in the object list to the object that is at the bottom. It is easy to exert control on this filter even if you do not remember the stacking order of your objects. All you need to do is to send one object to the back and one to the front.

In Illustration 11.2, I created a random composition. I will show you one way to use the result. I frequently like to use luck or controlled randomness in my designs. I like the spontaneity and freshness of the unplanned work, and it can lead me down unexpected paths, especially when I am freely creating or just playing. I hope that you will also find the technique useful.

PROJECT Illustration 11.2: Fish On The Move

1. Create a new document in Illustrator.

2. Choose the Pencil tool and draw a very simple fish shape (even if you have no drawing ability, you should be able to do this). Fill the fish with yellow. Figure 11.6 shows the fish shape that I used. Remember, you can refine the shape just by drawing near the selected shape.

Figure 11.6
A simple fish shape is very easy to draw.

Figure 11.7
Now the fish has an eye and a long strand of seaweed.

3. Draw a small circle for the eye (color it dark green) and a wiggly, closed shape for a piece of seaweed. Make the seaweed a lighter green. Figure 11.7 shows my interpretation.

4. Select the entire image (Mac: Command+A; Windows: Ctrl+A). Drag the selected shapes (i.e., the fish, its eye, and the seaweed) onto the Brushes palette. Create a Scatter brush. Figure 11.8 shows the suggested settings for the new brush (although your size percentage will depend on the size of your original).

5. Group the original fish and cut it. Create a new document and paste the fish into it—just in case you need it again. Save the image. You may close the original fish image now if you want. However, you need to leave open the image in which you defined the fish brush. That is the document in which you need to work.

6. Select the Paintbrush tool. Set the fill to None and the stroke color to yellow (although the stroke color doesn't matter in this instance). Draw random paths covering the image area of the document (see Figure 11.9).

7. Select the entire image (Mac: Command+A; Windows: Ctrl+A). Choose Object|Expand. Expand the fill, the stroke, and the object as shown in Figure 11.10.

8. Select the entire image (Mac: Command+A; Windows: Ctrl+A). Ungroup (Mac: Shift+Command+G; Windows: Shift+Ctrl+G) two times. This removes all the groupings in the file.

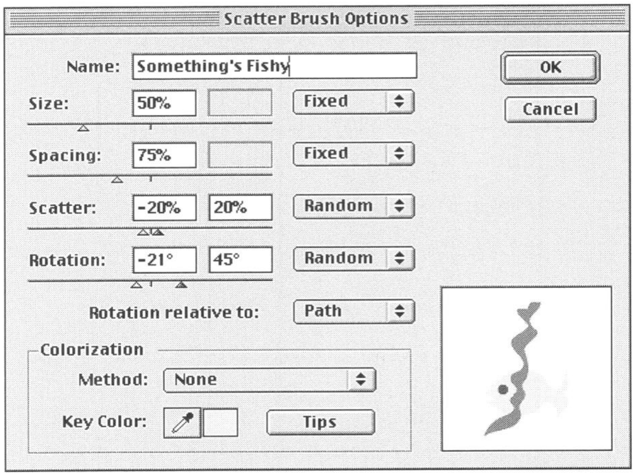

Figure 11.8
Use these settings for your fish brush.

Scatter Brush Options		
Name: Something's Fishy		OK
Size: 50% — Fixed		Cancel
Spacing: 75% — Fixed		
Scatter: -20% 20% Random		
Rotation: -21° 45° Random		
Rotation relative to: Path		
Colorization		
Method: None		
Key Color: — Tips		

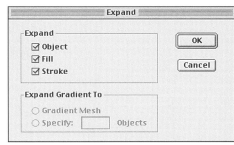

Figure 11.9 (left)
The fish brush is used to draw multiple paths in the image.

Figure 11.10 (right)
The Expand dialog box.

9. Each former stroke now contains a fish, an eye, a piece of seaweed, and the original unfilled, unstroked stroke. You need to remove all the invisible strokes from the image. Select one of the strokes. Figure 11.11 shows what this looks like. Choose Edit|Select|Same Fill Color. Figure 11.12 shows that only the invisible strokes are selected. This is an easy way to find them all. Press Delete or Backspace to remove all the selected strokes.

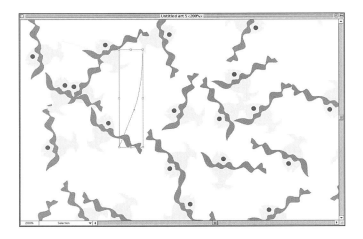

Figure 11.11
One selected invisible stroke.

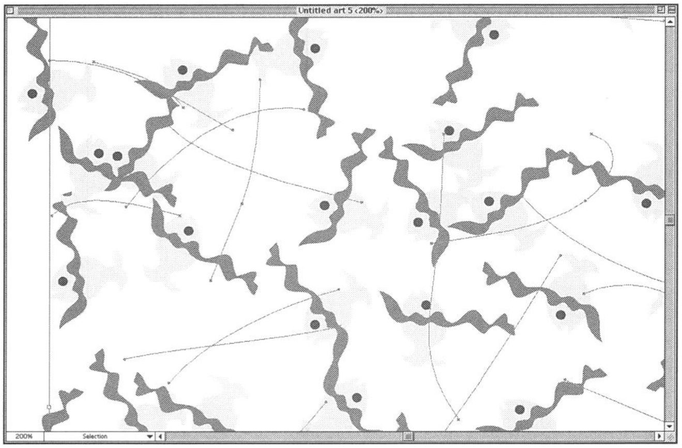

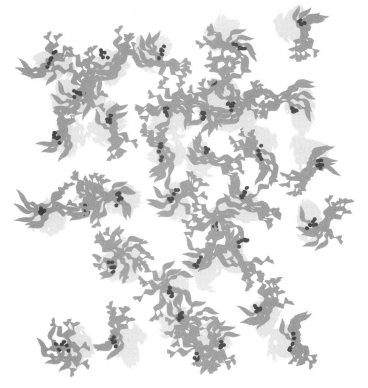

Figure 11.12

The Select|Same Fill Color command chooses only the invisible strokes that originally formed the spines of the scatter brush strokes.

10. Select the entire image (Mac: Command+A; Windows: Ctrl+A). Select Object|Transform|Transform Each. Figure 11.13 shows the settings that I used in the Transform Each dialog box. Make sure that you select the Random checkbox and then click on Copy. Repeat the transformation five more times (Mac: Command+D; Windows Ctrl+D). Figure 11.14 shows the result.

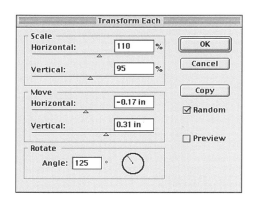

Figure 11.13 (left)

The Transform Each command allows you to randomize the transformation of each object in an image.

Figure 11.14 (right)

The result is an impressionistic "stamping" of the original design.

11. Now the project gets interesting. It is time to select colors. Select a shape and give it a bright yellow fill. Send it to the back (Object|Arrange|Send To Back). Select another shape and fill it with CMYK: 64, 84, 5, 0. Bring it to the front (Object|Arrange|Bring To Front).

12. Select the entire image (Mac: Command+A; Windows: Ctrl+A). Choose Filter|Colors|Blend Front To Back.

13. Now you need to rearrange the shapes. Leave the entire image selected. Click on the Horizontal Align Center command and then on the Vertical Align Center command in the Align palette. You now have a clump of fish, fish eyes, and seaweed in the center of the image.

14. Randomly select objects and drag them to all the areas of the image. Figure 11.15 shows the slight scattering that is needed.

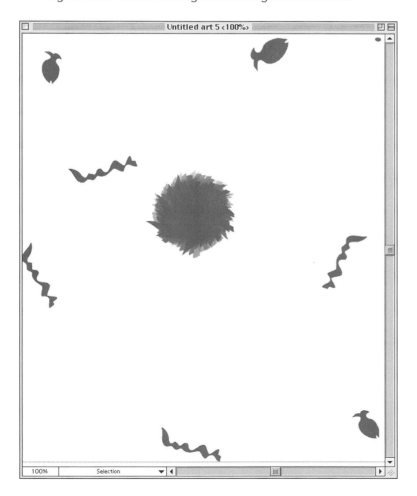

Figure 11.15

After the objects are centrally aligned, drag some of them at random to far-flung corners of your image.

15. Select the entire image (Mac: Command+A; Windows: Ctrl+A). Click on the Vertical Distribute Center command and then on the Horizontal Distribute Center command in the Align palette. This will probably create clumps of shapes similar to those shown in Figure 11.16.

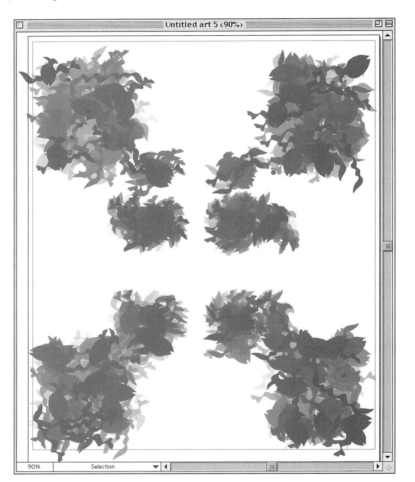

Figure 11.16

The Distribute commands move the fish into regular clumps.

16. You need to rearrange the fish into a less regular pattern. I do this by selecting random fish and moving them and then selecting only some of the areas of the image. If you then click on one or more of the Align or Distribute commands, you can move the shapes over different areas. This is a very iterative process and can continue for a while (until you like the result). I move shapes and select areas and use the Align and/or Distribute commands repeatedly. Figure 11.17 shows a result that is more visually pleasing than the one in Figure 11.16. It took an additional 50 to 75 steps to get the image to this point.

Figure 11.17
A more random arrangement can be created by repeated moving, selecting areas, and using the Align and Distribute commands.

17. As a final "arranging" step, choose Object|Transform|Transform Each and use the settings in Figure 11.18 (or your own). Do not click on Copy. Figure 11.19 shows the result. Notice that the fish "eyes" have now been transformed into confetti.

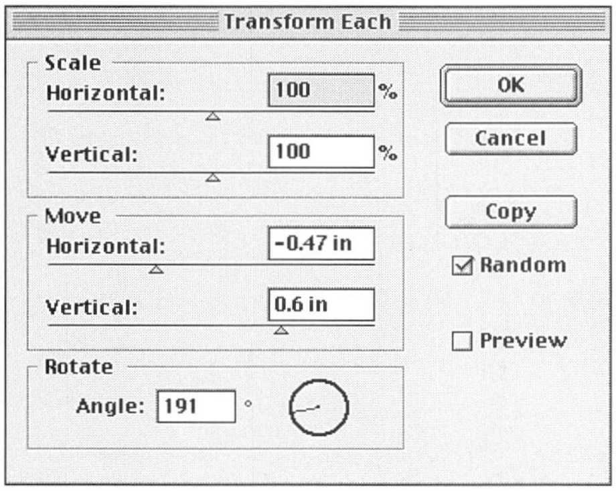

Figure 11.18
Use the Transform Each command for a last touch of randomness.

Figure 11.19
The fish are now more evenly and randomly distributed.

18. Select the shapes that lie outside the printable areas of the document and drag them back into the image.

19. Select the entire image (Mac: Command+A; Windows: Ctrl+A). Click on Stroke and change the stroke color to black. Set the width of the stroke to 1 pt. Figure 11.20 shows the result.

20. Choose the Rectangle tool and create a rectangle the size of the image. Set the stroke to None. Click on the Gradient palette and drag the purple to the start color of the gradient. Drag black to the end color. Choose a Linear gradient with an Angle of 0.

21. Choose Object|Rasterize. Rasterize to RGB mode at 72 ppi. Apply the Ocean Ripples filter (Filter|Distort|Ocean Ripples) with a Ripple Size of 2 and a Ripple Magnitude of 20 as shown in Figure 11.21.

22. Choose Filter|Colors|Convert to CMYK. Many of the filters for raster objects do not work in CMYK mode. Ocean Ripples happens to

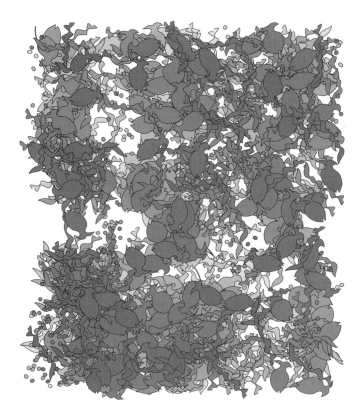

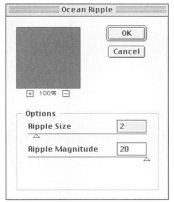

work, but if you are not sure which filter you want to apply, it's easier to rasterize to RGB and then convert to CMYK afterward. (Besides, this chapter is about color changes, so you may as well practice the filter!)

23. Finally, choose Object|Arrange|Send To Back.

24. Hmmm. Purple on purple is a bit dull. Perhaps the colors need to be rearranged so that the yellow objects are closer to the top. Although it would be lovely if there were a command to invert the stacking order of the objects, I have not found one. However, you can reapply the Blend Front To Back filter. Select a yellow fish and bring it to the front. Select the darkest purple fish (or apply the dark purple to the fish) and send it to the back. Select the raster background and send it to the back so that it is truly the bottom object. Now select the entire image (Mac: Command+A; Windows: Ctrl+A). Press Shift and click on the raster object to deselect it. Choose Filter|Colors|Blend Front To Back. Now the yellow fish are closer to the surface.

25. Save the image at this point. Figure 11.22 shows the fish in grayscale. Export a copy of your fish as FISHY.PSD in Photoshop 5 format. You do not need to save it in layers.

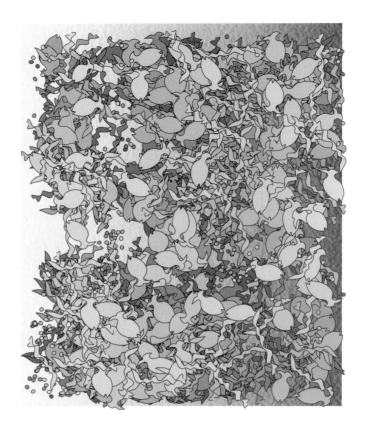

Figure 11.22
The fish image is complete.

Now that you have created this random image, two questions remain: How can you vary the effect, and what can you do with it? Let's tackle each question in turn.

Fish Variations

In the fish tale that you just created, the fish eyes turned into confetti. If you prefer fish eyes, the easiest way to keep them in place is to create a compound path when you first create the fish. The steps are simple: Create the fish, create the eye, and choose Object|Compound Paths|Make. The eye is knocked out of the fish. You can then proceed as you did in Illustration 11.2, but all the fish retain their eyes (of course, you get no sprinkles—unless you also create a tiny circle or two to live near the seaweed). The only problem with this variation is that the Blend Front To Back command doesn't work on compound paths (as you discovered in Illustration 11.1). So you can do it, but you have to release the compound paths before you apply the color filter and then individually re-create each compound path. If you have the patience, it is possible to do this, but I'm not sure that it's worth the effort.

Another variation to try is to use different ranges of color for the fish, the seaweed, and the confetti. This is fairly easy to do. Before you apply the Blend Front To Back color filter, select a fish, change its fill, and send it to the back. Select another fish, change its fill color, and bring it to the front. Select one of the remaining fish. Choose Edit|Select|Same Fill Color. All the remaining fish are selected. With Shift pressed, add the top and bottom fish to the selection, and then choose Filter|Colors|Blend Front To Back. Do the same thing with the seaweed.

To distribute the colors more randomly in levels, you can apply the color filters to the untransformed objects (prior to Step 10 of Illustration 11.2). When you then execute Step 10, your colors are already blended, and you will get multiple levels of color blend. Figure 11.23 shows another fish image done using combinations of all the variations. You can also see it in the Color Studio.

Using The Multi-Fish Image

You can take the fish that you have color-blended and embed them in a raster image in Photoshop. Here's how.

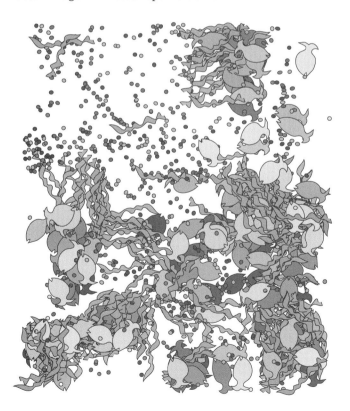

Figure 11.23
The fish and the seaweed are colored separately and the colors applied before the objects are transformed and copied.

PROJECT Illustration 11.3: Fishing In Photoshop

1. Open the image SHARKREEF.PSD in Photoshop. This image is cobbled together from two of my photos. I also included an alpha channel in the image.

2. Also in Photoshop, open either your version of FISHY.PSD or my version on the CD-ROM enclosed in this book.

3. Load the Alpha 1 channel (Mac: Command+Option+4; Windows: Ctrl+Alt+4). This is the outline of a fish. Using the Rectangle tool (not the Move tool), drag the selection outline into the FISHY.PSD image. This moves *only* the marquee and none of the shark. Position the marquee where it encloses an attractive area of the fish image. Press Shift+Command+Option (Mac) or Shift+Ctrl+Alt (Windows) and drag the marquee (this time with the fish inside of it) back into the SHARKREEF.PSD image. It should appear exactly over the marquee area in your image. The drag-and-drop operation creates Layer 1. Figure 11.24 shows the image covered with a fishy shark.

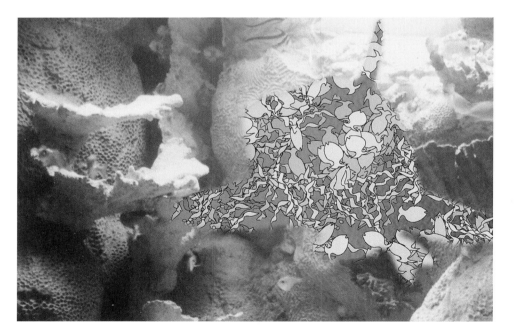

Figure 11.24
The marquee shape is now filled with part of your fishy image.

4. Make the Background layer active. Load the Alpha 1 channel again. Create a New Layer Via Copy from the selection (Mac: Command+J; Windows: Ctrl+J). This creates Layer 2. Drag the thumbnail for Layer 2 above the thumbnail for Layer 1 in the Layers palette.

5. Apply the Emboss filter (Filter|Stylize|Emboss) with the settings shown in Figure 11.25. Desaturate the image (Mac: Shift+Command+U; Windows: Shift+Ctrl+U). Change the Apply mode to Hard Light.

6. Make a new layer (click on the New Layer icon at the bottom of the Layers palette). This is Layer 3. Drag Layer 3 so that it is between Layer 1 and the Background layer in the Layers palette. Fill the layer (Mac: Option+Delete; Windows: Alt+Backspace) with a purple selected from your FISHY image. I used RGB: 114, 70, 120. Change the Apply mode to Color.

7. Make the Background layer active. Double-click on the thumbnail for the Background layer. This brings up the Make Layer dialog box (see Figure 11.26). Accept the default name of Layer 0 and set the Opacity to 48%. Click on OK.

8. Create a new Background layer (Layer|New|Background). Figure 11.27 shows the Layers palette at this point.

9. Drag the thumbnail for Layer 0 to the New Layer icon in the Layers palette. This creates Layer 0 copy. Choose Filter|Stylize|Emboss and use the settings shown in Figure 11.28 (the same angle and pixel height but a different Emboss amount from Step 5). Change the Apply mode to Hard Light. Leave the layer Opacity at 48%. Figure 11.29 shows the image at this point.

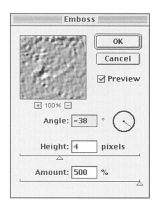

Figure 11.25
The Emboss filter is used to add texture to the fish-covered shark.

Figure 11.26
Changing the Background layer to Layer 0 allows you to adjust its opacity.

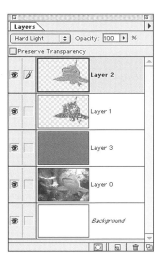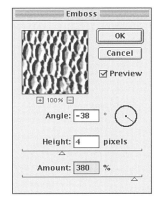

Figure 11.27 (left)
The Layers palette, after you add a new Background layer.

Figure 11.28 (right)
These Emboss filter settings are used to add texture to the entire base image.

Figure 11.29

The image looks as though it's starting to become a unified piece of work.

10. Make Layer 0 the active layer. Press Option (Mac) or Alt (Windows) and click on the Eye icon next to Layer 0 to hide all the layers but Layer 0. Select the Lasso tool and trace around the eye of the shark. Press Option (Mac) or Alt (Windows) and click to reveal all the layers again. Create a New Layer Via Copy from the selection (Mac: Command+J; Windows: Ctrl+J). This is Layer 4. Drag its thumbnail to the top of the Layer stack and change the layer Opacity to 100%.

11. Make the FISHY.PSD image active. Using the Move tool, drag the FISHY.PSD image and drop it into your working image. Position it so that the bottom left of the FISHY.PSD image is in the bottom left of the working image.

12. Use the Free Transform command (Mac: Command+T; Windows: Ctrl+T) and drag the control handle of the bounding box on the right over to the right edge of the image. This scales Layer 5 dispro-portionately, but the distortion isn't really obvious. It also manages to make it the same size as your working image. Figure 11.30 shows Layer 5 being transformed.

13. Drag the thumbnail for Layer 5 until it is above the Background layer but below Layer 0 in your image. Change the layer Opacity to 70%. Now there are fish swimming in the coral background. Figure 11.31 shows the current image.

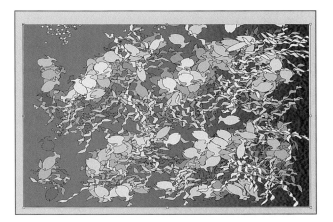

Figure 11.30
The Free Transform command is used to scale the FISHY image so that it covers the entire working image.

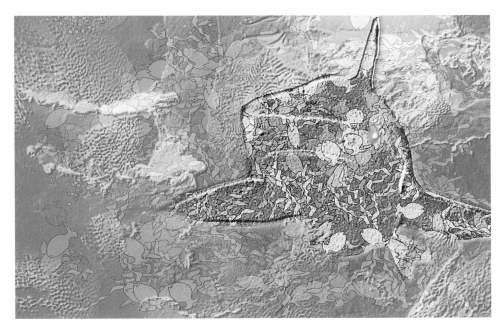

Figure 11.31
The shark still looks a bit flat.

14. Finally, you need to add shadow to give the shark more depth. Make Layer 1 the active layer. Drag the thumbnail for Layer 1 to the New Layer icon at the bottom of the Layers palette to make a copy of the layer. Drag the thumbnail for the Layer 1 copy *below* that of Layer 1 in the Layer stack.

15. Your foreground color swatch in the Toolbox should still contain the color that you used as a fill for Layer 3. Make the Color palette visible if it is not and change the slider to CMYK. Even though you are working in RGB color, by using the CMYK slider you can easily darken the color in your foreground color swatch. Add a little black to the purple foreground color to darken it.

Figure 11.32

The settings used in the Gaussian Blur filter to make a shadow.

16. Fill Layer 1 copy with the foreground color with Preserve Transparency On (Mac: Shift+Option+Delete; Windows: Shift+Alt+Backspace). Change the Apply mode to Multiply and the Opacity to 78%.

17. Use the Move tool to move the shadow substantially to the left and down. Apply a Gaussian Blur filter (Filter|Blur|Gaussian) with an amount of 18.0 (see Figure 11.32).

18. Finally, make Layer 1 active. Change the Opacity to 87%. Figure 11.33 shows the finished image in grayscale. You can see it in color in the Color Studio.

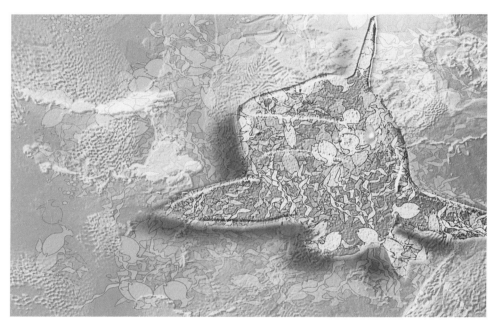

Figure 11.33

The finished image.

Mixing A Logo

The Pathfinder palette contains two commands that are useful for creating color effects. To see them, however, you need to select Show Options on the Pathfinder Palette menu bar. The Hard Mix and Soft Mix commands are then available.

Adobe Illustrator doesn't allow transparency (for this, you need a third-party filter). The Hard Mix and Soft Mix commands allow you to approximate transparency. These commands do not work on compound paths and do not work on gradient fills or gradient mesh objects. If you can tolerate the limitations, they are a good addition to your arsenal of tools.

The Mix Hard command keeps the highest percentage of each color component. For example, if you mix an object that contains CMYK:

100, 0, 0, 0 (cyan) with an object that contains CMYK: 0, 0, 100, 0 (yellow), you will get CMYK 100, 0, 100, 0 (green). You have no options on the Hard Mix command.

The Soft Mix command allows you to select the percentage of mix. Using the previous example with a 50 percent mixture of the colors, you would get CMYK 33.33, 0, 66.67, 0. The math is not intuitive, but the color is a soft green, which is about what you would visually expect to see.

One advantage of using either command is that you can create a harmonious range of colors regardless of the original colors in the image. The next illustration shows the use of these commands in a technique that I frequently use when I am asked to develop an abstract logo design for a company. The technique is very simple, but the effectiveness comes from the design principle of repeating elements and shapes. The unity of the colors comes from the use of the Hard Mix or the Soft Mix command.

PROJECT Illustration 11.4: Steven's Study Skills

You are going to develop a logo for a company called Steven's Study Skills.

1. Create a new document (Mac: Command+N; Windows: Ctrl+N) in Illustrator.

2. Using the Ellipse tool, draw an oval shape that resembles a head (see Figure 11.34).

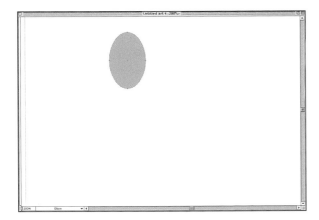 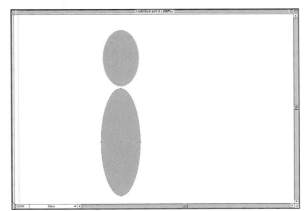

Figure 11.34 (left)
A vaguely head-shaped object is the start of a logo.

Figure 11.35 (right)
An elongated oval can suggest a body.

3. With the same tool, draw an elongated body as shown in Figure 11.35.

4. Draw another oval to resemble an arm and then use the Rectangle tool to draw two shapes that look like books. Figure 11.36 shows this stage completed.

5. Change the fill color for the shapes to whatever you want them to be. I used a different color for each shape and chose them randomly.

6. Select the entire image (Mac: Command+A; Windows: Ctrl+A). Choose Object|Transform|Transform Each. Figure 11.37 shows the settings that I used. Select the Random checkbox. Don't click on Copy. Repeat the transformation (Mac: Command+D; Windows: Ctrl+D) at least four times (or more times) until you are happy with the result. The shapes should take on a more abstract look.

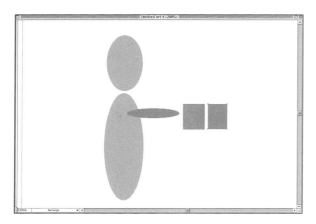

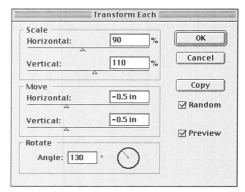

Figure 11.36 (left)

With simple shapes, you can create a very abstract student.

Figure 11.37 (right)

After you specify parameters in the Transform Each dialog box, you can choose to have Illustrator apply them randomly to each object.

7. Rearrange the altered shapes so that they are vaguely in their starting positions. You may rotate them as you feel necessary. Figure 11.38 shows my arrangement.

8. Select the Rectangle tool. Set your fill color to a bright red. Drag a rectangle so that it extends fairly far out on the left but does not quite cover the original shapes on the top, bottom, or right side. Figure 11.39 shows the position of the rectangle.

9. Click on Mix Soft in the Pathfinder palette. Choose 50% for the mix rate. Figure 11.40 shows the selected mixed item. Illustrator creates the illusion of a blended, transparent shape by taking the individual shapes and merging them into one grouped object that might also contain compound paths.

10. Ungroup (Mac: Shift+Command+G; Windows: Shift+Ctrl+G). Select the red rectangle and move it slightly to the left. With the Rotate tool, rotate the shape about four degrees. Figure 11.41 shows the final spacing I used.

11. Add the logo text "Steven's Study Skills". I used FFJustlefthand Caps at 72 points with 64-point leading. Any handwriting font should

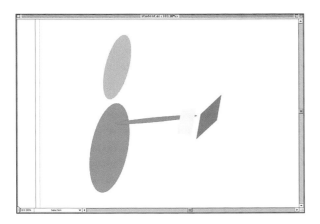

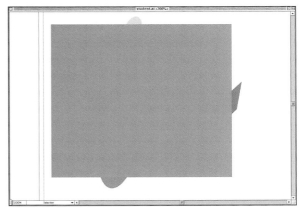

work as well. If you like the logo text, I have included an outline version of this text as LOGOTYPE.AI on the CD-ROM that accompanies this book. Figure 11.42 shows the logo with type added.

The design principle behind this exercise is that you create a negative shape that mimics the positive one formed by the original objects. Rotating this shape adds interest to the design but is not necessary. You can vary this technique in many ways. One is to take the completed logo into

Figure 11.38 (left)

After you transform the shapes, you can rearrange them as you like.

Figure 11.39 (right)

The rectangle should not quite cover the entire set of underlying shapes.

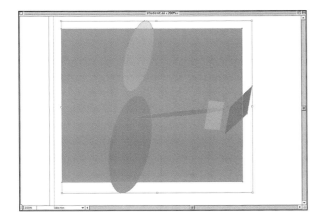

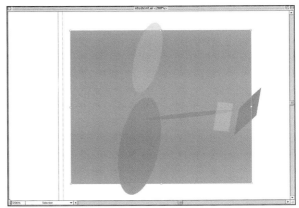

Photoshop. Figure 11.43 shows my additions to the logo in Photoshop. I created three layers in Illustrator. The bottom layer contains the red background shape, the top layer contains the text, and the middle layer holds all the other objects. I exported the image as a layered Photoshop 5 file. In Photoshop, I added 200 pixels to the canvas size in each dimension and created a white Background layer. I applied the KPT Texture Explorer filter to the red background shape, using Procedural + "Glue" in the Filter dialog box. I then used Fade Filter (Mac: Shift+Command+F; Windows: Shift+Ctrl+F) to change Apply mode to Multiply and the Opacity to 25%

Figure 11.40 (left)

Illustrator creates a mix by grouping the objects together into one.

Figure 11.41 (right)

Rotating and moving the background shape creates an attractive negative shape in the logo.

Figure 11.42 (left)
Lettering added to the logo completes the design.

Figure 11.43 (right)
You can use Photoshop to enhance your logo design.

so that there is a much gentler mix of original and KPT filter. I used the Inner Bevel filter in the Eye Candy 3 filter set to create the bevels on the student shape. I then cut Opacity of the red layer to 91% to slightly soften the color. The altered image looks more impressive, but I also like the stark colors and flat texture of the original.

Transparent Objects

Both Freehand and CorelDRAW have permitted transparent objects, and Illustrator users have been drooling in envy. Adobe, however, didn't implement transparency in this version of Illustrator. A small filter company, HotDoor, has released an inexpensive plug-in ($39.00) to add some transparent effects to Illustrator.

Figure 11.44
The HotDoor Transparency filter adds a simple palette to Illustrator.

Transparency needs to be installed in the Plug-Ins directory of Illustrator. Figure 11.44 shows the simple palette that the filter adds to Illustrator. The plug-in is simple to use, as its palette suggests. You may designate any simple shape (i.e., one without blends, gradients, gradient meshes, masks, or compound paths) as a transparent object. If you select Live Update, the transparent effect immediately recalculates when you move the object. The transparent object can be moved on top of almost any type of object (pattern, gradient mesh, and so on) and will show the effect of transparency. You cannot see the result over a bitmapped object or over any of Illustrator's brush effects. A major difference between this filter and Freehand's Lens fill is that the portion of the shape that is not over another shape keeps 100 percent opacity. In Freehand, if the object is supposed to be 50 percent transparent, it will contain 50 percent of its color mixed with white when it isn't on top of anything else. Reducing opacity throughout the object is far more realistic than the way it is done in the Transparency filter. Illustration 11.5 presents a notebook cover that shows the features of HotDoor's filter. In the Illustration, I assume that you have the full version of the filter. If you don't own this filter, you may download the latest version of the demo from **www.hotdoor.com** or use the one on this book's CD-ROM. It's a crippled demo and doesn't allow you to change transparency levels or to perform a live update. If you like the filter, you can purchase it online.

PROJECT Illustration 11.5: Stephanie's Notebook

1. In Illustrator, create a new document.

2. Scatter some random shapes over the document (see Figure 11.45). Fill one or more of them with a pattern, making sure that at least one shape overlaps another. Add a wide stroke to several of the shapes.

Figure 11.45
Create some random rectangles.

3. Select the shape that overlaps another. In the Transparency palette, click on the Make button. Select the Live Update checkbox. Drag the transparency slider from 25% to 75% to see how the color on top reacts. Figure 11.46 shows a transparency on the left of 25% and on the right of 75%. The transparency of 25% results in a darker overlap because the shape is more opaque at 25% transparency than it is at 75%. (Remember that the terminology is backward to the way transparency works in Photoshop, where you are actually setting *opacity*). Notice the unnatural way in which the black shape on top stays black when it isn't over the other object. Figure 11.47 shows the way the shapes would look in Freehand (or in reality).

4. Draw a large rectangle on top of the shapes so that it intersects with most of them. Figure 11.48 shows my image at this stage.

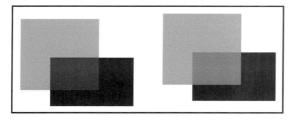

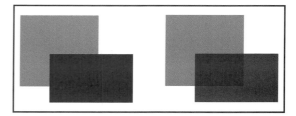

Figure 11.46 (left)
The top shape has 25% transparency on the left and 75% transparency on the right.

Figure 11.47 (right)
You can make the entire object look transparent if you place a white shape behind the transparent object and send it to the back.

5. Select the rectangle and click on Make in the Transparency palette.

6. Type the name of your favorite student in 18-point text. Select the text object and choose Text|Create Outlines. Group the text outlines (Mac: Command+G; Windows: Ctrl+G). Drag the group onto the Brushes palette and define the brush as an Art brush as shown in Figure 11.49. Use the Paintbrush tool to randomly paint the name across the shapes. As long as these brush strokes stay on top of the transparency lens, all is well. If you place one of them behind a lens, you will no longer see the portion under the lens object.

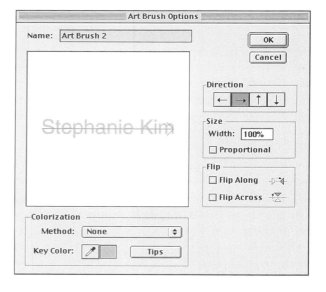

Figure 11.48 (left)
Drawing the final rectangle.

Figure 11.49 (right)
Create an Art brush from your favorite student's name.

7. Finally, to keep the transparency accurate, draw a large white rectangle that covers everything in the image and then send it to the back. Figure 11.50 shows the final image.

When you are ready to save the image, you need to select the entire image and click on Expand in the Transparency palette. This "sets" the transparent effect so that it is changed into "native" Illustrator shapes, masks, and compound paths. Otherwise, you would not be able to take the file to a service bureau or share it with someone who does not own the filter.

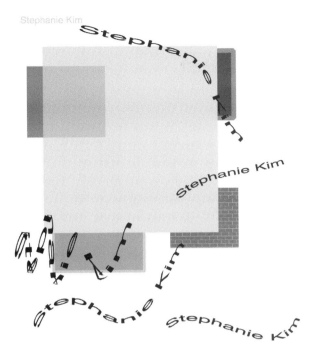

Figure 11.50
Stephanie's completed notebook cover.

Moving On

In this chapter, you explored some of the color effects that are possible using Illustrator 8. Although this version has fewer of them than does Illustrator 6 or 7 (because of the absence of several filter add-on sets), you can still create interesting color changes. You can use the Blend Horizontal and Blend Vertical filters in a variety of ways. You can explore the Saturate, Adjust Colors, and Invert filters. You can use the Mix Hard or Mix Soft command, and you can even add transparency effects using the HotDoor Transparency filter.

In Chapter 12, Michael will show you a variety of effects and techniques that can be used to create images for the Web.

WEB GRAPHICS

12

T. MICHAEL CLARK

If you design graphics, chances are that you've needed to design Web graphics. Chances are that you've also been faced with a whole new set of questions and problems— questions such as which file format to use and problems such as color palettes. This chapter answers these questions, helps you solve these problems, and more.

File Formats

One of the first questions you might have when designing graphics for the Web is which file format to use. Although you're currently limited to only two formats, JPEG and GIF, the question is not a simple one to answer. I expect that, as an illustrator, however, you should have an idea of the basics of both the JPEG and GIF file formats. To that end, I'll not present you with the basic concepts such as color palettes, compression, or the various options that Illustrator makes available. Rather, I'll present a short description of GIF versus JPEG and then move on to how you can create some cool effects for the Web, using Illustrator.

GIF Or JPEG?

Even though you're limited to only the two formats, it's still a tough question: Which one should you use for your Web images? A good rule of thumb is that simple images with few colors and hard edges would be better off as GIFs. You can really squeeze these down in size by using, for example, limited palettes. Even fairly large (in terms of dimensions) images can be compressed to manageable file sizes if the number of colors is limited.

Images with gradients and gradient meshes might be better off as JPEGs. Images that contain this type of information will actually compress better than images with solid colors when using the JPEG format. Make sure that any type is still readable. This can be problematic with smaller type and high compression.

If you're not sure which format is best, try experimenting. Set aside a temporary folder and save the image, using the different methods and setting the various options. After you've saved a couple of instances of the image, use your favorite image browser to display the images. Look at the quality, then check the sizes of each instance to see how the file size compares to the compression method and the quality. Eventually, you'll be able to save an image with the correct method just by using your experience and judgment.

Previewing JPEG And GIF Files

Wouldn't it be nice if you could preview the effects of saving your images as JPEGs or GIFs in realtime? Well, you can if you own a program such as Adobe's ImageReady.

With ImageReady, you can open files saved in a variety of formats, including Illustrator AI and Photoshop PSD files. After opening a file saved directly from Illustrator, you can open another window and preview the effects of saving the file in either JPEG or GIF (see Figure 12.1).

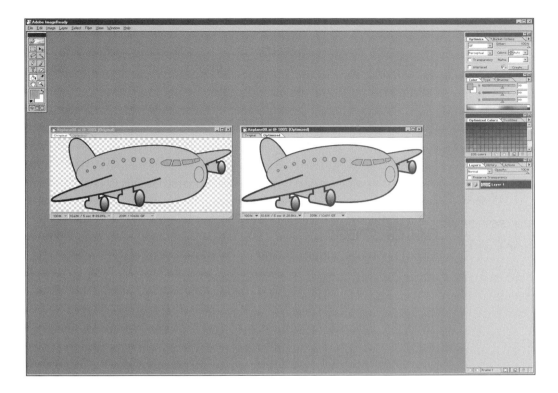

Figure 12.1
Adobe's ImageReady.

You'll get realtime feedback as you set the different options. You can go from JPEG to GIF, set the number of colors, select a palette, choose a method, and more. In short, you can set any or all of the options that we've looked at so far and all the changes will be immediately obvious to you.

If you spend even a small part of your working day developing graphics for the Web, ImageReady is an indispensable tool.

A Final Word On Web Graphics Formats

Before I move to the next section, I should state that I firmly believe that more and more users are surfing the Web with machines capable of displaying more than 256 colors and that they are doing so with their browsers set to 800×600 rather than 640×480. Some designers might argue this with this, but the results of surveys I've read in the last year or so support this new reality.

The bottom line: You should design for what you want or for what the client is paying for. If the client insists that you use Web-safe colors, use them, whether or not I think it's necessary.

Buttons

Hardly any images are more important on a Web page than the buttons. The buttons not only enable your readers to navigate around your Web site, they can actually convey a feeling of what the Web site is all about.

Using Illustrator, you can draw every type of button conceivable, from simple and iconic to more dramatic 3D buttons. Buttons can be made completely from type or from symbols. You're limited only by your imagination when it comes to creating buttons.

Iconic Buttons

Icons, done properly, can make great buttons. However, you must make sure that icons aren't too ambiguous, especially if you don't have any text accompanying the buttons. A button as simple as the one seen in Figure 12.2 is a good choice.

You can be fairly sure that most surfers will recognize the fact that clicking on this simple button will return them to your home page. The icon shown in Figure 12.2 was drawn by combining several simple shapes. You could embellish on this simple button by adding a gradient-filled circle behind the original shape, as shown in Figure 12.3.

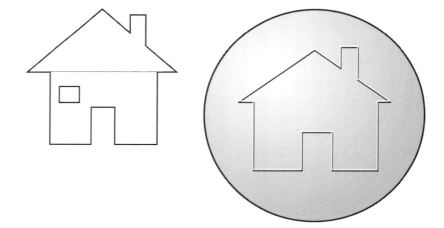

Figure 12.2 (left)
A simple iconic Home button.

Figure 12.3 (right)
Embellishing the Home icon.

Along with adding the gradient-filled circle, I copied and pasted the original shape. I set both shapes to a fill of None and set the stroke of one to white and the other to black. Lining them up gives the impression of the home being "raised." This simple techniques adds some depth to the otherwise simple icon. You can use this and other techniques (many of which are included in this book) to give some simple shapes a bit of flavor.

Shaped Buttons

I mentioned previously that buttons are ubiquitous. This is borne out by the fact that Adobe has included a set of button actions on the Application CD-ROM. The Buttons actions aren't installed by default, but you can find them on the Application CD-ROM in the Illustrator Extras|Action Sets folder.

To use these actions, follow the steps I outlined in Chapter 6 to load them into Illustrator. Figure 12.4 shows all 14 buttons that you can create using the Buttons action.

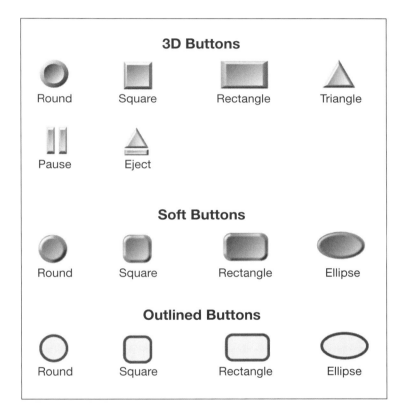

Figure 12.4

All the buttons that you can create using the Buttons action are included on the Illustrator Application CD-ROM.

Illustration 12.1 shows how to create a similar button on your own. In this Illustration, the button has a shape (pill) other than the ones included in the action set.

Illustration 12.1: Creating A Pill-Shaped Button

1. Open a new file in Illustrator.

2. Turn on the grid (Choose View|Show Grid).

3. Select the Rectangle tool.

4. Choose View|Snap To Grid.

5. Set the default Fill to white and the Stroke to black.

6. Draw a rectangle that stretches horizontally between four of the darker lines in the grid and between two of the vertical lines (i.e., the rectangle should be about three inches long and one inch high).

7. Select the Ellipse tool.

8. Place the mouse cursor in the middle of the line that makes up the left side of the rectangle that you just created.

9. Click and drag the mouse to get the ellipse started. With the ellipse shape started, holding down both keys (Mac: Shift+Option; Windows: Shift+Alt), draw the ellipse from the center out to ensure that it's a circle rather than an ellipse. Having the Snap To Grid option on will make it easy to draw the ellipse or circle to the same size as the height of the rectangle.

10. Repeat the process on the right side of the rectangle.

11. You should now have a rectangle and two circles as seen in Figure 12.5.

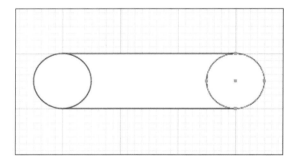

Figure 12.5
The beginnings of a pill-shaped button.

12. Select the Selection tool and draw a marquee around all three objects.

13. Choose Window|Show Pathfinder.

14. Click on the Unite icon to combine all three shapes into one.

You now have the basic pill shape. You can leave it this way and simply add text. If you want, however, you can add some dimension to the button by following the remainder of these steps.

15. Make sure that the button is still selected. If it's not, select the shape by using the Selection tool.

16. Choose Object|Path|Offset Path and set the options in the Offset Path dialog box as follows: Set the Offset to 0.139 inches, the Joins to Bevel, and the Miter Limit to 0.056 inches. These settings assume that the pill shape you created is the same size as the one described in the preceding steps. If you're working on a different-size button, play around with the settings. The object is to get a button that has a larger button behind it.

17. After creating the offset, the larger button will still be selected.

18. Click on the Gradient icon to fill the larger button with a gradient. I used a simple black-to-white linear gradient and set the Angle to 45 degrees.

19. Select the inside shape and fill it with the same gradient but set the Angle to 135 degrees. You might want to (as I have) also set the Stroke of the inner shape to None (see Figure 12.6).

Figure 12.6
The final pill-shaped 3D button created in Illustrator.

Filling the shape with the opposing gradients is what gives it the appearance of depth. This technique can be used on many different shapes. All that's left is to add some text and save as either a Web-ready GIF or a JPG. (Hint: With the gradient, JPG might be a better choice.)

Although they're not really buttons in the strictest sense, image maps are navigational aids that function similarly to buttons. In the next section, I show you how to create image maps in Illustrator 8. As a bonus, Illustrator will even create the HTML code for the image map.

Image Maps

Image maps are images that have more than one "hot spot"—that is, a selectable area contained within the image. Much as an image can be a button that, when clicked by Web surfers, will direct them to another page, an image map can function as many buttons that can direct the surfer to different pages. To put it another way, an image map is an image that contains areas that function the same way as buttons work.

Until recently, constructing an image map was rather tedious. I spent the better part of an afternoon (on more than one occasion) specifying coordinates by using the Info palette in Photoshop, jotting them down on small scraps of paper, and then transferring them to HTML, only to find that I had made a typo somewhere. No more!

With Illustrator 8, it's incredibly easy to spec out portions of an image to use in an image map. However, even better is the fact that Illustrator will write the HTML for the image map! That's right, when you add hot spots, or links, to an object in an Illustrator image, most of the work needed to turn the image into an image map will be done for you.

Follow along with Illustration 12.2 to see how easily you can construct an image map in Illustrator.

PROJECT Illustration 12.2: Image Mapping: Here There Be Buttons

1. Open a new file and draw some buttons, using any technique you want.

   Figure 12.7 shows a series of simple objects drawn using some of the techniques described throughout this book. For example, you can refer to Chapter 10 to see how the type was laid out on the path around the arrows.

Figure 12.7
A series of objects to be used as buttons in an image map.

2. To set up an object as a hot spot in the image map, simply select it by using the Selection tool and enter the URL (Universal Resource Locator—usually an address that points to a Web page, such as **www.grafx-design.com**) in the space provided in the Info palette (see Figure 12.8).

   You can see in Figure 12.8 that the Home button, or icon, is selected and that I entered "home.html" as the URL for this object.

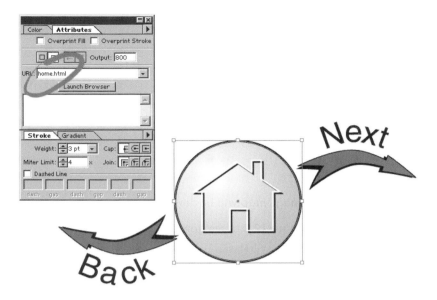

Figure 12.8
The URL for this object points to the home page.

3. To complete the image map, select each object you want to use as a link and enter the appropriate URL in the Info palette.

   This image was designed to be used as a navigation tool through a series of tutorials (here, the reader would be reading the second tutorial). Given that, the Back button should point to a page named something like tutorial01.html, and the Next button should point to tutorial03.html.

   To set the Back object (the arrow and the type), I grouped together the two objects and then, after selecting them with the Selection tool, I entered "tutorial01.html" as the URL. I repeated the process for the Next type and arrow and pointed them at tutorial03.html.

4. To save the image and the image map HTML, export your image as either a JPEG or a GIF.

5. In the JPEG Options (or GIF89a Options) dialog box, select the Image Map checkbox (see Figure 12.9).

6. Leave the Anchor as the default. This option defaults to the name of the file.

7. Use Explorer (or the equivalent on a Mac) to open the folder to which you exported the image map file. You should see an HTML file with the same name. For example, I named the exported JPEG file "Navigate.jpg", which created a file named Navigate.html.

8. Double-click on the file's icon, and it should open in your Web browser (see Figure 12.10).

Note: *Today, most browsers recognize client-side (HTML) image maps, and you should choose this option. A discussion of client-side versus server-side image maps is beyond the scope of this book. For more information about this subject, consult any good book on HTML.*

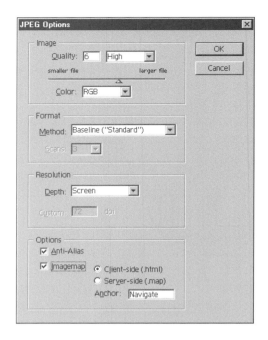

Figure 12.9
Exporting an image as an image map.

9. To add the HTML (see Listing 12.1 for an Illustrator 8 code example) of the new image map to the HTML of an existing Web page, open the source of the new file from within your browser (in Netscape, choose View|Page Source), select the text of the image map (see the following code), and choose Edit|Copy.

Listing 12.1   HTML code generated by Illustrator for a client-side image map.

```
<!-- Begin Adobe Illustrator 7.0 generated client-side imagemap -->
<map name="Navigate">
<area shape="rect" coords="259,79,400,195" href="home.html">
<area shape="rect" coords="262,82,403,198" href="home.html">
<area shape="rect" coords="377,0,658,136" href="tutorial03.html">
<area shape="rect" coords="423,63,611,128"
href="tutorial03.html">
<area shape="rect" coords="0,197,281,333" href="tutorial01.html">
<area shape="rect" coords="50,168,238,233"
href="tutorial01.html">
<area shape="rect" coords="232,47,431,246" href="home.html">
</map>

<img src="Navigate.jpg" usemap="#Navigate">

<!-- End Adobe Illustrator 7.0 generated Netscape imagemap -->
```

10. Open the HTML page to which you want to add the image map in your favorite HTML editor and choose Edit|Paste.

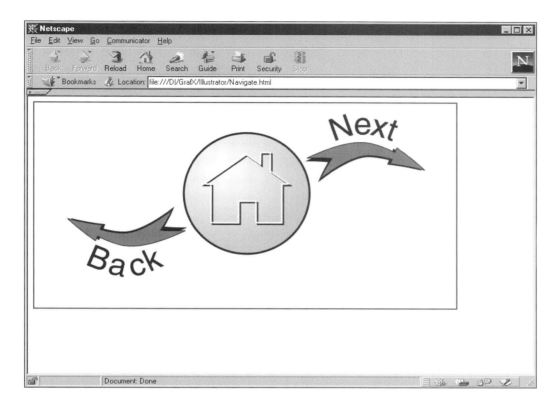

Figure 12.10
The image map file opened in Netscape.

You might have to do a little editing, but all the grunt work (i.e., creating the coordinates and so on) has been done. It's as simple as that! No matter how complex the image and the objects you want to use as links, the process is the same, as follows:

1. Select an object.

2. Enter a URL in the Info palette

3. Repeat for all links.

4. Export the image as a Web-ready JPEG or GIF.

5. Copy and paste the HTML generated by Illustrator into an existing HTML document.

Animated GIFs

I must admit that, at first, I shunned the idea of animated GIFs. This probably is because the first ones I saw on the Web were, well, cheesy. However, I've come to see that they can be a great medium that you can use to express all kinds of ideas.

Although Illustrator can't construct the final animation file, it's a great program when you are creating the frames needed for an animated GIF.

Because Illustrator files consist of vector-based objects that can easily be rearranged, these files are perfect for the basis of an animation file.

After putting together the images for the separate frames of your animation, you need a program that will put together the frames to form the animated GIF. Here are a few programs I recommend, along with the URLs (Web addresses) where you can download evaluation demos:

- Adobe's ImageReady, **www.adobe.com**

- Auto F/X's Universal Animator, **www.autofx.com**

- Gamani's Gif Movie Gear, **www.gamani.com**

- Ulead's GIF Animator, **www.ulead.com**

Undoubtedly, many more similar programs are available, but these are the ones I've tried. For the purposes of this exercise, I'll use Gamani's GIF Movie Gear.

I'll show you the basic steps needed to construct a simple GIF animation (I believe that simpler is always better). You can expand on the process, using Illustrator and the GIF animation software of your choice, to construct animations that are as complex as you want.

Although you won't be able to see the final animation on these pages, I'll include the Illustrator file and the resulting animated GIF file on this book's companion CD-ROM so that you can dissect them at your leisure to see what I did.

The animation I decided to create in Illustration 12.3 will be (or could be, anyway) used as a Web page advertising banner to advertise this *Illustrator 8 f/x and design* book. As I said, I prefer simple, and this is a simple animation. I wanted something maybe a little humorous that would do a good job of getting across the message that this book is about creating special effects.

The idea that I came up with is completely constructed with type. To get started, I drew up a quick storyboard and brainstormed the type that I wanted. I decided on two main frames. The first says "Need to add some Pizazz to your graphics?" and the second says "You should read Illustrator 8 f/x."

The humor comes in as, after holding on the second frame for a couple of seconds, the subsequent frames show the letters "f/x" toppling over and bouncing. Sound difficult? Not really, when you use Illustrator to construct the frames. It will be fairly easy to follow along with this Illustration. If you're ready, let's get started.

PROJECT ## Illustration 12.3: Lights! Computer! Action!

1. Open a new Illustrator file.

2. Set the Fill as the default white and Stroke to black.

3. Create a rectangle using the Rectangle tool.

4. Edit|Copy and Edit|Paste the rectangle so that you have two rectangles for the two main frames.

5. Enter the type for the first frame using the Type tool over the first rectangle (see Figure 12.11).

Need to add some Pizazz to your graphics?

Figure 12.11
The first frame of the animated GIF.

Note how I changed the color of the word, "Pizazz". I used the technique that's described in Chapter 10, where all the techniques (only a couple) used in this example are described.

6. Using the Type tool, enter all the text for the second frame except for the final "f/x". This will be done differently because it's the only part of the image that will be animated.

7. Select the Ellipse tool and create a small circle to the right of the type that you created in Step 6.

8. Select the Path Type tool and enter "f/x" on the circle. You should now have the two main frames as seen in Figure 12.12.

Need to add some Pizazz to your graphics?

You should read Illustrator 8

Figure 12.12
The two main frames of the animation.

Because I'm using GIF Movie Gear in this Illustration, (mainly because it can import Photoshop files), I will copy the images into Photoshop. I could save each image as a separate GIF, but having all the frames stored in a PSD file makes it easier for the way that I work.

9. If you have Photoshop, select the first frame that you created and choose Edit|Copy, then open Photoshop and choose File|New (leave

the default settings in the dialog box) and Edit|Paste. The frame will come in as a new layer over the background layer. If you're using some other program to construct the animation, you should read the manual to find out how to import the separate images as frames. For example, ImageReady will allow you to import the file directly from Illustrator or even to open an Illustrator format file, which other programs won't do.

10. Use the Selection tool to place the "f/x" at the top of the circular path (see Figure 12.13).

You should read Illustrator 8

11. Copy this frame into your animation program or into Photoshop (as I've done) in preparation for creating the animation. You should now have two complete frames.

12. Use the Selection Tool to move the "f/x" type down the circular path in a clockwise direction and copy and paste the entire second frame into your animation program as the third frame.

13. Continue this process, moving the type a little further along, until the "x" is parallel to the horizon (see Figure 12.14).

You should read Illustrator 8

This ended up being frame 5 in my file. The first frame is the type with the question, the second is the frame with the book title and the "f/x" straight up. The third, fourth, and fifth frames have the "f/x" moving clockwise in increments, as if on a pivot, until it falls over. Figure 12.15 shows all the frames needed to construct the animated GIF file.

Remember that I want the "f/x" to bounce. Because it will do so along the same line that it fell through, it will be a simple matter to copy and paste the needed frames in the animation software. Figure 12.16 shows all the frames in place for the final animation with GIF Movie Gear. Frame 6 is a copy of frame 4, and frame 7 is a copy of frame 5. Copying and pasting these two additional frames will make the "f/x" appear to bounce once after hitting bottom.

Need to add some Pizazz to your graphics?

You should read Illustrator 8

You should read Illustrator 8

You should read Illustrator 8

You should read Illustrator 8

Figure 12.15
All the frames needed to construct the animation.

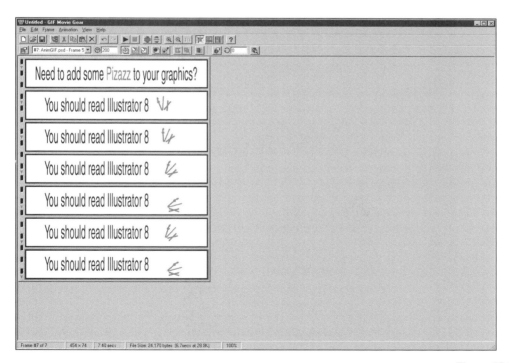

Figure 12.16
The final animation sequence is ready.

With all the frames in place, only a couple of steps remain.

14. Set the timing of the first frame to 3 seconds (refer to the manual that ships with your animation program).

15. Set the timing of the second frame and the last frame to 2 seconds.

16. Set the timing of the third, fourth, fifth, and sixth frames to 1/10 of a second.

With this timing, the first frame will display long enough for the reader to read the question. The second and the last frames will display long enough for the letters "f/x" to sit still for a moment, and the in-between frames will move quickly enough that it appears as if the "f/x" has toppled over.

I include the layered PSD file and the final GIF animation file on the CD-ROM that accompanies this book so that you can further dissect them to see how I created the effect.

After you have all the frames in place and have set up the timing for each one, all that's left is to optimize the file, save it, place it on your Web server, and point to it from within an HTML file. Read the documentation that came with your GIF animation software for instructions on how to optimize the file before saving it.

Using all the options available in GIF Movie Gear, I could squeeze the final animation down to a mere 9.2Kb. That means that this file will download, with a 56K modem, in only 3.5 seconds—not bad for a seven-frame animated banner (and you will have an efficient Web page design provided, of course, that you don't overload the rest of the page with large images).

I encourage you to play around with some of the techniques in the other chapters of this book to see whether you can come up with ideas for GIF animations. They can be quite fun to create and even more fun to watch.

Moving On

This chapter wraps up the book. I hope you have as much fun reading about and trying the techniques that Sherry and I presented in this book as we had in creating and writing about them.

If you are that rare breed that has read this book from Chapter 1 to here, your next step is to use the techniques you learned as a springboard to push your own creativity to new heights. If, like most artists, you are picking and choosing subjects that most interest you, make sure that you browse chapters that may not seem interesting to you—you'll probably learn something that you didn't realize you can do with Illustrator, and you'll be that much richer for the effort.

Also, don't forget to check out David Xenakis' great prepress appendix. We'll see you on the Net.

APPENDIX:
TRAPPING IN
ILLUSTRATOR

DAVID XENAKIS

Trapping is one of the most misunderstood and confusing topics in the world of digital prepress. This confusion has nothing to do with the difficulty of the subject. Instead, it has to do with the color printing process—where the trapping is crucial— taking place at some remove in time and space from the construction of the digital files that require trapping.

Many users of graphics software are brilliant in the execution of their software but have no clear idea how a press actually transforms the digital files into ink on paper. Illustrator is one of the few programs in which trapping is almost completely a manual process, and so it is a good program for learning to grasp what trapping is and how it works.

General Trapping Information

A color press isn't really a single press. It is a set of presses, all hooked together with rollers that draw a sheet of paper from one to the next. Each press puts down a single ink color. The completed piece is ejected from the press with all inks present. We can show the process graphically in four steps in Figure A.1.

Imagine that the press is to print the graphic using three colors: red, blue, and black. First, the dark border is placed on the blank sheet (a). The red horizontal bars are added next (b). Finally, the dark blue area surrounding the star is added (c) to give the completed flag (d). Note that the light stripes and the star are simply places where there is an absence of ink.

If all goes well and the press keeps the sheet in perfect registration as it moves through the inking stations, the printed figure will look perfect. However, the world—and most presses—are imperfect: As the sheet emerges from the other end of the press, you are likely to find that instead of the color boundaries matching one another perfectly, you have a sheet that has printed as shown in Figure A.2.

The red ink of the horizontal bands has slightly shifted so that it doesn't line up with the black rule. The blue rectangle containing the star has also shifted so that it doesn't line up with anything. What should have been a simple job is now an embarrassing and amateurish piece of printing.

What happened? As the sheet of paper was drawn through each station of the press, it moved slightly off track. As a result, the inks did not go onto the sheet in the proper places. Was the press or the pressperson at fault? Neither. Misregistration is a fact of life. Presses are typically engineered with incredible precision. However, it's a mechanical impossibility that a sheet of paper—which is not without imperfections and small variations in its thickness—can move through the press, sometimes at a speed of 300 feet per minute, without some misregistration.

Trapping is the means of fixing the visual problem presented by misregistration. Trapping does not prevent the problem; it simply disguises it. When a graphic document containing multiple inks is trapped, the color boundaries are moved so that the colors overlap one another.

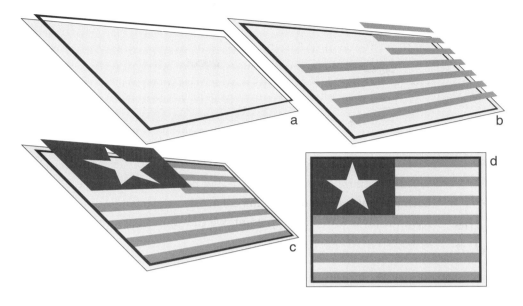

Figure A.1 (top)
Three inks are printed, one at a time, to make the completed figure.

Figure A.2 (bottom)
When the press sheet doesn't move perfectly through the press, the boundaries between adjoining colors show gaps and areas of overlap.

The amount of the overlap, or trap, is the amount of misregistration predicted for a given press.

Don't be intimidated by the idea of trapping. The principles involved are fairly straightforward.

Simple Trapping: Principle 1

If two adjoining color areas utilize significant percentages of the same ink, no trap is required. For example, the object in Figure A.3 shows a yellow oval atop a red rectangle. Working on the computer, you are likely to think of red and yellow as separate colors. However, remember when working with process colors that red is a *built color*, that is, a color made up of two or more inks printing atop one another and is composed of 100% magenta and 100% yellow. Yellow is, then, a common color, used both in the oval shape and in the darker area surrounding the oval. No trapping is required.

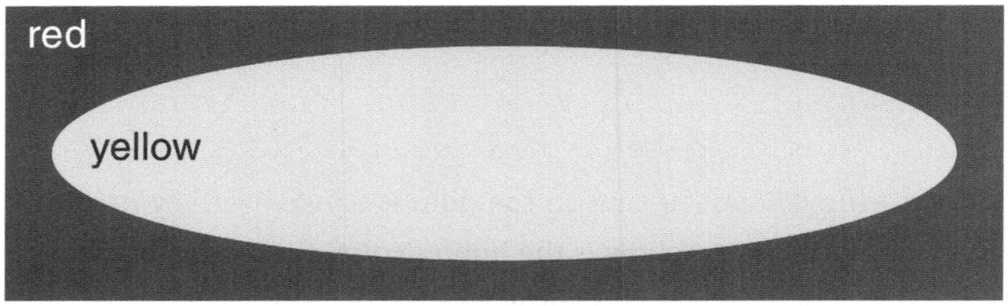

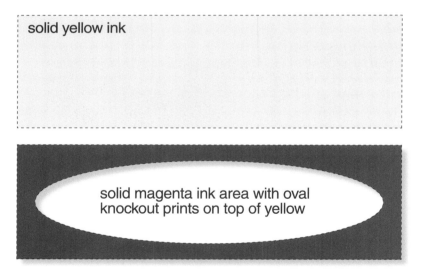

red

yellow

Figure A.3
When two adjoining colors share a substantial percentage of the same ink (in this case, yellow), no trapping is needed.

Take a moment to make sure that you understand this point completely. It will save you a lot of work! Think through the process as you look at Figure A.4. Remember that red is not, in process color work, a single ink color. In this case, a magenta area is put down on top of an all-over yellow area. Any misregistration of the two inks will be visible only along the outer edges of the figure, where it isn't really noticeable.

solid yellow ink

solid magenta ink area with oval knockout prints on top of yellow

Figure A.4
Yellow (upper) and magenta (lower) inks printing one atop each other give the yellow oval surrounded by red.

Simple Trapping: Principle 2

When any two colors meet each other, first you need to determine which is the lighter of the two colors. The lighter of the colors should overlap into the area of the darker color. Get this one straight: You'll see why later.

Because the lighter color must extend into the area of the darker color, it's fortuitous that Illustrator furnishes a perfect way to extend the color of an object past its boundaries: the stroke. It also furnishes a way to ensure that the two colors in the overlapping area mix with each other rather than butting up against each other: overprint.

Imagine that the three star shapes in Figure A.5 are 100% yellow on a 100% cyan background. Yellow is the lighter color. The dotted lines in the figure represent the paths. The star on the left has no stroke. The middle star is stroked with yellow. The one on the right is stroked with cyan. Both strokes are set to overprint. Yellow is the lighter color. Following the previously stated rule, the middle star is correctly trapped. The right-hand shape is included to illustrate one of the reasons for the rule: By using cyan as the trapping color, the darker color enters into the boundaries of the light-colored object and shrinks it.

The opposite case is shown in Figure A.6: yellow background and cyan stars. The original is on the left. The center star is stroked with yellow (the lighter color extends into darker color), whereas the right-hand star is stroked with cyan. As in the previous figure, the dotted line represents the original path. Notice that the left-hand star is larger than the other two.

Although these two cases might not seem serious, imagine the consequence of incorrect color movement when it comes to type. Imagine, for example, cyan text on a yellow background as shown in Figure A.7. Note that the dotted lines indicate the original letter shapes. The word on the right is stroked with yellow, the one on the left with cyan. The right-hand

WHICH COLOR IS LIGHTER?

Sometimes it's difficult to tell which of two colors is the lighter; for example, 100% cyan and 100% magenta are vibrant colors, but it isn't easily apparent which of the two is darker. If you are unsure, here's a quick way to tell. First, select objects filled with the colors in question and then choose Filter | Colors | Convert To Grayscale. After you see the colors in grayscale, you'll have no doubt about the relative darkness and lightness of the two. Use the Undo command to restore the objects to their former colors.

Figure A.5 (top)
Three yellow stars on a cyan background. The center star is correctly trapped because it extends the yellow color into the darker surrounding color. The right-hand star is incorrectly trapped: The darker color spreads into the boundaries of the lighter object and visually shrinks it.

Figure A.6 (bottom)
Three cyan stars on a yellow background. The center star is correctly trapped because it extends the yellow, lighter background color into the darker object color. The right-hand star is incorrectly trapped: The darker object color spreads into the boundaries of the lighter background and visually enlarges the star shape.

OVERPRINT AND KNOCKOUT

These two terms mean precisely what they seem to mean. If an object is set to overprint, the colored inks that compose it will print on top of any other object beneath it in the Illustrator stacking order. If an object is not set to overprint, it knocks out its own shape from any objects beneath it. Illustrator allows you to set the overprint status for both the stroke and the fill of an object. You set these attributes by clicking on the appropriate checkboxes on the Attributes palette (choose Window | Show Attributes).

letters have apparently been bolded. Although this might not cause serious problems with large-scale display type, the consequences are more serious when smaller letters are used. Book text seems to become bold text, and in some cases the counters will begin to fill in.

Simple Trapping: Principle 3

Where colors overlap, it is usually a good idea to cut back on the amount of ink *of the spreading color* in the zone of the trap so that it is as inconspicuous as possible.

This process is best illustrated in Figure A.8 by showing the familiar stars as cyan shapes on a magenta background. Cyan and magenta are strong colors. When they are used together, they form a deep, rich blue. If the stroke is set to the same 100% value as the fill, the star is shown with a dark outline. This will be the case whether the color of the overprinted stroke is cyan or magenta.

If you change the ink percentages (Figure A.9) to 100% magenta and 40% cyan, the object is still correctly trapped, but the visually intrusive dark line has been made relatively inconspicuous.

How Much Trap Is Enough?

The amount of trap is a figure that can be furnished by your print house. The figure varies widely, but a fairly good general figure is about .004 inches. Expressed in points, that number is .288. Remember that because the stroke width is on both the inside and the outside of the object, the stroke weight needed for a trap of .004 inches is .576 points (2×.288).

Figure A.7

The importance of correct color movement in trapping is shown by these two words. The left-hand word has been correctly trapped. Incorrect trapping on the right has resulted in bolded letters, which, in the case of smaller letterforms, might result in filled counters.

Figure A.8

Two strong colors can blend in the zone of trap to form a dark outline around the shape. This outline might not be objectionable but can be visually intrusive.

Figure A.9
If the ink percentages are changed—in this case, the trapping color has been reduced by 60%—the trap is much less visible.

Common Trapping Situations: Closed Shapes Placed On Other Closed Shapes

The two possibilities for this situation are illustrated by the drawing in Figure A.10. The examples in the upper part of the figure are easy to figure out. A simple stroke (with overprint) in the lighter color will handle the job. This will increase the size of the protruding light object on the right, but not by any amount that will be visually significant. (Note that the stroke for the upper-right object must be identical to the color of the object: If the ink percentages are reduced, they will show as lighter-colored lines wherever the shape extends past the darker background.) The trap for the right-hand object in the lower set is also straightforward: Add an overprinting stroke in the lighter color.

Figure A.10
The most common trapping situation involves closed shapes that sit atop other closed shapes.

The left-hand shape in the lower part of the figure provides a more challenging problem. The obvious solution, a stroke in the lighter color, will not work because then the stroke will be visible when it is outside the light-colored back rectangle. The easiest way to deal with this is the *oh, what the heck* solution: Stroke (with overprint) the dark object in its own color. A few other solutions might work better in critical situations. Here is an example:

1. Click on the dark rectangle to select it. Press Command+C (Mac) or Ctrl+C (Windows) to copy it.

2. With the object still selected, choose Object|Path|Slice. The drawing will change from the way it is shown in Figure A.10 to a solid-colored background object that has been carved into three pieces (Figure A.11a).

3. Click on the shape that was behind the original dark rectangle and delete it (Figure A.11b). Use the Paste In Back command (Mac: Command+B; Windows: Ctrl+B). This places the dark rectangle back into the drawing and in a new place in the stacking order, *behind* the two light-colored rectangles.

4. Select the two light-colored rectangles and apply an overprinted stroke in the light color (Figure A.12a). This stroke will overlap the large dark rectangle (because the two rectangles are now in front of it in the stacking order) only where the two smaller light-colored rectangles touch it. Note that this procedure slightly enlarges the two light-colored rectangles.

5. A more elegant way to construct the trap for the dark rectangle, and one that doesn't enlarge the background shapes except in the trap zone, is to click (with the Direct Select cursor and Shift held down) on the two boundary lines next to the dark rectangle. The arrows in Figure A.12b show these lines. Copy and use the Paste In Front command (Mac: Command+F; Windows: Ctrl+F). These two line segments should be set to no fill and with a stroke—overprinting—of the same color as the light background shape. This is a much more satisfactory way to construct the trap.

Common Trapping Situations: Shapes Where The Stroke Requires Trapping To An Adjacent Color

In the example shown in Figure A.13, the lighter object is a single stroked line (no fill) sitting atop a darker background. The stroke cannot be spread along its edges, and overprinting will simply merge the entire object with the background color. To solve this problem, you can follow either of two procedures. Both work perfectly, and the method you use depends on circumstances and your own preferences.

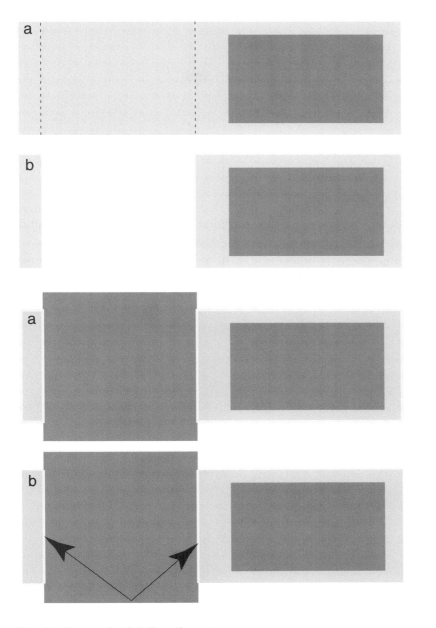

Figure A.11

The Slice command carves the background rectangle into three pieces (a). The area originally behind the dark rectangle can now be deleted (b).

Figure A.12

After you use the Paste In Back command, the dark rectangle is restored to the drawing. The light rectangles, now in front of the dark shape, can be stroked—overprinted—to make the trap (a). A better trap (b) is made by copying the two line segments of the light rectangles, pasting them in front, and stroking them—overprinting—with the light color.

For the first method, follow these steps:

1. Select the stroke and copy it (be sure that overprint is not turned on). Use the Paste In Back command.

2. Change the width of the pasted path to be .576 points wider than the original stroke and set it to overprint. The example in Figure A.14 shows the new stroke as slightly lighter than the original for the sake of clarifying the process. When finished, only the light-colored edge area overprints, making the edge of the upper stroked path trap perfectly to the background.

For the second method, follow these steps:

1. Click on the stroke to select it.

2. From the Object menu, select Path|Outline Path. This operation converts the stroked path to a filled object of the same shape. Assign an appropriately colored overprinting stroke to the new object.

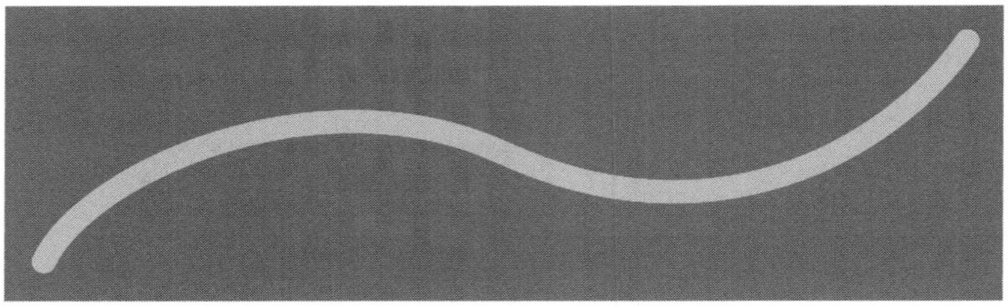

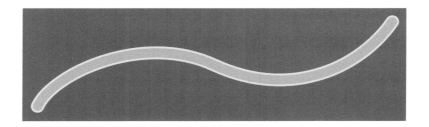

Figure A.13

Here is a special trapping problem. The light object is simply a stroked path. Setting the path to overprint will simply merge the color with that of the background.

Figure A.14

Using the Paste In Back command, a slightly heavier stroke can be set to overprint. The upper path does not overprint, leaving the zone of trap as just the narrow light area surrounding it. Alternately, the path can be outlined (Object|Path|Outline Path) and an overprinting stroke applied to the resulting filled object.

Another trapping problem involving strokes is shown in Figure A.15. We will assume that this figure shows a worst-case configuration in which all three of the tones are separate inks and the dark borders around the uppermost rectangles are strokes. As you can see, a number of trap problems exist here because the edge of each stroke must trap (or be trapped by) the colors on the inside and the outside of the path.

The problems can be satisfactorily handled by using the Outline Path command. Select each of the four shapes and run the filter. Before deselecting, press Command+Shift+G (Mac) or Ctrl+Shift+G (Windows) to ungroup the objects, as the filter groups all the objects. This filter outlines the edges of the stroke and converts it to a narrow compound object perfectly positioned where the stroke originally existed. This compound shape is now filled with the original stroke color.

Begin with the two upper rectangles. Use the Direct Select cursor with Option (Mac) or Alt (Windows) held down. Carefully click the inner edge of the two compound frames. Copy and Paste In Front. An exact copy of the inner edge of the frame is now sitting atop the inner edge of the frame. Change its paint attributes to no fill and an overprinted stroke of the light color (Figure A.16). Perform the same operations with the outer

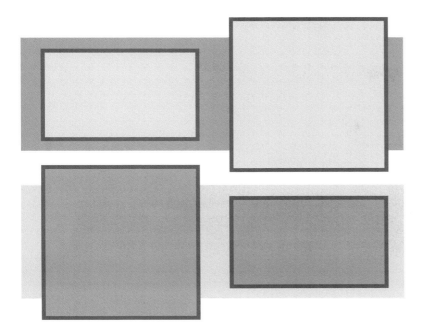

Figure A.15
Strokes provide a further element of complexity in this figure: The stroke colors must be trapped to the colors on each side of the stroke.

frame of the upper-left rectangle and use the medium color (overprinting) for its stroke. Treat the lower-right rectangle in the same way.

For the larger rectangles, click on and copy the outer paths. Use these paths with the Slice command to create separate shapes of the backmost objects as described previously. Apply the strokes to these objects. The diagram shows the strokes applied (in lighter colors to make the trap relationships clear).

Figure A.16
After using the Outline Path command on all the strokes, the individual pieces—inside and outside—of the compound shapes can be used for trapping purposes.

Common Trapping Situations: Trapping Gradients

The example in Figure A.17 shows a gradient sitting on top of a solid color. The background could also be another gradient. This configuration could be handled in several ways. Because the background is a solid color and is generally lighter than the gradient object, it should be possible to place an overprinting stroke of the background color on the gradient object.

If the background object is filled with a different gradient, you can still make a satisfactory trap. First assign an overprinting stroke of some color (it doesn't matter what color because it will change) to the oval object (Figure A.18). With the oval shape selected, use the Outline Path. The shape created by this filter, as mentioned previously, is a narrow compound path that can now be filled with the same gradient as the oval (or with the gradient of the background) and with the fill set to overprint. If the two gradients are not aligned (Figure A.19), select the small compound shape and the large oval gradient shape and use the Gradient Vector tool on both (Figure A.20).

If the background of the shape contains a gradient, it might be necessary that the inner should be trapped with the background gradient attribute. Assign the compound shape (overprinting fill) the background gradient. If needed, select the narrow shape and the background and re-vector the gradients to cause them to align.

Common Trapping Situations: Trapping With Patterns And Brushes

When using patterns and brushes, options for choke and spread are limited.

However, patterns and brushes must be trapped in the same way any other Illustrator objects are trapped. The fundamental rule for pattern tiles and brushes that trap is that the trapping must be done on the original artwork before the pattern or brush is created. Otherwise, you will have to use the Expand command and individually trap each of the hundreds of objects laid down by your use of pattern fills and brushes. The

Figure A.17
Trapping a gradient is fairly simple if the gradient object sits atop a solid-colored background. Simply stroke the gradient object—overprinting—with the background color.

Figure A.18
First assign an overprinting stroke to the gradient shape. It makes no difference what color you use because it will change later.

Figure A.19
After using the Outline Path command, fill the narrow compound shape with the same gradient (with the fill overprinting).

Figure A.20
If the two gradients don't line up (as they do not in Figure A.19), select both the compound shape and the gradient behind it and use the Gradient Vector tool to make them match.

drawing in Figure A.21 shows four patterns and illustrates what is required for the patterns to trap correctly.

The pattern fill in Figure A.21a is a tile containing colored circular shapes and a bounding rectangle of no stroke or fill. If the pattern-filled shape always sits on an area of the background color—in this case, the color is lighter—it's simple to define a new pattern on the basis of this one. In the new pattern, each of the small circles is stroked (with overprint) in the background color. If the background is darker than the circles, the stroke is the same color as the circles.

This does not solve the problem of using the pattern when it might be placed on several different background colors and sometimes several

CHOKE AND SPREAD

Choke and *spread* are traditional trapping terms. If the surrounding color is brought inside the object to be trapped, the object is described as choked. If the color of the object is brought outside into the background, that color is described as spreading. Two other more colloquial trapping terms in general use are *skinny* (meaning the same as choke) and *fatty* (meaning the same as spread).

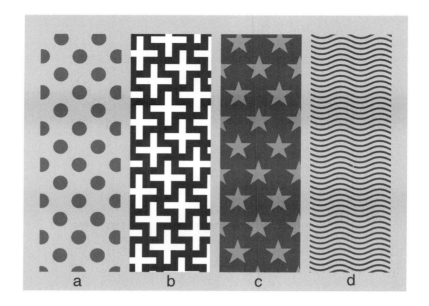

Figure A.21

This drawing illustrates four possible problems that confront an Illustrator user whenever tiled patterns or brushes are used in the artwork.

REDUCING THE SIZE OF THE CIRCLES

A quick and precise way to reduce the size of circles—and other objects that will be enlarged by a trapping stroke—is to assign the stroke and then to use the Outline Path command. After using the command, look at the shape in Artwork mode. You'll see three concentric paths. Use the Direct Select cursor to select and delete the outer two paths, retaining the inner path. This inner path, the one you didn't delete, is now filled with the original color and the overprinting stroke is added to it. With the stroke added to it, it is exactly the same size as the original circle.

colors at the same time. In such a case, the only solution is to give the small circles in the tile an overprinting stroke in the same color as the circles. Note that in this case, color reduction is not a good idea because the scaled-back color would show up as a lighter stroke if the pattern were used on areas where there is no background color.

Trapping this pattern in this way is not a very satisfactory solution because it makes the small circles larger than they were in the original pattern. To compensate for this, you can make the original circles slightly smaller so that the addition of the stroke spreads them to the desired size (see the tip in this section, "Reducing The Size Of The Circles").

The pattern shown in Figure A.21b is a set of black shapes on a lighter color. If the color is white, the black shapes need no modification. If the lighter color is not white, the black shapes should be stroked (with overprint) with the lighter color.

Trapping the objects of the pattern tile does not solve the trapping of the pattern-filled object to the color behind it. In the example shown, stroking (with overprint) the patterned object with the color of the large background rectangle could solve the problem. If the rectangle extends beyond the boundary of the background rectangle, it must be treated as shown in Figures A.12 and A.22, with the resulting background pieces stroked with their own color. When the procedure has gotten to this point, use the Direct Select cursor to click on the ends of the light rectangular background shapes (dotted line). Copy these lines and use the Paste In Front command.

Figure A.22
Even though the rectangle on the left has been filled with a pattern, it can still be used to slice a segment out of the background shape.

Figure A.23
Selecting the line segments at the inner ends of the background rectangle, copying, and using the Paste in Front command allows the addition of two trapping strokes.

Change the paint style of these lines to overprinted stroke in the color of the light rectangles (shown lighter in Figure A.23 for clarity).

The pattern in Figure A.21c has the same trapping challenges except that the star shapes are light on a darker background. You would trap the stars to the background before defining the pattern or brush and then trap the edge of the pattern-filled rectangle separately.

The pattern in Figure A.21d presents the most difficult challenge of the samples presented. The pattern is made up of simple stroked paths, and the solution is really not very complex. The pattern tile used is shown in Figure A.24 with the bounding rectangle—no stroke or fill—delineated as a dotted line. When using a pattern of this sort, the only practical way to trap is to spread the strokes used for the tiles. The reasons are the same as for the example shown in Figures A.13 and A.14. Spreading the strokes is accomplished by using the previously described procedure. The trapped tile would appear as shown in Figure A.25: Wider, overprinting strokes underlie narrower, non-overprinting strokes in position on top.

Common Trapping Situations: Anti-Trapping When Using Rich Black

Often, when large areas of black are used in a design, it is common practice to mix percentages of other process colors with the black to increase its density. For example, the reversed letters in Figure A.26a might be

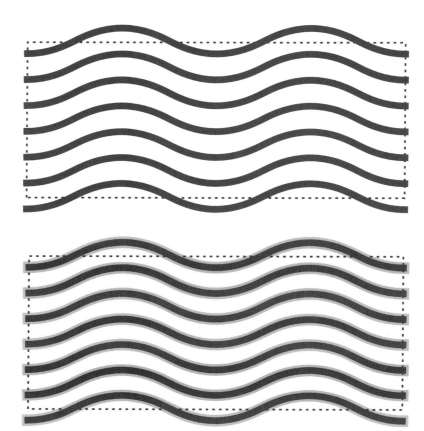

Figure A.24
This pattern tile is composed of stroked lines.

Figure A.25
Before defining the pattern tile, slightly wider lines set to overprint are placed beneath the original tile's lines.

surrounded by an ink area composed of 100% black and 100% cyan. This would give the black a density and richness that it would otherwise lack. The problem is that misregistration on press can cause the identical areas of cyan and black to print slightly out of alignment (Figure A.26b). If that happens, you see the edges of the black area and edges of the letter fringed with cyan dots. This is an aspect of trapping that requires some thought because, to prevent the fringing effect, you will need to add black strokes that *do not overprint*.

Begin with the white letters. Select, copy, and hide them. Use the Paste In Back command to place new copies of the letters in the identical position. Stroke these letters with a width appropriate for a trap. Make the stroke solid black (only black) and be sure to *not* set the stroke to overprint. Use the Show All command to bring the original letter back. You won't be able to see any difference to your artwork, but you will have accomplished a minor piece of magic. Surrounding the white letters will be a narrow zone of just black ink that will knock out the cyan component of the surrounding color and drive the cyan away from the edges of the letters. (The effect is shown in Figure A.26c with the edges of the letters lightened to make them easy to see.)

Figure A.26
When using a black composed of more than one ink, it is a good idea to surround any reversed areas and the edges of the rich black shapes with zones of pure black set not to overprint (c). This prevents the extra ink from fringing the white areas.

Next, select and copy the black background rectangle. Use the Paste In Front command to place a new rectangle precisely over the first. Set the new rectangle's fill to be None and use the same stroke of black as you used for the letters. Be sure this stroke does not overprint.

If you want to get really fussy about the whole thing, use the Paste In Front command once more, sweep select all the rectangles and letters, and use the Objects|Masks|Make command. This will eliminate the outer edge of the stroke around the background rectangle and leave the artwork exactly the same size it was when you began.

The Illustrator Trap Command

Sometimes, Illustrator objects can be subjected to Illustrator's own Trap command (Figure A.27). The cases in which this command works are limited, but it does perform superbly.

The Trap command can be used only with color-filled objects atop (completely or partially) other color-filled objects. The items to be trapped cannot have strokes, nor can they be filled with patterns or gradients.

```
┌─────────────── Pathfinder Trap ───────────────┐
│ ┌─ Setting ──────────────────┐  ┌──────────┐  │
│ │  Thicknes [0.25]  points   │  │    OK    │  │
│ │  Height/Widt [100]  %      │  ├──────────┤  │
│ │  Tint       [40]  %        │  │  Cancel  │  │
│ │                            │  ├──────────┤  │
│ ├─ Option ───────────────────┤  │ Defaults │  │
│ │  ☐ Traps with Process Col  │  └──────────┘  │
│ │  ☐ Reverse Trap            │               │
│ └────────────────────────────┘               │
└───────────────────────────────────────────────┘
```

Figure A.27
The Illustrator Trap dialog box.

Notice that this dialog box allows you to reduce the color in the trap zone. It can also convert the traps of spot colors to process colors as well as make the trapping color spread the wrong direction (Reverse Trap). With respect to the former, I cannot image a case in which this might prove necessary. However, the option has been present as long as the Trap command has been a part of Illustrator.

When the Trap command (click on the right-hand icon in the bottom section of the Pathfinder palette) is used on a set of shapes such as those shown in Figure A.28, a small filled object is created that perfectly extends the light color into the darker color. This shape is created without the necessity of slicing the background object. This preserves the original artwork's editability and is a valuable feature of the Trap command. When you need to use the Trap command, it is a good idea to convert any strokes to filled compound objects by using the Outline Path command. The Trap command will then function very well indeed.

Figure A.28

The Pathfinder's Trap command constructs small overlaid shapes filled with the appropriate color needed to trap the two objects.

INDEX

A

Actions, 238
  creating, 242-259
  debugging with stops, 246
  dry-running of, 242-243
  installed, 240-241
  interactivity in, 247-252
  randomness in, 252-259
  recording, 239-240, 244-245
  saving, 247
  signing, 246-247
  to-do lists for, 248
Actions palette, 4, 8-13, 238-259
  actions installed with, 240-241
  controls in, 238-240
  default mode of, 238-239
  extra actions with, 241
  menu for, 240
Add Anchor Points command, 263
Adjust Colors dialog box, 250, 257
Adobe Streamline program, creating vector images with, 86
.AI format, saving images in, 19
Aliased images, 37-38
Align palette, 69-70, 277, 281, 428-429
Angel fish image, 147-148
Angel image, 127
Angles
  of gradient ramps, 155
  with radial gradients, 156-160
  of rotation, 363, 364
Animated GIF files, 457-462
Anti-aliasing of images, 35-38, 47-51
  in copying and pasting, 55-56
  drag-and-drop operation and, 56-57
  Place command and, 51
  scaling and rotating and, 54-55
  text in, 58
Anti-alias PostScript checkbox, 35
Anti-trapping, in color printing, 477-479
Apple image, 398-399
Apply mode, 73
Area Type tool, 397-399
Around and around image, 421-423
Arrow tool, 102
Art brushes, 342, 343-344, 345

abstract strokes with, 366-368, 369
changing type shapes with, 405-410
expanded definitions of, 368-374
Pattern and Scatter brushes versus, 354-366
with type, 405, 406
Art Brush Options dialog box, 359, 364, 373, 406, 444
Artistic Sample dialog box, 318
Astronomy image, 292-293
Attribute Setting command, in Actions palette, 251
Attributes palette, 249, 468
Auto-Select Layer checkbox, 61
Auto Trace Tolerance settings, 15-16
Auto Trace tool, 13-16
  creating vector images with, 86
Averaging of points, problems with, 205

B

Background, shadowing and, 131-132
Background patterns, creating, 63-72
Beveled Frame action, 8
"BigStroke" Art brush, 368, 369
Bitmap dialog box, 117
Bitmapped images
  creating gradient meshes from, 178-182
  with raster image-editing programs, 32-33
  tracing, 18-23
Bitmap textures, 113-118
  adjusting densities of, 118
Black and white images, 91
Black areas, anti-trapping with, 477-479
Blatner, David, 62
Blend Front To Back filter, 423, 433
Blend Horizontal filter, 421-423
Blend Options dialog box, 141, 245, 411
Blends, 138-155
  basic, 138-143
  clip art, 147-149
  with colors, 420-423, 432-433
  irregular line, 145-147
  linear, 143-155
  linear landscapes with, 149-152
  live, 138-140
  manipulating objects in, 143
  reversing directions of, 142
  reversing front to back in, 142

H

I

COLOPHON

From start to finish, Coriolis Group designed *Illustrator 8 f/x and design* with the creative professional in mind.

The cover was created on a Power Macintosh using QuarkXPress 3.3, Adobe Photoshop 5.0.2, Alien Skin Black Box 2 filters, and the Trajan and Futura font families. It was printed using four-color process, metallic silver ink and spot UV coating.

The interior layout was also produced on a Power Macintosh with Adobe PageMaker 6.52, Microsoft Word 98, Adobe Photoshop 5.0.2 and Adobe Illustrator 8. The body text is Stone Informal, heads are Avenir Black, and Chapter titles are Copperplate 31ab.

Illustrator 8 f/x and design was printed by Courier Stoughton, Inc. of Stoughton, Mass.

WHAT'S ON THE CD-ROM

The *Illustrator f/x and design* companion CD-ROM contains elements specifically selected to enhance the usefulness of this book, including:

- *Practice files for the book's projects*—You can adapt these real-world images based on your needs; they are yours to use freely:
  - All the starting images needed to work the Illustrations in the book. For each chapter, these are to be found in the CHxxSTART folder.
  - Key finished projects in editable and layered formats so that you can explore the construction of the final images. These are located in the CHxxEND folder for each chapter.
- Trial version of Illustrator 8 for Mac and Windows.
- Try-out versions of Extensis Vector Tools, Extensis PhotoGraphics, and Extensis PhotoTools, which will work either in Illustrator (Vector Tools, minus the toolbars and Tips) or Photoshop.
- *Demo versions of AutoF/X filters:*
  - Photo/Graphic Edges—This plug-in applies stylized edge effects to any image.
  - Typo/Graphic Edges—This Photoshop plug-in enables you to add edge effects to type and clip art.
  - Universal Animator—This product allows you to create animations in any application and to view the animations as they are created.
  - Page/Edges—a Photoshop border-generating filter to create cool page edges.
- Demo versions of Hot Door Transparency (live transparent lens effects) and Hot Door Cad Tools filters (drafting and dimensioning plug-in tools for Illustrator 7, Mac and Windows).
- *RAYflect plug-in demos (Photoshop plug-ins accessible in Illustrator):*
  - Four Seasons—The powerful, photorealistic sky and atmosphere generator extension for Photoshop.
  - PhotoTracer—Allows every level of user to easily and quickly add 3D graphics to their creations.
- Demo Versions of Vertigo3D filters—Creates and manipulates 3D text right from your keyboard.
- Demo Version (Macintosh only) of ZaxWerks 3D Invigorator (Windows to be available Summer '99).
- Clip art samples from Direct Imagination and Ultimate Symbol.

System Requirements

Macintosh

Software:

- Your operating system must be System 7 or higher (OS 8+ recommended).
- A full copy of Adobe Illustrator 8 is needed to complete the projects included in this book.
- A full copy of Adobe Photoshop 5 is needed to complete some of the projects in this book.

Hardware:

- Adobe Illustrator 8 requires a PowerPC processor.
- At least 32MB of RAM.
- A color monitor (16 million colors) is recommended.

PC

Software:

- Your operating system must be Windows 95, 98, NT4 or higher.
- A full copy of Adobe Illustrator 8 is needed to complete the projects included in this book.
- A full copy of Adobe Photoshop 5 is needed to complete some of the projects in this book.

Hardware:

- A Pentium processor is strongly recommended.
- At least 32MB of RAM.
- A color monitor (16 million colors) is recommended.